Student Study Guide

to accompany Petrucci's
General Chemistry

Student Study Guide

to accompany Petrucci's
General Chemistry

Fifth Edition

Robert K. Wismer

Millersville University
Millersville, PA

MACMILLAN PUBLISHING COMPANY
New York

COLLIER MACMILLAN PUBLISHERS
London

Macmillan Publishing Company
866 Third Avenue, New York, New York 10022

Collier Macmillan Canada, Inc.

ISBN 0-02-394792-6

Printing: 1 2 3 4 5 6 7 8 Year: 9 0 1 2 3 4 5 6 7 8

ACKNOWLEDGEMENTS

Any book is the product of many individuals and this one is also. Many improvements have resulted from their efforts; the errors that remain are mine. I owe special thanks to Professor Ralph H. Petrucci for his frequent comments during the writing of this edition and his extensive advice on the previous ones.

The students of Introductory Chemistry I and II at Millersville University have used this book for eight years. They have not hesitated to correct errors, criticize poor expression, and offer helpful suggestions. Several students from other institutions have written suggestions and criticisms. Lynn A. Wainwright deserves particular recognition for she proofread the entire text and the answers.

The Chemistry Department faculty at Millersville University have been very supportive throughout the writing process, especially Professors Lewis M. Bass, Donald W. Gauntlett, Thomas G. Greco, Jan M. Shepherd, and Gerald S. Weiss—all fellow lecturers of Introductory Chemistry.

Robert McConnin, chemistry editor at Macmillan Publishing Company, supervised this book. Elisabeth Belfer of college production at Macmillan has made the entire writing process as painless as possible with her advice and encouragement throughout the task of producing a final version. Any beauty in the printed pages herein is largely the result of her guidance.

This book was written with Microsoft Word on a Macintosh Plus computer, using MacBillboard and Thunderscan for most pictures. Dr. Jan M. Shepherd assisted in producing many chemical structures with ChemDraw. My thanks to Dave Measel and Russ Baker at Microsoft for their technical advice.

But I still owe my largest debt of gratitude to my wife Debbie and our children Michael, Mary, and two-year old Karen. They have understood a husband and father who always is thinking of "the book," and have constantly loved, supported, and encouraged him.

Robert K. Wismer

Contents

How to Use This Book

The purpose of this study guide is to help you learn chemistry. Students of chemistry have a wide variety of backgrounds and different reasons for taking chemistry. A study guide is designed primarily for those students who have difficulty learning a subject. But there is something of value here for all—because all of us have trouble at some point in learning new material. Because of this, you may not find all of the material useful to you. However, each suggestion is worth an honest effort on your part. It is the result of several years of experience helping students with what many of them consider a very difficult subject—chemistry.

Chemistry is both an art and a science. In fact, often students complain that "chemistry has too many rules," and "there are too many exceptions to the rules." Thus, its study requires a different approach than does the study of many other subjects. It is not sufficient to read the text once, twice, or even several times. In fact, if you follow that approach, you will spend a great deal of time on chemistry but probably will not learn it. And it is essential that you learn chemistry well, particularly the basics. Chemistry is a vertical discipline; it builds on itself. And just as you would have great difficulty with arithmetic if you had never learned the "seven times table," so too chemistry will more than live up to its brutal reputation for you unless you learn its fundamentals well.

Learning chemistry requires that you do three things: write, practice, and test yourself. Writing helps you firm up your ideas and thus makes you aware of what you really know and what you don't know. Francis Bacon stated it well: "Reading maketh a full man, conference a ready man, and writing an exact man."

Practice in the study of chemistry means that you practice solving problems. You certainly would not expect to be able to play a difficult piano composition without first practicing it. Neither can you expect to be able to solve problems without practice, lots of practice.

Finally, you need to test yourself. This has two purposes. First, it helps you find out how well you have learned what you have studied. Second, you should feel more at ease during class examinations, since you have examined yourself already.

This study guide contains material to help you with these three areas of study. Although the study guide can be used with any textbook, it is designed for use with *General Chemistry: Principles and Modern Applications,* fifth edition, by Ralph H. Petrucci (1989, Macmillan Publishing Co., Inc.) This book is referred to throughout the study guide as *the text*. Each chapter in the study guide contains the same material as the text chapter of the same number.

The study guide is organized around learning objectives, which emphasize the fundamentals of each chapter. Throughout the study guide, each objective is referred to by chapter number and objective number; thus objective 14-5, refers to objective 5 in Chapter 14. Occasonally a figure or a table in *the text* is referred to in the review of an objective. All of these figures and tables are reprinted in Appendix II of this study guide.

Organization by objectives divides the material of each chapter into more easily comprehended parts. Each chapter begins with short essays on each objective, grouped under "Chapter Objectives." These essays explain the material, describe how to work problems, and give detailed examples of worked-out problems. The material in the essays is taken from the author's classroom experience and include many analogies, examples, and illustrations. The analogies are often fanciful, sometimes ridiculous. This is necessary to have them illustrate the intended point and it does make them more memorable. Sometimes topics are explained from a different viewpoint from that of the text. There is no contradition here but mutual reinforcement. If you have trouble with a topic, a different viewpoint may be enough to make it clear. Even if you have no difficulty, the alternate view may highlight aspects that had not occurred to you and thus increase your appreciation of the topic.

For objectives involving memorization, and especially for learning new terms, 3 × 5 cards are very useful. Write the term on one side of the card and your version of the definition on the other side. You can carry the cards with you and study them during spare moments: waiting in line, before classes start, and so on. In this way, you profitably use the many small—and otherwise wasted—periods of time throughout the day. In addition, you can shuffle the cards to make sure you know the terms in any order, pick out the hard one and concentrate only on those, and turn the cards over to look at the definition and then recall the term.

Most objectives require learning a new concept or how to solve a new type of problem. For those involving a new concept, first read the essay carefully and then summarize the concept in your words. Since a 3 × 5 card is too small for most concepts, this summary should be written in your lecture notebook (see "How to Take Good Notes"). Notice that both definitions and new concepts should be stated in your words. Copying from a book will help you very little. Putting the definition or concepts in your words ensures that you understand it. This is what you want. After all, *you* are the one who has to learn it and be able to use it.

Each problem-solving objective includes a detailed example of the type of problem being considered. The "tricks of the trade" are included in these examples, and often one type of problem is solved by two or more methods so that you will understand better how the answer is obtained. Sometimes there will be but one example, solved by a method different from that of the text. Keep in mind that there are many different ways to solve a problem and choose the one that is easiest to you. When a method of solving a problem is broken down into several steps, each step is explained.

You gain practice in problem solving by doing, so the "Chapter Objectives" section is followed by "Drill Problems." The problems are grouped by objective, and their presence is indicated in the essay section by a star (*) in the margin by the objective number. Occasionally there are more than 26 drill problems for a given objective, and thus they could not all be referred to with the letters of the alphabet. In these cases, those Greek alphabet capital letters which are different from Latin capital letters, are used. (The Scandinavian letter Ø also is occasionally used.) Greek letters often are used in scientific equations, and you thus may become familiar with some of them. The complete Greek alphabet is given in the table that follows.

The Greek Alphabet (capital letter first in each case, followed by lower case letter, and name)

Α α alpha	Β β beta	Γ γ gamma	Δ δ delta	Ε ε epsilon	Ζ ζ zeta
Η η eta	Θ θ theta	Ι ι iota	Κ κ kappa	Λ λ lambda	Μ μ mu
Ν ν nu	Ξ ξ xi	Ο ο omicron	Π π pi	Ρ ρ rho	Σ σ sigma
Τ τ tau	Υ υ upsilon	Φ φ phi	Χ χ chi	Ψ ψ psi	Ω ω omega

The drill problems are straightforward questions based on each objective. By working them, you are practicing your mastery of that objective (rather like practicing scales or chords when learning to play the piano). If you drill yourself and become confident in these basic techniques, solving problems in which several are combined will be much easier. There should be enough drill problems for each objective for you to practice until you feel confident in solving that type of problem. Answers to all drill problems are given in Appendix I. The drill problems all use actual data, not "made-up" numbers. This should give you a better feeling for the "real world" of chemistry.

When you can solve the drill problems, go on to the Exercises in the text. The drill problems give practice in the techniques of solving problems. The exercises in the text require that you be able to use these techniques in solving more involved problems.

For testing yourself, you will find in each chapter four "Quizzes" and a "Sample Test." The quizzes are designed to be a quick way to check your knowledge. There are four so that you can check you progress at several points: before you intensively study the chapter, during your study , and just before a class examination. The tests are included to give you some idea of the kinds of questions you might encounter on a class examination. The test questions differ in style within each chapter and between chapters. Different instructors use different style questions and knowing many styles can be of great help on examinations. Answers to the quizzes are in Appendix I, as are detailed solutions to the tests.

All problems have been twice solved by the author. The entire book has been class tested with a 25¢ reward offered for the first notification of each error that was found. Nonetheless, errors may still be present. The author would very much appreciate notification of such errors (The rewards—now 50¢—still are being given!) and any suggestions for improvement. All such suggestions regarding the previous editions have been carefully considered and most have been included in this edition.

Here's hoping your study of chemistry is both successful and enjoyable!

How to Take Good Notes

There are several steps involved in producing a good notebook—one that will help you in your study of a subject. Although what follows is written for a chemistry lecture notebook, it can be adapted with small changes for the lecture notebook of any subject.

The first step is prior preparation. Before going to lecture, carefully read the chapter summary. Then skim the material in the text that will be covered. Do not read in detail. You are trying to get an overview of what is to be taught so that you understand how each topic relates to the others. Write a brief outline in your lecture notebook. You should also write out questions about points that seem unclear. Finally, review the notes you took during the previous lecture. In this way, you will understand how each point is related to the whole subject.

During lecture, take notes on only one side of each page. You will need the other side later. Many professors write all important material on the board and, of course, you will copy this in your notebook. Think about your brief outline and your questions as you take notes, and answer those questions that you can. Leave a blank line after each major topic in your lecture notes so that you can more easily see the main points when you review your notes.

Immediately after lecture, review the notes you have taken. Correct the items that are wrong, add material you remember, and emphasize important points. This should take five minutes at most and it is absolutely essential that you do it immediately, while the lecture is fresh in your mind. The longer you delay, the less value this process will have.

The evening after lecture, read the portions of the text and the study guide that were covered in lecture. Use the blank sides of the notebook pages to write clarifications and any questions that you might have. Often, just the process of writing down a question will clarify the concept enough that you will understand it.[†]

Now try to work the problems. Start with the drill problems and work some from each objective. If you can't solve a certain type of problem, reread the essay on that objective and the appropriate part of the text. If you still can't solve it, write down a specific question in your notebook. Try to identify which part of the problem you can not solve and solve the rest of the problem or at least detail how you will solve it.

Ask other students in the class for answers to your questions. Sometimes they will understand concepts which you do not; often you will be able to help them. But do not just work problems to get the right answer. Make sure that you understand the technique well enought so that you can easily solve other problems of the same type by yourself.

Finally, ask your instructor for help with your unanswered questions. By working on your own and writing down your questions, you will know exactly what to ask and you will understand the answer much better. Do not wait until the day before the exam to question your instructor. Ask daily as new questions arise. Write the answers to your questions in your notebook. Also write a solved example of each type of problem that gave you trouble.

Now you have a notebook that really is helpful. It not only contains your lecture notes but also has questions and answers on every point that gave you trouble. Such a notebook is invaluable when you study for class exams and also when, in a later course, you want to refresh your knowledge.

† See "How to Study in College," 2nd ed., by Walter Pauk, pp. 125-175 (Houghton Mifflin Co., 1974). Or consult the staff at the learning center on your campus.

Figures and Tables in Petrucci's *General Chemistry* that are referenced in the Study Guide.

SG page	F.or T.			Caption of Figure or Table
13	Fig. 2-12		5/e	A mass spectrometer {INCLUDE CAPTION}
84	Fig. 8-5	b	SG	The electromagnetic spectrum
87	Fig. 8-17	b	SG	Energy-level diagram for the hydrogen atom
97	Fig. 9-1	e	1c	An illustration of the periodic law—atomic volume as a function of atomic number
97	Fig. 9-8	a	n	Covalent radii of atoms
97	Fig. 9-13	a	n	First ionization energies as a function of atomic number
137	Fig. 12-11	a	n	Vapor pressure curves of several liquids
138	Fig. 12-20	c	1o	Phase diagram for carbon dioxide
138	Fig. 12-21	c	1o	Phase diagram for water {INCLUDE CAPTION}
146	Fig. 13-15	a	n	Vapor pressure lowering by a nonvolatile solute
149	Tbl. 13-3	c	1o	Some common types of colloids
156	Fig. 14-23	a	n	The Frasch process for mining sulfur
158	Tbl. 9-8		5/e	Some selected reactions of the halogen elements
202,03	Tbl. 17-2	d	d	Ionization constants for some weak acids and weak bases in water at 25°C
202	Tbl. 17-3		5/e	Ionization constants of some common polyprotic acids
222,30	Tbl. 19-1		13c	Solubility product constants at 25°C
227	Tbl. 19-2	a	n	Formation constants for some complex ions
246	Tbl. 21-1	e	1c	Some selected standard electrode potentials
265,71	Fig. 23-1	a	n	Standard electrode potential diagram for chlorine
265,71	Fig. 23-6	a	n	Electrode potential diagram for oxygen
265,71	Fig. 23-7	a	n	Electrode potential diagram for sulfur
265,71	Fig. 23-10	a	n	Electrode potential diagram for nitrogen
267	Tbl. 23-7	d	2d	Solubilities of some metal sulfides
268	Tbl. 23-12	a	n	Preparation of oxides of nitrogen
277	Fig. 24-3	a	n	Electrode potential diagram for vanadium
277	Fig. 24-8	a	n	Electrode potential diagram for manganese
279	Fig. 24-15	e	1c	Qualitative analysis of cation group 3 *(caption)*
281	Fig. 5-12	b	SG	A qualitative analysis scheme for cations
299	Tbl. 26-6	a	n	Units of radiation dosage
314	Tbl. 10-4	a	n	Some synthetic carbon-chain polymers
318	Tbl. 28-3	a	n	Some common amino acids
318	Fig. 28-20	a	n	Hydrolysis products of nucleic acids
319	Tbl. 28-1	a	n	Some common fatty acids
320	Fig. 28-8	a	n	Some common disaccharides
320	Fig. 28-9	a	n	Two common polysaccharides
324	Fig. 28-12	e	4c	An alpha helix—secondary structure of a protein *(paste 4 =)*
325	Fig. 28-21	a	n	A portion of a nucleic acid chain
325	Fig. 28-22	a	n	DNA model

The number given in the left-most column is the page number of the referencee in the 1989 edition of the Study Guide. The Figure or Table number given in the next column is from the 5th edition of Petrucci's *General Chemistry* (*the text*), with the source depending on the chapter in *the text.*

3. For chapters 1-10 of *the text*——uncorrected page proofs.
3. For chapters 11-16 of *the text*——copies of Dr. Petrucci's corrected galleys.
3. For chapters 17-28 of *the text*——uncorrected galleys.

These figure and table numbers should not be followed blindly but rather the captions given in the right hand column should be checked in each instance, as figure and table numbers may well change between the source and the corrected page proofs. If changes in the figure or table numbers in the text occur, appropriate changes should also be made on the corresponding page in the Study Guide.

1 Matter—Its Properties and Measurement

CHAPTER OBJECTIVES

0. Define and use the terms listed in "Some New Terms."

This should be an objective of every chapter. You should write each term and its definition on a 3 × 5 card. A very serious mistake is to copy definitions from the book. The definitions should be in *your* words so they make sense to *you*. After all, *you* will be the one using these terms in the future. Consider the term: "heterogeneous mixture." The definition given is that such a mixture separates "into physically distinct regions of differing properties." How might this be restated? Perhaps as: "A mixture that, by itself, separates into parts that I can easily tell apart." Now think of several examples. Fruit salad is one. You easily can distinguish the pears from the juice, the pineapple, and the grapes. Another example is Italian salad dressing that has been shaken. On standing, the oil separates from the vinegar and the herbs are clearly seen as separate pieces. Finally, use the term in a sentence. "Hash is a heterogeneous mixture because I can see two parts: meat and potatoes."

1. Use the terms *element, compound, homogeneous mixture,* and *heterogeneous mixture,* to describe common materials.

This is an extension of objective 0. Not only should you know definitions of terms, but you should be able to apply these definitions. Pick everyday items, classify them, and write your choices on paper. You should choose simple items: the graphite in your pencil, wood, a penny, a stone, aluminum foil, air, vinegar, water, milk, glass, and so on.

2. Write the names and chemical symbols of the more common elements, including the first forty.

This is a memorization task: to know the symbol that goes with the name and vice versa. "Common" means different things to different people. A reasonable goal would be those elements with atomic number from 1 (H) to 40 (Zr), along with Ag, Cd, Sn, Sb, Te, I, Xe, Cs, Ba, La, Ce, W, Pt, Au, Hg, Pb, Bi, Ac, and U. The atomic number is the whole number in each box in the periodic table, often given the symbol Z. This is a 3 × 5 card project with the name on one side and the symbol on the other.

3. Distinguish between physical and chemical properties and simple physical and chemical changes.

Choose a few simple items and write facts you know about each one. Then decide which property is chemical (depends on reacting the item with another or producing a new substance) and which is physical (does not so depend). For example wood is brown or tan (physical), breaks more easily with than across the grain (physical), floats in water (physical), burns in air (chemical), blackens in sulfuric acid (chemical), and is somewhat flexible (physical). Note also that chemical changes produce new substances while physical ones do not.

4. Describe the principal features of the scientific method and its limitations.

Start by writing a summary of the scientific method, using the terms *hypothesis, law, theory,* and *experiment.* Now construct an example of the scientific method. A fanciful one follows.

Law: Nordic children have fair skin, blond hair, and blue eyes.

Hypotheses: Trolls eat dark–haired, –eyed, or –skinned children (mythological).

The lack of sun prevents the skin, eyes, and hair of young children from "tanning" (environmental). Children inherit these traits from their parents (genetic).

Experiments: The environmental hypothesis can be disproved by moving Nordic children to sunnier climates. Their hair, eyes, and skin do not darken.

The principle known as "Occam's razor" (the explanation that makes the fewest assumptions is the best) is used to discard the mythological hypothesis. After all, trolls are imaginary—they never have been observed.

Just because the environmental and mythological hypotheses are invalid, does not prove the genetic one. Experiments are needed to test it, such as observing the children produced by one Nordic parent and a Mediterranean parent. But no amount of testing "proves" a hypothesis. A hypothesis always *may* fail a future experimental test. Then a new explanation for the law must be devised.

What are the limitations of the scientific method? One criticism is that it is too rigid. It does not permit conclusions to be reached from insufficient data—a process that often is a mark of genius. Can you think of others? You develop a better appreciation for any method or theory if you know what it can and what it cannot do.

4a. Know common units in the English system, and the relationships between them.

Common English units, their abbreviations, and the relationships between them follow. They should be memorized (a 3 × 5 card project) if you do not know them already.

Volume measure:	gallon (gal) = 4 quarts (qt)
	pint (pt) = 16 fluid ounces (fl oz) = 2 cups (c)
	quart (qt) = 2 pt = 32 fl oz
	tablespoon (T) = 3 teaspoons (tsp) = 1/2 fl oz
Linear measure:	yard (yd) = 3 feet (ft) = 36 inches (in.)
	mile (mi) = 1760 yd = 5280 ft
	foot = 12 in.
Mass measure:	pound (lb or #) = 16 ounces (oz)
	ton = short ton = 2000 lb

5. For the metric system, state the basic units of mass, length, and volume, and the common prefixes.

There are three units to be learned: gram (g) for mass, meter (m) for length, and liter (L) for volume. The following prefixes should be memorized (a 3 × 5 card project).

mega	(M)	1,000,000 or 10^{+6}	one million of
kilo	(k)	1000 or 10^{+3}	one thousand of
deci	(d)	1/10	one tenth of
centi	(c)	1/100 or 10^{-2}	one hundredth of
milli	(m)	1/1000 or 10^{-3}	one thousandth of
micro	(μ)	1/1,000,000 or 10^{-6}	one millionth of
nano	(n)	1/1,000,000,000 or 10^{-9}	one billionth of

You should also know how to combine the prefixes and the units and be able to write the abbreviation of the resulting unit. For example, mL is the abbreviation for milliliter, 1/1000 liter, and Mg is the abbreviation for megagram, 10^6 gram (in addition to being the chemical symbol for the element magnesium).

6. State the relationships between English and metric units.

This is a 3 × 5 card project. There is a vast number of relationships, a number that can be reduced to only three if you know and use the interrelationships within each system (objectives 1-4a and 1-5). These three can be one from each column below.

Mass	*Length*	*Volume*
kg = 2.20 lb	m = 39.37 in.	0.946 L = qt
453.6 g = lb	2.540 cm = in.	L = 1.057 qt
28.35 g = oz	30.48 cm = ft	29.6 mL = fl oz

7. Describe the relationship between SI and metric units.

At this point, you only need to be aware that SI uses the kilogram (1000 g), meter (100 cm), and cubic meter (m^3 = 1000 L) as basic units. Contrast these with the basic units of the metric system. The liter is being replaced in SI by the cubic decimeter (dm^3).

* **8. Determine the number of significant figures in a numerical calculation.**

* **9. Express the result of a calculation with the appropriate number of significant figures.**

Here is another 3 × 5 card project. Write down, in your own words, the five rules given *in the text* for determining significant figures and the rules for addition, subtraction, multiplication, and division. Write one rule per card and memorize them. You might want to add the rounding rule: "When rounding a number ending with 5, round up or down so that the digit before the 5 becomes or remains an even number." Practice your understanding of these rules by working through some of the drill problems. Rewrite the rules if necessary to make them more useful to you.

* **10. Express numbers in scientific notation.**

This is very closely related to being able to work with significant figures. A pair of rules might be the following.
a. A number in scientific notation is written as a many–digit number, with only one significant figure to the left of the decimal point. This many–digit number is multiplied by 10 to some power.
b. The power of 10 is determined by starting with the original number, the number being converted to scientific notation. Count the number of places that the decimal point must be moved in order to have only one significant figure to the left of the decimal point. For each place that the decimal point is moved left, the power of 10 is increased by 1. For each place that the decimal point is moved right, the power of 10 is decreased by 1.

EXAMPLE 1-1 Express each of the following numbers in scientific notation: (a) 1742, (b) 0.0008649, (c) 114.692×10^6.

(a) $1742 = 1.742 \times 10^3$ Since the decimal point was moved 3 places to the left to transform 1742 to 1.742, the power of 10 is 3.
(b) $0.0008649 = 8.649 \times 10^{-4}$ Since the decimal point was moved four places to the right to transform 0.0008649 to 8.649, the power of 10 is –4.
(c) $114.692 \times 10^6 = 1.14692 \times 10^8$ Since the decimal point was moved 2 places left to transform 114.692 to 1.14692, the power of 10 is 6 + 2 = 8.

* **11. Write a conversion factor from a relationship between two quantities, and use conversion factors to solve problems.**

Probably the most powerful problem–solving method you can learn is the conversion factor method, also known by other names such as the factor units method. The method is valuable not only in general chemistry but in any numerical problem–solving course. There are many ways to use the method. One of these follows.

(1) Identify what is wanted and what units it should have. Write this down.
(2) Identify what is given, what you start with. Write this down with its given units.
(3) Multiply what is given by one or more conversion factors until you have changed the units into those of the answer.

EXAMPLE 1-2 How many teaspoons are there in 5.00 gallons?

Apply the steps in the order given above.
(1) ? teaspoons (tsp) (2) 5.00 gallons (gal)
(3) $5.00\ \text{gal} \times \dfrac{4\ \text{qt}}{1\ \text{gal}} \times \dfrac{32\ \text{fl oz}}{1\ \text{qt}} \times \dfrac{3\ \text{tsp}}{0.5\ \text{fl oz}} = 3.84 \times 10^3\ \text{tsp}$

There is no requirement that you remember the factors from right to left. You might have remembered 3 tsp = 0.5 fl oz and have started with that, building the string of factors outward from the center.

EXAMPLE 1-3 How many gallons are in 1.00 cubic foot (ft^3)?

There is no simple relationship between gallons and cubic feet in the English system, but there is a relationship between volume and cubic length in the metric system (L = 1000 cm^3). Thus, the steps in the solution are as follows.

(1) ? gal (2) 1.00 ft^3

(3) $1.00\ ft^3 \times \left(\frac{12\ in.}{1\ ft}\right)^3 \times \left(\frac{2.54\ cm}{1\ in.}\right)^3 \times \frac{1\ L}{1000\ cm^3} \times \frac{1\ qt}{0.946\ L} \times \frac{1\ gal}{4\ qt} = 7.48\ gal$

In this example, we started in the center (L = 1000 cm^3) of the string of conversion factors and worked outward. Note also that two relationships (12 in. = ft and 2.54 cm = in.) are cubed to get the needed cubic units of length. Finally, recognize that it is essential that you memorize the relationships in the English system, those in the metric system, and those between the two systems. In this way you have a sufficient pool of relationships to use in constructing conversion factors.

We should note that there is another method for solving problems—ratios or proportions. To convert 5.00 in. to cm using this technique, one sets up the ratio

1.00 in. is to 2.54 cm as 5.00 in. is to ? cm

or 1.00 in. : 2.54 cm :: 5.00 in. : ? cm

or $\frac{1.00\ in.}{2.54\ cm} = \frac{5.00\ in.}{?\ cm}$

A little algebra gives the following equation.

$$?\ cm = 5.00\ in. \times \frac{2.54\ cm}{1.00\ in} = 12.7\ cm$$

which is the conversion factor method. Therefore, the two methods *are* equivalent. But the ratio technique is cumbersome (a new ratio is needed for each change of units), it does not allow you to work backward or from the middle in solving the problem, and it makes squaring or cubing units very difficult.

* **12. Express and use density in the form of conversion factors.**

Density is both a physical property of a substance and the means of interconverting mass and volume of that substance. The defining equation ($d = m/V$) has three variables: density, mass, and volume. There are only three general types of problems involving density, illustrated in the following three examples.

EXAMPLE 1-4 (Given mass and volume, determine density.) A 732 g object occupying 242 mL has what density? Simply apply the definition to determine density.

$$\text{density} = \frac{\text{mass}}{\text{volume}} = \frac{732\ g}{242\ mL} = 3.02\ g/mL$$

EXAMPLE 1-5 (Given volume and density, determine mass.) A 27.4 mL gold (19.3 g/mL) object has what mass? The conversion factor method is used here.

(1) ? g Au (2) 27.4 mL Au (3) $27.4\ mL \times \frac{19.3\ g\ Au}{1\ mL} = 529\ g\ Au$

EXAMPLE 1-6 (Given mass and density, determine volume.) A 75.2 g piece of zirconium (6.42 g/mL) has what volume? Again the conversion factor method is used.

(1) ? mL Zr (2) 75.2 g Zr (3) $75.2\ g\ Zr \times \frac{1\ mL\ Zr}{6.49\ g} = 11.6\ mL\ Zr$

* **13. Express and use percent composition in terms of conversion factors.**

Percent means parts per hundred. Thus, 40.0% C in acetic acid means 40.0 g C in 100.0 g of acetic acid. (40.0% C = 40.0 g C / 100.0 g acetic acid) Again, three types of problems can be expected. Can you identify all three types? One of them is illustrated in the following example.

EXAMPLE 1-7 What mass of acetic acid contains 247 g C?

We proceed via the conversion factor method.

(1) ? g acetic acid (2) 247 g C (3) $247\ g\ C \times \frac{100.0\ g\ acetic\ acid}{40.0\ g\ C} = 618\ g\ acetic\ acid$

* **14. Be able to convert between Fahrenheit and Celsius temperatures.**

You need to memorize and be able to use two relationships.

$°C = (°F - 32) \times \frac{5}{9}$ and $°F = (°C \times \frac{9}{5}) + 32$

EXAMPLE 1-8 Perform the following temperature conversions.

(a) 112°F = ? °C; (b) 25°C = ? °F; (c) –30°C = ? °F

(a) $°C = (112°F - 32)(\frac{5}{9}) = 80.(\frac{5}{9}) = 44°C$

(b) $°F = (25°C \times \frac{9}{5}) + 32 = 45 + 32 = 77°F$

(c) $°F = (-30°C \times \frac{9}{5}) + 32 = -54 + 32 = -22°F$.

The following list may help you develop a useful "feel" for various Celsius temperatures. Notice that a 5°C temperature rise corresponds to a 9°F increase.

0°C = 32°F	water freezes, winter day
5°C = 41°F	early spring morning
10°C = 50°F	early spring afternoon
15°C = 59°F	mid spring afternoon
20°C = 68°F	heated home in winter
25°C = 77°F	cooled home in summer
30°C = 86°F	summer afternoon
35°C = 95°F	hot summer day
37°C = 98.6°C	normal body temperature

* **15. Solve algebraic equations that arise in the course of working chemistry problems.**

Solving an algebraic equation generally means obtaining a new equation, with the symbol for one variable isolated on one side (generally the left–hand side) and the remainder of the equation on the other side. We then say that we have "solved the equation" for the variable that stands alone. (Of course, that variable does not appear on the other side of the equation.) A very powerful method of solving equations is to perform the same operation on both sides of the equation. This is done several times with different operations, until the symbol for one variable *is* isolated. Keep in mind the following list of things that you can do to both sides of any equation.

a. Add the same thing to both sides.
b. Subtract the same thing from both sides.
c. Multiply both sides by the same thing.
d. Divide both sides by the same thing.
e. Raise both sides to the same power. This includes the power –1, so both sides can be inverted or flipped over. It also includes fractional powers, so that you can take the square root, for instance, of both sides.

Often solving an equation will require a lot of trial and error. *Do not become discouraged!* The reason why some people can solve algebraic equations more quickly than you is that they have had more practice. Study the example and practice on the drill problems.

EXAMPLE 1-9 Solve the following equation for P.

$(P + an^2/V^2)(V - nb) = nRT.$

First divide each side by $(V - nb)$.

$(P + an^2/V^2) = nRT/(V - nb)$

Then subtract an^2/V^2 from each side.

$P = [nRT/(V - nb)] - (an^2/V^2)$

DRILL PROBLEMS

8. (1) How many significant figures are in each of the following numbers?

A. 1837 B. 3.14145×10^4 C. 6005 D. 0.08206
E. 0.000014 F. 302400 G. 632 H. 8.732
I. 14.163000 J. 14.000 K. 19.7324 L. 302400.0
M. 149356 N. 205.8 O. 0.0019872 P. 8.7300
Q. 1900 R. 20000 S. 426.1 T. 1200.43
U. 1900.00 V. 0.00743 W. 6000 X. 60.0

(2) Round off each of the following numbers to four significant figures.

Y. 6.16782 Z. 213.25 Γ. 1200.43 Δ. 3135.69
Θ. 6.19648 Λ. 14.163000 Ξ. 0.0022457 Π. 152.00
Σ. 0.0019872 Υ. 302400 Φ. 14.16300 Ψ. 3.14145×10^4

9. (1) Multiplication and division, following the rules for significant figures.

A. 1.86/3.14 B. $(6.6262 \times 10^{-27})(2567)$
C. (37.2)(1.5) D. (200)(87.45)
E. (998)(32.157)/36 F. 4.51545/0.15
G. $(5.00 \times 10^2)/36.72$ H. 374.1/1800
I. 31.11/2.04 J. $(1.40 \times 10^8)(37.842)/147.3562$

(2) Addition and subtraction, following the rules for significant figures.

K. 104 + 37.2 – 18.57 L. 87.6 – 0.005
M. 6.23 + 915 – 1012.7 N. 87.9 + 11.3 + 9.6
O. 36.516 + 0.00258 – 32.157 P. $6.47 \times 10^2 + 4.2 \times 10^1 + 6.8$
Q. 4.30 + 29.1 + 100.3452 R. 204.5 – 96.5 – 32.1

(3) Combined operations, following the rules for significant figures.

S. (94.3)(12) – 7.62 + 300.0 T. $(5.19 \times 10^{-2} + 1.83)(2.19 \times 10^2)$
U. $0.318 + 1.6 \times 10^{-2}$ (6.40/12.1) –2.19 V. (3.18)(2.4)/1.92) – 0.17
W. 1.32 + 0.06 + (28/43)

10. Express each number in scientific notation.

A. 0.00374 B. 1200 C. 4063.89
D. 175.1×10^3 E. 6460.4×10^7 F. 0.06627×10^{-25}
G. 9475×10^{-6} H. 0.00374×10^7 I. 0.0000142×10^1
J. 17645 K. 212,000,000 L. 0.00266
M. 843×10^5 N. 94.00×10^6 O. 0.0004963×10^{-4}
P. 843.214×10^{-3} Q. 0.00212×10^{12} R. 0.00839×10^2
S. 894.13 T. 0.000 000 831 4 U. 49000.6
V. 9204×10^5 W. 87012×10^{23} X. 0.001413×10^{-4}
Y. 17645×10^{-15} Z. 0.00266×10^3 Γ. $0.000\,000\,831\,4 \times 10^2$

11. Perform the following conversions of units.

(1) One or two conversion factors per problem.

A. 7.3 ft = ___in. B. 6.40 qt = ___fl oz C. 12750 yd = ___mi
D. 37.54 oz = ___lb E. 16.54 cm = ___mm F. 0.0374 m = ___µm
G. 146 cm^3 = ___L H. 37.54 L = ___m^3 I. 27.2 kg = ___mg
J. 206 g = ___lb K. 250 cm^3 = ___fl oz L. 7.30 ft = ___m
M. 12.0 fl oz = ___L N. 187 lb = ___kg O. 8.5 in. = ___cm
P. 1.00 gal = ___L Q. 1.00 cup = ___cm^3 R. 0.500 oz = ___g
S. 127 lb = ___kg T. 22.4 L = ___gal U. 28.3 g = ___oz
V. 100 cm^3 = ___fl oz W. 15.0 m = ___yd X. 5.15 ft = ___cm

(2) With units raised to higher powers than unity.

Y. 100 yd^3 = ___m^3 Z. 1.00 m^3 = ___yd^3 Γ. 1.00 ft^3 = ___$in.^3$
Δ. 1.00 $in.^2$ = ___cm^2 Θ. 1.00 ft^2 = ___cm^2 Λ. 1.00 yd^3 = ___gal

Ξ. 55.0 gal = ___ ft^3 Π. 250 $in.^3$ = ___ L

12. Perform the following conversions. You may need the following densities: copper, 8.92 g/cm^3; ethanol, 0.789 g/cm^3; gold, 19.3 g/cm^3; iron, 7.86 g/cm^3; lead, 11.34 g/cm^3; table salt, 2.17 g/cm^3; water, 1.00 g/cm^3.
A. 5.00 cm^3 water = ___ g = ___ oz
B. 5.00 cm^3 gold = ___ g = ___ oz
C. 5.00 cm^3 Pb = ___ g = ___ oz
D. 250 cm^3 ethanol = ___ g
E. 250 cm^3 water = ___ g
F. 50.0 g Pb = ___ cm^3
G. 50.0 g Cu = ___ cm^3
H. 1.00 lb Fe = ___ cm^3
I. 1.00 lb Au = ___ cm^3
J. 4.00 lb salt = ___ cm^3
K. 1.00 qt water = ___ lb
L. 1.00 qt ethanol = ___ g
M. 5.00 lb Fe = ___ $in.^3$
N. 5.00 lb Au = ___ $in.^3$

13. Answer the following problems based on percent composition.
A. A copper penny has a mass of 3.015 g and contains 95.0% Cu. What is the mass of copper present?
B. An automobile weighs 1.00 ton and contains 13% Al and 75% Fe. What is the mass of Al present. What is the volume of Fe if Fe has a density of 7.86 g/cm^3?
C. Air contains 78% N_2 and 21% O_2. A house 40.0 ft × 30.0 ft × 14.0 ft contains how many liters of nitrogen? How many liters of air must one have in order to have 680 L of oxygen?
D. 50.0 lb of tin is sufficient to produce how many pounds of soft solder, which contains 70.0% Sn and 30.0% Pb?
E. Liquid bleach contains 5.0% sodium hypochlorite (NaOCl). How many pounds of bleach can be made from 1.00 lb of NaOCl? How many gallons is this, if bleach has a density of 1.04 g/mL?
F. Soft whiskey is 42.0% ethanol and 58.0% water by volume. It has a density of 0.933 g/mL. How many quarts of soft whiskey contain 1.00 lb of ethanol?
G. A certain vinegar is 5.00% acetic acid by mass and has a density of 1.007 g/mL. What volume in liters of this vinegar contains 100.0 g of acetic acid?
H. A solution of 36.00% (by mass) sulfuric acid in water has a density of 1.271 g/mL. How many grams of sulfuric acid are needed to make 5.00 L of this solution?
I. A copper penny has a mass of 3.015 g and contains 95.0% Cu. If Cu costs 80¢ per pound, what is the value of the copper in 100 pennies? How many pennies contain one dollar's worth of copper?
J. The compound silver nitrate contains 63.5% Ag. If Ag costs $12.00/oz, what is the value of the silver in 125 g of silver nitrate? What mass of silver nitrate in grams contains a dollar's worth of Ag?

14. Convert the Celsius temperatures to Fahrenheit, and the Fahrenheit temperatures to Celsius.
A. 32°F = ___ °C
B. 100°F = ___ °C
C. 98.6°C = ___ °F
D. 25°F = ___ °C
E. 100°C = ___ °F
F. 4.0°C = ___ °F
G. −14°F = ___ °C
H. 500°F = ___ °C
I. −60°C = ___ °F
J. −273.15°C = ___ °F

15. Solve the following equations for the indicated variable.
A. $69.72 = 68.9257\,x + 70.9249\,(1 - x)$ for x
B. $V = (4/3)\pi r^3$ for r
C. $14 = x/(1 - x)$ for x
D $13.6\,h = 1.15\,H$ for h
E $pvt = PVT$ for T
F $PV = nRT$ for R
G $PV = nRT$ for the ratio n/V
H $PV = mRT/\mathfrak{M}$ for $\mathfrak{M}$
I $PV = mRT/\mathfrak{M}$ for the ratio m/V
J $(1/2)\mathfrak{M}u^2 = (3/2)RT$ for u
K $\nu = Rc[(1/4) - (1/n^2)]$ for n
L $\lambda = h/m\nu$ for ν
M $16 = x^2/(1 - x)^2$ for x

QUIZZES

Each of the following short quizzes is designed to test your mastery of the objectives of this chapter. Limit yourself to 20 minutes for each one. The questions are multiple choice; choose the best answer for each question.

Quiz A

1. A 6.0 in.–ruler is how long in centimeters? (a) 6.0/2.54; (b) 6.0/39.37; (c) (6.0)(36)(39.37); (d) (6.0)(2.54); (e) none of these.
2. What is the mass in pounds of a small child who weighs 23.0 kg? (a) (23.0)(1000/454); (b) (23.0/2.20); (c) (23.0/1000)454 (d) (23.0)(1000)(454);(e) none of these.
3. 742 mL is how many fluid ounces? (a) 742/946; (b) 742/[(946)(32)]; (c) (742/32)(946); (d) (742)(946)(32); (e) none of these.
4. An object weighs 946 g and occupies 241 mL. What is its density? (a) 241/946; (b) 241 + 946; (c) (241)(946); (d) 1/[(241)(946)]; (e) none of these.
5. Italian salad dressing is an example of (a) an element; (b) a compound; (c) a homogeneous mixture; (d) a heterogeneous mixture; (e) none of these.
6. The density of margarine in 0.96 g/mL. What is the mass of 250 mL? (a) 1/[(0.96)(250)]; (b) 0.96/250; (c) 250/0.96; (d) (250)(0.96); (e) none of these.
7. 88 ft/s is how fast in meters/min? (a) (88/12)(2.54/100)(60); (b) (88)(12)(2.54)(60/100); (c) (88)(12)(100/2.54)(60); (d) (88/12)(100/2.54)/60; (e) none of these.
8. Matter which cannot be further separated by physical means but which can be separated by chemical means is called a(n) (a) element; (b) compound; (c) homogeneous mixture; (d) heterogeneous mixture; (e) none of these.
9. A concise summary of many observations is called a(n) (a) law; (b) theory; (c) experiment; (d) hypothesis; (e) none of these.
10. 32°C is equivalent to (a) 0°F; (b) 58°F; (c) 18°F; (d) 50°F; (e) none of these.
11. Phosphorus, sodium, and helium, respectively, have the symbols (a) Ph, Na, H; (b) P, So, He; (c) P, Na, He; (d) Ps, Na, He; (e) none of these.
12. Which answer has the correct number of significant figures? (a) (14.7 + 24.312) × 87.27 = 3402 (b) (58 + 18 + 51)/3.000 = 36; (c) [97.2/(114 – 37)] = 1.26; (d) (1.172 – 0.4963)(4.194) = 2.83; (e) none of these.

Quiz B

1. The density of gold is 19.3 g/mL. What is the volume in mL of 231 g of Au? (a) 231/19.3; (b) 19.3/231; (c) (19.3)(231); (d) 1[(19.3)(231)]; (e) none of these.
2. What is the volume in milliliters of one cup? (a) (1/2)(946/2); (b) (1)(2)(946)(2); (c) (1)(2)(946/2); (d) (1/2)(946) (e) none of these.
3. A 6-ft person is how many centimeters tall? (a) (6)(12)(2.54); (b) (6/12)(2.54); (c) (6)(2.54/12); (d) (6)(12/2.54); (e) none of these.
4. One acre equals how many square meters (640 acres = 1 mi^2)? (a) (1.0)(640)/[(1760)(36)(0.0254)]: (b) (1.0/640)/[(1760)(36)(0.0254)]; (c) $(1.0/640)/[(1760)(36)(0.0254)]^2$; (d) $(1.0)[(1760)(36)(0.0254)]^2$; (e) none of these.
5. Matter that cannot be further broken down by chemical means is called a(n) (a) element; (b) compound; (c) homogeneous mixture; (d) heterogeneous mixture; (e) none of these.
6. "A less dense solid will flo..t in a more dense liquid" is an example of a(n) (a) hypothesis; (b) theory; (c) experiment; (d) law; (e) none of these.
7. An object weighs 314 g and occupies 432 mL. What is its density? (a) (314)(432); (b) 432/314; (c) 314/432; (d) 1/[(314)(432)]; (e) none of these.
8. 200 lb equals how many kilograms? (a) (200/454)(1000) (b) (200)(2.20); (c) (200)(454/1000); (d) (200)(1000/454); (e) none of these.
9. Sound is an example of a(n) (a) element; (b) compound; (c) homogeneous mixture; (d) heterogeneous mixture; (e) none of these;
10. Carbon, fluorine, and lithium, respectively, have the symbols (a) C, F, Li; (b) Ca, Fl; Li; (c) C, Fl, Li;

(d) Cr, F, Li; (e) none of these.

11. Which answer has the correct number of significant figures? (a) (37 + 54 + 42)/3.000 = 44.3; (b) (726 + 835)/24.0 = 65.04; (c) (202 + 74)/73.1 = 3.8; (d) (1.43 + 7.26 + 9.1)/14.23 = 1.250; (e) none of these.
12. 432°C is equivalent to (a) 720°F; (b) 272°F; (c) 835°F; (d) 810°F; (e) none of these.

Quiz C

1. A procedure designed to test the truth or the validity of an explanation about many observations is called a(n) (a) law; (b) theory; (c) experiment; (d) hypothesis; (e) none of these.
2. An object has a mass of 764 g and occupies 265 mL. What is its density? (a) 265/764; (b) 764/265; (c) (265)(764); (d) 1/[(764)(265)]; (e) none of these.
3. What is the volume in quarts of a 250-mL beaker? (a) (250)(1000)(1.057); (b) (250/1000)(0.946); (c) (250/1000)(1.057); (d) (250)(1000/1.057); (e) none of these.
4. What is the mass in kilograms of an object that weighs 29.0 lb? (a) (29.0/454)(1000); (b) (29.0)(2.20); (c) (29.0)(1000/454); (d) (29.0)(454/1000); (e) none of these.
5. If a room is 10.0 m wide, its width in feet is (a) (10.0)(100/2.54); (b) (10.0/1000)(2.54); (c) (10.0)(1200/2.54); (d) (10.0)(100)(2.54)(12); (e) none of these.
6. The density of copper is 8.92 g/cm^3. What is the volume of 651 g? (a) (651)(8.92); (b) 8.92/651; (c) 651/8.92; (d) 651 + 892; (e) none of these.
7. 14.7 lb/in.^{2} equals how many kg/cm^2? (a) 14.7(0.454)(2.54)^2; (b) (14.7)(454)(2.54)^2; (c) (14.7/2.20)(2.54); (d) (14.7)(2.20)/(2.54)^2; (e) none of these.
8. Red food coloring is an example of a(n) (a) element; (b) compound; (c) homogeneous mixture; (d) heterogeneous mixture; (e) none of these.
9. Matter which can be separated by physical means but *not* by chemical means is called a(n) (a) element; (b) compound; (c) homogeneous mixture; (d) heterogeneous mixture; (e) none of these.
10. Bromine, magnesium, and beryllium, respectively, have the symbols (a) B, Ma, Be; (b) Br, Mg, B; (c) Br, Mn, Be; (d) Br, Mn, B; (e) none of these.
11. Which answer has the correct number of significant figures? (a) (107.4 – 17.2)/9.164 = 9.843; (b) 9.843 × (44.7 – 65.2) = 1082; (c) (1.172 + 0.4963)(3.145) = 5.25; (d) (651 + 8.92)/14.174 = 46.558; (e) none of these.
12. 450°F is equivalent to (a) 250°C; (b) 232 °C; (c) 218°C; (d) 778°C; (e) none of these.

Quiz D

1. An object weighs 876 g and occupies 923 mL. What is its density? (a) 923/876; (b) 876/923; (c) (923)(876); (d) 923 + 876; (e) none of these.
2. An automobile weighs 2750 pounds. What is its mass in Mg? (a) (2750/2.2)(1000); (b) (2750)(2.20)(1000) (c) (2750)(2.2)/1000; (d) (2750/2.20)/1000; (e) none of these.
3. 100 yards is how long in meters (36 in. = 1 yd)? (a) (100)(36)(2.54); (b) (100)(36)(2.54/100) (c) (100/36)(100/2.54); (d) (100/36)(2.54/100) (e) none of these.
4. What is the volume in milliliters of 27.5 fl oz (32 fl oz = 1 qt)? (a) (37.5)(32/946); (b) (37.5/32)(946); (c) (37.5)(32)(946); (d) 37.5/[(32)(946)]; (e) none of these.
5. 13.6 g/cm^3 equals how many lb/qt? (a) (13.6/454)(1000/0.946); (b) (13.6)(454/1000)(1.057); (c) (13.6)(454)(1.057/1000); (d) (13.6/454)(1000/1.057); (e) none of these.
6. Gold is a(n) (a) element; (b) compound; (c) homogeneous mixture; (d) heterogeneous mixture (e) none of these.
7. Matter which can be further separated by physical means but which appear uniform to the eye is called a(n) (a) element; (b) compound; (c) homogeneous mixture; (d) heterogeneous mixture; (e) none of these.
8. An explanation of many observations, which has been proven true by many tests is called a(n) (a) law; (b) theory; (c) experiment; (d) hypothesis; (e) none of these.
9. The density of silver is 10.5 g/mL. What is the volume of 475 g? (a) 475/10.5; (b) 10.5/475; (c) 1/[(10.5)(475)]; (d) (475)(10.5) (e) none of these.
10. Hydrogen, potassium, and chlorine, respectively, have the symbols (a) H, Po, Cl; (b) Hy, K, C; (c) H, K, Cl; (d) H, P, Ch; (e) none of these.
11. Which answer has the correct number of significant figures? (a) (10.46 + 1.7394)/12.2 = 1.000;

(b) (14.234 + 0.005169)/0.1243 = 115; (c) (0.723 + 9.746)/15.493 = 0.6757;
(d) (300.4 + 0.00216)/9.18745 = 32.697; (e) none of these.

12. 98.6°F is equivalent to (a) 54.7°C; (b) 37.0°C; (c) 120°C; (d) 86.8°C; (e) none of these.

SAMPLE TEST

This test is designed to take 30 minutes. Treat it as if it were an in–class examination. Only allow yourself a periodic table for reference and a calculator. Express your answers with the correct number of significant figures and with proper units.

1. A 10-karat gold ring contains 41.7% gold and weighs 10.72 g. If gold sells for $652.50/oz, what is the value of the gold in the ring?
2. In the production of ammonia, the following processes occur.
   a. *Air* is liquefied at low temperature and high *pressure*.
   b. The *temperature* of the *liquid air* is raised gradually until the oxygen boils off. Essentially pure *liquid nitrogen* remains.
   c. *Natural gas* is reacted with *steam* to produce *carbon dioxide* and *hydrogen*. The proportions of the products vary with the composition of the natural gas.
   d. Hydrogen and nitrogen (both gases) are combined at high temperatures and pressures. They react to form *ammonia gas*. Some unreacted hydrogen and nitrogen remain.
   e. The gases are cooled until the ammonia liquefies. Then the still gaseous hydrogen and nitrogen are recirculated to react again.

   Each of the steps described above involves a physical change or a chemical reaction. Indicate which occurs in each step and briefly explain why you chose as you did.

   Each of the *italicized* words or phrases in the descriptions above refers to an element, a compound, a mixture, or none of these. Identify each one and explain your choice.
3. Perform the following conversions
   a. 12.5 lb/in^2 = ____oz/cm^2 = ____kg/m^2
   b. 77.4°F = ____°C
   c. 5.2 m^3 = ____L = ____qt

2 Atoms and the Atomic Theory

CHAPTER OBJECTIVES

* **1. State and apply the laws of conservation of mass, definite composition, and multiple proportions.**

The first goal is to memorize your versions of these three laws: (a) mass is neither created nor destroyed; (b) a compound has the same composition, no matter what its source; and (c) the ratio of the masses of one element combined (in different compounds) with a constant mass of another is a small whole number. Application involves recognizing when the laws are violated and when they are obeyed. For example (for the law of conservation of mass), 4.00 g of methane plus 16.00 g of oxygen produce 11.00 g of carbon dioxide and what mass of water? If water and carbon dioxide are the only products, 9.00 g of water must be produced. This is because there are 20.00 g of reactants (methane and oxygen) and there must be the same mass of products.

EXAMPLE 2-1 (The law of constant composition) A sample of methane taken from a mine (mine gas) was found to contain only C and H with 3.00 g C for every 1.00 g of H. How much hydrogen should be present in 100.0 g of methane taken from a natural gas well?

Note that 4.00 g of the mine gas contains 3.00 g C and 1.00 g H. These proportions should be the same for the natural gas methane. Thus, we can use the conversion factor method.

(1) ? g H (2) 100.0 g methane (3) $100.0 \text{ g methane} \times \frac{1.00 \text{ g H}}{4.00 \text{ g methane}} = 25.0 \text{ g H}$

EXAMPLE 2-2 (The law of multiple proportions) Four compounds contain hydrogen, chlorine, and oxygen in the following proportions. Demonstrate that these compounds obey the law of multiple proportions.

Hypocholorous acid:	% H = 1.91%	% Cl = 67.7%	% O = 30.5%
Chlorous acid:	= 1.46%	= 51.9%	= 46.8%
Chloric acid:	= 1.18%	= 42.0%	= 56.8%
Perchloric acid:	= 0.99%	= 35.3%	= 63.6%

(We discuss how to obtain these percentages in objective 3-5.) Consider a fixed mass, such as 10.0 g, of chlorine (any fixed mass will do). The combining masses of oxygen in order are

$10.0 \text{ g Cl} \times (30.5 \text{ g O}/67.7 \text{ g Cl}) = 4.51 \text{ g O}$ $\quad 4.51/4.51 = 1.00$
$10.0 \text{ g Cl} \times (46.8 \text{ g O}/51.9 \text{ g Cl}) = 9.02 \text{ g O}$ $\quad 9.02/4.51 = 2.00$
$10.0 \text{ g Cl} \times (56.8 \text{ g O}/42.0 \text{ g Cl}) = 13.5 \text{ g O}$ $\quad 13.5/4.51 = 3.00$
$10.0 \text{ g Cl} \times (63.6 \text{ g O}/35.3 \text{ g Cl}) = 18.0 \text{ g O}$ $\quad 18.0/4.51 = 4.00$

If each combining mass is divided by 4.51 g, they are seen to be related as 1:2:3:4. In other words, they are related as small whole numbers.

Notice that percents are converted to masses in grams here, as was done in objective 1-13. These are the masses present in 100.0 g of each compound. Although there is nothing special about 100.0 g, choosing that mass makes the calculation simply a matter of replacing "%" with "g."

2. State the basic assumptions of Dalton's atomic theory.

You should write down and memorize your versions of Dalton's assumptions. Shortened versions are (a) Elements are composed of indestructible atoms. (b) Atoms of a given element are alike; atoms differ between elements. (c) Atoms unite in small whole number ratios to form compounds.

3. List some of the characteristic properties of cathode rays and of canal (anode) rays.

A brief list of cathode ray properties is (a) straight–line travel; (b) travel from cathode when current flows; (c) deflected as if negatively charged; (d) properties are independent of current source, tube material, cathode material, and the gas that filled the tube; (e) invisible, but give off light when they strike the glass; and (f) have mass.

Canal rays (positive rays or anode rays) form when electrons (cathode rays) knock electrons out of neutral atoms. Canal rays are thus positive ions or cations (cat'-eye-uns). Their properties are (a) straight–line travel; (b) travel toward cathode when current flows; (c) deflected as if positively charged; (d) properties depend on the gas that filled the tube but *not* on other factors; and (e) have mass but are invisible until they strike certain other objects. Canal rays are not fundamental particles since their properties are not the same under all circumstances but depend on the gas that originally filled the tube. The properties of these two types of rays lead to the idea that the atom is composed of positive and negative parts, and the negative parts (the electrons) are easily removed. You should memorize your versions of the properties of these two types of rays.

4. Describe the production of x rays, the phenomenon of radioactivity, and the characteristics of α, β, and γ radiation.

X rays are given off when cathode rays (electrons) are stopped by an object. X rays penetrate matter easily and have no mass or charge. The degree to which they penetrate matter depends on both their energy and the type of matter, principally its density.

Becquerel discovered radioactivity by chance. He observed that certain materials produce penetrating radiation that is unaffected by attempts to enhance or diminish it. Radioactivity is a characteristic property of certain elements. α (alpha) particles are $^4He^{2+}$ cations. They are massive, have poor penetrating power, but high ionizing power. β (beta) particles are electrons. They have moderate penetrating and ionizing powers. When α or β particles are emitted by an atom, the atom changes to one of a different element (as is described in objective 25-1). γ (gamma) rays are similar to x rays, but of very high energy. They have a high penetrating power, but poor ionizing power. Except for carrying away energy, they leave unchanged the atom from which they are emitted.

5. Describe Thompson's *e/m* experiment, Millikan's oil drop experiment (to measure the charge on the electron), and Rutherford's gold–foil experiment (to establish the existence of the atomic nucleus).

In Thompson's experiment, a beam of electrons of known energy is deflected by a magnetic field—the beam curves. The radius of the curve depends on: the energy of the electrons, the mass of the electrons, and the strength of the magnetic field. The energy of the electrons in turn depends on their charge. Thompson knew the magnetic field strength but neither the mass nor the charge of the electron. To make the experiment clearer, consider an experiment very similar in principle. Attach a weight to a spring and, holding the other end of the spring, whirl the weight about you over your head. The distance between your head and the mass (the radius of the curve) depends on three things, as shown in Table 2-1.

TABLE 2-1 Relation Between Thomson's Experiment and Its Analog

Depends on	To make radius of curve greater	Analog in Thomson's experiment
mass of weight	increase weight	electron mass
speed of weight	increase speed	electron energy
stiffness of spring	use a slacker spring	magnetic field strength

Millikan's experiment is designed to measure the fundamental unit of charge. Oil drops are charged with x rays and allowed to fall between two charged plates. The charge on each drop can be found by varying the voltage between the plates until the electrical force felt by the charged drop just balances the force of gravity. Then the drop stands still. The charges on the drops always are found to be an exact multiple of 1.602×10^{-19} coulomb.

Geiger and Marsden beamed α particles at thin gold foil. Most of the particles passed straight through, but some were deflected, a few by great angles. The experiment is similar to shooting BBs through a hole into a hot air balloon, in the center of which is a metal object. If enough BBs are shot, the size and shape of the metal object can be deduced from the pattern produced by the BBs striking the inner surface of the balloon. Rutherford interpreted these data as evidence for a nuclear atom. *Your* descriptions of these three experiments should be committed to memory.

6. State the features of Rutherford's nuclear atom and how it differs from Thompson's model of the atom.

The Rutherford model of the atom consists of a very small nucleus (about 10^{-13} cm in diameter), which contains all of the positive charge and more than 99.9% of the mass, surrounded by a tenuous cloud of electrons. The electrons possess all of the negative charge, less than 0.1% of the mass, and fill most of the space in the atom (which is about 10^{-8} cm in diameter).

Thompson's model lacks the nuclear kernel. The negative electrons are embedded in a uniform sphere of positive charge, rather like plums in a pudding. The model is sometimes called the "plum pudding" model.

7. Perform simple calculations involving the masses and charges of the proton, neutron, and electron.

These calculations generally involve determining charge–to–mass ratios. In order to do them successfully, you need to have memorized the charge and mass of the proton (charge = +1, mass = 1.0073 amu), neutron (charge = 0, mass = 1.0087 amu), and electron (charge = –1, mass = 0.00055 amu). The unit of charge is that of the electron: 1.602×10^{-19} coulomb. The atomic mass unit, amu (sometimes called the dalton, d), equals 1.66×10^{-24} g. Note that the mass of the electron is about 1/1830 that of the proton.

* **8. List the numbers of protons, neutrons, and electrons present in atoms and ions, using the symbolism $^{A}_{Z}X$.**

The complete symbol for an atom or ion consists of the elemental symbol surrounded by subscripts and superscripts. Two of the four "corners" thus created are used for an atomic species and the third for an ionic species. (The fourth is used when writing formulas of compounds.)

(1) The leading superscript (upper left) is the mass number. This is also the number of nucleons; a nucleon is a proton or a neutron. Older books may show the mass number as a trailing superscript (upper right), as this was correct until about 1960. Thus ^{235}U was U^{235}. Often one encounters U-235 instead of ^{235}U, especially in newspapers and magazines.

(2) The leading subscript (lower left) is the atomic number. It actually is unnecessary since the elemental symbol determines the atomic number, but often it is included for emphasis.

(3) The trailing superscript (upper right) is the charge or the number of protons (atomic number) minus the number of electrons. The sign (+ or –) always is included. The number is zero for a neutral atom, but the zero is written only for emphasis.

For example, $^{19}F^{-}$ has an atomic number of 9 (9 protons), a mass number of 19 (19 nucleons = 9 protons + x neutrons and thus x = 10 neutrons), and a charge of –1 (9 protons – y electrons = –1 where y = 10 electrons).

* **9. Describe how atomic mass ratios are determined by mass spectrometry and use these ratios to determine relative atomic masses.**

The mass spectrometer is an elegant version of Thomson's experiment for determining the charge–to–mass ratio of electrons, as described in objective 2-5. This apparatus was modified to determine the properties of canal rays by the addition of a velocity selector (the electric and magnetic fields just after the collimator) to ensure that all ions are of the same velocity. The instrument (shown in *Figure 2-12 in the text*) determines ratios of atomic masses from which one computes relative atomic masses. For example, if naturally occurring carbon were analyzed, the ratio of the mass of carbon-13 to that of carbon-12 would be 1.083613. The relative mass of carbon-13 then is

$$\text{mass of } ^{13}C = 1.083613 \times 12.00000 = 13.00335$$

* **10. Calculate the atomic weight of an element from the known masses and relative abundances of its naturally occurring isotopes.**

The atomic weight of an element is the decimal number appearing in the element's block in the periodic table. For example, the atomic weight of carbon is 12.01115. This atomic weight represents an average atomic mass. It can be determined from the combining weights of elements in chemical reactions. These combining weights represent the average mass of many atoms of the same element. The individual atoms are of different masses; all of the stable isotopes are present.

An example should make the concept and the calculation of atomic weights clearer. Suppose the average weight of an athlete in an 80-member track team is needed. The team consists of 52 runners weighing 67 kg each and 28 shot–putters weighing 110 kg each. The coach obtains the total team weight as follows.

52 runners × 67 kg/runner =	3.5 Mg
28 shot–putters × 110 kg/shot–putter =	3.1 Mg
Total =	6.6 Mg

(Remember: Mg = 10^6 g = 1000 kg.) and the average weight of an athlete as

$$\frac{3.5\text{ Mg} + 3.1\text{ Mg}}{80\text{ athetes}} = \frac{6.6\text{ Mg}}{80\text{ athletes}} = 82\text{ kg/athlete}$$

Any of the following expressions also may be used to calculate the average weight.

$$\frac{[52\text{ runners} \times 67\text{ kg/runner}] + [28\text{ shot–putters} \times 110\text{ kg/shot–putter}]}{80\text{ athletes}} = 82\text{ kg}$$

$$\left(\frac{52\text{ runners}}{80\text{ athletes}} \times \frac{67\text{ kg}}{\text{runner}}\right) + \left(\frac{28\text{ shot–putters}}{80\text{ athletes}} \times \frac{110\text{ kg}}{\text{shot–putter}}\right) = 82\text{ kg}$$

$$\left(\frac{0.65\text{ runner}}{\text{athlete}} \times \frac{67\text{ kg}}{\text{runner}}\right) + \left(\frac{0.35\text{ shot–putter}}{\text{athlete}} \times \frac{110\text{ kg}}{\text{shot–putter}}\right) = 82\text{ kg}$$

The last expression is of the same type as that which is used to determine average atomic weight. We use the fraction of atoms (athletes) of each isotope (runner or shot–putter) times the mass of that isotope to find the average mass of all the atoms (the average mass of an athlete).

* **11. Obtain and use relationships between the mole, the Avogadro constant (Avogadro's number), and the molar mass of an element.**

The atomic weight is the nonintegral number appearing with each element in the periodic table. In Chapter 2 we thought of the atomic weight as the average weight of an atom of a particular element. However, it is very inconvenient (some might say impossible) to work with individual atoms. Therefore, chemists have chosen a larger quantity—Avogadro's number—of atoms to work with. If the atomic weight of an element is expressed in grams, that mass—the molar mass—will contain Avogadro's number of atoms of that element.

Avogadro's number is truly huge—6.022×10^{23}. The number is so large that if one million workers attempted to move 6.022×10^{23} grains of sand with shovels, the task would require 450 years, assuming 10 shovelfuls per worker each minute and 250 million grains of sand (about 15 pounds) per shovelful. Thus, instead of counting atoms, we weigh out an amount of material and use the mass of Avogadro's number of atoms (the molar mass) to determine how many atoms are present.

Usually we are not concerned with how many atoms are present. We simply need to know that there are enough atoms of each element for a particular chemical reaction. Hence, we speak of moles of atoms. A mole of an element contains Avogadro's number of atoms. To make the concept clearer, imagine that a packer of machinery must put 500 bolts into each box. For each bolt there must be one nut and two washers, for a total of 500 nuts and 1000 washers. Counting these out would take quite a long time. Knowing the masses of 100 pieces of each type (100 bolts = 99.1 g, 100 nuts = 42.7 g, 100 washers = 41.2 g) makes the job faster and just as accurate as counting. In fact, the instructions to the packer might be phrased: "Put in five dops of bolts (one dop weighs 99.1 g)." ["Dop" is a made–up unit that contains 100 pieces.] Now the packer is not concerned with the number of pieces in a "dop." In similar fashion, chemists are not concerned very often with the number of atoms in a mole. Moles are convenient amounts of material to use—much more so than individual atoms.

The abbreviation for mole is mol. Other symbols such as m, $\overline{\text{m}}$, and M are incorrect and should not be used as abbreviations for mole, for they will cause confusion.

One mole of an element (a) contains Avogadro's number of atoms; (b) contains 6.022×10^{23} atoms; and (c) has a mass in grams equal to the molar mass of the element. Each of these relations can be used to construct a conversion factor. Sample factors are

$$\frac{6.022 \times 10^{23} \text{ atoms}}{\text{mole}} \qquad \frac{6.022 \times 10^{23} \text{ Mg atoms}}{24.305 \text{ g Mg}} \qquad \frac{24.305 \text{ g Mg}}{\text{mol Mg}}$$

EXAMPLE 2-3 87.4 g Mn is equivalent to how many moles of Mn?

(1) ? moles Mn (2) 87.4 g Mn (3) $87.4 \text{ g Mn} \times \frac{\text{mol Mn}}{54.9 \text{ g Mn}} = 1.59 \text{ mol Mn}$

EXAMPLE 2-4 47.2 g Cs contains how many atoms?

(1) ? Cs atoms (2) 47.2 g Cs (3) $47.2 \text{ g Cs} \times \frac{6.022 \times 10^{23} \text{ Cs atoms}}{132.9 \text{ Cs}} = 2.14 \times 10^{23} \text{ Cs atoms}$

EXAMPLE 2-5 How many moles are present in 5.00×10^9 atoms? (Notice that we do not have to specify which element since there is the same number of atoms—Avogadro's number—in a mole of any element.)

(1) ? moles (2) 5.00×10^9 atoms (3) $5.00 \times 10^9 \text{ atoms} \times \frac{\text{mole of atoms}}{6.022 \times 10^{23} \text{ atoms}} = 8.30 \times 10^{-15} \text{ mol}$

DRILL PROBLEMS

1. (1) Law of Conservation of Mass.

A. If 3.41 g of hydrogen sulfide is combined with 4.80 g of oxygen, what should the products weigh? If the products are 21.9% water, what mass of water is produced?

B. 7.95 g of copper(II) oxide is combined with 0.20 g of hydrogen. What is the mass of products? The products are 77.9% Cu. What mass of Cu is produced? What is the percentage of Cu in copper(II) oxide?

C. Sugar, a carbohydrate, can be decomposed by heating into only carbon and water vapor. If 18.0 g of sugar is heated, the carbon that remains has a mass of 7.2 g. What mass of water vapor is driven off? What is the percentage of water in sugar?

D. When reacted with carbon at high temperatures, iron(III) oxide produces solely molten iron and carbon dioxide. 3.192 g of iron(III) oxide and 0.36 g of carbon produce 1.32 g of carbon dioxide. What mass of molten iron is produced? What is the percent of iron in iron(III) oxide? What mass of carbon dioxide is produced when 1.00 g of molten iron is made? How many tons of carbon dioxide are produced when 1.00 ton of molten iron is made?

E. When 9.56 g of copper(II) sulfide is heated in the presence of 4.80 g of oxygen, the products contain 55.4% copper(II) oxide. What is the total mass of products? What mass of copper(II) oxide is produced?

F. When 18.0 g of wood is burned in oxygen, the products weigh 37.2 g and contain 29.0% water. What mass of water is produced? What mass of oxygen is needed? What mass of oxygen is needed to burn 4.00 lb of wood?

(2) Law of Constant Composition.

G. Sulfur dioxide, produced by reacting copper(II) sulfide with oxygen, contains 50.1% S. If 57.2 g of sulfur dioxide is made by *burning sulfur in oxygen,* what mass of sulfur was burned? What mass of oxygen was used?

H. Carbon dioxide, produced by reacting iron(III) oxide with carbon, contains 27.3% carbon. What mass of carbon must be *burned in oxygen* to produce 100.0 g of carbon dioxide? What mass of oxygen is needed?

I. Water, produced by reacting hydrogen with oxygen, contains 11.1% H. What mass of hydrogen must be *reacted with hot copper(II) oxide* to produce 75.0 g of water? What mass of oxygen is present in 75.0 g of water?

J. When calcium carbonate (limestone) is heated, the calcium oxide (lime) produced contains 28.5% oxygen. What mass of oxygen is needed to produce 74.2 g of lime when *Ca is reacted directly with oxygen* (lime is the only product)? What mass of Ca is needed?

K. When pure table salt is obtained from sea water, it contains 39.3% Na. 14.0 g of chlorine *reacted with excess Na* produces table salt as the only product. What mass of table salt is produced? What mass of Na will react?

L. Ammonia, produced by reacting magnesium nitride with water, contains 17.6% H. *The direct reaction of hydrogen and nitrogen* produces only ammonia. What mass of ammonia can be made from 32.1 g of hydrogen? What mass of nitrogen is needed?

(3) Law of Multiple Proportions

Each problem below gives the percents of elements in several compounds. Demonstrate that the law of multiple proportions is valid in each case.

M.	carbon monoxide	42.9% C	57.1% O	
	carbon suboxide	52.9% C	47.1% O	
	carbon dioxide	27.3% C	72.7% O	
N.	potassium oxide	83.0% K	17.0% O	
	potassium peroxide	71.0% K	29.0% O	
	potassium trioxide	62.0% K	38.0% O	
O.	acetylene	7.7% H	92.3% C	
	methane	25.0% H	75.0% C	
	ethane	20.0% H	80.0% C	
	ethene	14.3% H	85.7% C	
P.	chlorine monoxide	18.4% O	81.6% Cl	
	chlorine dioxide	31.1% O	68.9% Cl	
	chlorine heptoxide	61.2% O	38.8% Cl	
Q.	sulfur monochloride	52.5% Cl	47.4% S	
	sulfur dichloride	68.9% Cl	31.1% S	
	sulfur tetrachloride	81.6% Cl	18.4% S	
R.	ammonia	82.4% N	17.6% H	
	hydrazine	87.5% N	12.5% H	
	hydrazoic acid	97.7% N	2.3% H	
	diimine	93.3% N	6.7% H	
S.	orthophosphoric acid	3.1% H	31.6% P	65.3% O
	phosphorous acid	3.7% H	37.8% P	58.5% O
	hypophosphorous acid	4.5% H	47.0% P	48.5% O
T.	sulfurous acid	2.4% H	39.1% S	58.5% O
	sulfuric acid	2.0% H	32.7% S	65.2% O
	peroxomonosulfuric acid	1.8% H	28.1% S	71.0% O

8. Fill in the blanks in the table below. Some of the combinations are impossible. If so, write IMPOSSIBLE across the line. The first line is completed as an example.

	Symbol	*Ionic charge*	*Mass number*	*Atomic number*	*No. of electrons*	*No. of neutrons*
	$^{122}Sn^{2+}$	+2	122	50	48	72
A.	______	______	81	35	36	______
B.	______	+3	59	______	______	32
C.	$^{43}Ca^{2+}$	______	______	______	______	______
D.	______	−3	______	7	______	8
E.	______	0	20	10	______	______
F.	______	______	127	53	54	______
G.	______	+1	23	______	______	12
H.	$^{192}Os^{4+}$	______	______	______	______	______
I.	______	+3	26	______	______	30
J.	______	+2	52	24	______	______
K.	______	______	60	27	25	______
L.	______	−1	17	______	______	8
M.	$^{80}Se^{2-}$	______	______	______	______	______
N.	______	−4	14	______	______	8
O.	______	+4	118	50	______	______

9. Fill in the blanks in the table below. The first line is completed as an example.

	Mass of first isotope	*Ratio of masses:, 1st to 2nd isotope*	*Mass of second isotope*
	13.00335	1.083613	12.00000
A.	190.9609	0.7745905	______
B.	77.9204	______	69.9243

C.	70.9249	______	14.0067
D.	68.9257	______	18.9984
E.	45.9537	0.283281	______
F.	______	4.80336	52.9407
G.	______	4.98818	49.9461
H.	______	1.50951	105.907
I.	63.9280	______	6.01888
J.	53.9389	1.29638	______
K.	77.9204	1.43613	______
L.	57.9353	______	150.923

10. Fill in the blanks in the table below. Make use of the periodic table to fill in the first two columns. Assume that each of these elements has only two isotopes.

	Element		*Isotope A*		*Isotope B*	
	Symbol	*Atomic weight*	*Atomic weight*	*%Abundance*	*Atomic weight*	*% Abundance*
A.	______	______	6.015	7.42	7.016	______
B.	B	______	10.013	______	11.009	______
C.	______	12.011	12.000	98.89	______	______
D.	______	______	19.992	90.2	21.991	______
E.	______	63.54	62.930	______	64.928	______
F.	Cl	______	34.969	75.53	______	______
G.	______	39.102	38.964	93.1	______	______
H.	______	______	68.926	60.4	70.935	______
I.	Br	______	78.918	______	80.916	______
J.	Rb	______	84.912	72.15	______	______

11. Fill in the blanks in the table below. The first line is done as an example. Use the periodic table when needed.

	Element	*Atomic weight*	*Mass of element,g*	*Amount, mol*	*No. of atoms*
	Cs	132.9	47.2	0.355	2.14×10^{23}
A.	H	______	______	0.412	______
B.	______	32.1	87.4	______	______
C.	O	______	______	1.59	______
D.	C	______	______	______	3.02×10^{23}
E.	______	35.5	______	0.0427	______
F.	______	14.0	2.14	______	______
G.	Mg	______	99.1	______	______
H.	______	31.0	______	______	3.01×10^{24}
I.	Br	______	______	6.12×10^{-5}	______
J.	K	______	______	______	8.69×10^{19}
K.	Be	______	______	5.02×10^{10}	______
L.	______	19.0	0.000302	______	______

QUIZZES (20 minutes each) Select the best answer to each question.

Quiz A

1. There are two stable isotopes of chlorine: Cl-35 = 34.9689 amu (75.5 3%) *and* Cl-37 = 36.9659 amu. What is the atomic weight of chlorine? (a) (0.7553)(34.9689) + (0.2447)(36.9659); (b) (34.9689 + 36.9659)/2; (c) (0.7553)(36.9659) + (0.2447)(34.9689); (d) (75.53)(34.9689) + (24.47)(36.9659); (e) none of these.
2. $^{35}Cl^-$, ^{40}Ar, and $^{39}K^+$ all have the same (a) mass number; (b) atomic number; (c) number of electrons; (d) number of neutrons; (e) none of these.
3. When 16.0 g of methane is burned in 64.0 g of oxygen, 44.0 g of carbon dioxide and 36.0 g of water are produced. This is an example of the law of (a) conservation of mass; (b) multiple proportions; (c) constant composition; (d) relative proportions; (e) none of these.

4. Isotopes always have the same (a) mass number; (b) atomic number; (c) number of electrons; (d) number of neutrons; (e) none of these.
5. A subatomic particle that has a very small mass and a negative charge is called (a) a proton; (b) a neutron; (c) an electron; (d) an isotope; (e) none of these.
6. According to experimental evidence, cathode rays do *not* possess which property? (a) negative charge; (b) fundamental particle; (c) radioactivity; (d) mass; (e) they possess all these properties.
7. The combination that led to the determination of the charge on the electron was (a) Dalton–oil drop; (b) Rutherford–gold foil; (c) Millikan–oil drop; (d) Thomson–magnetic field; (e) Becquerel–gold foil.
8. Dalton's assumptions included (a) isotopes; (b) the nuclear atom; (c) indestructible atoms; (d) electrons; (e) none of these.
9. 1.60×10^{22} Cu atoms (a) is 0.0531 mol Cu; (b) is 1.69 g Cu; (c) contains 2.32×10^{23} protons; (d) contains 4.64×10^{23} neutrons; (e) none of these.

Quiz B

1. There are two stable isotopes of silver: Ag-107 = 106.9041 amu (51.82%) *and* Ag-108 = 107.9047 amu. What is the atomic weight of silver? (a) (0.5182)(108.9047) + (0.4818)(106.9041); (b) (108.9047/0.5182) + (106.9041/0.4818); (c) (0.5182)(106.9041) + (0.4818)(108.9047); (d) (106.9041 + 108.9047)/2; (e) none of these.
2. When two carbon oxide samples are analyzed, one contains 36.0 g of C and 32.0 g of O, while the other contains 12.0 g of C *in* 28.0 g of the oxide. This is an example of the law of (a) conservation of mass; (b) multiple proportions; (c) constant composition; (d) fixed percentages; (e) none of these.
3. $^{19}F^-$, ^{20}Ne, and $^{24}Mg^{2+}$ all have the same (a) mass number; (b) atomic number; (c) number of electrons; (d) neutron number; (e) none of these.
4. A subatomic particle that has the same mass as the hydrogen nucleus and a positive charge is called (a) a proton; (b) a neutron; (c) an electron; (d) an isotope; (e) none of these;
5. Two atoms with the same number of neutrons are called (a) protons; (b) daltons; (c) electrons; (d) isotopes; (e) none of these.
6. A fundamental particle of the atom is the (a) alpha particle; (b) beta particle; (c) x ray; (d) canal ray; (e) none of these.
7. The combination that lead to the proposal that the atom has a nucleus was (a) Rutherford–oil drop; (b) Dalton–plum pudding; (c) Thomson–curved electron beam; (d) Millikan–oil drop; (e) Becquerel–radioactivity.
8. Canal rays are the same as (a) positive ions; (b) negative ions; (c) electrons; (d) neutrons; (e) none of these.
9. There is 4.024 mol of protons (a) in 0.1006 mol of Zr; (b) in 0.9177 g of Zr; (c) associated with 2.423×10^{24} electrons in Zr atoms; (d) associated with 9.693×10^{25} neutrons in Zr atoms; (e) none of these.

Quiz C

1. There are two stable isotopes of gallium: Ga-69 = 68.9257 amu (60.4%) *and* Ga-71 = 70.9249 amu. What is the atomic weight of gallium? (a) (0.604)(68.9257) + (0.396)(70.9249); (b) (68.9257 + 70.9294)/2; (c) (0.396)(68.9257) + (0.396)(70.9249); (d) (60.4)(68.9257) + (39.6)(70.9249); (e) none of these.
2. Both HCl and HBr are strong acids with similar properties. It also is true that most chlorine compounds are similar chemically to the corresponding bromine compounds. This is an example of the law of (a) conservation of mass; (b) multiple proportions; (c) constant composition; (d) average reactivity; (e) none of these.
3. The mass of an element, expressed in a scale where carbon-12 has a mass of 12.00000, is called (a) mass number; (b) atomic weight; (c) natural abundance; (d) isotope mass; (e) none of these.
4. An entity that contains only protons and neutrons is called (a) an atom; (b) a molecule; (c) an ion; (d) a nucleus; (e) none of these.
5. $^{40}Ca^{2+}$, $^{39}K^+$, and $^{41}Sc^+$ all have the same (a) number of electrons; (b) mass number; (c) atomic number; (d) number of neutrons; (e) none of these.
6. All of the following scientists contributed to determining the *structure* of the atom *except* (a) Thomson; (b) Rutherford; (c) Millikan; (d) Dalton; (e) Becquerel.
7. Which of the following is *not* a property of canal rays? (a) straight–line travel; (b) fundamental particle; (c) mass; (d) have positive charge; (e) depend on the gas in the tube.
8. Which of the following is *not* a fundamental particle of the atom? (a) proton; (b) neutron; (c) beta particle; (d) alpha particle; (e) none, all are fundamental particles.

9. 91.84 g of Ti (as) is 4.175 mol of Ti; (b) contains 5.531 × 10^{25} Ti atoms; (c) contains 1.155 × 10^{24} protons; (d) contains 2.541 × 10^{25} electrons; (e) none of these.

Quiz D

1. There are two stable isotopes of europium. The element has an atomic weight of 151.96, and one isotope has a mass of 150.9196 amu and a percent abundance of 47.820%. What is the isotopic mass of the other isotope in amu? (a) 152.92; (b) 153.00; (c) 153.09; (d) 149.97; (c) none of these within 0.02 amu.
2. When decomposed chemically, a 73.0 g sample of HCl produced 71.0 g Cl_2 and 2.0 g H_2, while a 34.0 g sample of H_2S produces 32.0 g S and 2.0 g H_2. This is an example of the law of (a) conservation of mass; (b) multiple proportions; (c) constant composition (d) hydrogen conservation; (e) none of these.
3. A subatomic particle that has about the same mass as the hydrogen atom and a negative charge is called (a) a proton; (b) a neutron; (c) an electron; (d) an isotope; (e) none of these.
4. ^{104}Pd, ^{105}Pd, and ^{108}Pd all have the same (a) mass number; (b) atomic number; (c) number of electrons; (d) number of neutrons; (e) none of these.
5. "Compounds are composed of atoms combined in small whole number ratios and all atoms of the same element have the same weight" is an example of (a) a law; (b) a theory; (c) an experiment; (d) an observation; (e) a hypothesis.
6. Choose the incorrect pair. (a) alpha–radioactivity; (b) Rutherford–nuclear atom; (c) Millikan–electric charge; (d) Dalton–isotope; (e) cathode rays–electrons.
7. Which of the following is a form of radioactivity under a different name? (a) proton; (b) neutron; (c) isotope; (d) hydrogen nucleus; (e) none of these.
8. The mass of an atom is largely determined by the number of (a) protons; (b) neutrons; (c) electrons; (d) isotopes; (e) nucleons.
9. 1.774 mol of Ca (a) contains 2.100 × 10^{25} Ca atoms; (b) weighs 35.48 g; (c) contains 70.96 mol of protons; (d) contains 1.068 × 10^{24} protons; (e) none of these.

SAMPLE TEST (30 minutes)

1. An attempt was made to determine the atomic weight of element X. X forms a compound with oxygen that contains 46.7% X and has the formula XO. Oxygen's atomic weight is 16.00 amu. What is the atomic weight of X?
2. Match the terms in the left–hand column one–for–one with those in the right–hand column by writing the correct letter in each blank.

a. same atomic number	____ Millikan
b. alpha particle	____ Rutherford
c. proton number	____ Thomson
d. atomic theory	____ isotope
e. x ray	____ atomic number
f. electron charge	____ mass number
g. nuclear atom	____ Dalton
h. proton	____ fundamental particle
i. isotope	____ radioactivity
j. nucleon number	____ cathode tube emission

3. A certain element contains one atom of mass 10.013 amu for every four atoms of mass 11.009 amu. Compute the atomic weight of this element.

3 Chemical Compounds

CHAPTER OBJECTIVES

1. Distinguish between a mole of atoms and a mole of molecules.

A molecule is simply a cluster of atoms bound together. In our nut–bolt–washer analogy of objective 2-11, a bolt with two washers on it and the nut screwed on is similar to a molecule (Fig. 3-1). If these parts were preassembled, the packers would weigh out "dops" of assemblies (224.2 g/"dop"). Similarly, chemists find it convenient to work with moles of molecules, since molecules are "preassembled" groups of atoms.

Notice that some elements do not exist as individual atoms or as a large mass of somewhat attracted atoms, but rather as distinct molecules. You should firmly memorize the following forms: H_2, F_2, Cl_2, Br_2, I_2, O_2, S_8, N_2, and P_4.

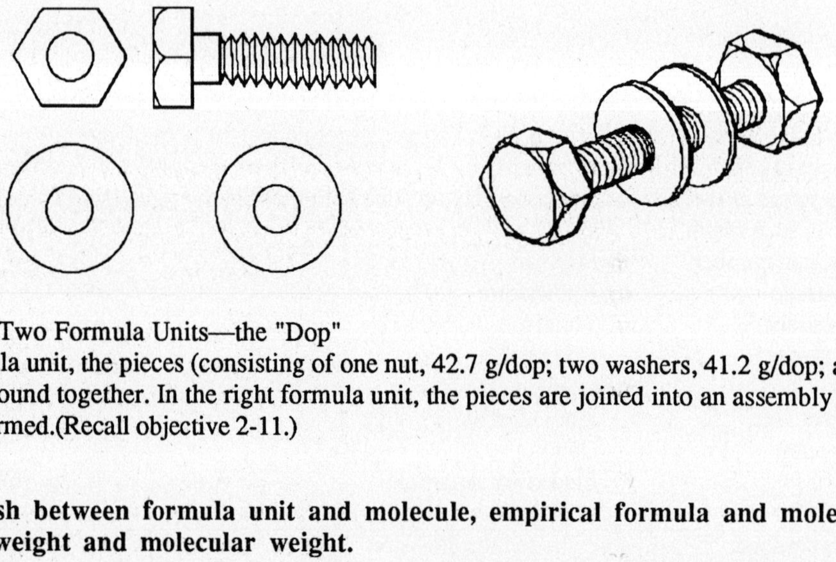

FIGURE 3-1 Two Formula Units—the "Dop"
In the left formula unit, the pieces (consisting of one nut, 42.7 g/dop; two washers, 41.2 g/dop; and one bolt, 99.1 g/dop) are not bound together. In the right formula unit, the pieces are joined into an assembly (224.2 g/dop)—a "molecule" is formed.(Recall objective 2-11.)

* 2. Distinguish between formula unit and molecule, empirical formula and molecular formula, and formula weight and molecular weight.

As we have noted, a molecule is a group of atoms bound together. The atoms in a formula unit are not necessarily bound together. In our nut–bolt–washer analogy, a formula unit consists of a nut, a bolt, and two washers, whether or not they are bound together. The grouping of 1 nut + 1 bolt + 2 washers is the smallest collection of pieces that contains the pieces in the same ratio they are in when packed for shipment. Likewise, the formula unit is a small collection of atoms that contains the atoms in the same ratio they have in the compound.

A molecular formula counts the actual number of atoms of each type in the compound. Thus, the molecular formula of hydrogen peroxide is H_2O_2, and each molecule contains four atoms: two H and two O. An empirical formula expresses the *smallest* combining ratio of the atoms in the compound. For hydrogen peroxide the empirical formula is HO. To change the molecular formula into an empirical one, divide the subscripts by the largest whole number that will go evenly into all of them. In the case of H_2O_2 this number is 2. Some examples follow.

formula is HO. To change the molecular formula into an empirical one, divide the subscripts by the largest whole number that will go evenly into all of them. In the case of H_2O_2 this number is 2. Some examples follow.

Compound name	*Molecular formula*	*Divide by*	*Empirical formula*
hydrazine	N_2H_4	2	NH_2
propene	C_3H_6	3	CH_2
diborane	B_2H_6	2	BH_3
acetic acid	$C_2H_4O_2$	2	CH_2O

Just as the atomic weight in grams is the mass of a mole of atoms, so the formula weight in grams is the mass of a mole of formula units, and the molecular weight in grams is the mass of a mole of molecules. It is proper to refer to any one of these three masses as the molar mass ($\mathcal{M}$). To determine the molar mass (sometimes called the mole weight), one starts with the chemical formula of the compound. Each subscript is interpreted as the number of moles of each element in a mole of the compound. Thus, a mole of N_2O_5 contains two moles of nitrogen and five moles of oxygen. Remember that a mole of an element has a mass equal to the atomic weight of that element in grams.

EXAMPLE 3-1 Determine the molar mass of CsF.

$$1 \text{ mol Cs} \times \frac{132.9 \text{ g Cs}}{\text{mol Cs}} = 132.9 \text{ g Cs}$$

$$1 \text{ mol F} \times \frac{19.0 \text{ g F}}{\text{mol F}} = \underline{19.0 \text{ g F}}$$

$$151.9 \text{ g/mol CsF}$$

EXAMPLE 3-2 Determine the molar mass of $HC_2H_3O_2$.

$$4 \text{ mol H} \times 1.01 \text{ g H/mol H} = 4.04 \text{ g H}$$
$$2 \text{ mol C} \times 12.0 \text{ g C/mol C} = 24.0 \text{ g C}$$
$$2 \text{ mol O} \times 16.0 \text{ g O/mol O} = \underline{32.0 \text{ g O}}$$
$$60.0 \text{ g/mol } HC_2H_3O_2$$

EXAMPLE 3-3 Determine the molar mass of $Al_2(SO_4)_3$.

A mole of the compound contains two moles of aluminum ions and three moles of sulfate ions. Since *each* moles of sulfate ions contains one mole of S and four moles of O, three moles of sulfate ions contain three moles of S and twelve moles of O (3 moles of sulfate ions × 4 moles O/ mol sulfate ions). Notice in the calculation below that *atomic*, rather than *ionic* masses are used, since there are as many electrons "missing" from the cations as are present in excess among the anions.

$$2 \text{ mol Al} \times 27.0 \text{ g Al/mol Al} = 54.0 \text{ g Al}$$
$$3 \text{ mol S} \times 32.1 \text{ g S/mol S} = 96.3 \text{ g S}$$
$$12 \text{ mol O} \times 16.00 \text{ g O/mol O} = \underline{192.0 \text{ g O}}$$
$$342.3 \text{ g/mol } Al_2(SO_4)_3$$

*** 3. Calculate the number of atoms, ions, formula units, or molecules in a substance from a given mass or vice versa.**

Once we can determine molar masses (objective 3-2), we can find the amount in moles of a given mass of substance. We must know something about the substance before we can state whether or not these are moles of molecules. Only covalent compounds exist as molecules. All binary (two–element) compounds of two nonmetals are covalent. (See objective 3-12 for definitions of nonmetals and metals.) Thus HCl, H_2O, NH_3, and PCl_5 are covalent compounds and composed of molecules.

All binary compounds composed of a metal and a nonmetal are ionic. Thus NaCl, $FeCl_3$, and K_2S are ionic and do not contain molecules, but rather consist of ions. A group of ions identical in number and type to that given by the formula is a formula unit and not a molecule. A Na^+ cation and a Cl^- anion compose the formula unit of NaCl, while the formula unit of $FeCl_3$ consists of one Fe^{3+} cation and three Cl^- anions.

EXAMPLE 3-4 11.34 g of In_2S_3 contains (a) how many moles of the compound? (b) how many formula units of the compound? (c) how many molecules? (d) how many In^{3+} ions?

The molar mass of In_2S_3 is determined first.

2 mol In × 114.8 g In/mol In = 229.6 g In

3 mol S × 32.1 g S/mol S = 96.3 g S

325.9 g/mol In_2S_3

The various parts of the problem can be solved with the conversion factor method.

(a) (1) ? mol In_2S_3 (2) 11.34 g In_2S_3

$$(3)\ 11.35\text{ g } In_2S_3 \times \frac{\text{mol } In_2S_3}{325.9\text{ g } In_2S_3} = 0.03480\text{ mol } In_2S_3$$

(b) (1) ? formula units In_2S_3 (2) 0.03480 mol In_2S_3

$$(3)\ 0.03480\text{ mol } In_2S_3 \times \frac{6.022 \times 10^{23}\text{ f.u. } In_2S_3}{\text{mol } In_2S_3} = 2.096 \times 10^{22}\ In_2S_3\text{ f. u.}$$

(c) There are no molecules, since In_2S_3 is not a covalent compound.

(d) (1) ? In^{3+} ions (2) 11.34 g In_2S_3

$$(3)\ 11.34\text{ g } In_2S_3 \times \frac{\text{mol } In_2S_3}{325.9\text{ g } In_2S_3} \times \frac{6.022 \times 10^{23}\text{ f.u. } In_2S_3}{\text{mol } In_2S_3} \times \frac{2\ In^{3+}\text{ ions}}{\text{f.u. } In_2S_3}$$

$$= 4.191 \times 10^{22}\ In^{3+}\text{ ions}$$

* 4. Use chemical formulas as a source of conversion factors for stoichiometric calculations.

A chemical formula is not only "a recipe for something you wouldn't want to eat," as a sixth grader said, but also contains many relationships. $CuSO_4{\cdot}5H_2O$ contains 1 copper(II) ion, 1 sulfate ion, and 5 water molecules *or* 1 Cu atom, 1 S atom, 9 O atoms, and 10 H atoms. A mole of $CuSO_4{\cdot}5H_2O$ contains 1 mol of copper(II) ions, 1 mol of sulfate ions, and 5 mols of water molecules *or* 1 mol Cu atoms, 1 mol S atoms, 9 mol S atoms, and 10 mol H atoms. We already have used these relationships in obtaining the molar mass of a compound. The molar mass of $CuSO_4{\cdot}5H_2O$ is 249.6 g/mol. We can use the above relationships in other ways, as Example 3-5 illustrates.

EXAMPLE 3-5 How many moles of oxygen atoms are in 29.95 g $CuSO_4{\cdot}5H_2O$?

$$29.95\text{ g cmpd} \times \frac{\text{mol cmpd}}{249.6\text{ g cmpd}} \times \frac{9\text{ mol O atoms}}{\text{mol cmpd}} = 10.8\text{ mol O atoms}$$

* 5. Use a compound's formula to determine its percent composition.

The percent of one element in a compound is given by the following relationship. Both masses are determined in the course of computing the molar mass of the compound.

$$\frac{\text{mass of element}}{\text{mass of compound}} \times 100 = \text{percent of element}$$

EXAMPLE 3-6 What is the percent H in $(NH_4)_3PO_4$?

First compute the molar mass.

3 mol N × 14.0 g N/mol N = 42.0 g N

12 mol H × 1.01 g H/mol H = 12.1 g H

1 mol P × 31.0 g P/mol P = 31.0 g P

4 mol O × 16.0 g O/mol O= 64.0 g O

149.1 g/mol $(NH_4)_3PO_4$

Then use the masses computed in the course of determining the molar mass to calculate the percent by mass of the desired element, hydrogen.

$$\%H = \frac{12.1\text{ g H}}{149.1\text{ g } (NH_4)_3PO_4} \times 100 = 8.12\%\text{ H}$$

* 6. Use the percent composition of a compound to determine its empirical formula.

The percent composition of a compound gives the combining *masses* of the elements. The empirical formula gives the combining number of atoms or *moles of atoms* in the compound. Obtaining the empirical formula involves first determining how many moles of each element are present. These numbers of moles then are reduced to the simplest integers.

EXAMPLE 3-7 A compound contains 27.3% C and 72.7% O. What is its empirical formula?

First find the amount in moles of each element.

$$27.3 \text{ g C} \times \frac{\text{mol C}}{12.0 \text{ g C}} = 2.28 \text{ mol C}$$

$$72.7 \text{ g O} \times \frac{\text{mol O}}{16.0 \text{ g O}} = 4.54 \text{ mol O}$$

This gives $C_{2.28}H_{4.54}$ as the formula, but we wish to have whole numbers as subscripts. One way to achieve this is to divide all mole numbers by the smallest. Thus 2.20 mol C and 4.54 mol O becomes (2.28/2.28 =) 1.00 mol C and (4.54/2.28 =) 1.99 mol O. As a result we obtain the formula $CO_{1.99}$ or CO_2.

EXAMPLE 3-8 A.compound contains 47.0% C, 5.9% H, and 47.0% O. What is its empirical formula?

First find the amount of each element in moles contained in 100.0 g of the compound.

$$47.0 \text{ g C} \times \frac{\text{mol C}}{12.0 \text{ g}} = 3.92 \text{ mol C} \qquad \frac{3.92}{2.94} = 1.34 \text{ mol C}$$

$$5.9 \text{ g H} \times \frac{\text{mol H}}{1.0 \text{ g}} = 5.9 \text{ mol H} \quad \textit{then} \quad \frac{5.9}{2.94} = 2.0 \text{ mol H}$$

$$47.0 \text{ g O} \times \frac{\text{mol O}}{16.0 \text{ g}} = 2.94 \text{ mol O} \qquad \frac{2.94}{2.94} = 1.00 \text{ mol O}$$

We round any of the final mole numbers that is within 0.05 of a whole number. Here, we recognize that 0.34 is about 1/3. Thus, to obtain whole numbers, we multiply each final mole number by 3.

$1.34 \text{ mol C} \times 3 = 4.02 \text{ mol C}$ (rounded to 4.00 mol C)

$2.0 \text{ mol H} \times 3 = 6.0 \text{ mol H}$

$1.00 \text{ mol O} \times 3 = 3.00 \text{ mol O}$

The empirical formula is $C_4H_6O_3$

* **7. Use the masses of CO_2 and H_2O produced from the complete combustion of a compound containing only C, H, and O to determine the percent composition of the compound.**

The combustion analysis of a compound containing only C, H, and O requires the chemist to weigh the compound, and then weigh the two products of combustion: carbon dioxide and water. *All* of the carbon in the compound is in the carbon dioxide and all of the compound's hydrogen is in the water. The masses of these two elements in the products are computed first. The mass of oxygen then is computed by subtraction, if necessary. The percent of oxygen is determined by difference because the oxygen in the products is from two sources: the compound that was burned and the oxygen in which it was burned. This, of course, is not the case with either the carbon or the hydrogen.

EXAMPLE 3-9 4.27 g of a C–H–O compound produces 10.78 g CO_2 and 1.89 g H_2O when burned in air. What is the percent composition of the compound?

The masses of C and H are found as follows.

$$\text{mass C} = 10.78 \text{ g } CO_2 \times \frac{\text{mol } CO_2}{44.01 \text{ g } CO_2} \times \frac{\text{mol C}}{\text{mol } CO_2} \times \frac{12.01 \text{ g C}}{\text{mol C}} = 2.942 \text{ g C}$$

$$\text{mass H} = 1.89 \text{ g } H_2O \times \frac{\text{mol } H_2O}{18.0 \text{ g } H_2O} \times \frac{2 \text{ mol H}}{\text{mol } H_2O} \times \frac{1.01 \text{ g H}}{\text{mol H}} = \underline{0.212 \text{ g H}}$$

3.154 g C, H

The mass of oxygen then is found by difference,

$$\text{mass O} = 4.27 \text{ g compound} - 3.154 \text{ g C \& H} = 1.12 \text{ g O}$$

and the percent composition for each element then is determined.

$$\%O = \frac{1.12 \text{ g O}}{4.27 \text{ g cmpd}} \times 100 = 26.2\% \text{ O}$$

$$\%H = \frac{0.212 \text{ g H}}{4.27 \text{ g cmpd}} \times 100 = 4.96\% \text{ H}$$

$$\%C = \frac{2.94 \text{ g O}}{4.27 \text{ g cmpd}} \times 100 = 68.9\% \text{ C}$$

*** 8. Use the results of precipitation analysis to determine the percent of a compound in a mixture or of an element in a compound.**

A very powerful technique is that of isolating one element or compound within a solid that settles to the bottom of a liquid solution. Such a solid is called a precipitate. Precipitates are separated from the solution, dried to a constant composition, and weighed. The following example gives a sample of the type of calculations required.

EXAMPLE 3-10 4.21 g of a silver–containing alloy is dissolved in nitric acid. Sodium chloride solution is added in slight excess and a silver chloride precipitate is obtained. The solid is washed, dried, and weighed. Its mass is 15.32 g. Determine the percent of silver in the alloy.

$$15.32 \text{ g AgCl} \times \frac{\text{mol AgCl}}{143.32 \text{ g}} \times \frac{\text{mol Ag}}{\text{mol AgCl}} \times \frac{107.87 \text{ g Ag}}{\text{mol Ag}} = 11.53 \text{ g Ag}$$

$$\%\text{Ag in alloy} = \frac{\text{mass Ag}}{\text{mass of alloy}} \times 100 = \frac{11.53 \text{ g Ag}}{14.21 \text{ g alloy}} \times 100 = 81.14\% \text{ Ag}$$

*** 9. Apply precipitation analysis to determine atomic weights.**

Dalton's most serious handicap in determining atomic weight was that he did not know the formulas of the compounds that he analyzed. But if the chemical formula is known, the determination of atomic weight is straightforward.

EXAMPLE 3-11 The sulfur present in a 11.78-g sample of M_2S is precipitated as 17.27 g of $BaSO_4$. What is the atomic weight of element M?

We first compute the mass of S in the $BaSO_4$ (and thus also the mass of S in M_2S).

$$\text{g S} = 17.27 \text{ g BaSO}_4 \times \frac{\text{mol BaSO}_4}{233.40 \text{ g BaSO}_4} \times \frac{\text{mol S}}{\text{mol BaSO}_4} \times \frac{32.06 \text{ g S}}{\text{mol S}} = 2.372 \text{ g}$$

Hence, in the 11.78 g of M_2S, there are 11.78 g M_2S – 2.372 g S = 9.41 g M. Now we can determine the number of moles of S.

$$\text{mol S} = 2.372 \text{ g S} \times \frac{\text{mol S}}{32.06 \text{ g S}} = 0.07399 \text{ mol S}$$

and, since 1 mol S unites with 2 mol M in M_2S, the 11.78 g of M_2S contains

$$0.07399 \text{ mol S} \times \frac{2\text{mol M}}{\text{mol S}} = 0.1480 \text{ mol M}$$

Now we compute the molar mass of element M.

$$\text{molar mass} = \frac{\text{mass M}}{\text{mol M}} = \frac{9.41 \text{ g M}}{0.1480 \text{ mol M}} = 63.5 \text{ g/mol M}$$

The atomic weight of M is 63.5

*** 10. Know and apply the conventions used in determining oxidation states.**

TABLE 3-1 Determination of Oxidation States

	Method of applying the rules
1.	Apply the rules from the top to the bottom of the list.
2.	Search the list to find a rule that fits. Apply it.
3.	Then start again at the top of the list to find the next rule that fits.
	Oxidation state rules
a.	The O.S. (oxidation state) of all uncombined elements = 0.
b.	The sum of the O.S. in a compound = 0.
c.	The sum of the O.S. in an ion = ionic charge.
d.	Alkali metals (family 1A) have O.S. = +1.
e.	Alkaline earths (2A) have O.S. = +2.
f.	H has O.S. = +1.
g.	O has O.S. = –2.
h.	F, Cl, Br, I (in order) have O.S. = –1.
i.	S, Se, Te (in order) have O.S. = –2.
j.	N, P (in order) have O.S. = –3.
k.	Others have O.S. = group number, or O.S. = group number–8, with the more electronegative element having the negative oxidation state.

Because oxidation state is a formal rather than an experimental concept, it is possible to devise a rigid set of rules that work in all but the most unusual circumstances. One set of rules, slightly different from that *of the text,* is given in Table 3-1.

In Example 3-12, the oxidation states are determined in the order in which they are listed; the letter of the applicable rule is given in parentheses before the oxidation state.

EXAMPLE 3-12 Determine the oxidation state of each element in F_2, O_2^{2-}, Na_2S, CaC_2, HCl, HFO, ClF_3, CS_2, FeN, SiC, and SO_3^{2-}.

F_2	(a) F = 0		
O_2^{2-}	(c) O = −1		
Na_2S	(d) Na = +1	(b) S = −2	
CaC_2	(e) Ca = +2	(b) C = −1	
HCl	(f) H = +1	(b) Cl = −1	
HFO	(f) H = +1	(g) O = −2	(b) F = +1
ClF_3	(h) F = −1	(b) Cl = +3	
CS_2	(i) S = −2	(b) C = +4	
FeN	(j) N = −3	(b) Fe = +3	
SiC	(k) C = −4	(b) Si = +4	
SO_3^{2-}	(g) O = −2	(b) S = + 4	

* **11. Know the names, formulas, and charges of the ions in Tables 3-2 and 3-3 and be able to write formulas and names of the compounds formed from these ions.**

The first goal is straightforward and is a 3 × 5 card project. Write the name of the ion on one side and the symbol and charge on the other side of the card. Then practice until you can recall one while looking at the other. It may seem early to learn chemical nomenclature and how to write formulas. But these form the basis of the language of chemistry. Chemists speak to each other with the names of compounds, but chemical formulas are most useful for calculations. Thus, to understand chemists you must have a firm grasp of nomenclature and to work problems you need to be good at writing formulas. Far too many students find chemistry difficult simply because they have not mastered nomenclature and formula writing.

The lists of ions presented in Tables 3-2 and 3-3 may seem rather extensive, but they contain practically all of the ions you are likely to encounter in general chemistry. To help you memorize charges, note how often the charge on a cation equals its group number in the periodic table, and how often an anion's charge equals its group number minus 8. (Some of the polyatomic ions in Table 3-3 are named systematically. See objective 3-13.)

Naming an ionic compound is surprisingly simple. Write down the name of the cation (positive ion), follow it with a space, and then the name of the anion (negative ion). As examples, consider iron(III) chloride, sodium nitrate, and mercury(I) phosphate.

TABLE 3-2 Names, Formulas, and Charges of Some Common Ions

H^+	hydrogen	Mn^{2+}	manganese(II)	Cr^{2+}	chromium(II) *or* chromous
Li^+	lithium	Ni^{2+}	nickel(II)	Cr^{3+}	chromium(III) *or* chromic
Na^+	sodium	As^{3+}	arsenic(III)	Fe^{2+}	iron(II) *or* ferrous
K^+	potassium	Zn^{2+}	zinc	Fe^{3+}	iron(III) *or* ferric
Rb^+	rubidium	Cd^{2+}	cadmium	Co^{2+}	cobalt(II) *or* cobaltous
Cs^+	cesium	Ag^+	silver	Co^{3+}	cobalt(III) *or* cobaltic
Mg^{2+}	magnesium	Au^+	gold(I) *or* aurous	Cu^+	copper(I) *or* cuprous
Ca^{2+}	calcium	Au^{3+}	gold(III) *or* auric	Cu^{2+}	copper(II) *or* cupric
Sr^{2+}	strontium	Sn^{2+}	tin(II) *or* stannous	Hg_2^{2+}	mercury(I) *or* mercurous
Ba^{2+}	barium	Sn^{4+}	tin(IV) *or* stannic	Hg^{2+}	mercury(I) *or* mercuric
Al^{3+}	aluminum	Pb^{2+}	lead(II) *or* plumbous	Tl^+	thallium(I) *or* thallous
		Pb^{4+}	lead(IV) *or* plumbic	Tl^{3+}	thallium(III) *or* thallic
H^-	hydride	F^-	fluoride	Cl^-	chloride
Br^-	bromide	I^-	iodide	N^{3-}	nitride
O^{2-}	oxide	S^{2-}	sulfide	Se^{2-}	selenide

TABLE 3-3 Names, Formulas, and Charges of Some Common Polyatomic Ions

NH_4^+	ammonium	SO_4^{2-}	sulfate	FO^-	hypofluorite
$C_2H_3O_2^-$	acetate	HSO_4^-	hydrogen sulfate	ClO^-	hypochlorite
CO_3^{2-}	carbonate	SO_3^{2-}	sulfite	ClO_2^-	chlorite
HCO_3^-	hydrogen carbonate	HSO_3^-	hydrogen sulfite	ClO_3^-	chlorate
$C_2O_4^{2-}$	oxalate	$S_2O_3^{2-}$	thiosulfate	ClO_4^-	perchlorate
CN^-	cyanide	HS^-	hydrogen sulfide	BrO^-	hypobromite
OCN^-	cyanate	OH^-	hydroxide	BrO_3^-	bromate
SCN^-	thiocyanate	O_2^{2-}	peroxide	BrO_4^-	perbromate
NO_2^-	nitrite	CrO_4^{2-}	chromate	IO^-	hypoiodite
NO_3^-	nitrate	$Cr_2O_7^{2-}$	dichromate	IO_3^-	iodate
PO_4^{3-}	phosphate	MnO_4^-	permanganate	IO_4^-	periodate
HPO_4^{2-}	hydrogen phosphate	MnO_4^{2-}	manganate		
$H_2PO_4^-$	dihydrogen phosphate				

Writing formulas from names is not quite so simple. The formula contains more than just the symbols for the cation and the anion. The cation and anion symbols are multiplied so that the total charge from the cations just balances the total charge of the anions. The total cation charge plus the total anion charge thus equals zero. Fortunately, there are not very many ways to combine cations and anions. A look at Tables 3-2 and 3-3 reveals that cations have charges of +1 (like Li^+), +2 (Ca^{2+}), and +3 (Al^{3+}). Cations also exist with higher charges, but +4 (Ce^{4+}) is the highest positive charge you are likely to encounter. Anions seem to come in only three varieties: –1 (like F^- or MnO_4^-), –2 (S^{2-} or CrO_4^{2-}), and –3 (PO_4^{3-}). Thus, there are only twelve (4 × 3) different common types of ionic compounds.

The easiest type of formula to write occurs in those compounds in which the anion and cation charges are equal but of opposite signs. Sodium chloride is composed of Na^+ and Cl^- ions and the total anion and cation charges sum to zero if we take one cation and one anion: NaCl. In like manner, magnesium sulfate is composed of Mg^{2+} and SO_4^{2-} ions and has the formula $MgSO_4$. Finally, aluminum phosphate is $AlPO_4$. These compounds lie along the diagonal in Table 3-4 [from the top line of column (1) to the fourth line of column (3)].

TABLE 3-4 Formulas for twelve common types of ionic compounds
Examples are given in brackets

		Anions					
		(1)		(2)		(3)	
Cations		X^-	[F^-]	X^{2-}	[SO_4^{2-}]	X^{3-}	[PO_4^{3-}]
M^+	[NH_4^+]	MX	[NH_4F]	M_2X	[$(NH_4)_2SO_4$]	M_3X	[$(NH_4)_3PO_4$]
M^{2+}	[Ca^{2+}]	MX_2	[CaF_2]	MX	[$CaSO_4$]	M_3X_2	[$Ca_3(PO_4)_2$]
M^{3+}	[Al^{3+}]	MX_3	[AlF_3]	M_2X_3	[$Al_2(SO_4)_3$]	MX	[$AlPO_4$]
M^{4+}	[Ce^{4+}]	MX_4	[CaF_4]	MX_2	[$Ce(SO_4)_2$]	M_3X_4	[$Ce_3(PO_4)_4$]

The next easiest type of compound has one ion with a charge of +1 or –1. The charge on the other ion indicates how many of the +1 or –1 ions are needed. For example, sodium phosphate is composed of Na^+ cations (charge = +1) and PO_4^{3-} anions (charge = –3) To balance the –3 charge of the phosphate ion requires a total cation charge of +3, which is supplied by 3 Na^+ cations. Compounds of this general type lie across the top line and down column (1) of Table 3-4.

The remaining compounds in the table are similar to aluminum sulfate, which is composed of Al^{3+} and SO_4^{2-} ions. One method of finding the number of each type of ion in the formula follows.

(1) Use as many anions as the value of the cation charge. In this case, the charge of Al^{3+} is +3, so three anions are needed.

(2) Use as many cations as the negative of the anion charge. Sulfate's charge is –2, so two cations are needed.

(3) If the two numbers you have produced (number of cations and number of anions) both can be divided evenly by the same number, perform this division before your write down the formula.

An example will make the use of step (3) clearer.

EXAMPLE 3-13 What is the formula of cerium(IV) oxalate?

The formula for cerium(IV) ion is Ce^{4+} and that for oxalate ion is $C_2O_4^{2-}$.
(1) Ce^{4+} has a charge = + 4. Thus, use four $C_2O_4^{2-}$ anions.
(2) $C_2O_4^{2-}$ has charge = −2. Thus, use two Ce^{4+} cations.
(3) But 4 and 2 both can be divided evenly by 2.
This yields two $C_2O_4^{2-}$ anions and one Ce^{4+} cation.
The formula of cerium(IV) oxalate is $Ce(C_2O_4)_2$

Remember: *Simply because a compound's name and formula can be written does not mean that the compound exists.* (The ability to name something does not guarantee that it exists. After all, the headless horseman doesn't exist, does he?)

*** 12. Be able to write formulas and names of simple binary covalent compounds and of binary acids.**

As the text states, covalent compounds are formed between nonmetallic elements. The easiest way to identify the nonmetals uses the periodic table. The nonmetals are the elements in the area of the table above and to the right of a zigzag line running from between B and Al to between Po and At. All other elements are metals.

Although there are a few exceptions (such as water, H_2O; ammonia, NH_3; and methane, CH_4) , the names of the binary covalent compounds are obtained from the names of the two elements. The elements are named in the same order as they appear in the formula. The first element name is unchanged; the ending of the second becomes "-ide." The element names have prefixes depending on the subscript of that element in the formula, except that the prefix mono- (meaning one of) is rarely used for the first element in the formula. The other prefixes are: di = 2, tri = 3, tetra = 4, penta = 5, hexa = 6, hepta = 7, and octa = 8.

Yet another method of naming binary covalent compounds is the Stock system. This system uses the oxidation state of the first element, in parentheses following its name. Thus, it is similar to binary ionic compound nomenclature. For example, N_2O_3 is nitrogen(III) oxide. The Stock system is becoming more prevalent, but it still is awkward in naming some compounds. For example, nitrogen(IV) oxide can be either NO_2 or N_2O_4.

Binary acids consist of hydrogen and a nonmetal. HCl and H_2S both are binary acids. The name of a binary acid has the prefix "hydro-" and the suffix "-ic" surrounding the root name of the element. Thus, HCl is hydrochloric acid and H_2S is hydrosulfuric acid. These binary acid names are used when the compounds are dissolved in water, that is, in aqueous solution. When they are present as pure compound (usually gases), their names follow the conventions for ionic compounds: hydrogen chloride and hydrogen sulfide.

*** 13. Use oxidation states to name oxoacids and oxoanions.**

Oxidation states are helpful in balancing some chemical equations (see objective 4-14), and in naming many compounds. In addition to the oxidation state nomenclature that we have already studied, oxidations states are helpful in naming oxoacids and oxoanions. The general formula of an oxoanion is XO_n^{m-} where X is the central atom, n the number of oxygens, and $m-$ the charge on the anion. For a given representative element X, the oxidation state of X varies between oxoanions in steps of 2, as shown in Table 3-5 for chlorine oxoanions.

TABLE 3-3 Salts and acids of chlorine oxoanions

Oxidation state	*Salt*	*Example*	*Acid*	*Example*
+1	hypo- -ite	NaClO sodium hypochlorite	hypo- -ous	HClO hypochlorous acid
+3	-ite	$NaClO_2$ sodium chlorite	-ous	$HClO_2$ chlorous acid
+5	-ate	$NaClO_3$ sodium chlorate	-ic	$HClO_3$ chloric acid
+7	per- -ate	$NaClO_4$ sodium perchlorate	per- -ic	$HClO_4$ perchloric acid

All oxoanions of the same family with the same oxidation state have similar names. Thus, $KBrO_3$ is potassium bromate and HFO is hypofluorous acid. Another generality, for the representative elements, is that the *-ate* anion and *-ic* acid endings are used when the oxidation state of the central atom equals the periodic table family number. The only exceptions to this occur in the halogens, where the *-ate* and *-ic* endings correspond to a +5 oxidation state and in the noble gases where they correspond to +6. Thus, H_2SO_4 is sulfuric acid, since sulfur is in family 6A and has an oxidation state of +6 in this compound. Likewise, $Na_2B_4O_7$ has boron (a member of family 3A) in an oxidation state of +3 and is sodium borate. Notice that the formulas of the compounds do not determine the names, only the oxidation state of the central atom. H_3PO_4, phosphor*ic* acid, and HNO_3, nitr*ic* acid, both have central atoms from family 5A, both with oxidation states of +5.

* **14. Be able to use the formulas for hydrates and coordination compounds in the same ways as those of simpler compounds**

In hydrates, water is associated with another compound. An example is $CaSO_4{\cdot}2H_2O$, calcium sulfate dihydrate, commonly called gypsum. Some or all of the associated water, known as water of hydration, can be removed by heating. For example, at 128°C, $1\frac{1}{2}$ mol H_2O is removed from $CaSO_4{\cdot}2H_2O$, producing $CaSO_4{\cdot}\frac{1}{2}H_2O$ (also written as $2CaSO_4{\cdot}H_2O$), calcium sulfate hemihydrate, or plaster of Paris. And at 163°C, $CaSO_4{\cdot}\frac{1}{2}H_2O$ becomes $CaSO_4$, the anhydrous (literally "without water") compound. Interpreting the formula of a hydrate follows the procedure given in objective 3-4, where the example is $CuSO_4{\cdot}5H_2O$, copper(II) sulfate pentahydrate.

Coordinate compounds are similar to simpler compounds, except that either the anion or the cation (or both) is a complex ion. An example is $Co_3[Fe(CN)_6]_2$; the formula of the complex ion is enclosed in brackets, [ ]. Each mole of $Co_3[Fe(CN)_6]_2$ contains

3 mol Co^{2+} cations and 2 mol $[Fe(CN)_6]^{3-}$ anions *or*

3 mol Co, 2 mol Fe, 12 mol C, and 12 mol N

The nomenclature of complex ions is discussed in objective 25-3.

DRILL PROBLEMS

2. (1) Determine the molar mass of each compound below.

A. Hg_2Cl_2 B. NaCl C. $SnCl_4$ D. $Ca(HS)_2$ E. $NaHCO_3$
F. $Al(NO_3)_3$ G BaCl$_2$ H. H_2O_2 I. HCN J. $C_2Cl_6S_2$
K. $(NH_4)_2CrO_4$ L. B_2H_6

(2) Fill in the blanks in the table below.

Molecular formula	*Molar mass*	*Mass of compound*	*Moles of compound*
M. $(NH_4)_2CrO_4$	_____	174 g	_____
N. Li_2SO_3	_____	_____g	2.47
O. BaCl	_____	8.21 oz	_____
P. AgClO	_____	_____g	0.0694
Q. $FeCl_3$	_____	0.124 lb	_____
R. H_2O_2	_____	5.00 lb	_____
S. NaOH	_____	_____g	1.47
T. $Mg(CN)_2$	_____	_____g	0.00836
U. $Hg(NO_3)_2$	_____	4.12 g	_____
V. CuCl	_____	_____g	7.97
W. $NaNO_3$	_____	1.43 lb	_____
X. $HBrO_4$	_____	_____g	2.18

3. & 4. Use the concept of a mole of substance to answer the following questions.

A. How many C atoms are in 14.7 g $NaHCO_3$?
B. What is the mass of 1.42×10^{23} NH_3 molecules?
C. What is the mass of 64.2 mol $CsIO_3$?
D. How many O atoms are in 74.2 g $Al_2(SO_4)_3$?
E. How many molecules are in 163.7 g HCl?
F. How many atoms are in 82.3 g H_2O_2?
G. What is the mass of 7.42×10^{27} H_2SO_4 molecules?

H. What is the mass of 27.32 mol $(NH_4)_2Cr_2O_7$?
I. How many atoms are in 64.2 g NH_4MnO_4?
J. What mass of $BaCl_2$ contains 23.1×10^{19} ions?
K. How many O atoms are in 2.68 lb of $NaClO_3$?
L. In $C_4H_{10}O$, how many moles of C are present with 18.4 mol H?
M. In NH_4CN, how many moles of H are present with 0.125 mol C?
N. In $K_4Fe(CN)_6$, how many moles of C are present with 74.2 mol K?
O. In $Al_2(SO_4)_3$, how many moles of Al are present with 16.2 mol O?

5. & 6. Fill in the blanks in the table below. The first line is an example.

	Empirical formula	*"Empirical weight"*	*Elements and percent by weight*					
	NH_4Cl	53.5	N	26.2	H	7.5	Cl	66.4
A.	_____	_____	H	2.44	S	39.02	O	_____
B.	NaOH	_____	Na	_____	O	_____	H	_____
C.	_____	_____	Mg	_____	C	14.2	O	56.9
D.	$Al(NO_3)_3$	_____	Al	_____	N	_____	O	_____
E.	_____	_____	Li	6.5	Cl	_____	O	60.2
F.	ZnC_2O_4	_____	Zn	_____	C	_____	O	_____
G.	_____	_____	K	_____	Mn	_____	O	_____
H.	$Mg(CN)_2$	_____	Mg	_____	C	_____	N	_____
I.	_____	_____	Cu	51.4	C	_____	O	_____
J.	H_3PO_4	_____	H	_____	P	_____	O	_____
K.	_____	_____	Hg	_____	Br	35.0	O	21.0

7. Each line below gives the mass of a carbon–hydrogen–oxygen compound that has been burned completely in oxygen to give the specified masses of CO_2 and H_2O. Based on these data, determine the empirical formula of each compound.

	Mass of compound	*Mass of CO_2*	*Mass of H_2O*
A.	1.320 g	1.489 g	0.914 g
B.	74.6 g	226.3 g	115.8 g
C.	8.14 g	11.19 g	9.16 g
D.	0.0152 g	0.0514 g	0.0105 g
E.	0.943 g	2.412 g	0.987 g
F.	11.7 g	22.4 g	13.7 g
G.	19.4 g	28.5 g	11.6 g
H.	7.32 g	11.84 g	5.81 g
I.	0.0255 g	0.0244 g	0.0100 g
J.	0.556 g	0.388 g	0.159 g
K.	146 g	222 g	45 g
L.	92.4 g	317.6 g	52.0 g

9. Fill in the blanks in the table below. The first line is done as an example.

	Element being determined	*Sample mass, g*	*Precipitate*		*% of element in sample*
			Formula	*Mass, g*	
	Ag	14.21	AgCl	15.32	82.14
A.	Ba	13.20	$BaSO_4$	15.14	_____
B.	Mg	96.41	MgO	17.46	_____
C.	Pb	9.915	$PbCrO_4$	11.321	_____
D.	Ca	82.44	$Ca_3(PO_4)_2$	1.42	_____
E.	Al	19.00	Al_2O_3	10.56	_____
F.	Cr	14.32	Cr_2O_3	5.17	_____
G.	Li	0.847	Li_3PO_4	0.201	_____
H.	F	37.32	CaF_2	40.03	_____
I.	S.	1.755	$BaSO_4$	1.972	_____
J.	C	1.650	$CaCO_3$	13.124	_____

9. The mass and formula of a compound are given below, in the first two columns. The compound is dissolved in water and the anion of the compound reacts with another cation to form a precipitate of known identity. The for-

mula and the mass of this precipitate are given in the second two columns. Use these data to determine the atomic weight of the metallic element (that of the cation) present in the original compound. (Hint: Recall Example 3-11.)

	Original compound		*Precipitate*		*Atomic weight*
	Formula	*Mass, g*	*Formula*	*Mass, g*	*of M*
A.	MS	44.72	$BaSO_4$	114.99	_____
B.	MCl_3	13.27	AgCl	35.17	_____
C.	MCl_2	6.34	AgCl	21.00	_____
D.	MF	13.62	CaF_2	12.66	_____
E.	$M(SO_4)_2$	20.10	$PbSO_4$	36.70	_____
F.	$M_2(SO_4)_3$	59.87	$BaSO_4$	122.53	_____
G.	M_2CO_3	24.55	$CaCO_3$	17.76	_____
H.	$M_2(CrO_4)_2$	54.39	$PbCrO_4$	82.87	_____

10. Give the oxidation state of each element in each compound in the drill problems for objective 3-2.

11. Give the name or formula, as appropriate, of each of the following compounds.

A. NaCl B. $Al_2(SO_4)_3$ C. $SrCl_2$ D. $CrHPO_4$ E. $Na_2S_2O_3$
F. CuCl G. $FePO_4$ H. Li_2SO_3 I. $(NH_4)_2CrO_4$ J. $KHSO_4$
K. $Ca(C_2H_3O_2)_2$ L. $Cu(HS)_2$ M. $MgCO_3$ N. $Cr(CN)_3$ O. $BaSO_4$
P. potassium permanganate Q. sodium acetate R. iron(III) chloride
S. strontium nitrate T. silver chlorate U. copper(I) oxide
V. mercury(I) nitrate W. iron(III) nitrate X. strontium bicarbonate
Y. zinc phosphate Z. potassium chromate Γ. magnesium cyanide
Δ. aluminum oxide Θ. sodium hydroxide Λ. calcium hypochlorite

12. Give the name and formula, as appropriate, of each compound.

A. H_2O B. H_2S C. CO_2 D. N_2O_3 E. P_2O_5
F. ICl_5 G. NCl_3 H. SCl_4 I. ClO_2 J. SO_3
K. nitrogen monoxide L. diphosphorus pentoxide M. ammonia
N. hydrogen fluoride O. silicon tetrafluoride P. xenon(VI) fluoride
Q. phosphorus tribromide R. silicon(IV) sulfide S. methane
T. boron trichloride U. nitrogen(III) oxide V. antimony(V) sulfide

13. For each compound below, supply the correct name or formula, and give the oxidation state of each element.

A. $NaClO_4$ B. selenic acid C. $(NH_4)_2S$
D. $CuIO_3$ E. H_2CO_3 F. phosphoric acid
G. sodium borate H. $NaNO_3$ I. LiClO
J. hyposulfurous acid K. $AlPO_4$ L. boric acid
M. $HClO_4$ N. nitrous acid O. magnesium chlorite
P. $ScAlO_3$ Q. Mg_2GeO_4 R. NH_4NO_3
S. HFO T. hyponitrous acid U. H_4TeO_4

14. Correctly interpret the formula of each hydrate or coordination compound to answer the following questions.

A. 186 g of $CoCl_2 \cdot 6H_2O$ contains how many moles of H?
B. There is 21.7 g of O in how many moles of $Tl(NO_3)_3 \cdot 3H_2O$?
C. 14.2 g of $Na_3[Au(S_2O_3)_2]$ contains how many moles of S?
D. There is 1.48 g of H in how many moles of $[Ag(NH_3)_2]_2SO_4$?
E. 57.2 g of $Rb_2S \cdot 4H_2O$ contains how many moles of O?
F. There is 7.62 moles of O in how many grams of $3K_2S_2O_3 \cdot 5H_2O$?
G. 91.6 g of $K_2[Pt(SCN)_6]$ contains how many moles of C?
H. There is 34.0 g of N in how many grams of $[Cr(NH_3)_6]_2(SO_4)_3$?
I. There are how many moles of O in 67.9 g of $Co(IO_3)_2 \cdot 6H_2O$?
J. 1.88 moles of $CrSO_4 \cdot 7H_2O$ contains how many grams of O?
K. 44.3 g of $NH_4[Au(CN)_2]$ contains how many moles of N?
L. There is 3.11 mol of S in how many moles of $Al_2[Pt(SCN)_6]_3$?

QUIZZES (20 minutes each.) Choose the best answer for each question or fill in the blank.

Quiz A

1. A compound contains 37% N and 63% O. Its empirical formula is (a) NO_2; (b) NO; (c) N_3O_2; (d) N_2O_3; (e) none of these.
2. The mole weight of $CuSO_4$ (Cu = 63.5, S = 32.1, O = 16.0 g/mol) is (a) 63.5 + 32.1 + 16.0; (b) (2 × 63.5) + 32.1 + 16.0; (c) 63.5 + 32.1 + (16.0 × 4); (d) (2 × 63.5) + 32.1 + (4 × 16.0); (e) none of these.
3. What is the mass fraction of oxygen in $C_6H_{12}O_6$ (C = 12.0, H = 1.01, O = 16.0 g/mol)? (a) 16.0/(12.0 + 1.01 + 16.0); (b) 16.0[12.0 + (2 × 1.01) + (6 × 16.0)]; (c) 16.0/[(6 × 12.0) + (12 × 1.01) + (6 × 16.0)]; (d) (6 × 16.0)/[(12.0 + (2 × 1.01) + 16.0]; (e) none of these.
4. 34.1 is the mole weight of which compound(H = 1.0, Si = 28.1, P = 31.0, S = 32.1, Cl = 35.5 g/mol)? (a) HCl; (b) SiH_4; (c) H_2S; (d) PH_3; (e) none of these.
5. How many molecules are contained in 1.00 g H_2O (H = 1.0, O = 16.0 g/mol; $N = 6.022 \times 10^{23}$)? (a) (1.0/16.0)*N*; (b) (1.0/17.0)*N*; (c) (18.0/1.0)*N*; (d) (17.0/1.0)*N*; (e) none of these.
6. How many moles of atoms are contained in 15.0 g H_2SO_4 (H = 1.0, S = 32.1, O = 16.0 g/mol)? (a) 15.0/98.1; (b) 7 × 15.0/98.1; (c) 15.0/49.1; (d) 7 × 15.0/49.1; (e) none of these.
7. The oxidation state of N in HNO_3 is (a) +1; (b) +3; (c) +5; (d) −1; (e) none of these.
8. The formula of iron(III) phosphate is (a) $Fe_2(PO_4)_3$; (b) $IrPO_4$; (c) $FePO_3$; (d) $FePO_4$; (e) none of these.
9. The name of $Sr(HCO_3)_2$ is (a) strontium oxalate; (b) strontium carbonate; (c) sodium bicarbonate; (d) strontium bicarbonate; (e) none of these.
10. The name of P_2O_5 is (a) potassium peroxide; (b) phosphorus(V) oxide; (c) phosphorus peroxide; (d) dipotassium pentoxide; (e) none of these.
11. Potassium perbromate has a formula of (a) KBr; (b) K_2BrO_4; (c) $KBrO_4$; (d) $KBrO_2$; (e) none of these.

Quiz B

1. A compound contains 46.7% N and 53.3 % O. What is the empirical formula of this compound? (a) NO; (b) NO_2: (c) N_2O_3; (d) N_2O; (e) none of these.
2. 97 g/mol is the mole weight (H = 1.0, Br = 79.9, S = 32.1, O = 16.0, P = 31.0 g/mol) to *two significant digits* of (a) HBrO; (b) H_3PO_4; (c) H_2SO_4; (d) $H_2S_2O_2$; (e) none of these
3. The mass fraction of carbon in Na_2S (Na = 23.0, S = 32.1 g/mol) is (a) 12.0/[23.0 + (32.1 × 2)]; (b) 12.0/32.1; (c) 32.1/[(23.0 × 2) + 12.0]; (d) 32.1/(32.1 + 23.0); (e) none of these.
4. The mole weight of $Mg_3(PO_4)_2$(Mg = 24.3, P = 31.0, O = 16.0 g/mol) is (a) 24.3 + 31.0 + 16.0; (b) 24.3 + 31.0 + (6 × 16.0); (c) (24.3 × 3) + 31.0 + (16.0 × 4); (d) 24.3 + (31.0 + 16.0 × 4)] × 2); (e) none of these.
5. How many atoms are contained in 95.0 g Fe_2O_3 (Fe = 55.8, O = 16.0 g/mol; $N = 6.022 \times 10^{23}$)? (a) (2 × 95.0)/71.8; (b) [(5 × 95.0)/71.8]*N*; (c) (95.0/159.6)*N*; (d) [(5 × 95.0)/159.6]*N*; (e) none of these.
6. How many moles of formula units are contained in 45.1 g Al_2O_3 (Al = 27.0, O = 16.0 g/mol)? (a) 45.1/43.0; (b) 45.1/102.0; (c) (5 × 45.1)/43.0; (d) (5 × 45.1)/102.0; (e) none of these.
7. Mn has an oxidation state of +4 in (a) $MnCl_2$; (b) MnO_2; (c) $MnSO_4$; (d) $KMnO_4$; (e) none of these.
8. The formula of mercury(I) cyanide is (a) $Hg(CN)_2$: (b) HgCN; (c) MgCN; (d)$Hg_2(CN)_2$; (e) none of these.
9. The name of $CrSO_3$ is (a) cerium sulfite; (b) chromium(II) sulfite; (c) chromium(III) sulfate; (d) chromium(II) sulfate; (e) none of these.
10. The formula of nitrogen(III) oxide is ______________________________
11. The formula of sodium chlorate is (a) $NaClO_4$; (b) $NaClO_3$; (c) $NaClO_2$; (d) NaClO; (e) NaCl.

Quiz C

1. A compound contains 1.8% H, 42.0% O; and 56.0% S. What is the empirical formula of the compound (H = 1.0, O = 16.0, S = 32.1 g/mol)? (a) H_2SO_4; (b) H_2SO_3; (c) HSO; (d) $H_2S_2O_3$; (e) none of these.
2. 51.7 is the mole weight of which compound (Li = 6.9; F = 19.0; Na = 23.0; P = 31.0; S = 32.1; Cl = 35.5; K = 39.1 g/mol)? (a) NaCl; (b) KF; (c) MgS; (d) Li_3P; (e) none of these.
3. What is the mass fraction of Mg in $Mg(ClO_3)_2$ (O = 16.0, Mg = 24.3, Cl = 35.5 g/mol)? (a) 24.3/(24.3 + 35.5 + 16.0); (b) 24.3/[24.3 + 35.5 + (3 × 16.0)]; (c) 24.3/[24.3 + (2 × 35.5) + 2 × 16.0)]; (d) 24.3/[24.3 + (2 × 35.5) + (6 × 16.0)]; (e) none of these.

4. The formula weight of $Al(ClO_4)_3$ (O = 16.0, Al = 27.0, Cl = 35.5 g/mol) is (a) 27.0 + [3 $\times$ (35.5 + 16.0)]; (b) 27.0 + (3 $\times$ 35.5) + (12 $\times$ 16.0); (c) 27.0 + 35.5 + 16.0; (d) 27.0 + [(35.5 + 16.0) $\times$ 3]; (e) none of these.
5. How many atoms of all kinds are contained in 42.1 g of CaC_2 (C = 12.0, Ca = 40.1 g/mol; N = 6.022 $\times$ 10^{23})? (a) (42.1/52.1)(2) N; (b) (42.1/52.1)(3) N; (c) (42.1/64.1)(3) N; (d) (42.1/64.1)(2) N; (e) none of these.
6. How many oxygen atoms are contained in 63.5 g of Cl_2O_7 (O = 16.0, Cl = 35.5 g/mol; N = 6.022 $\times$ 10^{23})? (a) (63.5/183) N; (b) [(7 $\times$ 63.5)/167] N; (c) [(7 $\times$ 63.5)/183] N; (d) [(7 $\times$ 63.5)/167] N; (e) none of these.
7. Iodine has an oxidation state of +7 in (a) HIO; (b) $NaIO_2$; (c) $Mg(IO_3)_2$; (d) $Al(IO_4)_3$; (e) KI.
8. The name of $FePO_4$ is (a) fermium phosphate; (b) iron(II) phosphite; (c) iron(III) phosphate; (d) iron(II) phosphate; (e) none of these.
9. The formula for magnesium chlorate is (a) $Mg(ClO_3)_2$; (b) $Mn(ClO_3)_2$; (c)$MnClO_2$; (d) $MgClO_4$; (e) none of these.
10. The name of Cl_2O_7 is ________________
11. The name of which compound ends with "ite"? (a) $NaHCO_3$; (b) HNO_2; (c) $NaNO_3$; (d) $AlPO_3$; (e) HFO.

Quiz D

1. What is the mass fraction of carbon in CH_3CN (H = 1.0, C = 12.0, N = 14.0)? (a) (2 $\times$ 12.0)/[(2 $\times$ 12.0) + (3 $\times$ 1.0) + 14.0] (b) 12.0/(12.0 + 1.0 + 14.0); (c) 12.0/[12.0 + (3 $\times$ 1.0) + 14.0]; (d) 12.0/[12.0 + (3 $\times$ 1.0) + 12.0 + 14.0]; (e) none of these.
2. The mole weight of $Na_2S_2O_3$ (O = 16.0, S = 23.0, S = 32.1 g/mol) is (a) [2 $\times$ (23.0 + 32.1 + 16.0) + 16.0; (b) 23.0 + 32.1 + 16.0; (c) 2 $\times$ (23.0 + 32.1 + 16.0); (d) (2 $\times$ 23.0) + 32.1 + (3 $\times$ 16.0); (d) none of these.
3. A compound contains 46.7%O, 51.8% Cl, and 1.5% H (H = 1.0, O = 16.0; Cl = 35.5 g/mol). This compound's empirical formula is (a) HClO; (b) $HClO_2$; (c) $HClO_3$; (d) $HClO_4$; (e) none of these.
4. 102.5 is the mole weight of (O = 16.0, Na = 23.0, Cl = 35.5 g/mol) (a) NaCl; (b) NaClO; (c) $NaClO_2$; (d) $NaClO_3$; (e) none of these.
5. How many hydrogen atoms are contained in 14.8 g of C_3H_8 (H = 1.0, C = 12.0; N = 6.022 $\times$ 10^{23})? (a) (14.8/13.0)(3) N; (b) (14.8/13.0)(8) N; (c) (14.8/44.0) N; (d) (14.8/44.0) N; (e) none of these.
6. 8.75 g of $AlCl_3$ contains how many moles of ions (Al = 27.0, Cl = 35.5 g/mol)? (a) 8.75/(27.0 + 35.5); (b) 8.75/[(3 $\times$ 27.0) + 35.5]; (c) 8.75 $\times$ (27.0 + 35.5); (d) 8.75/[27.0 + (3 $\times$ 35.5)]; (e) none of these.
7. The oxidation state of Mn in $MgMnO_4$ is (a) +2; (b) +7; (c) +6; (d) +4; (e) + 3.
8. The name of Hg_2SO_3 is (a) mercury(I) sulfite; (b) mercury(II) sulfite; (c) mercury(I) sulfate; (d) mercury(II) sulfate; (e) none of these.
9. The formula of copper(II) nitrite is (a) $CuNO_2$; (b) $Cu(NO_3)_2$; (c) $Co(NO_2)_2$; (d) $Cu(NO_2)_2$; (e) Cu_2NO_2.
10. The formula of phosphorus(III) oxide is __________.
11. The name of which compound ends with "ate"? (a) HIO_4; (b) Na_4SiO_3; (c) $KClO_2$; (d) HFO; (e) NO_2.

SAMPLE TEST (30 minutes)

1. A synthetic food company is attempting to make synthetic alcohol. Before the product is tested on laboratory animals, it is chemically analyzed. In determining the compound's empirical formula, a chemist burns 9.72 g of the compound in oxygen. The products are 16.88 g CO_2 and 9.21 g H_2O. Determine the empirical formula of the compound, assuming that it contains only the elements H, C, and O.
2. Iron is important in the body primarily because it is present in red blood cells and acts to carry oxygen to the various organs. Without oxygen, these organs will die. There are about 2.6 $\times$ 10^{13} red blood cells in the blood of an adult human, and the blood contains a total of 2.9 g of iron. How many.iron atoms are there in each red blood cell?
3. Give the name or formula, as appropriate, of each compound and the oxidation state of each atom.

a. copper(II) sulfate	b. $Na_2S_2O_3$	c. magnesium chlorite
d. Na_3BO_4	e. ammonium oxide	f. NI_3
g. Ca_2XeO_4	h. barium peroxide	i. antimony(III) sulfide

4 Chemical Reactions

CHAPTER OBJECTIVES

* **1. Write word equations and symbolic equations for chemical reactions.**

Chemical equations are summaries of experimental results. Consider the description: "A solution of lead(II) nitrate is added to a solution of potassium chromate. A bright yellow precipitate forms, that later is identified as lead(II) chromate. Evaporation of the remaining solution produces crystals of potassium nitrate." The word equation for this reaction is

lead(II) nitrate solution + potassium chromate solution

Even more compact is the symbolic equation.

$$Pb(NO_3)_2(aq) + K_2CrO_4(aq) \longrightarrow PbCrO_4(s) + 2\ KNO_3(aq)$$

In the symbolic equation, the substances on the left are known as reactants; those on the right are the products. The form of each substance is indicated in parentheses: (g) for a gas, (l) for a liquid, (s) for a solid, and (aq) for a substance dissolved in water—an aqueous solution. Be careful not to confuse (aq) with (l). NaCl(aq) means sodium chloride—table salt—dissolved in water. This is salt water, a mixture that can exist at room temperature. NaCl(l), on the other hand, means molten sodium chloride, which requires a temperature of at least 801°C to achieve.

You may need to review nomenclature (objectives 3-11 through 3-13) in order to transform a word equation into a symbolic equation. Some nomenclature problems are included for review in the Drill Problems. The solubility rules of objective 5-3 will enable you to predict when an insoluble compound—designated (s)—is formed in an aqueous solution.

* **2. Balance chemical equations by inspection.**

An unbalanced chemical equation gives the identities of the reactants and the products but not how much of each is consumed or produced. For that, we need the balanced chemical equation. Balancing by inspection means that one multiplies each species by a whole number—called the stoichiometric coefficient—placed before the chemical formula of that species. These numbers are chosen so that there are the same number of atoms of each element on both sides of the balanced chemical equation. The whole numbers inserted are called stoichiometric coefficients. *Do not change the chemical formulas of the reactants or products!*

EXAMPLE 4-1 Balance the expression $H_2(g) + O_2(g) \longrightarrow H_2O(l)$

The formula of water shows that hydrogen and oxygen unite in a ratio of 2:1, but the left–hand side of the expression shows one mole of H_2 uniting with one mole of O_2. The first step in balancing the expression is to multiply $H_2(g)$ by 2—that is, assign $H_2(g)$ the stoichiometric coefficient of 2. This produces

$$2\ H_2(g) + O_2(g) \longrightarrow H_2O(l)$$

Counting atoms, we see 4 H and 2 O on the left–hand side and 2 H and 1 O on the right–hand side. So we multiply $H_2O(l)$ by 2. That is, 2 becomes the stoichiometric coefficient of $H_2O(l)$.

$$2\ H_2(g) + O_2(g) \longrightarrow 2\ H_2O(l)$$

Alternatively, we could have noticed that in the original unbalanced expression there are 2 O atoms on the left–hand side and only 1 O atom on the right–hand side. To balance this, 2 becomes the stoichiometric coefficient of water, producing

$$H_2(g) + O_2(g) \longrightarrow 2\ H_2O(l)$$

Then, we would have observed that there are 2 H atoms on the left–hand side and 4 H atoms on the right–hand side. This would have lead to 2 as the stoichiometric coefficient for $H_2(g)$, as before. We also could have incorrectly "balanced " this expression by changing H_2O to H_2O_2. This is incorrect because the formula of the product was changed.

EXAMPLE 4-2 Balance the expression

$$Pb(NO_3)_2(aq) + K_2CrO_4(aq) \longrightarrow PbCrO_4(s) + KNO_3(aq)$$

This expression is easy to balance if we recognize that only ionic compounds are involved and count the number of *ions* on each side. There are the same number of Pb^{2+} and CrO_4^{2-} ions on each side, but there are 2 K^+ and 2 NO_3^- ions on the left–hand side and only one of each on the right–hand side. To make the number of K^+ and NO_3^- ions the same on each side, we merely need to assign 2 as the stoichiometric coefficient of $KNO_3(aq)$.

$$Pb(NO_3)_2(aq) + K_2CrO_4(aq) \longrightarrow PbCrO_4(s) + 2\ KNO_3(aq)$$

EXAMPLE 4-3 Balance the expression $NH_3(g) + O_2(g) \longrightarrow NO(g) + H_2O(l)$

We first notice that N and H are combined in a ratio of 1:3 in NH_3 on the left–hand side and thus they must be produced in that ratio on the right–hand side. A complication arises because of H being present in H_2O where there are 2 H atoms in the formula. This requires an even number of H on the left–hand side of the expression. Thus the coefficient of NH_3 is an even number; we begin with 2.

$$2\ NH_3(g) + O_2(g) \longrightarrow NO(g) + H_2O(l)$$

We now balance N and H atoms with stoichiometric coefficients for NO(g) and $H_2O(l)$.

$$2\ NH_3(g) + O_2(g) \longrightarrow NO(g) + 3\ H_2O(l)$$

There are now two O atoms on the left–hand side and 5 O atoms on the right–hand side. We could use $\frac{5}{2}$ as the coefficient of $O_2(g)$,

$$2\ NH_3(g) + \tfrac{5}{2}\ O_2(g) \longrightarrow 2\ NO(g) + 3\ H_2O(l)$$

but it is preferable to use whole numbers (integers) as stoichiometric coefficients. Thus we multiply all coefficients by 2.

$$4\ NH_3(g) + 5\ O_2(g) \longrightarrow 4\ NO(g) + 6\ H_2O(l)$$

There are only two rules to follow when balancing by inspection.

(1) *Never* change the formulas of the compounds.

(2) Make sure there are the same number of atoms of each element on both sides of the equation.

* 2a. Balance "ionic equations" by inspection.

The technique of balancing is the same as in objective 4-2, but there is one additional rule.

(3) Make sure that the total charge on the left–hand side equals the charge on the right–hand side.

EXAMPLE 4-4 Balance the expression $Al + Cu^{2+} \longrightarrow Al^{3+} + Cu$

There is one atom of each element on each side. But the charge on the left is +2 while that on the right is +3. In the balanced equation, atomic balance is maintained by using the same coefficient for both forms of aluminum (Al and Al^{3+}), and likewise for both forms of copper (Cu^{2+} and Cu).

$$2\ Al + 3\ Cu^{2+} \longrightarrow 2\ Al^{3+} + 3\ Cu$$

Note that the coefficient for copper (3) and the charge of aluminum (+3) are the same.

* 3. Predict the combustion products of carbon–hydrogen and carbon–hydrogen–oxygen compounds and write a balanced equation.

Combustion means that the compound burns in oxygen, $O_2(g)$. When C–H–O or C–H compounds burn completely in oxygen, the *only* products formed are $CO_2(g)$ and $H_2O(l)$. The resulting equation then is balanced by inspection.

EXAMPLE 4-5 Write the balanced equation for the combustion of C_3H_7OH (l).

First write the unbalanced expression.

$$C_3H_7OH(l) + O_2(g) \longrightarrow CO_2(g) + H_2O(l)$$

Then balance by inspection (objective 4-2).

$$2\ C_3H_7OH(l) + 9\ O_2(g) \longrightarrow 6\ CO_2(g) + 8\ H_2O(l)$$

* **4. Predict the products of some simple types of chemical reactions: combination, decomposition, displacement, and metathesis reactions.**

A major accomplishment of chemistry is the prediction of reaction products. The number and type of products depend in the type of the reaction. In one system of classification, there are four simple types of reactions: combination, decomposition, displacement, and metathesis or double displacement.

Combination: In a combination reaction, two substances unite to produce a single substance. The two substances may be two elements

$$C(s) + O_2(g) \longrightarrow CO_2(g)$$

$$Cu(s) + S(s) \longrightarrow CuS(s)$$

or two compounds

$$SO_3(g) + H_2O(l) \longrightarrow H_2SO_4(l)$$

$$CaO(s) + H_2O(l) \longrightarrow Ca(OH)_2(s)$$

$$Li_2O(s) + CO_2(g) \longrightarrow Li_2CO_3(s)$$

or an element and a compound.

$$2\ FeCl_2(s) + Cl_2(g) \longrightarrow 2\ FeCl_3(s)$$

$$2\ Cu_2O(s) + O_2(g) \longrightarrow 4\ CuO(s)$$

Notice that the products are composed of ions that you already have learned (Cu^{2+}, S^{2-}, Ca^{2+}, OH^-, Li^+, CO_3^{2-}, Fe^{3+}, Cl^-, O^{2-}) or are compounds that should be familiar (CO_2, NH_3, H_2SO_4).

Decomposition: In decomposition reactions, a single compound produces two or more substances. These substances may be compounds

$$CaCO_3 \xrightarrow{\text{heat}} CaO(s) + CO_2(g)$$

$$NH_4HCO_3(s) \xrightarrow{\text{heat}} NH_3(g) + H_2O(l) + CO_2(g)$$

or they may be elements

$$2\ HgO(s) \xrightarrow{\text{heat}} 2\ Hg(l) + O_2(g)$$

$$MgCl_2(l) \xrightarrow{\text{heat \& electricity}} Mg(l) + Cl_2(g)$$

or they may be elements and compounds.

$$2\ KClO_3(s) \xrightarrow{\text{heat}} 2\ KCl(s) + 3\ O_2(g)$$

$$2\ NaNO_3(s) \xrightarrow{\text{heat}} 2\ NaNO_2(s) + O_2(g)$$

Again, notice that the compounds produced are either familiar (NH_3, CO_2, H_2O) or are composed of ions that you have memorized (Ca^{2+}, O^{2-}, K^+, Cl^-, Na^+, NO_2^-) Notice also that heat or electricity is normally required to decompose compounds. Few compounds decompose spontaneously. Finally, notice that this is the only one of the four types of reaction with but one reactant.

Displacement: In a displacement reaction, normally one element reacts with a compound to take the place of one of the elements in that compound. Sometimes, however, one compound displaces another from a third compound. Some examples follow.

$$Zn(s) + CuSO_4(aq) \longrightarrow ZnSO_4(aq) + Cu(s)$$

$2\ Na(s) + 2\ H_2O(l) \longrightarrow 2\ NaOH(aq) + H_2(g)$

$Mg(s) + 2\ HCl(aq) \longrightarrow MgCl_2(aq) + H_2(g)$

$3\ H_2(g) + Fe_2O_3(s) \xrightarrow{heat} 3\ H_2O(g) + 2\ Fe(s)$

$Cl_2(g) + 2\ KI(aq) \longrightarrow 2\ KCl(aq) + I_2(s)$

$CO_2(g) + Ca(OH)_2(s) \longrightarrow CaCO_3(s) + H_2O(g)$

In each equation above, the "displacing element" (or compound) is written first and the element (or compound) that is displaced is written last.

Metathesis: Metathesis or double displacement reactions occur when two ionic compounds "change partners" or swap ions.

$BaCl_2(aq) + K_2SO_4(aq) \longrightarrow BaSO_4(s) + 2\ KCl(aq)$

$NaCN(aq) + HCl(aq) \longrightarrow HCN(g) + NaCl(aq)$

$2\ KOH(aq) + H_2SO_4(aq) \longrightarrow 2\ H_2O(l) + K_2SO_4(aq)$

$MgCO_3(s) + 2\ HCl(aq) \longrightarrow H_2O(l) + CO_2(g) + MgCl_2(aq)$

$Na_2SO_3(aq) + 2\ HI(aq) \longrightarrow H_2O(l) + SO_2(g) + 2\ NaI(aq)$

In the first three equations above, it is clear that the cations have exchanged anions. For example, Ba^{2+} is associated with Cl^- in in the reactants of the first equation and with SO_4^{2-} in the products. This exchange is not so evident in the last two equations until we recognize that the first two products in each case are the result of decomposition reactions. ["$H_2CO_3(aq)$" is enclosed in quotes because the evidence for the existence of this compound is rather poor.

"$H_2CO_3(aq)$" $\longrightarrow H_2O(l) + CO_2(g)$

$H_2SO_3(aq) = SO_2{\cdot}H_2O \longrightarrow H_2O(l) + SO_2(g)$

* **5. Derive from balanced chemical equations conversion factors for use in stoichiometric calculations.**

A chemical equation can be interpreted either in terms of atoms, ions, molecules, and formula units or in terms of moles of these entities. Thus, there are two ways of interpreting the equation

$2\ Na(s) + 2\ H_2O(l) \longrightarrow 2\ NaOH(aq) + H_2(g)$

The first interpretation is that 2 sodium atoms react with two water molecules to produce 2 formula units of sodium hydroxide (or 2 sodium ions and 2 hydroxide ions) and 1 hydrogen molecule. The second is that two moles of Na(s) react with 2 moles of $H_2O(l)$ to produce two moles of NaOH(aq) and 1 mole of $H_2(g)$.Therefore, if we are given the number of moles of reactant or product and the balanced chemical equation, we can determine the amounts of all other species involved.

EXAMPLE 4-6 Determine the amounts of Sr(s) and $H_2O(l)$ that reacted and the amount of $Sr(OH)_2(aq)$ produced when 3.4 mol $H_2(g)$ is produced by the reaction

$Sr(s) + 2\ H_2O(l) \longrightarrow Sr(OH)_2(aq) + H_2(g)$

The balanced equation states that 1 mol $Sr(OH)_2(aq)$ is produced for every 1 mol $H_2(g)$ produced.

$$3.4\ \text{mol}\ H_2(g) \times \frac{1\ \text{mol}\ Sr(OH)_2(aq)}{1\ \text{mol}\ H_2(g)} = 3.4\ \text{mol}\ Sr(OH)_2(g)$$

Note carefully that 1 mol $H_2(g)$ is not *equal* to 1 mol $Sr(OH)_2(aq)$. They are equivalent to each other in that they are produced together. In like fashion, we determine the amounts of reactants consumed.

$$3.4\ \text{mol}\ H_2(g) \times \frac{1\ \text{mol}\ Sr(s)}{1\ \text{mol}\ H_2(g)} = 3.4\ \text{mol}\ Sr(s)$$

$$3.4\ \text{mol}\ H_2(g) \times \frac{2\ \text{mol}\ H_2O(l)}{1\ \text{mol}\ H_2(g)} = 6.8\ \text{mol}\ H_2O(l)$$

Note that the stoichiometric coefficient for each species is used in the conversion factor. It is essential that you thoroughly master this technique. All of the material in this chapter either leads up to or depends on this objective.

* **5a. Solve problems based on balanced chemical equations with quantities given or sought in a variety of units.**

This objective combines the previous objective with the techniques of using conversion factors that you mastered when studying Chapters 1 and 3. You may wish to review the conversion factor method (objective 1-11). One or more of the conversion factors will be obtained from a balanced chemical equation.

EXAMPLE 4-7 Under normal conditions, hydrogen gas has a density of 0.0824 g/L. What volume of hydrogen gas in liters is needed to produce 2.50 lb of iron from $Fe_2O_3(s)$? Water is the other product of the reaction.

$$Fe_2O_3(s) + H_2(g) \longrightarrow Fe(s) + H_2O(l)$$

Review objective 4-1 if you have trouble writing the unbalanced expression. The expression then is balanced by inspection (objective 4-2).

$$Fe_2O_3(s) + 3\ H_2(g) \longrightarrow 2\ Fe(s) + 3\ H_2O(l)$$

and we now can apply the conversion factor method.

(1) ? L $H_2(g)$ (2) 2.50 lb Fe(s)

(3) $2.50 \text{ lb Fe} \times \frac{454 \text{ g Fe}}{\text{lb}} \times \frac{\text{mol Fe}}{55.8 \text{ g}} \times \frac{3 \text{ mol } H_2}{2 \text{ mol Fe}} \times \frac{2.02\text{g } H_2}{\text{mol } H_2} \times \frac{\text{L } H_2(g)}{0.0824\text{g } H_2}$

$= 748 \text{ L } H_2$

Note in Example 4-7 that *only one* conversion factor (3 mol H_2/2 mol Fe) is obtained from the balanced chemical reaction. It is very unusual that more than one conversion factor will be taken from a given chemical equation. Also notice that moles (not grams or liters) of different substances are related. Therefore, when you wish to convert from one substance to another, you must convert to (and then from) moles.

* **6. Define the terms associated with solutions including molar concentration units, compute concentrations and solution volumes.**

The first goal is to define in your words and commit to memory the terms: solute, solution, solvent, concentration, aqueous solution, and molarity. You should also know the common symbol for molarity, brackets around the chemical formula of the solute. Thus, [NaCl] means "the molar concentration of sodium chloride." The definition of molarity (= moles of solute/liters of solution) suggests three types of problems, illustrated by the next three examples.

EXAMPLE 4-8 (Determine molarity given moles of solute and volume of solution.) 205 mL of solution contains 4.1 g NaCl. What is the concentration of the solution in moles per liter?

We divide the *amount* of solute (in moles) by the *volume* of solution (in liters).

$$[\text{NaCl}] = \frac{4.1 \text{ g NaCl} \times (\text{mol NaCl}/58.5 \text{ g NaCl})}{205 \text{ mL solution} \times \text{L}/1000 \text{ mL}} = \frac{0.070 \text{ mol NaCl}}{0.205 \text{ L soln}} = 0.34 \text{ M}$$

EXAMPLE 4-9 (Determine moles of solute given solution volume and concentration.) 114 mL of 0.357 M $Al(NO_3)_3$ contains what amount of solute in moles?

The concentration is a conversion factor.

(1) ? mol $Al(NO_3)_3$ (2) 114 mL solution

(3) $114 \text{ mL soln} \times \frac{\text{L}}{1000 \text{ mL}} \times \frac{0.357 \text{ mol } Al(NO_3)_3}{\text{L soln}} = 0.0407 \text{ mol } Al(NO_3)_3$

EXAMPLE 4-10 (Determine volume of solution given solution concentration and moles of solute.) What volume in mL of 0.628 M $AgNO_3$ solution contains 55.1 g $AgNO_3$?

(1) ? mL solution (2) 55.1 g $AgNO_3$

(3) $55.1 \text{ g } AgNO_3 \times \frac{\text{mol } AgNO_3}{169.9 \text{ g}} \times \frac{\text{L soln}}{0.628 \text{ mol } AgNO_3} \times \frac{1000 \text{ mL}}{\text{L}} = 516 \text{ mL}$

* **7. Solve dilution problems and those involving the mixing of two solutions.**

There is yet another type of problem, which concerns the dilution of solutions. We first determine the amount of solute in the original solution with the technique of Example 4-9, and then compute the molarity with the new solution volume and the method of Example 4-8.

EXAMPLE 4-11 54.0 mL of water is mixed with 114 mL of 0.357 M $NaNO_3$ solution.What is the final concentration of sodium nitrate, $[NaNO_3]$?

$$114\text{ mL soln} \times \frac{1}{1000\text{ mL}} \times \frac{0.357\text{ mol NaNO}_3}{\text{L soln}} = 0.0407\text{ mol NaNO}_3$$

Total solution volume = 54.0 mL + 114 mL = 168 mL.

$$[NaNO_3] = \frac{\text{mol NaNO}_3}{\text{L solution}} = \frac{0.0407\text{ mol NaNO}_3}{168\text{ mL} \times (\text{L}/1000\text{ mL})} = 0.242\text{ M}$$

Sometimes an alternate definition of molarity (= millimoles of solute / milliliters of solution) is more convenient because most solution volumes are expressed in milliliters. The solution for Example 4-11 then becomes

$$114\text{ mL soln} \times \frac{0.357\text{ mmol NaNO}_3}{\text{mL soln}} = 40.7\text{ mmol NaNO}_3$$

$$[NaNO_3] = \frac{40.7\text{ mmol NaNO}_3}{168\text{ mL}} = 0.242\text{ M}$$

When two solutions are mixed, the concentration of a solute (or solute ion) that is *not* present in both solutions is determined with the method of Example 4-11. If a solute (or a solute ion) is present in both solutions, the amount of solute in each solution must be determined before its concentration in the final solution can be calculated, as shown in Example 4-12.

EXAMPLE 4-12 54.0 mL of 0.628 M $NaNO_3$ is mixed with 114 mL of 0.357 M $NaNO_3$ solution.What is the final concentration of sodium nitrate?

As in Example 4-11, the total solution volume is 168 mL.
First the amount of nitrate ion in each of the two solutions is determined.

$$54.0\text{ mL} \times \frac{0.628\text{ mmol NaNO}_3}{\text{mL soln}} = 33.9\text{ mmol NaNO}_3$$

$$114\text{ mL} \times \frac{0.357\text{ mmol NaNO}_3}{\text{mL soln}} = 40.7\text{ mmol NaNO}_3$$

$$[NaNO_3] = \frac{(33.9 + 40.7)\text{ mmol NO}_3^-}{168\text{ mL soln}} = 0.444\text{ M}$$

* **8. Solve stoichiometry problems when either the reactants or the products are species in solution and concentration and volume data are given.**

This objective is a combination of objectives 4-5 and 4-7. It therefore only requires the techniques you have mastered already. Apply these techniques to somewhat more involved problems. Be sure that you have worked some of the drill problems of these two previous objectives before you attempt the drill problems of this objective. Remember that, when changing from one substance to another, you must convert to (and from) moles.

* **9. Determine the reactant(s) in excess, the limiting reagent, and the amounts of products obtained in a chemical reaction.**

Limiting reagent problems, as they commonly are called, are based on the fact that reactants combine to yield definite amounts of products. If there is an excess of one reactant, that excess will undergo no chemical change. The reactant that determines the final amount of product is the limiting reagent. Chemical reactions are different in this respect from cookie recipes. Using 12 oz of chocolate chips, in a recipe calling for 16 oz, only causes the cookies to have a scant number of chocolate chips.

Chemical reactions are more similar to automobile production. If a manufacturer has 5000 spark plugs, only 1250 four–cylinder cars can be produced. Likewise, 4000 wheels limits production to 1000 cars. It makes no difference that there are 9000 engines and 4000 car bodies on hand. Limiting reagent problems often are most easily solved by determining the amount of each product that can be produced from each reactant. The limiting reagent is the one that produces the least amount of product.

$$5000 \text{ spark plugs} \times \frac{1 \text{ car}}{4 \text{ spark plugs}} = 1250 \text{ cars}$$

$$4000 \text{ wheels} \times \frac{1 \text{ car}}{4 \text{ wheels}} = 1000 \text{ cars}$$

$$3000 \text{ bodies} \times \frac{1 \text{ car}}{1 \text{ body}} = 3000 \text{ cars}$$

$$9000 \text{ engines} \times \frac{1 \text{ car}}{1 \text{ engine}} = 9000 \text{ cars}$$

Hence, only 1000 cars can be produced. Notice the limiting reagent is *not* the smallest number of parts (3000 bodies), but the number of parts that produces the smallest number of cars.

EXAMPLE 4-13 1.50 g of Mg reacts with 342 mL of 0.350 M HCl solution to produce what mass in grams of $MgCl_2$? What mass of the excess reactant is left over?

The reaction is a displacement (See objective 4-4).

$$Mg(s) + 2\,HCl(aq) \longrightarrow MgCl_2(aq) + H_2(g)$$

$$1.50 \text{ g Mg} \times \frac{\text{mol Mg}}{24.3 \text{ g Mg}} \times \frac{\text{mol } MgCl_2}{\text{mol Mg}} = 0.0617 \text{ mol } MgCl_2$$

$$342 \text{ mL soln} \times \frac{\text{L}}{1000 \text{ mL}} \times \frac{0.350 \text{ mol HCl}}{\text{L soln}} \times \frac{\text{mol } MgCl_2}{2 \text{ mol HCl}} = 0.0599 \text{ mol } MgCl_2$$

Thus, HCl is the limiting reagent, and only 0.0599 mol $MgCl_2$ is produced. We convert this amount to a mass in grams.

$$0.0599 \text{ mol } MgCl_2 \times \frac{95.3 \text{ g } MgCl_2}{\text{mol } MgCl_2} = 5.71 \text{ g } MgCl_2$$

To determine the mass of Mg (the excess reactant) left over, first compute the mass of Mg that reacted and then subtract that mass from the 1.50 g provided.

$$342 \text{ mL soln} \times \frac{\text{L}}{1000 \text{ mL}} \times \frac{0.350 \text{ mol HCl}}{\text{L soln}} \times \frac{\text{mol Mg}}{2 \text{ mol HCl}} \times \frac{24.3 \text{ g Mg}}{\text{mol Mg}} = 1.45 \text{ g Mg}$$

Mass in excess = 1.50 g − 1.45 g = 0.05 g Mg.

* **10. Define the terms actual yield, theoretical yield, and percent yield and compute these quantities for a given reaction.**

The theoretical yield of a reaction is the computed quantity of product based on the quantities of reagents used. This quantity may have to be determined by solving a limiting reagent problem. The actual yield is the quantity of product actually obtained (often called the experimental yield). The percent yield is calculated from their ratio.

$$\text{percent yield} = \frac{\text{actual yield}}{\text{theoretical yield}} \times 100$$

EXAMPLE 4-14 The reaction $N_2 + 3\,H_2 \longrightarrow 2\,NH_3$ has a percent yield of 62.1% under certain conditions. (a) What mass of NH_3 is produced from 74.8 g N_2? (b) What mass of H_2 is required to produce 36.5 g NH_3?

We use the conversion factor method in each case, labeling the actual yield as "g NH_3 produced" and the theoretical yield as "g NH_3 calculated."

$$\text{(a) g } NH_3 = 74.8 \text{ g } N_2 \times \frac{1 \text{ mol } N_2}{28.0 \text{ g } N_2} \times \frac{2 \text{ mol } NH_3}{1 \text{ mol } N_2} \times \frac{17.0 \text{ g } NH_3}{1 \text{ mol } NH_3}$$

$$\times \frac{62.1 \text{ g } NH_3 \text{ produced}}{100.0 \text{ g } NH_3 \text{ calculated}} = 56.4 \text{ g } NH_3 \text{ produced}$$

$$\text{(b) g } H_2 = 36.5 \text{ g } NH_3 \text{ produced} \times \frac{100.0 \text{ g } NH_3 \text{ calculated}}{62.1 \text{ g } NH_3 \text{ produced}} \times \frac{1 \text{ mol } NH_3}{17.0 \text{ g } NH_3}$$

$$\times \frac{3 \text{ mol } H_2}{2 \text{ mol } NH_3} \times \frac{2.02 \text{ g } H_2}{1 \text{ mol } H_2} = 10.5 \text{ g } H_2 \text{ required}$$

Notice that the correct use of percent yield as a conversion factor requires careful labeling of both numerator and denominator. Often problems involving consecutive reactions (objective 4-13) will be complicated by including the percent yield of each reaction.

*** 11. Compute the amount of product produced or reactant consumed by two or more simultaneous reactions.**

Simultaneous reactions either consume an identical reactant or yield an identical product. One is asked to determine the amount of the common species involved. The techniques of solving stoichiometric problems (objective 4-7) are used for each reactant in turn and the results are added.

EXAMPLE 4-15 What amount of $AgNO_3$ is consumed when 406 mL of a solution that is both 0.100 M in KCl and 0.250 M in Na_2CrO_4 reacts completely?

Both reactions are metatheses.

$$KCl(aq) + AgNO_3(aq) \longrightarrow AgCl(s) + KNO_3(aq)$$

$$Na_2CrO_4(aq) + AgNO_3 \longrightarrow Ag_2CrO_4(aq) + 2\ NaNO_3(aq)$$

The amount of $AgNO_3$ required to react with the KCl present is

$$406\text{ mL} \times \frac{\text{L}}{1000\text{ mL}} \times \frac{0.100\text{ mol KCl}}{\text{L soln}} \times \frac{\text{mol AgNO}_3}{\text{mol KCl}} = 0.0406\text{ mol AgNO}_3$$

The amount of $AgNO_3$ required to react with the Na_2CrO_4 present is

$$406\text{ mL} \times \frac{\text{L}}{1000\text{ mL}} \times \frac{0.250\text{ mol Na}_2\text{CrO}_4}{\text{L soln}} \times \frac{2\text{ mol AgNO}_3}{\text{mol Na}_2\text{CrO}_4} = 0.203\text{ mol AgNO}_3$$

Total $AgNO_3$ = 0.0406 mol + 0.203 mol = 0.244 mol.

*** 12. Compute the amount of product produced by two or more consecutive reactions.**

Consecutive reactions yield products from reactants in a series of steps rather than in a single reaction. Thus the solution of a stoichiometric problem will contain not just one conversion factor obtained from a chemical reactions, but several—one from each step. An example of a series of reactions is the Solvay process for making sodium carbonate. It consists of six steps.

$$CaCO_3(s) \xrightarrow{\text{heat}} CaO(s) + CO_2(g) \qquad [1]$$

$$CaO(s) + H_2O(l) \longrightarrow Ca(OH)_2(aq) \qquad [2]$$

$$Ca(OH)_2(aq) + 2\ NH_4Cl(aq) \longrightarrow 2\ NH_3(g) + 2\ H_2O(l) + CaCl_2(aq) \qquad [3]$$

$$NH_3(g) + CO_2(g) + H_2O(l) \longrightarrow NH_4HCO_3(aq) \qquad [4]$$

$$NaCl(aq) + NH_4HCO_3(aq) \longrightarrow NaHCO_3(s) + NH_4Cl(aq) \qquad [5]$$

$$2\ NaHCO_3(s) \xrightarrow{\text{heat}} Na_2CO_3(s) + H_2O(l) + CO_2(g) \qquad [6]$$

EXAMPLE 4-16 What mass of sodium carbonate in grams can be produced from 64.0 g of $CaCO_3(s)$ using the Solvay process?

$$64.0\text{ g CaCO}_3 \times \frac{\text{mol CaCO}_3}{100.0\text{ g}} \times \frac{\text{mol CO}_2}{\text{mol CaCO}_3} \times \frac{\text{mol NH}_4\text{HCO}_3}{\text{mol CO}_2} \times \frac{\text{mol NaHCO}_3}{\text{mol NH}_4\text{HCO}_3}$$

$$\times \frac{\text{mol Na}_2\text{CO}_3}{2\text{ mol NaHCO}_3} \times \frac{106.0\text{ g Na}_2\text{CO}_3}{\text{mol Na}_2\text{CO}_3} = 33.9\text{ g Na}_2\text{CO}_3$$

The principal difficulty in solving this type of problem is in tracing the product (or a part thereof) through the series of reactions. In this case the element carbon was used. Sometimes this tracing is more easily done if one works backward from the products. Notice that each chemical equation furnishes only one conversion factor. In Example 4-16, only conversion factors from equations [1], [4], [5], and [6] are needed.

* **13. Determine the overall reaction for a process consisting of several steps.**

The easiest way to find the overall reaction is by combining the chemical reactions as if they were algebraic equations, eliminating those species that appear on both sides. The equations are manipulated so that as many species cancel as possible. For the Solvay process, equations [1] through [6] of objective 4-13, we proceed as follows.

[6] + 2 × [5] = [7]

$$\cancel{2\,NaHCO_3} + 2\,NaCl + 2\,NH_4HCO_3 \longrightarrow Na_2CO_3 + H_2O + CO_2 + \cancel{2\,NaHCO_3} + 2\,NH_4Cl$$

$$2\,NaCl + 2\,NH_4HCO_3 \longrightarrow Na_2CO_3 + H_2O + CO_2 + 2\,NH_4Cl \qquad [7]$$

[7] + 2 × [4] = [8]

$$2\,NaCl + \cancel{2\,NH_4HCO_3} + 2\,NH_3 + \cancel{2}\,CO_2 + \cancel{2}\,H_2O \longrightarrow Na_2CO_3 + \cancel{H_2O} + \cancel{CO_2} + 2\,NH_4Cl + \cancel{2\,NH_4HCO_3}$$

$$2\,NaCl + 2\,NH_3 + CO_2 + H_2O \longrightarrow Na_2CO_3 + 2\,NH_4Cl \qquad [8]$$

[8] + [3] = [9]

$$2\,NaCl + \cancel{2\,NH_3} + CO_2 + \cancel{H_2O} + Ca(OH)_2 + \cancel{2\,NH_4Cl} \longrightarrow Na_2CO_3 + \cancel{2\,NH_4Cl} + \cancel{2\,NH_3} + \cancel{2}\,H_2O + CaCl_2$$

$$2\,NaCl + CO_2 + Ca(OH)_2 \longrightarrow Na_2CO_3 + H_2O + CaCl_2$$

[1] + [2] = [10]

$$CaCO_3 + \cancel{CaO} + H_2O \longrightarrow \cancel{CaO} + CO_2 + Ca(OH)_2 \qquad [10]$$

[9] + [10] = overall reaction = [11]

$$2\,NaCl + \cancel{CO_2} + \cancel{Ca(OH)_2} + CaCO_3 + \cancel{H_2O} \longrightarrow Na_2CO_3 + \cancel{H_2O} + CaCl_2 + \cancel{CO_2} + \cancel{Ca(OH)_2}$$

$$2\,NaCl + CaCO_3 \longrightarrow Na_2CO_3 + CaCl_2 \qquad [11]$$

DRILL PROBLEMS

1. Write a balanced equation for each of the following word equations.

A. aluminum + magnesium oxide ⟶ magnesium + aluminum oxide
B. aluminum chloride + sodium hydroxide ⟶ aluminum hydroxide + sodium chloride
C. silver nitrate + sodium phosphate ⟶ silver phosphate + sodium nitrate
D. chlorine + potassium iodide ⟶ potassium chloride + iodine
E. ferric sulfate + calcium hydroxide ⟶ ferric hydroxide + calcium sulfate
F. barium nitrate + ammonium carbonate ⟶ ammonium nitrate + barium carbonate
G. potassium hydroxide + sulfuric acid ⟶ potassium sulfate + water
H. ferrous hydroxide + hydrochloric acid ⟶ ferrous chloride + water
I. cupric sulfate + hydrogen sulfide ⟶ cupric sulfide + sulfuric acid
J. silver nitrate + potassium chromate ⟶ silver chromate + potassium nitrate

1a. (1) Give the correct name of each of the following compounds.

A. $K_2Cr_2O_7$	B. $AgBrO_3$	C. $Mg(HSO_3)_2$	D. KIO
E. $NaClO_4$	F. KCNS	G. $Pb(CO_3)_2$	H. CuC_2O_4
I. K_2O	J. Tl_2S	K. $Sn(MnO_4)_2$	L. $Mn(IO_3)_2$
M. $Ni(CN)_2$	N. $Mn(CNO)_2$	O. $CdSO_4$	P. $MnSO_3$
Q. $(NH_4)_2Cr_2O_7$	R. $CoMnO_4$	S. AgCNS	T. Li_2O_2
U. $Hg_2(C_2H_3O_2)_2$	V. Tl_2S_3	W. $Ag_2Cr_2O_7$	X. $Au_2(SO_4)_3$
Y. $Mg(IO_3)_2$	Z. $Cd(HSO_3)_2$		

(2) Give the chemical formula of each of the following compounds.

A. strontium peroxide	B. manganese(II) sulfite	C. lead(II) oxalate
D. rubidium bromate	E. magnesium cyanate	F. sodium bisulfide

G. calcium hypoiodite H. iron(II) peroxide I. silver sulfide
J. barium iodate K. cadmium sulfate L. potassium manganate
M. arsenic(III) chloride N. chromium(II) bromite O. tin(IV) carbonate
P. arsenic(III) sulfide Q. manganese(II) bicarbonate
R. lead(IV) carbonate S. gold(I) dichromate T. potassium oxalate
U. manganese(II) thiosulfate V. hydrogen peroxide W. gold(III) iodate
X. sodium cyanide Y. copper(I) cyanate Z. stannous fluoride

2. Balance the following equations by inspection.

A. $P + O_2 \longrightarrow P_2O_5$
B. $Na + O_2 \longrightarrow Na_2O_2$
C. $Al + HCl \longrightarrow AlCl_3 + H_2$
D. $Ca + H_2O \longrightarrow Ca(OH)_2 + H_2$
E. $FeCl_3 + Ca(OH)_2 \rightarrow Fe(OH)_3 + CaCl_2$
F. $Al + N_2 \longrightarrow AlN$
G. $HCl + Fe_2O_3 \longrightarrow FeCl_3 + H_2O$
H. $Cl_2 + H_2O \longrightarrow HCl + HClO_3$
I. $Al(OH)_3 + HCl \longrightarrow AlCl_3 + H_2O$
J. $CaSO_3 + H_2SO_4 \rightarrow CaSO_4 + H_2O + SO_2$
K. $NaCl + H_2SO_4 \longrightarrow Na_2SO_4 + HCl$
L. $Pb(NO_3)_2 \longrightarrow PbO + NO + O_2$
M. $HNO_2 \longrightarrow HNO_3 + NO + H_2O$
N. $Ca(OH)_2 + H_3PO_4 \rightarrow Ca_3(PO_4)_2 + H_2O$
O. $SiF_4 + H_2O \longrightarrow HF + SiO_2$

2a. Determine the proper stoichiometric coefficients to balance the following ionic equations.

A. $Cu + Fe^{3+} \longrightarrow Cu^{2+} + Fe$
B. $Zn + H^+ \longrightarrow Zn^{2+} + H_2$
C. $H^+ + S^{2-} \longrightarrow H_2S(g)$
D. $Pb^{2+} + Hg^{2+} \longrightarrow Pb^{4+} + Hg_2^{2+}$
E. $Fe^{3+} + Sn^{2+} \longrightarrow Fe^{2+} + Sn^{4+}$
F. $H^+ + N^{3-} \longrightarrow NH_3$
G. $CN^- + Fe^{2+} \longrightarrow Fe(CN)_6^{4-}$
H. $Bi^{3+} + S^{2-} \longrightarrow Bi_2S_3$

3. Write the balanced equation for the combustion of each of the following compounds.

A. methanol, CH_3OH B. benzene, C_6H_6 C. pentane, C_5H_{12}
D. sucrose, $C_{12}H_{22}O_{11}$ E. carbon monoxide, CO F. acetic acid, $HC_2H_3O_2$
G. naphthalene, $C_{10}H_8$ G. ninhydrin, $C_9H_6O_4$ H. benzoic acid, C_6H_5COOH
J. testosterone, $C_{19}H_{28}O_2$

4. Classify each of the following reactions as combination, decomposition, displacement, or metathesis. Predict the products. Balance the resulting equation.

A. $Al + Br_2 \longrightarrow$ B. $Al + Fe_2O_3 \longrightarrow$ C. $AlCl_3 + KOH \longrightarrow$
D. $BaCl_2 + Na_2CO_3 \longrightarrow$ E. $BaCl_2 + AgNO_3 \longrightarrow$ F. $BaO + SO_3 \longrightarrow$
G. $Ba(OH)_2 + "H_2CO_3" \rightarrow$ H. $Br_2 + H_2 \longrightarrow$ I. $Cd(NO_3)_2 + (NH_4)_2S \rightarrow$
J. $Ca + H_2O \longrightarrow$ K. $HCl + Al \longrightarrow$ L. $H_2O + Li_2O \longrightarrow$
M. $Ba(OH)_2 + H_3PO_4 \longrightarrow$ N. $BaO + H_2O \longrightarrow$ O. $CaO + HCl \longrightarrow$
P. $CaO + CO_2 \longrightarrow$ Q. $Ca(OH)_2 + HNO_3 \longrightarrow$ R. $CaO + HNO_3 \longrightarrow$
S. $NH_3(g) + HCl(g) \longrightarrow$ T. $SO_2 + O_2 \longrightarrow$ U. $AgNO_3 + Cu \longrightarrow$
V. $Cl_2 + FeCl_2 \longrightarrow$ W. $Cl_2 + KI \longrightarrow$ X. $N_2 + H_2 \longrightarrow$
Y. $Al(OH)_3 + H_2SO_4 \longrightarrow$ Z. $CuBr + Br_2 \longrightarrow$

5. & 5a. Beneath a reactant or a product of each balanced equation below is given the quantity of material produced or consumed. Densities are given at right. Fill in the blanks under each equation.

	$Fe(OH)_3(s)$ +	$3\ HCl(aq)$ $\longrightarrow$	$FeCl_3(s)$ +	$3\ H_2O$	Density
A.	34.5 g	____g	____g	____cm³	H_2O 1.00 g/cm³
B.	____g		21.2 cm³	____g	$FeCl_3$ 2.90 g/cm³
C.	____g	18.2 g	____g		

	$C_5H_{12}(g)$ +	$8\ O_2(g)$ $\longrightarrow$	$5\ CO_2(g)$ +	$6\ H_2O(l)$	
D.	14.3 g	____g	____g	____cm³	H_2O 1.00 g/cm³
E.	____g		1.26 L	____g	CO_2 1.96 g/L
F.	____g	1.46 g	____L	____cm³	

	$N_2(g)$ +	$3\ H_2(g)$ $\longrightarrow$	$2\ NH_3(g)$	
G.	1.46 g	____g	____g	NH_3 0.625 g/L
H.	____g	____g	21.4 L	
I.	____g	19.4 g	____L	

	$6\ HCl(g)$ +	$2\ HNO_3(aq)$ $\longrightarrow$	$4\ H_2O(g)$ +	$2\ NO(g)$ +	$3\ Cl_2(g)$	
J.	18.4 g	____g	____g	____L		NO 1.34 g/L
K	14.2 L	____g	____g	____g		HCl 1.63 g/L
L.	____L	____g	____L	19.4 g		

	$3\ Cu(s)$ +	$8\ HNO_3(aq) \longrightarrow$	$3\ Cu(NO_3)_2(aq)$ +	$2\ NO(g)$ +	$4\ H_2O(l)$
M.	3.14 g	____g		____g	
N.	____g		____g	2.16 L	
O.	19.4 cm³	____g	____g	____L	
P.	____cm³		21.2 g	____L	____g

NO 1.34 g/L
Cu 8.96 g/cm³

6. Each line in the table below represents a different solution. Fill in the blanks, based on the numbers that are given.

	Solute		*Solution*		
	Identity	*Mass, g*	*Moles*	*Volume, mL*	*Concentration, M*
A.	$HC_2H_3O_2$	____	____	250.0	3.414
B.	HCl	____	1.270	400.0	____
C.	$Pb(NO_3)_2$	____	0.221	____	1.104
D.	BaCl2	____	____	275.0	0.888
E.	NaCl	14.3	____	____	1.462
F.	KNO_3	49.7	____	174.0	____
G.	$KHCO_3$	____	____	200.0	1.064
H.	KBr	13.75	____	350.0	____
I.	HNO_3	____	12.4	____	6.00
J.	CsCl	____	0.200	____	1.207
K.	H_2SO_4	____	____	400.0	13.026
L.	H_3PO_4	21.9	____	300.0	____

7. In each line in the table below, two solutions (or a solution and water, a solvent) are mixed to produce a new solution. Fill in the blanks based on the numbers given.

	Solution A			*Solution B*			*Final Solution*		
	Solute	*M*	*Volume*	*Solute*	*M*	*Volume*	*Solute*	*M*	*Volume*
A.	KNO_3	0.250	120 mL	water		450 mL	KNO_3	____	____
B.	HCl	0.250	____mL	water		____mL	HCl	0.100	600 mL
C.	$MgSO_4$	1.74	87.5 mL	water		____mL	$MgSO_4$	1.00	____mL
D.	NaCl	1.40	120 mL	water		500 mL	NaCl	____	____mL
E.	$MgSO_4$	2.40	____mL	water		____mL	Mg^{2+}	0.250	300 mL
F.	Na_2SO_4	3.00	140 mL	water		____mL	Na_2SO_4	0.50	____mL
G.	K_2SO_4	0.350	120 mL	water		600 mL	K_2SO_4	____	____mL
H.	AlI_3	0.750	____mL	water		____mL	AlI_3	0.250	400 mL
I.	KNO_3	0.642	100 mL	NaCl	0.342	200 mL	NaCl	____	____mL
J.	Na_2SO_4	0.578	50. mL	KCl	0.618	____mL	Na_2SO_4	0.214	____mL
K.	$NaNO_3$	0.745	100 mL	$NaNO_3$	0.312	200 mL	$NaNO_3$	____	____mL
L.	$MgSO_4$	0.134	____mL	$MgSO_4$	0.817	____mL	$MgSO_4$	0.400	500 mL
M.	HCl	____	____mL	HCl	1.00	300 mL	HCl	1.45	600 mL
N.	$NaNO_3$	1.50	200 mL	$NaNO_3$	4.00	300 mL	$NaNO_3$	____	____mL

8. Beneath a reactant or a product in each equation below is given either the mass of the species or the volume and concentration of the species in solution. Fill in the blanks with the data provided. To simplify the calculations, assume that any water produced does not dilute the final solution.

	$3\ AgNO_3(aq)$ +	$Na_3PO_4(aq) \longrightarrow$	$Ag_3PO_4(s)$ +	$3\ NaNO_3(aq)$
A.	25.0 mL,1.20 M		____g	100 mL, ____M
B.		____mL, 1.00 M	13.2 g	200 mL, ____M
C.	____mL, 2.00 M		____g	150 mL, 1.75 M
D.		84.0 mL, ____M	6.25 g	____mL, 1.40 M

	$Mg(CN)_2(s)$ +	$H_2SO_4(aq) \longrightarrow$	$2\ HCN(g)$ +	$MgSO_4(aq)$
E.	13.4 g		____g	____mL,1.00 M
F.	____g	____mL,6.00 M	9.47 g	
G.	____g	172 mL, 3.50 M	____g	172 mL, ____M
H.	____g		____g	200 mL,1.75 M

	$Al_2O_3(s)$ +	$3\ H_2SO_4(aq)$ ⟶	$Al_2(SO_4)_3(aq)$ +	$3\ H_2O(l)$
I.	2.31 g	____mL, 6.00 M		____g
J.	____g		174 mL, 3.12 M	____g
K.		196 mL, 1.00 M	196 mL, ____M	____g
L.	____g	____mL, 1.00 M		7.42 g

	$2\ KOH(aq)$ +	$CO_2(g)$ ⟶	$K_2CO_3(aq)$ +	$H_2O(l)$
M.	274 mL, ____M	____g	274 mL, ____M	17.4 g
N.	194 mL, 1.00 M	____g	____mL, 0.50 M	____g
O.	142 mL, ____ M	1.34 g		____g
P.		____g	86.5 mL, 1.00 M	____g

	$2\ KOH(aq)$ +	$H_2SO_3(aq)$ ⟶	$K_2SO_3(aq)$ +	$2\ H_2O(l)$
Q.	182 mL, 1.00 M	50.0 mL, ____M	____mL, 0.750 M	____g
R.	____mL, 1.32 M		143 mL, 1.00 M	____g
S.		____mL, 4.50 M	____mL, 3.21 M	14.3 g

10. Below each equation is given the mass of each reactant. Determine the mass in grams of the product listed first. Also determine the mass in grams of each reactant that is left over.

A. $3\ AgNO_3(aq) + Na_3PO_4(aq) \longrightarrow Ag_3PO_4(s) + 3\ NaNO_3(aq)$
102 g 25.4 g

B. $Ba(OH)_2(aq) + H_2SO_4(aq) \longrightarrow 2\ H_2O(l) + BaSO_4(s)$
105 g 83.1 g

C. $4\ C(s) + BaSO_4(s) \longrightarrow 4\ CO(g) + BaS(s)$
12.4 g 13.2 g

D. $Zn(s) + 2\ HCl(aq) \longrightarrow H_2(g) + ZnCl_2(aq)$
7.15 g 14.2 g

E. $Mg(s) + S(s) \longrightarrow MgS(s)$
21.3 g 19.4 g

F. $Cu_2O(s) + 2\ HCl(g) \longrightarrow 2\ CuCl(s) + H_2O(g)$
49.7 g 32.1 g

G. $4\ NaOH(aq) + SiO_2(s) \longrightarrow 2\ H_2O(l) + Na_4SiO_4(s)$
146 g 84.3 g

H. $2\ Al(OH)_3(s) + 3\ H_2SO_4(aq) \longrightarrow 6\ H_2O(l) + Al_2(SO_4)_3(aq)$
242 g 127 g

I. $SO_3(g) + 2\ HNO_3(aq) \longrightarrow H_2SO_4(aq) + N_2O_5(aq)$
47.2 g 195 g

J. $6\ HCl(aq) + Fe_2O_3(s) \longrightarrow 2\ FeCl_3(aq) + 3\ H_2O(l)$
143 g 204 g

K. $2\ HNO_3(aq) + 3\ H_2S(g) \longrightarrow 4\ H_2O(l) + 2\ NO(g) + 3\ S(s)$
123 g 164 g

L. $Na_2TeO_4(s) + 4\ NaI(s) + 6\ HCl(aq) \longrightarrow 6\ NaCl(s) + Te(s) + 2\ I_2(s) + 3\ H_2O(l)$
174 g 157.1 g 63.2 g

M. $4\ Zn(s) + H_3AsO_4(aq) + 8\ HCl(aq) \longrightarrow 4\ ZnCl_2(aq) + AsH_3(g) + 4\ H_2O(l)$
105 g 194 g 128 g

N. $2\ KNO_3 + 4\ H_2SO_4 + 3\ Hg \longrightarrow K_2SO_4 + 3\ HgSO_4 + 4\ H_2O + 2\ NO$
48.7 g 102 g 193 g

O. $3\ K_2Cr_2O_7 + 2\ C_2H_3OCl + 12\ H_2SO_4 \longrightarrow Cl_2 + 6\ CrSO_4 + 4\ CO_2 + 15\ H_2O$
247 g 92.1 g 374 g

10. In each line below each reaction, determine the mass of reactant, the mass of product, or the percent yield as requested.

	$P_4(s) + 6\ Cl_2(g)$ ⟶	$4\ PCl_3(g)$	
A.	24.6 g	100.0 g	____%
B.	19.3 g	____g	74.2%
C.	____g	81.2 g	63.0%

$$CaCO_3(s) \xrightarrow{\text{heat}} CaO(s) + CO_2(g)$$

D.	87.6 g	30.7 g	___%
E.	143 g	___ g	74.1%
F.	___ g	45.3 g	82.4%

$$4\,NH_3(g) + 5\,O_2(g) \longrightarrow 4\,NO(g) + 6\,H_2O(l)$$

G.	48.2 g	30.7 g	___%
H.	___ g	137 g	73.5%
I.	14.6 g	___ g	89.7%

$$Fe_2O_3(s) + 3\,H_2(g) \xrightarrow{\text{heat}} 2\,Fe(s) + 3\,H_2O(l)$$

J.	64.7 g	14.0 g	___%
K.	18.9 g	___ g	94.6%
L.	___ g	14.3 g	79.6%

$$2\,HNO_3(aq) + 3\,H_2S(g) \longrightarrow 4\,H_2O(g) + 2\,NO(g) + 3\,S(s)$$

M.	124 g	85.1 g	___%
N.	8.43 g	___ g	93.1%
O.	___ g	1.74 g	90.2%

11. Answer each question based on the reactions that follow it.
   A. 14.5 g of a $AgNO_3$–$Pb(NO_3)_2$ mixture that is 74.0% $AgNO_3$ consumes how many grams of NaCl? ...how many mL of 0.500 M NaCl(aq)?
      $$AgNO_3(aq) + NaCl(aq) \longrightarrow AgCl(s) + NaNO_3(aq)$$
      $$Pb(NO_3)_2(aq) + 2\,NaCl(aq) \longrightarrow PbCl_2(s) + 2\,NaNO_3(aq)$$
   B. 17.4 g of a $CaCO_3$–$NaHCO_3$ mixture that is 45.0% $CaCO_3$ produces what mass of $CO_2(g)$? ...consumes how many mL of 0.100 M HCl?
      $$CaCO_3(s) + 2\,HCl(aq) \longrightarrow CaCl_2(aq) + H_2O(l) + CO_2(g)$$
      $$NaHCO_3(s) + HCl(aq) \longrightarrow NaCl(aq) + H_2O(l) + CO_2(g)$$
   C. 74.2 g of a KOH–LiOH mixture that is 17.2% KOH consumes what mass of $CO_2(g)$? ...produces what total mass of solid?
      $$2\,KOH(s) + CO_2(g) \longrightarrow K_2CO_3(s) + H_2O(l)$$
      $$2\,LiOH(s) + CO_2(g) \longrightarrow Li_2CO_3(s) + H_2O(l)$$
   D. 347 mL of a solution with $[H_2SO_4] = 0.640$ M and $[HCl] = 0.935$ M consumes what volume of 0.623 M KOH? ...produces what mass of $H_2O(l)$?
      $$H_2SO_4(aq) + 2\,KOH(aq) \longrightarrow K_2SO_4(aq) + 2\,H_2O(l)$$
      $$HCl(aq) + KOH(aq) \longrightarrow KCl(aq) + H_2O(l)$$
   E. 9.52 g of a Mg–Fe mixture containing 17.4% Mg produces what mass of $H_2(g)$? ...consumes what volume of 0.200 M HCl?
      $$Mg(s) + 2\,HCl(aq) \longrightarrow MgCl_2(aq) + H_2(g)$$
      $$2\,Fe(s) + 6\,HCl(aq) \longrightarrow 2\,FeCl_3(aq) + 3\,H_2(g)$$
   F. 84.6 g of a ZnO–Al_2O_3 mixture containing 43.6% ZnO produces what mass of $H_2O(l)$? ...consumes what volume of 3.47 M H_2SO_4?
      $$ZnO(s) + H_2SO_4(aq) \longrightarrow ZnSO_4(aq) + H_2O(l)$$
      $$Al_2O_3(s) + 3\,H_2SO_4(aq) \longrightarrow Al_2(SO_4)_3(aq) + 3\,H_2O(l)$$
   G. 17.3 g of a Na_3PO_4–Na_2CrO_4 mixture that contains 26.1 % Na_2CrO_4 produces what mass of $NaNO_3(aq)$, once the solvent water is evaporated? ...what mass of silver–containing solid?
      $$3\,AgNO_3(aq) + Na_3PO_4(aq) \longrightarrow Ag_3PO_4(s) + 3\,NaNO_3(aq)$$
      $$2\,AgNO_3(aq) + Na_2CrO_4(aq) \longrightarrow Ag_2CrO_4(s) + 2\,NaNO_3(aq)$$

12. Answer each question based on the series of reactions that follows it.
   A. 19.4 g of CuS(s) produces what mass of H_2SO_4?
   B. 27.5 g of H_2SO_4 requires what mass of $O_2(g)$ for its production?
      $$2\,CuS(s) + 3\,O_2(g) \longrightarrow 2\,CuO(s) + 2\,SO_2(g)$$
      $$2\,SO_2(g) + O_2(g) \longrightarrow 2\,SO_3(g)$$
      $$SO_3(g) + H_2O(l) \longrightarrow H_2SO_4(aq)$$

C. 83.1 g of Cu(s) requires what mass of NaCl(s)?
D. 34.2 g of $H_2O(l)$ requires what volume of 12.5 M H_2SO_4?
E. 28.2 g of NaCl produces what mass of copper?
$2\ NaCl(s) + H_2SO_4(aq) \longrightarrow Na_2SO_4(aq)\ 2\ HCl(g)$
$6\ HCl(g) + 2\ Fe(s) \longrightarrow 2\ FeCl_3(aq) + 3\ H_2(g)$
$CuO(s) + H_2(g) \longrightarrow Cu(s) + H_2O(l)$
F. 85.1 g N_2 produces what mass of HNO_3? Do not recycle by–product NO(g).
G. 14.3 g HNO_3 requires what mass of $H_2(l)$?
$N_2(g) + 3\ H_2(g) \longrightarrow 2\ NH_3(g)$
$4\ NH_3(g) + 5\ O_2(g) \longrightarrow 4\ NO(g) + 6\ H_2O(g)$
$2\ NO(g) + O_2(g) \longrightarrow 2\ NO_2(g)$
$3\ NO_2(g) + H_2O(l) \longrightarrow 2\ HNO_3(aq) + NO(g)$
H. 19.5 g $CaCO_3(s)$ produces what mass of $CaCl_2$?
I. 73.9 g $CaCO_3(s)$ produces what mass of $NH_3(g)$?
J. 43.1 g $H_2O(l)$ requires what mass of $CaCO_3(s)$?
$CaCO_3(s) \longrightarrow CaO(s) + CO_2(g)$
$CaO(s) + H_2O(l) \longrightarrow Ca(OH)_2(s)$
$Ca(OH)_2(aq) + 2\ NH_4Cl(aq) \longrightarrow CaCl_2(aq) + 2\ NH_3(g) + 2\ H_2O(l)$
K. 21.7 g $P_4(s)$ produces what mass of AgCl(s)?
L. 34.2 g $Cl_2(g)$ produces what volume of 3.40 M HNO_3?
M. 27.4 g AgCl(s) requires what mass of $P_4(s)$? ...of $Cl_2(g)$?
$P_4(s) + 3\ Cl_2(g) \longrightarrow 4\ PCl_3(g)$
$PCl_3(g) + 3\ H_2O(l) \longrightarrow H_3PO_4(aq) + 3\ HCl(aq)$
$HCl(aq) + AgNO_3(aq) \longrightarrow AgCl(s) + HNO_3(aq)$

13. Write the overall reaction for each of the five series of reactions given in the drill problems for objective 4-12.

QUIZZES (20 minutes each). Choose the best answer or fill in the blanks for each question. Nomenclature questions are included for review.

Quiz A

1. The name of $NaNO_2$ is ________________.
2. The formula of sulfuric acid is __________.
3. $AlPO_4 + Ca(OH)_2 \longrightarrow Al(OH)_3 + Ca_3(PO_4)_3$ When this equation is balanced with the smallest whole number coefficients, the coefficient of $Al(OH)_3$ is (a) 4; (b) 8; (c) 5; (d) 7; (e) none of these.
4. $Fe_2O_3 + C \longrightarrow Fe + CO$ When this equation is balanced with the smallest whole number coefficients, the sum of all the coefficients is (a) 4; (b) 8; (c) 5; (d) 7; (e) none of these.
5. $PCl_5 + 4\ H_2O \longrightarrow H_3PO_4 + 5\ HCl$ 12.0 g PCl_5 (208.5 g/mol) produces how many grams of HCl (36.5 g/mol)? (a) 12.0(36.5 × 5)/208.5; (b) 12.0(26.5/208.5); (c) 12.0 × 36.5/208.5; (d) 12.0/208.5 (e) none of these.
6. $Fe_2O_3 + 3\ H_2 \longrightarrow 2\ Fe + 3\ H_2O$ 15.0 g Fe_2O_3 (159.0 g/mol) and 3.1 g H_2 (2.0 g/mol) produce how many grams of H_2O (18.0 g/mol)? (a) 15.0(18.0/159.6) (b) 15.0(18.0/159.6) × 3; (c) 3.1(18.0/2.0) × 3;
(d) 3.1(18.0/2.0); (e) none of these.
7. Complete and balance $Na_2S(s) + HCl(aq) \longrightarrow$ ________________
8. How many mL of 14.5 M HNO_3 must be diluted to produce 5.00 L of 1.00 M solution? (a) 5000(14.5/1.00); (b) 5000(1.00/14.5); (c) 5.00(1.00/14.5); (d) 5.00(14.5/1.00); (e) none of these.
9. 56.7 g of sodium bicarbonate (84.0 g/mol) is enough solute for how many mL of 2.77 M solution? (a) (56.7/84.0)/2.77; (b) (84.0)(56.7)(2.77); (c) (84.0/56.7)(1000/2.77); (d) (56.7/84.0)(1000/2.77); (e) none of these.
10. $N_2(g) + 3\ H_2(g) \longrightarrow 2\ NH_3(g)$ has an 85% yield. What mass in grams of $NH_3(g)$ are produced from 12.0 g $N_2(g)$? (a) (12.0/28.0)(2)(17.0)(0.85); (b) (12.0/14.0)(2)(17.0/0.85); (c) (12.0/28.0)(2)(17.0/0.85); (d) (12.0/14.0)(2)(0.85)(17.0); (e) none of these.

Quiz B

1. The name of $Mg(ClO_3)_2$ is ______________________.
2. The formula of sodium hypofluorite is ________________.
3. $NO_2 + H_2 \longrightarrow NH_3 + H_2O$ When this equation is balanced with the smallest whole number coefficients, the coefficient of nitrogen(IV) oxide is (a) 1; (b) 2; (c) 3; (d) 4; (e) none of these.
4. $H_2S + O_2 \longrightarrow H_2O + SO_3$ When this equation is balanced with the smallest whole number coefficients, the sum of the coefficients is (a) 4; (b) 5; (c) 6; (d) 8; (e) none of these.
5. How many moles of SO_2 are produced when 72.0 g H_2O is produced by the process $2\ H_2S + 3\ O_2 \longrightarrow 2\ H_2O + 2\ SO_2$ (H = 1.0, O = 16.0, S = 32.1 g/mol)? (a) (72.0/17.0)(3/2); (b) (72.0/18.0); (c) (72.0/18.0)(2/3); (d) (72.0)(18.0); (e) none of these.
6. $2\ Al + Fe_2O_3 \longrightarrow 2\ Fe + Fe_2O_3$ 2.5 g Al (27.0 g/mol) and 7.2 g Fe_2O_3 (158.9 g/mol) produce how many grams of Fe (55.8 g/mol)? (a) 2.5(55.9/27.0); (b) 7.2(55.9/158.9); (c) 2.5(55.9 × 2)/(27.0 × 2); (d) (7.2)(55.9 × 2)/158.9; (e) none of these.
7. Complete and balance $Al_2O_3(aq) + H_2SO_4(aq) \longrightarrow$ ______________________.
8. 67.5 mL of a 3.15 M solution is diluted to 2.50 L with water. The molarity of the resulting solution is (a) (3.15)(67.5)(2500); (b) (3.15/67.5)/2500; (c) 3.15(67.5/2500); (d) 3.15(2500/67.5); (e) none of these.
9. 151.6 mL of a 0.45 M solution of iron(II) chloride contains how many grams of solute? (a) (151.6)(0.45)(126.8); (b) (151.6/1000)(0.45)(146.1); (c) (151.6/0.45)(126.8); (d) (151.6/0.45)(146.1); (e) none of these.
10. $CO(g) + 2\ H_2(g) \longrightarrow CH_3OH(l)$ has a 72.1 % yield. 14.0 g $CH_3OH(l)$ requires what mass of hydrogen? (a) (14.0/32.0)(2)(2.0/72.1); (b) (14.0/32.0)(2)(1.0/72.1); (c) (14.0/32.0)(2)(2.0/0.721); (d) (14.0/32.0)(2)(2.0)(0.721); (e) none of these.

Quiz C

1. The name of $RbBrO_4$ is ______________________.
2. The formula of nitric acid is ______________________.
3. $H_3PO_4 + NaOH \longrightarrow H_2O + Na_3PO_4$ When this equation is balanced with the smallest whole number coefficients, the coefficient for water is (a) 1; (b) 2; (c) 3; (d) 4; (e) none of these.
4. $AgNO_3 + CuCl_2 \longrightarrow Cu(NO_3)_2 + AgCl$ When this equation is balanced with the smallest whole number coefficients, the sum of all the coefficients is (a) 8; (b) 5; (c) 6; (d) 4; (e) none of these.
5. $2\ C_3H_7OH + 9\ O_2 \longrightarrow 6\ CO_2 + 8\ H_2O$ 47.2 g CO_2 (44.0 g/mol) requires what mass in grams of C_3H_7OH (60.0 g/mol)? (a) 47.2[(6 × 44.0)/(2 × 60.0)]; (b) 47.2[(2 × 60.0)/(6 × 44.0)]; (c) 47.2(60.0/44.0); (d) 47.2(44.0/60.0); (e) none of these.
6. $Mg_3N_2 + 6\ H_2O \longrightarrow 3\ Mg(OH)_2 + 2\ NH_3$ 2.5 mol Mg_3N_2 (100.9 g/mol) and 6.0 mol H_2O (18.0 g/mol) produce how many grams of NH_3 (17.0 g/mol)? (a) (6.0/18.0)(2 × 17.0); (b) 2 × 17.0; (c) 2.5(2 × 17.0); (d) (2.5 × 100.9)(2 × 17.0); (e) none of these.
7. Complete and balance $Ba(OH)_2(aq) + H_2SO_4(aq) \longrightarrow$ ______________________.
8. If 85.6 mL of a 6.75 M solution is diluted to 6.20 L with water, what is the concentration of the final solution? (a) 6.75(85.6/6.20); (b) 6.75(6.20/85.6); (c) 6.75(6200/85.6); (d) 6.75(85.6/6200); (e) none of these.
9. 433 mL of a 0.675 M solution of sodium sulfite contains how many grams of solute? (a) (433/1000)(0.675)(126.1); (b) (433/1000)(0.675)(142.1); (c) (433)(0.675)(103.1); (d) (433)(0.675)(126.1); (e) none of these.
10. $2\ SO_2(g) + O_2(g) \longrightarrow 2\ SO_3(g)$ has a 93.7% yield. What mass in grams of $SO_3(g)$ is produced from 6.75 g of $SO_2(g)$? (a) (6.75/64.1)(80.1/93.7); (b) (6.75/64.1)(80.1/0.937); (c) 6.75/0.937; (d) (6.75/64.1)(80.1)(93.7); (e) none of these.

Quiz D

1. The name of H_2SO_4 is ______________________.
2. The formula of sodium bicarbonate is ______________________.
3. $PCl_3 + H_2O \longrightarrow HCl + H_3PO_3$ When this equation is balanced with the smallest whole number coefficients, the coefficient of hydrochloric acid is (a) 6; (b) 2; (c) 3; (d) 4; (e) none of these.
4. $SCl_4 + H_2O \longrightarrow H_2SO_3 + HCl$ When this equation is balanced with the smallest whole number coefficients, the sum of the coefficients is (a) 4; (b) 6; (c) 8; (d) 10; (e) none of these.

5. How many moles of NH_3 are required to produce 256 g N_2H_4 via the process. $4\ NH_3 + Cl_2 \longrightarrow N_2H_4 + 2\ NH_4Cl$ (Cl = 35.5, N = 14.0, H = 1.0 g/mol)? (a) (256/16)(4); (b) (256/32)(2); (c) (256/32)(4); (d) (256/32)/4; (e) none of these.
6. $2\ Fe_2O_3 + 3\ C \longrightarrow 4\ Fe + 3\ CO_2$ 18.0 g Fe_2O_3 (159.7 g/mol) and 2.5 g C (12.0 g/mol) produce what mass in grams of Fe (55.8 g/mol)? (a) 18.0(55.8/159.7); (b) 18.0(4 × 55.9)/(2 × 159.7); (c) 2.5(55.8/12.0); (d) 2.5(55.8 × 4)/(3 × 12.0); (e) none of these.
7. Complete and balance $CaCO_3(s) + HCl(aq) \longrightarrow$ ________________.
8. How many mL of 17.0 M HCl solution is needed to produce 2.0 L of a 3.00 M HCl solution? (a) (2000.17.0)/3.00; (b) 2.000(17.0)(3.00); (c) 2000(17.0/3.00); (d) 2000(3.00/17.0); (e) none of these.
9. 81.2 mL of a 14.0 M solution of nitric acid contains how many grams of solute? (a) (81.2/1000)(14.0)(81.0); (b) (81.2/1000)(14.0)(63.0); (c) (81.2)(14.0)(47.0); (d) (81.2)(14.0)(65.0); (e) none of these.
10. $2\ P(s) + 5\ Cl_2(g) \longrightarrow 2\ PCl_5(g)$ has a 89.0% yield. 19.2 g $PCl_5(g)$ requires how many grams of $Cl_2(g)$? (a) (19.2/208.5)(5/2)(71.0/0.890); (b) (19.2/66.5)(5/2)(71.0/0.890); (c) (19.2/208.5)(71.0)(0.890); (d) (19.2/66.5)(5/2)(71.0)(0.890); (e) none of these.

SAMPLE TEST (30 minutes)

1. Give the name or formula, as appropriate, of each of the following compounds.
   a. $Pb(CNS)_2$ b. lead(IV) thiosulfate c. arsenic(III) sulfide
   d. $Mn(BrO)_2$ e. HNO_2 f. periodic acid
   g. stannous sulfide h. ammonium bicarbonate
2. What volume in mL of 0.160 M KNO_3 must be mixed with 200.0 mL of 0.240 M K_2SO_4 to produce a solution having $[K^+] = 0.400$ M?
3. Complete and balance the following equations.
   a. $BaCl_2 + AgNO_3 \longrightarrow$ b. $Ca(OH)_2 \xrightarrow{heat}$ c. $NH_3 + HCl \longrightarrow$
   d. $Zn + CuSO_4 \longrightarrow$ e. $C_2H_6 + O_2 \longrightarrow$
4. Consider the reaction $CuO(s) + 2\ HCl(aq) \longrightarrow CuCl_2(aq) + H_2O(l)$. If 55.2 g CuO is added to 158 mL of 2.70 M HCl, what mass of H_2O is formed?
5. 26.4 g of a NaOH–CaO mixture that contains 40.0% CaO reacts with HCl according to the equations
   $NaOH(s) + HCl(aq) \longrightarrow NaCl(aq) + H_2O(l)$
   $CaO(s) + 2\ HCl(aq) \longrightarrow CaCl_2(aq) + H_2O(l)$
   All of the mixture reacts and there is no HCl left over. The resulting solution is evaporated to dryness. What is the weight of solid obtained?

5 Introduction to Reactions in Aqueous Solutions

CHAPTER OBJECTIVES

* **1. Identify compounds according to whether they are non-, weak, or strong electrolytes; strong or weak acids or bases; or salts.**

As a general rule, ionic compounds are completely dissociated into ions when they dissolve in water. That is, ionic compounds are strong electrolytes. These soluble ionic compounds also are known as salts. Recall that an ionic compound consists of a cation and an anion. Hence, a thorough mastery of the common ions listed in Tables 3-2 and 3-3 will make it easy for you to recognize ionic compounds. If the ions of the compound are not common ones, you still can recognize an ionic compound as consisting of a metal and a nonmetal. (Recall objective 3-12, to distinguish between a metal and a nonmetal.)

Also, it is easy to remember that practically all covalent compounds are nonelectrolytes. Of course, covalent compounds contain only nonmetal atoms. (Ammonium ion is the only common cation that contains no metal atoms.)

There are two large classes of covalent compounds that are electrolytes: acids andbases. Strong acids and strong bases are completely dissociated into ions in aqueous solution; they are strong electrolytes. It is essential that you memorize the common strong acids and bases, listed in Table 5-1. The periodic table can help. The strong bases are the hydroxides of the Group 1A and 2A elements with the exceptions of $Be(OH)_2$ and $Mg(OH)_2$. Three of the strong acids are hydrogen halides. HF is a weak acid.

TABLE 5-1
Common Strong Acids and Bases

Acids	*Bases*
HCl	LiOH
HBr	NaOH
HI	KOH
HNO_3	RbOH
$HClO_3$	CsOH
H_2SO_4	$Ca(OH)_2$
	$Sr(OH)_2$
	$Ba(OH)_2$

The formulas of all acids,weak as well as strong, begins with H, and their names end with the word "acid." (Later, when you study organic chemistry, you will note that the formulas of organic acids end with —COOH. Thus, we shall write the formula of acetic acid as $HC_2H_3O_2$, but organic chemists write it as CH_3COOH.) Of course, if an acid is not strong (listed in Table 5-1), it must be weak, and thus only partially dissociated in aqueous solution. Weak bases are harder to identify. Most of them are based on ammonia, NH_3. They are produced by replacing one of the H's with a group of atoms. Thus, CH_3NH_2, where —CH_3 replaces —H, is a weak base (methylamine).

*** 1a. Be able to write the names and formulas of acids.**

Notice that acids are named differently than salts. HNO_3 is not "hydrogen nitrate" but nitric acid. In order to form the name of the acid, one begins with the name of the anion. Remove the anion name's suffix (the last three letters in each case) and replace it with the suffix of the acid as illustrated in Table 5-2. In the case of binary acids (those composed of hydrogen and only one other element) the prefix hydro– also is added.

TABLE 5-2 Nomenclature of Acids

		Examples	
Anion suffix	*Acid suffix*	*Anion name, formula*	*Acid name, formula*
–ite	–ous	nitrite, NO_2^-	nitrous acid, HNO_2
–ate	–ic	oxalate, $C_2O_4^{2-}$	oxalic acid, $H_2C_2O_4$
–ide	(hydro–) –ic	iodide, I^-	hydroiodic acid, HI

*** 2. Relate the molarities of ions in solutions to the concentrations of strong electrolytes from which they are derived.**

Determining the concentrations of ions in aqueous solutions is similar to determining the concentration of the solute. (Recall objectives 4-6 and 4-7.) There is one added feature—each mole of solute may produce more than one mole of cations or anions. The calculation thus involves an additional conversion factor, derived from the chemical formula of the solute.

EXAMPLE 5-1 How many millimoles of nitrate ion are present in 114 mL of 0.357 M $Al(NO_3)_3$ solution?

This is similar to Example 4-9, with the addition of the conversion factor mentioned above.

$$114 \text{ mL soln} \times \frac{0.357 \text{ mmol Al(NO}_3)_3}{\text{mL soln}} \times \frac{3 \text{ mmol NO}_3^-}{\text{mmol Al(NO}_3)_3} = 122 \text{ mmol NO}_3^-$$

Of course, problems similar to Examples 4-8, 4-10, 4-11, and 4-12 also can involve ionic solutes. Perhaps the most involved one of these is the situation where two solutions, which have one ion in common, are mixed.

EXAMPLE 5-2 54.0 mL of 0.314 M $Mg(NO_3)_2$ is mixed with 114 mL of 0.357 M $Al(NO_3)_3$ solution. What is the final concentration of nitrate ion, $[NO_3^-]$?

Total solution volume = 54.0 mL + 114 mL = 168 mL.
First the amount of nitrate ion in each of the two solutions is determined.

$$54.0 \text{ mL} \times \frac{0.628 \text{ mmol Mg(NO}_3)_2}{\text{mL soln}} \times \frac{2 \text{ mmol NO}_3^-}{\text{mmol Mg(NO}_3)_2} = 33.9 \text{ mmol NO}_3^-$$

$$114 \text{ mL} \times \frac{0.357 \text{ mmol Al(NO}_3)_3}{\text{mL soln}} \times \frac{3 \text{ mmol NO}_3^-}{\text{mmol Al(NO}_3)_3} = 122 \text{ mmol NO}_3^-$$

$$[NO_3^-] = \frac{(33.9 + 122) \text{ mmol NO}_3^-}{168 \text{ mL soln}} = 0.928 \text{ M}$$

*** 3. State general rules that apply to the aqueous solubilities of ionic compounds, and write net ionic equations based on these solubility rules.**

Solubility rules enable us to determine whether a precipitation reaction will occur. In most precipitation reactions that we will study, aqueous solutions of two soluble compounds are mixed, and their ions rearrange into those of two new compounds, one of which is insoluble. This is the description of a metathesis reaction (recall objective 4-4) in which one of the products is insoluble. An example is the reaction of $Ba(NO_3)_3(aq)$ and $Al_2(SO_4)_3(aq)$, equation [1].

$$3 \text{ Ba(NO}_3)_2\text{(aq)} + \text{Al}_2\text{(SO}_4)_3\text{(aq)} \longrightarrow 3 \text{ BaSO}_4\text{(s)} + 2 \text{ Al(NO}_3)_3\text{(aq)} \quad [1]$$

The solubility rules in the text can be restated in a fashion that is perhaps more memorable. As you learn these rules, remember three points. First, the rules are organized by the anions of the compounds. Second, the compounds of one group of anions generally are soluble, while those of another group generally are insoluble. Third, most of the rules have exceptions, even though they deal only with common compounds. The restatement of the rules follows, with some additional anions (written in *italics*) included for completeness.

1. The common compounds of ammonium ion (NH_4^+) and the alkali metal (periodic group 1A) cations are **soluble.**
2. Most of the common compounds of the following anions are **soluble.**
   a. The common compounds of nitrate (NO_3^-), chlorate (ClO_3^-), perchlorate (ClO_4^-), acetate ($C_2H_3O_2^-$), and *permanganate* (MnO_4^-), *nitrite* (NO_2^-) ions are **soluble.** Exceptions are moderately soluble $AgC_2H_3O_2$ and insoluble $AgNO_2$.
   b. The common compounds of chloride (Cl^-), bromide (Br^-), and iodide (I^-) ions are **soluble.** Exceptions are the insoluble compounds of these anions with Pb^{2+}, Hg_2^{2+}, and Ag^+.
   c. The common compounds of sulfate (SO_4^{2-}) and *thiosulfate* ($S_2O_3^{2-}$) ions are **soluble.** Exceptions are the insoluble compounds of these anions with Sr^{2+}, Ba^{2+}, Pb^{2+}, and Hg_2^{2+}, and their moderately soluble compounds with Ca^{2+} and Ag^+.
3. Most of the common compounds of the following anions are **insoluble.** Exceptions, of course, are the compounds that they form with ammonium ion and the alkali metal cations.
   a. The common compounds of the carbonate (CO_3^{2-}), chromate (CrO_4^{2-}), phosphate (PO_4^{3-}), *sulfite* (SO_3^{2-}), and *oxalate* ($C_2O_4^{2-}$) ions are **insoluble.** An exception is soluble $Fe_3(C_2O_4)_3$.
   b. The common compounds of the hydroxide (OH^-) and *oxide* (O^{2-}) ions are **insoluble.** Exceptions are their moderately soluble compounds with Ca^{2+}, Sr^{2+}, and Ba^{2+}.
   c. The common compounds of the sulfide ion (S^{2-}) are **insoluble.** Exceptions are the soluble sulfides of the alkaline earth (group 2A) cations.

Ionic equations were first discussed in objective 4-2a. To produce a net ionic equation, one can start with a non-ionic equation, one written in terms of compounds. Consider the reaction of aqueous solutions of $Al_2(SO_4)_3$ and $Ba(NO_3)_2$.The products of the reaction are predicted by switching the partners of each ion. (Thus, this is a metathesis reaction.) They are $BaSO_4$ and $Al(NO_3)_3$. The solubility rules (specifically, rules 2.c. and 2.a.) indicate that $BaSO_4$ is insoluble and $Al(NO_3)_3$ is soluble. The unbalanced chemical expression then is.

$$Al_2(SO_4)_3(aq) + Ba(NO_3)_2(aq) \longrightarrow BaSO_4(s) + Al(NO_3)_3(aq) \quad [2]$$

We now take advantage of a simple generality: *Ionic compounds dissociate completely into ions when they dissolve in water.*The ions are written for each compound that is designated (aq), and we have

$$Al^{3+}(aq) + SO_4^{2-}(aq) + Ba^{2+}(aq) + SO_4^{2-}(aq) \longrightarrow BaSO_4(s) + Al^{3+}(aq) + NO_3^-(aq) \quad [3]$$

[Be careful to not break into ions those compounds that are designated (s), (l), or (g) when writing net ionic equations.] Notice that the ions are written without stoichiometric coefficients. To obtain the net ionic expression, eliminate those ions that appear on both sides of the expression. These ions are known as spectator ions.

$$SO_4^{2-}(aq) + Ba^{2+}(aq) \longrightarrow BaSO_4(s) \quad [4]$$

If necessary, balance this expression to obtain the net ionic equation. The net ionic equation is more general than any specific equation. In this case, for example, it tells us that a $BaSO_4(s)$ precipitate will form when any soluble barium compound is mixed with a soluble sulfate compound in aqueous solution.

* **4. Write net ionic equations for neutralization reactions and for reactions that result in the dissolving (dissolution) of a water-soluble substance or the evolution of a gas.**

In neutralization reactions, an acid and a base react to form a salt and water. An example is the reaction of sodium hydroxide with sulfuric acid.

$$2\ NaOH(aq) + H_2SO_4(aq) \longrightarrow Na_2SO_4(aq) + 2\ H_2O(l) \quad [5]$$

The net ionic equation for this neutralization reaction, and for all others, is the reaction of hydroxide ion with hydrogen ion. (We often write hydrogen ion as H^+ for simplicity, even though we recognize that it exists in aqueous solution as H_3O^+.)

$$OH^-(aq) + H^+(aq) \longrightarrow H_2O \quad [6]$$

Recognize that sometimes acid-base reactions may occur between an acid and an oxide (of a metal) or between a base and an oxide (of a nonmetal). The following two equations are examples.

$$H_2SO_4(aq) + CaO(s) \longrightarrow CaSO_4(aq) + H_2O \quad [7]$$

$$2\ NaOH(aq) + SO_2(aq) \longrightarrow Na_2SO_3(aq) + H_2O \quad [8]$$

The net ionic equation for the dissolving of a water soluble substance has that substance as the reactant, and its dissolved form as the products. In the case of strong electrolytes, the products are ions, as in the case of the dissolution of $Al_2(SO_4)_3$. (Note that we used this net ionic equation, without the stoichiometric coefficients, in obtaining equation [3] from equation [2].)

$$Al_2(SO_4)_3(s) \xrightarrow{H_2O} 2\ Al^{3+}(aq) + 3\ SO_4^{2-}(aq) \qquad [9]$$

A limited number of gases are produced from aqueous solutions, most through the actions of acids and bases. Two of these, sulfur dioxide and carbon dioxide, were described in objective 4-4. These are formed by the reaction of an acid on sulfites, and carbonates, respectively.

$$2\ H^+(aq) + SO_3^{2-}(aq) \longrightarrow H_2SO_3(aq) = SO_2{\cdot}H_2O \longrightarrow H_2O(l) + SO_2(g) \qquad [10]$$

$$2\ H^+(aq) + CO_3^{2-}(aq) \longrightarrow \text{"}H_2CO_3(aq)\text{"} \longrightarrow H_2O(l) + CO_2(g) \qquad [11]$$

Sulfur dioxide gas also is formed by the reaction of an acid on thiosulfates. Hydrogen sulfide gas is formed by the reaction of an acid on sulfides. And ammonia gas is formed by the reaction of a base [OH^-, such as from NaOH(aq)] on ammonium ion. This, in fact, is the basis of a qualitative test for ammonium ion.

$$2\ H^+(aq) + S_2O_3^{2+}(aq) \longrightarrow H_2S_2O_3(aq) \longrightarrow SO_2(g) + H_2O(l) + S(s) \qquad [12]$$

$$2\ H^+(aq) + S^{2-}(aq) \longrightarrow H_2S(aq) \qquad [13]$$

$$OH^-(aq) + NH_4^+(aq) \longrightarrow NH_3(g) + H_2O(l) \qquad [14]$$

Sulfur dioxide, carbon dioxide, and hydrogen sulfide, also are produced by reaction with the bisulfite, bicarbonate, and bisulfide ions, respectively. Some gases other than these four are produced, many as the result of oxidation-reduction equations.

*** 5. State and apply the conditions under which reactions between ions in aqueous solution may go to completion.**

Reactions between ions in aqueous solution will go to completion if some of the ions are removed from solution. This will occur when the reaction forms one of three types of substance.

- a precipitate, that is, a compound that is insoluble in the solution
- a gas that escapes from the solution
- a nonelectrolyte, such as water, that results from the union of ions

This method of prediction is straightforward to apply when both reactants are strong electrolytes. Sometimes, however, substances of these types are present on both sides of the chemical equation. We then are unable to predict whether the reaction will go to completion or not. Such is the case with the reaction between cobalt(II) sulfate and hydrogen sulfide.

$$CoSO_4(aq) + H_2S(aq) \longrightarrow CoS(s) + H_2SO_4(aq) \qquad [15]$$

We recognize cobalt(II) sulfide as a precipitate, but we also note that H_2S normally appears as a gas escaping from solution. In this and similar cases we will be content for now to state that the reaction proceeds to a state of equilibrium.

*** 6. Recognize an oxidation-reduction reaction by changes in oxidation state and identify the oxidizing and reducing agents in an oxidation–reduction reaction.**

An oxidation–reduction reaction or a *redox* reaction has occurred when some of the atoms change oxidation state in going from reactants to products. (Objective 3-11 discusses how to assign oxidation states.) An atom is *oxidized* when its oxidation state increases. An atom is *reduced* when its oxidation state decreases. Oxidation and reduction *always* occur together in a reaction, *never* one without the other.

An oxidizing agent (or oxidant) accepts electrons from the substance that is being oxidized. A reducing agent (or reductant) gives electrons to the substance that is being reduced. Thus, the oxidizing agent itself is reduced and the reducing agent is oxidized. (If this seems confusing, think of the following. If you give money to someone, you are that person's enriching agent; yet you are impoverished by the transaction.) An individual atom is *not* the oxidizing or reducing agent but rather the agent is the compound or ion that contains the atom that changes oxidation state. In the equation in objective 4-14, $HNO_3(aq)$ or $NO_3^-(aq)$ is the oxidizing agent, and Cu(s) is the reducing agent.

* **7. Balance oxidation-reduction reactions by the oxidation-state change method.**

The oxidation–state change method is a fast way to balance oxidation–reduction reactions. Its steps are illustrated in balancing the oxidation of copper metal by nitric acid.

$$Cu(s) + HNO_3(aq) \longrightarrow Cu(NO_3)_2(aq) + NO(g) + H_2O(l)$$

1. Write the oxidation state of each element below its symbol.

$$\underset{0}{Cu} + \underset{+1\ +5\ -2}{HNO_3} \longrightarrow \underset{+2\ \ +5\ -2}{Cu(NO_3)_2} + \underset{+2\ -2}{NO} + \underset{+1\ -2}{H_2O}$$

2. Determine which element is oxidized (increases in oxidation state) and connect its reactant and product forms with an arrow above the equation. Write the increase in oxidation state above the arrow. Similarly identify the reduced element and write its decrease above the arrow.

+2 (Cu → Cu)

$$\underset{0}{Cu} + \underset{+1\ +5\ -2}{HNO_3} \longrightarrow \underset{+2\ \ +5\ -2}{Cu(NO_3)_2} + \underset{+2\ -2}{NO} + \underset{+1\ -2}{H_2O}$$

−3 (N in HNO_3 → N in NO)

3. Multiply the oxidation-state increase and decrease by whole numbers so that they have the same magnitude (or absolute value). The whole number multiplier is the stoichiometric coefficient for the compound (or element) at each end of the arrow.

3(+2)

$$3\,\underset{0}{Cu} + 2\,\underset{+1\ +5\ -2}{HNO_3} \longrightarrow 3\,\underset{+2\ \ +5\ -2}{Cu(NO_3)_2} + 2\,\underset{+2\ -2}{NO} + \underset{+1\ -2}{H_2O}$$

2(−3)

4. Balance the rest of the equation by inspection. Do not change the coefficients you already have determined unless the atoms in that compound appear on the other side of the equation with two oxidation states. In this case, the coefficients of Cu, $Cu(NO_3)_2$, and NO are not changed. But HNO_3 contains the nitrogen atom that appears on the product side with two oxidation states (+5 in HNO_3 and +2 in NO). Since there are 8 N atoms on the right-hand side, 8 is the coefficient of HNO_3.

$$3\,Cu(s) + 8\,HNO_3(aq) \longrightarrow 3\,Cu(NO_3)_2(aq) + 2\,NO(g) + 4\,H_2O(l)$$

* **8. Separate an oxidation-reduction equation into half equations; complete and balance the half equations; and recombine them into a balanced net oxidation-reduction equation.**

This method of balancing redox equations is more time-consuming that the oxidation-state change method of objective 5-7. However, it has two advantages. First, does not rely on your ability to determine oxidation states, a fact that many beginners appreciate. Second, it tends to be a bit more reliable than the oxidation-state change method. And finally, it produces two balanced half equations that can be used to determine the standard cell voltage of the reaction (objective 20-3). We illustrate the method by balancing the following redox reaction.

$$HCl(aq) + MnO_2(s) \longrightarrow Cl_2(g) + H_2O(l) + MnCl_2(aq)$$

1. Identify the substances that are oxidized or reduced, separate them into ions (if that is how they exist in the reaction mixture), and write two *couples*.A couple consists of the oxidized and the reduced species of an element separated by an arrow or a bar (|), with the reactant written first. In this reaction, chlorine changes oxidation state from being combined in HCl(aq) to being uncombined in $Cl_2(g)$. Manganese changes oxidation state from $MnO_2(s)$ to $MnCl_2(aq)$. Chlorine is present as $Cl^-(aq)$ in HCl(aq) and manganese as $Mn^{2+}(aq)$ in $MnCl_2(aq)$. Because $MnO_2(s)$ is a solid, it is not separated into ions.

$$Cl^-(aq) \longrightarrow Cl_2(g) \quad \textit{or} \quad Cl^-(aq)|Cl_2(g)$$

$$MnO_2(s) \longrightarrow Mn^{2+}(aq) \quad \textit{or} \quad MnO_2(s)|Mn^{2+}(aq)$$

2. Next, balance all atoms except H and O.

$$2\,Cl^-(aq) \longrightarrow Cl_2(g)$$

3. Now, use H_2O and $H^+(aq)$ to balance H and O atoms. First, add one H_2O for each O atom needed. Then add one $H^+(aq)$ for each H atom needed.

$$MnO_2(s) \longrightarrow Mn^{2+}(aq) + 2\ H_2O$$

$$MnO_2(s) + 4\ H^+(aq) \longrightarrow Mn^{2+}(aq) + 2\ H_2O$$

4. Balance charge by adding electrons to reactants or products as needed.

$$2\ Cl^-(aq) \longrightarrow Cl_2(g) + 2\ e^- \quad [16]$$

$$2\ e^- + 4\ H^+(aq) + MnO_2(s) \longrightarrow Mn^{2+}(aq) + 2\ H_2O \quad [17]$$

These are the two balanced half equations. In equation [16], electrons are produced; this is an oxidation. Chloride ion is oxidized to elemental chlorine and electrons are produced. In equation [17] electrons are consumed; this is a reduction. Manganese(IV) in $MnO_2(s)$ is reduced to manganese(II), $Mn^{2+}(aq)$.

5. Now, multiply the two half equations by whole numbers so that the total number of electrons produced in the oxidation equals the total number consumed in the reduction. Add the resulting (multiplied) half equations, and eliminate substances that appear on both sides. This is the net ionic equation.

$$2\ Cl^-(aq) + MnO_2(s) + 4\ H^+(aq) \longrightarrow Cl_2(g) + Mn^{2+}(aq) + 2\ H_2O$$

6. Finally, add spectator ions as needed to produce the original substances. In this case, we add 2 $Cl^-(aq)$ to each side of the equation.

$$MnO_2(s) + 4\ HCl(aq) \longrightarrow Cl_2(g) + MnCl_2(aq) + 2\ H_2O$$

In alkaline solution, a step is added after step 3. The purpose is to eliminate the $H^+(aq)$, which appears only in extremely small concentrations in alkaline solutions. Consider the following reaction.

$$KMnO_4(aq) + H_2S(g) \longrightarrow MnO_2(s) + S(s) + KOH(aq) + H_2O$$

1. & 2. Write half equations, and balance all atoms except H and O.

$$MnO_4^-(aq) \longrightarrow MnO_2(s)$$

$$H_2S(g) \longrightarrow S(s)$$

3. Balance O and H by adding H_2O and H^+.

$$MnO_4^-(aq) + 4\ H^+(aq) \longrightarrow MnO_2(s) + 2\ H_2O$$

$$H_2S(g) \longrightarrow S(s) + 2\ H^+(aq)$$

3a. Now add, to *both sides* of a half equation, a number of $OH^-(aq)$ equal to the number of $H^+(aq)$ present in that reaction.

$$MnO_4^-(aq) + 4\ H^+(aq) + 4\ OH^-(aq) \longrightarrow MnO_2(s) + 2\ H_2O + 4\ OH^-(aq)$$

$$H_2S(g) + 2\ OH^-(aq) \longrightarrow S(s) + 2\ H^+(aq) + 2\ OH^-(aq)$$

And now, when $H^+(aq)$ and $OH^-(aq)$ appear one the same side of a half equation, combine these two ions to form H_2O and then eliminate H_2O if it appears on both sides of the half equation.

$$MnO_4^-(aq) + 2\ H_2O \longrightarrow MnO_2(s) + 4\ OH^-(aq)$$

$$H_2S(g) + 2\ OH^-(aq) \longrightarrow S(s) + 2\ H_2O$$

4. Balance charge to produce two balanced half equations.

$$3\ e^- + MnO_4^-(aq) + 2\ H_2O \longrightarrow MnO_2(s) + 4\ OH^-(aq) \quad [18]$$

$$H_2S(g) + 2\ OH^-(aq) \longrightarrow S(s) + 2\ H_2O + 2\ e^- \quad [19]$$

5. Equation [18] is multiplied by 2, and equation [19] by 3, and $OH^-(aq)$ and H_2O are eliminated when they appear on both sides of the equation.to produce the balanced net ionic equation.

$$2\ MnO_4^-(aq) + 3\ H_2S(g) + 4\ H_2O + 6\ OH^-(aq) \longrightarrow 2\ MnO_2(s) + 8\ OH^-(aq) + 3\ S(s) + 6\ H_2O$$

Then, $OH^-(aq)$ and H_2O are eliminated when they appear on both sides of the equation to produce the balanced net ionic equation.

$$2\ MnO_4^-(aq) + 3\ H_2S(g) \longrightarrow 2\ MnO_2(s) + 2\ OH^-(aq) + 3\ S(s) + 2\ H_2O$$

6. Two $K^+(aq)$ are added as spectator ions to each side to obtain the balanced oxidation-reduction equation.

$$2\ KMnO_4(aq) + 3\ H_2S(g) \longrightarrow 2\ MnO_2(s) + 2\ KOH(aq) + 3\ S(s) + 2\ H_2O$$

The procedure outlined here differs somewhat from that of the text in that $OH^-(aq)$ are added to the half equations rather than two the net ionic equation. This is done to make the half equations ones that can be used to determine the reaction voltage.

*** 9. Extend the stoichiometric methods of Chapter 4 to reactions involving precipitation, neutralization, or oxidation-reduction.**

Just because reactions are given different names, does not mean that the techniques for dealing with them have changed. In fact, practically all precipitation, gas-evolution, and acid-base reactions are metathesis reactions. And many decomposition, displacement, and combination reactions are oxidation-reduction reactions. Of course, some reactions do not neatly fit into any category. Consider the reaction of solid lithium hydroxide with water.

$$Li_2O(s) + H_2O(l) \longrightarrow 2\ LiOH(aq) \qquad [20]$$

This reaction clearly is a combination reaction; but it does not fit into the categories of precipitation, acid-base, or oxidation-reduction. Another example is the reaction of chlorine with water.

$$3\ Cl_2(aq) + 3\ H_2O(l) \longrightarrow 5\ HCl(aq) + HClO_3(aq) \qquad [21]$$

This is clearly an oxidation-reduction reaction; and yet it cannot be classified as a combination, decomposition, displacement, or metathesis reaction. The drill problems for this objective provide practice in classifying reactions by each of the two schemes.

*** 10. Be familiar with the technique of titration—how the experiment is performed, what data are collected, why an indicator is used, and how to use the data.**

Titrations involve reactions in solution that go to completion. Titrant, a solution of known concentration, is added to a solution of known volume until the endpoint of the titration is reached. The titrant is added in such a way that its total volume is known precisely. Often an indicator is added so that a color change occurs when the endpoint is reached The data are (a) the chemical reaction that occurs, called the titration reaction; (b) the volumes of the solution being titrated and of the titrant; and (c) the concentration of the titrant. From these data, one determines the concentration of the titrated solution or the amount of titrated material present. Thus, the drill problems of this objective are merely applications of objective 4–8.

11. Describe the qualitative analysis scheme for cations and the way in which the types of reactions studied in this chapter enter into the scheme.

In the cation qualitative analysis scheme, cations are first separated into groups with precipitation reactions.

1. Ag^+, Hg_2^{2+}, Pb^{2+} precipitate as chlorides from acidic solution.
2. Hg^{2+}, Pb^{2+}, Bi^{3+}, Cu^{2+}, Cd^{2+}, As^{3+}, Sn^{4+}, and Sb^{3+} precipitate as sulfides from acidic solution.
3. Mn^{2+}, Fe^{2+}, Ni^{2+}, Co^{2+}, and Zn^{2+} precipitate as sulfides from alkaline solution; and Fe^{3+}, Al^{3+}, and Cr^{3+} precipitate as hydroxides from that same alkaline solution.
4. Ca^{2+}, Sr^{2+}, and Ba^{2+} precipitate as carbonated from ammoniacal solution.
5. Mg^{2+}, NH_4^+, K^+, and Na^+ remain in solution after the above treatments.

Even though you are not expected to know the details of the cation qualitative analysis scheme at this point, you should realize that the scheme is consistent with principles studied up to now. For example, the separation of cations into groups, described above, can be explained by reference to the solubility rules (objective 5-3).

*** 12. Draw conclusions about the presence or absence of ions in an unknown from experimental observations.**

The formation of a precipitate or the evolution of a gas can provide evidence that is sufficient to enable you to determine which ions are present and which ones are absent in an ion. If the smell of the gas is known, further conclusions can be drawn. $CO_2(g)$ is odorless, $H_2S(g)$ has the odor of rotten eggs, $SO_2(g)$ has the sharp odor of burning sulfur (the smell of a burning match), and $NH_3(aq)$ has the characteristic odor of ammonia (smelling salts or ammonia cleaning solution).

EXAMPLE 5-3 A solid is known to be $BaSO_4$, $Ba(NO_3)_2$, $CaCO_3$, or Na_2SO_3. $Na_2SO_4(aq)$ is added to its aqueous solution and no precipitate forms. When dilute HCl(aq) is added to the unknown solid, a gas forms and a sharp odor is observed. What is the identity of the unknown?

The solid cannot be $BaSO_4$ because it is soluble in water. Likewise, it cannot be $Ba(NO_3)_2$ because no precipitate (of $BaSO_4$) forms when a solution of sulfate ion is added. The formation of a gas also eliminates $BaSO_4$ and $Ba(NO_3)_2$, since neither anion evolves a gas on the addition of acid. The sharp odor is that of $SO_2(g)$, which is produced from the treatment of sulfite ion with an acid. Thus, the unknown solid is Na_2SO_3.

DRILL PROBLEMS

1. State whether each of the following compounds is a non-, weak, or strong electrolyte. If relevant also state whether each is a weak or strong acid or base or a salt.
A. HCl B. H_2O C. NH_3 D. NaCl E. HNO_3 F. $HC_2H_3O_2$
G. $BaCl_2$ H. SO_2 I. $MnBr_2$ J. KNO_3 K. $MgSO_4$ L. NaOH
M. FeS_2O_3 N. HCN O. CH_4 P. $RbMnO_4$ Q. Na_2CO_3 R. $Ni(NO_3)_2$
S. NH_4Cl T. NH_2OH U. H_2O_2 V. SrI_2 W. K_2S X. HF
Y. $AgClO_3$ Z. Na_3PO_4 Γ. CH_3OH Δ. $CuSO_4$ Θ. K_3PO_4 Λ. $H_2C_2O_4$
Ξ. PCl_3 Π. SCl_4 Σ. $CaCl_2$ Υ. $AuClO_3$ Φ. $CoSO_4$ Ψ. CO_2

1a. Give the name or formula as appropriate of each of the following acids.
A. H_2SO_4 B. nitric acid C. sulfuric acid D. $HC_2H_3O_2$
E. acetic acid F. HCl G. H_3PO_4 H. $HClO_4$
Uncommon Acids I. HIO_3 J. hydrofluoric acid
K. nitrous acid L. HFO M. $H_2C_2O_4$
N. periodic acid O. H_2MnO_4 P. hypobromous acid
Q. hydrocyanic acid R. HIO S. carbonic acid
T. permanganic acid U. perchloric acid V. $HClO_2$

2. (1) Determine the indicated concentration in each of the following solutions.
A. $[NO_3^-]$ in 2.16 M $Al(NO_3)_3$ B. $[SO_4^{2-}]$ in 1.22 M Na_2SO_4
C. $[Fe^{3+}]$ in 0.233 M $Fe_2(C_2O_4)_3$ D. $[Na^+]$ in 0.210 M Na_3PO_4
E. $[Br^-]$ in 2.40 M $MgBr_2$ F. $[Cl^-]$ in 0.743 M $FeCl_3$
G. $[Mg(ClO_3)_2]$ when $[ClO_3^-] = 0.538$ M H. $[Hg_2(NO_3)_2]$ when $[NO_3^-] = 0.428$ M
I. $[Al_2(SO_4)_3]$ when $[Al^{3+}] = 0.338$ M J. $[BaBr_2]$ when $[Br^-] = 0.406$ M
K. $[Fe(NO_3)_3]$ when $[NO_3^-] = 0.972$ M L. $[PO_4^{3-}]$ when $[K_3PO_4] = 0.665$ M

(2) In each line in the table below, two solutions (or a solution and water, a solvent) are mixed to produce a new solution. Fill in the blanks based on the numbers given.

	Solution A			*Solution B*			*Final Solution*		
	Solute	*M*	*Volume*	*Solute*	*M*	*Volume*	*Solute*	*M*	*Volume*
M.	$MgSO_4$	2.40	___mL	water		___mL	Mg^{2+}	0.250	300 mL
N.	Na_2SO_4	3.00	140 mL	water		___mL	SO_4^{2-}	0.50	___mL
O.	K_2SO_4	0.350	120 mL	water		600 mL	K^+	___	___mL
P.	$Al(NO_3)_3$	0.750	___mL	water		___mL	NO_3^-	0.250	400 mL
Q.	KNO_3	0.642	100 mL	NaCl	0.342	200 mL	Na^+	___	___mL
R.	$NaNO_3$	0.745	100	NaCl	0.312	200 mL	Na^+	___	___mL
S.	$MgSO_4$	0.134	___mL	Na_2SO_4	0.817	___mL	SO_4^{2-}	0.400	500 mL
T.	KCl	___	___mL	HCl	1.00	300 mL	Cl^-	1.45	600 mL
U.	NaCl	1.50	200 mL	$MgCl_2$	2.00	300 mL	Cl^-	___	___mL

3. Use the solubility rules to determine whether each of the following compounds is soluble or insoluble in water.
A. Ag_2CrO_4 B. CdS C. $(NH_4)_2S$ D. Cr_2O_3 E. $AlCl_3$ F. $Pb(NO_3)_2$
G. K_3PO_4 H. $Sr(OH)_2$ I. $FeCl_3$ J. CaS K. $BaCO_3$ L. $Cr_2(SO_4)_3$
M. $AgNO_2$ N. Hg_2Cl_2 O. $CdSO_4$ P. $ZnCl_2$ Q. $MgSO_4$ R. $Al(NO_3)_3$
S. Rb_2SO_3 T. $Zn(OH)_2$ U. $FeCl_3$ V. SrS W. $CuSO_4$ X. $Hg_6(PO_4)_2$
Y. $MnBr_2$ Z. CaO Γ. Ag_2SO_4 Δ. NiS_2O_3 Θ. $FeCrO_4$ Λ. $Sr(NO_2)_2$
Ξ. $AgClO_3$ Π. CaC_2O_4 Σ. $Ba(OH)_2$ Υ. $AlPO_4$ Φ. Bi_2S_3 Ψ. $(NH_4)_2CO_3$

3. & 4. Write each of the following as a net ionic expression, then balance the result. All substances are dissolved in aqueous solution unless otherwise indicated, with three exceptions. Water, of course, is a liquid. Insoluble compounds are determined by application of the solubility rules. And sulfur dioxide, carbon dioxide, hydrogen sulfide, and ammonia are gases.

A. $HCl + Pb(NO_3)_2 \rightarrow PbCl_2 + HNO_3$
B. $Ca(OH)_2 + K_3PO_4 \rightarrow Ca_3(PO_4)_2 + KOH$
C. $BaCl_2 + K_2CO_3 \longrightarrow BaCO_3 + KCl$
D. $NaCl + H_3PO_4 \longrightarrow Na_3PO_4 + HCl(g)$
E. $KOH + H_2SO_4 \longrightarrow K_2SO_4 + H_2O$
F. $CaS + HCl \longrightarrow CaCl_2 + H_2S$
G. $BaI_2 + Na_2SO_4 \longrightarrow BaSO_4 + NaI$
H. $AgNO_3 + K_2CrO_4 \rightarrow Ag_2CrO_4 + KNO_3$
I. $Na_2CO_3 + HCl \rightarrow NaCl + H_2O + CO_2$
J. $NH_4Br + KOH \rightarrow NH_3 + H_2O + KBr$
K. $CdSO_4 + K_2S \longrightarrow CdS + K_2SO_4$
L. $Mg(OH)_2 + H_3PO_4 \rightarrow H_2O + Mg_3(PO_4)_2$
M. $FeCl_3 + NaOH \longrightarrow Fe(OH)_3 + NaCl$
N. $AgNO_3 + K_3PO_4 \rightarrow Ag_3PO_4 + KNO_3$
O. $H_3PO_4 + Sr(OH)_2 \rightarrow Sr_3(PO_4)_2 + H_2O$
P. $Al(OH)_3 + HCl \longrightarrow AlCl_3 + H_2O$

5. Complete and balance each of the following equations. Determine whether you expect each reaction to go to completion and state why.

A. $Hg_2(NO_3)_2(aq) + H_3PO_4(aq) \longrightarrow$
B. $Pb(NO_3)_2(aq) + HCl(aq) \longrightarrow$
C. $Sr(NO_3)_2(aq) + H_2SO_4(aq) \longrightarrow$
D. $RbOH(aq) + H_2SO_3(aq) \longrightarrow$
E. $AsCl_3(aq) + H_2S(aq) \longrightarrow$
F. $MnCl_2(aq) + H_2S(aq) \longrightarrow$
G. $Ba(OH)_2(aq) + H_2SO_4(aq) \longrightarrow$
H. $Zn(OH)_2(aq) + HCl(aq) \longrightarrow$
I. $Al_2O_3(aq) + HNO_3(aq) \longrightarrow$
J. $Cu_2O(s) + HCl(aq) \longrightarrow$
K. $Cr_2O_3(s) + H_2SO_4(aq) \longrightarrow$
L. $KOH(aq) + CO_2(g) \longrightarrow$
M. $CuSO_4(aq) + H_2S(aq) \longrightarrow$
N. $ZnCl_2(aq) + H_2S(aq) \longrightarrow$
O. $ZnO(s) + H_3PO_4(aq) \longrightarrow$
P. $Al_2O_3(s) + H_2SO_4(aq) \longrightarrow$
Q. $HCl(aq) + Na_2S(s) \longrightarrow$
R. $K_2CO_3(aq) + H_2SO_4(aq) \longrightarrow$
S. $NH_4Cl(aq) + NaOH(aq) \longrightarrow$
T. $CuSO_3(s) + HBr(aq) \longrightarrow$

6. Identify the oxidizing agent and the reducing agent in each equation in the drill problems of objective 5–7.

7. Balance each of the following equations with the oxidation state change method.

A. $Ag(s) + HNO_3(aq) \longrightarrow AgNO_3(aq) + H_2O(l) + NO(g)$
B. $HCl(aq) + PbO_2(s) \longrightarrow H_2O(l) + Cl_2(l) + PbCl_2(s)$
C. $Al(s) + HBr(aq) \longrightarrow AlBr_3(aq) + H_2(g)$
D. $BaSO_4(s) + C(s) \longrightarrow BaS(s) + CO(g)$
E. $Br_2(l) + H_2O(l) + SO_2(g) \longrightarrow HBr(aq) + H_2SO_4(aq)$
F. $Br_2(l) + KOH(aq) \longrightarrow KBr(aq) + KBrO_3(aq) + H_2O(l)$
G. $C(s) + HNO_3(aq) \longrightarrow CO_2(g) + H_2O(l) + NO_2(g)$
H. $Ca_3(PO_3)_2 + C \longrightarrow Ca_3(PO_4)_2 + P_4$
I. $ClO_2(g) + H_2O_2(aq) + KOH(aq) \longrightarrow H_2O(l) + KClO_2(aq) + O_2(g)$
J. $Cu(s) + HNO_3(aq) \longrightarrow Cu(NO_3)_2(aq) + NO_2(g) + H_2O(l)$
K. $HClO_3(aq) \longrightarrow HClO_4(aq) + ClO_2(g) + H_2O(l)$
L. $H_3AsO_3(aq) + H_2O(l) + I_2(s) \longrightarrow HI(aq) + H_3AsO_4(aq)$
M. $Cr(OH)_3(s) + Na_2O_2aq) \longrightarrow Na_2CrO_4(aq) + NaOH(aq) + H_2O(l)$
N. $FeI_2(aq) + H_2SO_4(aq) \longrightarrow Fe_2(SO_4)_3(aq) + I_2(s) + SO_2(g) + H_2O(l)$
O. $FeS(s) + HNO_3(aq) \longrightarrow Fe(NO_3)_3(aq) + S(s) + NO_2(g) + H_2O(l)$

8. (1) Write the balanced half equation for each of the following couples, in acidic solution, unless indicated (B) as being in alkaline solution.

A. $Ag(s)|Ag^+(aq)$
B. $HNO_3(aq)|NO(g)$
C. $HCl(aq)|Cl_2(g)$
D. $PbO_2(s)|Pb^{2+}(aq)$
E. $Br_2(l)|BrO_3^-(aq)$
F. $Br_2(l)|BrO_3^-(aq)$ (B)
G. $HNO_3(aq)|NO_2(g)$
H. $H_2O_2(aq)|O_2(g)$ (B)
I. $HClO_3(aq)|ClO_2(g)$
J. $HClO_3(aq)|HClO_4(aq)$
K. $Cr(OH)_3(s)|CrO_4^{2-}$ (B)
L. $O_2^{2-}(aq)|OH^-(aq)$ (B)
M. $H_2SO_4(aq)|SO_2(g)$
N. $NO_2^-(aq)|NO_3^-(aq)$ (B)
O. $MnO_4^{2-}(aq)|MnO_2(s)$ (B)

(2) Balance each of the following oxidation-reduction equations by the ion-electron method.

A. $CuSO_4(aq) + H_2O_2(aq) \longrightarrow Cu(s) + O_2(g) + H_2SO_4(aq)$
B. $Mg(s) + AgNO_3(aq) \longrightarrow Ag(s) + Mg(NO_3)_2(aq)$
C. $Al(s) + H_2SO_4(aq) \longrightarrow Al_2(SO_4)_3(aq) + H_2(g)$
D. $Fe_2(SO_4)_3(aq) + Pb(s) \longrightarrow PbSO_4(s) + Fe(s)$
E. $NaI(aq) + F_2(g) \longrightarrow I_2(s) + NaF(aq)$
F. $H_2(g) + KClO(aq) \longrightarrow KCl(aq) + H_2O(l)$, in alkaline solution

G. $MnO_2(s) + HCl(aq) \longrightarrow MnCl_2(aq) + H_2O(l) + Cl_2(g)$
H. $I_2(s) + H_2O(l) + Br_2(l) \longrightarrow HBr(aq) + HIO_3(aq)$
I. $Ag(s) + HNO_3(aq) \longrightarrow AgNO_3(aq) + NO(g) + H_2O(l)$
J. $S(s) + H_2O(l) + Pb(NO_3)_2(aq) \longrightarrow Pb(s) + H_2SO_3(aq) + HNO_3(aq)$
K. $O_2(g) + H_2O(l) \longrightarrow O_3(g) + H_2O_2(aq)$
L. $H_2S(g) + Br_2(l) \longrightarrow S(s) + HBr(aq)$
M. $SnSO_4(aq) + FeSO_4(aq) \longrightarrow Sn(s) + Fe_2(SO_4)_3(aq)$
N. $H_2SO_4(aq) + S(s) + H_2O(l) \longrightarrow H_2SO_3(aq)$
O. $I_2(s) + H_2O(l) \longrightarrow HI(aq) + HIO_3(aq)$

9. Balance each of the reactions of the Drill Problems of Objectives 4-2 and 4-4 in two ways;
- as combination (c), decomposition (de), displacement (di), or metathesis (m); or
- as precipitation (p), gas-evolution (g), acid-base (a), or oxidation-reduction (o) reactions.

Use ? if the reaction does not fit into one of these categories.

10. Determine the concentration of the solution being titrated or the purity of the solid in each reaction that follows.

A. $HCl(aq) + NaOH(aq) \longrightarrow NaCl(aq) + H_2O(l)$
18.7 mL of 01.00 M NaOH titrates 25.0 mL of solution with [HCl] = ____M.

B. $HNO_3(aq) + KOH(aq) \longrightarrow KNO_3(aq) + H_2O(l)$
26.42 mL of 0.125 M KOH titrates 20.00 mL of solution with $[HNO_3]$ = ____M.

C. $HCl(aq) + KOH(aq) \longrightarrow KCl(aq) + H_2O(l)$
17.42 mL of 1.43 M HCl titrates 20.00 mL of solution with [KOH] = ____M.

D. $HNO_3(aq) + NaOH(aq) \longrightarrow NaNO_3(aq) + H_2O(l)$
19.74 mL of 1.50 M HNO_3 titrates 25.00 mL of solution with [NaOH] = ____M.

E. $H_2SO_4(aq) + 2\ KOH(aq) \longrightarrow K_2SO_4(aq) + 2\ H_2O(l)$
20.41 mL of 0.100 M KOH titrates 20.00 mL of solution with $[H_2SO_4]$ = ____M.

F. $2\ HNO_3(aq) + CaO(s) \longrightarrow Ca(NO_3)_2(aq) + H_2O(l)$
1.75 g of pure CaO(s) reacts with 20.13 mL of solution with $[HNO_3]$ = ____M.

G. $2\ HCl(aq) + Mg(OH)_2(aq) \longrightarrow MgCl_2(aq) + 2\ H_2O(l)$
19.40 mL of 0.100 M HCl titrates 25.00 mL of solution with $[Mg(OH)_2]$ = ____M.

H. $CaCO_3(s) + 2\ HCl(aq) \longrightarrow CaCl_2(aq) + H_2O(l) + CO_2(g)$
19.41 mL of 1.00 M HCl titrates 5.00 g of $CaCO_3$-containing solid. $CaCO_3$ = ____%.

I. $AgNO_3(aq) + NaCl(aq) \longrightarrow AgCl(s) + NaNO_3(aq)$
17.42 mL of 0.100 M $AgNO_3$ titrates 25.00 mL of solution with [NaCl] = ____M.

J. $2\ AgNO_3(aq) + K_2CrO_4(aq) \longrightarrow 2\ KNO_3(aq) + Ag_2CrO_4(s)$
19.21 mL of 0.500 M K_2CrO_4 titrates 50.00 mL of solution with $[AgNO_3]$ = ____M.

K. $Sr(NO_3)_2(aq) + Na_2SO_4(aq) \longrightarrow 2\ NaNO_3(aq) + SrSO_4(s)$
25.00 mL of 0.100 M $Sr(NO_3)_2$ titrates 27.20 mL of soln. with $[Na_2SO_4]$ = ____M.

L. $Al_2O_3(s) + 6\ HNO_3(aq) \longrightarrow 2\ Al(NO_3)_3(aq) + 6\ H_2O(l)$
24.31 mL of 0.100 M HNO_3 titrates 0.100 g Al_2O_3-containing solid. Al_2O_3 = ____%

M. $Na_2CO_3(s) + 2\ HCl(aq) \longrightarrow 2\ NaCl(aq) + H_2O(l) + CO_2(g)$
13.41 mL of 0.25 M HCl titrates 4.00 g Na_2CO_3-containing solid. Na_2CO_3 = ____%

N. $Mg(CN)_2(s) + H_2SO_4(l) \longrightarrow MgSO_4(aq) + 2\ HCN(g)$
17.21 mL of 0.500 M H_2SO_4 titrates 1.05 g $Mg(CN)_2$-containing solid. $Mg(CN)_2$ = ____%

O. $KCNS(aq) + AgNO_3(aq) \longrightarrow AgCNS(s) + KNO_3(aq)$
14.32 mL of 0.200 M $AgNO_3$ titrates 20.00 mL of solution with [KCNS] = ____M.

12. Identify the unknown.in each of the following cases from the list of three compounds given, based on the information supplied.

A. $AgNO_3$, NH_4NO_3, $PbSO_3$. When NaOH(aq) is added, a gas is evolved that has a sharp odor.
B. $BaCl_2$, $FeSO_4$, $MgCO_3$. A precipitate forms when Na_2S(aq) is added to a solution of the unknown.
C. AgI, KCl, $FeBr_2$. No precipitate forms when Na_2CrO_4(aq) is added to a solution of the unknown.
D. $AgClO_3$, $Ba(NO_3)_2$, $FeSO_4$. A precipitate forms when NaCl(aq) is added to a solution of the unknown.
E. $BaCrO_4$, $MgSO_4$, $Sr(NO_3)_2$. A precipitate forms when a small amount of NaOH(aq) is added to a solution of the unknown.
F. $PbSO_3$, Na_2CO_3, FeS. When HCl(aq) is added, a gas with a rotten-egg odor is evolved.
G. $FeSO_4$, $MgSO_4$, $PbSO_4$. A precipitate forms when Na_2CrO_4(aq) is added to a solution of the unknown.
H. $CoCl_2$, Hg_2Br_2, $BaCl_2$. No precipitate forms when Na_2SO_4(aq) is added to a solution of the unknown.

I. $Cu(OH)_2$, $AlPO_4$, MgS. No precipitate forms when $NaNO_3$(aq) is added to a solution of the unknown.
J. $(NH_4)_2S$, $CaCO_3$, $NaC_2H_3O_2$. No gas is evolved when HCl(aq) is added.
K. $BaBr_2$, $MgSO_4$, Na_2CO_3. A precipitate forms when $Co(ClO_3)_2$(aq) is added to a solution of the unknown.
L. BaS, $FeSO_3$, $MgCO_3$. An odorless gas is evolved when HCl(aq) is added.

QUIZZES (20 minutes each) Choose the best answer to each question or complete and balance the equation.

Quiz A

1. Which of the following consists of a nonelectrolyte, a strong electrolyte, and a base? (a) H_2O, NH_3, NaOH; (b) KCl, $BaBr_2$, Na_2SO_3; (c) $MnBr_2$, CH_4, H_2O; (d) CO_2, $MgSO_4$, KOH; (e) none of these.
2. $2\ KClO_3 \xrightarrow{heat}$? When this reaction is completed, it is which of the following types? (a) precipitation; (b) gas evolution; (c) acid-base; (d) oxidation reduction; (e) none of these.
3. Which of the following consists of two insoluble compounds and a soluble one? (a) $HgBr_2$, Ag_2CrO_4, $Co(ClO_4)_2$; (b) $CoSO_4$, CuCl, $AgNO_3$; (c) $PbSO_4$, $Ca(ClO_3)_2$, BaS; (d) $CuCl_2$, $Ba(C_2H_3O_2)_2$, Na_2CO_3; (e) none of these.
4. 11.2 mL of 0.100 M NaOH is needed to titrate 25.00 mL of HCl(aq). [HCl] = (a) 0.448 M; (b) 2.23 M; (c) 0.223 M; (b) 0.112 M; (e) none of these.
5. Which of the following represents an oxidant followed by a reductant, for the equations of questions 7 and 8 of this quiz? (a) I_2, $HClO_3$; (b) $HClO_3$, I_2; (c) I^-, $Cr_2O_7^{2-}$; (d) H^+, I^-; (e) none of these.
6. What is $[Al^{3+}]$ in an $Al_2(SO_4)_3$ solution that has $[SO_4^{2-}]$ = 0.342 M? (a) 0.228 M; (b) 0.513 M; (c) 0.342 M; (d) 0.114 M; (e) none of these.
7. When the equation $HClO_3 + H_2O + I_2 \longrightarrow HIO_3 + HCl$ is balanced with the smallest integer coefficients, the coefficient of chloric acid is (a) 10; (b) 12; (c) 6; (d) 8; (e) none of these.
8. $Cr_2O_7^{2+} + I^- + H^+ \longrightarrow Cr^{3+} + I_2 + H_2O$ When the smallest whole number coefficients are used, the correct coefficient for I^- in the balanced equation is (a) 2; (b) 3; (c) 6; (d) 14; (e) none
9. Complete and balance: $AgClO_3(aq) + H_2SO_4(aq) \longrightarrow$ ________
10. Complete and balance: $LiOH(aq) + H_2C_2O_4(aq) \longrightarrow$ ________

Quiz B

1. Which of the following consists of a strong electrolyte, an acid, and a base? (a) KCl, $CoSO_4$, CH_4; (b) HCl, $HC_2H_3O_2$, NH_3; (c) $Mg(NO_3)_2$, H_2SO_4, $Ca(OH)_2$; (d) SO_2, HF, HCl; (e) none of these.
2. $AgNO_3(aq) + H_2SO_4(aq) \longrightarrow$? When this reaction is completed, it is which of the following types? (a) precipitation; (b) gas evolution; (c) acid-base; (d) oxidation-reduction; (e) none of these.
3. Which of the following consists of two soluble compounds and an insoluble one? (a) $HgBr_2$, $MnSO_4$, $Na_2C_2O_4$; (b) $Na_2S_2O_3$, NH_4Cl, CoI_2; (c) MnS, $Cu(OH)_2$, Al_2O_3; (d) $Mn(ClO_3)_2$, $Ca(MnO_4)_2$, Hg_2SO_4; (e) none of these.
4. 19.4 mL of 2.00 M HNO_3 is needed to titrate 25.00 mL of LiOH(aq). [LiOH] = (a) 0.776 M; (b) 1.55 M; (c) 3.10 M; (d) 1.29 M; (e) none of these.
5. Which of the following represents an oxidant followed by a reductant, for the equations of questions 7 and 8 of this quiz? (a) KNO_2, $KMnO_4$; (b) $KMnO_4$, KNO_2; (c) NO_3^-, H_2O; (d) NO, NO_3^-; (e) none of these.
6. What is $[NO_3^-]$ in a $Ca(NO_3)_2$ solution that has $[Ca^{2+}]$ = 0.840 M? (a) 0.840 M; (b) 1.64 M; (c) 0.420 M; (d) 1.26 M; (e) none of these.
7. When the equation $NO_3^- + H_2O \longrightarrow NO + OH^-$ is balanced with the smallest whole number coefficients, the coefficient of hydroxide ion is (a) 3; (b) 2; (c) 4; (d) 1; (e) none of these.
8. When the equation $KMnO_4 + KNO_2 + H_2O \longrightarrow MnO_2 + KNO_3 + KOH$ is balanced with the smallest whole number coefficients, the coefficient of potassium nitrate is (a) 1; (b) 2; (c) 3; (d) 4; (e) 6.
9. Complete and balance: $MnSO_3(aq) + HCl(aq) \longrightarrow$ ________
10. Complete and balance: $ZnSO_4(aq) + BaS(aq) \longrightarrow$ ________

Quiz C

1. Which of the following consists of a nonelectrolyte, a weak electrolyte, and a strong electrolyte? (a) CO_2, NaCl, $MnSO_4$; (b) H_2SO_4, $HC_2H_3O_2$, CuCl; (c) SO_2; HF, $FeSO_4$; (d) $Ba(ClO_3)_2$, $K_2S_2O_3$, $NaMnO_4$; (e) none of these.
2. $MnSO_3(s) + HCl(aq) \longrightarrow$? When this reaction is completed, it is which of the following types?

(a) precipitation; (b) gas evolution; (c) acid-base; (d) oxidation-reduction; (e) none of these.

3. Which of the following consists of three insoluble compounds? (a) FeC_2O_4, SrS, $AgC_2H_3O_2$; (b) $Ba(OH)_2$, MnS_2O_3, $Sr(ClO_3)_2$; (c) $CrCl_3$, $SnSO_4$, $(NH_4)_2S$; (d) $ZnSO_4$, AgI, CdS; (e) none of these.
4. 21.6 mL of 0.500 M CsOH is needed to titrate 20.00 mL of $HClO_3$(aq). $[HClO_3]$ = (a) 0.270 M; (b) 0.540 M; (c) 1.08 M; (d) 0.926 M; (e) none of these.
5. Which of the following represents an oxidant followed by a reductant, for the equations of questions 7 and 8 of this quiz? (a) Cs_2MnO_4, H_2O; (b) H_2O, Cs_2MnO_4; (c) H_2S, Br_2; (d) Br_2, H_2S; (e) none of these.
6. What is $[Na^+]$ in a $Na_3(PO_4)_3$ solution that has $[PO_4^{3-}] = 0.444$ M? (a) 0.111 M; (b) 0.148 M; (c) 0.592 M; (d) 1.33 M; (e) none of these.
7. When the equation $Br_2 + H_2S \longrightarrow S + HBr$ is balanced with the smallest whole number coefficients, the sum of the coefficients is (a)7; (b) 8; (c) 4; (d) 5; (e) 6.
8. When the equation $Cs_2MnO_4 + H_2O \longrightarrow CsMnO_4 + CsOH + MnO_2$ is balanced with the smallest whole number coefficients, the coefficient of MnO_2 is (a) 1; (b) 2; (c) 3; (d) 4; (e) none of these.
9. Complete and balance: $FeS(s) + H_2SO_4(aq) \longrightarrow$ _________
10. Complete and balance: $Al(OH)_3(s) + H_2SO_4(aq) \longrightarrow$ _________

Quiz D

1. Which of the following consists of a weak acid, a weak base, and a salt? (a) HCl, NH_3, Na_2SO_4; (b) HNO_2, NH_3, NH_4NO_2; (c) HCl, $Ca(OH)_2$, $CaSO_4$; (d) HNO_2, $HMnO_4$, Cs_2CrO_4; (e) none of these.
2. $NaI(aq) + Cl_2(aq) \longrightarrow$? When this reaction is completed, it is which of the following types? (a) precipitation; (b) gas evolution; (c) acid-base; (d) oxidation-reduction; (e) none of these.
3. Which of the following consists of two insoluble compounds and a soluble one? (a) CuS, CaS, K_2S; (b) $PbSO_4$, MnS; $Co(NO_2)_2$; (c) $AgClO_4$, $Mn(C_2H_3O_2)_2$, CaO; (d) $BaBr_2$, $Ba(BrO_3)_2$, AgBr; (e) none of these.
4. 31.2 mL of 0.200 M HI(aq) is needed to titrate 25.00 mL of KOH(aq). [KOH] = (a) 1.25 M; (b) 0.801 M; (c) 0.245 M; (d) 4.08 M; (e) none of these.
5. Which of the following represents an oxidant followed by a reductant, for the equations of questions 7 and 8 of this quiz? (a) Cl^-, MnO_4^-; (b) MnO_4^-, H^+; (c) H_2O_2, KOH; (d) ClO_2, H_2O_2; (e) none of these.
6. What is $[Na^+]$ in a Na_2SO_4 solution that has $[SO_4^{2-}] = 0.344$ M? (a) 0.0860 M; (b) 0.172 M; (c) 0.258 M; (d) 0.688 M; (e) none of these.
7. When the equation $MnO_4^- + Cl^- + H^+ \longrightarrow Mn^{2+} + Cl_2 + H_2O$ is balanced, the sum of the smallest whole number coefficients is (a) 21; (b) 24; (c) 38; (d) 43; (e) 51.
8. When the equation $ClO_2 + H_2O_2 + KOH \longrightarrow H_2O + KClO_2 + O_2$ is balanced with the smallest integer coefficients, the coefficient of water is (a) 3; (b) 5; (c) 2; (d) 4; (e) none of these.
9. Complete and balance: $BaCO_3(s) + HNO_3(aq) \longrightarrow$ _________
10. Complete and balance: $Ba(OH)_2(aq) + H_2SO_4(aq) \longrightarrow$ _________

SAMPLE TEST (20 minutes)

1. Complete and balance the following equations, and identify each one in two ways: as combination, decomposition, displacement, or metathesis; and as precipitation, oxidation-reduction, gas-evolution, or acid-base.
   a. $Sr(NO_3)_2(aq) + H_2SO_4(aq) \longrightarrow$
   b. $Na_2CO_3(s) + HCl(aq) \longrightarrow$
   c. $Zn(s) + CuSO_4(aq) \longrightarrow$
   d. $CaO(s) + HNO_3(aq) \longrightarrow$
   e. $NaI(aq) + Cl_2(aq) \longrightarrow$
   f. $BaBr_2(aq) + AgNO_3(aq) \longrightarrow$
   g. $CaO(s) + CO_2(g) \longrightarrow$
   h. $Al(s) + Cl_2(g) \longrightarrow$
2. Balance each of the following oxidation-reduction reactions.
   a. $Ag(s) + HNO_3(aq) \longrightarrow AgNO_3(aq) + NO(g) + H_2O(l)$
   b. $Cl_2(aq) + KOH(aq) \longrightarrow KCl(aq) + KClO(aq) + H_2O(l)$
   c. $Fe_2(SO_4)_3(aq) + SnSO_4(aq) \longrightarrow Sn(SO_4)_2(aq) + FeSO_4(aq)$
   d. $K_2C_2O_4(aq) + KMnO_4(aq) + H_2SO_4(aq) \longrightarrow MnSO_4(aq) + K_2SO_4(aq) + CO_2(g) + H_2O(l)$
3. An unknown, which is a pure compound, forms a solution when water is added. A precipitate forms when $Ba(NO_3)_2(aq)$ is added to this solution, but not when $Cu(NO_3)_2(aq)$ is added. When $HC_2H_3O_2(aq)$ is added to the unknown solid, no gas is produced. When NaOH(aq) is added to the solid, a gas with a pungent odor is produced. What is the identity of the unknown?

6 Gases

CHAPTER OBJECTIVES

* **1. Be able to convert between the common units of pressure.**

Your goal is to firmly commit the following expressions to memory and use them in problems. (Notice that there is no abbreviation for torr; t or T is incorrect.) The problems are largely of the conversion factor type.

1.000 atmosphere (atm) = 760 mmHg = 760 torr
= 101,325 N/m^2 = 101,325 pascal (Pa)
= 14.7 $lb/in.^2$
1000 pascal = 1.000 kilopascal (kPa)

2. Explain the operation of a mercury barometer, an open-end manometer, and a closed-end manometer and be able to use the data obtained with these devices.

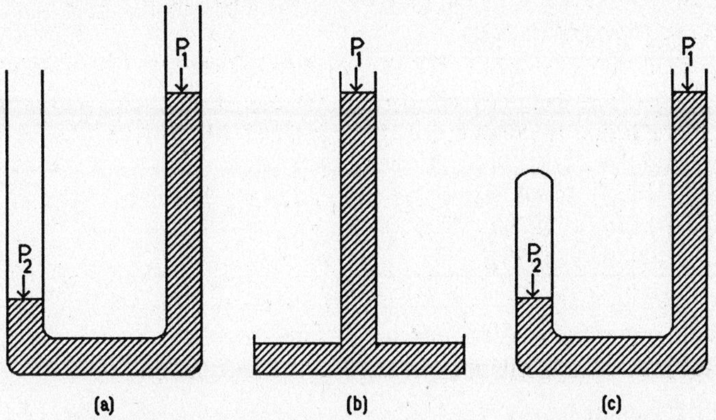

FIGURE 6-1 Liquid-Filled Pressure-Measuring Devices
(a) General Principle; Open-end Manometer. (b) Barometer. (c) Closed-End Manometer.

All liquid-filled pressure-measuring devices operate on the same principle. A liquid is confined in a U-shaped tube, as shown in Figure 6-1a. If the pressure is different on the two ends of the liquid column—that is, if P_1 differs from P_2—the liquid levels in the two arms of the tube are different. The liquid level is higher on the low pressure end of the tube. (Think of sucking on a soda straw to remember which side is higher; the side with lower pressure.) In all parts of Figure 6-1 P_1 is a lower pressure than P_2.The difference in liquid levels (h) is proportional to the pressure difference.

$$\Delta P = P_2 - P_1 \propto h \quad \text{or} \quad \Delta P = gdh \qquad [1]$$

In equation [1], g is the gravitational constant, 9.80 m/s^2 or 980 cm/s^2, and d is the density of the liquid. Normally, we know one of the pressures (P_1 or P_2) and the density of the liquid, are given or can measure a value of h, and wish to find the other pressure. There is a simpler method of finding ΔP than equation [1]. It depends on the fact that a denser liquid produces a smaller level difference. In fact, for two liquids, A and B,

$$h_A d_A = h_B d_B \quad or \quad h_A = h_B d_B / d_A \qquad [2]$$

Frequently, mercury is used as liquid A (d_A = 13.6 g/cm^3) because 760 mmHg = 1.000 atm.

Frequently, mercury is used as liquid A (d_A = 13.6 g/cm³) because 760 mmHg = 1.000 atm.

EXAMPLE 6-1 An open-end manometer is filled with dibutyl phthalate (DBP, d = 1.047 g/cm³). If the level difference is 625 mm, what is the pressure difference in atmospheres?

$$\Delta P = h_{Hg} = 625 \text{ mm DBP} \times \frac{1.047 \text{ g/cm}^3}{13.6 \text{ g/cm}^3}$$

$$= 48.1 \text{ mmHg} \times \frac{1 \text{ atm}}{760 \text{ mmHg}} = 0.0633 \text{ atm}$$

In a closed-end manometer and in a barometer, the pressure in the low pressure end (the closed end) is almost zero. It equals the vapor pressure of the liquid being used. At 25°C, the vapor pressure of mercury is 6×10^{-6} atm, that of glycerol is 6×10^{-7} atm, and that of dibutyl phthalate is 2×10^{-7} atm. See objectives 13-2 through 13-5 for a discussion of vapor pressure.

* **3. Learn Boyle's law both mathematically and graphically and be able to use it in calculations.**

Robert Boyle observed that for a fixed amount of gas at constant temperature the product of volume and pressure is constant. The condition of fixed amount means that the apparatus must not leak and also that no physical or chemical change that produce or consume a gas may occur within the container. There are no units specified for pressure or temperature in Boyle's law. Any units are satisfactory, but they must be consistent: all pressures must be measured in the same units (for example, inches of mercury, in. Hg), and so must be all volumes (for example, in.³). Graphically, the law is demonstrated when one plots $1/V$ vs. P (Figure 6-2a) or $1/P$ vs. V (Figure 6-2b) and obtains a straight line. Other plots give curved lines. The data for Figure 6-2, given in Table 6-1, are part of Boyle's original data.

TABLE 6-1 Robert Boyle's data of pressure and volume

P, in. Hg	V, in.³	$P \times V$	$1/P$	$1/V$
33.50	10.50	352.8	0.02985	0.09524
39.25	9.00	353	0.02548	0.111
54.31	6.50	353	0.01841	0.154
74.13	4.75	353	0.01346	0.211
117.56	3.00	353	0.0085063	0.333

Problems based on Boyle's law, sometimes called pressure-volume problems, supply an initial pressure and volume, and a final pressure or volume. A convenient expression to use is equation [3].

$$P_iV_i = P_fV_f \quad (i = \text{initial}; f = \text{final}) \qquad [3]$$

EXAMPLE 6-2 When measured at 1.25 atm, a gas occupies 247 cm³. What is its volume in liters at a pressure of 76.5 mmHg?

Since the two pressures are given in different units, we convert one of them.

$$1.25 \text{ atm} \times \frac{760 \text{ mmHg}}{\text{atm}} = 950 \text{ mmHg} = P_i$$

Then equation [3] gives the final volume.

$$950 \text{ mmHg} \times 247 \text{ cm}^3 = 76.5 \text{ mmHg} \times V_f$$

$$3.07 \times 10^3 \text{ cm}^3 \times (\text{L}/1000 \text{ cm}^3) = 3.07 \text{ L}$$

* **4. Learn Charles's law both mathematically and graphically and be able to use it in calculations.**

Charles observed that for a fixed amount of gas at constant pressure the ratio of volume to temperature is constant. Mathematically, this law can be expressed in three ways, as shown in equation [4].

$$V = \text{constant} \times T \quad or \quad \frac{V}{T} = \text{constant} \quad or \quad \frac{V_i}{T_i} = \frac{V_f}{T_f} \qquad [4]$$

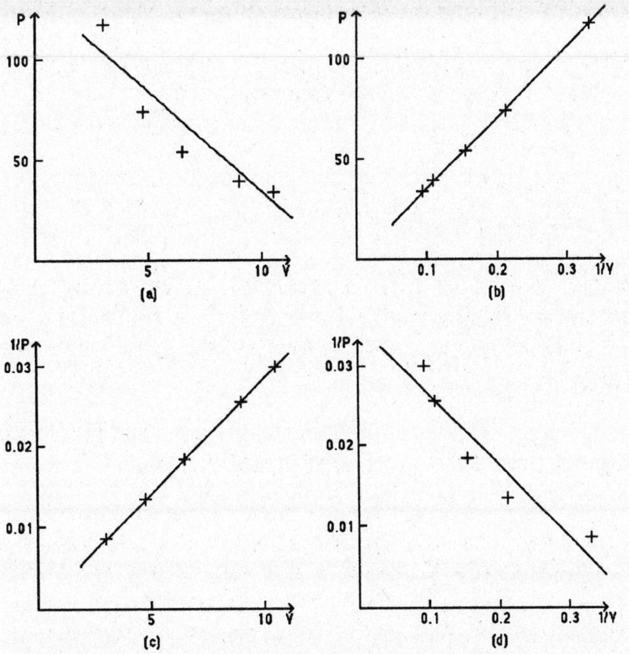

FIGURE 6-2 Graphs of Boyle's Pressure-Volume Data
In each graph, the solid line is the best straight line (estimated visually) through the data points.
(a) P vs. V, (b) P vs. $1/V$, (c) $1/P$ vs. V, (d) $1/P$ vs. $1/V$

Graphically, data points fall along a straight line when volume is plotted against temperature. The only requirement for *graphical* data is that all volumes must have the same units and all temperatures must be measured on the same scale (either K or °C or °F). When solving Charles's law *problems,* however, one *must* express temperatures on an absolute scale, such as kelvins.

EXAMPLE 6-3 At 0.0°C, a gas has a volume of 22.414 L. What is its volume at 25.0°C?

Both temperatures must be expressed on the kelvin scale before Charles's law can be used.

$T_i = 0.0 + 273.15 = 273.2\text{ K} \quad T_f = 25.0 + 273.15 = 298.2\text{ K}$

Then application of Charles's law, equation [4], gives the final volume.

$$\frac{22.414\text{ L}}{273.2\text{ K}} = \frac{V_f}{298.2\text{ K}}$$

$$V_f = 22.414\text{ L} \times \frac{298.2\text{ K}}{273.2\text{ K}} = 24.47\text{ L}$$

5. Discuss the significance of the absolute zero of temperature and be able to convert between Celsius and kelvin temperatures.

When Charles's law ($V = \text{constant} \times T$) is carefully examined, one concludes that a gas will occupy no volume once its kelvin temperature has been lowered to zero. Charles's law ($V/T = \text{constant}$) is not valid on the Celsius scale. For example, consider the data of Example 6-3.

$$\frac{V_i}{T_i} = \frac{22.414\text{ L}}{0°\text{C}} = \infty \quad \text{but} \quad \frac{V_f}{T_f} = \frac{24.47\text{ L}}{25.0°\text{C}} = 0.979\text{ L/°C}$$

Charles's law *is* valid on the Kelvin scale.

$$\frac{V_i}{T_i} = \frac{22.414\text{ L}}{273.2\text{ K}} = 0.08204\text{ L/K} \quad \text{but} \quad \frac{V_f}{T_f} = \frac{24.47\text{ L}}{298.2\text{ K}} = 0.08206\text{ L/K}$$

The relationship to be memorized is $T(\text{K}) = t(°\text{C}) + 273.15$. Notice that the degree sign (°) is not used in expressing temperatures on the kelvin scale, which are represented by a capital T. Celsius temperatures, formerly called centigrade temperatures, are symbolized by a lower case t. It is very important to remember that *all problems involving gas laws must use kelvin temperatures.*

6. State what is meant by STP and the STP molar volume, and be able to use the latter in calculations.

STP is the abbreviation for Standard Temperature and Pressure: $T = 273.15$ K and $P = 1.000$ atm. The volume occupied by one mole of an ideal gas at STP is 22.414 L. This volume is known as the STP molar volume.

EXAMPLE 6-4 What volume of $H_2(g)$, measured at STP, is needed to reduce 36.1 g WO_3 according to

$$WO_3(s) + 3\ H_2(g) \xrightarrow{\text{heat}} W(s) + 3\ H_2O(l)$$

The problem is solved with the conversion factor method.

$$36.1 \text{ g } WO_3 \times \frac{\text{mol } WO_3}{231.8 \text{ g } WO_3} \times \frac{3 \text{ mol } H_2}{\text{mol } WO_3} \times \frac{22.4 \text{ L } H_2}{\text{mol } H_2} = 10.5 \text{ L } H_2$$

7. State and be able to use Avogadro's law.

Avogadro's law states that equal volumes of gases at the same temperature and pressure contain equal numbers of molecules (or of moles) of the gases, even if the gases are different. It is crucial to compare the gases at the *same temperature and pressure*; otherwise the law is invalid. Avogadro's law is expressed mathematically in equation [5].

$$V = n \times \text{constant} \quad \text{or} \quad \frac{V}{n} = \text{constant} \quad \text{or} \quad \frac{V_1}{n_1} = \frac{V_2}{n_2} \qquad [5]$$

* **8. Solve for one of *P, V, n,* or *T* when given values of the other three for an ideal gas.**

This objective involves the proper use of the ideal gas equation, $PV = nRT$. Since R has the units of L-atmospheres/mole-kelvins, the pressure must be expressed in atmospheres, the volume in liters, the amount of gas in moles, and the temperature in kelvins.

EXAMPLE 6-5 What is the volume of 3.04 moles of gas confined at 741 mmHg and 14.0°F?

We first express the pressure in atmospheres and the temperature in kelvins.

$$P = 741 \text{ mmHg} \times \frac{1.00 \text{ atm}}{760 \text{ mmHg}} = 0.975 \text{ atm}$$

$$°C = (°F - 32) \times \tfrac{5}{9} = (14.0 - 32.0)(\tfrac{5}{9}) = -10.0°C$$

$$T = -10.0 + 273.15 = 263.2 \text{ K}$$

We then solve the ideal gas equation for V. (Notice that the units in the expression cancel appropriately.)

$$V = \frac{nRT}{P} = \frac{(3.04 \text{ mol})(0.0821 \text{ L atm mol}^{-1} \text{ K}^{-1})(263.2\text{K})}{0.975 \text{ atm}} = 67.4 \text{ L}$$

* **9. Obtain the value of one final variable (*P, V, n,* or *T*), given the values of the other final variables and of all the initial variables, excluding those that remain unchanged, in the ideal gas law.**

This problem often is called an "initial-final problem." It depends on a rearranged version of the ideal gas law, $R = PV/nT$, which produces equation [6].

$$\frac{P_iV_i}{n_iT_i} = \frac{P_fV_f}{n_fT_f} \quad (i = \text{initial}, f = \text{final}) \qquad [6]$$

Any of the simple gas laws (Boyle's, Charles's, Guy-Lussac's, Avogadro's) can be produced from this equation. Equation [6] also can be used to solve problems for which there is no "law."

EXAMPLE 6-6 2.40 mol of an ideas gas is confined to a 26.4 L vessel at 250.0 K. If an additional 2.00 mol of gas is added to the vessel and the pressure does not vary, what is the final temperature?

Notice that this problem does not explicitly state that the volume of the container does not vary. It could be a balloon, for instance. However, the final volume is not given and thus one *assumes* that the volume is constant. You often will have to make such an assumption to solve a problem. Both volume and pressure are fixed and thus $V_i = V_f$ and $P_i = P_f$. We use equation [6].

$$\frac{P_iV_i}{n_iT_i} = \frac{P_iV_i}{n_fT_f} \quad \text{or} \quad n_iT_i = n_fT_f$$

We are given the following data.

$n_i = 2.40$ mol $\quad T_i = 250.0$ K

$n_f = n_i + 2.0 = 2.40$ mol $+ 2.00$ mol $= 4.20$ mol

which produces

$(2.40 \text{ mol})(250.0\text{K}) = (4.40 \text{ mol})T_f$

$$T_f = \frac{(2.40 \text{ mol})(250.0 \text{ K})}{4.40 \text{ mol}} = 136.4 \text{ K}$$

Notice that the equation produced to solve this problem, $n_iT_i = n_fT_f$, is not one of the classical gas laws.

We conclude with a word of caution. Frequently it is stated that this kind of problem can be solved with an equation of the type: $T_f = T_i \times$ (mole ratio). The "mole ratio" is the quotient of the two numbers of moles (4.40 and 2.40). But is the ratio 4.40/2.40 or 2.40/4.40? In order to confine more moles in the same volume, at the same pressure, the temperature must be reduced, and thus the correct ratio is 2.40/4.40. This kind of intuitive or "gut" feeling about how gases behave comes with experience, which many general chemistry students lack. Thus you may get the ratio upside down, especially in the heat of an examination. Starting with equation [6] is safer.

* **10. Use the alternate version of the ideal gas law for calculating molar masses of gases and determining gas densities.**

An alternate form of the ideal gas law is given in equation [7].

$$PV = \frac{m}{\mathcal{M}} RT \qquad [7]$$

Equation [7] is obtained from the ideal gas law by substituting the expression

$$n = m/\mathcal{M} \qquad [8]$$

where m = the mass of the gas and $\mathcal{M}$ = the molar mass of the gas. It is convenient, but not essential, that you know the combined equation [7]. The two parts, the ideal gas law and equation [8], work just as well, as Example 6-7 illustrates. You should use the technique that you find easier.

EXAMPLE 6-7 What is the molar mass of a gas if 11.770 g has a volume of 5.00 L at 30°C and 742 mmHg?

With the combined equation [7]:

$P = 742 \text{ mmHg} \times 1 \text{ atm}/760 \text{ mmHg} = 0.976$ atm

$T = 30°\text{C} + 273.15 = 303$ K $\quad m = 11.770$ g $\quad V = 5.00$ L

$$PV = \frac{m}{\mathcal{M}} RT$$

$$(0.976 \text{ atm})(5.00\text{L}) = \frac{11.770 \text{ g}}{\mathcal{M}}\left(0.0821 \frac{\text{L atm}}{\text{mol K}}\right)(303 \text{ K})$$

$\mathcal{M} = 60.0$ g/mol

With the ideal gas equation and equation [8]:

$(0.976 \text{ atm})(5.00 \text{ L}) = n(0.0821 \text{ L atm mol}^{-1} \text{ K}^{-1})(303 \text{ K})$

$n = 0.196 \text{ mol} = m/\mathcal{M} = 11.770 \text{ g}/\mathcal{M}$

$\mathcal{M} = 11.770 \text{ g}/0.196 \text{ mol} = 60.0$ g/mol

* **11. Combine gas molar masses with empirical formulas to determine molecular formulas.**

Percent composition data only enable us to determine the empirical formula of a compound. But several compounds may have the same empirical formula. The molar mass of a compound allows us to determine the molecular formula of a compound if we know its empirical formula. Thus, if the gas of Example 6-7 has an empirical formula of CH_2O, its molecular formula will be an integral multiple of its molecular formula: CH_2O, $C_2H_4O_2$, $C_3H_6O_3$, etc. The "empirical molar mass" is the mass of 1 mol C, 2 mol H, and 1 mol O or 12.0 g + 2 × 1.0 g + 16.0 g = 30.0 g/mol. Since the molar mass obtained in Example 6-7 (60.0 g/mol) is twice this empirical molar mass, the molecular formula must be twice the molecular formula, or $C_2H_4O_2$.

* **12. Solve stoichiometry problems involving gases.**

A large variety of stoichiometry problems use the ideal gas law either to determine the amount of reactant or product or to determine the volume (or occasionally the pressure) of a reactant or product. If you can solve stoichiometry problems, and those problems based on the ideal gas law, these combined problems should give you little trouble. However, in problems involving the law of combining volumes, the data supplied often are insufficient to allow you to compute the number of moles. Remember that equal volumes of gases at the same temperature and pressure contain equal numbers of moles (Avogadro's law). Thus, when temperature and pressure are fixed, gas volumes can be used in the same manner as numbers of moles.

EXAMPLE 6-8 3.04 L of $O_2(g)$ and excess $SO_2(g)$ both at 397 K and 0.351 atm combine to produce what volume of SO_2? $2\,SO_2(g) + O_2(g) \longrightarrow 2\,SO_3(g)$

We assume that the pressure and temperature remain fixed. Thus, we have

$$3.04\text{ L } O_2 \times \frac{2\text{ L } SO_3(g)}{1\text{ L } O_2(g)} = 6.08\text{ L } SO_3(g)$$

* **13. Solve problems involving mixtures of gases with (a) the ideal gas law, (b) Dalton's law of partial pressures, or (c) Amagat's law of partial volumes.**

Since ideal gas molecules do not attract or repel each other, a mixture of several gases displays properties that are the sums of the individual gas properties. Each gas in the mixture acts as if the other gases were not present, unless a chemical reaction occurs. Thus, the total pressure is the sum of the partial pressures. (The partial pressure is the pressure that the individual gas would exert if it were confined in the same volume at the same temperature as the mixture.) In like fashion, the total gas volume is the sum of the partial volumes. (The partial volume is the volume an individual gas would occupy if confined at the same temperature and pressure as the mixture.) This additive feature of pressures (Dalton's law) and volumes (Amagat's law) is valid only for ideal gases. For other states of matter—solids and liquids—only masses and amounts (that is, moles) can be summed to obtain the total mass and the total amount of the mixture.

EXAMPLE 6-9 2.46 L of argon at 0.320 atm and 400 K is combined with 4.12 L of neon at 0.440 atm and 400 K in a rigid 5.00 L container at 400 K. Determine (a) the total final pressure, (b) the partial pressure of argon in the 5.00 L container, and (c) the partial volume of neon in the 5.00 L container.

We first compute the amount of each gas and the total moles of gas placed in the 5.00 L container.

$$n_{Ar} = \frac{PV}{RT} = \frac{(0.320\text{ atm})(2.46\text{ L})}{(0.0821\text{ L atm mol}^{-1}\text{ K}^{-1})(400\text{ K})} = 0.0240\text{ mol}$$

$$n_{Ne} = \frac{(0.440\text{ atm})(4.12\text{ L})}{(0.0821\text{ L atm mol}^{-1}\text{ K}^{-1})(400\text{ K})} = 0.0552\text{ mol}$$

$$n_{total} = n_{Ar} + n_{Ne} = 0.0240\text{ mol} + 0.0552\text{ mol} = 0.792\text{ mol}$$

(a) We compute the final total pressure with the ideal gas law.

$$P_f = \frac{nRT}{V} = \frac{(0.0792)(0.0821\text{ L atm mol}^{-1}\text{ K}^{-1})(400\text{ K})}{5.00\text{ L}} = 0.520\text{ atm}$$

(b) The pressure of argon in the 5.00 L container can be computed in two ways.

(1) $$P_{Ar} = \frac{n_{Ar}RT}{V} = \frac{(0.0240\text{ mol})(0.0821\text{ L atm mol}^{-1}\text{ K}^{-1})(400\text{ K})}{5.00\text{ L}} = 0.158\text{ atm}$$

(2) $$P_{Ar} = \frac{n_{Ar}}{n_{total}} P_{total} = \frac{0.0240\text{ mol}}{0.0792\text{ mol}} \times 0.520\text{ atm} = 0.158\text{ atm}$$

(c) The partial volume of neon is computed from its mole fraction (objective 13-2b).

$$V_{Ne} = \chi_{Ne} V_{total} = \frac{n_{Ne}}{n_{total}} V_{total} = \frac{0.0552\text{ mol}}{0.0792\text{ mol}} \times 5.00\text{ L} = 3.48\text{ L}$$

TABLE 6-2 Vapor pressure of water at various temperatures

t, °C	*P*, mmHg	*t*, °C	*P*, mmHg	*t*, °C	*P*, mmHg	*t*, °C	*P*, mmHg
13.0	1.2	19.0	16.5	25.0	23.8	31.0	33.7
14.0	2.0	20.0	17.5	26.0	25.2	32.0	35.7
15.0	2.8	21.0	18.7	27.0	26.7	33.0	37.7
16.0	3.6	22.0	19.8	28.0	28.3	34.0	39.9
17.0	4.5	23.0	21.1	29.0	30.0	35.0	42.2
18.0	5.5	24.0	22.4	30.0	31.0	36.0	44.6

14. Compute the pressure of gases collected over water.

When a gas is collected over water, it is mixed with water vapor. Thus, the total pressure is the sum of the pressure of the gas collected and the vapor pressure of water.

$$P_{\text{total}} = P_{\text{gas}} + P_{\text{water}} \qquad [9]$$

The vapor pressure of water varies with temperature. Several values are given in Table 6-2.

15. State the postulates and the basic mathematical relationships of the kinetic molecular theory of gases.

The postulates are a set of assumptions about the structure of gases and how gas molecules interact. One form of these assumptions follows.

1. Gases are composed of molecules.
2. Gas molecules have no volume—they are point masses.
3. Gas molecules neither attract nor repel each other—there are no forces between them.
4. Gas molecules are in constant, random motion.

The kinetic energy of an average ideal gas molecule is given by two equations, [10] and [11], where $k = R/N = 1.38 \times 10^{-23}$ J molecule^{-1} K^{-1}. k is known as Boltzmann's constant.

$$\overline{e_k} = \frac{1}{2} m \overline{u^2} \qquad [10]$$

$$\overline{e_k} = \frac{3}{2} kT \qquad [11]$$

16. Deduce the simple gas laws from the kinetic molecular theory.

Let us consider Boyle's law as an example. One of the consequences of random motion of molecules is that molecules continually are hitting the sides of the container. The average molecular speed, and thus the average force exerted by each collision with the wall, depends only on the temperature (equation [12], below). If the gas now is compressed into a smaller volume, each collision still has the same force but there are many more collisions. In fact, if the volume is halved, there are twice as many collisions and the pressure will double. That the should hit each side twice as often. You should write down the same type of explanation for each of the other simple gas laws: Charles's, Avogadro's, and Dalton's.

* **17. Compute molecular velocities and know and apply Graham's law.**

When the two expressions for average kinetic energy, equations [10] and [11] of objective 6-12, are combined and solved for $\sqrt{\overline{u^2}} = u_{\text{rms}}$, one obtains

$$u_{\text{rms}} = \sqrt{\frac{3kT}{m}} = \sqrt{\frac{3RT}{\mathcal{M}}} \qquad [12]$$

where m = molecular mass (kg/molecule), $\mathcal{M}$ = molar mass (kg/mole of molecules), and $R = 8.314$ J mol^{-1} K^{-1}, giving a velocity in m/s. Equation [12] is true for all ideal gases, specifically for two different gases at the same temperature.

$$u_{\text{rms,A}} = \sqrt{\frac{3RT}{\mathcal{M}_{\text{A}}}} \quad \text{and} \quad u_{\text{rms,B}} = \sqrt{\frac{3RT}{\mathcal{M}_{\text{B}}}} \qquad [13]$$

$$\frac{u_{\text{rms,A}}}{u_{\text{rms,B}}} = \sqrt{\frac{\mathcal{M}_{\text{B}}}{\mathcal{M}_{\text{A}}}} \qquad [14]$$

Equation [14], obtained from the kinetic molecular theory, is similar to Graham's law, which was experimentally obtained. An example of Graham's law is the rate at which a helium-filled toy balloon deflates—far faster than a similar balloon filled with air. This occurs because the lower-molar-mass helium effuses more rapidly through the small pores in the balloon than the high-molar-mass oxygen and nitrogen of air.

EXAMPLE 6-10 Determine the average velocity of a xenon molecule at 37°C (310K)

$$u_{\text{rms}} = \sqrt{\frac{3 \times 8.314 \text{ J mol}^{-1} \text{ K}^{-1} \times 310 \text{ K}}{0.1313 \text{ kg/mol}}} = 243 \text{ m/s}$$

Note that the unit joule is equal to kg $m^2 s^{-2}$.

EXAMPLE 6-11 How much faster than air does helium effuse through a pinhole? The molar mass of air is 29.0 g/mol.

$$\frac{\text{rate of He}}{\text{rate of air}} = \frac{u_{He}}{u_{air}} = \sqrt{\frac{\mathcal{M}_{air}}{\mathcal{M}_{He}}} = \sqrt{\frac{29.0 \text{ g/mol}}{4.00 \text{ g/mol}}} = 2.69$$

Helium effuses 2.69 times as fast as air.

18. Explain why real gases differ from ideal gases and how the differences lead to the van der Waals equation. Know under what conditions gases are most nearly ideal.

Two assumptions of the kinetic molecular theory clearly are incorrect: gas molecules *do* have volume and they *do* attract each other. If these assumptions were true, liquids would have no volume because the molecules composing them would be volumeless. Furthermore, liquids would spontaneously and completely vaporize as no forces would act to hold the molecules together. The ideal gas equation can be modified to account for these two inconsistencies. The volume in the ideal gas equation is the free volume—the space within which molecules are free to move. In a real gas the molecules take up, or exclude, some of the volume of the container.

free volume = container volume – excluded volume

or $V_{free} = V - nb$ (b = excluded volume per mole of gas) [15]

The ideal gas pressure assumes the molecules do not attract each other. But because there is an attraction, the molecules slow down as they approach the wall. They are attracted back into the body of the gas by other molecules. Hence, they strike the wall less hard and the measured pressure, P_{meas}, is less than the ideal pressure would be. The corrected pressure, P_{corr}, approximates what the ideal gas pressure would be.

$$P_{corr} = P_{meas} + \frac{an^2}{V^2} \qquad [16]$$

The constant a is multiplied by the square of the concentration of the gas, n/V. The concentration is squared because molecule A attracts molecule B, but molecule B also attracts molecule A. The result of these two modifications (equations [15] and [16]) is the van der Waals equation.

$$\left(P + \frac{an^2}{V^2}\right)(V - nb) = nRT \qquad [17]$$

The constants a and b principally depend on the gas being considered. To a small extent, they also depend on pressure and temperature. The volume correction ($-nb$) will be insignificant if it is small compared to the volume of the container. For a given quantity of the gas, the container volume is large when the pressure is low and the temperature is high. A large container volume also means that the concentration is low and thus the pressure correction ($+an^2/V^2$) will be small. Hence a gas will behave most like an ideal gas at low pressure and high temperature.

DRILL PROBLEMS

1. Fill in the blanks in each line below.
   A. 30.0 atm = ___lb/in.^{2} = ___kPa
   B. 142 mmHg = ___lb/in.^{2} = ___Pa
   C. 242 lb/in.^{2} = ___atm = kPa
   D. 5.00×10^2 atm = ___torr = ___Pa
   E. 5.00×10^2 mmHg = ___atm = ___Pa
   F. 5.00×10^2 Pa = ___ atm = ___torr
   G. 5.00×10^2 lb/in.^{2} = ___atm = ___ Pa
   H. 3.52 kPa = ___lb/in.^{2} = ___mmHg

3. Fill in the blanks in the table below. Neither temperature nor amount changes.

	P_i	V_i	P_f	V_f
A.	732 mmHg	6.49 L	1.02 atm	___L
B.	30.0 mmHg	35.2 L	___atm	35.2 mL

C.	9.75 Pa	1.46 L	923 mmHg	____L
D.	104 kPa	22.4 L	____atm	1.00 L
E.	247 kPa	0.972 L	143 torr	____L
F.	1.43 atm	0.947 qt	____atm	4.06 qt
G.	65.0 lb/in.^{2}	342 ft^3	14.7 lb/in^2	____ft^3
H.	175.5 kPa	12.4 m^3	____kPa	8.46 m^3
I.	1.11 atm	85.2 L	0.506 atm	____L
J.	4.11 atm	24.2 m^3	____mmHg	100.0 m^3

4. Fill in the blanks in the table below. Neither pressure nor amount changes.

	V_i	T_i	V_f	T_f
A.	6.49 L	273 K	____L	94.2°C
B.	0.947 qt	446 K	3.37 qt	____K
C.	85.2 L	200. K	____L	400. K
D.	24.2 m^3	37.0°C	19.4 m^3	____°C
E.	22.4 L	273 K	____L	298 K
F.	22.4 L	273 K	100.0 L	____K
G.	14.7 L	50.6°C	____L	300. K
H.	342 ft^3	300. K	67.4 ft^3	____°C

8. Fill in the blanks in the table below.

	P	V	n	T
A.	1.23 atm	____L	2.50 mol	298 K
B.	1.00 atm	22.4 L	1.46 mol	____K
C.	2.64 atm	1.23 L	____mol	143 K
D.	____atm	9.61 L	8.41 mol	312 K
E.	746 mmHg	____qt	1.00 mol	87.6°C
F.	1024 kPa	14.7 L	9.00 mol	____°C
G.	5.17 atm	94.2 L	____mol	25.0°C
H.	1726 kPa	____L	4.40 mol	37.0°C
I.	____torr	173 qt	1.26 mol	−40.0°C
J.	1.00 atm	____L	12.4 g CO_2	94.2°C
K.	7.43 lb/in.^{2}	91.6 L	____g H_2O	293°F
L.	____lb/in.^{2}	123 gal	9.63 g Ne	31°F
M.	1.46 kPa	84.1 m^3	143 g He	____°C
N.	9.47 atm	____L	22.0 g Xe	14.6°C
O.	522 mmHg	19.2 L	____g CH_4	83.1°C
P.	____torr	245 L	19.6 g N_2	250.0 K

9. (1) Fill in the blank in each line below. All numbers have the units given in the column headings unless otherwise specified. All substances are ideal gases. The number of moles does not change within each line.

	P_i, atm	V_i, L	T_i, K	P_f, atm	V_f, L	T_f, K
A.	1.42	7.26	274	3.74	2.16	____
B.	0.926	24.7	80.1°C	1.42	____	481
C.	1.43	7.32 qt	91.7 °C	____	3.17	260.
D.	4.60 torr	9.32	____	8.23	0.164	272.°C
E.	2.72	____	403	0.247	10.3	52.4°F
F.	____	82.1 gal	96.4°C	1.79	47.2	294
G.	829 torr	500 mL	73.1°F	0.725	1.42	____
H.	32.1	16.4	24.7°C	7.15	____	578
I.	466 torr	8.83	98.6°F	____	8.92 qt	78.0°F
J.	1776 torr	68.4	____	1086 torr	54.3	347
K.	5.13	____	671°C	2.62	9.13	403
L.	____	1.34	72.4°C	1.46	8.71	272

(2) Fill in the blanks in each line below.

M.	T_i = 296 K	n_i = 14.3 mol	T_f = 123°C	n_f = ____mol
N.	P_i = 14.3 atm	n_i = 1.23 mol	P_f = ____atm	n_f = 2.16 mol
O.	V_i = 22.4 L	n_i = 1.00 mol	V_f = 14.2 L	n_f = ____mol

P. $T_i = 500.$ K	$n_i = 2.74$ mol	$T_f =$ ____°C	$n_f = 1.43$ mol
Q. $P_i = 21.2$ atm	$n_i = 5.00$ mol	$P_f = 12.3$ atm	$n_f =$ ____mol
R. $V_i = 19.4$ L	$n_i = 1.32$ mol	$V_f =$ ____L	$n_f = 4.61$ mol

10. Fill in the blanks in each line that follows. The numbers have the units given in the column headings, unless otherwise specified.

	V, L	P, atm	n, mol	$\mathfrak{M}$, g/mol	T, K	*mass*, g	*density*, g/L
A.	25.0	2.00	____	____	27°C	____	1.25
B.	____	10.4	2.32	____	165	75.0	____
C.	72.1	46.8 torr	1.76	____	____	34.2	____
D.	34.6	____	0.832	17.4	–14°C	____	____
E.	10.4	1364 kPa	1.54	27.3	____	____	____
F.	134	72.1	____	79.2	279	____	____
G.	____	7.15	5.32	34.1	217	____	____
H.	____	3.12	1.91	____	36.0°C	76.4	____
I.	____	5.27	____	80.9	302	417	____
J.	____	____	2.73	75.2	463	____	2.14
K.	____	____	1.26	____	275	104	0.312
L.	____	____	____	44.0	350.	396	4.16
M.	____	34.6	____	____	46.3°C	4.76	54.0
N.	75.2	0.268	____	____	450.	____	0.178
O.	3.74	24.2	5.441	____	____	____	145
P.	7.42	1.74	____	8.42	____	____	87.4

11. The mass and the empirical formula of each gas used in part (1) of the drill problems of objective 6-9 are given below. Determine the molecular formula of each gas.

A. 19.3 g CH_2	B. 101 g CHF	C. 73.3 g CF	D. 1.75 g NH_2
E. 9.16 g CH_2	F. 168 g O	G. 1.17 g CH	H. 991 g NO_2
I. 9.48 g C_2H_2N	J. 158 g CHO	K. 43.4 g CH_2O	L. 17.1 g CH_2O

12. (1) Balance each equation if necessary. Then fill in the blanks in each line. All gases are measured at the same temperature and pressure.

$$C_5H_{12}(g) + O_2(g) \longrightarrow CO_2(g) + H_2O(g)$$

	$C_5H_{12}(g)$	$O_2(g)$	$CO_2(g)$	$H_2O(g)$
A.	7.41 L	____L	____L	
B.		____L	87.9 L	____L
C.	6.75 L	73.1 L	____L	____L

$$N_2(g) + H_2(g) \longrightarrow NH_3(g)$$

	$N_2(g)$	$H_2(g)$	$NH_3(g)$
D.	27.2 L	____L	____L
E.	8.63 L	25.2 L	____L
F.	____L	____L	21.4 L

$$6\,HCl(g) + 2\,HNO_3(g) \longrightarrow 4\,H_2O(g) + 2\,NO(g) + 3\,Cl_2(g)$$

	$HCl(g)$	$HNO_3(g)$	$H_2O(g)$	$NO(g)$	$Cl_2(g)$
G.	9.14 L		____L	____L	
H.	____L		2.17 L	____L	____L
I.	1.46 L	0.521 L	____L	____L	

$$4\,NH_3(g) + 5\,O_2(g) \xrightarrow{Fe} 4\,NO(g) + 6\,H_2O(g)$$

	$NH_3(g)$	$O_2(g)$	$NO(g)$	$H_2O(g)$
J.	4.61 L		____L	____L
K.	36.0 L	42.1 L	____L	____L
L.		____L	2.74 L	____L

$$4\,NH_3(g) + 3\,O_2(g) \longrightarrow 2\,N_2(g) + 6\,H_2O(g)$$

	$NH_3(g)$	$O_2(g)$	$N_2(g)$	$H_2O(g)$
M.	____L	____L	19.2 L	
N.	24.9 L	18.6 L	____L	____L
O.	____L	____L	____L	14.6 L

(2) A given quantity of a C–H–O compound is burned in $O_2(g)$. Both the products and the reactants are at the pressure and the temperature specified. Fill in the blank in each line.

P. 14.2 g $C_2H_6(g) \longrightarrow$ ____L $CO_2(g)$ 1.94 atm, 50.°C

Q. 42.1 L $C_3H_8(g)$ + 19.2 g $O_2 \longrightarrow$ ____L H_2O 0.746 atm, 428 K

R. 7.07 g $C_2H_5OH(l)$ + ____L $O_2(g) \longrightarrow$ ____L $CO_2(g)$ 0.203 atm, 540. K

S. 1.43 L $C_3H_8(g)$ + ____ L $O_2(g) \longrightarrow$ ____L $CO_2(g)$ 0.761 atm, 397 K

T. 8.16 L $H_2CO(g)$ + ___g $O_2(g) \longrightarrow$ ___g $H_2O(l)$ + ___L $CO_2(g)$ 0.331 atm, 100°C
U. 1.43 g $C_6H_6(l)$ + 9.14 L $O_2(g) \longrightarrow$ ___L $H_2O(g)$ 0.621 atm, 150.°C
V. 12.3 L $CH_4(g)$ + 123 L $O_2(g) \longrightarrow$ ___L $CO_2(g)$ 0.714 atm, 200.°C
W. 1.13 g $C_2H_6O_2(l)$ + 4.37 L $O_2(g) \longrightarrow$ ___L $CO_2(g)$ 0.506 atm, 350 K
X. 9.06 g $C_3H_8O(l) \longrightarrow$ L $CO_2(g)$ 1.23 atm, 74°F
Y. 8.32 g $C_6H_{14}(l) \longrightarrow$ ___L $CO_2(g)$ + ___g $H_2O(l)$ 790 torr, 50.°C

13. Gas *A* is mixed with Gas *B* to produce the mixture of gases. All gases are ideal, and no chemical reactions occur. Fill in the blank in each line below.

	Gas A			Gas B			Mixture		
	P, atm	*V*, L	*T*, K	*P*, atm	*V*, L	*T*, K	*P*, atm	*V*, L	*T*, K
A.	1.04	27.2	298	0.96	21.2	304	1.00	___	298
B.	2.14	7.86	250.	1.32	9.42	273	___	11.6	260.
C.	1.21	8.17	308	1.41	16.2	308	1.31	25.0	___
D.	1.73	3.12	298	___	2.14	298	204	6.15	298
E.	1.96	74.2	401	1.96	4.01	___	1.96	80.0	374
F.	1.43	2.13	298	___	2.13	298	7.42	2.13	250.
G.	2.17	14.6	294	2.17	16.4	306	4.34	___	300.
H.	3.00	17.2	286	3.00	17.2	2.90	___	17.2	294
I.	1.62	9.46	271	2.61	8.46	271	1.23	17.9	___

17. Apply the kinetic molecular theory and Graham's law.
   A. How fast is the average molecule of each of the following gases traveling at 300K? He, Ne, H_2, N_2, O_2, CO_2, Ar, CH_4, NH_3, H_2O?
   B. At what temperature is the average speed of each of the following molecules equal to 300. m/s? He, Ne, H_2, N_2, O_2, CO_2, Ar, CH_4, NH_3, H_2O?
   C. How much faster does He effuse than N_2 at 300 K?
   D. How much faster does N_2 effuse than O_2 at 1.50 atm?
   E. If 1.25 mol He effuses in 8.00 hr, how long is needed for 1.25 mol of Ar to effuse? Both gases are at 74°C.
   F. If 0.230 mol N_2 effuses in 12.4 hr, how long is needed for 12.4 mol of H_2O to effuse? Both gases are at 4.71 atm.

QUIZZES (20 minutes each) Choose the best answer for each question.

Quiz A

1. STP refers to which one of the following conditions? (a) 1.00 atm, undefined temperature; (b) 1.00 atm, 0°C; (c) 760 torr, 298 K; (d) 100 torr, 0°C; (e) none of these.
2. The kinetic energy of a gas is directly proportional to its (a) temperature; (b) pressure; (c) volume; (d) composition; (e) none of these.
3. A mathematical statement of Boyle's law is (a) V/T = constant; (b) V/n = constant; (c) P/T = constant; (d) PV = constant; (e) none of these.
4. At 1000.°C the pressure of a gas is 5020. kPa. What is its pressure in kPa at 3450.°C? (a) 5020(1273/3723); (b) 5020(3723/1273); (c) 5020(3450/1000); (d) 5020(1000/3450); (e) none of these.
5. At 850. kPa and 252 K a gas occupies 17.9 mL. What is its kelvin temperature at 722 kPa and a volume of 625 mL? (a) 252(722/850)(625/17.9); (b) 252(722/850)(17.9/625); (c) 252(850/722)(17.9/625); (d) 252(850/722)(625/17.9); (e) none of these.
6. 4.0 L of argon at 600. mmHg and 2.0 L of xenon at 300. mmHg are put in a 6.00 L container. The final pressure is (a) 3000. mmHg; (b) 480. mmHg; (c) 500. mmHg; 450 mmHg; (e) none of these.
7. 2.30 mol of gas occupy a volume of 62 mL at 6.14 atm and what kelvin temperature? (a) (6.14)(62)/[(2.30)(0.0821)]; (b) (6.14)(2.30)(62)/(0.0821); (c) (6.14)(0.062)/[(2.30)(0.0821)]; (d) (6.14)(0.062)(0.0821)/(62); (e) none of these.
8. 85.1 g of gas occupies 120. L at 0.224 atm and 450.°C. Its mole weight is (a) 85.1(0.0821/0.224)(450/120); (b) 85.1(0.224/0.0821)(120/450); (c) 85.1(0.224/0.0821)(120/723); (d) 85.1(0.0821/0.224)(120/723); (e) none of these.

9. In the reaction $4\ NH_3(g) + Cl_2(g) \longrightarrow N_2H_4(g) + 2\ NH_4Cl(s)$, how many liters of ammonia will combine with 14.0 L of chlorine? (a) (14.0/22.4); (b) (14.0)(22.4)(4); (c) (14.0)(4); (d) (14.0/22.4)(1/4); (e) none of these.
10. A gas has a temperature of 200. K. What is its kelvin temperature if the average kinetic energy of the molecules doubles? (a) $\sqrt{2}$(200 K); (b) 200 K; (c) (2)(200 K); (d) (3/2)(200 K); (e) none of these.

Quiz B

1. The postulates of the kinetic molecular theory of gases include all those that follow *except* (a) no forces exist between molecules; (b) molecules are point masses; (c) molecules are repelled by the walls of the container; (d) molecules are in random motion; (e) all are postulates.
2. The mutual attraction of gas molecules is an important aspect of (a) Avogadro's law; (b) Dalton's law; (c) Graham's law; (d) van der Waals's theory; (e) none of these.
3. Subtracting the vapor pressure of water from the total pressure of a gas collected over water is an example of the application of (a) Avogadro's law; (b) Dalton's law; (c) Graham's law; (d) van der Waals's theory; (e) none of these.
4. A gas has a volume of 16.0 L at 37°C. What is its volume in liters at 82°C? (a) 16.0(82/37); (b) 16.0(37/82); (c) 16.0(310/355); (d) 16.0(355/310); (e) none of these.
5. At 1.70 atm and 27°C a gas occupies 14.0 L. What is its volume in L at 1.05 atm and –3°C? (a) 14.0(1.70/1.05)(270/300); (b) 14.0(1.05/1.70)(270/300); (c) 14.0(1.05/1.70)(300/270); (d) 14.0(1.70/1.05)(300/270); (e) none of these.
6. 1.0 L of hydrogen at 700. kPa and 2.0 L of nitrogen at 1000. kPa are put in a 3.0 L-container. The final pressure is (a) 900. kPa; (b) 800. kPa; (c) 850. kPa; (d) 750. kPa; (e) none of these.
7. How many moles of gas occupy 3.14 L at 272 K and 0.122 atm? (a) (0.112)(272)/[(0.0821)(3.14)]; (b) (0.0821)(272)/[(0.112)(3.14)]; (c) (0.112)(3.14)/[(0.0821)(272)]; (d) (0.0821)(3.14)/[(0.112)(272)]; (e) none of these.
8. 22.4 g of gas occupies 15.2 L at 77°C and 3.21 atm. Its mole weight is (a) 22.4(0.0821/3.21)(350/15.2); (b) 22.4(0.0821/3.21)(77/15.2); (c) 22.4)(0.0821/3.21)(15.2/350); (d) 22.4(3.21/0.0821)(350/15.2); (e) none of these.
9. In the reaction $2\ H_2S(g) + 3\ O_2(g) \longrightarrow 2\ H_2O(l) + 2\ SO_2(g)$, how many liters of $O_2(g)$ are needed to produce 16.2 L of SO_2? (a) (16.2)(2/3); (b) (16.2)(3/2); (c) 16.2; (d) (16.2)(22.4)(3/2); (e) none of these;
10. The rate of effusion of a gas (38 g/mol) is 3.14 L/s. What is the rate in L/s for a second gas (28 g/mol) (a) 3.14(38/28); (b) 3.14 (28/38); (c) 3.14$\sqrt{38/28}$; (d) 3.14$\sqrt{28/38}$; (e) none of these.

Quiz C

1. If the kelvin temperature of a gas is doubled, then (a) the average molecule moves twice as fast; (b) every molecule moves twice as fast; (c) every molecule has twice the kinetic energy; (d) the average molecule has twice the kinetic energy; (e) none of these.
2. The ideal gas constant has the value of 0.08206. The units of this number are (a) L atm mol^{-1} K^{-1}; (b) L torr mol^{-1} K^{-1}; (c) mL torr mol^{-1} K^{-1}; (d) mL atm mol^{-1} K^{-1}; (e) none of these.
3. $PV = nRT$ is a statement of (a) Avogadro's law; (b) Dalton's law; (c) Graham's law; (d) van der Waals's theory; (e) none of these.
4. At 5.12 atm the temperature of a gas is 47°C. What is the Celsius temperature at 7.14 atm? (a) 320(7.14/5.12) – 273; (b) 320(5.12/7.14) – 273; (c) 47(5.12/7.14) – 273; (d) 47(7.14/5.12) + 273; (e) none of these.
5. At 250. torr and 150.°C a gas occupies 75.2 L. What is its volume in liters at 520. torr and 275°C? (a)75.2(250/520)(275/150); (b) 75.2(250/520)(548/423); (c) 75.2(520/250)(548/423); (d) 75.2(520/250)(275/150); (e) none of these.
6. 3.0 L of neon at 640. kPa and 2.0 L of xenon at 250. kPa are put in a 5.0 L container. The final pressure is (a) 445 kPa; (b) 484 kPa (c) 2420 kPa; (d) 406 kPa; (e) none of these.
7. 2.24 mol of gas exerts what pressure in atm at 25°C and a volume of 72 L? (a) (2.24)(0.0821)(25/72); (b) (2.24)(0.0821)(72/298); (c) (2.24)(72)(298/0.0821); (d) (25)(72)(0.0821/2.24);
8. 22.4 g of gas occupies 22.4 L at 273 K and 2.24 atm. Its mole weight is (a) 22.4; (b) 22.4(2.24/0.0821)(273/22.4); (c) 22.4(0.0821/2.24)(273/22.4); (d) 22.4(0.0821/2.24)(546/22.4); (e) none of these.
9. In the reaction $3\ H_2S(g) + 2\ HNO_3 \longrightarrow 3\ S + 2\ NO(g) + 4\ H_2O$, how many liters of $H_2S(g)$ are needed to produce 75.1 L of NO(g)? (a) 75.1(2/3); (b) 75.1(3/2); (c) 75.1/22.4)(3/2); (d) 75.1; (e) none of these.

10. 3.12 mol of Ne (20.2 g/mol) effuses through a pinhole in 14.2 min. How many minutes are needed for 3.12 mol of methane (CH_4, 16.0 g/mol) to effuse through the same pinhole? (a) 14.2; (b) 14.2(20.2/16.0); (c) $14.2\sqrt{20.2/16.0}$; (d) $14.2\sqrt{16.0/20.2}$; (e) none of these.

Quiz D

1. Pressure is the result of (a) molecular speed only; (b) molecular speed and mass; (c) molecular mass only; (d) repulsive forces between molecules; (e) attractive forces between molecules.
2. Gases approach ideal behavior most closely under what set of conditions? (a) high temperature, low pressure; (b) low temperature, low pressure; (c) high temperature, high pressure; (d) low temperature, high pressure; (e) no such simple statement can be made.
3. Equal volume of two gases at the same temperature and pressure have (a) equal weights; (b) Avogadro's number of molecules; (c) identical chemical compositions; (d) the same number of molecules; (e) none of these.
4. At 946 mmHg the volume of a gas is 2.54 L. What is its volume in liters at 454 mmHg? (a) 2.54(454/946)(760); (b) 2.54(454/946); (c) 2.54(946/454)(760); (d) 2.54(946/454); (e) none of these.
5. At 8.59 atm and 25°C a gas occupies 16.2 L. What is its pressure in atm at 421 K and a volume of 14.0 L? (a) 8.59(421/25)(16.2/14.0); (b) 8.59(25/421)(14.0/16.2); (c) 8.59(298/421)(14.0/16.2); (d) 8.59(421/298)(16.2/14.0); (e) none of these.
6. 2.0 L of argon at 400 kPa and 3.0 L of neon at 600 kPa are put in a 5.0 L container. The final pressure is (a) 500 kPa; (b) 520 kPa; (c) 580 kPa; (d) 2600 kPa; (e) none of these.
7. 0.82 mol of gas exerts what pressure in torr at 23°C and a volume of 72 L? (a) (0.82)(0.0821)(296)(760/72); (b) (0.82)(0.0821)(296/72); (c) (0.82)(0.0821)(72/296); (d) (0.82)(0.0821)(296)/[(72)(760)]; (e) none of these.
8. 374 g of gas occupies 82.4 L at 12.4 atm and 750 K. Its mole weight is (a) 374(0.0821/12.4)(750/82.4); (b) 374(0.0821/12.4)(82.4/750); (c) 374(12.4/0.0821)(82.4/750); (d) 374(12.4/0.0821)(750/82.4); (e) none of these.
9. In the reaction $4\,NH_3(g) + 5\,O_2(g) \longrightarrow 4\,NO(g) + 6\,H_2O$, how many milliliters of O_2 are needed to form 200 mL of NO? (a) 200(4/5); (b) 200(4/6); (c) 200(5/6); (d) 200(6/4); (e) none of these.
10. 2.45 mL of methane (CH_4 16.0 g/mol) effuses through a pinhole in 1.25 min. What is the mole weight (in g/mol) of a second gas is 2.45 mL of it (measured at the same T and P) requires 2.61 min to effuse through the same pinhole? (a) $16.0(2.61/1.25)^2$; (b) 16.0(2.61/1.25); (c) $16.0\sqrt{1.25/2.61}$; (d) $16.0\sqrt{2.61/1.25}$; (e) none of these.

SAMPLE TEST (20 minutes)

1. Convert the units of each of the following properties as indicated.
   a. Pressure: 26.4 lb/in.^{2} = ____atm = ____kPa   b. Volume: 2.0 ft^3 = ____m^3 = ____L
   c. Temperature: 92.0°F = ____°C = ____K
2. A gas that has the pressure, volume, and temperature given in the previous problem has a mass of 1.305 lb. What is the mole weight, in g/mol, of this gas?
3. What is the numerical value of the ideal gas constant R when it has the units: mL mmHg mmol^{-1} K^{-1}?
4. Two gases are to be mixed into a 5.000 L container at 291.0 K. Gas A originally is confined in 14.20 L at 1.067 atm and 303.1 K. Gas B originally is confined in 1.251 L at 26.42 atm and 327.5 K. Once the two gases are mixed in the 5.000-L container,
   a. What is the total pressure?   b. What is the partial pressure of Gas A?
   c. What is the partial pressure of Gas B?
5. 1 mol of helium and 1 mole of neon are mixed in a container.
   a. Which gas has the larger mass?
   b. Which gas has the greater average molecular speed?
   c. Which type of molecule strikes the wall of the container more frequently?
   d. Which gas exerts the larger pressure?

7 Thermochemistry

CHAPTER OBJECTIVES

1. Distinguish between heat and work.

Work moves an object through a distance and heat raises its temperature. Although work can be transformed completely into heat, heat cannot be transformed completely into work. One difference between the two is that work is an organized form of energy and can move large objects. Heat is more random and mainly is able to make small particles—atoms and molecules—move more rapidly. In addition, heat and work only appear during energy changes. We readily can measure energy—as heat or as work—when it moves from one place to another, but we cannot measure energy in place. Thus, we do not speak of absolute energy, but rather of energy changes.

* 2. Use specific heat to determine temperature changes and quantities of heat.

Specific heat is the quantity of heat needed to raise the temperature of one gram of substance by one degree Celsius. Specific heat depends on the nature of the substance being heated. The specific heat of water, for example, is greater than that of iron. Specific heat has the units of joules per gram per degree Celsius (J g^{-1} $°C^{-1}$) or calories per gram per degree Celsius (cal g^{-1} $°C^{-1}$). The quantity of heat absorbed by an object that undergoes a temperature change is given by equation[1].

heat absorbed = (mass)(specific heat)(temperature change)

$$q = (m)(\text{sp. ht.})(\Delta T) \qquad [1]$$

The molar heat capacity (C_p) is the quantity of heat needed to raise the temperature of one mole of a substance by one degree Celsius. C_p has the units of J mol^{-1} $°C^{-1}$ or cal mol^{-1} $°C^{-1}$. Thus the heat absorbed by an object also is given by equation [2].

$$q = n\, C_p\, \Delta T \qquad (n = \text{number of moles of substance}) \qquad [2]$$

There is a standard convention for using the symbol Δ (delta). It represents the final value minus the initial value of a property: $\Delta T = T_f - T_i$; $\Delta n = n_f - n_i$; $\Delta V = V_f - V_i$; and so forth.

EXAMPLE 7-1 What quantity of heat is needed to raise the temperature of 500.0 g Cu (C_p = 24.5 J mol^{-1} $°C^{-1}$) from 20.0°C to 50.0°C?

$$\text{heat} = n\, C_p\, \Delta T = 500.0 \text{ g} \times \frac{\text{mol Cu}}{63.54 \text{ g}} \times 24.5 \frac{\text{J}}{\text{mol °C}}(50.0°\text{C} - 20.0°\text{C})$$

$$= 5.78 \times 10^3 \text{ J} = 5.78 \text{ kJ}$$

One can combine equation [1] or [2] with the law of conservation of energy to determine either specific heats of materials or final temperatures of mixtures.

EXAMPLE 7-2 150.0 g H_2O (sp. ht. = 4.184 J g^{-1} $°C^{-1}$) at 20.0°C is mixed with 100.0 g of V (sp. ht. = 0.502 J g^{-1} $°C^{-1}$) at 100.0°C. What is the final temperature of the mixture?

We use the law of conservation of energy first. (The negative sign is inserted before the heat gained by vanadium because the metal actually loses heat in going to a lower temperature.)

$$-\text{heat gained by V} = \text{heat gained by } H_2O$$

$$-(100.0 \text{ g})(0.502 \frac{\text{J}}{\text{g °C}})(T_f - 100.0°\text{C}) = (150.0 \text{ g})(4.184 \frac{\text{J}}{\text{g °C}})(T_f - 20.0°\text{C})$$

$$-50.2\, T_f + 5.02 \times 10^3 = 627.6\, T_f - 1.26 \times 10^4$$

$$1.76 \times 10^4 = 677.8\, T_f \quad or \quad T_f = 26.0°\text{C}$$

EXAMPLE 7-3 140.0 g H_2O at 25.0°C is mixed with 100.0 g of metal at 100.0°C. The final temperature of the mixture is 29.6°C. What is the specific heat of the metal?

–heat gained by the metal = heat lost by the water

$$-(100.0\text{g})(\text{sp. ht.})(29.6°\text{C} - 100.0°\text{C}) = (140.0 \text{ g})(4.184 \text{ J g}^{-1} \text{ °C}^{-1})(29.6°\text{C} - 25.0°\text{C})$$

$$(7040 \text{ g°C})(\text{sp. ht.}) = 2694 \text{ J} \quad or \quad \text{sp. ht.} = 0.38 \text{ J g}^{-1} \text{ °C}^{-1}$$

* 3. Interconvert joules and calories. Apply the first law of thermodynamics: $\Delta E = q + w$.

It seems obvious to a casual observer that heat and work are different. Heat is always associated with changes either in temperature or in the state of matter, as from a solid to a liquid. On the other hand, work is associated with motion. Actually, heat and work are simply two different ways in which energy changes appear. Nevertheless, both the English and the metric systems have two different units of energy—one for heat and the other for work.

calorie: the heat needed to raise the temperature of one gram of water from 14.5°C to 15.5°C. (The Calorie, written with a capital C, is a nutritional Calorie, equal to 1000 calories.)

Btu (British thermal unit): the heat needed to raise the temperature of one pound of water one degree Fahrenheit at or near 39.1°F.

Joule: the work done when a force of one newton acts through a distance of one meter.

foot-pound: the work done when a force of 1 pound acts through a distance of one foot.

The calorie now is defined as one calorie = 4.184 Joule. The joule is an SI unit; the plan is to phase out the calorie as a unit of energy.

EXAMPLE 7-4 Determine the number of (a) Joules in 267.2 cal; (b) calories in 794.2 J.

The conversion factor method is used in each part.

a. 265.2 cal × (4.184 J/cal) = 1117 J

b. 791.2 J × (cal/4.184 J) = 189.1 cal

The ability to phrase heat and work in the same units is needed when we use the first law of thermodynamics: $\Delta E = q + w$. In this statement of the first law, both heat and work represent energy; flowing into the system. q is the heat absorbed by the system, and w is the work done on the system. The first law can be thought of as an energy bookkeeping equation. It operates in a fashion similar to determining the assets of a business, such as a gold bullion merchant. The assets may be in either cash (heat) or gold (work). The two assets may be interconverted, and either cash or gold may be lost or gained by the merchant. The comparison fails in that energy systems do not make a profit (like a shrewd merchant), the "rate of exchange" between heat and work is fixed, and in energy systems we do not know the value of E (the total assets) and can only determine its change.

EXAMPLE 7-5 A system absorbs 189.1 cal of heat and does 1117 J of work. What is the value of ΔE for this system, in J?

The energy values are those given in Example 7-4. In this case we must be careful to note that work is negative; work is not done on the system, but rather done by the system.

$\Delta E = q + w = 791.2 \text{ J} - 1117 \text{ J} = -326 \text{ J}$

4. State the meaning of the concept "state function," especially as demonstrated by enthalpy.

The concept of a state function is one of the most powerful in thermodynamics. A state function, or state property, only depends on where the system starts (its initial state) and where it ends (its final state). A state function does *not* depend on how the change in the system occurred (the pathway). For example, the distance from Los Angeles to New York is a state function. Yet the actual mileage traveled depends on the route. Thus, the mileage traveled is a path function. Both heat and work are path functions although they may equal state functions if the path is specified. An important path is one of constant pressure, a condition frequently encountered in the laboratory. The heat evolved at constant pressure equals the state function, change in enthalpy of the system.

* **5. Calculate the heat of a reaction at constant volume, q_V, using bomb calorimetry data.**

The heat given off by a reaction occurring in a bomb calorimeter increases the temperature of both the bomb and the water that surrounds it.

$$q_V = -\text{ heat gained by water} - \text{heat gained by bomb}$$

$$= -(\text{mass } H_2O)(\text{sp. ht.})(\Delta T) - (C_B)(\Delta T) \quad [3]$$

C_B is the heat capacity of the bomb—the quantity of heat needed to raise the temperature of the bomb by 1°C. The change in the state function internal energy, $\Delta E = q_V$, the heat produced at constant volume. Notice that a path function (heat) has been restricted by specifying its pathway (one of constant volume) so that it equals a state function.

Most often, combustion reactions are those run in a bomb calorimeter. In fact, there are extensive tabulations of heats of combustion. In order to relate these to enthalpy changes (see objective 6), you need to know the products of the combustion reaction. Some of the usual conventions are that $CO_2(g)$ is formed from compounds that contain C, $H_2O(l)$ from those containing H, $N_2(g)$ from those containing N, and $SO_2(g)$ from those contining S. In addition, be aware that sometimes heats of combustion are tabulated as positive numbers, even though the enthalpy change is negative; most combustion reactions are exothermic. Consult the legend of the table you are using to see what conventions are used.

6. Explain the purpose of enthalpy change (ΔH).

The enthalpy change for a process, ΔH, is obtained by measuring the heat given off at constant pressure, q_P. Since enthalpy is a state function, the value of ΔH depends only on the initial and final states of the system for a given process. ΔH does *not* depend on how the change from initial to final state occurred. Note also that the condition of constant pressure is the same as that under which we typically study chemical reactions—in a reaction vessel open to the (constant-pressure) atmosphere. Thus, for chemical reactions ΔH provides a readily measured and useful state function.

* **7. Explain how to use a Styrofoam coffee cup calorimeter and interpret the data obtained.**

A nest of Styrofoam coffee cups with a cardboard lid is almost a perfectly adiabatic container—very little heat is lost or gained by the system inside the cups. All the heat produced or absorbed by any process (including a chemical reaction) occurring in the calorimeter produces a temperature change of the system. The process occurs at constant pressure and thus we measure $q_P = \Delta H$.

$$\text{heat absorbed} = (\text{mass})(\text{sp. ht.})(\Delta T) \quad [4]$$

EXAMPLE 7-6 50.0 mL each of NaOH(aq) and HCl(aq), both at a concentration of 1.00 M and a temperature of 20.5°C, are mixed in a Styrofoam cup calorimeter. The final temperature of the system is 27.3°C. The resulting 0.500 M NaCl(aq) has a density of 1.02 g/mL and a specific heat of 4.02 J g^{-1} °C^{-1}. Compute ΔH in kJ/mol for the reaction.

$$NaOH(aq) + HCl(aq) \longrightarrow NaCl(aq) + H_2O(l)$$

$$q_P = [(1.02 \text{ g/mL})(100.0 \text{ mL})](4.02 \text{ J g}^{-1} \text{ °C}^{-1})(27.3\text{°C} - 20.5\text{°C}) = 2.79 \times 10^3 \text{ J}$$

This is the heat absorbed by the solution. The heat produced by the reaction is the negative of this value.

$$q_{P,\text{rxn}} = -2.79 \times 10^3 \text{ J}$$

The amount of NaCl then is computed.

$$\text{no. mol NaCl} = 0.1000 \text{ L soln} \left(\frac{0.500 \text{ mol NaCl}}{\text{L soln}}\right) = 0.0500 \text{ mol NaCl}$$

Finally, we compute ΔH.

$$\Delta H = \frac{q_{P\ \text{rxn}}}{\text{no. mol NaCl}} = \frac{-2.79 \times 10^3 \text{ kJ}}{0.0500 \text{ mol NaCl}} = -55.7 \times 10^3 \text{ J/mol} = -55.7 \text{ kJ/mol}$$

* **8. Apply Hess's law of constant heat summation.**

It is not necessary to actually *run* a chemical reaction in order to determine how much heat it will produce or consume. We merely combine (on paper) the heats of a series of consecutive reactions whose overall result is the same as the reaction we wish to study. On paper, we start with the same reactants (initial state) and end with the same products (final state) but by (perhaps) a different pathway.

EXAMPLE 7-7 What is $\Delta H°$ for the reaction C(graphite) + 2 S(rhombic) $\longrightarrow$ $CS_2(l)$?

The following thermochemical reactions are provided.

(*a*) C(graphite) + $O_2(g) \longrightarrow CO_2(g)$ $\Delta H° = -393.5$ kJ/mol
(*b*) S(rhombic) + $O_2(g) \longrightarrow SO_2(g)$ $\Delta H° = -296.9$ kJ/mol
(*c*) $CS_2(l) + 3\,O_2(g) \longrightarrow CO_2(g) + 2\,SO_2(g)$ $\Delta H° = -1075.2$ kJ/mol

We combine the given reactions: (*a*) + 2 (*b*) – (*c*). Whatever we do to the reactions (multiply them by a constant or change their sign—reverse them) we also do to the associated $\Delta H°$s

(*a*) C(graphite) + $O_2(g) \longrightarrow CO_2(g)$ $\Delta H° = -393.5$ kJ/mol
2(*b*) 2 S(rhombic) + $2\,O_2(g) \longrightarrow 2\,SO_2(g)$ $\Delta H° = -593.8$ kJ/mol
–(*c*) $CO_2(g) + 2\,SO_2(g) \longrightarrow CS_2(l) + 3\,O_2(g)$ $\Delta H° = +1075.2$ kJ/mol

C(graphite) + 2 S(rhombic) $\longrightarrow CS_2(l)$ $\Delta H = +87.9$ kJ/mol

*** 9. State the definitions of "standard state" and "standard formation reaction" and write the standard formation reaction for any substance.**

To use Hess's law, all reactants and products must be under the same conditions. Consider the $CO_2(g)$ produced by reaction (*a*) in objective 7-8 and consumed by reaction –(*c*). If reaction (*a*) produces $CO_2(g)$ at 298 K and 1.00 atm and reaction –(*c*) consumes $CO_2(g)$ at 323 K and 4.00 atm, then energy is needed to heat and compress the gas. On the other hand, if both reactants and products are in a well-defined state at one temperature, no energy is needed for heating and compression (or produced by cooling and expansion). The *standard state* is a pressure of one atmosphere or, for a solute in solution, a concentration of one mole per liter of solution. The degree sign (°) on $\Delta H°$ indicates that the molar enthalpy change is measured with reactants and products in their standard states. Note that the standard state does not mention the temperature. Temperature should be written as a subscript in kelvins, $\Delta H°_{298}$. Often, however, the subscript is omitted and 298 K is assumed.

The standard state also does not mention anything about the form of a substance. We can correctly refer to the standard state of each of the two allotropes of oxygen: $O_2(g)$ and $O_3(g)$ or ozone. Also, we can refer to the standard state of steam or water vapor, $H_2O(g)$; of ice, $H_2O(s)$; and of liquid water, $H_2O(l)$. It is most helpful to speak of all substances as being formed from one set of reactants. Since all matter is composed of atoms, we pick the *most stable form* (the one of lowest energy) of each element. These ideas are contained in the concept of the standard formation reaction: the reaction that produces one mole of a substance in its standard state from the elements in their standard states and most stable forms. The stable forms of the elements at 25°C (298.15 K) are given in Table 7-1. The standard formation reactions for $C_2H_5OH(l)$, $P_2O_5(s)$, and $CaCO_3(s)$ follow.

$$2\,C(graphite) + 3\,H_2(g) + \tfrac{1}{2}O_2 \longrightarrow C_2H_5OH(l)$$

$$2\,P(s, white) + \tfrac{5}{2}O_2(g) \longrightarrow P_2O_5(s)$$

$$Ca(s) + C(graphite) + \tfrac{1}{2}O_2(g) \longrightarrow CaCO_3(s)$$

TABLE 7-1 Most Stable Forms of Elements at 25°C

H_2 — gas	He — gas	C — graphite	Sb — crystalline, III
N_2 — gas	O_2 — gas	F_2 — gas	Hg —liquid
Ne — gas	P — white solid	S — rhombic solid	I_2 — solid
Cl_2 — gas	Ar — gas	Mn — crystalline, α	U — crystalline, III
As — gray solid	Se — gray solid	Br_2 — liquid	Xe — gas
Kr — gas	Cd — crystalline, α	Sn — white solid	
All other elements — crystalline solid (c) or (s)			

The standard molar enthalpy of formation is the heat evolved at constant pressure (the enthalpy change) when the standard formation reaction occurs. But what if the reaction does not occur as written? We already have answered this question in Example 7-7. The reaction

$$C(graphite) + 2\,S(rhombic) \longrightarrow CS_2(l)$$

does not occur, and it is the formation reaction for $CS_2(l)$. In Example 7-7, we determined $\Delta H°_f$, the standard enthalpy of formation for $CS_2(l)$, by combining the $\Delta H°$ values of several reactions that do occur.

*** 10. Apply Hess's law in the special case of standard formation reactions, that is, compute ΔH°_{rxn} from ΔH°_f values.**

Standard formation reactions greatly simplify the application of Hess's law.

EXAMPLE 7-8 What is ΔH°_{rxn} in kJ/mol for the oxidation of ammonia in the presence of a platinum catalyst?

$$4\,NH_3(g) + 5\,O_2(g) \xrightarrow{Pt} 4\,NO(g) + 6\,H_2O(g)$$

$$\Delta H^\circ_f \quad -46.19 \qquad 0.00 \qquad +90.37 \qquad -285.85 \text{ kJ/mol}$$

The ΔH°_fs that are given are equivalent to the following thermochemical equations

(d) $\frac{1}{2}N_2(g) + \frac{3}{2}H_2(g) \longrightarrow NH_3$ $\Delta H^\circ = -46.19$ kJ/mol

(e) $O_2(g) \longrightarrow O_2(g)$ $\Delta H^\circ = 0.00$ kJ/mol

(f) $\frac{1}{2}N_2(g) + \frac{1}{2}O_2(g) \longrightarrow NO(g)$ $\Delta H^\circ = +90.37$ kJ/mol

(g) $\frac{1}{2}O_2(g) + H_2(g) \longrightarrow H_2O(l)$ $\Delta H^\circ = -285.85$ kJ/mol

Notice that ΔH°_f for $O_2(g)$ is 0.00 kJ/mol since the product of the formation reaction is an element in its most stable form. [In contrast, $\Delta H^\circ_f = 142.3$ kJ/mol for ozone, $O_3(g)$.] Equations *(d)* through *(g)* can be combined to yield the equation for the catalytic oxidation of ammonia: $-4(d) - 5(e) + 4(f) + 6(g)$.

$-4\,(d)$	$4\,NH_3(g) \longrightarrow 2\,N_2(g) + 6\,H_2(g)$	$\Delta H^\circ = -184.8$ kJ/mol
$-5\,(e)$	$5\,O_2(g) \longrightarrow 5.O_2(g)$	$\Delta H^\circ = 0.00$ kJ/mol
$+4\,(f)$	$2\,N_2(g) + 2\,O_2(g) \longrightarrow 4\,NO(g)$	$\Delta H^\circ = +316.5$ kJ/mol
$+6\,(g)$	$3\,O_2(g) + 6\,H_2(g) \longrightarrow 3\,H_2O(l)$	$\Delta H^\circ = -1715.1$ kJ/mol
	$4\,NH_3(g) + 5\,O_2(g) \longrightarrow 4\,NO(g) + 6\,H_2O(l)$	$\Delta H^\circ = -1168.8$ kJ/mol

Notice that the numbers we used to multiply equations (*d*) through (*g*) are the stoichiometric coefficients in the original equation! In order to determine ΔH°_{rxn} for a reaction, we first add the standard formation enthalpies for the products, each one multiplied by the stoichiometric coefficient which that substance has in the balanced chemical equation. From this sum of product ΔH°_f values, we subtract a similarly constructed sum of reactant ΔH°_f values

$$\Delta H^\circ_{rxn} = \sum_{products} \nu_{prod}\, \Delta H^\circ_{f,prod} - \sum_{reactants} \nu_{reac}\, \Delta H^\circ_{f,reac} \qquad [5]$$

In equation [5], ν_{prod} and ν_{reac} represent the stoichiometric coefficients in the balanced chemical equation. For the calculation of Example 7-8. equation [5] becomes

$$\Delta H^\circ_{rxn} = 4\,\Delta H^\circ_f[NO(g)] + 6\,\Delta H^\circ_f[H_2O(l)] - 4\,\Delta H^\circ_f[NH_3(g)] - 5\,\Delta H^\circ_f[O_2(g)]$$
$$= 4(90.37) + 6(-285.85) - 4\,(-46.19) - 5\,(0.00) \text{ kJ/mol}$$
$$= -1168.8 \text{ kJ/mol}$$

DRILL PROBLEMS

2. (1) Fill in the blanks in each line that follows. The numbers have the units given in the column headings unless otherwise specified.

	q, J	m, g	sp. ht. , J g^{-1} °C^{-1}	t_i, °C	t_f, °C
A.	640	124	0.931	25.0	____
B.	____	75.2	0.587	20.0	30.0
C.	374	103	____	19.5	25.4
D.	1300	____	0.863	17.2	30.1
E.	194 cal	125	0.750	21.2	____
F.	____	142	1.086	66.1°F	52.3°F
G.	876	4.32 oz	____	22.1	30.4
H.	444	____	0.0953 cal g^{-1} °C^{-1}	19.4	26.5
	q, J	n, mol	C_p, J mol^{-1} °C^{-1}	t_i, °C	t_f, °C
I.	1962	14.0	72.0	19.3	____
J.	83.5	____	32.1	20.4	22.1
K.	572	1.65	____	21.4	35.7
L.	____	4.32	26.1	19.7	17.3
M.	1043	5.72	56.2	16.8	____

(2) Fill in the blank in each line that follows. Substance *A* is mixed with substance *B*. The mixture has the temperature given in the right-hand column. No heat is lost from or gained by the system.

	Substance A			Substance B			Mixture
	m, g	Sp. ht. J g^{-1} °C^{-1}	*t*, °C	*m*, g	*Sp. ht.* J g^{-1} °C^{-1}	*t*, °C	*t*, °C
N.	52.0	4.18	25.0	194	____	35.0	31.2
O.	124	4.60	20.0	10.2	0.412	75.2	____
P.	76.4	4.31	15.0	24.2	1.326	____	27.3
Q.	142	3.90	24.7	____	0.435	68.4	37.4
R.	7.51	0.682	94.1	____	3.66	27.2	39.0
S.	9.64	1.293	85.7	274	____	12.8	25.9
T.	7.28	3.60	106.0	15.4	4.18	____	82.0
U.	16.3	1.188	23.1	75.6	3.90	47.5	____
V.	12.3	1.686	64.7	105	3.06	19.4	____

3. (1) Convert the following values that are in calories to Joules, and those that are in Joules to calories.
A. 17.64 J B. 572.3 cal C. 2455 J D. 542.0 cal E. 312.4 J F. 73.20 cal
G. 627.6 J H. 1.264 cal I. 6773 J J. 2.600 cal K. 4002 J L. 34.51 cal

(2) Determine the value of ΔE, in J or kJ, of the system for each of the following situations.
M. 87.5 cal of heat absorbed and 434 J of work done by the system.
N. 96.2 cal of heat given off and 804 J of work done by system.
O. 134 cal of heat given off and 506 J of work done on the system.
P. 222 cal of heat absorbed and 734 J of work done on the system.
Q. 987 cal of heat absorbed and 2.17 kJ of work done by the system.
R. 455 cal of heat given off and 1.98 kJ of work done by the system.
S. 884 cal of heat given off and 3.02 kJ of work done on the system.
T. 976 cal of heat absorbed and 1.201 kJ of work done on the system.
U. 434 cal of heat given off and 1.502 kJ of work done on the system
V. 303 cal of heat absorbed and 863 J of work done on the system
W. 155 cal of heat given off and 733 J of work done by the system
X. 87.6 cal of heat absorbed and 95.0 J of work done by the system

5. Fill in the blank in each line below. *m* = the mass of substance burned in the calorimeter and q_v is the heat of combustion in kJ/g of that substance.

	m, g	q_v, kJ/g	m_{H_2O}, g	C_B, kJ/°C	t_f	t_i
A.	0.750	26.45	1043.3	____	24.1°C	20.0°C
B.	0.932	____	1032.1	1.614	26.9°C	20.5°C
C.	0.516	17.29	974.1	1.312	____	19.6°C
D.	____	34.09	836.4	1.435	77.9°F	68.1°F
E.	3.000	14.25	1017.0	____	25.9°C	17.3°C
F.	0.816	____	894.3	1.826	22.1°C	15.1°C
G.	____	40.56	1055.2	1.414	26.4°C	18.8°C
H.	1.024	13.08	1000.0	1.732	____°F	59.1°F

7. Fill in the blank in each line below. Assume that the solution formed has a specific heat of 4.00 J g^{-1} °C^{-1}.

	Solute	*Solute* mass, g	*Water* mass, g	t_f, °C	t_i, °C	$\Delta H^\circ_{solution}$ kJ/mol of solute
A.	$KClO_3$	9.04	91.20	22.4	29.7	____
B.	$NaNO_3$	8.61	157.98	____	18.6	−21.38
C.	NH_4Cl	18.31	____	17.1	22.0	+15.15
D.	NH_4NO_3	15.27	473.15	22.7	25.1	____
E.	NaOH	11.73	143.06	____	20.0	−42.89
F.	NaCl	11.98	104.32	____	20.1	+ 3.88
G.	RbI	74.24	569.87	25.0	28.4	____

8. For each lettered part, compute $\Delta H°$ in kJ/mol for the first reaction by appropriately combining the $\Delta H°$s of the remaining reactions.

	$\Delta H°$, kJ/mol
A. $CO_2(g) + H_2(g) \longrightarrow H_2O(l) + CO(g)$	
$H_2(g) + \frac{1}{2}O_2(g) \longrightarrow H_2O(l)$	– 285.9
$CO(g) + \frac{1}{2}O_2(g) \longrightarrow CO_2(g)$	– 393.5
B. $2\ C_2H_5OH(l) + O_2(g) \longrightarrow 2\ C_2H_4O(l) + 2\ H_2O(l)$	
$C_2H_5OH(l) + 3\ O_2(g) \longrightarrow 2\ CO_2(g) + 3\ H_2O(l)$	– 1370.7
$C_2H_4O(l) + \frac{5}{2}O_2(g) \longrightarrow 2\ CO_2(g) + 2\ H_2O(g)$	– 1167.3
C. $2\ C(gr) + \frac{5}{2}H_2(g) + \frac{1}{2}Cl_2(g) \longrightarrow C_2H_5Cl(g)$	
$H_2(g) + \frac{1}{2}O_2(g) \longrightarrow H_2O(l)$	– 285.9
$C(gr) + O_2(g) \longrightarrow CO_2(g)$	– 393.5
$C_2H_4(g) + HCl(g) \longrightarrow C_2H_5Cl(g)$	– 72.0
$C_2H_4(g) + 3\ O_2(g) \longrightarrow 2\ CO_2(g) + 2\ H_2O(l)$	– 1410.8
$H_2(g) + Cl_2(g) \longrightarrow 2\ HCl(g)$	– 184.6
D. $Na(s) + \frac{1}{2}Cl_2(g) \longrightarrow NaCl(s)$	
$2\ Na(s) + 2\ HCl(g) \longrightarrow 2\ NaCl(s) + H_2(g)$	– 637.4
$H_2(g) + Cl_2(g) \longrightarrow 2\ HCl(g)$	– 184.6
E. $2\ XO_2(s) + CO(g) \longrightarrow X_2O_3(s) + CO_2(s)$ (X is a metallic element)	
$XO_2(s) + CO(g) \longrightarrow XO(s) + CO_2(g)$	– 83.7
$X_3O_4(s) + CO(g) \longrightarrow 3\ XO(s) + CO_2(g)$	+ 25.1
$3\ X_2O_3(s) + CO(g) \longrightarrow 2\ X_3O_4(s) + CO_2(g)$	– 50.2
F. $2\ MnO_2(s) + CO(g) \longrightarrow Mn_2O_3(s) + CO_2(s)$	
$MnO_2(s) + CO(g) \longrightarrow MnO(s) + CO_2(g)$	– 150.6
$Mn_3O_4(s) + CO(g) \longrightarrow 3\ MnO(s) + CO_2(g)$	– 54.4
$3\ Mn_2O_3(s) + CO(g) \longrightarrow 2\ Mn_3O_4(s) + CO_2(g)$	– 142.3
G. $3\ V_2O_3(s) \longrightarrow V_2O_5(s) + 4\ VO(s)$	
$2\ V(s) + \frac{3}{2}O_2(g) \longrightarrow V_2O_3(s)$	– 1213
$2\ V(s) + \frac{5}{2}O_2(g) \longrightarrow V_2O_5(s)$	– 1561
$V(s) + \frac{1}{2}O_2(g) \longrightarrow VO(s)$	– 418
$V(s) + O_2(g) \longrightarrow VO_2(s)$	– 720
H. $V_2O_5(s) + 3\ V(s) \longrightarrow 5\ VO(s)$	
Data as in part G.	
I. $Fe_2O_3(s) + 3\ CO(g) \longrightarrow 2\ Fe(s) + 3\ CO_2(g)$	
$Fe_2O_3(s) + CO(g) \longrightarrow 2\ FeO(s) + CO_2(g)$	– 2.9
$Fe(s) + CO_2(g) \longrightarrow FeO(s) + CO(g)$	+ 11.3
J. $2\ C(gr) + H_2(g) \longrightarrow C_2H_2(g)$	
$CaO(s) + H_2O(l) \longrightarrow Ca(OH)_2(s)$	– 65.3
$CaO(s) + 3\ C(gr) \longrightarrow CaC_2(s) + CO(g)$	+ 462.3
$CaC_2(s) + 2\ H_2O(l) \longrightarrow Ca(OH)_2(s) + C_2H_2(g)$	– 125.5
$2\ C(gr) + O_2(g) \longrightarrow 2\ CO(g)$	– 220.9
$2\ H_2O(l) \longrightarrow 2\ H_2(g) + O_2(g)$	+ 571.5
K. $Cu(s) + \frac{1}{2}O_2(g) \longrightarrow CuO(s)$	
$Cu_2O(s) + \frac{1}{2}O_2(g) \longrightarrow 2\ CuO(s)$	– 143.9
$CuO(s) + Cu(s) \longrightarrow Cu_2O(s)$	– 11.3

9. Write the standard formation reaction for each substance.

A. $C_2H_5OH(l)$	B. $(NH_4)_2SbCl_5(s)$	C. $Hg_2Cl_2(s)$	D. $H_3PO_4(s)$
E. $NaIO_3(s)$	F. $SnF_2(s)$	G. $XeO_3(s)$	H. $HNO_3(l)$
I. $KBrO_3(s)$	J. $ICl_3(g)$		

10. Use the enthalpies of formation (all in kJ/mol) in Table 17-1 (in Chapter 17 of this text) to determine the enthalpy change of each reaction written below.
A. $CaCO_3(s) \longrightarrow CaO(s) + CO_2(g)$
B. $NH_3(g) + HCl(g) \longrightarrow NH_4Cl(s)$
C. $NH_3(g) + H_2O(l) + CO_2(g) \longrightarrow NH_4HCO_3(s)$
D. $CaO(s) + 2\ HCl(g) \longrightarrow CaCl_2(s) + H_2O(l)$
E. $Li_2CO_3(s) \longrightarrow Li_2O(s) + CO_2(g)$
F. $Li_2O(s) + H_2O(l) \longrightarrow 2\ LiOH(s)$
G. $4\ NH_3(g) + 3\ O_2(g) \longrightarrow 2\ N_2(g) + 6\ H_2O(l)$
H. $4\ NH_3(g) + 5\ O_2(g) \longrightarrow 4\ NO(g) + 6H_2O(l)$
I. $2\ NO(g) + O_2(g) \longrightarrow 2\ NO_2(g)$
J. $3\ NO_2(g) + H_2O(l) \longrightarrow 2\ HNO_3(l) + NO(g)$
K. $2\ HNO_3(l) + Li_2O(s) \longrightarrow 2\ LiNO_3(s) + H_2O(l)$
L. $Li_2O(s) + 2\ HCl(g) \longrightarrow LiCl(s) + H_2O(l)$
M. $LiOH(s) + HCl(g) \longrightarrow LiCl(s) + H_2O(l)$
N. $4\ LiNO_3(s) \longrightarrow 2\ Li_2O(s) + 4\ NO_2(g) + O_2(g)$
O. $CaO(s) + H_2O(l) \longrightarrow Ca(OH)_2(s)$

QUIZZES (20 minutes each) Choose the best answer for each question. For problems with a numerical answer, an answer is correct if it is within 1% of the correctly computed value.

Quiz A

1. The standard state of a substance is the (a) pure form at 1 atm; (b) most stable form at 25°C and 1 atm; (c) most stable form at 0°C; (d) pure gaseous form at 25°C; (e) none of these.
2. 75.0 g of water at 10°C is mixed with 125.0 g of water at 50°C. The final temperature of the mixture is (a) 30°C; (b) 40°C; (c) 35°C; (d) 25°C; (e) none of these.
3. $2\ Mg(s) + O_2(g) \longrightarrow 2\ MgO(s)$ $\Delta H° = -1203$ kJ/mol O_2 The heat given off in kJ/mol when 1.0 g of MgO(s) is formed is (a) 29.9; (b) 1203; (c) 601.7; (d) 14.9; (e) none of these.
4. 10.0 g of H_2O (sp. ht. = 4.184 J g^{-1} °C^{-1}) cools from 60.°C to 30.°C. How many joules of heat are lost to the surroundings? (a) 1393; (b) 62.8; (c) 12.6; (d) 1255; (e) none of these.
5. Which is the formation reaction for NO(g)? (a) $\frac{1}{2} N_2(g) + \frac{1}{3} O_3(g) \longrightarrow NO(g)$; (b) $N(g) + O(g) \longrightarrow NO(g)$; (c) $NO_2(g) \longrightarrow NO(g) + \frac{1}{2} O_2(g)$; (d) $N_2O_3(g) \longrightarrow NO(g) + NO_2(g)$; (e) none of these.
6. $Sn(s) + 2\ Cl_2(g) \longrightarrow SnCl_4(s)$ $\Delta H° = -545.2$ kJ/mol
$SnCl_2(s) + Cl_2(g) \longrightarrow SnCl_4(s)$ $\Delta H° = -195.4$ kJ/mol
What is $\Delta H°$ in kJ/mol for $Sn(s) + Cl_2(g) \longrightarrow SnCl_2(g)$? (a) 349.8; (b) –349.8; (c) 740.6; (d) 195.4; (e) none of these.
7. $C_2H_6(g) + \frac{7}{2} O_2(g) \longrightarrow 2\ CO_2(g) + 3\ H_2O(g)$ $\Delta H° = -1541$ kJ/mol
$H_2(g) + \frac{1}{2} O_2(g) \longrightarrow H_2O(l)$ $\Delta H° = -286$ kJ/mol
$C(gr) + O_2(g) \longrightarrow CO_2(g)$ $\Delta H° = -393$ kJ/mol
The value of ΔH° in kJ/mol for $2\ C(gr) + 3\ H_2(g) \longrightarrow C_2H_6(g)$ is (a) –1541; (b) 862; (c) 469; (d) 103; (e) none of these.
8. $SnO_2(s) + 2\ H_2(g) \longrightarrow Sn(s) + 2\ H_2O(l)$
$\Delta H°_f$ –580.7 0.00 0.00 –285.9 kJ/mol
$\Delta H°$ for this reaction in kJ/mol is (a) 294.8; (b) 8.9; (c) –294.8; (d) –571.8; (e) none of these.

Quiz B

1. Each of the following is a stable form of an element *except* (a) $O_2(g)$; (b) $F_2(g)$; (c) C(diamond); (d) $N_2(g)$; (e) all of these are stable.
2. 15.0 g of water at 100°C is mixed with 85.0 g of water at 20°C. The final temperature of the mixture is (a) 60°C; (b) 32°C; (c) 50°C; (d) 48°C; (e) none of these.
3. The molar heat of combustion of acetylene (26 g/mol) is –1301 kJ/mol. Combustion of 0.130 g of acetylene produces how many kilojoules of heat? (a) 50.05; (b) 384.9; (c) 6.51; (d) 169.1; (e) none of these.

4. 10.00 g of HCl solution (sp. ht. = 4.017 J g^{-1} °C^{-1}) is heated from 25.0°C to 40.0°C. How many joules of heat does this heating require? (a) 602.6; (b) 627.6; (c) 654.0; (d) 1004; (e) none of these.
5. Which is the formation reaction for $H_2O(l)$? (a) $H^+(aq) + OH^-(aq) \longrightarrow H_2O(l)$; (b) $H_2O(g) \longrightarrow H_2O(l)$; (c) $H_2(g) + \frac{1}{2} O_2(g) \longrightarrow H_2O(l)$; (d) $H_2(g) + {}^1/_3\, O_3(g) \longrightarrow H_2O(l)$; (e) none of these.
6. $MnO_2(s) \longrightarrow MnO(s) + \frac{1}{2} O_2(g)$ $\Delta H° = +136.0$ kJ/mol
   $MnO_2(s) + Mn(s) \longrightarrow 2\, MnO(s)$ $\Delta H° = -248.9$ kJ/mol
   For the reaction $Mn(s) + O_2(g) \longrightarrow MnO_2(s)$ the value of $\Delta H°$ in kJ/mol is (a) –112.9; (b) –384.9; (c) –520.9; (d) +361.9; (e) none of these.
7. $2\, ClF(g) + O_2(g) \longrightarrow Cl_2O(g) + F_2O(g)$ $\Delta H° = 167.4$ kJ/mol
   $2\, ClF_3(g) + 2\, O_2(g) \longrightarrow Cl_2O(g) + 3\, F_2O(g)$ $\Delta H° = 341.4$ kJ/mol
   For the reaction $ClF(g) + F_2(g) \longrightarrow ClF_3(g)$, the value of $\Delta H°$ in kJ/mol is (a) –217.5; (b) –435.0; (c) +223.6; (d) –130.2; (e) none of these.
8. $NH_3(g) + \frac{3}{4} O_2(g) \longrightarrow \frac{1}{2} N_2(g) + \frac{3}{2} H_2O(g)$
   $\Delta H°_f$ –46.0   0.00   0.00   –241.8 kJ/mol
   The molar enthalpy change for this reaction in kJ/mol is (a) –195.8; (b) +316.7; (c) –408.8; (d) –316.7; (e) none of these.

Quiz C

1. When an element is involved in a formation reaction, it does *not* have to be (a) pure; (b) at 1.00 M concentration; (c) at 1.00 atm pressure; (d) in its most stable form; (e) none of these.
2. 25.0 g of water at 50.0°C is mixed with 15.0 g of water at 30.0°C. The final temperature of the mixture is (a) 40.0°C; (b) 42.5°C; (c) 37.5°C; (d) 35.0°C; (e) none of these.
3. The molar enthalpy of combustion of $CS_2(l)$ is –897.46 kJ/mol. The products of combustion are $CO_2(g)$ and $SO_2(g)$. How many kJ of heat are given off when 6.41 g of $SO_2(g)$ is produced by the combustion of $CS_2(l)$? (a) 44.9; (b) 89.7; (c) 179; (d) 29.9; (e) none of these.
4. 1674 J of heat is absorbed by 25.0 mL of NaOH(aq) (d = 1.10 g/mL; sp. ht. = 4.10 J g^{-1} °C^{-1}). The temperature of the NaOH(aq) increases how many Celsius degrees? (a) 17.2; (b) 14.2; (c) 14.8; (d) 18.0; (e) none of these.
5. Which is the formation reaction for $CO_2(g)$? (a) $C(dia) + O_2(g) \longrightarrow CO_2(g)$; (b) $C(gr) + O_2(g) \longrightarrow CO_2(g)$ (c) $CO(g) + \frac{1}{2} O_2(g) \longrightarrow CO_2(g)$; (d) $CaCO_3(s) \longrightarrow CaO(s) + CO_2(g)$; (e) none of these.
6. $CH_4(g) + 2\, O_2(g) \longrightarrow CO_2(g) + 2\, H_2O(g)$ $\Delta H° = -890.4$ kJ/mol
   $CH_2O(g) + O_2(g) \longrightarrow CO_2(g) + H_2O(g)$ $\Delta H° = -563.5$ kJ/mol
   For the reaction $CH_4(g) + O_2(g) \longrightarrow CH_2O(g) + H_2O(g)$, the value of $\Delta H°$ in kJ/mol is (a) –236.6; (b) +118.3; (c) –326.9; (d) –1453.9; (e) none of these.
7. $2\, LiOH(s) \longrightarrow Li_2O(s) + H_2O(l)$ $\Delta H° = +379.1$ kJ/mol
   $LiH(s) + H_2O(l) \longrightarrow LiOH(s) + H_2(g)$ $\Delta H° = -110.9$ kJ/mol
   $2\, H_2(g) + O_2(g) \longrightarrow 2\, H_2O(l)$ $\Delta H° = -285.9$ kJ/mol
   For the reaction $2\, LiH(s) + O_2(g) \longrightarrow Li_2O(s) + H_2O(l)$, the value of $\Delta H°$ in kJ/mol is (a) –128.6; (b) –17.7; (c) 268.2; (d) 551.4; (e) none of these.
8. $2\, H_2S(g) + 3\, O_2(g) \longrightarrow 2\, H_2O(l) + 2\, SO_2(g)$
   $\Delta H°_f$ –20.1   0.00   –285.8   –297.1 kJ/mol
   For this reaction, what is the value of $\Delta H°$ in kJ/mol? (a) –562.8; (b) +562.8; (c) –3.14; (d) –477.4; (e) none of these.

Quiz D

1. The metric system unit of energy that was originally designed to measure heat is the (a) Joule; (b) Btu; (c) horse power; (d) calorie; (e) none of these.
2. 77.0 g of water at 95.0°C is mixed with 23.0 g of water at 5.0°C. The final temperature of the mixture is (a) 50.0°C; (b) 63.6°C; (c) 23.6°C; (d) 74.3°C; (e) none of these.
3. The molar enthalpy of combustion of $CH_3OH(l)$ is –726.51 kJ/mol. How much heat in kJ is given off when one mole $O_2(s)$ is used to burn $CH_3OH(l)$? (a) 726.51; (b) 363.26; (c) 1089.8; (d) 242.17; (e) none of these.
4. 40.0 g of solution has its temperature raised from 25.0°C to 50.0°C by the addition of 4067 J of heat. Its specific heat (in J g^{-1} °C^{-1}) is (a) 4.31; (b) 4.07; (c) 4.18; (d) 2.03; (e) none of these.

5. Which is the formation reaction for $SO_3(g)$? (a) S(monoclinic) + $\frac{3}{2}O_2(g) \longrightarrow SO_3(g)$; (b) $2\ SO_2(g) \longrightarrow SO_3(g) + SO(g)$; (c) $SO_2(g) + \frac{1}{2}O_2(g) \longrightarrow SO_3(g)$; (d) $H_2SO_4(l) \longrightarrow H_2O(l) + SO_3(g)$; (e) none of these.

6. $CO(g) + \frac{1}{2}O_2(g) \longrightarrow CO_2(g)$ $\Delta H° = -282.8$ kJ/mol
$C(gr) + O_2(g) \longrightarrow CO_2(g)$ $\Delta H° = -393.3$ kJ/mol
For the reaction $C(gr) + \frac{1}{2}O_2(g) \longrightarrow CO(g)$ the value of $\Delta H°$ in kJ/mol is (a) +282.8; (b) −676.1; (c) +110.5; (d) −110.5; (e) none of these.

7. $H_2S(g) + \frac{3}{2}O_2 \longrightarrow H_2O(l) + SO_2(g)$ $\Delta H° = -562.3$ kJ/mol
S(rhombic) + $O_2(g) \longrightarrow SO_2(g)$ $\Delta H° = -297.1$ kJ/mol
$H_2(g) + \frac{1}{2}O_2(g) \longrightarrow H_2O(l)$ $\Delta H° = -285.8$ kJ/mol
For the reaction $H_2(g)$ + S(rhombic) $\longrightarrow H_2S(s)$ the value of $\Delta H°$ in kJ/mol is (a) 562.3; (b) 265.3; (c) 276.6; (d) −20.6; (e) none of these.

8. $CH_3OH(l) + \frac{3}{2}O_2(g) \longrightarrow CO_2(g) + 2\ H_2O(l)$
$\Delta H°_f$ −238.5 0.00 −393.5 −285.3 kJ/mol
For this reaction, the value of $\Delta H°$ in kJ/mol is (a) −140.8; (b) −725.4; (c) +154.8; (d) −154.8; (e) none of these.

SAMPLE TEST (15 minutes)

1. Write the formation reaction for each of the following compounds.
a. $SnCl_4(s)$ b. $C_6H_5COOH(s)$ c. $COCl_2(g)$

2. Compute $\Delta H°_{rxn}$ for each of the following reactions. The value of $\Delta H°_f$ in kJ/mol is given below the formula of each species.
a. $SiO_2(s) + 4\ HF(g) \longrightarrow SiF_4(g) + 2\ H_2O(g)$
−859.4 −269 −1550 −285.9
b. $2\ CuS(s) + 3\ O_2(g) \longrightarrow 2\ CuO(s) + 2\ SO_2(g)$
−48.5 0.00 −155 −296.9

3. Each mole of LiCl produces 37.2 kJ of heat when it dissolves in water. What will be the final temperature when 5.00 g LiCl dissolves in 110.0 g of water at 20.00°C? The solution that is produced has a specific heat of 4.00 J g^{-1} °C^{-1}.

8 Electrons in Atoms

CHAPTER OBJECTIVES

* 1. **Apply the fundamental expression relating the frequency, wavelength, and velocity of electromagnetic radiation, with appropriate regard for units.**

Equation [1] is the relationship between the frequency of radiation ν (pronounced "nu") and its wavelength λ ("lambda"). The constant velocity of electromagnetic radiation has a value $c = 2.9979 \times 10^8$ m/s. We shall use the three significant figure value, $c = 3.00 \times 10^8$ m/s.

$$c = \lambda\nu = 3.00 \times 10^8 \text{ m/s} \qquad [1]$$

Frequency consistently is expressed in units of hertz (Hz). However, previous usage has resulted in many ways of writing hertz: /sec, /s, s^{-1}, cycles per second, cycles, and cps. For the units of wavelength the situation is not as simple. First, wavelengths often are not written in scientific notation as are frequencies, but rather are expressed in the most conveniently sized unit of length. For example, red light has a wavelength of 760 nm (7.60×10^{-7} m). Second, three non-SI units are or have been used for wavelengths. They are the micron (μ) = 10^{-6} m = micrometer (μm), the millimicron (mμ) = 10^{-9} m = nanometer (nm), and the Ångstrom (Å) = 10^{-10} m = 10^{-8} cm = 0.1 nm = 100 pm. To use equation [1], however, wavelengths must be expressed in meters.

EXAMPLE 8-1 A radio station broadcasts on a frequency of 95 kilocycles. What is the wavelength of this radiation in km?

95 kilocycles is the same as 95×10^3 Hz. Rearrangement of equation [1] produces $\lambda = c/\nu$.

$$\lambda = \frac{3.00 \times 10^8 \text{ m/s}}{95 \times 10^3 \text{ s}^{-1}} = 3.2 \times 10^3 \text{ m} \times \frac{\text{km}}{1000 \text{ m}} = 3.2 \text{ km}$$

EXAMPLE 8-2 Electromagnetic radiation with a wavelength of 3429 Å has what frequency?

$$\nu = \frac{c}{\lambda} = \frac{3.00 \times 10^8 \text{ m/s}}{(3429 \text{Å})(\text{m}/10^{10} \text{Å})} = 8.75 \times 10^{14} \text{ Hz}$$

* 2. **List the various types of radiation and their approximate wavelengths.**

This objective involves memorizing the various relationships shown in *Figure 8-5 in the text*. You should realize that the boundaries between the regions in the spectrum are not sharp divisions but rather gradual transitions. It is helpful to associate the various types of radiation with the changes in matter they produce or that produce them as shown in Table 8-1. These associations are mutual. For example, infrared radiation may cause changes in the vibrations of molecules when it strikes them, or a change in how a molecule vibrates may produce infrared radiation. In fact, this is how spectroscopy is used to identify molecules. Specific parts of molecules absorb light of specific wavelengths. By detecting this absorption, we often can identify a particular compound.

TABLE 8-1 Changes in Matter Associated with Various Types of Electromagnetic Radiation.

Type(s) of radiation	*Associated change in matter*
Cosmic rays, γ rays	Nuclear transformations (changes in the structure of the nucleus)
X rays	Inner electronic transitions in atoms
Ultraviolet rays, visible light	Outer electron transitions in atoms and in molecules
Infrared radiation	Vibration of atoms in molecules
Microwave radiation, radar	Rotation of molecules

3. Know how light is dispersed into a spectrum and the difference between continuous and line spectra.

When light passes through a prism, it is refracted or bent as it enters and as it leaves the prism. Different frequencies of light are refracted to different degrees. High-frequency (short-wavelength) light is refracted most and low-frequency (long-wavelength) light is refracted least. Thus, among the colors of visible light, red light is refracted least, next orange, then yellow, green, blue, indigo, and finally violet, which is refracted most (see Figure 8-1). (Note that the initial letters of the colors spell ROY G BIV.) You can remember which is bent the most by imagining that high-frequency light interacts the most frequently with the matter of the prism as it passes through and thus is refracted the most. This is only a memory aid, not what actually occurs; refraction occurs at the surfaces of the prism.

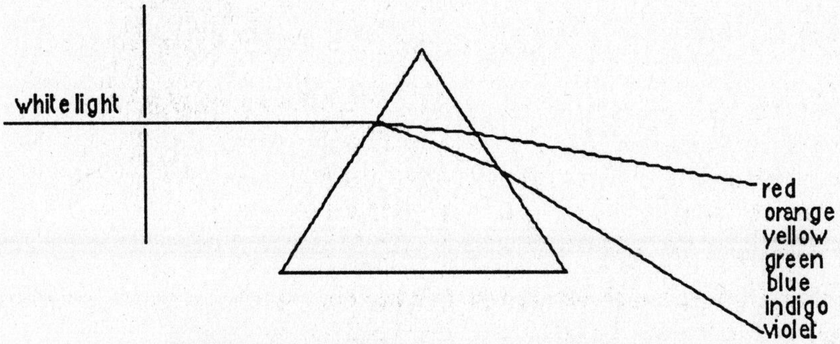

FIGURE 8-1 Dispersion of White Light by a Prism

When white light is dispersed by a prism, it produces a continuous spectrum. When an element is heated in a flame, and the resulting light is dispersed by a prism, the spectrum consists of several lines. The pattern of lines is characteristic of the element being heated. This line spectrum also is called an emission spectrum. A neon light is a common example of a heated element giving off light. Elements can be tentatively identified visually by the colors they emit when heated. Some of these flame colors are given in Table 8-2.

TABLE 8-2
Flame colors of some of the elements

Element	Color
Lithium	Carmine
Sodium	Orange-yellow
Potassium	Violet
Rubidium	Red
Cesium	Blue
Calcium	Orange-red
Strontium	Brick-red
Barium	Yellowish green
Copper	Azure blue
Lead	Light blue

*** 4. Use the Balmer equation to determine the wavelengths of lines in the spectrum of hydrogen.**

The Balmer equation is a specific example of the Rydberg equation, [2].

$$\frac{1}{\lambda} = R_H\left(\frac{1}{n^2} - \frac{1}{m^2}\right) \qquad [2]$$

The numbers n and m are whole numbers with m larger than n. $R_H = 109{,}678 \text{ cm}^{-1}$ The Balmer equation results when $n = 2$. The wavelengths calculated then are those of the Balmer series—a set of visible lines emitted by hydrogen. When $n = 1$, the Lyman series in the ultraviolet region is produced. The remaining series all are in the infrared: Paschen ($n = 3$), Brackett ($n = 4$), Pfund ($n = 5$), and Humphreys ($n = 6$). The Rydberg equation describes these lines. It is simply a summary of data, *not* based on a theoretical model.

EXAMPLE 8-3 What is the wavelength of the Lyman series line with $m = 2$? Is this visible light?

$$\frac{1}{\lambda} = 109{,}678 \text{ cm}^{-1}\left(\frac{1}{1^2} - \frac{1}{2^2}\right) = 82{,}258.5 \text{ cm}^{-1}$$

$\lambda = 1/(82258.5 \text{ cm}^{-1}) = 1.216 \times 10^{-5} \text{ cm} = 121.6 \text{ nm}$ (ultraviolet radiation)

*** 5. Know and be able to use Planck's equation.**

Planck's equation, $\varepsilon = h\nu$, enables us to relate the energy to the frequency of radiation.

EXAMPLE 8-4 What is the energy of a photon of light with a frequency of 110. Hz? A mole of such photons?

$\varepsilon = h\nu = (6.63 \times 10^{-34} \text{ J·s})(110. \times 10^6 \text{ Hz})$
$= 7.29 \times 10^{-26}$ J/photon
$E = N_A\varepsilon = (6.022 \times 10^{23} \text{ photons/mol})(7.29 \times 10^{-26} \text{ J/photon})$
$= 4.39 \times 10^{-2}$ J/mol photons

This is a truly small energy. Compare it with the 75.3 J required to raise the temperature of one mole (18.0 g) of water by 1.00°C.

Frequently $E = h\nu$ and $c = \lambda\nu$ are combined to produce equation [3], whose use is illustrated in Example 8-5.

$$E = hc/\lambda \qquad [3]$$

EXAMPLE 8-5 400. kJ is required to break a mole of N—H bonds. What is the wavelength of the radiation needed to break one N—H bond?

First we determine the energy needed to break one bond from $E = N_A\varepsilon$ or $\varepsilon = E/N_A$.
$\varepsilon = (400. \times 10^3 \text{ J/mol})/(6.022 \times 10^{23} \text{ bonds/mol}) = 6.64 \times 10^{-19}$ J/bond
Then we determine the wavelength from a rearranged version of equation [3].
$\lambda = hc/\varepsilon = (6.63 \times 10^{-34} \text{ J/s})(3.00 \times 10^8 \text{ m/s}) \div (6.64 \times 10^{-19} \text{ J})$
$= (3.00 \times 10^{-7} \text{ m})(10^9 \text{ nm/m}) = 300.$ nm

This radiation is just beyond the visible region, in the ultraviolet. This partly explains why some substances deteriorate when they are exposed to sunlight—the chemical bonds are disrupted.

6. Know and be able to apply Bohr's model of the hydrogen atom: the assumptions, the picture of the atom, the energy expression, and the energy-level diagram.

To explain the Rydberg equation, Bohr proposed a model of the hydrogen atom based on several assumptions.

1. The electron moves around the nucleus in one of several circular orbits. The electron does not spiral into the nucleus as classical physics requires.
2. In each orbit of radius r, the angular momentum of the electron, m_evr (where m_e and v are the electron mass and velocity) is restricted to values of $nh/2\pi$, where n is a whole number.

$$m_evr = nh/2\pi \qquad [4]$$

3. When an electron moves from one orbit to another, the energy difference ($\Delta\varepsilon$) between the two orbits appears as light emitted (if the second orbit is of lower energy) or absorbed (if the second orbit has higher energy) by the atom.

$$\Delta\varepsilon = h\nu \qquad [5]$$

After some algebraic manipulation, these equations, along with the principles of classical physics, give expressions for the energy of each orbit (ε_n) and the difference in energy between two orbits ($\Delta\varepsilon$).

$$\varepsilon_n = (2.181 \times 10^{-18}\ \text{J})/n^2 \qquad [6]$$

$$\Delta\varepsilon = (2.181 \times 10^{-18}\ \text{J})[(1/n^2) - (1/m^2)] \qquad [7]$$

The wavelength of light absorbed or emitted is given by the Rydberg equation, [2], which now is based on a theoretical model. When the energies determined from equation [6] are plotted, an energy-level diagram is produced (*Figure 8-17 in the text*). This diagram illustrates the various possible values of $\Delta\varepsilon$. Another aspect of each series in the hydrogen spectrum is the series limit. The energy levels get closer to each other as the quantum number n increases. Thus, the spectral lines also are closer in frequency.

7. Summarize de Broglie's and Heisenberg's ideas.

A serious problem with Bohr's model is why his second assumption, equation [4], is true. This was explained in 1924 by Louis de Broglie (and verified experimentally by Davisson and Germer in 1927), who stated that a particle of mass m and velocity v has a wavelength given by

$$\lambda = h/mv \qquad [8]$$

A whole number of wavelengths must fit in a circular orbit of the atom.

$$n\lambda = 2\pi r = \text{circumference} \qquad [9]$$

$$\text{or} \quad nh/mv = 2\pi r \quad \text{or} \quad nh/2\pi = mvr \qquad [10]$$

Equation [10] is Bohr's assumption (equation [4]). De Broglie's equation does much more than support Bohr's theory. It is a fundamental fact of nature. Particles moving at high velocities have wave properties. But even this statement is not complete. Particles *are* waves. It merely depends on how we observe them. In the same way, a woman can be both a mother to her children and a surgeon, depending on how (or when) we observe her. (Also, think of a coin twirling in the air. Is it heads or tails while it is in the air?) Another consequence of this wave-particle duality is the Heisenberg uncertainty principle.

$$\Delta x \Delta p > h/2p \quad (\text{where } p = mv = \text{momentum}) \qquad [11]$$

This uncertainty is a fundamental limitation of nature, not a result of the crudeness of measuring devices. This leads to a different picture of the electron in the atom. Bohr's definite orbits are replaced by orbitals—regions of space in which there is a reasonable chance (for example, better than 90%) of finding the electron. An orbital is somewhat similar to the boundaries of a city. We believe there is a good chance (or a high probability) of finding the population of the city somewhere within the city limits. In the same way there is a high probability of finding the electron within the boundaries of the orbital.

8. Explain the differences between Bohr's and Schroedinger's models of the atom.

Bohr treated the electron as a classical particle fixed in certain orbits defined by its wavelength. Schroedinger allows the electron to fully express its wave properties. Schroedinger's electron does not plod around an orbit but totally fills the space of the orbital. It does not fill this space as would a fly in a cage but rather as a gas fills a bottle. Yet, it does not fill the space by breaking into fragments or by making itself larger, but rather in the same way that a sound wave fills the room. Unfortunately the mathematics describing the Schroedinger atom are more complex than those needed by the Bohr model.

* **9. Know and be able to apply the quantum number relationships of wave mechanics.**

The *principal quantum number*, designated by n, can only be a whole number larger than zero.

$$n = 1, 2, 3, 4, 5, \ldots \qquad [12]$$

Electrons with the same principal quantum number are in the same *shell* or *level*. Often the shells are designated with capital letters.

Value of n:	1	2	3	4	5
Shell:	K	L	M	N	O

Thus all electrons with $n = 1$ are in the K shell. The average distance of the electron from the nucleus depends almost entirely on n. The *azimuthal quantum number, l* (also called the orbital quantum number) is a nonnegative whole number less than n.

$$l = 0, 1, 2, 3, ..., n - 1 \qquad [13]$$

Electrons with the same value of l are in the same *subshell* or *sublevel*. Subshells are designated by lower-case letters.

Value of l:	0	1	2	3	4	5	6	7
Subshell:	*s*	*p*	*d*	*f*	*g*	*h*	*i*	*k*

(The letters are the first letters of the words in: "Sober physicists don't find giraffes hiding in kitchens." This sentence is a convenient memory aid.) The energy of a many-electron atom depends on the sum $n + l$. Also, l specifies the shape of an orbital. The *magnetic quantum number*, m_l (or m, also called the orientation quantum number), ranges from $-l$ to $+l$ in whole number steps.

$$m_l = -l, -l+1, -l+2, ..., -1, 0, +1, ..., l-2, l-1, l \qquad [14]$$

Electrons with the same values of n, l, and m_l are in the same *orbital*. The orientation quantum number, m_l, specifies the general direction of the orbital (the direction in which it "points").

* 10. Know what an orbital is and sketch the appearance of *s*, *p*, and *d* orbitals.

An orbital is a region in space where there is a good chance or a high probability of finding an electron. All *s* orbitals are spherical in shape and increase in size as n increases. The *p* orbitals have a double squashed sphere shape (see Figure 8-2). Each squashed sphere is called a *lobe*, and the region of zero electron probability between the lobes is called a *node*. A *p* orbital corresponds to $l = 1$ and thus there are three *p* orbitals since m_l can equal −1, 0, or +1. Since $l = 2$ for *d* orbitals, there are five values for m_l (= −2, −1, 0, +1, +2) and thus five *d* orbitals. These are shown in Figure 8-3. Except for d_{z^2}, each orbital has four lobes and two nodes and looks rather like a four-leafed clover or four teardrops pointing to a common center.

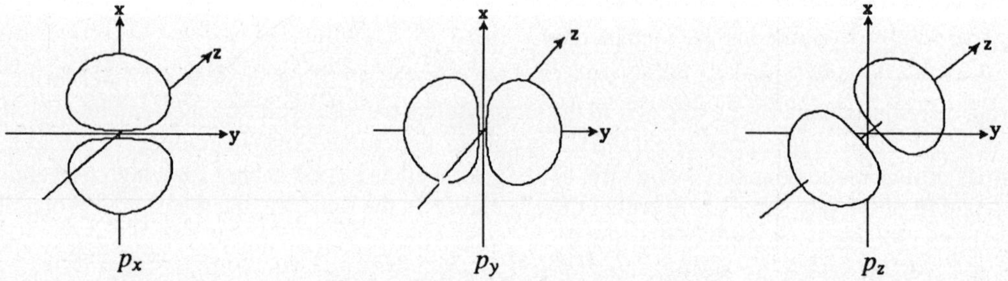

(a) The orbitals as they actually appear

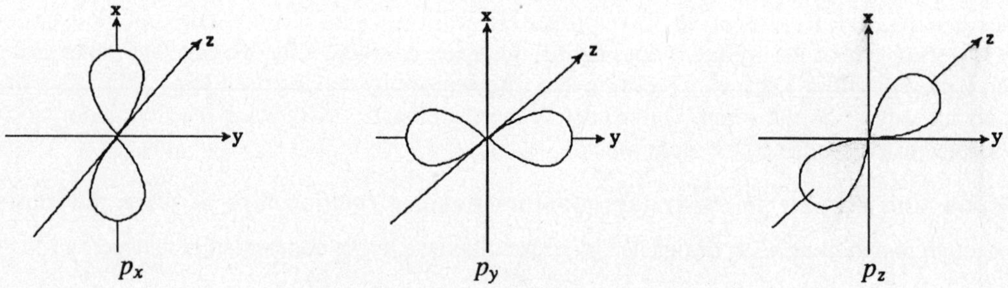

(b) The orbitals as they often are drawn

FIGURE 8-2 The three *p* orbitals

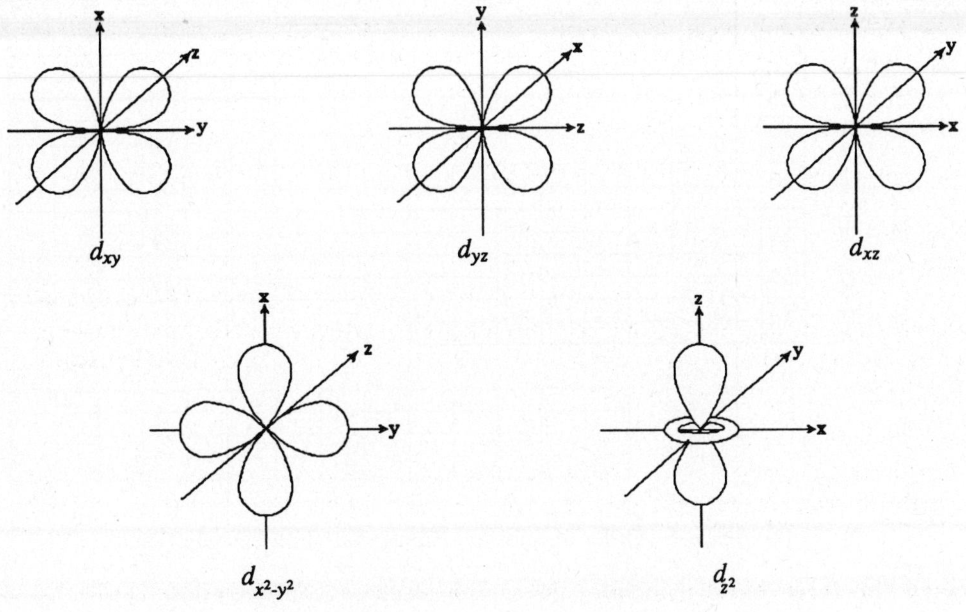

FIGURE 8-3 The five *d* orbitals

11. Know how orbital energies are modified when more than one electron is present in an atom.

For one-electron atoms and ions (H, He^+, Li^{2+}, and so forth), orbital energies depend on only the principal quantum number *n*. As other electrons are added, the principal levels split into sublevels of different energies. (As we shall see later in the text, sometimes when atoms combine, the orbitals within a subshell have different energies.) Listed in order of increasing energy, the orbitals are

$$1s,\ 2s,\ 2p,\ 3s,\ 3p,\ 4d,\ 5p,\ 6s,\ 4f,\ 5d,\ 6p,\ 7s,\ 5f,\ 6d,\ 7p \qquad [15]$$

There are two ways of remembering this order. One is based on Figure 8-4. If one follows the diagonal lines, the orbitals occur in the order given in expression [15]. A much more valuable way is based on the periodic table. First remember that each orbital can hold two electrons. Thus an *s* subshell with one orbital can hold two electrons, a *p* subshell with three orbitals can hold six electrons, a *d* subshell with five orbitals can hold ten electrons, and an *f* subshell with seven orbitals can hold fourteen electrons. Now look carefully at the periodic table in the front of this book. Notice that there is a region that is two columns wide. These columns, headed 1A and 2A, represent the filling of the *s* subshell. The six columns headed 3A, 4A, 5A, 6A, 7A, and 8A, represent the filling of the *p* subshell. The "waist" of the periodic table is ten columns wide—from 3B to 2B—and represents the filling of the five orbitals of the *d* subshell. Finally, the two long rows at the bottom are fourteen elements wide and represent the filling of the seven *f* orbitals. This order of filling is shown in Figure 8-5.

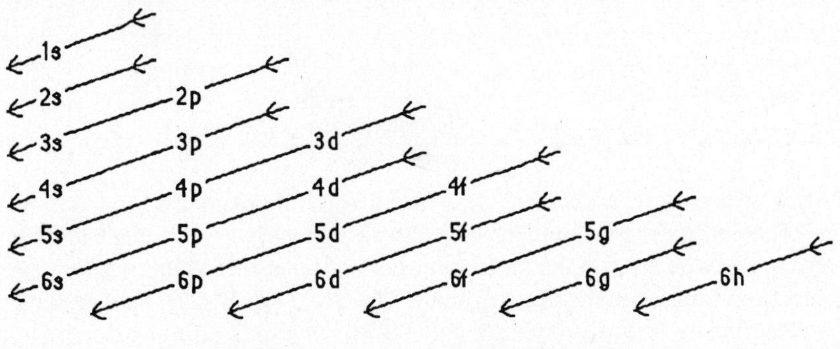

FIGURE 8-4 Order of Subshell Filling

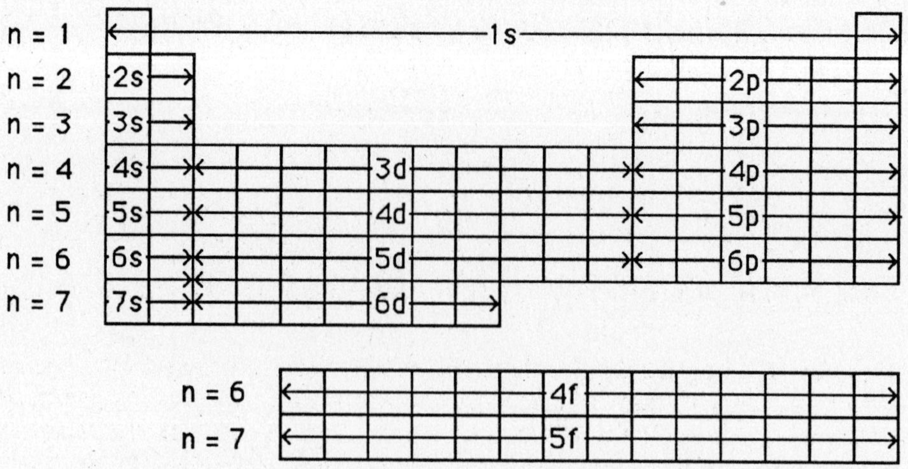

FIGURE 8-5 Order of Subshell Filling Based on Periodic Table

* **12. Learn the three basic principles governing electron configuration.**

1. *The order in which orbitals are filled.* The correct order is given in expression [15] and two ways of recalling it are discussed in objective 8-11.
2. *The Pauli exclusion principle and its implications.* The Pauli exclusion principle states that no two electrons in an atom may have the same set of four quantum numbers. An orbital contains electrons that have the same value of n, l, and m_l. Hence m_s (or s, the spin quantum number) must be different for the two electrons in an orbital. Note that m_s can have only two values: either $m_s = +\frac{1}{2}$ or $m_s = -\frac{1}{2}$.

   number of electrons/orbital = 2 [16]

   The number of orbitals in a subshell is governed by m_l, from $m_l = -l$ to $m_l = +l$ by whole number steps. There are $2l + 1$ values of m_l for a given value of l.

   number of orbitals/subshell = $2l + 1$ [17]

   number of electrons/subshell = $2(2l + 1)$ [18]

   The number of subshells in a shell is determined by the value of n. There are n subshells in each shell, from $l = 0$ to $l = n - 1$. In Table 8-3, the subshells of each shell are listed and the capacity of each subshell is given in parentheses.

   number of subshells/shell = n [19]

TABLE 8-3 Electrons in shells and subshells

Shell	Subshells (electron capacity)	Total electrons in each shell	
1	1*s*(2)	2	= 2
2	2*s*(2) 2*p*(6)	2 + 6	= 8
3	3*s*(2) 3*p*(6) 3*d*(10)	2 + 6 + 10	= 18
4	4*s*(2) 4*p*(6) 4*d*(10) 4*f*(14)	2 + 6 + 10 + 14	= 32
5	5*s*(2) 5*p*(6) 5*d*(10) 5*f*(14) 5*g*(18)	2 + 6 + 10 + 14 + 18	= 50

3. *Hund's rule of maximum multiplicity.* The multiplicity of an atom is related to its total spin or to the sum of the m_s quantum numbers. Hund's rule states that when electrons fill orbitals of the same energy, they do so with the same values of m_s if possible. Suppose, for example, that there are six electrons to place in the five d orbitals. The five d orbitals can be represented by five boxes and the electrons by arrows—pointing up(↑) if $m_s = +\frac{1}{2}$, and pointing down (↓) if $m_s = -\frac{1}{2}$. The three possibilities for six d electrons are summarized in Table 8-4. Hund's rule tells us to choose the last of these three.

TABLE 8-4 Possible Electron Configurations of Six *d* Electrons

Configuration	Number of electrons with $m_s = +\frac{1}{2}$	$m_s = -\frac{1}{2}$	Total spin
↑↓ ↑↓ ↑↓ □ □	3	3	0
↑↓ ↑↓ ↑ ↑ □	4	2	1
↑↓ ↑ ↑ ↑ ↑	5	1	2

* **13. Apply the Aufbau principle and write electron configurations with many different methods.**

With the Aufbau process, one builds the electron configuration of an atom upon the configuration of the previous atom (that of next lowest atomic number). There are many ways to write electron configurations. One is to give the four quantum numbers of the last electron added. Table 8-5 lists these values for the elements with atomic numbers up to $Z = 10$. This is a cumbersome technique and does not account for those cases (of which chromium is one) in which the quantum numbers of one of the previous electrons change. Note, in Table 8-5, that there is no good reason for assigning $m_s = +\frac{1}{2}$ before $m_s = -\frac{1}{2}$. If we would have assigned them in the opposite order, the configuration for H would be: $n = 1, l = 0, m_l = 0, m_s = -\frac{1}{2}$. You should be consistent: assigning either $m_s = +\frac{1}{2}$ or $m_s = -\frac{1}{2}$ first. There also is no preferred order of assigning m_l values. Table 8-5 shows them assigned in increasing order ($m_l = -1, 0, +1$) but they could have been assigned in decreasing order ($m_l = +1, 0, -1$). Again, either order is correct, as long as you are consistent.

TABLE 8-5 Quantum Numbers of the Last Electron Added for the First Ten Elements

Quantum number	H	He	Li	Be	B	C	N	O	F	Ne
n	1	1	2	2	2	2	2	2	2	2
l	0	0	0	0	1	1	1	1	1	1
m_l	0	0	0	0	−1	0	+1	−1	0	+1
m_s	$+\frac{1}{2}$	$-\frac{1}{2}$	$+\frac{1}{2}$	$-\frac{1}{2}$	$+\frac{1}{2}$	$+\frac{1}{2}$	$+\frac{1}{2}$	$-\frac{1}{2}$	$-\frac{1}{2}$	$-\frac{1}{2}$

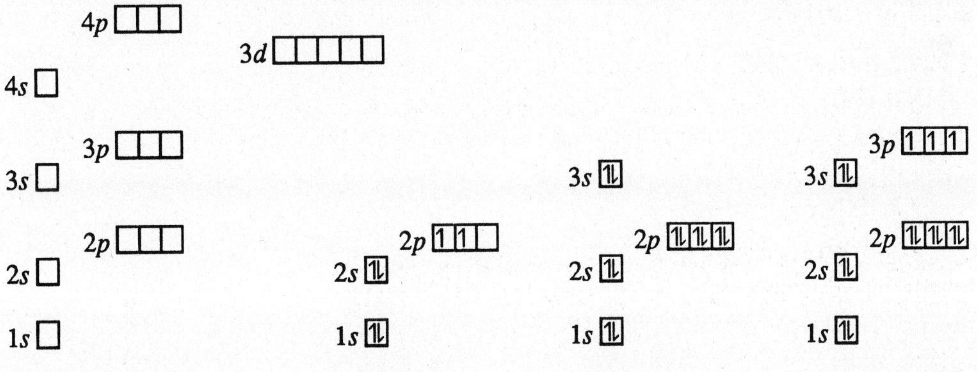

(a) Basic diagram (b) Carbon (c) Calcium (d) Phosphorus

FIGURE 8-6 Energy Level Diagrams

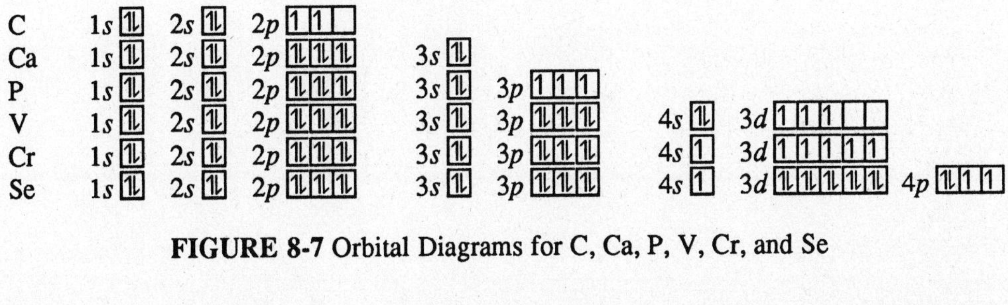

FIGURE 8-7 Orbital Diagrams for C, Ca, P, V, Cr, and Se

Electron configuration is written more compactly as an energy-level diagram, so called because the vertical direction indicates approximate orbital energy. The basic diagram is shown in Figure 8-6a. Each box represents an orbital to be filled with two electrons, drawn as arrows. The electron configurations of C, Ca, and P are shown in Figure 8-6. When energy-level diagrams are written on one line, as in Figure 8-7, they are called orbital diagrams.

The electron configuration of Cr is not built upon that of V. A useful generalization is that half-filled and filled subshells are unusually stable, and that if they can be obtained by "moving" only one electron between very closely spaced energy subshells, one should do so. The closely spaced subshells are: (4*s*, 3*d*), (5*s*, 4*d*), (6*s*, 4*f*, 5*d*), and (7*s*, 5*f*, 6*d*) This is an extension of Hund's rule. It works quite well for elements up to atomic number 60 but not very well after that.

TABLE 8-6 *spdf* Notation for C, Ca, P, V, Cr, Se

	Ordered by energy	Abbreviated	Ordered by n
C	$1s^2\,2s^2\,2p^2$	[He] $2s^2\,2p^2$	$1s^2\,2s^2\,2p^2$
Ca	$1s^2\,2s^2\,2p^6\,3s^2$	[Ne] $3s^2$	$1s^2\,2s^2\,2p^6\,3s^2$
P	$1s^2\,2s^2\,2p^6\,3s^2\,3p^3$	[Ne] $3s^2\,3p^3$	$1s^2\,2s^2\,2p^6\,3s^2\,3p^3$
V	$1s^2\,2s^2\,2p^6\,3s^2\,3p^6\,4s^2\,3d^3$	[Ar] $4s^2\,3d^3$	$1s^2\,2s^2\,2p^6\,3s^2\,3p^6\,3d^3\,4s^2$
Cr	$1s^2\,2s^2\,2p^6\,3s^2\,3p^6\,4s^1\,3d^5$	[Ar] $4s^1\,3d^5$	$1s^2\,2s^2\,2p^6\,3s^2\,3p^6\,3d^5\,4s^1$
Se	$1s^2\,2s^2\,2p^6\,3s^2\,3p^6\,4s^2\,3d^{10}\,4p^4$	[Ar] $4s^2\,3d^{10}\,4p^4$	$1s^2\,2s^2\,2p^6\,3s^2\,3p^6\,3d^{10}\,4s^2\,4p^4$

A more compact notation is *spdf* or spectroscopic notation. One writes the subshells in order of increasing energy, as in expression [15], and indicates the number of electrons in each subshell with a superscript. Sometimes the orbitals are written in order of increasing n value. Both methods are shown in Table 8-6. One often abbreviates the *spdf* notation by writing the symbol for the last noble gas in brackets to represent its electron configuration. But the *spdf* notation does not tell us how electrons are paired in the unfilled subshells. Often this information is expressed by writing the designation for the unfilled subshells, as shown in Table 8-7.

TABLE 8-7 Extended *spdf* notation

C	[He] $2s^2\,2p_x^1\,2p_y^1$
Ca	[Ne] $3s^2$
P	[Ne] $3s^2\,2p_x^1\,2p_y^1\,2p_z^1$
V	[Ar] $4s^2\,3d_{xy}^1\,3d_{xz}^1\,3d_{yz}^1$
Cr	[Ar] $4s^1\,3d_{xy}^1\,3d_{xz}^1\,3d_{yz}^1\,3d_{z^2}^1\,3d_{x^2-y^2}^1$
Se	[Ar] $4s^2\,3d^{10}\,4p_x^2\,4p_y^1\,4p_z^1$

DRILL PROBLEMS

1. & 2. Given the frequency (or wavelength), determine the wavelength (or frequency) of each of the following. Also classify each according to its region: gamma (γ) rays, x rays, ultraviolet (UV) radiation, visible light, infrared radiation, microwaves, radar waves, television (TV) waves, or radio waves.

A. $\nu = 2.7 \times 10^{14}$ Hz	B. $\lambda = 1.54$ Å	C. $\nu = 25$ MHz
D. $\lambda = 1.50$ km	E. $\nu = 1.75 \times 10^{22}$ Hz	F. $\lambda = 7256$ µm
G. $\nu = 51.5$ kHz	H. $\lambda = 124$ cm	I. $\nu = 7.42 \times 10^{11}$ MHz
J. $\lambda = 5.04 \times 10^{-5}$ cm	K. $\nu = 3.15 \times 10^{12}$ kHz	L. $\lambda = 12.45$ m
M. $\nu = 8.46 \times 10^{16}$ MHz	N. $\lambda = 1.359$ mm	O. $\nu = 5.02 \times 10^{11}$ kHz
P. $\lambda = 82.5$ nm	Q. $\nu = 14.6 \times 10^3$ MHz	R. $\lambda = 0.132$ pm
S. $\nu = 8.46 \times 10^{10}$ Hz	T. $\lambda = 4.40 \times 10^{-6}$ pm	U. $\nu = 525$ MHz

4. For the various lines of the hydrogen spectrum given below, determine n (the smaller integer), m the larger integer, or λ (the wavelength of the line). Also determine the series (Lyman, Balmer, Paschen, Brackett, Pfund, or Humphreys), or identify the region of the spectrum (cosmic, visible, ir, and so forth) to which the line belongs.

A. $m = 3, \lambda = 656.5$ nm B. $n = 1, \lambda = 102.6$ C. $n = 2, m = 10$
D. $n = 3, m = 4$ E. $m = 4, \lambda = 486.3$ nm F. $n = 2, \lambda = 434.2$ nm
G. $n = 4, m = 8$ H. $n = 6, m = 7$ I. $m = 6, \lambda = 7459.9$ nm
J. $n = 1, \lambda = 121.57$ K. $n = 2, m = 8$ L. $n = 1, m = 9$

5. Determine three of the following, given one of them: (1) frequency, ν; (2) wavelength, λ; (3) energy per photon, ε (in ergs for A through H and in joules for the remainder); and (4) energy per mole of photons, E (in kJ/mol)
A. $\lambda = 1.54$ Å B. $E = 418$ kJ/mol C. $\nu = 105$ MHz D. $\varepsilon = 1.07 \times 10^{-15}$ erg
E. $\lambda = 527$ nm F. $E = 28.5$ kJ/mol G. $\nu = 6.12 \times 10^{12}$ Hz H. $\varepsilon = 8.13 \times 10^{-20}$ erg
I. $\lambda = 1.34$ km J. $E = 526$ J/mol K. $\nu = 512$ kHz L. $\varepsilon = 3.14 \times 10^{-17}$ J
M. $\lambda = 6.15$ cm N. $E = 725$ J/mol O. $\nu = 34.2 \times 10^{15}$ Hz P. $\varepsilon = 1.24 \times 10^{13}$ J

9. (1) Give the letter designations, such as p_x, p_y, and p_z) and the set of magnetic quantum numbers (such as −1, 0, and +1) for the orbitals in each of the following subshells.
A. $1s$ B. $3p$ C. $4d$ D. $5s$ E. $4f$
F. $4p$ G. $3s$ H. $2p$ I. $5d$ J. $3d$

(2) Give the number and letter designations (such as $2s$ and $2p$), and the total number of electrons in each of the following shells.
K. *K* shell L. *M* shell M. 2nd shell N. $n = 4$ M. 5th level

10. Sketch each of the following sets of orbitals, each set on the same axes and to approximate scale.
A. $1s, 2s, 3s$ B. $2p_x, 2p_y$ C. $3d_{xy}, 3d_{x^2-y^2}$ D. $2p_x, 3p_x$
E. $3p_z, 3d_{z^2}$ F. $3d_{xz}, 3d_{z^2}$ G. $3s, 3p_x$ H. $3d_{xy}, 4d_{xy}$

12. The symbol for each element is followed by an electron configuration that violates one of the following principles: (1) the order of filling, or the Aufbau principle; (2) the Pauli exclusion principle; (3) Hund's rule of maximum multiplicity; (4) Hund's rule extended—stability of full and half-full subshells; (5) allowed quantum numbers (objective 8-10). State which principle is violated by each configuration and give the correct configuration for each atom.
A. N $1s^2\,2s^2\,2p_x^2\,2p_y^1$ B. Al $1s^2\,2s^2\,2p^6\,2d^3$ C. B $1s^2\,2s^3$
D. P $1s^2\,2s^2\,2p^6\,3p^5$ E. Cu [Ar] $4s^2\,3d^9$ F. Be $1s^2\,1p^2$
G. Mg [Ne] [⇈] H. C $1s^2\,2s^1\,2p_x^1\,2p_y^1\,2p_z^1$ I. V $1s^2\,2s^2\,2p^6\,3s^2\,3p^6\,3d^5$
J. S [Ne] $3s^2\,3p_x^2\,3p_y^2$ K. Mn [Ar] $4s^1\,3d^6$ L. N $1s^2\,1p^5$
M. Ag [Kr] $5s^2\,4d^9$ N. Ni [Kr] $5s^2\,4d_{xy}^2\,4d_{xz}^2\,4d_{yz}^2\,4d_{z^2}^2$
O. Na $1s^2\,1p^9$ P. Sc [Ne] $3s^2\,3p^6\,3d^3$ Q. Cl $1s^2\,1p^6\,2p^6\,3s^1$
R. C $1s^2\,2s^2\,2p_x^2$ S. Cl [Ne] [⇈] [⇈] [↿] T. B $1s^1\,2s^1\,2p_x^1\,2p_y^1\,2p_z^1$

In your own words, state each of the following principles or rules and give two examples illustrating the use of each.
U. Hund's rule extended V. Hund's rule of maximum multiplicity
W. The Aufbau principle X. Pauli exclusion principle.

13. (1) Give the symbol of the element of lowest atomic number that has the characteristics listed in each of the following parts. Also give the electron configuration of the element.
A. one electron with $m_l = 2$ B. two unpaired electrons
C. four pairs of electrons D. two electrons with $n = 3$ and $l = 2$
E. three electrons with $n = 3$ F. eleven p electrons
G. five d electrons H. three electrons with $m_l = 2$
I. five electrons with $m_l = +1$ J. five electrons with $m_l = 0$
K. five electrons with $m_s = \frac{1}{2}$

(2) The last few quantum numbers of the electrons or the last part of the electron configuration of different elements is given in each part below. Identify each element based on its electron configuration. (For example, boron might be: $n = 2, l = 1, m_l = -1, m_s = -\frac{1}{2}$; $n = 2, l = 0, m_l = 0, m_s = +\frac{1}{2}$)
L. $3s^2\,3p^4$ M. $n = 4, l = 0, m_l = 0, m_s = -\frac{1}{2}$; $n = 3, l = 2, m_l = -2, m_s = +\frac{1}{2}$
N. $5s^2\,4d^2$ O. $n = 5, l = 2, m_l = +2, m_s = -\frac{1}{2}$; $n = 6, l = 1, m_l = -1, m_s = +\frac{1}{2}$
P. $5d^1\,4f^1$ Q. $4d^5$ R. $3d^3$
S. $4p^4$ T. $n = 5, l = 0, m_l = 0, m_s = +\frac{1}{2}$; $n = 5, l = 0, m_l = 0, m_s = -\frac{1}{2}$

(3) Give the electron configuration of each of the following elements with both the abbreviated *spdf* notation and also an abbreviated orbital diagram.

U. N	V. S	W. Cl	X. Si	Y. Cu
Z. Na	Γ. Cs	Δ. Ce	Θ. Sc	Λ. O
Π. Fe	Σ. Sr			

Further Questions. (1) Use the Aufbau principle, the Pauli exclusion principle, and Hund's rule of maximum multiplicity to write reasonable electron configurations for atoms of the following elements.

A. K	B. C	C. Mn	D. Sc	E. Cr	F. Au
G. Si	H. Zn	I. Mg	J. Fe	K. P	L. Fe
M. U	N. Ne	O. Pr	P. Ag	Q. Al	R. Sn
S. Ge	T. Mo	U. Ga.	V. Ca	W. I	X. W
Y. B	Z. Cu				

(2) Compare these to the known configurations in the text. Which are different? Why?

QUIZZES (20 minutes each) Choose the best answer for each question.

Quiz A

1. The statement that one cannot simultaneously measure the speed and the position of the electron is the (a) Heisenberg uncertainty principle; (b) Pauli exclusion principle; (c) rule of maximum multiplicity; (e) none of these.
2. An experiment or effect that demonstrates that light is a particle is the (a) sun's spectrum; (b) destructive interference of light; (c) photoelectric effect; (d) electron microscope; (e) none of these.
3. The shape of a *p* orbital is (a) spherical; (b) similar to figure 8; (c) similar to four-leafed clover; (d) conical; (e) none of these.
4. Which series of subshells is arranged in order of increasing energy? (a) 6*s*, 4*f*, 5*d*, 6*p*; (b) 4*f*, 6*s*, 5*d*, 6*p*; (c) 5d, 4*f*, 6*s*, 6*p*; (d) 4*f*, 5*d*, 6*s*, 6*p*; (e) none of these.
5. The light with highest energy among those following is (a) television waves; (b) infrared (heat) radiation; (c) ultraviolet waves; (d) microwaves; (e) radio waves.
6. The radiation that follows that has the highest frequency has a wavelength of (a) 300 cm; (b) 200 nm; (c) 8.2 m; (d) 1.00 nm; (e) 7.31 μm.
7. A certain radiation has a frequency of $6.7 \times 10^{14}\ s^{-1}$. What is its wavelength in nanometers? (a) $(6.63 \times 10^{-34})(6.7 \times 10^{14})$; (b) $(3.0 \times 10^8)/[(6.7 \times 10^{14})(10^9)]$; (c) $(3.0 \times 10^8)(10^9)/(6.7 \times 10^{14})$; (d) $(6.7 \times 10^{14})(10^7)/(3.0 \times 10^8)$; (e) none of these.
8. The energy in joules of a photon of wavelength 1.23×10^{-5} m is (a) $(6.63 \times 10^{-34})(3.00 \times 10^8)/(1.23 \times 10^{-5})$; (b) $(6.63 \times 10^{-34})/(1.23 \times 10^{-5})$; (c) $(3.00 \times 10^8)/(1.23 \times 10^{-5})$; (d) $(1.23 \times 10^{-5})/(6.63 \times 10^{-34})$; (e) none of these.
9. The number of unpaired electrons in the outermost subshell of a Mg atom is (a) 0; (b) 1; (c) 2; (d) 3; (e) none of these.
10. $1s^2\,2s^2\,2p^6\,3s^2\,3p^6\,4s^2\,3d^3$ is the ground state electron configuration of (a) chromium; (b) vanadium; (c) scandium; (d) niobium; (e) none of these.
11. A sodium atom must gain or lose how many electrons to achieve an inert gas electron configuration? (a) gain 2; (b) gain 1; (c) lose 1; (d) lose 2; (e) none of these.
12. [Ar] $4s^2\,3d^1$ is the ground state electron configuration for (a) titanium; (b) zirconium; (c) vanadium; (d) calcium; (e) none of these.

Quiz B

1. What is an electron? (a) a wave; (b) a particle; (c) either, depending on how its is observed; (d) neither; (e) none of these.
2. The principle that is based on electrons attempting to be as far apart as possible is (a) Bohr theory; (b) Heisenberg principle; (c) exclusion principle; (d) Hund's rule; (e) none of these.
3. Which two orbitals are both located between the axes of a coordinate system, and not along the axes? (a) d_{xy}, d_{z^2}; (b) d_{yz}, p_x; (c) d_{xz}, p_y; (d) $d_{x^2-y^2}$, p_z; (e) none of these.
4. All of the terms that follow are the names of quantum numbers *except* (a) principal; (b) magnetic; (c) spin; (d) valence; (e) no choice is correct.

5. The radiation with the longest wavelength among those given is (a) infrared; (b) microwave; (c) radiowave; (d) x ray; (e) ultraviolet.
6. The radiation with the highest energy has a frequency of (a) 31.2 Hz; (b) 71.3 MHz; (c) 4.12 kHz; (d) 3.00×10^{10} Hz; (e) 296 kHz.
7. Radiation with a frequency of 3×10^{15} Hz has a wavelength of (a) 10 nm; (b) 100 nm; (c) 1000 nm; (d) 10,000 nm; (e) none of these.
8. Light with a frequency of 4.5×10^{10} s^{-1} has an energy in joules given by (a) $(6.63 \times 10^{-34})(3.0 \times 10^{8})/(4.5 \times 10^{10})$; (b) $(3.0 \times 10^{8})(6.63 \times 10^{-34})$; (c) $(4.5 \times 10^{10})(6.63 \times 10^{-34})$; (d) $(6.63 \times 10^{-34})/(4.5 \times 10^{10})$; (e) none of these.
9. The number of unpaired electrons in the outermost subshell of a carbon atom is (a) 0; (b) 1; (c) 2; (d) 3; (e) none of these.
10. The four quantum numbers that could identify the third 3*p* electron in sulfur are (a) $n = 3, l = 0, m_l = +1, m_s = +\frac{1}{2}$; (b) $n = 2, l = 2, m_l = -1, m_s = +\frac{1}{2}$; (c) $n = 3, l = 2, m_l = +1, m_s = -\frac{1}{2}$; (d) $n = 3, l = 1, m_l = -1, m_s = +\frac{1}{2}$; (e) none of these.
11. [Kr] $5s^2 4d^{10} 5p^5$ is the electron configuration of (a) Br; (b) I; (c) At; (d) Te; (e) none of these.
12. Which of the following represents the electron configuration of the element having the atomic number 17? (a) $1s^2 2p^8 3d^7$; (b) $1s^2 2s^8 3p^7$; (c) $1s^2 2p^2 2d^6 3f^7$; (d) $1s^2 2s^2 2p^6 3s^2 3p^5$; (e) none of these.

Quiz C

1. That the electron configuration of nitrogen is $1s^2 2s^2 2p_x^1 2p_y^1 2p_z^1$ rather than $1s^2 2s^2 2p_x^2 2p_y^1$ is postulated by the (a) Bohr-Sommerfeld theory; (b) the Pauli exclusion principle; (c) the Heisenberg uncertainty principle; (d) Hund's rule of maximum multiplictiy; (e) none of these.
2. In the equation $1/\lambda = R(1/2^2 - 1/n^2)$, *R* is known as the (a) ideal gas constant; (b) Boltzmann constant; (c) Rydberg constant; (d) Balmer constant; (e) none of these.
3. The shape of most *d* orbitals is (a) spherical; (b) figure 8; (c) figure 8 with a donut; (d) clover leaf; (e) none of these.
4. An electron in a 4*f* orbital has principal (*n*) and orbital (*l*) quantum numbers, respectively, of (a) 3 and 4; (b) 4 and 4; (d) 4 and 3; (e) none of these.
5. Among those that follow, the unit of frequency is (a) cm^{-1}; (b) s^{-1}; (c) Hz^{-1}; (d) pm; (e) none of these.
6. Of the following, the radiation that has the highest energy has a wavelength of (a) 560 nm; (b) 56.0 cm; (c) 5.60 μm; (d) 300 nm; (e) 56.0 nm.
7. The frequency of a microwave with a wavelength of 0.750 cm is (a) $0.750/3.00 \times 10^{10}$; (b) $6.63 \times 10^{-27}/0.750$; (c) $3.00 \times 10^{10}/0.750$; (d) $0.750/6.63 \times 10^{-27}$; (e) none of these.
8. Radiation with a frequency of of 3.3×10^{15} s^{-1} has what energy in joules per photon? (a) 2.0×10^{-49}; (b) 6.0×10^{-19}; (c) 3.0×10^{18}; (d) 2.2×10^{-18}; (e) none of these.
9. The number of unpaired electrons in the outermost subshell of a Cl atom is (a) 0; (b) 1; (c) 2; (d) 3; (e) none of these.
10. The four quantum numbers that could identify the *second* 2 *s* electron in the nitrogen atom are (a) $n = 1, l = 0, m_l = +1, m_s = +\frac{1}{2}$; (b) $n = 2, l = 1, m_l = 0, m_s = -\frac{1}{2}$; (c) $n = 2, l = 0, m_l = 0, m_s = -\frac{1}{2}$; (d) $n = 3, l = 0, m_l = 0, m_s = +\frac{1}{2}$; (e) none of these.
11. A Mg atom must gain or lose how many electrons to achieve an inert gas electron configuration? (a) lose 1; (b) lose 2; (c) lose 3; (d) gain 1; (e) none of these.
12. The ground state electron configuration of iron is (a) [Ar] $4s^2 3d^6$; (b) [Xe] $6s^2 4f^{14} 3d^7$; (c) [Ar] $3d^8$; (d) [Ar] $4s^1 3d^7$; (e) none of these.

Quiz D

1. In Bohr's atomic theory, when an electron moves from one energy level to another energy level more distant from the nucleus of the same atom (a) energy is emitted; (b) energy is absorbed; (c) no change in energy occurs; (d) light is given off; (e) none of these.
2. We employ several rules when determining electron configurations. Which of the following do we not use? (a) Aufbau principle; (b) Heisenberg principle; (c) Pauli principle; (d) Hund's rule; (e) none, we use all of these.
3. Which two orbitals are both located between the axes of a coordinate system, and not along the axes? (a) p_z, $d_{x^2-y^2}$; (b) d_{z^2}, d_{xy}; (c) d_{xy}, p_z; (d) d_{yz}, p_x; (e) none of these.

4. When an electron has a principal quantum number of 1 ($n = 1$), the orbital quantum number l must be (a) +1; (b) 0; (c) −1; (d) without restriction; (e) none of these.
5. The symbol λ is used in spectroscopy to represent (a) frequency; (b) energy; (c) speed; (d) wavelength; (e) none of these.
6. The light with the longest wavelength has which of the following frequencies? (a) 3.00×10^{13} Hz; (b) 4.12×10^5 Hz; (c) 8.50×10^{20} Hz; (d) 9.12×10^{12} Hz; (e) 3.12×10^9 Hz.
7. A radio wave has a frequency of 5.0 kHz. Its wavelength in meters is (a) $(3.0 \times 10^{10})/5.0$; (b) $(3.0 \times 10^{10})/(5.0 \times 10^3)$; (c) $(3.0 \times 10^8)/(5.0 \times 10^3)$; (d) $(3.0 \times 10^8)/5.0$; (e) none of these.
8. Radiation with a wavelength of 500. nm has an energy in joules of (a) $(6.63 \times 10^{-34})(3.00 \times 10^8)/(5.00 \times 10^{-7})$; (b) $(6.63 \times 10^{-34})(5.00 \times 10^{-5})/(3.00 \times 10^8)$; (c) $(6.63 \times 10^{-34})(3.00 \times 10^8)/500.$; (d) $(3.00 \times 10^8)(500.)/(6.63 \times 10^{-34})$; (e) none of these.
9. Unpaired electrons are found in the ground-state atoms of (a) Ca; (b) Ne; (c) Mg; (d) P; (e) none of these.
10. The four quantum numbers of the *last* electron of a calcium atom could be (a) $n = 4, l = 1, m_l = 0, m_s = +\frac{1}{2}$; (b) $n = 3, l = 0, m_l = 1, m_s = -\frac{1}{2}$; (c) $n = 4, l = 1, m_l = 0, m_s = -\frac{1}{2}$; (a) $n = 4, l = 0, m_l = 0, m_s = -\frac{1}{2}$; (e) none of these.
11. The number of electrons that a P atom must acquire to achieve a noble gas electron configuration is (a) 1; (b) 2; (c) 3; (d) 4; (e) none of these.
12. The ground-state electron configuration of silicon is (a) $1s^2\,2s^2\,2p^6\,3s^2\,3p_x^2$; (b) $1s^2\,2s^2\,2p^6\,3s^2\,3p_x^1, 3p_y^1$; (c) $1s^2\,2s^2\,2p^2$; (d) $1s^2\,2s^2\,2p^6\,3s^2\,3d^2$; (e) none of these.

SAMPLE TEST (20 minutes)

1. An FM station broadcasts with a 3.28 m wavelength. What is the energy (in kJ) per mole of photons emitted by this radio station?
2. Sketch the outline of each orbital on the axes that are drawn above the symbol for that orbital.

$3p_x$ $3d_{xy}$ $4s$

3. Give the full *spdf* electron configuration for each of the following atoms.
   a. P b. Mn c. Si
4. Give the abbreviated orbital diagram of the electron configurations of each of the following.
   a. Co b. Al c. O

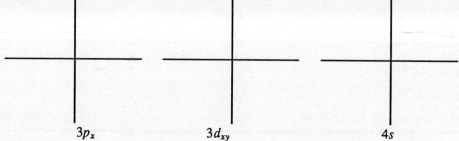

9 Atomic Properties and the Periodic Table

CHAPTER OBJECTIVES

1. Illustrate the periodic law with graphs of selected properties of the elements as a function of atomic number.

The periodic law states that the physical and chemical properties of the elements vary periodically with atomic number. This is shown in the graphs of atomic volume (*Figure 9-1 in the text*), covalent radius (*Figure 9-8 in the text*), ionization energy (*Figure 9-13 in the text*), and melting point (Figure 9-1) against atomic number. The same general pattern of peaks and valleys occurs in each graph.

2. Use the terms *periods, groups, families, representative elements,* and *transition elements* to describe individual elements and groupings of elements in the periodic table.

A *period* is a (horizontal) row in the periodic table. There are seven periods. A *group* or a *family* is a (vertical) column in the periodic table. Elements of the same family are called *cogeners*. For example, He, Ne, Ar, Kr, Xe, and Rn are cogeners. Several families have special names. Family 1A is the *alkali metals*. Family 2A is the *alkaline earth metals*. Family 7A is the *halogens* (meaning salt formers). Family 8A is the *noble gases*, the rare gases, or the inert gases. Family 6A is the *chalcogens* (meaning chalk formers). Family 1B is the *coinage metals*. Fe, Co, and Ni form the *iron triad*. Ru, Rh, Pd, Os, Ir, and Pt are the *noble metals*. The "B" families (the "waist" of the periodic table) are the *transition metals* or transition elements. Ce through Lu are the *lanthanide metals*, or the rare earth metals. Th through Lr are the *actinides*. The lanthanides and actinides together are the *inner transition elements*. The "A" families are the *representative elements*.

* **3. Use the periodic table to describe the Aufbau process; and explain the basic features of the electron configurations of the representative and transition elements, especially the number of valence electrons in each representative group.**

We used the periodic table to help remember the order of subshell filling (objective 8-11 and Figure 8-5). The electron configuration of an element determines its physical and chemical properties. An atom shows mainly its outer electrons—those in the shell of highest principal quantum number. Therefore, there is a striking difference in chemical and physical properties between elements in different representative families, since each family differs in the number of outer shell electrons. There is not so large a difference between transition families, because these families differ by inner shell electrons. The number of outer shell (that is, *s* and *p*) electrons of a representative element equals its family number. These outer shell electrons are called *valence electrons*.

EXAMPLE 9-1 What is the electron configuration of (a) the valence shell of elements in group 2A? (b) the element Ca?

(a) By reference to Figure 8-5, we see that each element in family 2A has two *s* electrons. The valence electron configuration is ns^2.
(b) Ca is in the fourth period; thus, its valence electron configuration is $4s^2$. Ca follows the noble gas, argon. Thus, the electron configuration of Ca is [Ar] $4s^2$.

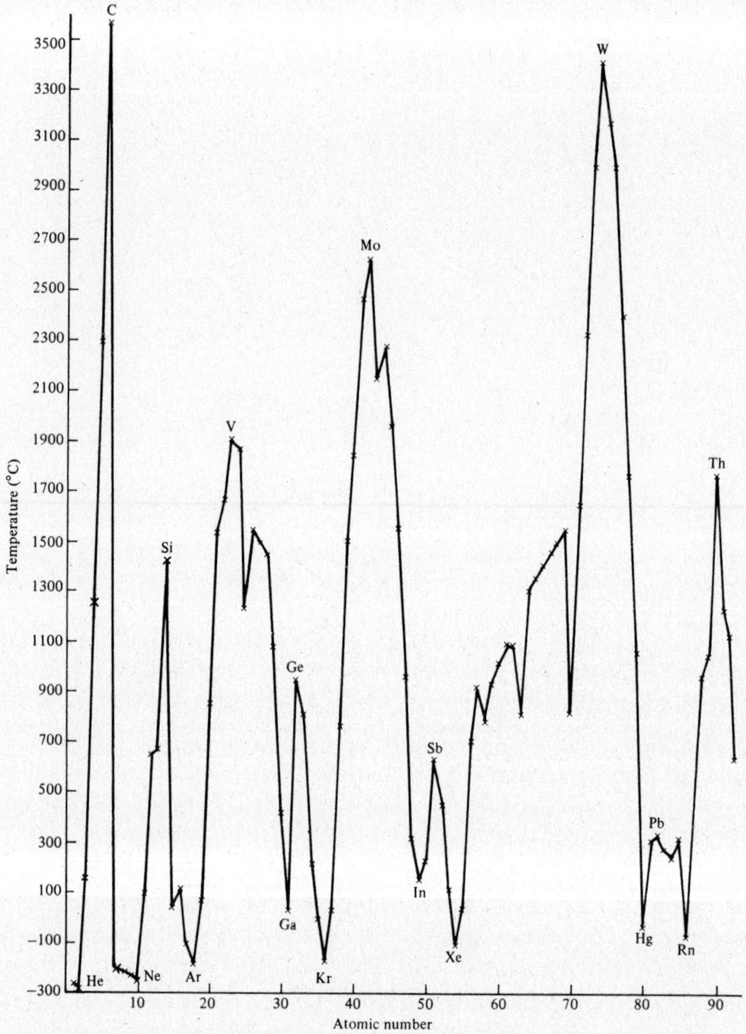

FIGURE 9-1 Melting Points of Elements vs. Atomic Number

4. Describe metals, nonmetals, metalloids, and noble gases in several ways and locate them in the periodic table.

Noble gases are the elements in group 8A. Nonmetals are in the triangular block in the upper right corner of the A groups: H, C, N, O, F, P, S, Cl, Se, Br, and I. Metalloids are in the two diagonal rows just below and to the left of the nonmetals: B, Si, Ge, As, Sb, Te, Po, and At. Metals are the remaining elements.

Physically, pure metals generally are malleable and ductile, melt at moderate temperatures, and conduct heat and electricity well. (These properties may change drastically when two metals are alloyed or mixed together.) Nonmetals tend to be insulators and form small molecules which condense poorly into solids; hence the solids often melt easily (except carbon with its high melting point). Chemically, metals lose electrons and form cations while nonmetals gain electrons and form anions. Metallic oxides form basic solutions with water while nonmetallic oxides form acidic aqueous solutions. Metals rarely combine chemically with each other. Nonmetals combine chemically to form definite compounds with nonmetallic properties. When metals combine with nonmetals, ionic compounds are formed. These compounds have high melting points and the melt conducts electricity while the solid does not.

Metalloids have physical and chemical properties that are between those of metals and those of nonmetals. For example, the aqueous solutions of their oxides often are amphoteric, meaning that they can react with either acids or bases.

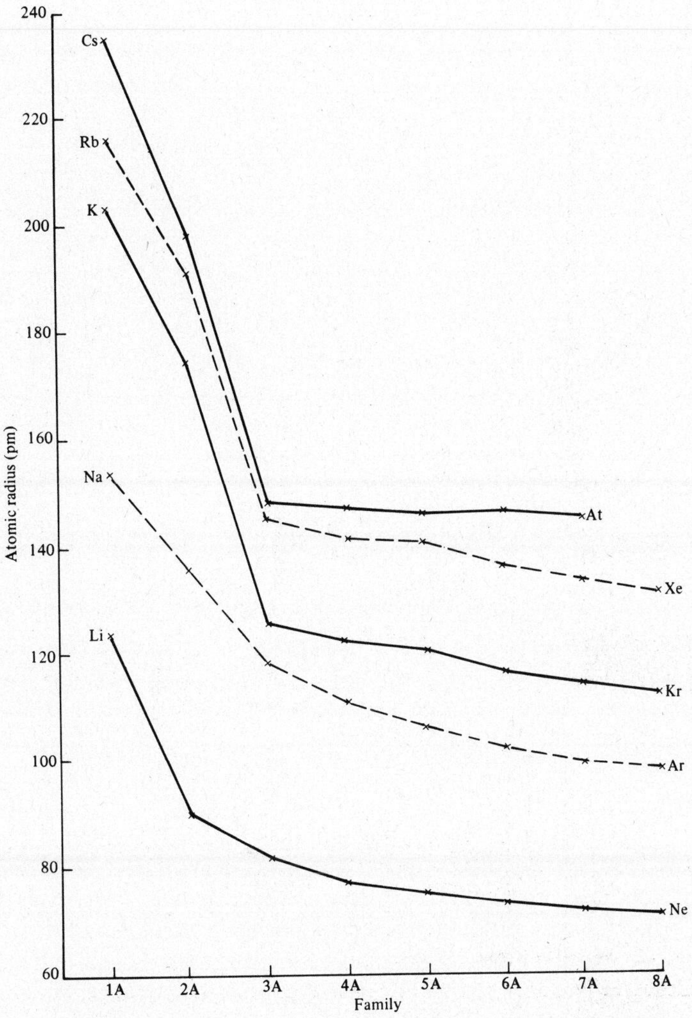

FIGURE 9-2 Atomic Radii of the Representative Elements

* **5. State the factors that influence atomic size; distinguish among covalent, ionic, metallic, and van der Waals radius; and describe the general trends in atomic size that occur within families and groups.**

Atoms are not small hard balls but indefinite spheres. Defining where the electron cloud of an atom ends is like trying to measure the extent of the metropolitan area of a city. (Such a nonbonded interaction defines the van der Waals radius.)But with both cities and atoms we can determine the center. For atoms, one half of the distance between two centers (one half of the internuclear distance) is called the atomic size. This idea is complicated by the fact that not all atoms are normally bound together in the same way. Some are bound by covalent bonds in molecules, some are attracted to each other in ionic crystals, and some are held in metallic crystals by the force of a "sea" of electrons acting as "glue" between cations. (The distance between nuclei in such a metallic crystal equals the metallic radius.)Fortunately, it is possible to form molecules of nearly every element (except the noble gases) in which two like atoms are held together by a single covalent bond. The covalent radius of these molecules often is called the atomic radius. Several family and periodic trends in atomic size follow.

1. Atomic size regularly increases from top to bottom (that is, "down") a family (see Figure 9-2). More electrons with the same outer configuration are present around the nucleus.

2. Atomic size gradually decreases from left to right in (that is, "across") a period of representative elements, as shown in Figure 9-2. The explanation is that within a family all electrons are being added to the same shell. It is helpful to think of these shells as similar to the layers of an onion. As each electron is added, the shell becomes more populated. But at the same time, protons are being added to the nucleus, making it more positively charged. Then the nucleus attracts the electrons more strongly. Thus the shell is pulled closer to the nucleus.

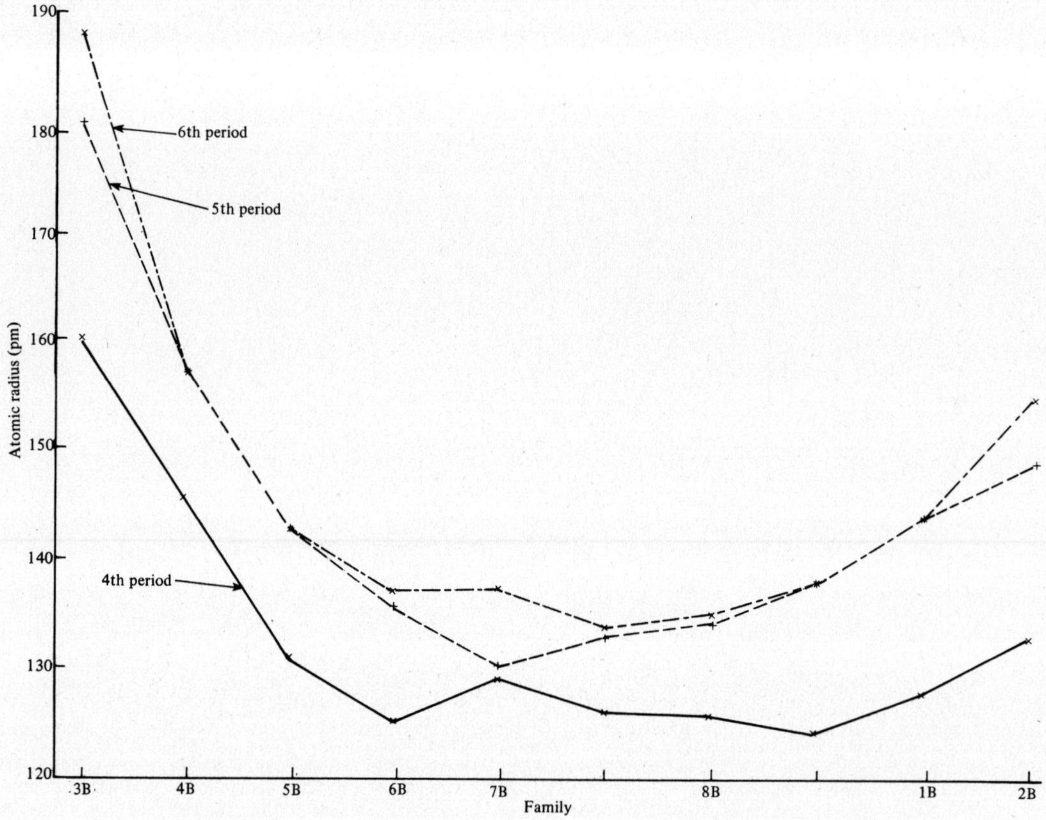

FIGURE 9-3 Atomic Radii of the Transition Elements

3. Across a transition series, size gradually but somewhat irregularly decreases then increases at the end of the series (see Figure 9-3). Within a transition series electrons are added to an inner (*d*) shell. This inner shell gradually will decrease in size, but since these *d* electrons are between the nucleus and the outer electrons, they "shield" the outer electrons from the nuclear charge. Thus, the outer shell should increase in size. The two trends oppose each other with the shielding effect initially being less important, but predominating at the end of the transition series.

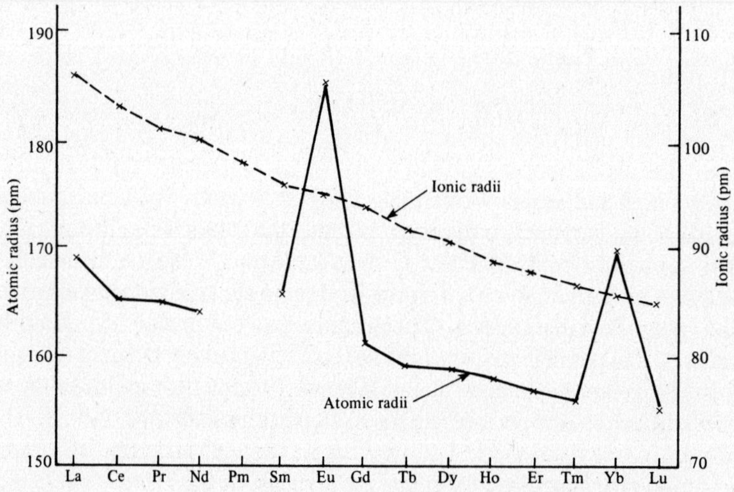

FIGURE 9-4 Atomic and Ionic Radii of the Lanthanide Elements

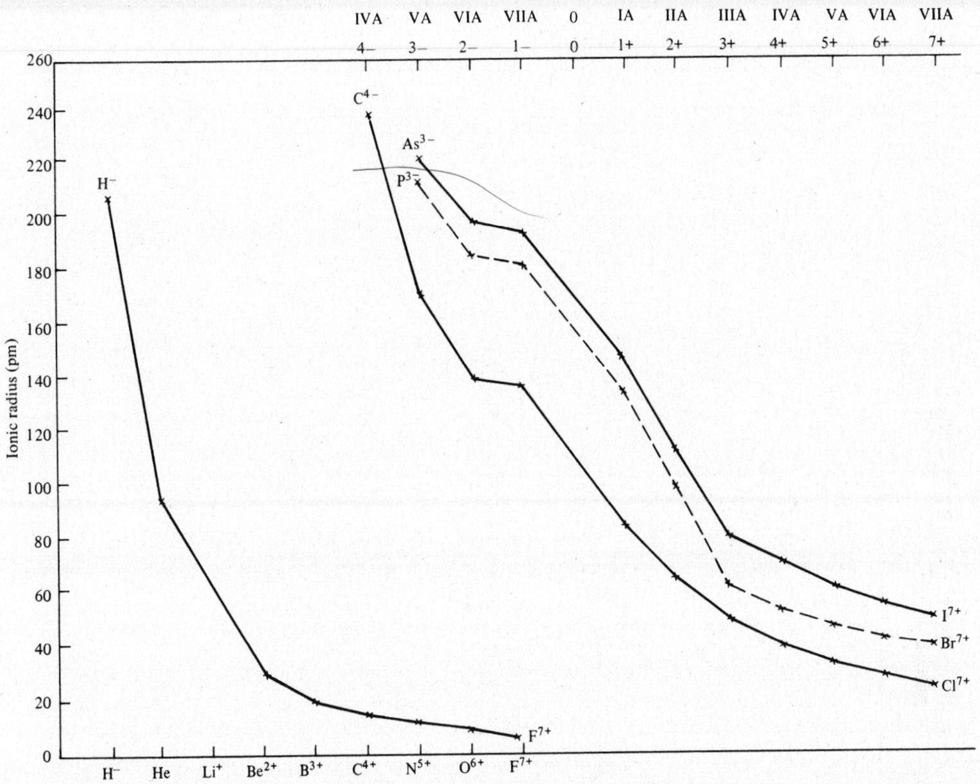

FIGURE 9-5 Ionic Radii of the Representative Elements (Isoelectronic Series)

4. Across the lanthanides, atomic radii gradually but irregularly decrease and the radii of the +3 cations steadily decrease (Figure 9-4). This steady decrease in size, called the lanthanide contraction, is due to the addition of electrons to a deep inner subshell (the 4*f* subshell). Note carefully the relative sizes of ions and atoms in Figure 9-4; the ions are very much smaller than the atoms. The lanthanide contraction helps to explain the similarity of chemical and physical properties between the transition elements of the fifth (Y to Cd) and sixth (La to Hg) periods. Because of the lanthanide contraction, atomic sizes are almost the same in these two transition periods (see Figure 9-3). Since the outer electron configurations are the same, the properties are very similar.

5. The sizes of isoelectronic ions decrease as the positive charge on the ions increase (see Figure 9-5). The explanation is that the larger nuclear charge exerts more pull on each electron and thus pulls these electrons in closer to the nucleus.

* **6. Define first, second, ... ionization energies; describe the factors that affect the magnitude of these ionization energies; and relate ionization energies to the location of elements in the periodic table.**

Ionization energy (or ionization potential, IP) is the energy required by the process $M(g) \longrightarrow M^+(g) + e^-(g)$. The ease of losing an electron is a measure of the ability of an element to act as a metal. Ionization energies are measured in kJ/mol, or in electron volts per atom (eV/atom), where 1 eV/atom = 96.49 kJ/mol.

1. Ionization energy decreases down a family of representative elements (see Figure 9-6). As atoms get larger, the outermost electrons are farther from the nucleus and hence are less strongly held.

2. Ionization energy gradually but irregularly increases across a representative period (see Figure 9-6). As atomic size decreases, electrons are more strongly attracted to the nucleus. Note that the trend is irregular in that it is harder than expected to remove electrons from s^2 and p^3 configurations and relatively easy to ionize p^1 and p^4 configurations. This demonstrates the unusual stability of full and half-full subshells.

3. Ionization energy gradually but irregularly increases across a transition period (see Figure 9-7). Again, as atomic size decreases, removing the outer electrons becomes more difficult.

4. There is no noticeable trend in ionization potentials of the lanthanides.

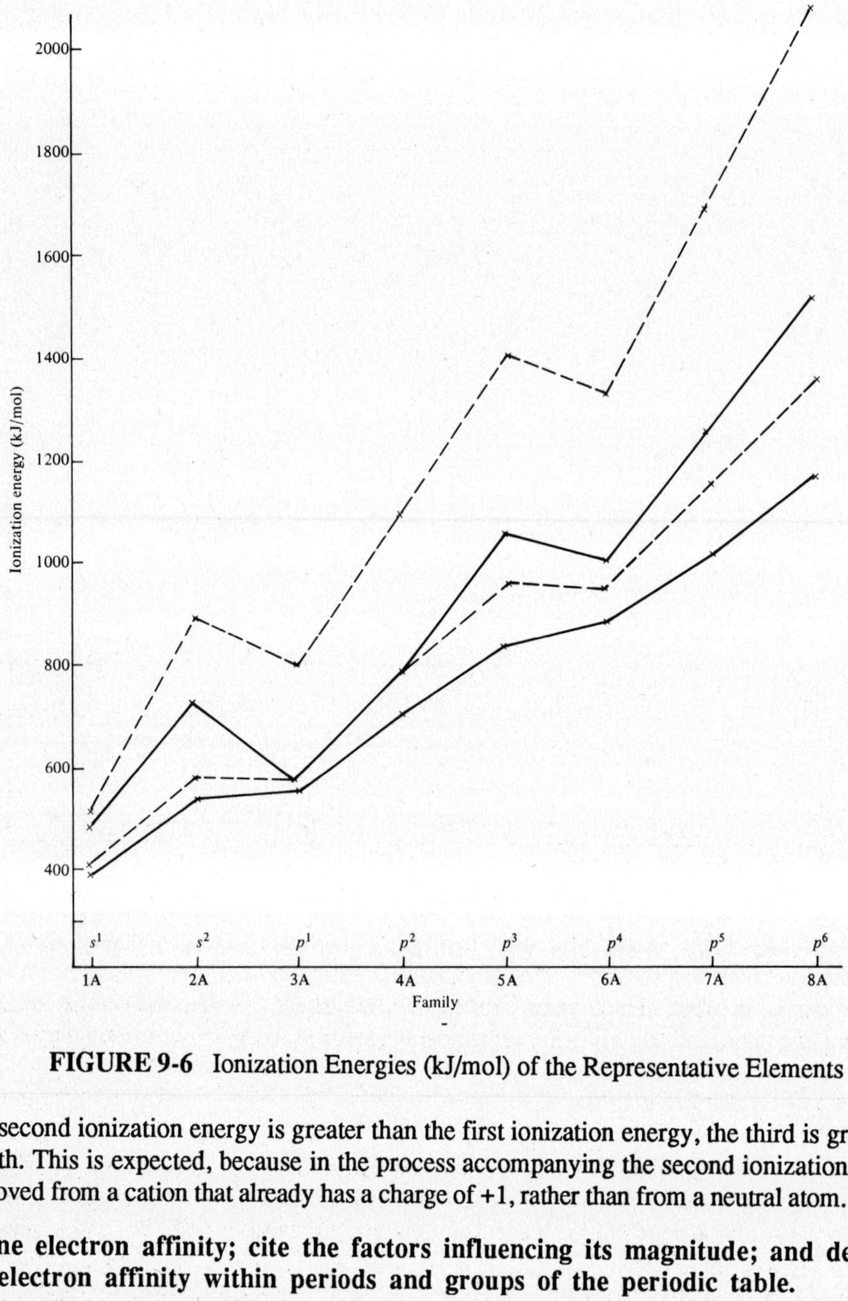

FIGURE 9-6 Ionization Energies (kJ/mol) of the Representative Elements

5. The second ionization energy is greater than the first ionization energy, the third is greater than the second, and so forth. This is expected, because in the process accompanying the second ionization energy an electron is being removed from a cation that already has a charge of +1, rather than from a neutral atom.

*** 7. Define electron affinity; cite the factors influencing its magnitude; and describe the variation of electron affinity within periods and groups of the periodic table.**

Electron affinity is the energy required by the process $X(g) + e^- \longrightarrow X^-(g)$. This is an indication of how nonmetallic an element is. Two trends are evident in Figure 9-8; electron affinity increases down a group and decreases gradually but irregularly across a period.

The increase in electron affinity down a group is explained by each atom being larger than the atom above it in the group. This means that the added electron is further away from the nucleus than in the smaller (and lighter) atom. With a large distance between the negatively charged electron and the positively charged nucleus, the force of attraction is relatively small. Thus, electron affinity increases—becomes more positive, indicating a less exothermic reaction—as one proceeds down a group.

The trend toward smaller (more exothermic) electron affinities across a family also is size-related. Recall that atoms become smaller across a period (Figure 9-2). The irregularities are due to adding an electron to a full subshell (an s^2 configuration) or a half-filled subshell (p^3). Because these two electron configurations are relatively stable, adding an electron to them is not very favorable energetically. In addition, adding an electron to an s^1, a p^2, or a p^5 electron will result in the creation of a full or half-filled subshell (s^2, p^3, or p^6, respectively). Creation of such a stable electron configuration is energetically favored—an exothermic process.

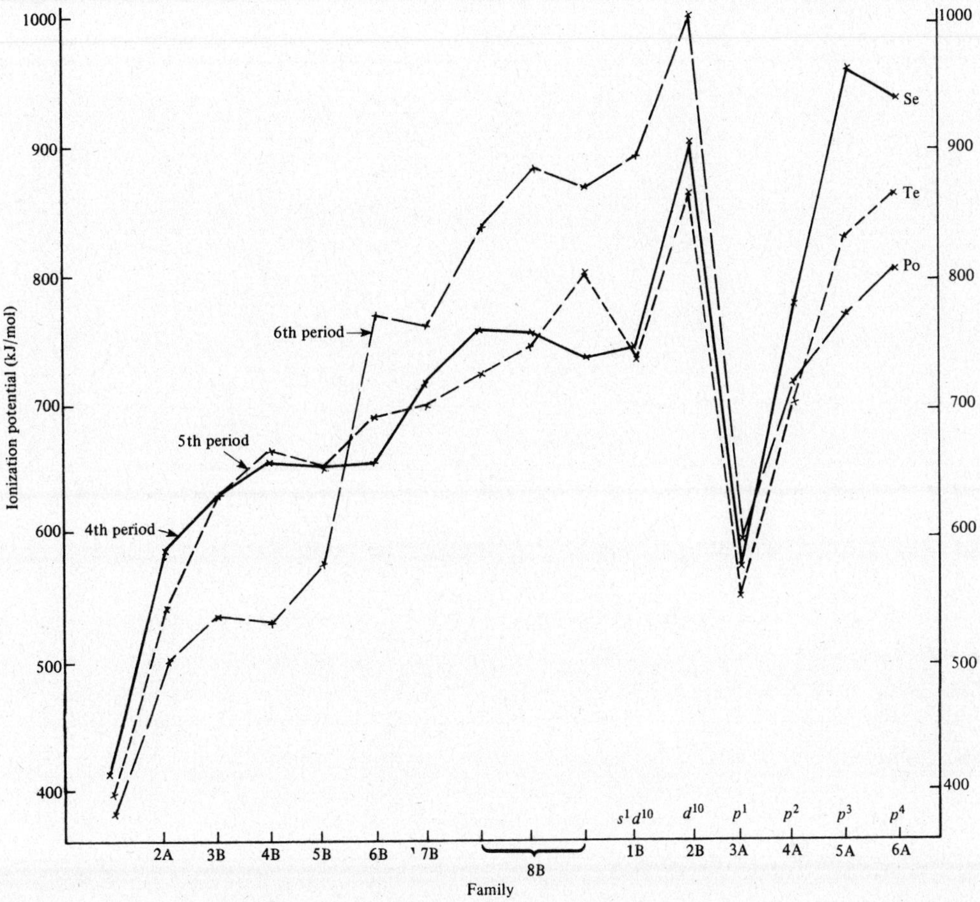

FIGURE 9-7 Ionization Potentials (kJ/mol) for the Elements of the Fourth, Fifth, and Sixth Periods.

Although the first electron affinity (*E A*) often is negative (an exothermic process), the second *E A* usually is positive. Adding an electron to an ion that has a charge of –1 is a process that is expected to require energy.

* **8. Define electronegativity, and use its value to assess the relative metallic/nonmetallic character of an element.**

Electronegativity is a combination of ionization potential and electron affinity—one number expressing how strongly an atom attracts electrons. Electronegativities are most often computed with a method devised by Linus Pauling (1901-); they are known as Pauling electronegativities. Again (see Table 9-1), two trends are evident: electronegativity increases across a period and decreases down a family. These trends are a result of the trends in ionization energies and electron affinities. Generally speaking, metals have electronegativity values below 2.0 and nonmetals have values above 2.0. Metalloids have electronegativity values of about 2.0.

TABLE 9-1 Pauling Electronegativities of Some Selected Elements

					H									
					2.20									
Li	Be									B	C	N	O	F
1.0	1.5									2.0	2.60	3.05	3.50	4.00
Na	Mg									Al	Si	P	S	Cl
0.9	1.2									1.5	1.90	2.15	2.60	3.15
K	Ca	Sc	Ti	V	...	Co	Ni	Cu	Zn	Ga	Ge	As	Se	Br
0.8	1.0	1.3	1.5	1.6		1.8	1.8	1.9	1.6	1.6	1.90	2.00	2.45	2.85
Rb	Sr	Y	Zr	Nb	...	Rh	Pd	Ag	Cd	In	Sn	Sb	Te	I
0.8	1.0	1.3	1.6	1.6		2.2	2.2	1.9	1.7	1.7	1.8	2.05	2.30	2.65
Cs	Ba	La	Hf	Ta	...	Ir	Pt	Au	Hg	Tl	Pb	Bi	Po	At
0.7	0.9	1.1	1.3	1.3		2.2	2.2	2.4	1.9	1.8	1.8	1.9	2.0	2.2

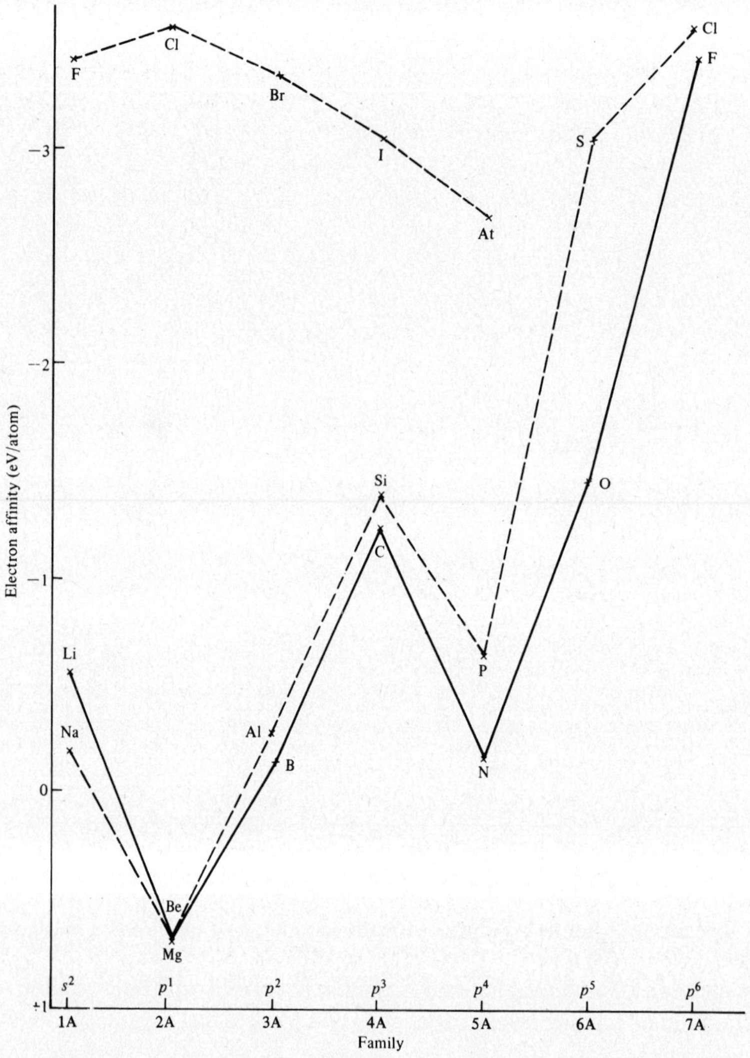

FIGURE 9-8 Electron Affinities: X(g) + e^- ⟶ X^-(g)

* **9. Relate the magnetic properties of an atom or ion to its electron configuration.**

An ion's electron configuration is determined from that of its atom. The electrons that are removed to form an ion are those of *highest principal quantum number*. Thus, when an iron atom ([Ar] $3d^6\ 4s^2$) forms an iron(II) ion, the two electrons lost are those of the 4*s* subshell, resulting in the electron configuration: [Ar] $3d^6\ 4s^0$. The resulting ion obeys Hund's rule. When all electrons in the atom or ion are paired, the atom or ion is repelled by a magnetic field and is thus diamagnetic. If there are unpaired electrons, the atom or ion is paramagnetic and is attracted into a magnetic field. The degree of attraction depends on the number of unpaired electrons in each atom or ion.

10. Use the periodic law, the periodic table, and trends in atomic properties to make predictions about the physical and chemical behavior of various elements.

In predicting metallic and nonmetallic character, ionization energies and electron affinities indicate that, when metallic atoms chemically combine with nonmetallic atoms, the metallic atoms will form cations and the nonmetallic atoms will form anions. For example, when sodium combines with chlorine, sodium anions and chloride anions—the constituents of the compound sodium chloride—are formed. Lewis theory, studied in Chapter 10, enables us to predict that a sodium cation should have a charge of +1 and the chloride anion should have a charge of −1.

DRILL PROBLEMS

3. (1) Give the general valence electron configuration for each of the following groups of elements. Use n to designate the principal quantum number. Use x as a superscript if the number of electrons in a subshell can have different values. For example, for the inner transition elements, one would write: $ns^2(n-1)d^1(n-2)f^x$

A. Transition element B. Alkaline earths C. Halogens
D. Group 4A elements E. Group 3A elements F. Group 6A elements
G. Alkali metals H. Noble gases I. Chalcogens
J. Group 3B elements K. Group 4B elements L. Group 5A elements

(2) Determine the valence electron configuration of each of the following atoms by referring only to the periodic table. Omit the principal quantum number. Thus, N would be s^2p^3. Also omit d electrons. They are not valence electrons.

M. P N. Si O. Ne P. Bi Q. Rb R. Se
S. In T. I U. Cs V. Xe W. Ba X. Ga
Y. Po Z. Pb Γ. Cl Δ. Be

(3) Refer only to the periodic table, and write the electron configuration of each element that follows. Use the abbreviated *spdf* notation in which Mn is [Ar] $4s^2\ 3d^5$. Use the extended Hund's rule where necessary. Check your predictions with the actual configurations *in the Appendix in the text.*

Θ. Cu Λ. Mo Ξ. Au Π. Eu Σ. La ϒ. Os
Φ. Fe Ψ. V Ω. Nb Ø. Y

5. Rewrite each of the lists that follow in order of increasing size (smallest to largest) of the species.

A. Ca, Sr, Mg, Be B. Te, Po, O, S C. Ar, He, Rn, Kr
D. Sn, Ge, Si, Pb E. Ga, In, B, Al F. B, Li, O, C
G. Bi, Cs, Tl, Ba H. Sn, In, Te, Rb I. K, Br, Ge, As
J. Mg, Al, Cl, P K. K, V, Ti, Fe L. Ba, Os, Hf, La
M. Ce^{3+}, W^{3+}, Tm^{3+}, Eu^{3+} N. Ac, W, U, Os O. P^{3-}, Ca^{2+}, Ar, Si^{4-}
P. O^{2-}, Na^+, C^{4-}, Ne Q. Sc^{3+}, Cl^-, K^+, S^{2-} R. Br^-, Ga^{5-}, Sr^{2+}, Y^{3+}
S. Si, Ge, O, Ne T. Mg^{2+}, F^-, Be^{2+}, C^{4-} U. K, Al, Na, N
V. N, Na^+, Li, Ne W. Tl, Ga, Ba, Ge

6. Rewrite each of the following lists in order of increasing value (smallest to largest) of the first ionization energy.

A. Rb, K, Na, Cs B. N, As, Sb, P C. Al, B, Tl, In
D. Be, Sr, Mg, Ca E. Na, Cl, Al, Ar F. Ca, Ge, Kr, As
G. Rb, Te, In, I H. N, Ne, Li, B I. F, Se, Ba, Ga
J. Sr, P, Sb, Ne K. Cs, Cl, As, Sn L. Al, In, Rb, F

7. Rewrite each of the following lists in order of increasing value (smallest or most negative, to largest) of electron affinity.

A. Br, Cl, F, I B. N, P, Bi, Sb C. Na, Rb, Li, K
D. K, Se, Ge, Br E. I, Sb, In, Rb F. F, P, In, Cs
G. Sb, Se, Cl, Tl H. Al, B, In, Tl I. As, Cl, In, Sr

8. (1) Rewrite each of the following lists in order of increasing value of electronegativity.

A. N, F, O, C B. K, Br, As, Ca C. F, Cl, I, Br
D. Rb, Cs, K, Li E. Rb, In, Ru, Se F. Li, Na, F, B
G. Cs, F, Cl, As H. Si, Ge, O, N I. Ba, Bi, Cs, Tl

(2) Rewrite each of the following lists in order of increasing metallic character, most nonmetallic to most metallic.

J. Rb, K, Na, Cs K. Cs, F, In, P L. As, Br, Ca, K
M. C, N, O, F N. Al, Cl, Mg, Na O. Ce, Fe, Ge, Se
P. Be, Ca, Mg, Sr Q. B, Be, C, Li R. Al, Ga, Rb, Sr

9. Rewrite each of the following lists of atoms and ions in order of increasing value of the paramagnetism of each atom or ion. When two species have the same value, write them together in parentheses, such as (He, Ne).

A. Ce^{3+}, K^+, C, P B. He, Cs^+, Sc, S C. B, Si, P, S
D. K, Ca, Sc, Ti E. V, Cr, Mn, Cu F. Zn, Ca, As, Ti
G. Na, Na^+, S, S^- H. Li, Be, B, C I. Fe^{2+}, Fe^{3+}, Cu^+, Cu^{2+}

QUIZZES (20 minutes each) Choose the best answer for each question.

Quiz A

1. The electrons lost when Fe ⟶ Fe^{2+} are (a) 4*f*; (b) 3*d*; (c) 4*s*; (d) 3*p*; (e) none of these.
2. A metalloid in the same periodic group as P is (a) N; (b) S; (c) Al; (d) As; (e) none of these.
3. An atom has the general ground-state *valence* electron configuration of ns^2np^5, where *n* is the principal quantum number. The element is (a) an alkali metal; (b) an alkaline earth metal; (c) a noble gas; (d) a halogen; (e) none of these.
4. If an element has some properties that are metallic and others that are nonmetallic, it is said to be (a) inert; (b) isoelectronic; (c) a metalloid; (d) an alloy; (e) none of these.
5. From left to right in a period of the periodic table, electronegativity (a) decreases, (b) increases; (c) remains the same; (d) shows irregular changes; (e) none of these.
6. In which of the following is there a consistent decrease in atomic radius as the atomic number increases? (a) halogens; (b) representative elements; (c) transition elements; (d) lanthanides; (e) none of these.
7. Of the following, the element with the smallest ionization potential is (a) Mg; (b) Na; (c) K; (d) Ca; (e) Cs.
8. An atom with an atomic radius greater than that of S is (a) O; (b) Cl; (c) Ca; (d) Li; (e) none of these.

Quiz B

1. The number of 3*d* electrons in a Co^{3+} ion is (a) 2; (b) 4; (c) 6; (d) 7; (e) none of these.
2. A nonmetal of group 5A is (a) Sb; (b) As; (c) N; (d) Bi; (e) none of these.
3. Which of these ions is the smallest? (a) O^{2-}; (b) F^-; (c) Na^+; (d) Mg^{2+}; (e) Al^{3+}.
4. The transition elements are all metallic because of their (a) *s* electrons; (b) lack of *s* electrons; (c) atomic size; (d) *d* electrons; (e) none of these.
5. Of the following elements, which possesses the lowest (most negative) electron affinity? (a) As; (b) O; (c) S; (d) Se; (e) Te.
6. Atoms in which either *s* or *p* electrons are being added are called (a) lanthanides; (b) metalloids; (c) representative elements; (d) transition elements; (e) none of these.
7. Going across a representative period from left to right, the radii of atoms (a) decrease; (b) increase; (c) remain constant; (d) show irregular patterns; (e) none of these.
8. The more metallic an element, the lower is its (a) ionization potential; (b) number of electrons; (c) electron affinity; (d) penetration of electrons.

Quiz C

1. The atom with an electronegativity greater than that of O is (a) H; (b) N; (c) S; (d) Rb; (e) none of these.
2. An atom with an atomic radius greater than that of Cs is (a) Ba; (b) Rb; (c) Sr; (d) Br; (e) none of these.
3. An element that is an example of a metalloid is (a) S; (b) Zn; (c) Ge; (d) Re; (e) none of these.
4. The most common ion of an atom with the electron configuration $1s^22s^22p^5$ would have the charge (a) +1; (b) –1; (c) +2; (d) 0; (e) none of these.
5. In the periodic table, the vertical (up and down) columns are called (a) periods; (b) transitions; (c) families; (d) metalloids; (e) none of these.
6. A series of atoms or ions that have the same electron configuration is said to be (a) degenerate; (b) isoelectronic; (c) amphoteric; (d) isotopic; (e) none of these.
7. A nonmetal will have a smaller ________ when compared to a metal of the same period. (a) atomic radius; (b) electronegativity; (c) ionization potential; (d) atomic weight; (e) none of these.
8. In the periodic table, in general, the electron affinity *increases* (a) top → bottom and right → left; (b) bottom → top and right → left; (c) top → bottom and left → right; (d) bottom → top and left → right; (e) none of these.

Quiz D

1. The species with a radius less than that of Ne is (a) Mg^{2+}; (b) F^-; (c) O^{2-}; (d) K^+; (e) none of these.
2. The representative elements are those that (a) are in the B groups; (b) fill *s* and *p* orbitals only; (c) fill *d* orbitals; (d) are metallic only; (e) none of these.
3. Which of the following elements would *not* be considered to be a metalloid? (a) Si; (b) As; (c) Ge; (d) Br; (e) Sb.

4. In general, the ionization potentials of elements decrease as one proceeds in the periodic table (a) bottom → top and right → left; (b) top → bottom and right → left; (c) bottom → top and left → right; (d) top → bottom and left → right; (e) none of these.
5. A metal will have a larger ________ when compared to a nonmetal of the same period. (a) ionization potential; (b) electron affinity; (c) number of valence electrons; (d) electronegativity; (e) none of these.
6. The electron lost when ionization occurs is the one with (a) highest principal quantum number; (b) lowest principal quantum number; (c) outer electron with highest orbital quantum number; (d) highest orbital quantum number; (e) none of these.
7. An element that has a large radius and a small ionization potential is likely to be a(n) (a) nonmetal; (b) metalloid; (c) metal; (d) inert gas; (e) none of these.
8. Of the following, which element possesses the largest ionization potential? (a) Mg; (b) Ca; (c) Sr; (d) Ba; (e) Ra.

SAMPLE TEST (15 minutes)

P_4	15	S_8	16	Cl_2	17
$3s^23p^3$	30.97	$3s^23p^3$	32.06	$3s^23p^5$	35.45
3 upe	106	2 upe	102	1 upe	99
17.0	2.1	15.5	2.5	18.7	3.0
11.0	0.741	10.4	0.732	13.0	0.485
As_4	33	Selenium	Z	Br_2	35
$4s^24p^3$	74.92	Config.	At.wt.	$4s^24p^5$	79.91
3 upe	120	No. upe	rad.	1 upe	114
13.1	2.0	*V*	E.N.	23.5	2.8
10.0	0.34	I.P.	sp.ht.	11.8	0.29
Sb_4	51	Te(metal)	52	I_2	53
$5s^25p^3$	121.75	$5s^25p^4$	127.60	$5s^25p^5$	126.90
3 upe	140	2 upe	136	1 upe	133
18.4	1.9	20.5	2.1	25.7	2.5
8.6	0.21	9.0	0.20	10.5	0.22

Suppose that the element selenium is yet to be discovered. Use the information in the partial periodic table above along with the principles of periodicity to predict the following properties of selenium.

1. Number of atoms per molecule.
2. Atomic number (*Z*).
3. Valence electron configuration (Config.).
4. Atomic weight (At.wt.).
5. Number of unpaired electrons in the isolated atom (No. upe)
6. Atomic radius, in picometers (rad.).
7. Atomic volume in ml/mol (*V*).
8. Electronegativity (E.N.).
9. First ionization potential in electron volts/atom (I.P.).
10. Specific heat in Joules g^{-1} $°C^{-1}$ (sp.ht.).

10 Chemical Bonding I: Basic Concepts

CHAPTER OBJECTIVES

1. State the basic assumptions of the Lewis theory.

Lewis theory assumes that in compounds an atom strives for as many paired electrons as possible, and then the electron configuration of the nearest noble gas. This second part is known as the *octet rule* since all noble gases except helium have eight valence electrons. An octet may obtained by losing, gaining, or sharing electrons. When electrons are lost, a cation is formed. When they are gained, an anion is formed. Cations and anions combine to form covalent compounds. When electrons are shared, a covalent bond is formed.

* **2. Relate the Lewis symbol for an element to its position in the periodic table.**

The valence electron configurations of all elements in a representative family are identical except for the value of the principal quantum number. The Lewis symbol is the symbol of the element surrounded by dots—one dot for each valence electron. See Table 10-1 where X is the symbol of the element. Note the following points.

1. The number of valence electrons of a representative element equals the family number.
2. There are four "sides" (top, bottom, left, and right) to a symbol. Each side can represent one valence orbital (s, p_x, p_y, p_z).
3. No given "side" represents any particular orbital. The following are correct Lewis symbols of oxygen.

   ·Ö̤ ·Ȯ̤ :Ȯ̤ :Ȯ: ·Ö: :Ö·

4. Pairs of electrons often are represented by a line rather than two dots. Either Ba: or Ba| can represent barium.

TABLE 10-1 Lewis Symbols for Representative Elements

Group or Family	Electron Configuration	Lewis Structures General	Examples			Ions
1A	s^1	X·	Na·			X^+
2A	s^2	X: *or* ·X·	Ba:	·Be·		X^{2+}
3A	s^2p^1	:X· *or* ·Ẋ·	·B:	·Ȧl·		X^{3+}
4A	s^2p^2	:Ẋ·	·C̈·	·Ṡi·		$[\lvert\overline{\underline{X}}\rvert]^{4-}$ *or* X^{4+}
5A	s^2p^3	:Ẋ·	·P·	·N·	·Ȧs:	$[\lvert\overline{X}\rvert]^{3-}$
6A	s^2p^4	:Ẋ:	:Ṡe·	·$\overline{\underline{O}}$·	·S̈·	$[\lvert\overline{\underline{X}}\rvert]^{2-}$
7A	s^2p^5	:Ẍ:	:C̈l:	$\lvert\overline{\underline{F}}$·	:Ï·	$[\lvert\overline{\underline{X}}\rvert]^{-}$
8A	s^2p^6	:Ẍ:	:Är:	$\lvert\overline{\underline{Xe}}\rvert$		

* **3. Write Lewis structures for simple ionic compounds.**

When electrons are gained or lost, the cations and anions shown in the last column of Table 10-1 are formed. Notice that anion Lewis symbols are surrounded by square brackets to separate the electron dots from the ionic charge. For example, the Lewis structure of calcium fluoride is depicted by

Ca^{2+} and two $[\overline{|\underline{F}|}]^-$

Square brackets also are used around the Lewis structures of polyatomic ions for clarity.

* **4. Use the Born-Fajans-Haber cycle to calculate lattice energies of ionic compounds from thermochemical, atomic, and molecular data.**

The Born-Fajans-Haber cycle divides the formation reaction [1] of an ionic solid into several steps, as shown in Figure 10-1 for $MgBr_2$.

$$Mg(s) + Br_2(l) \longrightarrow MgBr_2(s) \quad \Delta H^\circ_f = -517.6 \text{ kJ/mol} \qquad [1]$$

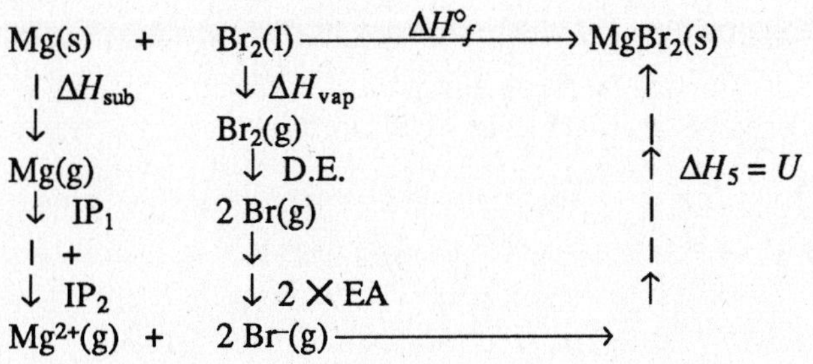

Figure 10-1 Born-Fajans-Haber cycle for $MgBr_2$

The energy of each step is found as follows.

1. The metallic solid is melted and then vaporized (or it is sublimed) and the heat of sublimation is measured: $\Delta H_1 = \Delta H_{sub}$

   $Mg(s) \longrightarrow Mg(g) \qquad \Delta H_1 = \Delta H_{sub} = +150.2$ kJ/mol

2. The metallic vapor is ionized to produce cations of the correct positive charge and the ionization potentials (IP) are measured: $\Delta H_2 = IP_1 + IP_2$ for Mg

   $Mg(g) \longrightarrow Mg^{2+}(g) + 2\,e^- \qquad \Delta H_2 = +2187.9$ kJ/mol

3. The nonmetal is vaporized if necessary and the heat of vaporization is measured: $\Delta H_3 = \Delta H_{vap}$ (or $= \Delta H_{sub}$ if the nonmetal is a solid).

   $Br_2(l) \longrightarrow Br_2(g) \qquad \Delta H_3 = +30.7$ kJ/mol

4. The nonmetal molecules are dissociated or broken apart into atoms, and the dissociation energy (D.E.) is measured: $\Delta H_4 =$ D.E.

   $Br_2(g) \longrightarrow 2\,Br(g) \qquad \Delta H_1 = +193.9$ kJ/mol

5. Electrons are added to the nonmetal atoms to produce anions of the correct negative charge and the electron affinity (EA) is measured: $\Delta H_5 =$ electron affinity $\times$ 2 (for 2 Br).

   $2\,Br(g) + 2\,e^- \longrightarrow 2\,Br^-(g) \qquad \Delta H_5 = -2710$ kJ/mol

6. The gaseous ions are combined and the lattice energy is calculated: $\Delta H_6 = U =$ lattice energy.

   $Mg^{2+}(g) + 2\,Br^-(g) \longrightarrow MgBr_2(s) \quad \Delta H_6 = ?$ kJ/mol

We have created two pathways for making $MgBr_2(s)$ from the elements: the formation reaction and steps (1) through (6). The energy for each pathway must be the same.

$$\Delta H_f^\circ = \Delta H_1 + \Delta H_2 + \Delta H_3 + \Delta H_4 + \Delta H_5 + \Delta H_6$$

$$= \Delta H_{sub} + (IP_1 + IP_2) + \Delta H_{vap} + \text{D.E.} + 2 \times \text{EA} + U \qquad [2]$$

−517.6 kJ/mol = +150.2 kJ/mol + (2187.9 kJ/mol) + 30.7 kJ/mol + 193.9 kJ/mol − 2710 kJ/mol + U

or $U = -367.9$ kJ/mol

Electron affinities are imprecise because they are hard to measure. If the lattice energy were known by some other technique, then the Born-Fajans-Haber cycle could be used to determine the electron affinity. In fact, the lattice energy of an ionic crystal can be *calculated* and the results agree well with those of the Born-Fajans-Haber cycle. The only data needed are the type of crystal structure and the ionic charges and radii. All of these data come from x-ray analysis of crystals (see objectives 12-12 through 12-14).

* 5. Describe the relationship between electronegativity difference and the percent ionic character of a bond.

The larger the electronegativity difference between two bonded atoms, the less covalent and the more ionic is the bond between them. An electronegativity difference of 0.5, as in S—Cl, produces about 6% ionic character. One of 1.0 (N—F) is about 22% ionic, one of 1.5 (C—F) about 43%, one of 2.0 (B—F) about 63%, one of 2.5 (Be—F) about 79%, and one of 3.0 (Li—F) is about 89% ionic. Often, in fact, it is possible to write both a covalent and an ionic Lewis structure for a molecule or an ion. The hydrogen halides (HF, HCl, HBr, and HI) are common examples. Others are drawn in Figure 10-2. The larger the electronegativity difference between the atoms, the greater the contribution of the ionic resonance structure. Resonance hybrids in general are more stable than a structure based on one resonance form would indicate. Hence, ionic resonance indicates a stronger bond, more difficult to break than a pure covalent bond.

$$H—\overset{\displaystyle ..}{\underset{\displaystyle ..}{C}}l| \qquad \left(\begin{array}{c} H \\ | \\ H—N—H \\ | \\ H \end{array}\right)^+ \qquad H—\overset{..}{\underset{..}{O}}—H \qquad |\overset{..}{\underset{..}{C}}l—N\_O|$$

$$H^+ \left[|\overset{..}{\underset{..}{C}}l|\right]^- \qquad \begin{array}{c} H \\ | \\ H—N| \\ | \\ H \end{array} + H^+ \qquad \left[H—\overset{..}{\underset{..}{O}}|\right]^- + H^+ \qquad \left[|\overset{..}{\underset{..}{C}}l|\right]^- + \left[|N≡O|\right]^+$$

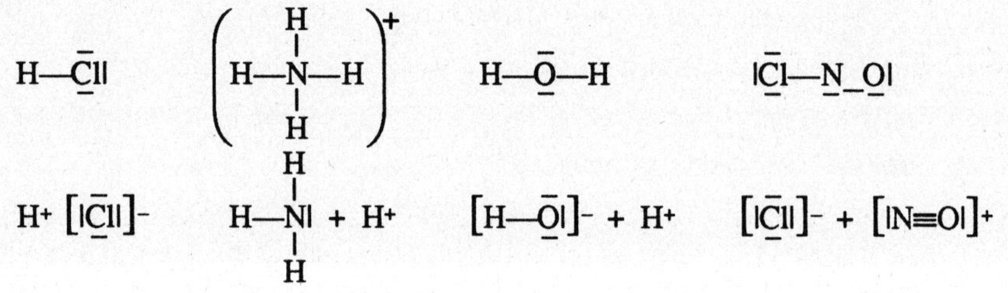

FIGURE 10-2 Some Ionic-Covalent Resonance Forms

* 6. Use the basic rules of Lewis theory to propose a plausible skeleton structure for a molecule and assign valence electrons to this structure.

Too often students believe that they *must* start with the Lewis symbols for the elements in order to draw Lewis structures. The guidelines that follow avoid that difficulty.

1. Count the total number of valence electrons in the species, remembering to add electrons for anions and subtract them for cations. The number of valence electrons of an element equals its family number.
2. Divide the total number of valence electrons by two to obtain the number of electron pairs.
3. Arrange the atomic symbols in the correct molecular skeleton. In some cases, this will be given to you. If it is not, the central atom often is written first. You may have to choose among several skeletons; use the octet rule and the formal charge on each atom to decide.
4. Place one pair of electrons between each pair of adjacent atoms. Then place the remaining electron pairs so that each atom except H has an octet of electrons, if possible.
5. If an atom lacks an octet of electrons, consider double and triple bonds between that atom and an adjacent atom. But remember that the halogens (F, Cl, Br, and I) rarely form double bonds and H never does.
6. A central atom may have an expanded octet (objective 10-10) but only if it is from the third period or higher.
7. As a first guess, try the following number of bonds around the atoms. One bond: H, F, Cl, Br, I. Two bonds: O, S, Se, Te, Be. Three bonds: N, P, As, Sb. Four bonds: C, Si, Ge.

EXAMPLE 10-1 Draw the Lewis structure of HClO.

(1) There are 1(H) + 7(Cl) + 6(O) = 14 valence electrons
(2) *or* 14/2 = 7 valence electron pairs.
(3) Possible skeletons: H—Cl—O, H—O—Cl, O—H—Cl

(4) H—$\overset{..}{\underset{..}{C}}$l—$\overset{..}{\underset{..}{O}}$| H—$\overset{..}{\underset{..}{O}}$—$\overset{..}{\underset{..}{C}}$l| |$\overset{..}{\underset{..}{O}}$—H—$\overset{..}{\underset{..}{C}}$l|

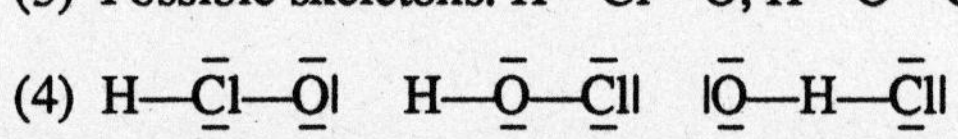

The third structure shows that we cannot have H as a central atom. H can have only one pair of electrons, not two. Based on guideline (7), H—$\overline{\underline{O}}$—$\overline{\underline{Cl}}$| is correct. We shall see later (Example 10-3) that formal charges make the same prediction.

EXAMPLE 10-2 Draw the Lewis structure of N_2O.

(1 & 2) 5(N) + 5 (N) + 6(O) = 16 valence electrons = 8 pairs.

(3) Two possible skeletons: N—N—O and N—O—N

(4) |$\overline{\underline{N}}$—N—$\overline{\underline{O}}$| and |$\overline{\underline{N}}$—O—$\overline{\underline{N}}$| are incorrect, since the central atom has no octet of electrons.

(5) There are many possibilities with multiple bonds. But based on guideline (7), none of these structures is correct. We need another criterion to decide between them. That criterion is formal charge (see Example 10-4).

N as central atom: |$\underline{N}$=N=$\underline{O}$| |N≡N—$\overline{\underline{O}}$| |$\overline{\underline{N}}$—N≡O|

O as central atom: |$\underline{N}$=O=$\underline{N}$| |N≡O—$\overline{\underline{N}}$|

* **7. Compute the formal charge on each atom in a Lewis structure; and use formal charges to determine which of several Lewis structures is the most plausible.**

To determine the formal charge (f.c.) on an atom, one uses equation [3] or equation [4].

f.c. = (group number) – (number of unshared electrons) – (number of shared electrons ÷ 2) [3]

Unshared electrons pairs are called lone pairs (l.p.) and shared electron pairs are called bond pairs (b.p.).

f.c. = (group number) – (2 × number of lone pairs) – (number of bond pairs) [4]

The most plausible structure is the one in which the formal charges are minimized, or the one in which the sum of the absolute values of the formal charges is the smallest. If formal charges remain in the species, the positive formal charges should be on the least electronegative atoms and the negative formal charges on the most electronegative atoms. Of course, there will always be some nonzero formal charges in an ion, since the sum of formal charges equals the charge on the ion (or zero for a molecule). Examples 10-3 and 10-4 illustrate the selection of the most plausible Lewis structure with the aid of formal charge.

EXAMPLE 10-3 Which of the structures of Example 10-1 is the most plausible?

The calculations (according to equation [2]) are shown below the symbol of each atom.

Lewis structure	H—$\overline{\underline{Cl}}$—$\overline{\underline{O}}$|	H—$\overline{\underline{O}}$—$\overline{\underline{Cl}}$|
group number	1 7 6	1 6 7
–2 × lone pairs	–0 –4 –6	–0 –4 –6
–bond pairs	–1 –2 –1	–1 –2 –1
formal charge	0 +1 –1	0 0 0

Thus, the formal charges make the same prediction (H—$\overline{\underline{O}}$—$\overline{\underline{Cl}}$|) as does guideline (7) of objective 10-4.

EXAMPLE 10-4 Which of the structures of Example 10-2 for N_2O is the most plausible?

Lewis structure	|$\underline{N}$=N=$\underline{O}$|	|N≡N—$\overline{\underline{O}}$|	|$\overline{\underline{N}}$—N≡O|
group number	5 5 6	5 5 6	5 5 6
–2 × lone pairs	–4 –0 –4	–2 –0 –6	–6 –0 –2
–bond pairs	–2 –4 –2	–3 –4 –1	–1 –4 –3
formal charge	–1 +1 0	0 +1 –1	–2 +1 +1

Lewis structure	|$\underline{N}$=O=$\underline{N}$|	|N≡O—$\overline{\underline{N}}$|
group number	5 6 5	5 6 5
–2 × lone pairs	–4 –0 –4	–2 –0 –6
–bond pairs	–2 –4 –2	–3 –4 –1
formal charge	–1 +2 –1	0 +2 –2

|N=N=O| and |N≡N—Ō| have the minimum total formal charge. Of these two, |N≡N—Ō| is the more plausible, as it has the negative formal charge (–1) on the more electronegative atom, oxygen.

* 8. Recognize situations when resonance occurs and draw plausible resonance structures.

Some compounds cannot be represented with a single Lewis structure. For these species, we write several equivalent structures, called resonance structures. For example, ozone (O_3) has 18 valence electrons or nine pairs. Two structures can be written:

$$|\bar{O}—O=O| \quad and \quad |O=O—\bar{O}|.$$

These two resonance structures have the following features, that are true for *all* resonance structures.

1. The number of bonds of each type (single, double, and triple) is the same in each resonance structure.
2. The molecular skeleton is the same in each structure.
3. The formal charge distribution is the same in each structure. Here, the central atom has a formal charge of +1, one terminal atom has a formal charge of 0 (the double-bonded one), and the other oxygen has a formal charge of –1.

All that is involved in creating one resonance structure from another is shifting a few pairs of electrons. Many students have the mistaken impression that the resonance hybrid (the molecule) "hops back and forth" between its various resonance forms. This is no more true than believing that a tangelo is sometimes an orange and other times a tangerine (or that you are sometimes your mother and at other times your father). There is nothing wrong with a resonance hybrid such as ozone (or a hybrid like a tangelo). Instead, the difficulty is with the Lewis theory itself; a single structure is inadequate. In the same way, a photograph is inadequate to fully represent a three-dimensional object such as a coffee cup; several photographs from different angles are needed. So also a resonance hybrid is depicted by several Lewis structures.

* 9. Draw Lewis structures for odd-electron and electron-deficient structures.

When the total number of valence electrons is odd, there is no way to pair all the electrons.

EXAMPLE 10-5 Draw the Lewis structure for ClO_2. Cl is the central atom.

$$7(Cl) + 2 \times 6(O) = 19 \text{ valence electrons } or \ 9\tfrac{1}{2} \text{ pairs.}$$

Possible Lewis structures and formal charge calculations follow.

Lewis structure	|Ō—Ċl—Ō|			·Ō—Ċl—Ō|		
group number	6	7	6	6	7	6
–2 × lone pairs	–6	–3	–6	–5	–4	–6
–bond pairs	–1	–2	–1	–1	–2	–1
formal charge	–1	+2	–1	0	+1	–1

Thus, the resonance hybrid (·Ō—Ċl—Ō| ←→ |Ō—Ċl—Ō·) is the best choice, based on formal charge. (Note that unpaired electrons rarely are used as bonding electrons in Lewis structures.)

Electron-deficient compounds occur when an atom (other than hydrogen) is surrounded by fewer than eight valence electrons. The atom is said to have an incomplete octet.

EXAMPLE 10-6 Draw the Lewis structure of BF_3.

$$3(B) + 3 \times 7(F) = 24 \text{ valence electrons } or \ 12 \text{ pairs}$$

$$|\bar{F}—B—\bar{F}| \quad or \quad |\bar{F}—B=\bar{F} \leftrightarrow |\bar{F}—B—\bar{F}| \leftrightarrow \bar{F}=B—\bar{F}|$$
$$\quad\quad |\bar{F}| \quad\quad\quad\quad |\bar{F}| \quad\quad\quad |\bar{F}| \quad\quad\quad\quad |\bar{F}|$$

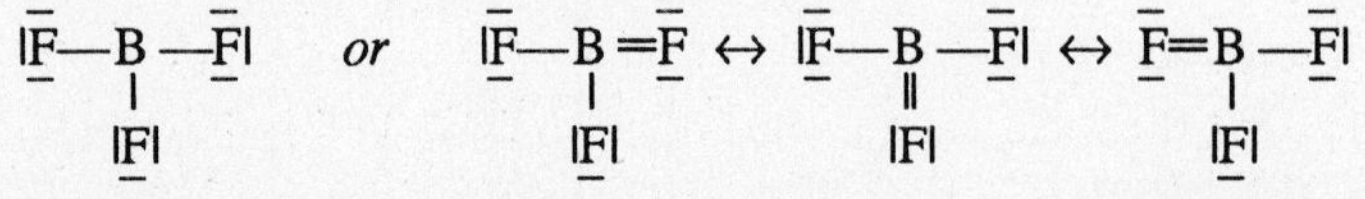

The structure at the left is electron deficient in that B is surrounded by only 6 electrons. Note that the formal charge on each atom in this structure is zero. In the resonance hybrid, there is no electron deficiency and the octet rule is obeyed. But B has a formal charge of −1 and the double-bonded F has a formal charge of +1. The true picture probably is a resonance hybrid of all four structures.

*** 10. State which elements can have expanded octets and be able to draw Lewis structures with expanded octets.**

For elements of the first and second periods, it is very difficult to place more than an octet around the atom mainly because it requires too much energy (see Figure 8-6 for the relative energy spacing between the $2p$ and $3s$ orbitals), but also because the central atom is very small (see Figure 9-2), leaving little room for more than four pairs of electrons. This is not the case for elements of the third period and higher and these elements can have expanded octets in Lewis structures.

EXAMPLE 10-7 Draw the Lewis structure of ICl_3. I is the central atom.

$7(I) + 3 \times 7(Cl) = 28$ valence electrons or 14 pairs

	I	each Cl
group number	7	7
−2 × lone pairs	−4	−6
−bond pairs	−3	−1
formal charge	0	0

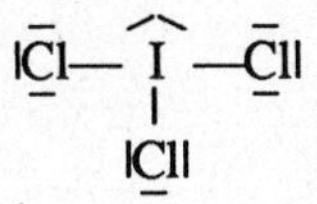

*** 11. Use bond distances to help in writing Lewis structures and bond energies to compute the enthalpy change of a reaction.**

Bond order (single, double, triple) lets us predict relative bond length and strength.

bond strength: single < double < triple
bond length: single > double > triple

This information can help us decide between various Lewis structures. In Example 10-6, we indicated the B—F to have some double-bond character. The bond length in BF_3 is somewhat shorter than an average B—F bond, confirming our prediction. Bond energies can be used to calculate an approximate $\Delta H°$ for a reaction involving reactants and products that all are gases. A fairly complete set of bond energies and lengths is given in Table 10-2.

TABLE 10-2
Bond Energies (kJ/mol of bonds broken) and Bond Lengths (pm, in parentheses)

Bond	I	Br	Cl	S	P	Si	F	O	N	C	H
H	291 (170)	364 (151)	431 (136)	339 (133)	318 (142)	293 (148)	565 (101)	464 (97)	389 (100)	414 (110)	435 (74)
C	238 (213)	276 (193)	330 (177)	259 (181)	264 (187)	289 (135)	490 (143)	360 (148)	305 (154)	347	
N	176	243 (214)	201 (197)	205 (166)	209 (160)	326 (174)	264 (137)	201 (119)	163 (145)		
O	201 (183)	218 (165)	205 (147)	268 (140)	360 (161)	368 (161)	184 (142)	138 (148)			
F	268 (176)	251 (163)	255 (159)	285 (154)	490 (156)	540 (128)	155				
Si	213 (243)	289 (217)	360 (201)	226 (214)	213	176 (233)					
P	213	272 (213)	318 (204)	230 (186)	172 (223)						
S	180	213 (227)	276 (199)	264 (204)							
Cl	209 (232)	218 (218)	243 (199)								
Br	180	192 (228)									
I	151 (266)										

P≡P	490
B—B	226 (177)
B—H	331 (116)

≡	C	N
N		946 (109)
C	837 (120)	891 (116)

=	C	O	N	S
S				427
N			418 (123)	485
O		498 (121)	607 (119)	331 (143)
C	611 (134)	707 (123)	615 (128)	531 (171)

EXAMPLE 10-8 Calculate the approximate enthalpy change for the uncatalyzed combustion of ammonia.

$$4 NH_3(g) + 3 O_2(g) \longrightarrow 2 N_2(g) + 6 H_2O(g)$$

First we draw the Lewis structures of all species.

$$4 \; H{-}\overset{\overset{\displaystyle H}{|}}{\underline{N}}{-}H + 3 \; |\underline{O}{=}\underline{O}| \longrightarrow 2 \; |\underline{N}{\equiv}\underline{N}| + 6 \; H{-}\overset{-}{\underline{O}}{-}H$$

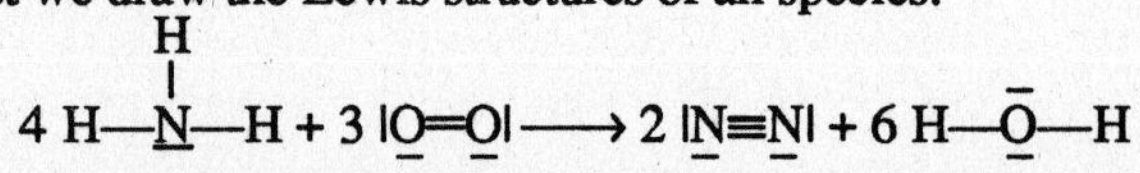

Then we determine how many bonds of each type are broken or formed and the energy required to break the bonds and produced by forming them.

Bonds broken	*Bonds formed*
12 N—H = 12 × 398 = 4668 kJ/mol	2 N≡N = 2 × 946 = 1892 kJ/mol
3 O=O = 3 × 498 = <u>1494 kJ/mol</u>	12 O—H = 12 × 464 = <u>5568 kJ/mol</u>
Total energy required = 6162 kJ/mol	Total energy produced = 7460 kJ/mol

Energy produced has a negative sign and energy required has a positive sign.

$$\Delta H° = \text{Net energy} = +6162 \text{ kJ/mol} - 7460 \text{ kJ/mol} = -1298 \text{ kJ/mol}$$

This value of $\Delta H°$ is approximate. The actual value, computed from $\Delta H°_f$ values, is −1266.25 kJ/mol. Because $\Delta H°$s computed from bond enthalpies often have an error of up to 10%, they can be used only for approximations.

* **12. Predict the electron-pair geometry and the molecular shape of a molecule or ion with VSEPR theory. Know that single bonds act similarly to multiple bonds in determining molecular shape.**

Lewis theory does not predict molecular shape. The valence-shell electron-pair repulsion (VSEPR) theory accounts for molecular shape. The basis of the theory is that pairs of electrons in the valence shell of an atom try to get as far apart as they can and still remain attached to the atom. VSEPR theory uses the following terms. AX_mE_n: the formula of a molecule or ion. A represents the *central atom*. X represents each of the *ligands*: atoms or groups of atoms bound to the central atom. E represents each of the *lone pairs*: unshared pairs of electrons on the central atom. m is the number of ligands or bonding pairs. A double bond or a triple bond is counted as one bonding pair. n is the number of lone pairs. A lone electron, such as that of N in NO_2, is counted as a lone pair. $m + n$ is the total number of electron pairs. This total determines the electron-pair geometry of the molecule or ion, as given in Table 10-3. The dashed lines in each sketch represent electron pairs pointing back, away from you, and the wedges represent electron pairs pointing forward, toward you. You may have trouble seeing these electron-pair geometries. Most beginners find a set of models helpful.

Although the total number of electron pairs determines the electron-pair geometry, we only can see the ligands. We see only the *effect* of the lone pairs. The possible VSEPR structures are given in Table 10-3 and drawn in Figure 10-2. AX_4E is an irregular tetrahedron rather than a trigonal pyramid because of the differences in the space occupied by types of electron pairs. Sizes decrease in the order given in expression [5], and repulsions between different types of electron pairs decrease in the order given in expression [6].

Sizes: lone pair > lone electron > triple bond > double bond > single bond   [5]

Repulsions: lone pair—lone pair > lone pair—bond pair > bond pair—bond pair   [6]

Since there is more room around the equator of a trigonal bipyramid than at its axes, the lone pair in AX_4E occupies the equatorial position as shown in Figure 10-3(a). You should make a model of each shape shown in Figure 10-3 so that you have a clear idea of each geometry.

TABLE 10-3 Electron-Pair Geometries

No. eln pairs	2	3	4	5	6
Electron-pair geometry	linear	equilateral triangle	tetrahedron	trigonal bipyramid	octahedron
Sketch	—A—	A	A	A	A
Bond angles	180°	120°	109.5°	120° (equatorial) 90° (axial)	90°

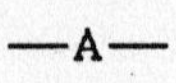

The application of VSEPR theory requires that you first draw the Lewis structure of the molecule or ion. Then, for each central atom, count the number of ligands and the number of lone pairs. From this information and Table 10-4 (which you should have memorized), name the molecular shape and draw a sketch of the molecule. The correct molecular shape is predicted even if you have not drawn the most plausible Lewis structure. Consider the three possibilities for N_2O of Example 10-4. All have two ligands and no lone pairs on the central nitrogen atom and all predict a linear structure.

TABLE 10-4 Summary of VSEPR Possibilities

		No. of Electron Pairs				
Formula	Example	Lone	Bonding	Total	Molecular shape	Electron-pair geometry
AX_2	$BeCl_2$	0	2	2	linear	linear
AX_3	BF_3	0	3	3	triangular	triangular
AX_2E	$SnCl_2$	1	2	"	bent	"
AX_4	CF_4	0	4	4	tetrahedron	tetrahedron
AX_3E	NF_3	1	3	"	triangular pyramid	"
AX_2E_2	OF_2	2	2	"	bent	"
AX_5	PCl_5	0	5	5	trigonal bipyramid	trigonal bipyramid
AX_4E	SF_4	1	4	"	irregular tetrahedron	" "
AX_3E_2	ClF_3	2	3	"	T-shaped	" "
AX_2E_3	XeF_2	3	2	"	linear	" "
AX_6	SF_6	0	6	6	octahedron	octahedron
AX_5E	BrF_5	1	5	"	square pyramid	"
AX_4E_2	XeF_4	2	4	"	square planar	"

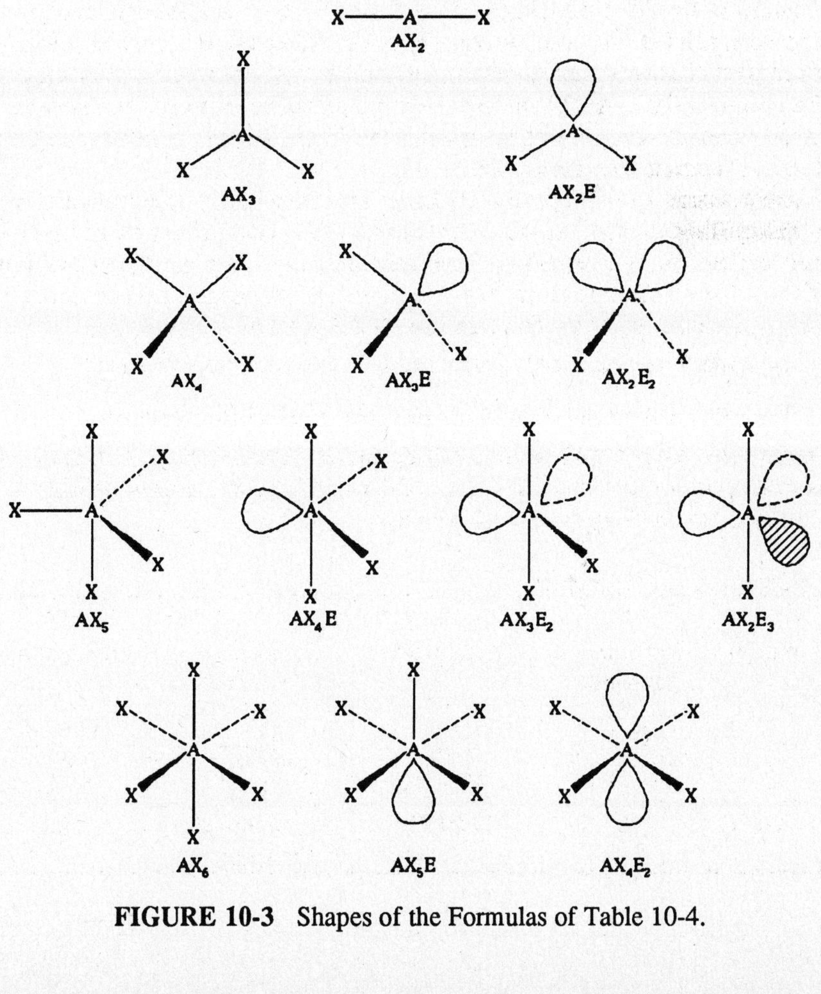

FIGURE 10-3 Shapes of the Formulas of Table 10-4.

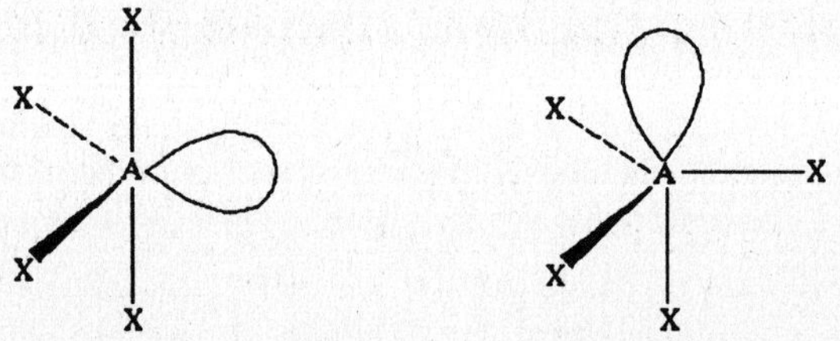

(a) Irregular Tetrahedron (b) Trigonal Pyramid

FIGURE 10-4 Possible Shapes of AX_4E Species.

* **13. Use electronegativities to determine if a bond is polar and use bond polarities and molecular shape to predict whether a molecule has a dipole moment.**

In a pure covalent bond, the bonding pair of electrons toms. In a polar bond , the bonding pair is shifted toward the more electronegative atom. Thus CO_2 and H_2O both possess polar bonds. A bond dipole is drawn as an arrow with a cross at the base (+⟶). The dipole points away from the more electropositive atom as shown in Figure 10-5. VSEPR theory predicts that CO_2 is linear and H_2O is bent. Since the two bond dipoles in CO_2 are equal and point in opposite directions, the cancel. CO_2 has no net dipole moment. But in the H_2O molecule, the bond dipoles do not point in exactly opposite directions and therefore they do not cancel. H_2O has a dipole moment.

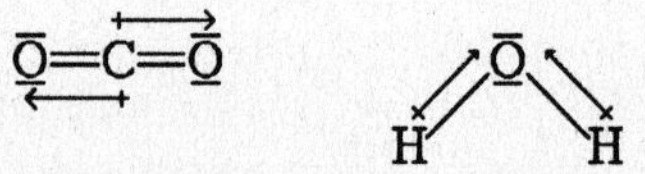

FIGURE 10-5 Bond Dipoles for CO_2 and H_2O

DRILL PROBLEMS

2. Give the Lewis structure for each of the following atoms. You may use a periodic table for assistance.
A. He B. Li C. Ne D. In E. I F. Sb
G. Sn H. Sr I. S J. F K. C L. Ca
M. K N. Al O. P P. Se

3. Give the Lewis structure of each of the following ionic species.
A. LiH B. Na_2S C. $CaBr_2$ D. SrO E. Mg_3N_2 F. Sc_2S_3
G. NaF H. TiO_2 I. MgS J. $LaCl_3$ K. K_2Se L. Mg_2C
M. $CeCl_4$ N. AlN O. Cs_2Te P. Rb_3P

6. Give the Lewis structure of each of the following covalently bonded species.
A. CH_4 B. H_2CO C. CH_3OH D. N_2H_4 E. OF_2
F. NF_3 G. SiO_4^{4-} H. NOCl I. $GeCl_4$ J. C_2H_4
K. $SiCl_4$ L. Cl_2CO M. $HClO_2$ N. HOCN O. HFO
P. HNO_2 Q. H_2O_2 R. C_2H_2 S. AsF_3 T. BrO_4^-
U. PO_4^{3-} V. $SeBr_2$ W. PO_3^{3-} X. BF_4^- Y. OF^-
Z. ClO_3^- Γ. HO_2^- Δ. CN^- Θ. HI Λ. CCl_2H_2
Π. H_2CO_2 Σ. C_2H_6 Φ. O_2^{2-}

7. Give at least two Lewis structures for each of the following species, evaluate the formal charge on each atom of each structure, and predict which structure is the most plausible.
A. HOCN B. HCN C. $HClO_3$ D. CO E. ONCl
F. CNS^- G. HCNS H. BCl_3 I. HON J. $BeCl_2$
K. $OSeF_2$ L. CN_2^{2-} M. HNO_2 N. $POCl_3$ O. ClCN
P. HN_3 Q. H_2CN_2 R. NO_2Cl

8. Give at least two resonance structures for each of the following resonance hybrids
A. HCO_2^- B. NO_2^- C. N_3^- D. SO_2 E. CO_3^{2-}
F. BO_3^{3-} G. NO_3^- H. PO_3^- I. C_2O_3 J. SO_3
K. $C_2O_4^{2-}$ L. O_3

9. Give the most plausible Lewis structure for each of the following species.
A. NO B. NO_2 C. $SnBr_2$ D. ClO_2 E. PbI_2
F. $BeCl_2$ G. PO_3 H. BeH_2 I. NCN^- J. BN
K. BI_3 L. CH_3 M. SO_2^+ N. BCl_3 O. O_2^-

10. Give the most likely Lewis structure for each of the following species.
A. BrF_3 B. $SbCl_5^{2-}$ C. BrO_3^- D. I_3^- E. PF_3Cl_2
F. $PbCl_2$ G. H_2SO_3 H. $SbCl_5$ I. ICl_4^- J. $SnCl_6^{2-}$
K. IF_5 L. SO_3^{2-} M. XeF_4 N. SeF_6 O. ArF_2
P. SCl_4 Q. ClF_3 R. PO_4^{3-} S. IS_2^- T. XeF_2
U. XeO_3

11. Use the bond enthalpies given in Table 10-2 to predict the value of $\Delta H°$ for each of the following reactions.
A. $CS_2(g) + 3\ O_2(g) \longrightarrow CO_2(g) + 2\ SO_2(g)$
B. $CH_4(g) + 2\ O_2(g) \longrightarrow CO_2(g) + 2\ H_2O(g)$
C. $H_2CO(g) + O_2(g) \longrightarrow CO_2(g) + H_2O(g)$
D. $PCl_3(g) + F_2(g) \longrightarrow PCl_3F_2(g)$
E. $I_2(g) + 3\ Cl_2(g) \longrightarrow 2\ ICl_3(g)$
F. $N_2(g) + 3\ H_2(g) \longrightarrow 2\ NH_3(g)$
G. $2\ N_2H_2(g) + O_2(g) \longrightarrow 2\ N_2(g) + 2\ H_2O(g)$
H. $4\ HCN(g) + 5\ O_2(g) \longrightarrow 2\ H_2O(g) + 2\ N_2(g) + 4\ CO_2(g)$
I. $CO_2(g) + 2\ Cl_2(g) \longrightarrow Cl_2CO(g) + Cl_2O(g)$
J. $SiH_4(g) + 4\ F_2(g) \longrightarrow SiF_4(g) + 4\ HF(g)$
K. $CO_2(g) + 2\ H_2(g) \longrightarrow CH_4O_2(g)$
L. $C_2H_6O(g) + 3\ O_2(g) \longrightarrow 2\ CO_2(g) + 3\ H_2O(g)$
M. $PCl_5(g) + 4\ H_2O(g) \longrightarrow H_3PO_4(g) + 5\ HCl(g)$ (Assume P=O is 710 kJ/mol)
N. $PCl_5(g) + H_2O(g) \longrightarrow POCl_3(g) + 2\ HCl(g)$
O. $HCNS(g) + H_2(g) \longrightarrow HCNO(g) + H_2S(g)$
P. $C_3H_8(g) + 5\ O_2(g) \longrightarrow 3\ CO_2(g) + 4\ H_2O(g)$

12. Give the electron-pair geometry, the molecular shape, and a perspective sketch (use solid, dashed, and wedged lines) of each of the following species.
A. BF_3 B. $GeCl_4$ C. HO_2^- D. BrF_3 E. $SbCl_5$
F. $PbBr_2$ G. H_2CO H. $BeCl_2$ I. KrF_4 J. ArF_2
K. SCl_4 L. NO_2^- M. NF_3 N. ClF_3 O. IF_5
P. PF_5 Q. ClO_4^- R. AsF_3 S. XeF_4 T. BCl_3
U. H_2S V. $POCl_3$ W. SF_4 X. CO_3^{2-} Y. NO_3^-
Z. N_2O Γ. BF_4^-

13. Which of the molecules in the question for objective 10-12 has a dipole moment and which way does the dipole point in each case?

QUIZZES (20 minutes each) Choose the best answer to each question.

Quiz A

1. Which of the following should be paramagnetic? (a) HCl; (b) CO_2; (c) PCl_3; (d) ClO_2; (e) none of these.
2. In which compound do two ions composed of nonmetal atoms combine to form an ionic compound? (a) NH_4NO_3; (b) $Al_2(SO_4)_3$; (c) Na_2SO_3; (d) $AlCl_3$; (e) none of these.
3. Which of the following molecules does *not* obey the octet rule? (a) HCN; (b) PF_3; (c) CS_2; (d) NO; (e) none of these.
4. Which molecule has no polar bonds? (a) H_2CO; (b) CCl_4; (c) OF_2; (d) N_2O; (e) none of these.
5. The incorrect Lewis structure of the following is (a) $\overline{\underline{O}}=C=\overline{\underline{O}}$; (b) H—B(—H)—H (H above B); (c) $|C=\overline{O}$; (d) $H-\overline{\underline{O}}-\overline{\underline{Cl}}|$; (e) none of these.
6. Of the following molecules, which contains the shortest bond distance? (a) Cl_2; (b) N_2; (c) O_2; (d) NH_3; (e) CO_2.
7. Which of the following compounds would be expected to have tetrahedral geometry? (a) AsF_3; (b) SF_4;

(c) $SiCl_4$; (d) XeF_4; (e) none of these.

8. According to VSEPR theory, the shape of ICl_4^- ion is (a) pyramidal; (b) linear; (c) tetrahedral; (d) square planar; (e) none of these.
9. Which of the following species has a tetrahedral shape? (a) ICl_4^-; (b) SF_4 ; (c) SiF_4; (d) XeF_4; (e) none of these.
10. Given the bond enthalpies I—Cl (209), H—H (435), H—Cl (341), H—I (297) in kJ/mol, compute $\Delta H°$ in kJ/mol for the reaction $ICl_3(g) + 2\ H_2(g) \longrightarrow HI(g) + 3\ HCl(g)$. (a) –93; (b) +264; (c) –523; (d) –84; (e) none of these within 2%.

Quiz B

1. Which of the following species should be paramagnetic? (a) NO^-; (b) NO; (c) N_2O; (d) NO_2; (e) none of these.
2. Which of the following exhibits ionic bonding? (a) BF_3; (b) AsH_3; (c) SO_3; (d) BaO; (e) none of these.
3. When a molecule should be represented by more than one electronic structure, it is best described in terms of (a) VSEPR theory; (b) hybridization; (c) resonance; (d) multiple covalent bonding; (e) none of these.
4. Which of the following molecules has polar bonds but is itself not polar? (a) CH_3Cl; (b) H_2O; (c) SCl_2; (d) Cl_2; (e) none of these.
5. The incorrect Lewis structure of the following is (a) |C≡O|; (b) $\overline{\underline{S}}=\overline{O}$; (c) $H—\overline{\underline{O}}—\overline{\underline{O}}—H$; (d) $|\overline{\underline{O}}—\overline{N}=\overline{O}|$; (e) none of these.
6. Which of the following rarely forms multiple bonds? (a) S; (b) C; (c) Cl; (d) O; (e) none of these.
7. Which of the following would be expected to be trigonal planar? (a) $H_2C{=}O$; (b) H_3O^+; (c) NH_3; (d) $:CH_3$; (e) none of these.
8. The electron-pair geometry of H_2O is (a) tetrahedral; (b) trigonal planar; (c) bent; (d) linear; (e) none of these.
9. Which of the following species has a triangular planar shape? (a) BF_3; (b) PF_3; (c) IF_3; (d) NH_3; (e) none of these.
10. Given the bond enthalpies C=O (707), O=O (498), H—O (464), C—H (414) in kJ/mol, compute $\Delta H°$ in kJ/mol for the reaction $CH_4(g) + 2\ O_2(g) \longrightarrow CO_2(g) + 2\ H_2O(g)$. (a) +618; (b) –519; (c) –618; (d) + 259; (e) none of these within 2%.

Quiz C

1. Which of the following is not paramagnetic? (a) NO; (b) CO; (c) NO_2; (d) BC; (e) none of these.
2. Which of the following exhibits covalent bonding? (a) NaF; (b) ICl; (c) K_2O; (d) SrI_2; (e) none of these.
3. Which of the following molecules is adequately represented by a single Lewis structure? (a) O_3; (b) NOCl; (c) SO_2; (d) N_2O; (e) none of these.
4. Which of the following does *not* have a molecular dipole moment? (a) HCl; (b) CO; (c) NCl_3; (d) BCl_3; (e) none of these.
5. As indicated by its Lewis structure, which of the following species would probably not exist as a stable molecule? (a) CH_3O; (b) CH_2O; (c) CH_2O_2; (d) C_2H_2; (e) none of these.
6. Which of the following atoms often is present in compounds that are electron deficient? (a) F; (b) O; (c) H; (d) B; (e) none of these.
7. The electron-pair geometry of NO_2^+ is (a) triangular; (b) bent; (c) tetrahedral; (d) octahedral; (e) none of these.
8. The molecular shape of PF_5 is (a) octahedral; (b) trigonal bipyramid; (c) irregular tetrahedron; (d) square pyramid; (e) none of these.
9. Which of the following species has a square planar shape? (a) ICl_4^-; (b) SF_4; (c) NH_4^+; (d) CH_2F_2; (e) none of these.
10. Given the bond enthalpies C—H (414), H—Cl (431), C—Cl (331), Cl—Cl (243) in kJ/mol, compute $\Delta H°$ in kJ/mol for the reaction $CH_4(g) + 4\ Cl_2(g) \longrightarrow CCl_4(g) + 4\ HCl(g)$. (a) +904; (b) +418; (c) –252; (d) –105; (e) none of these.

Quiz D

1. Which of the following molecules is not paramagnetic? (a) NCl; (b) NO_2; (c) BrCl; (d) O_2; (e) none of these.

2. The greater the electronegativity difference between two atoms in a bond, the greater the _____ of the bond, and the _____ it is. (a) covalent character, more stable; (b) ionic character, more stable; (c) ionic character, less stable; (d) covalent character; less stable; (e) none of these.
3. The most likely arrangement of atoms in S_2Cl_2 is (a) S—S—Cl—Cl; (b) S—Cl—S—Cl; (c) S—Cl—Cl—S; (d) Cl—S—S—Cl; (e) none of these.
4. Of the following, which has the largest dipole moment? (a) CH_4; (b) CH_3Cl; (c) BCl_3; (d) CO_2; (e) none of these.
5. As indicated by Lewis structures, which of the following species could probably not exist as a stable molecule? (a) NH_3; (b) N_2H_2; (c) N_2H_4; (d) N_2H_6; (e) none of these.
6. Which of the following contains the bond of largest energy? (a) H_2; (b) Cl_2; (c) CO_2; (d) N_2; (e) H_2O.
7. According to VSEPR theory, the molecular shape of the I_3^- ion is (a) trigonal planar; (b) linear; (c) trigonal bipyramid; (d) bent; (e) none of these.
8. Which of the following electron pairs exhibits the largest degree of repulsion? (a) lone pair-lone pair; (b) bond pair-bond pair; (c) lone pair-bond pair; (d) lone pair-unpaired; (e) bond pair-unpaired.
9. Which of the following species is T-shaped? (a) BF_3; (b) PCl_3; (c) CH_3^-; (d) ClF_3; (e) none of these.
10. Given the bond enthalpies H—O (464), Cl—Cl (243), Cl—O (205), H—Cl (431) in kJ/mol, compute $\Delta H°$ in kJ/mol for the reaction $H_2O(g) + Cl_2(g) \longrightarrow HOCl(g) + HCl(g)$. (a) –1100; (b) +393; (c) –393; (d) +71; (e) none of these within 2%.

SAMPLE TEST

1. Draw the best Lewis structure of each of the following species and indicate the formal charge of each atom of the structure. If the species is a resonance hybrid, be sure to draw all of the resonance forms.
   a. PO_3^- b. NO_2^+ c. CH_3Cl d. IF_3
2. For each of the following compounds give the names of the electron-pair geometry and the molecular shape. Sketch the molecule and indicate on the sketch the direction of the dipole moment of the molecule, if any. For sketches, use the "wedge and dash" convention.
   a. SiF_4 b. NF_3 c. SF_4 d. IF_5
3. Given the bond enthalpies H—O (464), H—C (414), C—C (347), C=O (707), O=O (498) in kJ/mol, determine which of the following compounds produces the most energy in kJ/g when it burns completely in oxygen: $CH_4(g)$, $CH_3OH(g)$, $H_2CO(g)$, $HCOOH(g)$. Is there any relationship between the oxidation state of carbon and the heat of combustion (in kJ/g or kJ/mol)?

11 Chemical Bonding II: Additional Aspects

CHAPTER OBJECTIVES

1. Explain the fundamental basis of valence bond theory; and explain why hybrid orbitals are often used to describe bonding in molecules rather than pure atomic orbitals.

Valence bond theory states that a bond forms when two atomic orbitals overlap. Each bond contains at most two electrons with their spins paired. A large amount of overlap means that a strong bond is formed. Atomic orbitals with directional character (those that "point" in a certain direction) overlap better than those with no directional character. For example, *p* orbitals overlap better than do *s* orbitals. We say they overlap more efficiently or there is a greater degree of overlap.

An attempt to explain the bonding in the CH_4 molecule provides an example of the need for hybridization. The promoted, or excited, configuration of C, given in expression [1], permits the formation of four C—H bonds.

C: [He] [↿] [↿][↿][↿] [1]

2*s* 2*p*

Each bond results from the overlap of a half-filled orbital on carbon (2*s* or 2*p*) with a half-filled orbital on hydrogen (1*s*). These bonds disagree in two ways with the experimental results for CH_4. Although the four bonds in CH_4 are the same, the bonds we have just formed differ from each other (three 1*s*-2*p* bonds and one 1*s*-2*s* bond). Also, the predicted bond angles are 90°, compared to observed bond angles of 109.5°. We can combine the four pure atomic orbitals to produce four equivalent hybrid orbitals. (Notice that hybridization is used *after* we know the bond angles, which VSEPR theory can predict.)

C: [He] [↿][↿][↿][↿] [2]

sp^3

* **2. Write hybridization schemes for the formation of sp, sp^2, sp^3, sp^3d, and sp^3d^2 hybrid orbitals; and predict the geometrical shapes of molecules in terms of the pure and hybrid orbitals used in bonding.**

In the process of hybridization, pure atomic orbitals are combined to produce hybrid orbitals that have the correct geometry for the resulting molecule. The hybridization of a species can be predicted by starting from the Lewis structure. There are as many hybrid orbitals as there are ligands plus lone pairs (that is, total electron pairs) on the central atom. The pure atomic orbitals are used in order of increasing energy. Thus, if the total number of electron pairs is 2, an *s* orbital and one *p* orbital are combined or hybridized to form two *sp* hybrid orbitals. If there are three electron pairs total, an *s* orbital and two *p* orbitals are hybridized to form three sp^2 hybrid orbitals. And if there are four electron pairs total, an *s* orbital and three *p* orbitals are hybridized to form four sp^3 hybrid orbitals. (Notice that the symbol for a hybrid orbital indicates which pure atomic orbitals have been combined to form it.) When hybrid orbitals are formed, those that are to hold lone pairs are full, while those that are to form bonds (or to hold bond pairs) usually have one electron. The exception is coordinate covalent bonds: formed from a full orbital (an electron pair) on one atom and an empty orbital on another. We can recognize coordinate covalent bonds because the atoms involved in them have nonzero formal charges in Lewis structures. Lewis structures and ground state and hybridized electron configurations for CH_4, BCl_3, BeF_2, NF_3, and Cl_2O are shown in Figure 11-1.

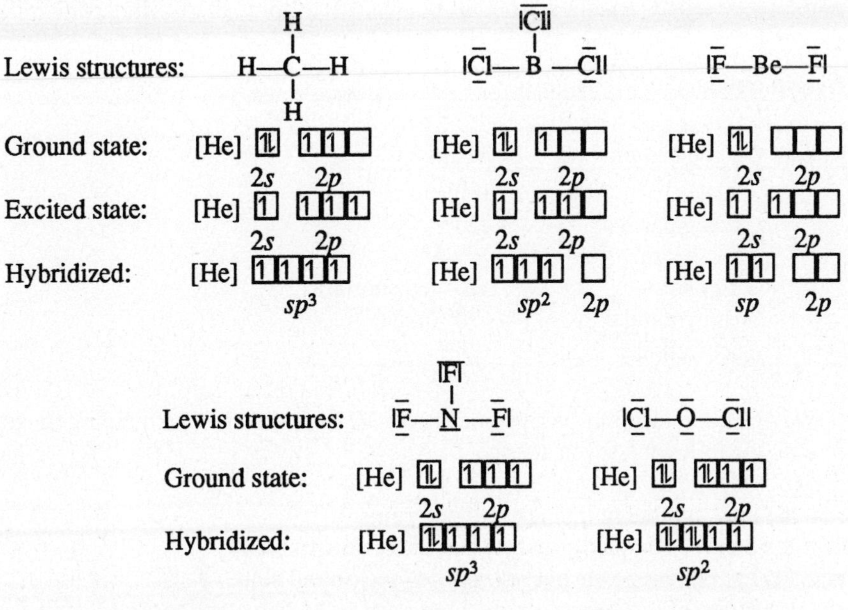

FIGURE 11-1 Bonding in CH_4, BCl_3, BeF_2, NF_3, and Cl_2O

The properties of hybrid orbitals depend on the properties of the pure atomic orbitals that were combined to form them. Consider direction. A *p* orbital points along a line. Thus, if an *s* orbital is combined with a *p* orbital, the two *sp* hybrids formed will point along a line. In similar fashion, sp^2 orbitals are confined to a plane (they must be flat), since any two *p* orbitals lie in a plane. And four sp^3 orbitals are three-dimensional.

When the number of ligands and bond pairs totals five or six, *d* orbitals must be included in the hybridization scheme, since there are only a total of four *s* and *p* orbitals in each shell. The Lewis structures and orbital diagrams for SF_4, ClF_3, XeF_4, and ClF_5 in Figure 11-2 illustrate the inclusion of *d* orbitals in hybridization schemes. Notice that the technique used is similar to that of Figure 11-1, except for including *d* orbitals.

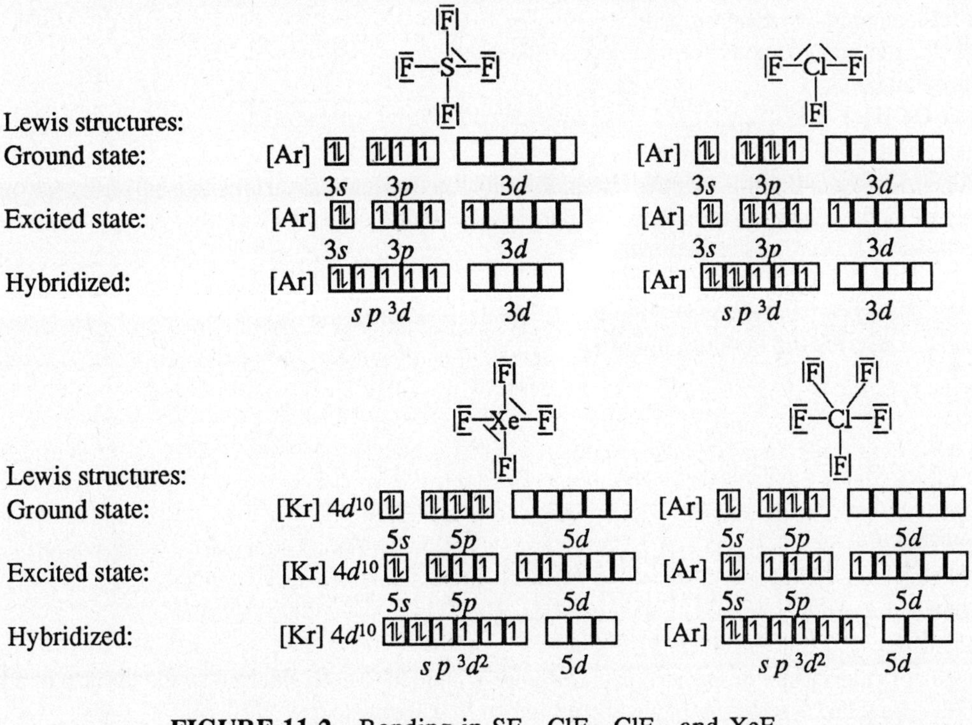

FIGURE 11-2 Bonding in SF_4, ClF_3, ClF_5, and XeF_4

*** 3. Use the relationship between VSEPR theory and valence bond theory to predict molecular geometries.**

TABLE 11-1 Geometry of Hybrid Orbitals

Number of ligands and lone pairs	Hybridization	Geometry of hybrid	Bond angles
2	*sp*	linear	180°
3	sp^2	triangular	120°
4	sp^3	tetrahedral	109.5°
5	sp^3d	trigonal bipyramidal	120° (equatorial) 90° (axial)
6	sp^3d^2	octahedral	90°

Table 11-1 is quite similar to Table 10-2, the electron-pair geometries of VSEPR theory. The bonding in single-bonded species can be predicted with the following seven steps.

1. Draw the most plausible Lewis structure for the species.
2. For each central atom, count the total number of ligands and lone pairs.
3. Use VSEPR theory to determine the electron-pair geometry and the molecular shape.
4. Use the information in Table 11-2 to determine the hybridization of the central atom.
5. Write the ground-state orbital diagram of the central atom (subtracting or adding electrons to account for the charge, if any, of the species) and the orbital diagram for the hybridization found in step 4.
6. Write the ground-state orbital diagram for each ligand. The orbital diagrams may have to be modified if a coordinate covalent bond is to be formed. The ligands need not be hybridized unless they also are central atoms.
7. Overlap two half-filled orbitals (one from the ligand and one from the central atom) for each bond, and describe the result.

EXAMPLE 11-1 Describe the bonding in OF_2.

We follow the steps outlined above.

(1) |F̄—Ō—F̄|
(2) 2 ligands + 2 lone pairs = 4 electron pairs.
(3) tetrahedral electron-pair geometry; bent molecular shape.
(4) sp^3 hybridization on O.
(5) ground state O: [He] 2*s* [↑↓] 2*p* [↑↓][↑][↑]   hybridized O: [He] sp^3 [↑↓][↑↓][↑][↑]
(6) ground state F: [He] 2*s* [↑↓] 2*p* [↑↓][↑↓][↑]
(7) Oxygen is bound to each fluorine by the overlap of a half-filled sp^3 orbital on O with a half-filled 2*p* orbital on F. The two lone pairs of oxygen are in sp^3 hybrid orbitals. Writing the two fluorines as F_A and F_B to tell them apart, the bonding scheme can be abbreviated

$O(sp^3)^1—F_A(2p)^1$ $O(sp^3)^1—F_B(2p)^1$

EXAMPLE 11-2 Describe the bonding in ClF_3.

(1) 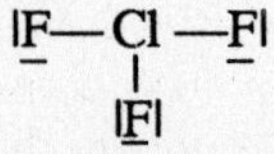

(2) 3 ligands + 2 lone pairs = 5 electron pairs.
(3) trigonal bipyramid electron-pair geometry; T-shaped molecular geometry.
(4) sp^3d hybridization
(5) ground state Cl: [Ne] 3*s* [↑↓] 3*p* [↑↓][↑↓][↑] 3*d* [ ][ ][ ][ ][ ]
sp^3d-hybridized Cl: [Ne] sp^3d [↑↓][↑↓][↑][↑][↑] 3*d* [ ][ ][ ][ ]
(6) ground state F: [He] 2*s* [↑↓] 2*p* [↑↓][↑↓][↑]

The two lone pairs of chlorine are in sp^3d orbitals. Chlorine is bonded to each fluorine by the overlap of a half-filled sp^3d hybrid orbital on Cl with a half-filled 2*p* orbital on F. Abbreviated:

$Cl(sp^3d)^1—F_a(2p)^1$ $Cl(sp^3d)^1—F_b(2p)^1$ $Cl(sp^3d)^1—F_c(2p)^1$

* **4. Describe multiple bonds between second period elements in terms of the overlap of *sp*, sp^2 and pure 2*p* orbitals to form σ bonds, and the sidewise overlap of *p* orbitals to form π bonds.**

A multiple bond consists of a σ bond and one or two π bonds. A π bond is formed by the side-to-side overlap of two *p* orbitals. σ bonds are formed by the following types of overlap: *s*-*s*, *s*-*p*, *p*-*p* (end-to-end), *s*-hybrid, *p*-hybrid, and hybrid-hybrid. σ bonds lie along the internuclear axis (the line joining the two atoms), whereas the two lobes of a π bond lie one either side of the axis. To achieve sidewise overlap, the p orbitals must, of course, be aligned. This means that the molecule is not free to rotate about a multiple bond as it is about a single bond. A π bond is weaker than a σ bond because its overlap is not as efficient. To account for π bonding, we add one more rule to the list first given in objective 11-4.

8. Overlap one half-filled *p* orbital on a ligand with one half-filled *p* orbital on the central atom to form each π bond.

EXAMPLE 11-3 Describe the bonding in ClNO.

(1) |C̲̅l—N̲=O̲| (2) 2 ligands + 1 lone pair = 3 electron pairs.
(3) triangular electron-pair geometry; bent molecular geometry. (4) sp^2 hybridization.
(5) ground state N: [He] 2*s* [↑↓] 2*p* [↑][↑][↑] sp^2-hybridized N: [He] sp^2 [↑↓][↑][↑] 2*p* [↑]
(6) ground state Cl: [Ne] 3*s* [↑↓] 3*p* [↑↓][↑↓][↑] ground state O: [He] 2*s* [↑↓] 2*p* [↑↓][↑][↑]
(7) Sigma bonding: $N(sp^2)2$: lone pair. $N(sp^2)^1$—$Cl(3p)^1$ $N(sp^2)^1$—$O(2p_y)^1$
(8) Pi bonding: $N(2p_z)^1$—$O(2p_z)^1$

The *y* and *z* subscripts on the 2*p* orbitals simply remind us not to use the same 2*p* orbital twice.

5. Propose plausible bonding schemes from Lewis structures or from experimental information about molecules (that is, bond lengths, bond angles, and so on).

To this point, we have used theoretical models (Lewis theory and VSEPR theory) to predict molecular structure. However, when experimental information disagrees with theory, we discard the theoretical model. H_2S is an example of the conflict between theory and experiment. VSEPR theory predicts a tetrahedral electron-pair geometry.

H—S̲̅—H [3]

This leads to sp^3 hybridization and a predicted H—S—H bond angle of 109.5°. The actual bond angle is 92°. It is very unlikely that the two lone pairs on S exert such a strong influence that they "close down" the bond angle from 109.5° to 92°. It is much more likely that hybridization does not occur and that the bonding is between half-filled 3*p* orbitals on S and half-filled 1*s* orbitals on H.

Do not become confused by this example of a conflict between theory and experiment. In practically all of the molecules and ions that you will encounter, VSEPR theory will predict the shape of the species. The information in Table 11-1 then can be used to specify the hybridization on each central atom.

6. Explain the fundamental basis of molecular orbital theory.

In molecular orbital theory, the atom centers are put in place and the electrons are "poured" around them. The electrons find the orbitals of lowest energy. These *molecular orbitals* extend over at least two atoms; they are not restricted to one atom as are atomic orbitals. Electrons in a bonding orbital spend most of their time between two nuclei and act rather like a "glue" holding the atoms together. Electrons in an antibonding orbital spend most of their time outside the region between the two nuclei. The nuclear charges are not screened from each other by the negative charge of the electrons and hence they repel each other.

7. Know that two atomic orbitals combine to form a bonding and an antibonding molecular orbital and sketch these molecular orbitals.

The bonding and antibonding molecular orbitals produced by the combination of two *s* atomic orbitals and of two *p* atomic orbitals are drawn in Figure 11-3. The asterisk (*) is used for an antibonding orbital and the *b* superscript (or sometimes no superscript) is used for a bonding orbital. The subscript (*s* or *p*) indicates the two atomic orbitals that were combined to form the molecular orbital.

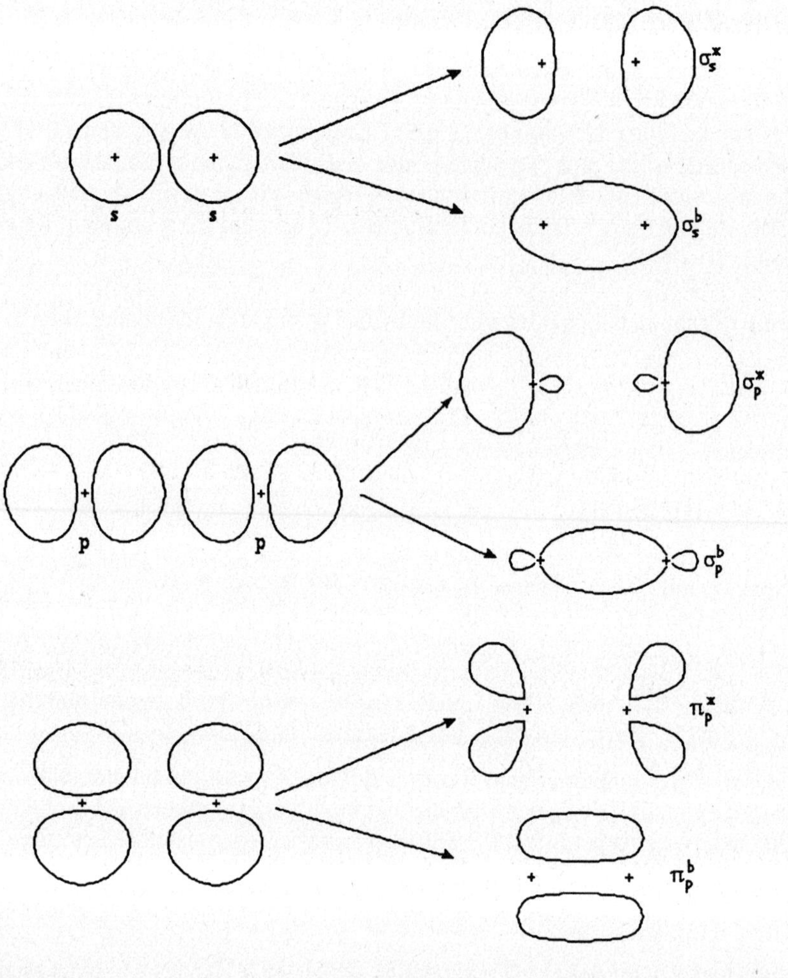

FIGURE 11-3 Molecular Orbitals Produced from *s* and *p* Atomic Orbitals

* **8. Assign probable electron configurations, determine bond orders, and predict magnetic properties of the diatomic molecules and ions of the first and second period elements.**

It is important to memorize the energy order of the molecular orbitals formed between two second period elements. This order is shown in the molecular orbital energy level diagrams of Figure 11-4 where it is applied to CN^-, BeC, and NO. Notice that the molecular orbitals are filled with valence electrons from lowest to highest energy. Electrons are not paired until all orbitals of the same energy are half filled. In this regard, filling molecular energy level diagrams obeys Hund's rule.

Molecular orbital theory defines bond order as follows.

Bond order = (number of bonding electrons − number of antibonding electrons) ÷ 2 [4]

Lewis theory and valence bond theory concepts equivalent to bond orders of 1, 2, and 3 are given in Table 11-2.

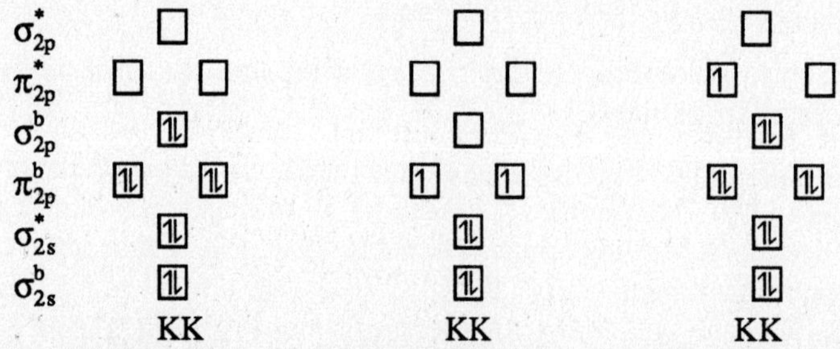

(a) CN^-, 10 v.e. (b) BeC, 6 v.e. (c) NO, 11 v.e.

FIGURE 11-4 Molecular Orbital Energy Level Diagrams for CN^-, BeC, and NO. (v.e. = valence electrons)

TABLE 11-2 Concepts Equivalent to Bond Order

Bond order	Lewis theory	Valence bond theory
1	single bond	one σ bond
2	double bond	one σ and one π bond
3	triple bond	one σ and two π bonds

EXAMPLE 11-5 Use the data of Figure 11-4 to determine the bond order of (a) CN^-, (b) BeC, (c) NO.

(a) CN^-: bond order = $(8 - 2) \div 2 = 3$
(b) BeC: bond order = $(4 - 2) \div 2 = 1$
(c) NO: bond order = $(8 - 3) \div 2 = 2.5$

A molecule with unpaired electrons is paramagnetic and one in which all electrons are paired is diamagnetic. Molecular orbital theory enables us to predict the number of unpaired electrons.

EXAMPLE 11-6 Use the data of Figure 11-4 to predict the number of unpaired electrons and the magnetic properties of (a) CN^-, (b) BeC, (c) NO.

(a) CN^-: all electrons are paired, diamagnetic
(b) BeC: two unpaired electrons, paramagnetic
(c) NO: one unpaired electron, paramagnetic.

When molecular orbital energy level diagrams are written on one line, they are called molecular orbital diagrams. These are more compact, and are shown in Figure 11-5.

CN^- KK σ^b_{2s} [↑↓] σ^*_{2s} [↑↓] π^b_{2p} [↑↓][↑↓] σ^b_{2p} [↑↓] π^*_{2p} [ ][ ] σ^*_{2p} [ ]
BeC KK σ^b_{2s} [↑↓] σ^*_{2s} [↑↓] π^b_{2p} [↑][↑] σ^b_{2p} [ ] π^*_{2p} [ ][ ] σ^*_{2p} [ ]
NO KK σ^b_{2s} [↑↓] σ^*_{2s} [↑↓] π^b_{2p} [↑↓][↑↓] σ^b_{2p} [↑↓] π^*_{2p} [↑][ ] σ^*_{2p} [ ]

FIGURE 11-5 Molecular Orbital Diagrams for CN^-, BeC, and NO

9. Describe the bonding in the benzene molecule (C_6H_6) through Lewis structures, valence bond theory, and molecular orbital theory.

The benzene molecule is a specific example of a resonance hybrid, which we consider in general in objectives 11-10 and 11-12. The delocalized *p* bonding consists of three filled bonding π orbitals and three empty antibonding π orbitals. The π bond order thus equals $(6 - 0) \div 2 = 3$, exactly as predicted by Lewis theory. The difference is that the π molecular orbitals are not localized between two carbon atoms but spread out over the entire molecule.

* **10. Explain why a single valence structure is inadequate to explain the bonding in a resonance hybrid.**

We considered resonance hybrids in objective 10-8. Here we extend these ideas by first considering the nitrite ion, NO_2^-, by the procedure outlined in objective 11-3. We write the two oxygens as O and Ø to distinguish them.

(1) [|O—N=Ø]⁻ ↔ [O=N—Ø|]⁻
resonance form a   resonance form b
(2) 2 ligands + 1 lone pair = 3 electron pairs.
(3) triangular electron-pair geometry; bent (or angular) molecular shape.
(4) sp^2 hybridization on N.
(5) ground state N: [He] 2*s* [↑↓] 2*p* [↑↓][↑][↑] sp^2-hybridized N: [He] sp^2 [↑↓][↑↓][↑] $2p_y$ [↑]
ground state O: [He] 2*s* [↑↓] 2*p* [↑↓][↑][↑]   modified O: [He] 2*s* [↑↓] 2*p* [↑↓][↑↓][ ]

The modified O is for a coordinate covalent bond; this is the single-bonded oxygen.

resonance form a	*resonance form b*
O: [He] 2*s* [↑↓] $2p_x$ [↑↓] $2p_y$ [↑↓] $2p_z$ [ ]	O: [He] 2*s* [↑↓] $2p_x$ [↑↓] $2p_y$ [↑] $2p_z$ [↑]
Ø: [He] 2*s* [↑↓] $2p_x$ [↑↓] $2p_y$ [↑] $2p_z$ [↑]	Ø: [He] 2*s* [↑↓] $2p_x$ [↑↓] $2p_y$ [↑↓] $2p_z$ [ ]

(7) Sigma bonding descriptions
resonance form *a*: $N(sp^2)^2$—$O(2p_z)^0$ $N(sp^2)^1$—$Ø(2p_z)^1$
resonance form *b*: $N(sp^2)^1$—$O(2p_z)^1$ $N(sp^2)^2$—$Ø(2p_z)^0$
Let us draw the $2p$ orbitals that are left, indicating the electrons in each with arrows. These are $2p_y$ orbitals. The two resonance forms are drawn in Figure 11-6. In resonance form *a* the π bond forms between N and Ø, and in resonance form *b* the π bond forms between N and O.
Thus we see that valence bond theory, strictly applied, still does not explain resonance structures. The problem is transferred from "lines between atoms" to pictures of *p* orbitals.

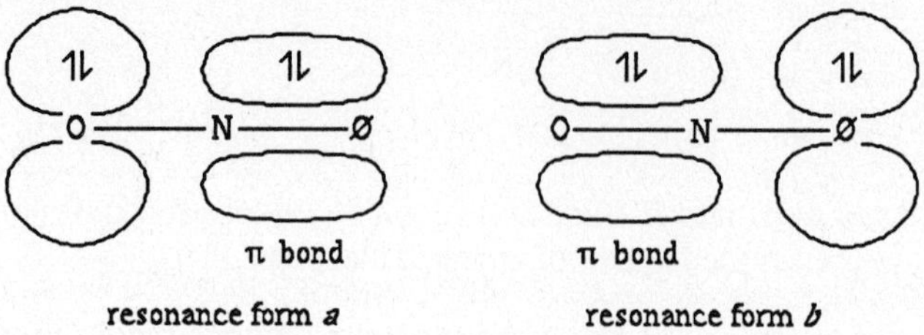

FIGURE 11-6 Valence Bond Pictures of the Two Resonance Forms of NO_2^-

11. Explain what is meant by a delocalized molecular orbital and how such orbitals are used to describe resonance hybrids.

Figure 11-6 does help us understand resonance hybrids. It reveals the very small difference between the two resonance structures for NO_2^-: the pairing of *p* electrons. Perhaps we should revise our concept of π bonding. Consider the following.

1. The electron is not a classical particle but has properties that require us to speak only of the probability of finding it in a small region. How can we demand that an electron remain in the small region of a π bond?
2. The kinetic energy of a particle is K.E. = $mv^2/2 = p^2/2m$, where $p = mv$. The deBroglie wavelength of a particle is $\lambda = h/p$. Thus $p = h/\lambda$ and K.E. = $h^2/2m\lambda^2$. The longer the wavelength of a particle, the lower will be its kinetic energy. The maximum wavelength of a particle is about the same size as the region in which it is confined. Thus, allowing electrons to move between more than two atoms should make a molecule of lower energy; one that is more stable. This is precisely what we have observed: a resonance hybrid is more stable than expected (that is, more stable than one resonance structure).

The extra stability possessed by a molecule due to resonance is known as the resonance stabilization energy. It occurs because the electrons are not restricted to one region; they are said to be *delocalized*. For NO_2^-, we would expect three π molecular orbitals extending over the entire molecule. There are three molecular orbitals because three $2p$ atomic orbitals contribute to the π bonding: one $2p_z$ orbital on each atom of the molecule ($2p_{z,O}$, $2p_{z,N}$, and $2p_{z,Ø}$). When molecular orbitals form, they usually divide into the same number of bonding as antibonding orbitals. If the original number of molecular orbitals is an odd number, there usually is a nonbonding molecular orbital. Hence, these three molecular orbitals in NO_2^- divide with one bonding, one nonbonding, and one antibonding. Since there are four π electrons in this ion, there are two π electrons in the lowest energy bonding MO and two in the nonbonding MO. The antibonding molecular orbital is empty. This yields a π bond order of 1, as depicted in the following molecular orbital energy level diagram for the π MO's of nitrite ion.

Antibonding π molecular orbital	π^*	□
Nonbonding π molecular orbital	π^n	⇅
Bonding π molecular orbital	π^b	⇅

FIGURE 11-7 Molecular Orbital Energy Level Diagram for the π Electrons of NO_2^-

12. State the special physical properties of metals and describe how the electron sea and band theories account for them.

The special physical properties of metals are that they (1) conduct electricity, (2) conduct heat, (3) are malleable and ductile, and (4) have luster. A metal's ability to conduct electricity (its conductance) decreases when the metal is heated or when it contains impurities. Both the electron sea and the band theories account for the ability to conduct heat and electricity and the ease of deformation by considering the valence electrons as relatively free to move throughout the crystal. The electrons conduct electricity and their vibrations conduct heat. Bonds between the centers of metal atoms break and re-form readily, accounting for the ease of deformation. The electron sea theory gives these electrons very little order. The band theory requires that they be in molecular orbitals extending over the entire crystal.

Luster is explained with the band theory by an electron absorbing light and going to a higher energy orbital. The light is re-emitted when the electron drops back down to a lower orbital. Because there are so many closely-spaced orbitals, all wavelengths of light are absorbed and re-emitted. Lower conductance at high temperatures is explained as being due to a disruption in the molecular orbitals because the atoms are vibrating (rather like walking across a stream on rocks during an earthquake). This also explains why small amount of impurities lower a metal's conductance (the rocks are at very different levels).

13. State how conductors, semiconductors, and insulators differ and explain how band theory explains these differences. Understand the meaning of *n*-type and *p*-type semiconductors and the impurities used to form each.

Conductors readily conduct an electrical current. Insulators do not unless a very high voltage is applied to them. Semiconductors conduct a current when only a moderate voltage is applied. The band theory explains these differences by noting that in insulators there is a wide energy gap (the band gap) between the full valence (or bonding) band and the empty conduction band. The band gap is small in semiconductors and there is no band gap (or a very small one) in conductors. A large band gap means that electrons must have a large voltage applied to promote them into the conduction band.

Insulators can be made into semiconductors by "doping" them with impurities. In one type, each impurity atom contains more valence electrons than does each atom of the insulator. Thus, due to the impurity there is a set of filled or partly filled orbitals within the band gap, as shown in Figure 11-8a. A moderate voltage excites these impurity electrons into the conduction band where they conduct a current. Thus a negative particle, the electron, conducts the current in *n*-type semiconductors.

In *p*-type semiconductors, each impurity atom contains fewer valence electrons than does each atom of the insulator. Hence, empty impurity orbitals are present in the band gap. A moderate voltage excites electrons from the full valence band into these empty orbitals, leaving "holes" behind in the valence band. These holes move in the opposite direction as the current flow. It seems as if the positive holes carry the current. A *p*-type semiconductor is shown in Figure 11-8b.

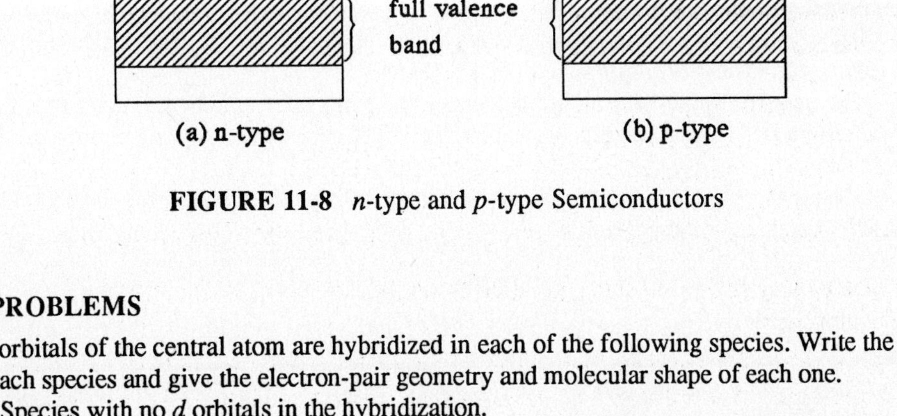

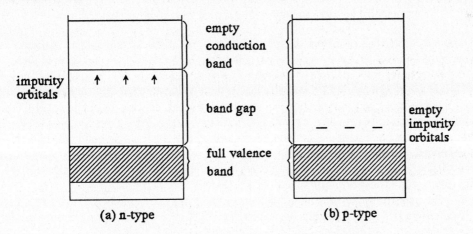

FIGURE 11-8 *n*-type and *p*-type Semiconductors

DRILL PROBLEMS

2. The orbitals of the central atom are hybridized in each of the following species. Write the hybridization scheme for each species and give the electron-pair geometry and molecular shape of each one.
   (1) Species with no *d* orbitals in the hybridization.

A. BeH_2	B. NH_4^+	C. $SnCl_2$	D. H_3O^+	E. PI_3
F. NH_2^-	G. SI_2	H. OF_2	I. BCl_3	J. CCl_2
K. CF_2Cl_2	L. BF_4^-	M. CH_3^-	N. SiH_4	O. CH_3^+

P. PCl_2^-

(2) Species with an expanded octet.

Q. SeF_2Cl_2 R. TeF_6 S. ClF_3 T. PCl_5 U. BrF_3
V. ArF_2 W. IF_5 X. ICl_2^- Y. $SbCl_5$ Z. I_3^-
Γ. PF_3Cl_2 Δ. BrF_5 Θ. KrF_4 Λ. $TeCl_4$ Ξ. $SbCl_5^{2-}$
Π. SF_4 Σ. ICl_4^- Υ. ICl_3 Φ. SeF_6 Ψ. XeF_4
Ω. $SnCl_6^{2-}$

3. Describe the bonding in each of the following species.

A. BeH_2 B. CH_3 C. KrF_4 D. BF_4^- E. $SnCl_2$
F. NH_4^+ G. ICl_4^- H. CH_3^+ I. SI_2 J. SeF_4
K. XeF_2 L. CCl_2 M. NH_2^- N. BrF_3 O. IF_5
P. ClO_2^- Q. BCl_3 R. $SbCl_5$ S. SeF_6 T. H_3O^+

4. Describe the bonding in each of the following species. The central atoms are underlined.

A. $N\underline{C}N^{2-}$ B. $H_2\underline{C}S$ C. $Cl_2\underline{CC}H_2$ D. $Cl_2\underline{C}S$ E. $S\underline{C}N^-$
F. $N\underline{N}O$ G. $\underline{C}O_2$ H. $H\underline{OC}N$ I. $H\underline{N}O$ J. $H\underline{C}N$
K. $H\underline{CC}H$ L. $\underline{N}O_2^+$ M. $\underline{C}OS$ N. $F\underline{P}O$ O. $H\underline{C}O\underline{O}H$

8. Write the simplified molecular orbital diagram of each of the following species. Based on that electron configuration, determine the number of (1) bonding and (2) antibonding electrons, (3) the bond order, the number of (4) σ and (5) π electrons, and (6) the number of unpaired electrons.

A. NO^+ B. O_2^{2-} C. N_2^- D. CN^+ E. BeO
F. BN G. Be_2^+ H. CN I. NeF J. F_2^-
K. O_2^- L. NO^- M. LiB

10. Give a complete description of the σ and π bonding in each of the following species. The central atom is underlined in each case.

A. $\underline{N}O_2$ B. $\underline{S}O_3$ C. $\underline{B}Cl_3$ D. $\underline{O}_3$ E. $H\underline{C}O_3^-$
F. $O\underline{C}F^-$ G. $\underline{C}O_3^{2-}$ H. $\underline{B}F_3$

QUIZZES (20 minutes each) Choose the best answer for each question.

Quiz A

1. The hybridization of N in NCl_2F is (a) *sp*; (b) sp^2; (c) sp^3; (d) not hybridized; (e) none of these.
2. The hybridization of P in PCl_5 is (a) not hybridized; (b) sp^3d^2; (c) sp^3d; (d) sp^3; (e) none of these.
3. Which of the following exhibits sp^2 hybridization? (a) $BeCl_2$; (b) BCl_3; (c) CCl_4; (d) NCl_3; (e) none of these.
4. Which of the following molecules has the shortest C—O bond length? (a) CO; (b) CO_2; (c) H_3COH; (d) H_2CO; (e) none of these.
5. A bond that has negligible electron density on the internuclear axis is a (a) single bond; (b) pi bond; (c) polar bond; (d) sigma bond; (e) none of these.
6. According to valence bond theory, bond strength is related to the degree of (a) orbital overlap; (b) covalent character; (c) resonance; (d) hybridization; (e) none of these.
7. A theory that can be used to explain why all bond distances and angles in a molecule such as methane are the same is (a) bond resonance; (b) delocalization of electrons; (c) bond polarities; (d) electronegativity; (e) bond hybridization.
8. Which of the following is predicted by molecular orbital theory to be paramagnetic? (a) CN^-; (b) CN^+; (c) NO^+; (d) LiB^{2+}; (e) none of these.

Quiz B

1. In the acetylene molecule, C_2H_2, the hybridization of each C atom is (a) *sp*; (b) sp^2; (c) sp^3; (d) d^2sp^3; (e) none of these.
2. Which of the following molecules displays sp^2 hybridization? (a) OF_2; (b) $F_2C{=}O$; (c) BeH_2; (d) NH_3; (e) none of these.
3. The hybridization of S in SF_4 is (a) not hybridized; (b) sp^3d; (c) sp^3d^2; (d) sp^3; (e) none of these.

4. The orientation in space (the geometry) of *sp*-hybridized orbitals is (a) perpendicular; (b) linear; (c) trigonal planar; (d) bent; (e) none of these.
5. If a double bond results from the sharing of four electrons, it consists of (a) 2 σ bonds; (a) 1 σ and 1 π bond; (c) 2 π bonds; (d) 1 π bond; (e) none of these.
6. Of the following, the molecule that has the shortest bond is (a) F_2; (b) N_2; (c) O_2; (d) HCl; (e) CO_2.
7. According to simple molecular orbital theory, the peroxide ion, O_2^{2-}, has (a) no unpaired electrons; (b) a bond order of 3/2; (c) its highest energy electrons in σ* orbitals; (d) no s electrons; (e) none of these.
8. The double covalent bond in ethylene, C_2H_4, (a) includes a lone pair of electrons on one of the carbon atoms; (b) is a sigma bond; (c) consists of one sigma and one pi bond; (d) is free to rotate because of delocalized electrons; (e) has a low electron density.

Quiz C

1. Which of the following molecules contains an sp^2-sp^2 sigma bond? (a) C_2H_4; (b) C_2H_6; (c) C_2H_2; (d) HCN; (e) CH_3OH.
2. The hybridization of central N in N_2O is (a) not hybridized; (b) *sp*; (c) sp^2; (d) sp^3; (e) none of these.
3. The hybridization of Xe in XeF_4 is (a) not hybridized; (b) sp^3; (c) dsp^3; (d) sp^3d^2; (e) none of these.
4. If the wave functions describing the 2*s* and two of the 2*p* orbitals are combined, the identical orbitals obtained form bond angles of (a) 90°; (b) 120°; (c) 180°; (d) 109.5°; (e) none of these.
5. A pi bond is (a) formed from two *s* orbitals; (b) formed from two *p* orbitals; (c) formed from an *s* and a *p* orbital; (d) stronger than a sigma bond; (e) none of these.
6. Generally, short bonds are (a) between highly electronegative atoms; (b) strong bonds; (c) polar bonds; (d) ionic bonds; (e) none of these.
7. The concept of an antibonding orbital is unique to the (a) valence bond theory; (b) theory of bond hybridization; (c) concept of resonance; (d) molecular orbital theory; (e) electron-pair repulsion theory.
8. A diatomic molecule has 15 electrons (including KK). Its bond order is (a) 2.5; (b) 2.0; (c) 1.5; (d) 1.0; (e) none of these.

Quiz D

1. Which of the following exhibits *sp* hybridization? (a) CO_2; (b) CCl_4; (c) H_2S; (d) NH_3; (e) none of these.
2. The hybridization of Cl in the perchlorate anion is (a) sp^3; (b) sp^2; (c) dsp^3; (d) not hybridized; (e) none of these.
3. The hybridization of Br in BrF_3 is (a) not hybridized; (b) sp^3d^2; (c) sp^3d; (d) sp^3; (e) none of these.
4. Which of the following does not use hybridized orbitals in its bonding? (a) CH_4; (b) CO_2; (c) H_2O; (d) HCl; (e) none of these.
5. A σ bond (a) cannot be formed by two *p* orbitals; (b) is weaker than a π bond; (c) has high electron probability between nuclei; (d) can only be formed from *s* orbitals; (e) none of these.
6. The shorter the internuclear distance between two bonded atoms, the (a) less stable the bond; (b) less multiple bonding character there is; (c) more energy is needed to break the bond; (d) more resonance there is; (e) none of these.
7. A pi bond is (a) stronger than a sigma bond; (b) concentrated along the bond axis; (c) a concept of bond hybridization theory only; (d) formed by the sidewise overlap of *p* orbitals; (e) none of these.
8. According to molecular orbital theory, NO has (not counting KK) (a) a bond order of $^3/_2$; (b) 4 antibonding electrons; (c) 6 sigma electrons; (d) 3 pi electrons; (e) none of these.

SAMPLE TEST (25 minutes)

1. Consider a molecule with the Lewis structure given below. For the central atoms C^a, C^b, and O, list the following.

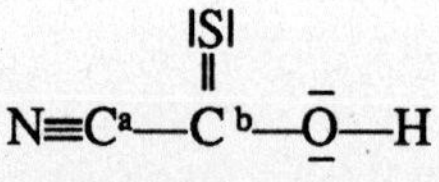

   a. Number of ligands
   b. Number of lone pairs
   c. Electron-pair geometry
   d. Molecular shape
   e. Orbitals in which the lone pairs (if any) are located
   f. Hybridization (if any)

   Now describe the bonding in the molecule by specifying which orbitals on each atom overlap to form the bonds in the molecule. Describe each bond separately.

   g. sigma bonds   h. pi bonds

   What are the (ideal) values of the following bond angles?

   i. N—C^a—C^b j. C^a—C^b—S
   k. C^a—C^b—O   l. C^b—O—H

2. Use the principles of molecular orbital theory to answer the following.
   a. What is the bond order of NO?
   b. Which of the ions, NO^+ or O_2^+, is more stable? Why?
   c. How many unpaired electrons are there in Ne_2^+, OF^+, CO?

12 Liquids, Solids, and Intermolecular Forces

CHAPTER OBJECTIVES

1. Explain surface tension and describe several phenomena based on it.

Energy is given off when atoms are attracted to each other. Forces of attraction exist between molecules and even between individual unbonded atoms. A molecule will have low energy if it is attracted to many other molecules. It then is more stable than a free molecule. A molecule in the body of a liquid is surrounded by and hence attracted to more molecules than is a molecule on the surface. We say that the surface molecule is less stable. A liquid thus becomes as stable as possible by decreasing its surface area. Hence, it resists attempts to increase its surface area and will form drops. When forces to another substance (adhesive forces) are stronger than forces within the liquid (cohesive forces), the liquid will wet that surface. Such a liquid also will rise in a capillary tube in a vain attempt to decrease the extra surface created by the liquid rising to meet the wall of the tube.

* **2. Describe vaporization, including its enthalpy, ΔH_{vap}, using the latter in calculations.**

Adding energy to a liquid causes some of the molecules to leave the surface of the liquid; they vaporize. The energy needed to vaporize a mole of the liquid is the molar enthalpy of vaporization, ΔH_{vap}. Vaporization always is an endothermic process and thus ΔH_{vap} always is positive. Molar enthalpies of vaporization of some common compounds are given in Table 12-1.

EXAMPLE 12-1 236.2 kJ of heat is added to 150. g of liquid water at 75.0°C. What is the final temperature and how much water remains as liquid?

First we determine the heat needed to raise the temperature of the water to 100.0°C.

heat = (moles)(specific heat)(temperature change)

$$= \left(150.\text{ g} \times \frac{\text{mol } H_2O}{18.0\text{ g}}\right) \times 98.7\text{ J mol}^{-1}\text{ K}^{-1} \times (100.0°\text{C} - 75.0°\text{C})$$

$$= 20.6 \times 10^3\text{ J} = 20.6\text{ kJ}$$

The heat remaining (236.2 kJ – 20.6 kJ = 215.6 kJ) is used to vaporize the water. We now determine the quantity of water that vaporizes.

$$n_{vap} = 215.6\text{ kJ} \times \frac{\text{mol } H_2O}{41.4\text{ kJ}} = 5.21\text{ mol } H_2O \times \frac{18.0\text{ g}}{\text{mol } H_2O} = 93.8\text{ g } H_2O$$

And then determine the mass of liquid water remaining, m_l.

$$m_l = 150.\text{ g liquid} - 93.8\text{ g vapor} = 56\text{ g liquid water remaining}$$

3. Explain what occurs when a liquid vaporizes in a closed container: the dynamic equilibrium between a liquid and its vapor.

When a liquid is placed in an open container, eventually it all vaporizes. If the container is closed, the vapor molecules cannot escape. The concentration of molecules in the vapor increases. It becomes more likely that some of these vapor molecules will condense. Soon molecules condense from the vapor just as rapidly as they vaporize; the vapor is saturated. From the ideal gas law ($PV = nRT$ or $P = RT \times n/V$), we see that the vapor concentration

(n/V) is proportional to its pressure. The pressure of the saturated vapor is known as the liquid's vapor pressure. When the vapor is saturated, the two processes of vaporization and condensation occur at the same speed.

rate of vaporization = rate of condensation [1]

The rate of vaporization depends on three factors: temperature, forces between molecules, and the surface area of the liquid. Increasing temperature causes liquid to vaporize more rapidly. Vaporization also occurs more rapidly from a large surface area than from a small one. Increasing the forces between molecules slows down the rate of vaporization. The rate of condensation depends on the pressure of the vapor and on the surface area. High pressure and large surface area both increase the rate of condensation. In the first case, there are more molecules present in the vapor that can condense and in the second case, the area that the molecules must hit in order to condense is larger.

TABLE 12-1 Heat Capacities and Molar Heats of Vaporization of Some Common Compounds

Formula	Name	ΔH_{vap}, kJ/mol	Heat capacities, J mol^{-1} K^{-1} Liquid	Vapor	Normal b. p., °C
NH_3	ammonia	31.2	76.1	44.4	−33.35
SO_2	sulfur dioxide	26.8	85.2	39.9	−10.0
HCN	hydrogen cyanide	30.7	70.7	36.0	26.0
$(C_2H_5)_2O$	diethyl ether	29.1	172.	108.	34.6
CS_2	carbon disulfide	28.4	76.1	45.6	46.3
$(CH_3)_2CO$	acetone	32.0	126.	75.3	56.2
Br_2	bromine	33.9	75.7	36.1	58.78
$CHCl_3$	chloroform	31.4	115.	65.7	61.2
CH_3OH	methanol	39.2	81.6	43.9	64.96
CCl_4	carbon tetrachloride	34.4	133.	83.3	76.8
C_2H_5OH	ethanol	40.5	113.	65.7	78.5
H_2O	water	44.0	75.3	36.	100.0
HCOOH	formic acid	41.4	98.7	45.2	100.7
CH_3COOH	acetic acid	39.7	123.	66.9	118.5

* **4. Determine vapor pressure experimentally, estimate values from tables or graphs, and predict whether vapor and/or liquid is present under specific conditions.**

In the transpiration method, a known volume of an inert gas is bubbled through a weighed liquid. Both liquid and gas are at the same temperature. The gas that leaves the liquid is saturated. The liquid then is reweighed. The loss in liquid mass divided by the liquid's mole weight is the number of moles of liquid (n_l) that has vaporized.

$$n_l = \frac{\text{liquid mass before bubbling} - \text{liquid mass after bubbling}}{\text{mole weight of liquid}} \quad [2]$$

The final mixture of gases is assumed to occupy the same volume as the inert gas does initially; only the pressure changes. Since both the temperature and the volume of the gas are known, the vapor pressure of the liquid (P_l) can be determined from the ideal gas law.

$$P_l = n_l RT/V \quad [3]$$

EXAMPLE 12-2 4.87 L of He is bubbled through isopropanol, $CH_3CHOHCH_3$, at 312.7 K. 1.50 g of liquid vaporizes. What is the vapor pressure of isopropanol (60.0 g/mol)at this temperature?

$$P_l = \frac{(1.50\ \text{g} \div 60.0\ \text{g/mol})(0.0821\ \text{L atm mol}^{-1}\ \text{K}^{-1})(312.7\ \text{K})}{4.87\ \text{L}} = 0.132\ \text{atm}$$

Determining the amount of liquid and vapor present in a container of known volume and temperature involves applying the ideal gas law to determine the moles of vapor present. We use the vapor pressure of the liquid as the pressure and the known volume and temperature. Thus, we can determine the number of moles of vapor. Any substance left over must be liquid and is assumed to be of negligible volume.

EXAMPLE 12-3 4.00 g of isopropanol is placed in a 10.0-L container at 312.7 K. What mass of liquid is present after equilibrium is achieved?

First determine the mass of isopropanol present as vapor.

$$m_{vap} = \frac{PV\mathcal{M}}{RT} = \frac{(0.132 \text{ atm})(10.0 \text{ L})(60.0 \text{ g/mol})}{(0.0821 \text{ L atm mol}^{-1} \text{ K}^{-1})(312.7 \text{ K})} = 3.08 \text{ g vapor}$$

Then the mass of liquid is determined by difference.

$$m_{liq} = 4.00 \text{ g total} - 3.08 \text{ g vapor} = 0.98 \text{ g liquid}$$

* **5. Explain how vapor pressure and temperature are related and the meaning of boiling. Also, be able to use the Clausius-Clapeyron equation.**

As the temperature increases, the vaporization rate increases and so does the vapor pressure. Vapor pressure is often tabulated or graphed against pressure. When the vapor pressure of the liquid reaches atmospheric pressure, bubbles of vapor form within the liquid and the liquid boils. This occurs at the boiling point, called the normal boiling point if the external pressure is 1.000 atm.

The Clausius-Clapeyron equation [4] is important for two reasons. First, it permits us to predict the vapor pressure of a liquid at several different temperatures. Second, it has the same form as two other important equations: the Arrhenius equation [5], which predicts the rate constant of a reaction at different temperatures (see objective 15-12), and the van't Hoff equation [6], which predicts the equilibrium constant of a chemical reaction at different temperatures (see objective 20-13).

$$\ln\frac{P_2}{P_1} = \frac{-\Delta H_{vap}}{R}\left(\frac{1}{T_2} - \frac{1}{T_1}\right) = 2.303 \log\frac{P_2}{P_1} \quad [4]$$

$$\ln\frac{k_2}{k_1} = \frac{-E_a}{R}\left(\frac{1}{T_2} - \frac{1}{T_1}\right) \quad [5]$$

$$\ln\frac{K_2}{K_1} = \frac{-\Delta H_{rxn}}{R}\left(\frac{1}{T_2} - \frac{1}{T_1}\right) \quad [6]$$

Thus, once we know how to use the Clausius-Clapeyron equation, we also will know how to use the other two. (Equations [5] and [6] are given just to show you the similarity of form. You do not have to learn them now.) There are five variables in the Clausius-Clapeyron equation: ΔH_{vap}, P_2, P_1, T_2, and T_1.

EXAMPLE 12-4 The vapor pressure of isopropyl alcohol is 100 mmHg at 39.5°C and 400 mmHg at 67.8°C. What is ΔH_{vap} of isopropyl alcohol?

Remember that temperature *must* be in kelvins.

$P_1 = 100$ mmHg, $T_1 = 312.7$ K, $P_2 = 400$ mmHg, $T_2 = 341.0$ K

$$\ln\frac{P_2}{P_1} = \frac{-\Delta H_{vap}}{R}\left(\frac{1}{T_2} - \frac{1}{T_1}\right)$$

$$\ln\frac{400 \text{ mmHg}}{100 \text{ mmHg}} = \frac{-\Delta H_{vap}}{8.314 \text{ J mol}^{-1} \text{ K}^{-1}}\left(\frac{1}{341.0 \text{ K}} - \frac{1}{312.7 \text{ K}}\right)$$

$\Delta H_{vap} = 43{,}430$ J/mol

EXAMPLE 12-5 What is the normal boiling point of isopropyl alcohol?

$P_1 = 100$ mmHg, $T_1 = 312.7$ K, $P_2 = 760$ mmHg, $T_2 = ?$ K

$$\ln\frac{760 \text{ mmHg}}{100 \text{ mmHg}} = \frac{43430 \text{ J mol}^{-1}}{8.314 \text{ J mol}^{-1} \text{ K}^{-1}}\left(\frac{1}{T_2} - \frac{1}{312.7 \text{ K}}\right)$$

$T_2 = 355.9$ K $= 82.7$°C

6. Describe the significance of the critical point.

Above the critical temperature a gas will not condense to a liquid no matter how much pressure is exerted. It will get denser and may become as dense as a liquid, but it will never show a meniscus. Similarly, above the critical pressure the familiar processes of condensation and vaporization do not occur not matter how the temperature changes. The critical temperature represents a very high kinetic energy of the molecules. Their speeds are so large that short-range intermolecular forces are not strong enough to hold molecules together.

* 7. Understand terms that apply to phase changes of solids.

The melting point and the freezing point are the same temperature. Melting also is called fusion and thus the molar heat absorbed on melting is called the heat of fusion, ΔH_{fus}. Solidification, the reverse of fusion, has an associated molar heat of solidification, ΔH_{sol}. ΔH_{sol} is negative since solidification is exothermic.

$$\Delta H_{fus} = -\Delta H_{sol} \qquad [7]$$

When a solid sublimes, it vaporizes without first becoming a liquid. The process of sublimation is the reverse of deposition (which is gas to solid). The molar enthalpy of sublimation, ΔH_{sub}, is positive; that is, sublimation is an endothermic process.

$$\Delta H_{dep} = -\Delta H_{sub} \qquad [8]$$

Sublimation also accomplishes the same phase change as melting and vaporization combined.

$$\Delta H_{sub} = \Delta H_{fus} + \Delta H_{vap} \qquad [9]$$

$$\Delta H_{dep} = \Delta H_{sol} + \Delta H_{cond} \qquad [10]$$

* 8. Interpret simple phase diagrams and use phase diagrams to predict changes that occur as a substance is heated, cooled, or undergoes a change in pressure.

A phase diagram summarizes the changes in phase of a given substance. The liquid-vapor curve is determined by the Clausius-Clapeyron equation. The points on the sublimation curve (the curve between solid and vapor) also come from the Clausius-Clapeyron equation, if ΔH_{sub} is used for ΔH_{vap}. The slope of the fusion line is given by equation [11].

$$\frac{\Delta P}{\Delta T} = \frac{\Delta H_{fus}}{T\,\Delta V_{fus}} \qquad [11]$$

where ΔV_{fus} is the volume increase of one mole on melting. The line between solid and liquid is straight enough to draw with a ruler. This line—the fusion curve—usually slopes upward and to the right (it has a positive slope) since ΔV_{fus} generally is positive. (Of course, ΔH_{fus} always is positive; fusion is an endothermic process.) In the case of water, however, ΔV_{fus} is negative (ice floats on water) and the fusion curve slopes upward and to the left (it has a negative slope). The point where the vapor pressure, sublimation, and fusion curves meet is the triple point where solid, liquid, and vapor exist together. You should remember that by convention, pressure is plotted vertically and temperature is plotted horizontally.

* 9. Describe the difference between intra- and intermolecular forces, distinguish the different types of forces between molecules, and explain how these forces influence molecular properties.

Intermolecular forces are those that exist between molecules, whereas intramolecular forces hold a molecule together. Strong intermolecular forces make a substance difficult to break apart, so it has relatively high melting and boiling points, heats of vaporization and fusion, and surface tension. Intermolecular forces are also called van der Waals forces and are of three types, of which hydrogen bonds (considered in objective 12-10) are the strongest.

The second type, and the weakest, is a dipole-dipole attraction. The negative end of one dipole is attracted to the positive end of another. The more polar are the molecules involved, the stronger are their forces. Dipole-dipole forces are relatively weak, ranging from 0.01 to 0.2 kJ/mol.

The third type of force occurs between an instantaneous dipole in a normally nonpolar molecule and the dipole that the instantaneous dipole creates or induces in another nonpolar molecule. These forces are called London forces or exchange forces. Temporary dipoles are more easily created in large molecules, especially those having atoms of high atomic number. In these atoms, the outer electrons are far from the nucleus and not held very firmly. Thus, large atoms are said to be very polarizable. This means that London forces are strong for molecules of high atomic weight. They ra ge from 0.4 to 10 kJ/mol.

10. State conditions that lead to hydrogen bonds, explain how hydrogen bonds differ from other types of intermolecular forces, and describe the effect of hydrogen bonds on physical properties.

Hydrogen bonds are formed when a hydrogen atom is bound to a F, O, or N atom and attracted to another F, O, or N atom. A Cl or S atom may substitute for either F, O, or N atom but the resulting hydrogen bond is not as strong. One explanation for H bonds is that the high electronegativity of F, O, or N pulls so much of the bond pair of electrons toward it that the hydrogen nucleus is almost a bare proton. Thus, a very intense dipole is created. It is doubtful that this is the full explanation, for the following reasons.

1. The hydrogen bond is very strong, ranging from 15 to 40 kJ/mol. Strengths of other intermolecular forces rarely exceed 10 kJ/mol.
2. It is common to see a fixed limit on the number of hydrogen bonds formed. For example, two hydrogen bonds form to oxygen in water.
3. Hydrogen bonds have a reasonably well defined length, 2.74 Å between oxygens in ice. Dipole-dipole attractions normally show a range of lengths.

The required distance and number of hydrogen bonds make the structure of ice more open than that of liquid water. Thus ice floats on water. The strength of the hydrogen bond results in unusually high melting and boiling points and heats of vaporization for substances comprised of small molecules.

11. Describe aspects of network covalent bonding: the conditions that lead to it, the properties of solids in which it occurs, and some examples of these solids.

A network covalent solid is held together by covalent bonds. In that respect, a crystal of such a solid is a single molecule. Covalent bonds range from 150 to 900 kJ/mol. Network covalent solids are formed between the nonmetals and the metalloids of families 3A, 4A, 5A, and 6A. Metals are held together by metallic bonds, described in objective 11-12. It is very difficult to break the strong covalent bonds that hold network covalent solids together. These solids have very high melting points (2000°C is not uncommon) and heats of fusion. They are very hard and make good abrasives. But they are brittle since once the bonds are broken they do not readily re-form.

*** 12. Predict relative lattice energies of ionic compounds and relate their magnitudes to the physical properties of those compounds (melting point, molar enthalpy of vaporization, and solubility in water).**

The lattice energy of an ionic compound is the energy of the reaction in which gaseous cations and gaseous anions unite to form one mole of the solid compound. Since this reaction releases energy, lattice energies are invariably exothermic (that is, they have a negative sign). Lattice energies can be determined from experimental data with the Born-Fajans-Haber cycle, as explained in objective 10-4. As also explained there, lattice energies also can be calculated from information obtained from the crystal structure of the compound. The calculations reveal that the lattice energy depends on terms that look like $Z_A Z_C/(r_A + r_C)$ where Z_A and Z_C are the anion and cation charge (such as −1, −2, +1, and +3) and r_A and r_C are the anion and cation radii. Thus, a compound that contains small, highly charged ions will have a large large lattice energy.

A large lattice energy means that the ions are firmly bound into the crystal. Hence, the crystal will be difficult to disrupt and will have a high melting point and a large heat of fusion. Because the ions also must be separated when the compound dissolves in water, we also would expect a large lattice energy to predict a low solubility in water. Here there are some exceptions because solubility also depends on the attractions between the ions and the water molecules. These ion-molecule attractions also are strong for small, highly charged molecules. Thus, although it may require a great deal of energy to separate the ions, most of that energy (if not more) may be released when the ions are hydrated (surrounded by water molecules).

13. Describe the three packing models for spheres and how x rays determine crystal structures.

The three ways to arrange atoms in solid metals are cubic close packed (also known as *abc*), hexagonal close packed (*aba*), and body-centered cubic. It helps a great deal to make models to see these structures. You can use Styrofoam balls touching each other and held together with glue. To see that the *abc* packing really is face-centered cubic, find three atoms on three different layers that touch each other and lie along a straight line. This is the diagonal of a face (corner, face center, corner). From this start you should be able to see the rest of the face-centered cube. In Figure 12-1, the atoms of a face-centered cube are labeled with the letters of the layer (*a*, *b*, or *c*) each is in.

$$n\lambda = 2d \sin\theta \qquad [12]$$

The Bragg equation [12] relates the spacing between layers of atoms (d) to the angle (θ) at which x rays of a certain wavelength (λ) are diffracted from the crystal. There are two types of layers in easily seen in Figure 12-1, the layers parallel to each face, and the *b* or *c* atom layers. Yet there are many other types of layers as well. These are hard to see in three dimensions unless you have a very good model. But you can see an arrangement of many rows in any orderly pattern: seat backs in a theater or tombstones in a military cemetery. X rays bounce off planes of atoms somewhat as light is reflected from a mirror. A plane of atoms is a good "mirror" for x rays if it contains many electrons. Thus if there are many atoms in a plane or if the atoms have high atomic numbers, the reflected x-ray beam will be strong. The spacings between layers and the strengths of the reflected x-ray beams are analyzed by computer to determine the structure of the crystal.

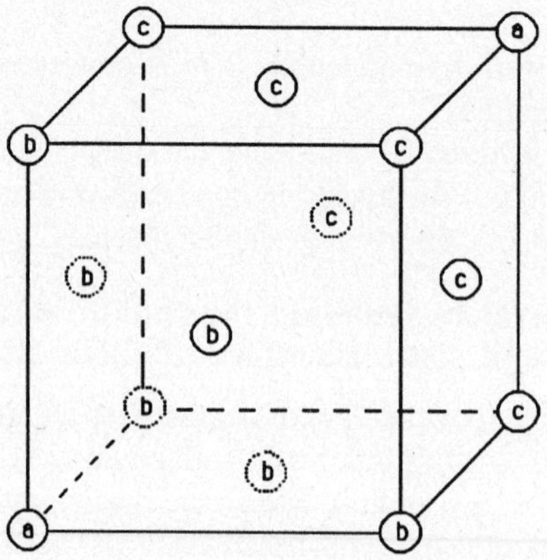

FIGURE 12-1 Close-packing Face-centered Cube

14. Explain what a unit cell is, use its dimensions in calculations, and analyze unit cells to determine crystal coordination numbers and chemical formulas.

FIGURE 12-2 Wall Paper Unit Cells and Formula Units

The smallest part of a crystal that will produce the entire crystal by simple repetition is called the unit cell. You can see two-dimensional unit cells in a wallpaper pattern (Figure 12-2). The dashed lines outline a "formula unit." Four of these formula units are needed in a unit cell since we cannot turn a unit cell around to repeat the pattern. The solid lines outline a unit cell. Calculations involving unit cells are straightforward if you have a clear picture of which atoms are touching. In a face-centered cubic unit cell, the atoms touch along the face diagonal. In a body-centered cubic unit cell, the atoms touch along the body diagonal. And in a simple cubic unit cell, the atoms touch along the edge. These ideas, along with a knowledge of geometry, allow one to use unit cell dimensions to calculate atomic radii. In addition, one can readily establish crystal coordination numbers in the same fashion, since one knows and can count all of the ions of opposite charge touching a given ion.

In order to determine den ities and chemical formulas, one should know how many unit cells "share" atoms on the outside. For a cubic unit cell, a face atom is shared by two unit cells, an edge atom by four, and a corner atom by eight.

15. Explain how liquid crystals differ from liquids and solids, and describe the basis for their use in various practical devices.

Liquid crystals flow as do liquids but reflect light in much the same way as crystalline solids. They are composed of long slender molecules that all are oriented in the same direction. This sameness of orientation, similar to spaghetti in a box, accounts for the optical properties. The freedom of the molecules to slide back and forth, over and under each other accounts for the fluid properties. The orientation is modified, and in some cases disrupted, by changes in temperature or pressure or by electrical fields. Hence, liquid crystals can be used in devices to detect very small variations of temperature and are employed to display images by varying the electrical field on a surface covered with a liquid crystal.

DRILL PROBLEMS

2. In each line below, the specified mass (m, g) of the indicated compound is heated from the initial temperature (t_i, °C) to the final temperature (t_f, °C) with the given amount of heat (q, kJ). Occasionally liquid and vapor are present together at the boiling point, in which case the mass of the liquid is given (m_l, g). Fill in the blanks in the table below. Data are given in Table 12-1.

Compound	*m*, g	t_i, °C	t_f, °C	*q*, kJ	m_l, g
A. methanol	200.	50.0	____	184.7	____
B. ethanol	204	20.2	100.0	____	0.00
C. sulfur dioxide	186	−51.0	−10.0	____	46
D. hydrogen cyanide	____	−14.0	36.5	20.20	0.00
E. acetic acid	400.	50.0	____	204	____
F. carbon disulfide	300.	25.0	90.	____	____
G. diethyl ether	55.0	25.0	____	1.43	____
H. carbon tetrachloride	____	25.0	76.8	16.4	14
I. chloroform	314	0.00	____	14.3	____
J. formic acid	87.0	41.0	126.0	____	0.00
K. acetone	435	8.2	56.2	____	100.
L. bromine	124	20.40	____	____	24
M. ammonia	100.	−52.6	____	8.17	____

4. (1) In each of the following transpiration experiments, a given volume of noble gas (V) is passed through a liquid at temperature t. The liquid loses m grams of liquid as vapor. Determine the vapor pressure in mmHg of each liquid.

Liquid	*V*, L	*t*, °C	*m*, g	*Liquid*	*V*, L	*t*, °C	*m*, g
A. acetone	214	−32.0	7.44	B. formic acid	86.7	20.0	7.64
C. diethyl ether	9.52	−10.0	4.94	D. chloroform	15.2	0.0	4.79
E. bromine	10.7	8.0	9.82	F. ethanol	83.1	−5.0	1.83
G. methanol	98.3	−43.0	0.241				

(2) Use the data of *Figure 12-11 in the text* to help determine how much liquid and how much vapor of each substance is present in each container of volume V and at temperature T.

Substance	*m*, g	*V*, L	*t*, °C	*Substance*	*m*, g	*V*, L	*t*, °C
H. diethyl ether	25.0	10.0	20	I. benzene	10.0	5.2	40
J. water	2.14	11.6	80	K. aniline	5.79	8.2	120
L. toluene	14.2	7.3	100	M. benzene	12.0	4.9	60
N. diethyl ether	12.0	6.1	30				

5. (1) The vapor pressure of several liquids at two temperatures follow. Determine the molar heat of vaporization of each liquid and its normal boiling point.

Liquid	P_1, mmHg	t_1, °C	P_2, mmHg	t_2, °C
A. BrF_5	10.0	−41.9	400.	+25.7
B. $COCl_2$	40.0	−50.3	100.	−35.6
C. H_2O_2	1.00	15.3	40.0	77.0
D. Si_2Cl_6	10.0	38.8	100	85.4
E. SO_2Cl_2	40.0	−1.0	100	17.8
F. SiH_3F	10.0	−141.2	400	−106.8
G. Rb	1.00	297	400	620

(2) Use the data of Table 12-1 to determine the vapor pressure at the given temperature, and the temperature needed to attain the given vapor pressure for each of the following liquids.

H. methanol, $t = 24.2°C, P = 15$ mmHg
I. bromine, $t = 75.8°C, P = 250$ mmHg
J. acetone, $t = 15.0°C, P = 30$ mmHg
K. $CHCl_3$, $t = 37.0°C, P = 300$ mmHg
L. SO_2, $t = -50.0°C, P = 500$ mmHg
M. HCOOH, $t = 25.0°C, P = 50$ mmHg
N. NH_3, $t = -50.0°C, P = 180$ mmHg
O. ethanol, $t = 50.0°C, P = 200$ mmHg

7. Fill in the blanks in the table below. All enthalpies are in kJ/mol

	Substance	ΔH_{fus}	ΔH_{vap}	ΔH_{sub}	ΔH_{sol}	ΔH_{cond}	ΔH_{dep}
A.	H_2O	6.01	44.0	____	____	____	____
B.	In	____	____	____	____	−232.8	−243.5
C.	LiI	____	176.9	____	____	____	−204.1
D.	NaCl	____	181.8	229.0	____	____	____
E.	LiBr	20.1	____	____	____	−158.6	____
F.	HCOOH	____	41.4	60.4	____	____	____
G.	BeO	____	____	725.1	−56.6	____	____

8. A. Given the following data for hydrogen, roughly sketch its phase diagram and label all regions. Solid vapor pressure at 10 K: 0.001 atm; normal melting point: 14.01 K; normal boiling point: 20.38 K; triple point: 13.97 K, 0.07 atm; critical point: 33.3 K, 12.8 atm.

B. Does H_2 freeze at 9.0 atm pressure? If so, what is the freezing point? If not, why not?
C. Does H_2 boil at 14.0 atm pressure? If so, what is the boiling point? If not, why not?
D. Does solid H_2 float on the liquids? Why or why not?
E. For hypothetical compound W, the vapor pressures of liquid and solid are as follows.

t, °C (solid)	−10	10	30	50	
P, atm (solid)	0.10	0.20	0.30	0.40	
t, °C (liquid)	50	150	200	250	300
P, atm (liquid)	0.40	0.64	0.76	0.88	1.00

The melting point of the solid varies with pressure as follows.

P, atm	0.40	1.00	1.60	2.20
t, °C	50	49.5	49.0	48.5

Above $t = 300°C$, liquid and vapor are indistinguishable. Sketch the phase diagram of compound W and label all points and regions.

F. Does compound W freeze at 9.00 atm pressure? If so, what is the freezing point? If not, why not?
G. Does compound W boil at 0.80 atm pressure? If so, estimate the boiling point. If not why not?
H. Does compound W solid float on the liquid? Why or why not?

Refer to *Figures 12-19 and 12-20 in the text* and the two phase diagrams you have just drawn in questions A. and E. to describe the phases changes that occur during each of the following changes in pressure or temperature. Give the approximate temperature and pressure at which each phase change occurs.

I. H_2O $P_1 = 2$ mmHg to $P_2 = 100$ mmHg at $T = 273.155$ K (fixed)
J. H_2O $T_1 = 253$ K to $T_2 = 423$ K at $P = 3.00$ atm (fixed)
K. CO_2 $T_1 = 194.7$ K to $T_2 = 473.2$ K at $P = 2.00$ atm (fixed)
L. CO_2 $T_1 = 194.7$ K to $T_2 = 473.2$ K at $P = 6.40$ atm (fixed)
M. CO_2 $T_1 = 173.2$ K to $T_2 = 555.1$ K at $P = 80.0$ atm (fixed)
N. H_2 $T_1 = 5.0$ K to $T_2 = 40.1$ K at $P = 1.00$ atm (fixed)
O. H_2 $P_1 = 0.001$ atm to $P_2 = 10.0$ atm at $T = 14.0$ K (fixed)
P. W $P_1 = 0.10$ atm to $P_2 = 2.00$ atm at $T = 322.4$ atm (fixed)
Q. W $T_1 = 273.0$ K to $T_2 = 573.2$ K at $P = 0.80$ atm (fixed)

9. For each of the following groups of substances, state which has the largest and which has the smallest value of the indicated property.

A. Boiling point: Xe, HI, H_2Te
B. Heat of vaporization: PF_3, NF_3, BCl_3
C. Vapor pressure at −25°C: HCl, HBr, HI
D. Vapor pressure at −10°C: CH_4, SiH_4, GeH_4
E. Boiling point: AlF_3, PF_3, ClF_3
F. Heat of vaporization: C_3H_8OH, C_2H_5OH, CH_3OH
G. Boiling point: NH_3, H_2O, HF

H. Heat of vaporization: Cl_2, Br_2, I_2
I. Vapor pressure at –10°C: CH_4, CF_4, CCl_4
J. Boiling point: CCl_3F, GeF_4, Br_2

12. For each of the following groups of compounds, state which has the largest and which has the smallest value of the indicated property. (Remember that lattice energies are exothermic!)
A. Lattice energy: MgS, SrS, CaS
B. Melting point: NaH, NaBr, NaCl
C. Solubility (M): $CaCl_2$, CaI_2, SrI_2
D. Melting point: MgS, Na_2S, Na_3P
E. Solubility (M): CaF_2, CaO, $CaCl_2$
F. Lattice energy: $MgCO_3$, Na_2CO_3, $Al_2(CO_3)_2$
G. Melting point: LiF, LiCl, LiI
H. Lattice energy: K_2S, Na_2S, MgS
I. Solubility (M): $Pb(NO_3)_2$, $Mg(NO_3)_2$, $Ca(NO_3)_2$
J. Melting point: RbCl, $InCl_3$, $SrCl_2$

QUIZZES (20 minutes each) Choose the best answer to each question.

Quiz A

1. The property of a liquid that does *not* depend on intermolecular forces is (a) surface tension; (b) boiling point; (c) vapor pressure; (d) heat of vaporization; (e) none of these.
2. 31.4 g of a solid requires 1.42 kJ of heat to melt. The solid has a mole weight of 154 g/mol., What is its heat of fusion in kJ/mol? (a) 31.4(1.42/154); (b) 154(1.42/31.4); (c) 31.4(1.42/154); (d) 1.42(1.54 × 31.4); (e) none of these.
3. In the phase diagram at right, point *A* is called the (a) triple point; (b) critical point; (c) melting point; (d) boiling point; (e) none of these.

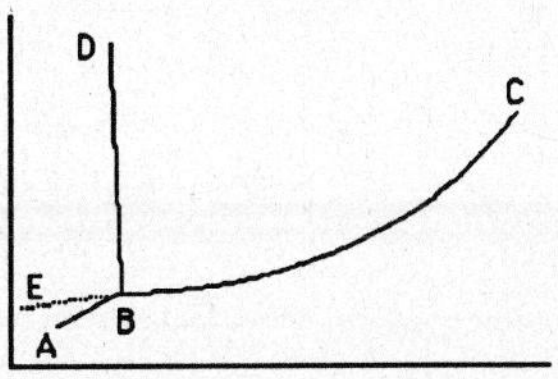

4. A liquid is in equilibrium with its vapor. If some of the vapor escapes, what is the immediate result? (a) vaporization rate decreases; (b) condensation rate decreases; (c) vaporization rate increases; (d) condensation rate increases; (e) none of these.
5. The process in which a solid is transformed into a vapor is called (a) vaporization; (b) sublimation; (c) gasification; (d) condensation; (e) none of these.
6. The vapor pressure of trichloroethene is given at four different temperatures. Which temperature is the normal boiling point? (a) 40 torr at 40.1°C; (b) 100 torr at 61.3°C; (c) 400 torr at 100.0°C; (d) 760 torr at 120.8°C; (e) none of these.
7. For anions of a fixed size, which type of hole is the smallest? (a) tetrahedral; (b) octahedral; (c) hexagonal; (d) cubic; (e) cannot be determined.
8. 3.42 g of a liquid of mole weight 100.0 g/mol vaporizes when 42.0 L of inert gas at 300 K is bubbled through. The vapor pressure of the liquid in mmHg is (a) (3.42/100.0)(0.0821 × 760); (b) (3.42/100.0)(8.314 × 760)(300/42.0); (c) (3.42 × 0.0821)(300/42.0); (d) insufficient information is given; (e) none of these.

Quiz B

1. A has a higher normal boiling point that B. This could be due to any of the following factors except the higher (a) mole weight of A; (b) vapor pressure of A; (c) intermolecular attractions of A; (d) vapor pressure of B; (e) none of these.
2. The liquid with the highest vapor pressure at –50°C is which of the following? (a) Xe; (b) Ne; (c) Cl_2; (d) F_2; (e) CCl_4.
3. The process in which a gas is transformed into a solid is called (a) vaporization; (b) condensation; (c) solidification; (d) deposition; (e) none of these.

4. In the phase diagram at right, point *H* represents the
(a) melting point;
(b) normal melting point;
(c) boiling point;
(d) normal boiling point;
(e) critical point.

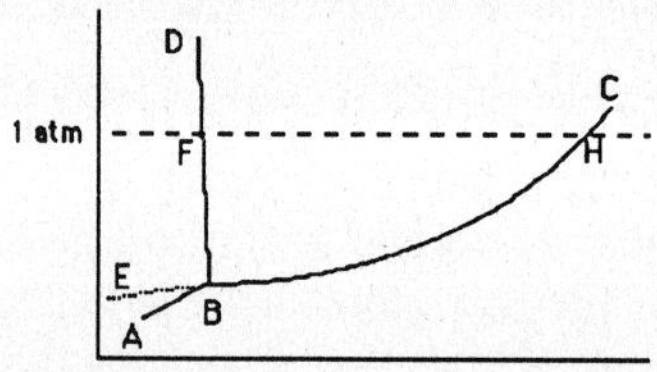

5. A liquid is in equilibrium with its vapor. A thin layer of a nonvolatile oil is poured on top of the liquid. What will now occur? (a) vaporization rate decreases; (b) condensation rate decreases; (c) vaporization rate increases; (d) condensation rate increases; (e) none of these.
6. The temperature at which a solid sublimes at 760 torr is called the (a) boiling point; (b) normal boiling point; (c) melting point; (d) normal melting point; (e) critical point.
7. A face-centered cubic cell contains all or part of how many different atoms? (a) 2; (b) 6; (c) 8; (d) 14; (e) none of these.
8. 105 kJ of heat vaporizes what mass of liquid if its mole weight is 86.0 g/mol and its molar heat of vaporization is 55.2 kJ/mol? (a) (105/55.2)/86.0; (b) 86.0(105/55.2); (c) 55.2(105/86.0); (d) (86.0/55.2)/105; (e) none of these.

Quiz C

1. The factor that has the largest effect on vapor pressure is (a) liquid surface area; (b) molecular dipole moment; (c) presence of hydrogen bonding; (d) liquid mole weight; (e) volume available for the vapor.
2. Which intermolecular force of the following is the strongest? (a) London; (b) hydrogen bonding; (c) dipole-dipole; (d) exchange; (e) all have the same strength.
3. In the phase diagram at right, point *C* is called the
(a) critical point;
(b) triple point;
(c) melting point;
(d) boiling point;
(e) none of these.

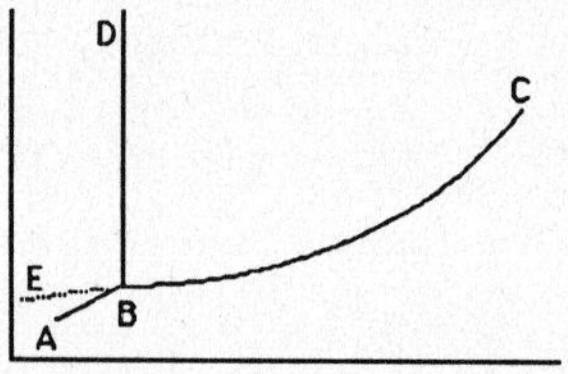

4. At the boiling point, the vapor pressure of a liquid is (a) 760 mmHg; (b) steadily decreasing; (c) constantly fluctuating; (d) atmospheric pressure; (e) none of these.
5. A liquid is in equilibrium with its vapor. Some vapor is allowed to escape. After equilibrium has be reestablished at the same temperature, it is true that the (a) vaporization rate is greater; (b) vaporization rate is less; (c) condensation rate is greater; (d) condensation rate is less; (e) none of these.
6. The process in which a liquid is changed into a gas is called (a) vaporization; (b) sublimation; (c) gasification; (d) condensation; (e) none of these.
7. 6.12 g of a solid with a mole weight of 79.0 g/mol is melted with 3.76 kJ of heat. The molar heat of fusion of the solid in kJ/mol is (a) 79.0(3.76/6.12); (b) (3.76/6.12)/79.0; (c) 6.12(3.76/79.0); (d) 3.76(6.12 × 79.0); (e) none of these.
8. The lowest melting solid of the following is (a) MgO; (b) NaCl; (c) CaS; (d) AlN; (e) Li_2O.

Quiz D

1. The mole weight of a compound directly affects all of the following *except* (a) vapor pressure of liquid; (b) vapor density; (c) solid density; (d) vapor pressure of solid; (e) none of these.
2. 4.25 kJ of heat is sufficient to melt 42 g of a solid of mole weight 210 g/mol. What is the heat of fusion of this solid in kJ/mol? (a) 42(4.25/210); (b) 210(4.25/42); (c) (4.25/210)/42; (d) 4.25(210 × 42); (e) none of these.
3. When the vapor pressure of a liquid equals atmospheric pressure, the temperature of the liquid equals (a) 100°C; (b) the boiling point; (c) the normal boiling point; (d) the vaporization point; (e) none of these.
4. The heat of sublimation equals the negative of the heat of (a) condensation; (b) vaporization; (c) solidification; (d) fusion; (e) none of these.
5. A liquid is in equilibrium with its vapor. Suddenly the pressure on the vapor is doubled. What will now occur? (a) vaporization rate increases; (b) condensation rate decreases; (c) condensation rate increases; (d) vaporization rate decreases; (e) none of these.
6. The highest melting solid of the following is (a) Mg_3N_2; (b) Ca_3P_2; (c) CaO; (d) NaCl; (e) $BaSO_4$.
7. Of the following, the liquid with the highest vapor pressure at 232 K is (a) H_2O; (b) NH_3; (c) CH_3OH; (d) CCl_3F; (e) CF_4.

8. Which of the following correctly labels regions A, B, and C in the phase diagram at right?
(a) A = solid, B = gas, C = liquid;
(b) A = gas, B = liquid, C = solid;
(c) A = gas, B = solid, C = liquid;
(d) A = liquid, B = solid, C = gas;
(e) none of these.

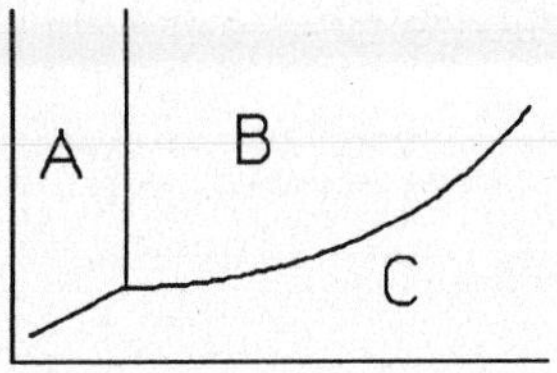

SAMPLE TEST (20 minutes)

1. The normal boiling point of acetone is 56.2°C and the molar heat of vaporization is 32.0 kJ/mol. At what temperature will acetone boil under a pressure of 50.0 mmHg?
2. Sketch a face-centered cubic cell.
3. Describe the processes that lead to the formation of exchange forces within molecules.
4. You have the following ions available:
   cations: Na^+, K^+, Ca^{2+}, Mg^{2+} anions: F^-, Br^-, O^{2-}, S^{2-}
   Which cation and which anion would you expect to combine to form the highest melting compound? Carefully explain your choice.

13 Solutions

CHAPTER OBJECTIVES

1. Explain how the relative forces between molecules predict whether an ideal solution, a nonideal solution, or a heterogeneous mixture will form.

1. In an ideal solution the two components act as if each were pure. The volume of the solution is the sum of the volumes of the two components. No heat is absorbed or given off when the solution forms. (The heat of solution is zero, $\Delta H_{soln} = 0$). When the forces between like molecules (A↔A and B↔B) are about the same as the forces between unlike molecules (A↔B), the solution formed is ideal. Often, however, the "like" forces (A↔A and B↔B) are not the same as the "unlike" forces (A↔B).
2. When the "unlike" forces are stronger, molecules in solution are more strongly attracted to each other than when they are pure. The two components will mix to form a solution and the extra energy is given off as heat: $\Delta H_{soln} < 0$.
3. When the "unlike" forces are weaker than the "like" ones, the molecules in solution are less strongly attracted to each other than when pure. Energy must be added during the solution process: $\Delta H_{soln} > 0$.
4. When the "like" forces are a great deal larger than the "unlike" forces, a solution does not form; a heterogeneous mixture is produced.

* **2. Know and be able to use percent concentration units: % (vol/vol), % (mass/mass), and % (mass/vol).**

Many different concentration units exist because of the many different ways in which solutions are used. Although chemists usually are concerned with amounts (moles) of substances, others often are interested in masses or volumes. Most concentration units express the quantity of solute in a certain quantity of solution. Then we can measure out solution and know how much solute we have. All types of concentration calculations will be much easier if you use the following general technique.

1. Write the definition of the concentration you are using. For example, the definition of % (mass/mass) is

$$\% \, (m/m) = \frac{\text{mass of solute}}{\text{mass of solution}} \times 100$$

2. Determine the quantity of solute given in the statement of the problem and use this as the numerator.
3. Determine the quantity of either solution or solvent (as required by the definition), and use this as the denominator.
4. Use the conversion factor method to separately change the units of numerator and denominator to agree with those given in the definition.

EXAMPLE 13-1 A solution of ethylene glycol in water contains 1.73 gal of ethylene glycol in 20.0 gal of solution. The densities are: solution, 1.0119 g/cm³; ethylene glycol, 1.1108 g/cm³; and water, 1.0000 g/cm³. The molar masses are: ethylene glycol, 62.07 g/mol; and water, 18.02 g/mol. Compute the (a) % (vol/vol), (b) % (mass/vol), and (c) % (mass/mass) of this solution.

(a) $$\frac{1.73 \text{ gal solute}}{20.0 \text{ gal solution}} \times 100 = 8.65\% \text{ (vol/vol)}$$

(b) $\frac{1.73 \text{ gal solute} \times 4 \text{ qt/gal} \times 946 \text{ cm}^3/\text{qt} \times 1.1108 \text{ g/cm}^3}{20.0 \text{ gal solution} \times 4 \text{ qt/gal} \times 946 \text{ cm}^3/\text{qt}} \times 100 = 9.61\% \text{ (m/v)}$

(c) $\frac{1.73 \text{ gal solute} \times 4 \text{ qt/gal} \times 946 \text{ cm}^3/\text{qt} \times 1.1108 \text{ g/cm}^3}{20.0 \text{ gal solution} \times 4 \text{ qt/gal} \times 946 \text{ cm}^3/\text{qt} \times 1.0119 \text{ g/cm}^3} \times 100 = 9.50\% \text{ (m/m)}$

* 3. **Know and be able to use the definitions of molarity and molality.**

Two concentration units frequently used by chemists are molarity and molality.

$$\text{molality } (m) = \frac{\text{moles of solute}}{\text{kilogram of solvent}} \qquad [1]$$

$$\text{molarity (M)} = \frac{\text{moles of solute}}{\text{liter of solution}} \qquad [2]$$

Molality is convenient for freezing-point depression and boiling-point elevation measurements as it does not change with temperature. Molarity does so change as solution volume generally expands as temperature increases. Molarity is valuable when one wishes to react two solutes. To obtain a certain amount of solute, in moles, one merely measures out a definite volume of solution.

EXAMPLE 13-2 What are the (a) molarity and (b) molality of the aqueous ethylene glycol solution of Example 13-1?

(a) $\text{M} = \frac{1.73 \text{ gal solute} \times 4 \text{ qt/gal} \times 946 \text{ cm}^3/\text{qt} \times 1.108 \text{ g/cm}^3 \times \text{mol}/62.07 \text{ g}}{20.0 \text{ gal solution} \times 4 \text{ qt/gal} \times 946 \text{ cm}^3/\text{qt} \times \text{L}/1000 \text{ cm}^3}$

$= \frac{117 \text{ mol solute}}{75.7 \text{ L solution}} = 1.55 \text{ M}$

(b) To determine molality, we need the mass of the solvent.

$\text{solution mass} = 75.7 \text{ L} \times \frac{1000 \text{ cm}^3}{\text{L}} \times \frac{1.0119 \text{ g}}{\text{cm}^3} \times \frac{\text{kg}}{1000 \text{ g}} = 76.6 \text{ kg soln}$

or $\text{solution mass} = 75.7 \text{ L} \times \frac{1.0119 \text{ kg}}{\text{L}} = 76.6 \text{ kg soln}$

$\text{solute mass} = 117 \text{ mol solute} \times \frac{62/07 \text{ g}}{\text{mol}} \times \frac{\text{kg}}{1000 \text{ g}} = 7.27 \text{ kg solute}$

solvent mass = solution mass – solute mass = 76.6 kg – 7.27 kg = 69.3 kg solvent

$\text{molality} = \frac{117 \text{ mol solute}}{69.3 \text{ kg solvent}} = 1.69\ m$

Notice in part (a), that the beginning units of solvent and solution volumes need not be the same, since both volumes are eventually converted to cm^3. Furthermore, as in this case, if the calculation is set up using the conversion factor method, unnecessary operations [such as 4 qt/gal in part (a)] stand out. Finally, take care not to confuse concentration units, such as molarity, with amounts of material (moles). Thinking of moles per liter as being the same as moles is similar to saying that the population of Australia is 5 persons, when actually there are 5 persons per square mile in this country of 16 million persons.

* 4. **Know and be able to use the definitions of mole fraction and mole percent.**

Mole fraction is defined as the moles of solute divided by the total moles of solution. If n_A is the symbol for the number of moles of substance A, then

$$\chi_A = \frac{n_A}{n_A + n_B + \ldots} = \text{mole fraction of A in solution} \qquad [3]$$

EXAMPLE 13-3 What is the mole fraction of ethylene glycol in the solution of Example 13-2?

$n_{\text{EtGly}} = 117 \text{ mol}$

$n_{H_2O} = 69.3 \text{ kg} \times \frac{1000 \text{ g}}{\text{kg}} \times \frac{\text{mol } H_2O}{18.02 \text{ g } H_2O} = 3846 \text{ mol } H_2O$

$\chi_{\text{EtGly}} = \frac{n_{\text{EtGly}}}{n_{\text{EtGly}} + n_{H_2O}} = \frac{117 \text{ mol}}{117 \text{ mol} + 3846 \text{ mol}} = 0.0296$

The mole percent is one hundred times the mole fraction.

mol % EtGly = 0.0296 × 100 = 2.96%

5. Distinguish among unsaturated, saturated, and supersaturated solutions and describe how a solute can be purified by recrystallization.

An unsaturated solution can dissolve more solute. A saturated solution holds the maximum amount of solute that it can at equilibrium. A supersaturated solution holds more than it can at equilibrium. Solute dissolves when it is added to an unsaturated solution. If solute is added to a saturated solution, the amount of undissolved solute will not change. If solute is added to a supersaturated solution, the amount of undissolved solute will increase.

To purify a solute by recrystallization, we lower the temperature of the solution containing that solute until the solute precipitates. Then we separate the crystallized solute from the saturated solution and dissolve it in pure fresh solvent. This process is repeated until pure solute is obtained.

*** 6. Apply Henry's law to calculations of gas solubility as a function of pressure.**

Henry's law states that the concentration (C) of gas dissolved in solution increases as the gas pressure (P) increases.

$$C = kP \qquad [4]$$

There is no agreement on the units to be used for P or for C. Thus, one must be careful that units are used correctly when solving Henry's law problems. Henry's law data are given in Table 13-1.

TABLE 13-1

Gas Solubilities per 100 cm^3 H_2O under 1.00 atm Gas Pressure

Gas	*t*, °C	*Gas Volume,* cm^3 at STP	*t*, °C	*Mass of Gas Dissolved*, mg
Ar	0	5.6	50	5.36
H_2	0	2.14	50	0.170
O_2	0	4.89	50	3.51
O_3	0	49	0	105
N_2	0	2.33	40	1.77
NO	0	7.34	60	3.17
Kr	0	11.0	50	17.5
CH_4	17	3.50	17	2.50

EXAMPLE 13-4 At 0°C, 49 cm^3 of ozone (O_3), measured at STP, dissolves in 100. cm^3 of water under a pressure of 1.00 atm. What mass in grams of ozone dissolves in 57.0 cm^3 of water under a pressure of 1750 mmHg?

First we compute the Henry's law constant for ozone in water at 0°C by applying equation [4].

$$\frac{49 \text{ cm}^3 \text{ at STP}}{100 \text{ cm}^3 \text{ H}_2\text{O}} \times \frac{\text{mol O}_3}{22414 \text{ cm}^3 \text{ at STP}} \times \frac{48.0 \text{ g O}_3}{100. \text{ cm}^3 \text{ H}_2\text{O}} = k \times 1.00 \text{ atm}$$

$$k = \frac{0.105 \text{ g O}_3}{100. \text{ cm}^3 \text{ H}_2\text{O·atm}}$$

Then we compute the concentration at 1750 mmHg.

$$C = \frac{0.105 \text{ g O}_3}{100. \text{ cm}^3 \text{ H}_2\text{O·atm}} \times 1750 \text{ mmHg} \times \frac{1 \text{ atm}}{760 \text{ mmHg}} = \frac{0.242 \text{ g O}_3}{100. \text{ cm}^3 \text{ H}_2\text{O}}$$

Finally we compute the mass of O_3 in 57.0 cm^3 of H_2O.

$$57.0 \text{ cm}^3 \text{ H}_2\text{O} \times \frac{0.242 \text{ g O}_3}{100. \text{ cm}^3 \text{ H}_2\text{O}} = 0.138 \text{ g O}_3$$

7. Describe the properties of solutions that are colligative properties.

Colligative properties depend on the *number* of particles (molecules or ions) of solute in a given quantity of solution.

1. *Vapor pressure lowering* obeys Raoult's law.

$$P_A = \chi_A P°_A \quad [5]$$

P_A = the vapor pressure of substance A above a solution in which χ_A is the mole fraction of A in solution. $P°_A$ is the vapor pressure of pure A.

2. *Boiling-point elevation.*

$$\Delta T_{bp} = K_b m \quad [6]$$

m is the molality of the solution, ΔT_{bp} is the increase in the boiling point, and K_b is the boiling point elevation constant. K_b depends only on the solvent.

3. *Freezing-point depression.*

$$\Delta T_{fp} = K_f m \quad [7]$$

m is the molality of the solution, ΔT_{fp} is the increase in the freezing point, and K_f is the freezing point depression constant. K_f depends only on the solvent.

4. *Osmotic pressure* obeys van't Hoff's law.

$$\pi = CRT \quad [8]$$

C is the molarity of the solution, T is its absolute temperature, and R is the ideal gas constant.

* 8. Apply Raoult's law. Describe the applications and the limitations of the law.

Raoult's law, equation [5], is true only for ideal solutions (objective 13-1, type 1). The two components of an ideal solution often are very similar chemically. For dilute *real* solutions, Raoult's law is valid for the solvent but not the solute.

EXAMPLE 13-5 At 25°C the vapor pressures of pure benzene and pure toluene are 95.1 mmHg and 28.4 mmHg, respectively. The total vapor pressure above a solution of these two liquids is 50.0 mmHg. What is the mole fraction of benzene in this solution?

The subscripts B and T stand for benzene and toluene. Raoult's law gives the vapor pressures.

$$P_T = \chi_T P°_T = \chi_T(28.4 \text{ mmHg}) \qquad P_B = \chi_B P°_B = \chi_B(95.1 \text{ mmHg})$$

Dalton's law produces the total vapor pressure.

$$P = P_1 + P_2 = 50.0 \text{ mmHg} = \chi_B\ (95.1 \text{ mmHg}) + \chi_T\ (28.4 \text{ mmHg})$$

Note that $\chi_B + \chi_T = 1$ or $\chi_T = 1 - \chi_B$.

$$P = 50.0 \text{ mmHg} = (1 - \chi_B)(28.4 \text{ mmHg}) + \chi_B\ (95.1 \text{ mmHg})$$

$$= 28.4 \text{ mmHg} - \chi_B\ (28.4 \text{ mmHg}) + \chi_B\ (95.1 \text{ mmHg})$$

$$\textit{or } \chi_B = (50.0 - 28.4) / (95.1 - 28.4) = 0.324$$

The partial pressure of benzene is given by

$$P_B = \chi_B P°_B = (0.324)(95.1 \text{ mmHg}) = 30.8 \text{ mmHg}$$

Then the mole fraction of benzene in the *vapor* is

$$y_B = 30.8 \text{ mmHg} / 50.0 \text{ mmHg} = 0.616$$

$\chi_B = 0.324$ is the mole fraction of benzene in the liquid and $y_B = 0.616$ is its mole fraction in the vapor. A plot of vapor pressure against mole fraction is called a vapor pressure diagram.

9. Explain how liquid-vapor equilibrium in non-ideal solutions differs from that in ideal solutions.

A solution that is not ideal does not obey Raoult's law. This can occur in two ways. If the vapor pressure of the solution is *less* than that predicted by Raoult's law, the solution is said to show a negative deviation from Raoult's law. This is due to the "unlike" forces (see objective 13-1) being stronger than the "like" forces and the molecules being more strongly held in the solution than they were in the pure components.

If a solution's vapor pressure is *greater* than that predicted by Raoult's law, the solution is said to show a positive deviation from Raoult's law. In this case, the "unlike" forces are weaker than the "like" forces and the molecules are not held in solution as strongly as they were in the pure components.

10. Describe how solution components can be separated by fractional distillation.

In fractional distillation, two components of a mixture are separated by repeatedly vaporizing and condensing portions of the mixture. Notice, from the answer to Example 13-5, that the vapor is richer or more concentrated in the more volatile component (the one with the higher vapor pressure). If this vapor is condensed and the resulting liquid is partially vaporized, the vapor that is produced will be richer yet in the more volatile component. This process can be extended through many cycles of partial vaporization followed by condensation. Eventually, the condensed vapor will be the pure liquid of the more volatile component. But what of the liquid that does not vaporize? Since the more volatile component has vaporized, the unvaporized liquid becomes more concentrated in the less volatile component. Given a sufficient number of cycles, the liquid left behind is pure and is the less volatile component.

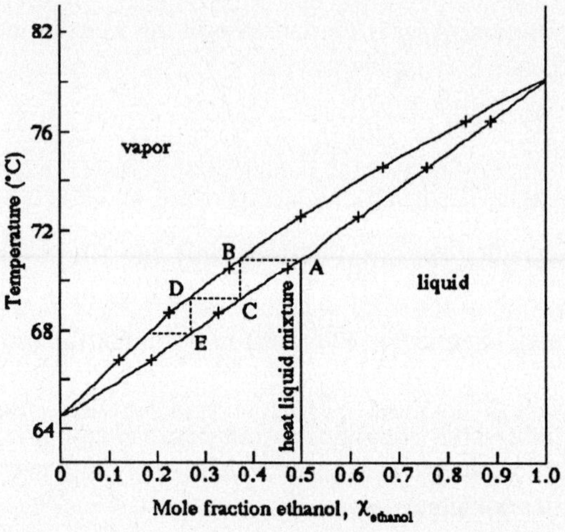

FIGURE 13-1 Boiling Point Diagram for the Methanol-Ethanol System

As an example of this process, consider Figure 13-1, a boiling point diagram for the ethanol-methanol system. Below the solid curved line, the entire system is liquid. Consider what occurs when a mixture with $\chi_{\text{ethanol}} = 0.50$ is heated (solid vertical line). At 71.5°C (point *A*), vapor appears with $y_{\text{ethanol}} = 0.38$. Note that the vapor (point *B*) is less concentrated in the less volatile component, ethanol. (The liquid and the vapor with which it is in equilibrium are connected by a dashed horizontal line.) This vapor condenses if the temperature is lowered to 69.8°C (point *C*). The vapor in equilibrium with this condensate (*D*) has $\chi_{\text{ethanol}} = 0.28$. An apparatus to achieve this separation is pictured in Figure 13-2. The mixture is heated in the round bottom flask at the bottom and the resulting vapors rise into the vertical column, which is loosely packed with an inert material, such as stainless steel turnings. The packing is cooler the farther it is from the heat source, and thus the vapor begins to condense. Points *A*, *C*, and *E* at the start of heating are indicated approximately on the column. Of course, the concentrated liquid that does not vaporize drips back down the column until it encounters a high enough temperature to vaporize. As time continues, the temperature of the flask must continually increase because its contents are becoming richer in the less volatile, higher boiling component (the more volatile one has vaporized).

* 11. Explain how vapor pressure lowering leads to boiling-point elevation and also to freezing-point depression. Use the equations for computing ΔT_{fp} and ΔT_{bp}.

According to Raoult's law (equation [5]), the vapor pressure of a component depends on its concentration. If the other component is nonvolatile, the vapor pressure of the solution will decrease as the concentration of the nonvolatile substance increases. Since the vapor pressure of the solution is lower at all temperatures, the liquid-vapor curve on a phase diagram, such as *Figure 13-15 in the text*, is lowered by the addition of a nonvolatile solute. In a similar fashion, if the solute does not freeze out with the solvent, the temperature at which the solution begins to freeze is lower than the freezing point of pure solvent. The constants K_b and K_f in equations [6] and [7] depend only on the solvent.

EXAMPLE 13-6 0.202 g of naphthalene (128 g/mol) lowers the freezing point of 10.453 g of cyclohexane by 3.08°C. 0.164 g of a solid unknown depressed the freezing point of 12.011 g of cyclohexane by 1.28°C. What is the molar mass of the unknown?

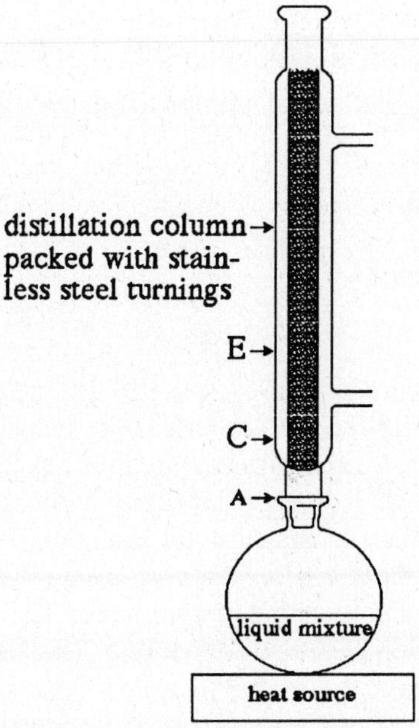

FIGURE 13-2 Fractional Distillation Apparatus

First compute the molality of the naphthalene solution and then K_f for cyclohexane.

$$\text{molality} = \frac{0.202 \text{ g naphthalene} \times 1 \text{ mol}/128 \text{ g}}{10.453 \text{ g cyclohexane} \times 1 \text{ kg}/1000 \text{ g}} = 0.151\ m$$

Then we compute the molality of the unknown solution, which we use to determine the amount of solute and its molar mass.

$$m = \frac{\Delta T_{fp}}{K_f} = \frac{1.28°\text{C}}{20.4°\text{C}/m} = 0.0627\ m$$

$$\text{moles solute} = 12.011 \text{ g solvent} \times \frac{\text{kg}}{1000 \text{ g}} \times \frac{0.0627 \text{ mol solute}}{\text{kg solvent}} = 7.54 \times 10^{-4} \text{ mol}$$

$$\text{molar mass} = \frac{0.164 \text{ g solute}}{7.54 \times 10^{-4} \text{ mol solute}} = 218 \text{ g/mol}$$

* **12. Describe the process of osmosis and use van't Hoff's law of osmotic pressure, equation [8].**

Osmosis occurs when two solutions of different concentrations are separated by a semipermeable membrane. If the membrane were not present, the solutions would mix, resulting in a solution of intermediate concentration. But only the solvent can flow through the membrane. The direction of flow is from the dilute to the concentrated solution, making it less concentrated. The pressure that must be exerted on the concentrated solution to halt the flow of solvent is the osmotic pressure. If we let C_c indicate the concentration of the concentrated solution and C_d that of the dilute one, equation [8] becomes

$$\pi = (C_c - C_d)\, RT \qquad [9]$$

EXAMPLE 13-7 An isotonic solution has a concentration of 0.308 M. What osmotic pressure develops when a living cell (which basically contains such an isotonic solution) is placed in a 0.104 M sugar solution at 37°C?

$$\pi = (0.308 - 0.104)\text{M} \times 0.0821 \text{ L atm mol}^{-1} \text{ K}^{-1} \times 310 \text{ K}$$
$$= 5.19 \text{ atm } (= 76.3 \text{ lb/in.}^2)$$

*** 13. Describe how the theory of electrolytic dissociation explains the behavior of aqueous solutions of strong, weak, and nonelectrolytes, including ionic concentrations and differences in the values of colligative properties from the value computed from equations [6] through [9].**

Colligative property data indicate that equations [6] through [9] should be modified as follows.

boiling-point elevation: $\Delta T_{bp} = iK_b m$ [10]

freezing-point depression: $\Delta T_{fp} = iK_f m$ [11]

osmotic pressure $\pi = iCRT$ [12]

$\pi = i(C_c - C_d)RT$ [13]

The value of i, the van't Hoff factor, depends on the type of solute and somewhat on its concentration but not on which colligative property is measured. When the electrical conductance of aqueous solutions is measured, the solutes can be grouped into three types. *Nonelectrolytes* produce aqueous solutions with a conductance equal to that of water. The van't Hoff factor equals 1. *Weak electrolytes* produce aqueous solutions with a slightly greater conductance than that of pure water. The van't Hoff factor is slightly larger than 1. Both the molar conductance and the van't Hoff factor increase as the solution becomes more dilute. *Strong electrolytes* produce aqueous solutions with a conductance much greater than that of pure water. The molar conductance of a strong electrolyte increases only very slightly as its solution becomes more dilute. The van't Hoff factor of its aqueous solution is a bit less than 2, 3, 4, or some whole number larger than 1, and increases slightly as the solution becomes more dilute. The molar conductances and van't Hoff factors for several concentrations of a strong and a weak electrolyte are given in Table 13-2.

TABLE 13-2 Variation of Molar Conductance (Λ)[a] and van't Hoff Factor (i) with Concentration for a Strong Electrolyte (HCl) and a Weak Electrolyte ($HC_2H_3O_2$)

Concentration,	*HCl*		*$HC_2H_3O_2$*	
mol/L	Λ	i	Λ	i
0.1	391	1.89	5.2	1.01
0.05	399	1.90	7.4	1.02
0.01	412	1.94	16.3	1.04
0.005	415	1.95	23.9	1.06
0.001	421	1.98	49.2	1.12
0.0005	423	2.00	67.7	1.17
0.0001	425	2.00	128	1.33

[a]The units of Λ are cm^2·mho/mol; i is unitless.

Svante Arrhenius pointed out the similarity of the conductance trend on dilution to that of the van't Hoff factor. He used this similarity to argue that ions must be present in solutions of electrolytes at all times, even when they are not conducting a current. Dilute solutions have higher van't Hoff factors and molar conductances because more ions are present, according to Arrhenius. Arrhenius believed that the larger amount of water present in dilute solutions of electrolytes causes more solute to dissociate into ions. Presently we believe that strong electrolytes dissociate *completely* into ions in aqueous solution. All ionic compounds are strong electrolytes. In addition, a few covalent compounds also are strong electrolytes, namely HCl, HBr, HI, HNO_3, H_2SO_4, $HClO_4$, and a few others.

EXAMPLE 13-7 50.0 mL of 1.20 M NaCl is mixed with 30.0 mL of 0.800 M $CaCl_2$. What is the concentration of each ion in the final solution?

final solution volume (V_f) = 50.0 mL + 30.0 mL = 80.0 mL

We first compute the concentration of NaCl in the final solution.

$$50.0 \text{ mL} \times \frac{1.20 \text{ mmol NaCl}}{\text{mL soln}} = 60.0 \text{ mmol NaCl}$$

$$[\text{NaCl}]_f = \frac{60.0 \text{ mmol NaCl}}{80.0 \text{ mL soln}} = 0.750 \text{ M}$$

The final $[CaCl_2]_f$ is computed in a different but equivalent way.

$$[CaCl_2]_f V_f = [CaCl_2]_i V_i$$

$$[CaCl_2]_f(80.0\ mL) = (0.800\ M)(30.0\ mL)$$

$$[CaCl_2]_f = \frac{(0.800\ M)(30.0\ mL)}{80.0\ mL} = 0.300\ M$$

The only source of Ca^{2+} ions is $CaCl_2$. Thus, $[Ca^{2+}] = [CaCl_2] = 0.300$ M. Likewise, the only source of Na^+ ions is NaCl and hence $[Na^+] = [NaCl] = 0.750$ M. But Cl^- ions are present in both solutes.

$$[Cl^-] = [Cl^- \text{ from } CaCl_2] + [Cl^- \text{ from } NaCl]$$

$$= 0.300\ M \times \frac{2\ mol\ Cl^-}{mol\ CaCl_2} + 0.750\ M \times \frac{1\ mol\ Cl^-}{mol\ NaCl} = 0.600\ M + 0.750\ M = 1.350\ M$$

14. Describe how interionic attractions in solution require modifications to Arrhenius's theory.

We stated in objective 13-13 that all strong electrolytes dissociate into ions or ionize completely in aqueous solution. Why then are the van't Hoff factors not precisely equal to integers and why do the van't Hoff factors decrease as the solution becomes more concentrated? The explanation given by Peter Debye and Erich Hückel is that positive and negative ions cluster around each other in aqueous solution. These clusters are more likely to form in concentrated solutions because the ions are closer together. In concentrated solutions, therefore, there are more clusters, fewer particles (since a cluster behaves in solution as if it were one particle), and the van't Hoff factor will be smaller than in dilute solutions.

15. Describe some of the properties of colloids, how colloidal dispersions differ from solutions and heterogeneous mixtures, and classify colloidal mixtures.

Colloidal mixtures or colloidal dispersions appear uniform—that is, like solutions—to the naked eye but often they are cloudy or foggy rather than clear. Colloidal dispersions often can be coagulated or separated by adding a solution of a strong electrolyte such as NaCl, NaOH, or HCl. This partly neutralizes the positive or negative charges that are present on the surface of each colloidal particle and that keeps it separated from its neighbors. The various types of colloidal dispersions are given in *Table 13-3 in the text.* These eight types, their names, and one or two examples of each should be memorized.

DRILL PROBLEMS

2. The aqueous solution described in each of the following lines is at 20°C. At 20°C, the density of water (the solvent) is 0.99823 g/cm³. The solutes all are liquids. Fill in the blanks in each line. The first line is completed as an example.

		Volume, cm^3		Mass, g		Density, g/cm^3		Percent composition		
	Solute formula	*Soln.*	*Solute*	*Solvent*	*Soln.*	*Soln*	*Solute*	*mass/mass*	*mass/vol*	*vol/vol*
	CH_3COOH	88.39	12.01	77.40	90.00	1.0182	1.0491	14.00	14.26	13.59
A.	CH_3COOH	___	___	___	31.16	1.0385	1.0491	30.00	___	___
B.	CH_3COCH_3	50.444	3.16	47.50	___	___	0.7908	___	___	___
C.	CH_3COCH_3	40.00	___	36.48	___	0.9967	0.7908	___	___	
D.	C_2H_5OH	134.51	___	___	___	0.8921	0.7893	___	53.53	___
E.	C_2H_5OH	___	___	___	6.87	0.9820	0.7893	10.00	___	___
F.	$(CH_2OH)_2$	59.441	4.87	54.60	___	___	1.1088	___	___	___
G.	$(CH_2OH)_2$	120.00	___	115.64	___	1.0038	1.1088	___	___	3.63
H.	HCOOH	137.994	___	___	___	1.0870	1.220	___	34.78	___
I.	HCOOH	___	___	___	38.11	1.0587	1.220	24.00	___	___
J.	$C_3H_5(OH)_3$	95.479	15.06	81.00	___	___	1.2613	___	___	___
K.	$C_3H_5(OH)_3$	75.00	___	43.72	___	1.1209	1.2613	___	___	42.65
L.	CH_3OH	83.673	___	___	___	0.8366	0.7914	___	70.27	___
M.	CH_3OH	___	___	___	62.56	0.8762	0.7914	68.00	___	___

3. Fill in the blanks on each line in the following table. All of these data are for actual aqueous solutions.

	Solute		*Solution*				
	Mol. wt. or Formula	*Number of moles*	*Volume,* mL	*Density* g/mL	*Molarity,* M	*Molality,* *m*	*% by weight*
A.	$HC_2H_3O_2$	___	250.0	1.0250	3.414	___	___
B.	___	___	500.0	1.020	1.334	___	7.00
C.	$BaCl_2$	___	275.0	___	0.888	___	16.00
D.	HCl	1.270	400.0	1.0544	___	___	___
E.	___	1.245	___	1.056	2.490	___	10.00
F.	$Pb(NO_3)_2$	0.221	___	1.3509	1.104	___	___
G.	___	0.289	300.0	1.023	___	___	4.00
H.	$KHCO_3$	___	200.0	___	1.064	___	10.00
I.	KBr	___	750.0	1.107	___	___	14.00
J.	___	___	340.0	1.0402	0.907	___	6.50
K.	$NaHCO_3$	___	270.0	1.0408	___	___	6.00
L.	K_2CrO_4	0.200	___	1.172	1.207	___	___
M.	$NaNO_3$	___	750.0	___	5.043	___	34.00
N.	H_2SO_4	___	400.0	1.681	13.026	___	___
O.	NaOH	6.00	___	1.4299	___	___	40.00
P.	___	___	300.0	1.1972	5.326	___	26.00
Q.	H_3PO_4	2.047	___	1.2536	5.117	___	___
R.	CsCl	___	240.0	1.107	___	___	13.00

4. Determine the mole fraction of the solute and the mole percent of the solvent for each solution of the drill problems of objective 13-3.

6. Use the data of Table 13-1 to determine the molality of each gas at each temperature under a pressure of 1.00 atm. The mass of 100. cm³ of water at each of the temperatures follows.

t, °C	0	17	40	50	60
mass, kg	0.1000	0.0999	0.0992	0.0988	0.0983

Compute the pressure or the molality of the gas in each solution described below.

A. Ar, $t = 0°C$, $P = 14.2$ atm
B. CH_4, $t = 17°C$, concn. = 0.0120 m
C. O_2, $t = 50°C$, $P = 160.$ mmHg
D. N_2, $t = 40°C$, concn. = 0.000314 m
E. NO, $t = 60°C$, $P = 1000.$ mmHg
F. O_3, $t = 0°C$, concn. = 0.100 m
G. Kr, $t = 50°C$, $P = 1000.$ mmHg
H. O_2, $t = 0°C$, concn. = 0.0100 m
I. H_2, $t = 50°C$, $P = 100.$ mmHg
J. H_2, $t = 50°C$, concn. = 0.0100 m

8. In each of the following ideal solutions, two liquids, A and B, are mixed together. The vapor pressures of the pure liquids are $P_A°$ and $P_B°$ and their mole fractions are χ_A and χ_B. The vapor pressure of the solution is P. Fill in the blanks in the table below.

	P_A, mmHg	χ_A	P_B, mmHg	χ_B	P, mmHg
A.	100.	0.75	50.	___	___
B.	180.	___	60.	___	125.
C.	___	0.25	20.	___	100.
D.	140.	0.30	___	___	220.
E.	___	___	160.	0.14	130.
F.	190.	___	270.	0.35	___
G.	120.	___	80.	___	85.
H.	220.	___	70.	___	100.

11. In each line that follows, the given mass of solute (in grams) is dissolved in the specified mass of solvent. The solvent has the boiling-point elevation constant K_b and the freezing-point depression constant K_f. t_b and t_f are the boiling and freezing points of the pure solvent. t_b' and t_f' are the boiling and freezing points of the solution. m is the molality of the solution and $\mathfrak{M}$ is the mole weight of the solute in grams/mole. Fill in the blanks in the following table.

	Solvent					Solute		Solution		
	mass, g	t_f, °C	t_b, °C	K_b, °C/*m*	K_f, °C/*m*	mass, g	𝔐, g/mol	*m*	t_b' °C	t_f' °C
A.	800.	16.62	117.9	3.07	3.90	17.2	___	0.201	118.52	___
B.	640.	5.51	80.09	2.53	___	19.4	140	___	___	4.45
C.	750.	5.68	210.81	5.24	___	___	126	0.244	___	3.97
D.	320.	43.02	181.75	___	7.40	7.62	___	0.284	182.76	___
E.	500.	0.00	100.00	0.512	1.86	12.7	78.2	___	___	–0.60
F.	600.	6.55	80.74	2.79	19.6	___	64.1	___	81.77	___

12. In the table that follows, each line represents a solution with osmotic pressure π (mmHg), concentration C (mol/L) and temperature t (°C). V is the volume of the solution (L) and 𝔐 is the mole weight of the solute (g/mol). Fill in the blanks in each line.

	Solution				Solute	
	π, mmHg	C, mol/L	t, °C	V, L	mass, g	𝔐, g/mol
A.	740.	___	25	0.250	___	75.2
B.	143	___	37	0.375	1.42	___
C.	___	___	18	0.800	3.14	146
D.	1476	___	20.	0.750	___	128
E.	476	___	23	0.240	7.51	___
F.	___	___	15	0.150	0.904	187
G.	1793	___	24	0.400	___	207
H.	500.	___	19	0.500	1.37	___
I.	___	___	30.	0.650	3.27	209
J.	473	___	23	0.450	___	150.
K.	760.	___	20.8	0.120	0.816	___
L.	___	___	26.8	0.350	2.06	314
M.	946	___	16.4	0.100	___	200.
N.	640.	___	19.9	0.175	0.943	___
O.	___	___	29.0	0.180	0.0843	187

13. Solution A is mixed with Solution B. Compute the concentration of each ion in the final solution.

	Solution A			Solution B		
	Volume	Solute	Molarity	Volume	Solute	Molarity
A.	100. mL	$LiNO_3$	0.241	240. mL	$Ca(NO_3)_2$	0.618
B.	200. mL	H_2SO_4	0.314	450. mL	HCl	0.114
C.	400. mL	NaBr	0.186	200. mL	$AlBr_3$	0.916
D.	400. mL	Na_3PO_4	0.126	150. mL	NaCl	0.224
E.	620. mL	HNO_3	1.07	180. mL	$NaNO_3$	2.03
F.	250. mL	$CaCl_2$	0.143	150. mL	$MgCl_2$	0.321
G.	150. mL	Na_2SO_4	0.187	350. mL	$Al_2(SO_4)_3$	0.103
H.	100. mL	Na_2SO_4	0.946	900. mL	Na_3PO_4	0.204
I.	176 mL	NaBr	0.700	224 mL	NaCl	0.600
J.	894 mL	LiCl	0.600	106 mL	Li_2SO_4	0.800

QUIZZES (20 minutes each) Choose the best answer for each question.

Quiz A

1. Moles of solute per mole of solvent is a definition of (a) mole fraction; (b) molarity; (c) molality; (d) solubility; (e) none of these.
2. A solution in which a relatively large amount of solute has been dissolved is called (a) unsaturated; (b) saturated; (c) supersaturated; (d) dilute; (e) none of these.
3. The concentration unit used in Raoult's law calculations is (a) mole fraction; (b) percent by weight; (d) molarity; (d) molality; (e) none of these.
4. 8.72 mL of a 14.0 M solution is diluted to 50.0 mL. What is the molarity of the resulting solution?

(a) 14.0(8.72/50.0); (b) (14.0/8.72)/50.0; (c) 14.0(50.0/8.72); (d) 14.0(50.0)(8.72); (e) none of these.

5. 141 mL of 0.175 M solution of silver nitrate contains how many grams of solute? (a) (141/1000)(0.175)(227.8); (b) (141)(0.175)(153.9); (c) (141/169.9)(0.175); (d) (141/1000)(0.175)(169.9); (e) none of these.
6. The normal boiling point of hexane is 68.7°C. What is the mole fraction of nonvolatile solute in a hexane solution that has a vapor pressure of 600 mmHg at this temperature? (a) 68.7/600; (b) 600/760; (c) 160/600; (d) 160/760; (e) none of these.
7. 0.23 mol of solute depresses the freezing point of 267 g of solvent by 6.15°C. What is the value of K_f for this solvent? (a) 6.15/(0.23/267); (b) 6.15/(0.23 × 0.267); (c) (6.15(0.23/0.267); (d) (6.15)(0.23)(0.267); (e) none of these.
8. 8.40 g (0.20 mol) NaF dissolves in 153 g (8.50 mol) of water. What is the mole fraction of NaF in the solution? (a) 0.20/(8.50 + 0.20); (b) 0.20/(153/1000); (c) [8.40/(153 + 8.40)](100); (d) (8.40/153)(100); (e) none of these.

Quiz B

1. Moles of solute per liter of solution is a definition of (a) mole fraction; (b) molarity; (c) molality; (d) solubility; (e) none of these.
2. A solution that has the capacity to dissolve more solute is called (a) unsaturated; (b) saturated; (c) supersaturated; (d) concentrated; (e) none of these.
3. Which of the following can be most directly linked to the lowering of vapor pressure by a nonvolatile solute? (a) osmotic pressure; (b) freezing-point depression; (c) boiling-point elevation; (d) solubility; (e) melting point depression.
4. 52.5 mL of a solution was diluted to 6.25 L and then had a concentration of 0.0316 M. What was the molarity of the original concentrated solution? (a) 0.316(52.5/6.25); (b) 0.0136(6.25/52.5); (c) 0.0316(52.5/6250); (d) 0.0136(52.5)(6.25); (e) none of these.
5. At 34.9°C ethanol has a vapor pressure of 100 mmHg. What is the mole fraction of nonvolatile solute present in an ethanol solution that has a vapor pressure of 92 mmHg at this temperature? (a) 34.9/100; (b) 92/34.9; (c) 100/92; (d) 92/100; (e) none of these.
6. 87.5 L of a 3.17 M solution of hydrochloric acid contains how many grams of solute? (a) (87.5/1000)(3.17)(36.5); (b) (87.5)(3.17)(52.5); (c) (87.5)(3.17)(36.5); (d) (87.5/1000)(3.17)(68.5); (e) none of these.
7. 17.4 L of a 0.623 M starch solution displays what osmotic pressure in atmospheres at 75°C? (a) (17.4)(0.623)(0.0821)(348); (b) (0.623)(0.0821)(75); (c) (17.4)(0.623)(75); (d) (17.4)(0.623/0.0821)(348); (e) none of these.
8. 23.4 g (0.40 mol) NaCl dissolves in 108 g (6.0 mol) of water. What is the molality of this solution? (a) 0.40/(0.40 + 6.0); (b) 0.40/(108/1000); (c) [23.4/23.4 + 108)](100); (d) (23.4/108)(100); (e) none of these.

Quiz C

1. Moles of solute per liter of solvent is a definition of (a) mole fraction; (b) molality; (c) molarity; (d) solubility; (e) none of these.
2. The following always form solutions (if they do not react chemically) (a) a gas in a liquid; (b) a solid in a liquid; (c) two liquids; (d) two gases; (e) none of these.
3. Which of the following is not a colligative property? (a) freezing-point depression; (b) boiling-point elevation; (c) osmotic pressure; (d) solubility; (e) none of these.
4. How many mL of 6.25 M solution can be produced from 57.5 mL of 16.4 M solution? (a) (57.5)(6.25)(16.4); (b) 57.5(6.25/16.4); (c) 57.5(16.4/6.25); (d) 16.4(6.25/57.5); (e) none of these.
5. 81.2 g of cesium chloride (168.4 g/mol) is enough solute for how many mL of 4.00 M solution? (a) (81.2/168.4)/4.00; (b) (81.2/168.4)(4.00/1000); (c) (81.2/4.00)(168.4)(1000); (d) (81.2)(4.00)(168.4/1000); (e) none of these.
6. The normal boiling point of acetic acid is 118.1°C. What is the vapor pressure in mmHg of a solution in which there is 0.35 mol of nonvolatile solute for every mole of acetic acid at this temperature? (a) 760(0.35/1.35); (b) 760(0.35/1.00); (c) 118.1/760; (d) 760/118.1; (e) none of these.
7. 0.25 mol of a nonvolatile, nonionic solute dissolved in 300 g of water (K_f = 1.86) will lower the freezing point how many degrees? (a) (1.86)(0.25)(1000/300); (b) (1.86/0.25)(300/1000); (c) (1.86)(0.25)(300/1000); (d) (0.25/1.86)(1000/300); (e) none of these.

8. 23.4 g (0.30 mol) of calcium fluoride dissolves in 189 g (10.5 mol) of water. What is the weight percent of calcium fluoride in this solution? (a) 0.30/(0.30 + 10.5); (b) 0.30/(189/1000); (c) [23.4/(189 + 23.4)](100); (d) (23.4/189)(100); (e) none of these.

Quiz D

1. Moles of solute per mole of solution is a definition of (a) percent by weight; (b) mole fraction; (c) molality; (d) molarity; (e) none of these.
2. A homogeneous mixture is a definition of all (a) solvents; (b) solvents; (c) solutions; (d) fluids; (e) none of these.
3. Colligative properties are similar in that they all (a) were discovered in college laboratories; (b) are due to solvent molecules linked together; (c) have no harmful side effects; (d) depend on the number of solute particles in solution; (e) none of these.
4. What volume (in liters) of 0.0325 M solution can be produced from 41.6 mL of 0.742 M solution? (a) 41.6(0.742/0.0325); (b) 41.6(0.0325/0.742); (c) (41.6/1000)(0.0325/0.742); (d) (1000/41.6)(0.742/0.0325); (e) none of these.
5. 864 mL of solution contains 143.9 g of sodium phosphate (164.0 g/mol). What is the molarity of the solution? (a) (164.0/143.9)(864/1000); (b) (143.9/164.0)(864/1000); (c) (143.9/164.0)(1000/864); (d) (164.0/143.9)(1000/864); (e) none of these.
6. 841 mL of a 0.625 M sugar solution displays what osmotic pressure in atmospheres at 100°C? (a) (0.625)(0.0821)(100); (b) (0.625)(0.841)(0.0821)(100); (c) (0.625)(0.841)(0.0821)(373); (d) (0.625)(0.0821)(373); (e) none of these.
7. 0.87 mol of solute depresses the freezing point of 742 g of solvent by 0.43 degrees. What is the value of K_f for this solvent? (a) (0.43)(0.87/742); (b) 0.43)(0.87)(0.742); (c) 0.43/(0.87/742); (d) 0.43/(0.87/0.742); (e) none of these.
8. 15.9 g (0.15 mol) of sodium carbonate dissolves in 81 g (4.50 mol) of water.What is the weight percent of sodium carbonate in this solution? (a) [15.9/(15.9 + 81)](100); (b) (15.9/81)(100); (c) 0.15/(81/1000); (d) 0.15/(0.15 + 4.50); (e) none of these.

SAMPLE TEST (20 minutes)

1. A solution contains 750. g of ethanol (46.0 g/mol) and 85.0 g of sucrose (180.0 g/mol). The volume of the solution is 810.0 mL. Determine the values of
   a. the density of the solution
   b. the percent of sucrose in the solution
   c. the mole fraction of sucrose
   d. the molality of the solution
   e. the molarity of the solution
2. What volume of ethylene glycol ($C_2H_6O_2$, 62.0 g/mol, density = 1.12 g/cm³) must be added to 20.0 L of water (H_2O, 18.0 g/mol, density = 1.00 g/cm³) to produce a solution freezing at 14.0°F? The freezing-point depression constant of water is 1.86°C/*m*.

14 An Introduction to Descriptive Chemistry: The First 20 Elements

CHAPTER OBJECTIVES

* **1. Describe and explain physical properties of a number of representative elements in terms of their atomic properties and positions in the periodic table.**

You should know the trends of physical and atomic properties and be able to predict relative values for different elements. These trends are summarized in Table 14-1. You should be able to apply them to all of the representative elements, not just the first 20.

TABLE 14-1
Periodic Trends of Physical and Atomic Properties of Elements

Property	*Behavior down a group*	*Behavior across a period*
ionization potential	decrease	increase
atomic or cationic radius	increase	decrease
hydration energy of cation	decrease	increase
density of solid	increase	increase
electrical conductivity for metals	decrease	increase
melting points of metals	decrease	increase
melting points of nonmetals	increase	decrease

EXAMPLE 14-1 Arrange the following elements in order of increasing melting point: Cl_2, Br_2, K, Mg, Ca.

Among the metals, we expect the melting point to increase from K to Ca to Mg. We know that both $Br_2(l)$ and $Cl_2(g)$ melt at lower temperatures than K. Thus, the expected order is: $Cl_2 < Br_2 < K < Ca < Mg$.

* **2. Use periodic relationships to predict certain properties of the elements and their simple compounds (such as melting and boiling point).**

If the values of a property are plotted against atomic number, a periodic trend often is evident. For example, the boiling points of a number of hydrides are: −87.7°C for PH_3, −88.5°C for GeH_4, −41.5°C for H_2Se, and −17.1°C for SbH_3. Suppose we wish to predict the boiling point of AsH_3. We plot boiling point against the atomic number of the non-hydrogen atom, either within a period (Figure 14-1a) or within a family (Figure 14-1b). The predicted boiling point for AsH_3 of −64°C from Figure 14-1a is not as good a prediction as the value of −52°C from Figure 14-1b, since the true value is −55°C. In general, the family or group trend is a better predictor than the period trend, since elemental properties vary gradually within a family but often change abruptly between groups. You should master this predictive skill for all representative elements, not just the first 20. Other properties that can be predicted are density (and hence atomic volume), ionization potential, heats of vaporization and fusion, atomic size, and, of course, formulas of compounds. For numerical properties, you will have to be supplied with sufficient data for surrounding elements.

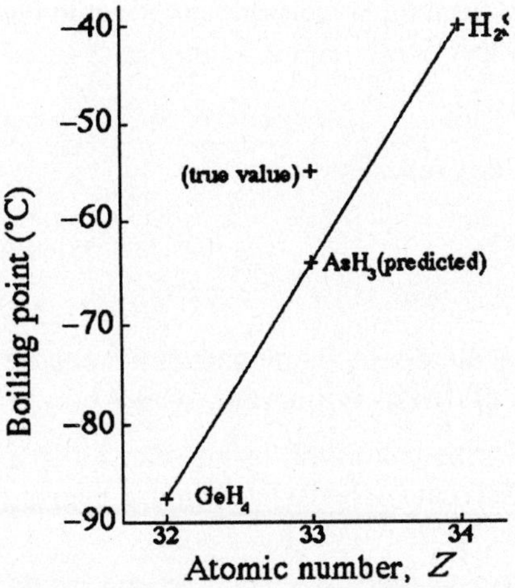

FIGURE 14-1a Boiling Point vs Atomic Number for GeH_4, AsH_3, and H_2Se.

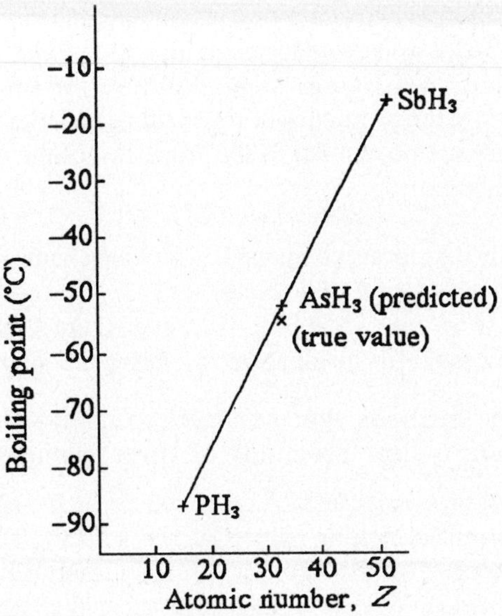

FIGURE 14-1b Boiling Point vs. Atomic Number for PH_3, AsH_3, and SbH_3.

3. Explain the origin of flame colors that are characteristic of some of the elements.

Flame colors of many elements are given in Table 8-2. These colors are emitted when an excited state gives off energy in returning to the ground state. The excited state is produced when the element is raised to a high temperature, as in a Bunsen burner flame. For the alkali metals, the excited state is produced by raising the outermost *s* electron to the *p* subshell. For example, an excited state of Na is $[Ne]3p^1$. For the alkaline earth metals, the excited states are energetic, and thus unstable, molecules, such as MgOH, CaOH, etc. In all cases the energy difference ΔE between the excited state energy (E^*) and the ground state energy (ΔE_0) determines the frequency and thus the color of the emitted light. (Violet light has high frequency while red light has low frequency.)

$$\Delta E = E^* - E_0 = h\nu \quad [1]$$

4. Cite examples of ways in which the first member of a group differs from other members of the same family, including a description of the "diagonal relationship."

The "diagonal relationship" refers to the similarity between an element of the second period and the element with one more valence electron but in the third period. Thus, one notes similarities between Li and Mg, Be and Al, etc. It is reasonable that the behavior of the second period element should be somewhat different from the other members of a family. For the second period element, the electron shell just below the valence shell has the configuration $(n-1)s^2$, while for the other elements in the family that configuration is $(n-1)s^2p^6$. Examples of diagonal behavior include the following.

1. Li (family 1A) reacts with N_2 directly to form a nitride, as does Mg (family 2A). The other alkali metals (2A) do not react directly with N_2.
2. Li_2CO_3 thermally decomposes to the metal oxide and CO_2(g), as do the alkaline earth (2A) carbonates. The other alkali metal (1A) carbonates are thermally stable. These relative thermal stabilities also are true of the hydroxides; NaOH and KOH are thermally stable while LiOH, $Mg(OH)_2$, and $Ca(OH)_2$ thermally decompose to produce the metal oxide and H_2O(g).
3. The fluorides, carbonates, and phosphates of Li (1A), Mg, and the other alkaline earth (2A) elements are relatively insoluble. The fluorides, carbonates, and phosphates of the other alkali metals are quite soluble.
4. Be (2A) and Al (3A) are unreactive toward both cold water and steam. Mg (2A) reacts with steam and Ca (2A) reacts with cold water.
5. BeO dissolves in alkaline solutions as does Al_2O_3. MgO and CaO do not so dissolve.
6. Molten BeF_2 and $BeCl_2$ (2A) are poor conductors of electricity as is also true of $AlCl_3$ (3A), but not of MgF_2, $MgCl_2$, CaF_2, and $CaCl_2$ (2A).

7. Boron (3A) has two allotropes: amorphous and crystalline; the latter is not quite as hard as diamond (4A).
8. Boron (3A) is a semiconductor, similar to Si (4A). Al (3A) is a metal.

In addition, the properties of a second period element often differ from other elements in the same family. However, these are not similar to the properties of the elements in the next family. Some examples of these differences include:

1. C (family 4A) forms compounds with long chains of atoms, unlike any other element.
2. N_2 (5A) possesses a triple bond.
3. HF is a weak acid while HCl, HBr, and HI are strong (all family 7A).
4. HF, H_2O, and NH_3 display strong hydrogen bonding. The hydrides of most other elements do not.

5. Describe methods that are used to obtain the first twenty elements from naturally occurring sources, some typical reactions of these elements, and some of their important uses.

Most metals of groups 1A and 2A are produced by the electrolysis of their molten (fused) halide salts. Often a flux is added to lower the melting point ($CaCl_2$ is added to NaCl). These metals cannot be isolated from aqueous solutions of their ions, since the free metals readily react with water to produce the metal ion and H_2(g).

$$2\ NaCl\ (l,\ with\ CaCl_2) \xrightarrow{electrolysis} 2\ Na(l) + Cl_2(g) \qquad [2]$$

$$2\ LiCl\ (l,\ with\ KCl) \longrightarrow 2\ Li(l) + Cl_2(g) \qquad [3]$$

$$CaCl_2\ (l,\ with\ CaF_2\ or\ KCl) \longrightarrow Ca(l) + Cl_2(g) \qquad [4]$$

$$MgCl_2 \longrightarrow Mg(l) + Cl_2(g) \qquad [5]$$

NaCl, LiCl, and $CaCl_2$ are mined from underground deposits. $MgCl_2$ is extracted from seawater by first adding $Ca(OH)_2$ to precipitate Mg^{2+} as $Mg(OH)_2$. Reaction of $Mg(OH)_2$ with HCl(aq) produces $MgCl_2$(aq), which is evaporated to dryness.

Several elements from among the first 20 occur as free elements in nature and they are "mined" from these sources. These include:

1. He occurs in natural gas from United States gas wells. It is separated by fractional distillation (see objective 13-10) of liquefied natural gas.
2. N_2, O_2, Ne, and Ar are present in air and are separated by fractional distillation of liquid air. The very small percentage of Ne (0.002 %) in air accounts for its relatively high cost compared to Ar (0.934 %).
3. Sulfur is mined from extensive underground deposits by the Frausch process, described in *Figure 14-23 in the text.*
4. Carbon is found naturally as diamonds and in coal. Heating coal at high temperature in the absence of air produces coke, almost pure carbon.

Several other elements are obtained by reacting their oxides or halides with a reducing agent at high temperatures. This reducing element may be an active metal

$$BeF_2(l) + Mg(l) \longrightarrow MgF_2(l) + Be(l) \qquad [6]$$

$$B_2O_3(s) + 3\ Mg(l) \longrightarrow 2\ B(amorphous,\ impure) + 3\ MgO(s) \qquad [7]$$

$$Na(g) + KCl(l) \longrightarrow K(g) + NaCl(l) \qquad [8]$$

or it may be carbon.

$$SiO_2(s) + 2\ C(s) \longrightarrow Si(l) + 2\ CO(g)\ [at\ 3000°C\ in\ an\ electric\ furnace] \qquad [9]$$

$$2\ Ca_3(PO_4)_2(s) + 10\ C(s) + 6\ SiO_2(s) \xrightarrow{1500°C} 6\ CaSiO_3(l) + 10\ CO(g) + P_4(g) \qquad [10]$$

$$2\ H_2O(g) + C(s) \longrightarrow CO_2(g) + H_2(g)\ \ [in\ two\ steps,\ with\ CO(g)\ as\ intermediate] \qquad [11]$$

The remaining elements are produced by various electrolytic processes.

$$2\ HF(l,in\ molten\ KHF_2) \longrightarrow H_2(g) + F_2(g) \qquad [12]$$

$$2\ Al_2O_3(l,\ in\ Na_3AlF_6) + 3\ C(s) \longrightarrow 4\ Al(l) + 3\ CO_2(g) \qquad [13]$$

$$2\ NaCl(aq) + H_2O(l) \longrightarrow 2\ NaOH(aq) + Cl_2(g) + H_2(g) \qquad [14]$$

The *reactions of* H_2 include: formation of covalent hydrides with elements of the carbon, nitrogen, oxygen, and halogen families (groups 4A, 5A, 6A, and 7A) and with Be, Mg, B, Al, and Ga (as in equation [15]); formation of ionic hydrides with the group 1A and 2A metals (see Table 14-2); and serving as a reducing agent for metal oxides (equation [16]).

$$N_2(g) + 3\ H_2(g) \rightleftharpoons 2\ NH_3(g) \qquad [15]$$

$$CuO(s) + H_2(g) \longrightarrow Cu(s) + H_2O(g) \qquad [16]$$

The *reactions of the alkali metals and the alkaline earth metals* are summarized in Table 14-2.

TABLE 14-2 Some Reactions of the Metals of Groups 1A and 2A

	Group 1A	*Group 2A*
X_2:	$2\ M(s) + X_2(g) \longrightarrow 2\ MX(s)$	$M(s) + X_2(g) \longrightarrow MX_2(s)$
O_2:	$4\ Li(s) + O_2(g) \longrightarrow 2\ Li_2O(s)$	$2\ M(s) + O_2(g) \longrightarrow 2\ MO(s)$
	$2\ Na(s) + O_2(g) \longrightarrow Na_2O_2(s)$	Ba forms BaO_2 above 500°C
	$M'(s) + O_2(g) \longrightarrow M'O_2(g)$	
	M' = K, Rb, Cs	
N_2:	$6\ Li(s) + N_2(g) \longrightarrow 2\ Li_3N(s)$	$3\ M(s) + N_2(g) \longrightarrow M_3N_2(s)$
H_2:	$2\ M(s) + H_2(g) \longrightarrow 2\ MH(s)$	$M'(s) + H_2(g) \longrightarrow M'H_2(s)$
		M' = Ca, Sr, Ba
H^+:	$2\ M(s) + 2\ H^+ \longrightarrow 2\ M^+ + H_2(g)$	$M(s) + 2\ H^+ \longrightarrow M^{2+} + H_2(g)$
H_2O:	$2\ M(s) + 2\ H_2O \longrightarrow 2\ M^+ + 2\ OH^- + H_2$	$M'(s) + 2\ H_2O \longrightarrow M'^{2+} + 2\ OH^- + H_2$
		M' = Ca, Sr, Ba
		$Mg(s) + H_2O \xrightarrow{\Delta} MgO(s) + H_2(g)$

Aluminum may be oxidized by strong acid (equation [17]), strong base (equation [18]), or, when finely divided, by $O_2(g)$ (equation [19]).

$$2\ Al(s) + 6\ H^+(aq) \longrightarrow 2\ Al^{3+}(aq) + 3\ H_2(g) \qquad [17]$$

$$2\ Al(s) + 2\ OH^-(aq) + 6\ H_2O(l) \longrightarrow 2\ Al(OH)^{4-} + 3\ H_2(g) \qquad [18]$$

$$4\ Al(s,powder) + 3\ O_2(g) \longrightarrow 2\ Al_2O_3(s) \qquad [19]$$

Carbon can be oxidized by either partial or complete combustion with oxygen (equations [20] and [21]). The element also serves as a reducing agent (equations [9] - [11]). Many carbon-containing compounds are synthesized from either the components of natural gas (equations [22] and [23]) or those of petroleum, rather than from the element.

$$C(s) + O_2(g) \longrightarrow CO_2(g) \qquad [20]$$

$$2\ C(s) + O_2(g) \longrightarrow 2\ CO(g) \qquad [21]$$

$$CH_4(g) + 4\ S(g) \longrightarrow CS_2(l) + 2\ H_2S(g) \qquad [21]$$

$$CH_4(g) + 4\ Cl_2(g) \longrightarrow CCl_4(l) + 4\ HCl(g) \qquad [23]$$

The important *reactions of* N_2 include its combination with $H_2(g)$ to form ammonia (equation [15]) and its combination with $O_2(g)$ at high temperatures to form oxides of nitrogen (equation [24]).

$$N_2(g) + x\ O_2(g) \longrightarrow 2\ NO_x(g) \qquad [24]$$

The important *reactions of P* include reaction with $O_2(g)$ to form $P_4O_6(s)$ or $P_4O_{10}(s)$, as in equation [25]; with a halogen ($X_2 = F_2$, Cl_2, Br_2, or I_2) to form $PX_3(g)$ or $PX_5(g)$, as in equation [26] (although PI_5 has not been synthesized); and with sulfur to form $P_4S_3(s)$ or $P_4S_{10}(s)$, as in equation [27]. In each of these cases, the oxidation state of P depends on the relative quantities of the reactants: smaller quantities of P yielding products of higher P oxidation state.

$P_4(s) + 5\ O_2(g) \longrightarrow P_4O_{10}(s)$ [25]

$P_4(s) + 10\ Cl_2(g) \longrightarrow 4\ PCl_5(g)$ [26]

$P_4(s) + 3\ S(s) \longrightarrow P_4S_3(s)$ [27]

The *reactions of O_2 and S* are confined mainly to the uses of these elements as oxidizing agents (Table 14-2, and equations [19] through [22], [24], [25], and [27] through [30]).

$S(s) + O_2(g) \longrightarrow SO_2(g)$ [28]

$2\ SO_2(g) + O_2(g) \longrightarrow 2\ SO_3(g)$, with $V_2O_5(s)$ as catalyst [29]

$2\ ZnS(s) + 3\ O_2(g) \longrightarrow 2\ ZnO(s) + 2\ SO_2(g)$ [30]

Halogens react with metals to produce ionic compounds (see Table 14-2) with the metal, as the cation, present in its highest common oxidation state. [For example, $FeCl_3(s)$ is formed by the reaction of Fe(s) with $Cl_2(g)$.] Their reactions with phosphorus is described in equation [26] and the preceding text. With sulfur, halogens produce SX_6, although only $F_2(g)$ produces this product without an excess of halogen. Finally, with $H_2(g)$, HX(g) is formed. All of these reactions are summarized in *Table 9-8 in the text.*

The important uses of the elements are given in Table 14-3.

TABLE 14-3 Uses of the Representative Elements and Some of Their Compounds

Substance	*Important or Unique Uses*
	Group 1A
H_2 gas	welding gas; production of $NH_3(g)$, synthetic methanol and hydrogenated vegetable oils.
Li metal	electrical batteries; future fusion fuel (objective 26-00).
Li_2CO_3	medication for gout and mental illness. *p*
$LiAlH_4$	reducing agent for organic reactions. [31]
Na metal	reducing agent for chemical manufacture; in sodium vapor lamps; future coolant in nuclear reactors.
NaOH	(caustic soda or soda lye) paper manufacture; aluminum ore purification; chemical manufacturing; petroleum refining; soap production. [14]
Na_2CO_3	(soda) glass; water softener in detergents. [obj. 4-13] *m*
Na_2SO_4	glass, soap, and paper pulp manufacture; dyeing textiles. [32]
$Na_5P_2O_{10}$	water softener in detergents. [33]
NaCl	chemical manufacturer; refrigeration; household use. *m*
$NaNO_3$	manufacture of fertilizers, KNO_3, and HNO_3. *m*
$NaHCO_3$	(bicarbonate of soda) stomach antacid; baking soda. [obj. 4-13]
NaCN	photographic process; in metal electroplating baths; rat poison. [34]
	Group 2A
Be metal	windows in x-ray tubes; in copper alloys that must be flexible and conduct electricity; in alloys for nonsparking safety tools; in aircraft alloys for lightness and strength.
Mg metal	lightweight, strong alloys often with Al; flashbulbs.
MgO	furnace liner as it is very high melting; antacid. [35]
$MgSO_4$	in sizing paper, cotton, and leather; epsom salts. *m*
	Group 3A
B	to harden steel; control rods for nuclear reactors.
B_4C_3	industrial abrasive. *e*
Borax	($Na_2B_4O_7 \cdot 10H_2O$) weak base; water softener. *m*
B_2O_3	Ingredient with SiO_2 of borosilicate glass (Pyrex). [36]
H_3BO_3	antiseptic; for fireproofing fabric. *m*
Al	Structural metal often alloyed with Cu, Mn, Si, Zn, and Fe; "silver" paint pigment; electric transmission lines; foil; in the thermite process ($Fe_2O_3 + 2\ Al \longrightarrow 2\ Fe + Al_2O_3$) for welding and incendiary bombs.
Al_2O_3	refractory (high temperature resistant) furnace liner; industrial abrasive; synthetic jewels. *m*
$Al_2(SO_4)_3$	for sizing paper. *a*
Alum	[$KAl(SO_4)_2 \cdot 12H_2O$] mordant to fix dye to cloth; to clarify water, size paper, fireproof fabric. [37]

	Group 4A
Si	semiconductor for microcircuitry
SiO_2	glass; Portland cement; sandpaper. *m*
Na_2SiO_3	(water glass) fabric and paper sizing; industrial glue; laundry detergent water softener (partly replacing $Na_5P_3O_{10}$). [38]
Silicates	water-softening equipment. *m*
SiC	Industrial abrasive. [39]
	Group 5A
N_2 gas	inert atmosphere for air-sensitive reactions and working active metals.
NH_3	refrigerant, fertilizer. [15]
N_2O	(laughing gas) inhalation anesthetic. [40]
HNO_3	to make nitrates for fertilizers, plastics, dyes, and explosives. (prob. 4-12-F)
$NaNO_2$	to make dyes; as a meat preservative. [41]
KNO_3	gunpowder ingredient. *p*
$AgNO_3$	photography. [42]
NH_4NO_3	fertilizer; explosive (amatol is a mixture of NH_4NO_3 and TNT, trinitrotoluene). *a*
NH_4Cl	as a flux to clean oxides from metal surfaces in galvanizing and soldering. *a*
P	red allotrope in the striking strip of safety matches; smoke bombs; tracer bullets; burned in air to make P_4O_{10}.
P_4S_3	in the tips of kitchen matches. [27]
P_4O_{10}	laboratory drying agent; with H_2O it forms H_3PO_4. [25]
$Ca(H_2PO_4)_2$	(triple superphosphate) fertilizer. [43]
H_3PO_4	fertilizer manufacture; in soft drinks; in dyeing. [44]
	Group 6A
O_2	steel refining; high temperature torches.
O_3	industrial oxidant; water purification. [formed by electric discharge through O_2]
Na_2O_2	paper and fabric bleach. *e*
H_2O_2	germicide; bleach. [45]
S	Manufacture of H_2SO_4; vulcanization of rubber; in fireworks, gunpowder, and matches.
H_2SO_4	principal product of the chemical industry; make fertilizers and plastics; clean steel. [pr 4-12-A]
SO_2	refrigerant; bleach; food preservative. *e*
Na_2SO_3	bleaches for natural fabrics and paper. *a*
$Na_2S_2O_3$	("hypo") photography; to remove excess Cl_2 used to bleach fabric and paper. [46]
	Group 7A
F_2	manufacture of fluorocarbons as lubricants and nonstick coatings (Teflon).
HF	etching glass ($SiO_2 + 4\ HF \longrightarrow SiF4 + 2\ H_2O$). [47]
NaF	insecticide; rat poison; very low concentrations in drinking water prevent tooth decay. *a*
Cl_2	water purification; plastic manufacture; solvent production; pulp and paper bleach; poison gas (bertholite) in World War I.
HCl	cleaning metals and masonry; removing boiler scale; refining some ores. *e*
$KClO_3$	oxidant in fireworks; weed killer. [48 & 49]
Ca(OCl)Cl	(bleaching powder) produces $Cl_2(g)$ when moistened. [50]
	Noble Gases
He	cryogenics; inert atmospheres; balloons, blimps; breathing mixture with O_2.
Ne	neon lamps.
Ar	inert atmospheres; to fill light bulbs.

TABLE 14-4 Preparation of Selected Compounds of the Representative Elements

$4\ LiH + AlCl_3 \longrightarrow LiAlH_4(s) + 3\ LiCl$ (in ether solution)	[31]
$2\ NaCl(s) + H_2SO_4(conc.) \longrightarrow 2\ HCl(g) + Na_2SO_4(aq)$	[32]
$2\ Na_2HPO_4(l) + NaH_2PO_4(l) \xrightarrow{\Delta} Na_5P_3O_{10}(l) + 2\ H_2O(g)$	[33]
$NaNH_2(s) + C(s) \xrightarrow{\Delta} NaCN(s) + H_2(g)$	[34]
$MgCO_3(s) \xrightarrow{\Delta} MgO(s) + CO_2(g)$	[35]
$2\ H_3BO_3(s) \xrightarrow{\Delta} B_2O_3(s) + 3\ H_2O(g)$	[36]
$K_2SO_4(aq) + Al_2(SO4)_3(aq) \longrightarrow 2\ KAl(SO_4)_3$ (evap. to crystals)	[37]
$SiO_2(s) + Na_2CO_3(s) \xrightarrow{\Delta} Na_2SiO_3(s) + CO_2(g)$	[38]
$SiO_2(s) + 3\ C(s) \xrightarrow{\Delta} SiC(s) + 2\ CO(g)$	[39]
$NH_4NO_3(s) \xrightarrow{\Delta} N_2O(g) + 2\ H_2O(g)$	[40]
$NaNO_3(l) + Pb(s) \xrightarrow{\Delta} PbO(s) + NaNO_2(s)$	[41]
$3\ Ag(s) + 4\ HNO_3(aq) \longrightarrow 3\ AgNO_3(aq) + NO(g) + 2\ H_2O(l)$	[42]
$[3Ca_3(PO_4)_2{\cdot}CaF_2] + 14\ H_3PO_4 + 10\ H_2O \longrightarrow 10\ [Ca(H_2PO_4)_2{\cdot}H_2O](s) + 2\ HF$	[43]
$Ca_3(PO_4)_2(s) + 3\ H_2SO_4(aq) + 6\ H_2O(l) \longrightarrow 2\ H_3PO_4(aq) + 3\ CaSO_4{\cdot}2\ H_2O(s)$	[44]
$BaO_2(s) + H_2SO_4(aq) \longrightarrow H_2O_2(aq) + BaSO_4(s)$	[45]
$Na_2SO_3(aq) + S(s) \longrightarrow Na_2S_2O_3(aq)$ (gentle heating needed)	[46]
$CaF_2(s) + H_2SO_4(conc.) \longrightarrow CaSO_4(s) + 2\ HF(g)$	[47]
$6\ Cl_2(g) + 6\ Ca(OH)_2(aq) \longrightarrow 5\ CaCl_2(aq) + Ca(ClO_3)_2(aq) + 6\ H_2O(l)$	[48]
$2\ KCl(aq) + Ca(ClO_3)_2 \longrightarrow 2\ KClO_3(s) + CaCl_2(aq)$ (at 100°C)	[49]
$Cl_2(g) + Ca(OH)_2(aq) \longrightarrow Ca(ClO)Cl(s) + H_2O(g)$	[50]

* **6. Name several important compounds of the first 20 elements, write equations for their preparation, and describe some of their uses.**

Uses of the first twenty elements and their compounds are given in Table 14-3. Within that table, *m* indicates that the substance is mined from natural sources, often followed by purification; *e* indicates that the compound is formed by direct combination of the elements, often at high temperature; *p* indicates that the compound is a precipitate formed when solutions of two more soluble compounds are mixed; and *a* indicates that the compound is formed by an acid-base reaction as is the case for $NH_4NO_3(s)$. Other methods of preparation are indicated in the table by equation numbers (surrounded by brackets, [ ]), or problem numbers (pr 4-12-F indicates the equations referred to by problem F in objective 12 of chapter 4). Several of the preparative equations for these compounds are given in Table 14-4.

* **7. Write equations for the reactions of various metals with water, nonoxidizing acids, and oxidizing acids.**

The equations for the reactions of the alkali metals and the alkaline earth metals with water and with nonoxidizing acids (H^+) are given in Table 14-2. Only these representative metals readily react with water. On the other hand, all representative metals react readily with an aqueous solution of H^+.

$Ca(s) + 2\ HCl(aq) \longrightarrow CaCl_2(aq) + H_2(g)$ [51]

The term "oxidizing acid" refers to an acid in which some element other than hydrogen is oxidized when the acid reacts. Nitric acid yields $NO(g)$, $N_2O(g)$, or $NH_4^+(aq)$, depending on the activity of the substance with which it reacts; more powerful reducing agents tend to give those products with N in a lower oxidation state. With active metals—primarily the alkali metals, the alkaline earth metals, and zinc—the product is $N_2O(g)$

$4\ Zn(s) + 10\ H^+ + 2\ NO_3^- \longrightarrow 4\ Zn^{2+} + 5\ H_2O + N_2O(g)$ [52]

unless the acid is quite dilute, in which case NH_4^+ is produced.

$4\ Zn(s) + 10\ H^+ + NO_3^- \longrightarrow 4\ Zn^{2+} + 3\ H_2O + NH_4^+$ [53]

Less active metals, and all nonmetals, yield NO(g) when they react with nitric acid.

$3\ Cu(s) + 8\ H^+ + 2\ NO_3^- \longrightarrow 3\ Cu^{2+} + 4\ H_2O + 2\ NO(g)$ [54]

8. Describe the different molecular or physical forms of helium, carbon, oxygen, sulfur, and phosphorus and the physical behavior associated with them.

Above its critical point of 5.3 K, *He* is gaseous, with a boiling point of 4.2 K. Lowering the temperature of liquid He at 1.00 atm pressure does not yield the solid, but rather a superfluid liquid phase, with zero viscosity and a high thermal conductivity. Solid He is produced by applying 25 atm pressure to the initial liquid phase.

At normal pressure and temperature, carbon and oxygen each exist in two allotropic forms. In the case of *carbon*, these two forms—graphite (2.267 g/cm³) and diamond (3.515 g/cm³)—are solids with different crystal structures arising from different types of bonding. In graphite, sp^2 orbitals bond carbon atoms together into sheets, with the remaining valence electron on each carbon atom located in a $2p$ orbital perpendicular to the plane of atoms. These $2p$ orbitals combine to form molecular orbitals that extend over the entire sheet. This bonding scheme is reflected in the physical properties of graphite: good electrical conductivity in the direction of the sheets but rather poor perpendicular to them (the *p* electrons in the molecular orbitals are free to move), and good lubricating properties (the sheets can slide across each other). Diamond, in contrast, is composed of sp^3-hybridized carbon atoms. The crystal thus is a three-dimensional network of covalently bonded carbon atoms. The physical properties of diamond—high melting point, extreme hardness, inability to conduct an electrical current—reflect the fact that the crystal is held together entirely by covalent bonds and there are no mobile electrons.

The two allotropes of *oxygen* are O_2, sometimes known as dioxygen, and O_3, ozone. O_3, produced from O_2 with either an electric discharge or ultraviolet light as an energy source, has a sharp, pungent odor and is an excellent oxidizing agent.

The two solid allotropes of *sulfur* are rhombic (2.08 g/cm³) and monoclinic (1.96 g/cm³). Rhombic crystals are somewhat squat, although we normally see this form as a powder—flowers of sulfur. Long, needle-shaped monoclinic crystals are unstable at room temperature, and slowly convert to rhombic crystals. The S_8 molecules of the solid allotropes also are present in the low temperature melt, which is relatively fluid. Higher temperatures produce S_8 and longer chains which entangle, producing a more viscous liquid. At still higher temperatures, the chains fragment, making the liquid less viscous.

Phosphorus is composed of tetrahedral P_4 molecules in white phosphorus, in the liquid, and in the vapor below 800°C. White phosphorus (1.828 g/cm³) is a highly reactive, waxy solid melting at 44.1°C. It burns spontaneously in air and is poisonous, very volatile, and soluble in $CS_2(l)$. Red phosphorus (2.34 g/cm³) is very much less reactive and less volatile than white phosphorus; it is insoluble in $CS_2(l)$ and nonpoisonous. Black phosphorus is formed white or red phosphorus is heated under pressure (200°C, 4000 atm) or in the presence of a catalyst. It is very unreactive, does not ignite in air below 400°C, is a semiconductor, and has a relatively high density (2.65 g/cm³). The structure consists of pleated sheets of phosphorus atoms and black phosphorus is somewhat flaky like graphite.

9. Explain the functions of O_3 and CO_2 in the heating of the atmosphere.

$O_3(g)$ is formed in the upper atmosphere when ultraviolet light ($\lambda = 260$ nm) is absorbed by O_2 molecules. The O_3 molecule also absorbs ultraviolet light ($\lambda = 210$ to 290 nm). It thus acts as a sunscreen, preventing this high energy (410 to 570 kJ/mol) radiation from reaching the earth's surface. When the radiation is absorbed, the O_3 molecule dissociates, and some of the energy of the radiation appears as the kinetic (heat) energy of the fragments. Thus, the upper atmosphere is warmer than would be the case if $O_3(g)$ were not present.

$CO_2(g)$ efficiently absorbs the infrared radiation ($\lambda = 5.4$ to 5.7 mm) that is emitted from the earth's surface. The excited CO_2 molecule that results then reradiates the infrared radiation: some back in the direction of earth and the rest into space. The net result is that much of the infrared radiation from the surface of the earth is reflected back and does not "lcak out" into space.

10. Discuss some of the factors involved in the formation of smog.

Photochemical smog depends on the presence in the atmosphere of nitrogen oxides, which are largely produced within internal combustion engines. The exhaust typically contains $NO(g)$ and $NO_2(g)$. $NO(g)$ readily reacts with $O_2(g)$ to form $NO_2(g)$. Sunlight dissociates $NO_2(g)$ into $NO(g)$ and $O(g)$. The oxygen atoms then readily unite with $O_2(g)$ to form $O_3(g)$, or react with gaseous hydrocarbons to produce a host of highly reactive radicals. (Radicals are atoms or molecules that have at least one unpaired electron.) The radicals, in turn, produce a wide variety of highly irritating or toxic compounds. Thus, photochemical smog depends on three ingredients: oxides of nitrogen, sunlight, and gaseous hydrocarbons.

Industrial smog, on the other hand, begins with the emission of $SO_2(g)$, mainly from coal-burning power plants and industries. Reaction of $SO_2(g)$ with $O_2(g)$ yields $SO_3(g)$ and further reaction with water produces a mist of H_2SO_4. This mist, combined with the dust and ash from the same coal fires, produces the irritating and low visibility atmosphere known as smog.

11. List some common air pollutants and describe their sources and methods used to control them.

The internal combustion engine produces four major air pollutants: nitrogen oxides, carbon monoxide, hydrocarbons, and lead. Redesign of automobile engines has reduced the amounts of these pollutants. Nitrogen oxides are produced when nitrogen and oxygen are heated at high pressures. Lowering the operating pressure (reducing the compression ratio) of the engine has reduced the quantity of nitrogen oxides. Carbon monoxide and hydrocarbon emissions have been reduced in two ways. First more air is mixed with the fuel (the engine runs "leaner"), giving more complete combustion. Second, a catalytic converter in the exhaust system encourages more complete combustion. Because the catalyst in the converter is "poisoned" (rendered ineffective) by lead, the compound tetraethyllead, $Pb(CH_2CH_3)_4$, must not be present in the fuel. [$Pb(CH_2CH_3)_4$ has been added to gasoline to increase its octane number, producing a higher grade fuel.] Since lead is no longer present in most gasoline, the quantity of lead pollution in the environment and the concentration of lead in the blood of Americans have decreased dramatically.

$SO_2(g)$ is produced in industrial operations, such as roasting metal sulfide ores, and in burning high sulfur coal. The problem is being attacked at two points. Methods are being developed to remove sulfur from coal before it is burned, or SO_2 is removed as it is produced. The fluidized bed method burns coal in the presence of $CaCO_3$ which absorbs the SO_2 (and oxygen) to yield gypsum, $CaSO_4$. Air is continuously passed through the combustion bed of finely ground coal and limestone to promote contact between the reactants. The air makes the bed look like a fluid; hence the name. Another method of removing $SO_2(g)$ is to bubble it through a basic solution such as $Ca(OH)_2(aq)$, in a process known as scrubbing. The main disadvantage is the resulting large quantity of $CaSO_4$ that must be disposed of.

DRILL PROBLEMS

1. Rewrite each of the following lists of species in order of increasing (smallest to largest) value of the indicated property. Base your predictions on what you have learned thus far.
   A. Ionization potential: Ca, Mg, Ba, C
   B. Atomic radius: K, Br, Ca, Se
   C. Ionic radius: Cs^+, Li^+, K^+, Be^{2+}
   D. Hydration energy: Cs^+, Al^{3+}, Mg^{2+}, Ca^{2+}
   E. Ionization potential: Cs, Al, F, Sr
   F. Density of solid: S, Mg, Na, I
   G. Electrical conductivity: Ca, Al, Cs, Rb
   H. Melting point: Al, Mg, Cs, Ca
   I. Melting point: Mg, Rb, Na, Cs
   J. Ionic radius: Al^{3+}, Li^+, Mg^{2+}, Si^{4+}
   K. Atomic radius: Br, I, Cs, Sr
   L. Hydration energy: Na^+, Al^{3+}, Mg^{2+}, K^+
   M. Electrical conductivity: Na, Cs, Rb, Li
   N. Melting point: Ca, Cs, Sr, Mg
   O. Density of solid: P, C, S, I

2. (1) Refer to a periodic table and use the data below to predict the values omitted from the tables that follow. Check your answers with data from a chemical handbook. How good are your predictions? (All temperatures are °C.)

Element	*Melting point*	*Boiling point*	*Critical temperature*	*Color*
F_2	–223	–188	A. _____	light yellow
Cl_2	–102	– 34.1	144	B. _____
Br_2	– 7.2	58.2	302	red-brown
I_2	114	185	553	deep violet
At_2	C. _____	D. _____	E. _____	F. _____

Substance	ΔH_{vap}, kJ/mol	*Boiling point*	*Melting point*	*Substance*	*Boiling point*	*Melting point*
CH_4	8.9	–164	–182.48	H_2O	J. _____	K. _____
SiH_4	12.5	–111.8	G. _____	H_2S	–60.8	–85.5
GeH_4	H. _____	– 88.5	–165	H_2Se	–41.5	–60.4
SnH_4	18.8	I. _____	–150	H_2Te	– 1.8	–48.9

	Substance					
	He	*Ne*	*Ar*	*Kr*	*Xe*	*Rn*
Melting point	–272.2	–284.6	–189.4	L. _____	–111.5	–71
Boiling point	–268.92	–246.02	–185.86	M. _____	–107.1	–65

(2) Use the data below and predict the melting point (°C), boiling point (°C), density of the element (g/cm³), and density of the oxide (g/cm³) for cesium

	Element								
	K	*Ca*	*Sc*	*Rb*	*Sr*	*Y*	*Ba*	*La*	*Cs*
Melting point	64	839	1539	39	769	1523	725	920	N. _____ °C
Boiling point	774	1484	2832	688	1384	3337	1640	3454	O. _____ °C
Density	0.86	1.54	2.99	1.53	2.6	4.34	3.51	6.19	P. _____ g/cm³
Oxide density	2.32	33.8	3.86	3.72	4.7	5.01	5.72	6.51	Q. _____ g/cm³

6. List at least three substances that have each of the following uses.

A. in structural alloys
B. as water softeners
C. in electrical devices
D. in soap production
E. as medications
F. as sizing for fabric or paper
G. as explosives
H. as a refrigerant
I. as fertilizers
J. for lining furnaces
K. in fireworks
L. as pigments
M. as reducing agents
N. in water purification
O. as oxidizing agents
P. as poisons
Q. to produce or use dyes
R. to resist or retard fire
S. in paper manufacture
T. in plaster, cement, or mortar
U. in glass
V. as abrasives
W. What does "sizing" mean?

6 & 7. Complete and balance the following equations. ("No reaction" may be a correct answer.)

A. potassium + water ⟶
B. aluminum + fluoride ⟶
C. calcium + water ⟶
D. magnesium + iodine ⟶
E. calcium oxide + water ⟶
F. sodium + oxygen ⟶
G. sodium + water ⟶
H. magnesium + hydrochloric acid ⟶
I. magnesium + hot water ⟶
J. sodium + nitrogen ⟶
K. calcium + oxygen ⟶
L. $NaF(aq) + Cl_2(g) \longrightarrow$
M. $H_2(g) + F_2(g) \longrightarrow$
N. $Cl_2(g) + H_2O(l) \longrightarrow$
O. $Mg(s) + H_2(g) \longrightarrow$
P. $Li(s) + N_2(g) \longrightarrow$
Q. $Cl_2(g) + S_8(s) \longrightarrow$
R. $Li(s) + O_2(g) \longrightarrow$
S. $Be(s) + H_2(g) \longrightarrow$
T. $Cl_2(g) + H_2S(g) \longrightarrow$
U. $F_2(g) + H_2(g) \longrightarrow$
V. $Na(s) + Cl_2(g) \longrightarrow$
W. $Li(s) + H_2(g) \longrightarrow$

QUIZZES (12 minutes each) Choose the best answer for each question.

Quiz A

1. Which element's flame test color is not due to an excited atom? (a) sodium; (b) lithium; (c) magnesium (d) potassium; (e) none of these.
2. Which two elements do not display a diagonal relationship? (a) Li and Mg; (b) Be and Al; (c) B and Si; (d) N and Si; (e) none of these.
3. Which has the highest electrical conductivity? (a) Al; (b) Na; (c) Cs; (d) Ba; (e) Ca.
4. Which has the highest atomic radius? (a) Rb; (b) Sr; (c) Mg; (d) Al; (e) Li.
5. Which is *not* produced with the use of electricity? (a) Na; (b) Cl_2; (c) Al; (d) Mg; (e) P_4.
6. Which is used as an antiseptic? (a) sodium chloride; (b) calcium carbonate; (c) boric acid; (d) ammonium nitrate; (e) dinitrogen oxide?
7. Which is *not* used to clean metals? (a) hydrochloric acid; (b) ammonium chloride; (c) potassium chlorate; (d) sulfuric acid; (e) all are used.
8. Which substance has the highest boiling point? (a) He; (b) Na; (c) CH_4; (d) SO_2; (e) CO_2.

Quiz B

1. Which of the following metals does not react readily with water? (a) Na; (b) Ca; (c) Mg; (d) Li; (e) K.
2. Which has the smallest density? (a) Li; (b) Cs; (c) Ba; (d) Mg; (e) Na.
3. Which has the greatest ionization potential? (a) Al; (b) Cs; (c) Ca; (d) Na; (e) Ba.
4. Which element has the highest melting point? (a) ozone; (b) sodium; (c) white phosphorus; (d) diamond; (e) sulfur.
5. Which is *not* found free in nature? (a) N_2; (b) O_2; (c) P_4; (d) S_8; (e) none of these.
6. Which is *not* used as an abrasive? (a) calcium oxide; (b) boron carbide; (c) aluminum oxide; (d) silicon dioxide; (e) silicon carbide.
7. Which is used in food? (a) phosphoric acid; (b) sodium fluoride; (c) hydrogen peroxide; (d) potassium chlorate; (e) silver nitrate.
8. The diagonal relationship is thought to be due to which of the following? (a) 2 K-shell electrons; (b) equal numbers of isotopes; (c) approximately equal atomic weights; (d) similar flame colors; (e) none of these.

Quiz C

1. Which has the lowest melting point? (a) Al; (b) Cs; (c) Li; (d) Mg; (e) Ba.
2. Which has the largest radius? (a) Ba; (b) Mg; (c) Be; (d) Li; (e) Al.
3. When Ca reacts with HNO_3(aq), one of the products is likely to be: (a) N_2O(g); (b) NO_2^-(aq); (c) NO(g); (d) NO_2(g); (e) H_2(g).
4. Which of the following is *not* a major pollutant? (a) NO_2(g); (b) CO(g); (c) HCN(g); (d) SO_2(g); (e) O_3(g).
5. Which element must be produced by electrolysis? (a) Cl_2; (b) Na; (c) F_2; (d) K; (e) none of these.
6. Which is used in the walls of houses? (a) calcium sulfate dihydrate; (b) calcium chloride; (c) aluminum oxide; (d) barium nitrate; (e) sodium silicate.
7. Which is *not* used in glass? (a) sodium carbonate; (b) boron trioxide; (c) potassium nitrate; (d) silicon dioxide; (e) none of these.
8. Which will ignite spontaneously in air? (a) sodium; (b) black phosphorus; (c) fluorine; (d) monoclinic sulfur; (e) none of these.

Quiz D

1. Which of the following produces more than one product when it reacts with Cl_2(g)? (a) Na; (b) Mg; (c) P; (d) H_2; (e) Be.
2. Which of the following melts at the highest temperature? (a) Cl_2; (b) F_2; (c) I_2; (d) O_2; (e) Br_2.
3. Which of the following properties decreases as one goes from left to right in a row of the periodic table? (a) density of solid; (b) ionization potential; (c) hydration energy of cation; (d) electrical conductivity for metals; (e) atomic radius.
4. Which of the following is *not* a pollutant produced by the internal combustion engine? (a) nitrogen oxides; (b) lead; (c) carbon monoxide; (d) hydrocarbon vapors; (e) sulfur oxides.
5. Carbon is used in the production of (a) K; (b) Ca; (c) Na; (d) Cl_2; (e) P_4.
6. Which is used as a water softener? (a) lithium carbonate; (b) sodium nitrate; (c) sodium carbonate; (d) potassium chloride; (e) hydrochloric acid.

7. Which is *not* used as a pigment? (a) barium sulfate; (b) aluminum; (c) calcium carbonate; (d) magnesium oxide; (e) none of these.
8. Which is *not* composed of small (less than ten atoms) molecules? (a) white phosphorus; (b) ozone; (c) monoclinic sulfur; (d) black phosphorus; (e) rhombic sulfur.

SYNTHESES

Rather than a sample test, this chapter concludes with twenty syntheses. You should give the reaction or series of reactions that will produce the desired end product from the list of reagents given. The product must be pure unless it is an aqueous solution. If the product is a solid, you must indicate how the solid is recovered from solution (evaporation, precipitation, or some other means). You may make any intermediates or by-products that you wish, but you must indicate how the by-products are separated from the desired end product. Use large-scale industrial processes if necessary. You also may use heat, light, and electricity as you need. You do not need to use all of the reagents listed.

1. $Na_2O_2(s)$ from Na(s), $O_2(g)$, and $H_2O(l)$
2. NaCl(s) from Na(s), $H_2O(l)$, and HCl(aq)
3. $Li_2CO_3(s)$ from Li(s), $H_2O(l)$, and $CO_2(g)$
4. $CaSO_4(s)$ from Ca(s), $Cl_2(g)$, $H_2O(l)$, and $H_2SO_4(aq)$
5. $Al_2(SO_4)_3$ from Al(s), $H_2SO_4(aq)$, and $O_2(g)$
6. $H_2(g)$ from Mg(s), $NH_3(g)$, and $H_2O(l)$
7. $Na_2CO_3(s)$ from $Na_2O(s)$ and $Li_2CO_3(s)$
8. $Ca(OH)_2(s)$ from $CaCO_3(s)$, $H_2O(l)$, $H_2(g)$, and $O_2(g)$
9. $CaCO_3(s)$ from Ca(s), $O_2(g)$, and C(s)
10. $CaSO_3(s)$ from Ca(s), $O_2(g)$, $H_2O(l)$, and S(s)
11. HCl(g) from Na(s), $Cl_2(g)$, and any strong acid (specify which one) except HCl.
12. H_3PO_4 from any elements; use no compounds.
13. $NH_3(g)$ from Li(s), $N_2(g)$, and $H_2O(l)$
14. $H_2(g)$ from Al(s), Ca(s), $NH_3(g)$, and $H_2O(l)$
15. $CaCO_3(s)$ from Ca(s), $O_2(g)$, $H_2O(l)$, and C(s)
16. H_2O_2 from $H_2SO_4(aq)$ and any elements
17. $LiNO_3(s)$ from $HNO_3(aq)$ and any elements
18. $H_2S(g)$ from Ca(s), S(s), and any strong acid.
19. HCl(g) from Mg(s), $Cl_2(g)$, and any strong acid (specify which one) except HCl.
20. $H_3PO_3(aq)$ from any elements; use no compounds.

15 Chemical Kinetics

CHAPTER OBJECTIVES

* 1. **Describe how the rate of a reaction is related to the rate of disappearance of a reactant or formation of a product.**

When a reaction goes faster, the reactants is used up more rapidly and the products are formed more rapidly. But how much more rapidly? Let us consider an example, the reaction of $C_2H_4Br_2$ with KI, equation [1].

$$C_2H_4Br_2 + 3\ KI \longrightarrow C_2H_4 + 2\ KBr + KI_3 \quad [1]$$

We can express the rate of a reaction as the rate of change of concentration of a reactant, Δ[reactant]/Δt, or of a product, Δ[product]/Δt. But we need to be more specific, for at a given reaction rate, KI is used up three times as fast as $C_2H_4Br_2$ and KBr is produced twice as fast as C_2H_4 or KI_3. By convention, the reaction rate is defined as in equation [2].

$$\text{rate of reaction} = -\Delta[\text{reactant}]/\ \nu_r\ \Delta t$$

$$= +\ \Delta[\text{product}]\ /\ \nu_p\ \Delta t \quad [2]$$

where ν_r is the stoichiometric coefficient of the reactant and ν_p is the stoichiometric coefficient of the product. The negative sign is placed before the change in the concentration of the reactant because its change always will be negative and we want the reaction rate to be positive. Thus, for the reaction between $C_2H_4Br_2$ and KI, we have

$$\text{rate of reaction} = -\ \Delta[C_2H_4Br_2]/\Delta t = -\ \Delta[KI]/3\Delta t$$

$$= \Delta[C_2H_4]/\Delta t = \Delta[KBr]/2\Delta t = \Delta[KI_3]/\Delta t \quad [3]$$

2. **Explain how to obtain the data needed for a kinetic study from the results of a simple chemical analysis.**

There are many methods of determining concentrations relatively rapidly, but two of the simplest are colorimetric and titrimetric. In reaction [1], only one species, KI_3, absorbs visible light. Thus, the concentration of KI_3 can be followed by recording the absorbance of visible light by the reaction mixture, since concentration and absorbance are directly related.

$$CH_3COOC_2H_5 + H_2O \longrightarrow CH_3COOH + C_2H_5OH \quad [4]$$

In the hydrolysis of ethyl acetate ($CH_3COOC_2H_5$), equation [4], neither the products nor the reactants absorb visible light. The concentration of one of the products, acetic acid (CH_3COOH), can be determined by titrating the reaction mixture. Thus, in either case, concentration vs. time data can be obtained readily.

* 3. **Establish the exact rate of a chemical reaction from the slope of a tangent line to the concentration vs. time graph. Also explain how to determine the initial rate.**

Although concentration vs. time data can be tabulated, it is most often presented graphically. Such a graph, known as a concentration vs. time curve, is presented in Figure 15-1 for the decomposition of HI(g) at 600 K, reaction [5].

$$2\ HI(g) \longrightarrow H_2(g) + I_2(g) \quad [5]$$

To determine the instantaneous rate from a concentration vs. time curve, a tangent line is drawn to the curve at the time for which one wishes to know the rate. This straight line should just touch the curve at the desired time and should "follow" the curve as far as possible. The slope of this line then is determined from two widely separated points on the line.

$$\text{slope} = \frac{\Delta[\text{graphed species}]}{\Delta t} = \frac{\text{concentration at time } t_1 - \text{concentration at time } t_2}{t_1 - t_2} \qquad [6]$$

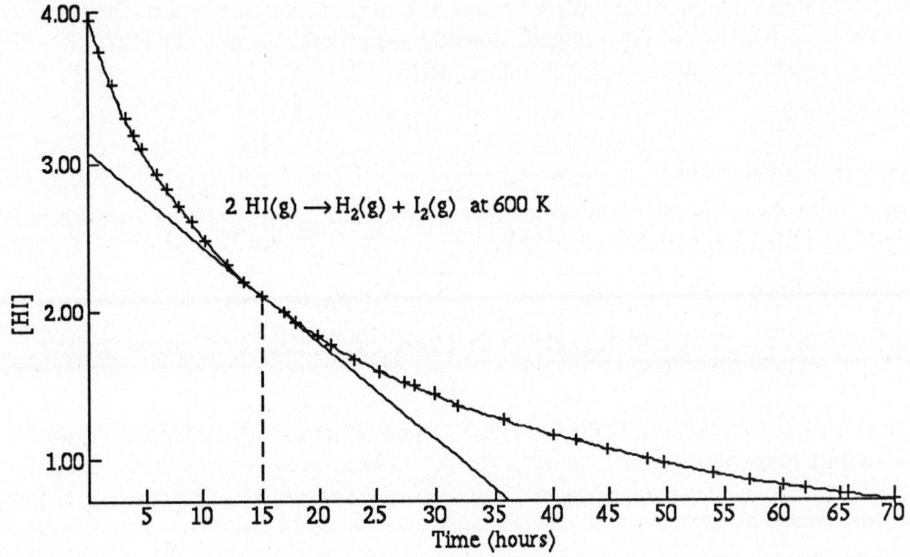

FIGURE 15-1 Concentration vs. Time Data for $2\,HI(g) \longrightarrow H_2(g) + I_2(g)$ at 600 K.

EXAMPLE 15-1 Use the data of Figure 15-1 to determine the rate of the decomposition of HI at 15 hours after the start of the reaction

First, determine (with the dashed vertical line) the point on the curve that represents 15 hours from the start of the reaction. Then draw the tangent to the curve at that point. (The tangent is drawn until it intersects the axes to make the slope easier to determine.) [HI] falls from 3.13 M to 1.00 M in the time interval from 0.00 hours to 32.3 hours. From this information, the slope of the tangent line can be determined.

$$\text{slope} = \frac{3.13\ \text{M} - 1.00\ \text{M}}{0.00\ \text{hr} - 32.3\ \text{hr}} = -0.0659\ \text{mol L}^{-1}\ \text{hr}^{-1}.$$

This slope represents the rate of change of [HI]. Thus, the rate of decomposition of HI is 0.0659 mol L^{-1} hr^{-1}.

One can also determine an instantaneous rate by calculation. There are two data points: the concentrations at two times very close to and on either side of the time at which the rate is to be determined. One computes the rate as one does the slope of the tangent line.

The initial rate of a reaction is a valuable sources of information about that reaction because it is the rate of the forward reaction only. It is not influenced by the reverse reaction, in which the products react to form the reactants, or by any other reactions that involve the products. This is because the products have not reached sufficient concentrations to react at appreciable rates. To determine the instantaneous value of the initial rate, the tangent line is drawn at the point where time, $t = 0$. For slow reactions—ones that are complete in minutes, rather than in seconds or even shorter—the initial rate is obtained by determining the change in the concentration of a product or a reactant during the first few seconds.

4. State the meaning of reaction order and use the rate law to determine the order of a reaction.

The law of mass action states that the rate of a reaction depends on the concentrations of its reactants. The reaction rate of equation [1] is given by

$$\text{rate of reaction} = k[C_2H_4Br_2]^x[KI]^y \qquad [7]$$

where x and y are determined by experiment. In this case, $x = 1$ and $y = 1$. Thus, this reaction is first order in $[C_2H_4Br_2]$ (since $x = 1$) and first order in [KI] (since $y = 1$). The reaction is said to be second order overall (since $x + y = 2$). We also say that this is a second-order reaction. The overall order is the sum of the orders of the various components.

* **5. Apply the method of initial rates to determine the rate law for a reaction.**

We reviewed the determination of initial rates by two methods in objective 15-3. Another general method is the study of clock reactions: those that produce a color change after a short period of time. The time period can be varied by changing the concentrations of the reactants. Consider the general reaction, equation [8], in which one of the products (species C) reacts to produce a color change, equation [9].

$$aA + bB \longrightarrow cC + dD \qquad [8]$$

$$C + \text{indicator} \longrightarrow \text{colored product} \qquad [9]$$

That same product, C, also reacts rapidly with yet another substance, E, as in equation [10]. Because of this reaction, C cannot produce a color until all of E is consumed.

$$C + E \longrightarrow \text{other products} \qquad [10]$$

A typical clock reaction uses I_2 as substance C, starch as the indicator, and $S_2O_3^{2-}$ as substance E. For this example, equations [8-10] become equations [11-13]

$$2\,I^- + H_2O_2 + 2\,H^+ \longrightarrow I_2 + 2\,H_2O \qquad [11]$$

$$I_2 + \text{starch} \longrightarrow \text{dark blue color} \qquad [12]$$

$$I_2 + S_2O_3^{2-} \longrightarrow 2\,I^- + 2\,S_4O_6^{2-} \qquad [13]$$

Reaction [12] does not occur until $S_2O_3^{2-}$ is used up. Such a small $[S_2O_3^{2-}]$ is present compared to $[I^-]$ and $[H_2O_2]$ that the changes in those two concentrations are negligible. Data from a typical series of experiments are given in Table 15-1 where the time is the period elapsed before the dark blue color appears.

TABLE 15-1 Initial Rates for: $2\,I^- + H_2O_2 + 2\,H^+ \longrightarrow I_2 + 2\,H_2O$

	Initial concentration, M			
	$[I^-]$	$[H_2O_2]$	*Time,* s	*Rate,* mol L^{-1} s^{-1}
1.	0.015	0.015	91	1.37×10^{-5}
2.	0.030	0.015	46	2.72×10^{-5}
3.	0.045	0.015	29	4.31×10^{-5}
4.	0.045	0.025	18	6.94×10^{-5}
5.	0.045	0.033	12	9.62×10^{-5}

$[S_2O_3^{2-}] = 0.0025$ M and $[H^+] = 1.7 \times 10^{-5}$ M throughout.

The rate of reaction [11] is given by equation [14].

$$\text{rate} = -\frac{\Delta[H_2O_2]}{\Delta t} = -\frac{\Delta[I^-]}{2\Delta t} = -\frac{\Delta[H^+]}{2\Delta t} = \frac{\Delta[S_2O_3^{2-}]}{2\Delta t} \qquad [14]$$

The last equality of equation [14] is established by the fact that the stoichiometry of reactions [11] and [13] show that the number of moles of I^- reacted equals the number of moles of $S_2O_3^{2-}$ consumed. Since all the $S_2O_3^{2-}$ is used up when the color appears, $\Delta[S_2O_3^{2-}] = 0.025$ M and Δt is the elapsed time from Table 15-1. These data can be used to determine the order of the reaction.

EXAMPLE 15-2 Use the data of Table 15-1 to determine the order of reaction [11].

$$\text{rate} = k[I^-]^x[H_2O_2]^y[H^+]^z$$

We cannot determine z since $[H^+]$ is constant throughout the reaction. The order of reaction with respect to a component cannot be determine unless the concentration of that component changes. Thus, we are dealing with the simplified rate law

$$\text{rate} = k'[I^-]^x[H_2O_2]^y \qquad [15]$$

Any two of the first three lines of data in Table 15-1 can be used to determine x. (They cannot be used to determine y because $[H_2O_2]$ is the same in all three cases.) From lines 1 and 2, we have, by substituting in equation [15],

$$1.37 \times 10^{-5}\ \text{mol L}^{-1}\ \text{s}^{-1} = k'(0.015\ \text{M})^x(0.015\ \text{M})^y$$

$$2.72 \times 10^{-5}\ \text{mol L}^{-1}\ \text{s}^{-1} = k'(0.030\ \text{M})^x(0.015\ \text{M})^y$$

Division of these last two equations one by the other gives

$$0.504 = (0.500)^x$$

The other terms cancel. Taking the logarithm of both sides yields

$$\log 0.504 = \log 0.500 \quad or \quad x = 0.298/0.301 = 0.990$$

In like fashion, we use lines 5 and 3 in Table 15-1 to determine y.

$$9.62 \times 10^{-5}\ \text{mol L}^{-1}\ \text{s}^{-1} = k'(0.045\ \text{M})^x(0.033\ \text{M})^y$$

$$4.31 \times 10^{-5}\ \text{mol L}^{-1}\ \text{s}^{-1} = k'(0.045\ \text{M})^x(0.015\ \text{M})^y$$

$$2.23 = (2.20)^y \quad or \quad y = 1.017$$

This reaction is first order in each of $[I^-]$ and $[H_2O_2]$ and second order overall.

* **6. Use the rate law and rate data to calculate a rate constant, k, or use the rate law and rate constant to calculate rate data.**

In Example 15-2, we determined, for reaction [11],

$$\text{rate} = k'[I^-]^1[H_2O_2]^1 \qquad [16]$$

EXAMPLE 15-3 What is the value of the specific rate constant, k', for reaction [11]?

We use line 3 of Table 15-1.
$4.31 \times 10^{-5}\ \text{mol L}^{-1}\ \text{s}^{-1} = k'\ (0.045\ \text{M})\ (0.015\ \text{M})$
$k' = 0.639\ \text{L mol}^{-1}\ \text{s}^{-1}$
$\text{rate} = (0.639\ \text{L mol}^{-1}\ \text{s}^{-1})\ [I^-]^1[H_2O_2]^1$

EXAMPLE 15-4 What is the rate of reaction [10] when $[I^-] = 0.020$ M and $[H_2O_2] = 0.080$ M?

$\text{rate} = (0.0639\ \text{L mol}^{-1}\ \text{s}^{-1})(0.020\ \text{M})(0.080\ \text{M})$
$= 1.02 \times 10^{-4}\ \text{mol L}^{-1}\ \text{s}^{-1}$

* **7. Establish, through rate data, equations, and graphs, whether a reaction is zero order, first order, or second order.**

We shall consider the general reaction $A \longrightarrow$ products for which rate $= -\Delta[A]/\Delta t$
Zero order: The rate law is $-\Delta[A]/\Delta t = k$. The concentration of A at time t is given by

$$[A]_t = [A]_0 - kt \qquad [17]$$

If we plot $[A]_t$ against time, the result is a straight line with a slope $= -k$. In addition, the rate of a zero-order reaction is constant.
First order: The rate law is $-\Delta[A]/\Delta t = k[A]$. The concentration of A at time t is given by

$$\log[A]_t - \log[A]_0 = \log([A]_t/[A]_0) = -kt/2.303 \qquad [18]$$

If we plot $\log [A]_t$ against time, the result is a straight line of slope $= -k/2.303$. In addition, the half-life of a first-order reaction is constant. (See objective 15-8.)
Second order: The rate law is $-\Delta[A]/\Delta t = k[A]^2$. The concentration of [A] at time t is given by

$$1/[A]_t = 1/[A]_0 + kt \qquad [19]$$

If we plot $1/[A]_t$ against time, the result is a straight line of slope $= k$.
Of course, for all three orders of reaction, the rate constant is constant. Three sets of data are given in Table 15-2 and each set is plotted in each of the three ways just described in Figure 15-2. Notice from Figure 15-2 that the reaction involving A is first order, that involving B is zero order, and that involving C is second order.

TABLE 15-2

Concentration-time data for the reactions: A ⟶ products, B ⟶ products, C ⟶ products

Time,s	[A]	log[A]	1/[A]	[B]	log[B]	1/[B]	[C]	log[C]	1/[C]
0	1.00	0.00	1.00	1.00	0.00	1.00	1.00	0.00	1.00
25	0.78	–0.108	1.28	0.75	–0.125	1.33	0.80	–0.097	1.25
50	0.61	–0.215	1.64	0.50	–0.301	2.00	0.67	–0.174	1.49
75	0.47	–0.328	2.13	0.25	–0.602	4.00	0.57	–0.244	1.75
100	0.37	–0.432	2.70	0 00	—	—	0.50	–0.301	2.00
150	0.22	–0.658	4.55				0.40	–0.398	2.50
200	0.14	–0.854	7.14				0.33	–0.481	3.03
250	0.08	–1.097	12.5				0.29	–0.538	3.45

EXAMPLE 15-5 Of the reactions for which data is given in Table 15-2, one is zero-order, one is first-order, and one is second-order. Without plotting data, determine the order of each.

Since [B] decreases by 0.25 M every 25 s, the rate is constant for the reaction B ⟶ products. Thus, this reaction is zero-order.

[A] drops to one-half of its initial value in slightly less than 75 seconds. In another 75 s (at $t = 150$ s), [A] drops again by a factor of one half. Finally, it seems as though [A] drops yet again by a factor of one half in another 75 s (at $t = 225$ s). Since this reaction has a constant half life, it must be first order.

By the process of elimination, the second-order reaction is C ⟶ products. We could also determine this fact by substituting several values of $1/[A]_t$ into equation [19] and noting that the resulting value of k is constant.

EXAMPLE 15-6 Use the graphs in Figure 15-2 to determine the rate constant for each of the reactions of Table 15-2.

Zero order: The slope of the straight-line graph ([B] vs. t) for the reaction B ⟶ products gives the value of the rate constant.

$$k = -\text{slope} = -\frac{[B]_2 - [B]_1}{t_2 - t_1} = -\frac{0.00 - 1.00}{100 - 0} = 1.00 \times 10^{-2}\ \text{mol L}^{-1}\ \text{s}^{-1}$$

First order: The slope of the straight-line graph (log [A] *vs. t*) for the reaction A ⟶ products equals the rate constant.

$$\text{slope} = \frac{\log[A]_2 - \log[A]_1}{t_2 - t_1} = \frac{-1.097 + 0.328}{250 - 75} = -4.39 \times 10^{-3}\ \text{s}^{-1}$$

$$k = -2.303 \times \text{slope} = 2.303 \times 4.39 \times 10^{-3}\ \text{s}^{-1} = 1.10 \times 10^{-2}$$

Second order: The slope of the straight-line graph (1/[C] vs. t) for the reaction C ⟶ products gives the value of the rate constant.

$$k = \text{slope} = \frac{1/[C]_2 - 1/[C]_1}{t_2 - t_1} = \frac{3.45 - 1.75}{250 - 75} = 9.71 \times 10^{-3}\ \text{L mol}^{-1}\ \text{s}^{-1}$$

* 8. Determine the half-life of a reaction that is zero-order, first-order, or second-order

The half-life of a reaction is the time needed for half of the reactant to be consumed. For a zero-order reaction, the half-life is $t_{1/2} = [A]_0/2k$. For a first-order reaction, the half-life is $t_{1/2} = 0.693/k$. And for a second-order reaction, the half-life is $t_{1/2} = 1/k[A]_0$.

EXAMPLE 15-7 Determine the initial value of the half-life for each of the reactions of Table 15-2.

Zero order: $$t_{1/2} = \frac{[B]_0}{2k} = \frac{1.00\ \text{mol L}^{-1}}{2 \times 1.00 \times 10^{-2}\ \text{mol L}^{-1}\ \text{s}^{-1}} = 50.0\ \text{s}$$

First order: $$t_{1/2} = \frac{0.693}{k} = \frac{0.693}{1.01 \times 10^{-2}\ \text{s}^{-1}} = 68.6\ \text{s}$$

Second order: $$t_{1/2} = \frac{1}{k[A]_0} = \frac{1}{9.71 \times 10^{-3}\ \text{L mol}^{-1}\ \text{s}^{-1} \times 1.00\ \text{mol L}^{-1}} = 103\ \text{s}$$

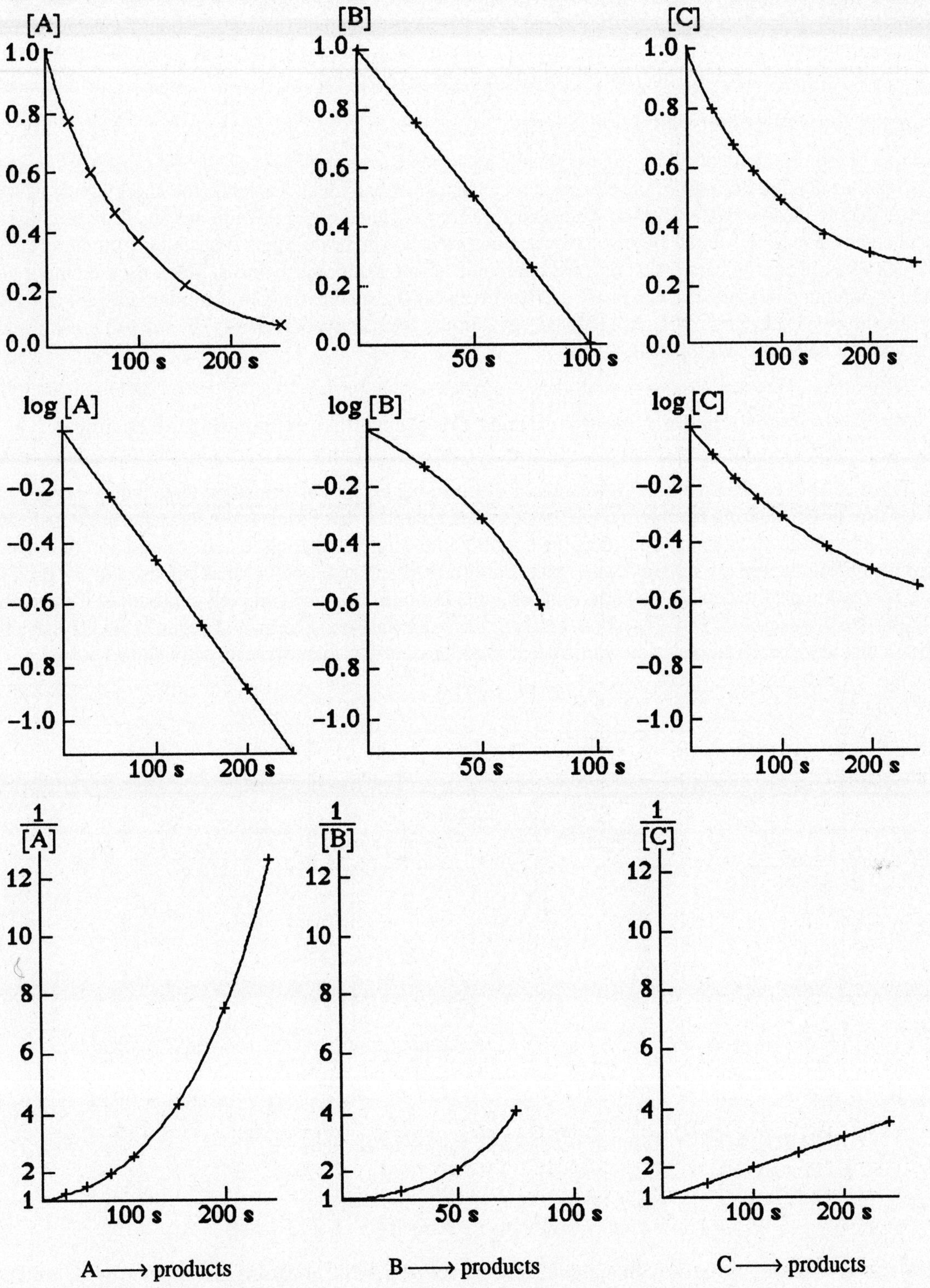

FIGURE 15-1 Graphs of the Data of Table 15-2

9. Describe the collision theory of reactions, stating the factors that affect collision frequency and those that lead to favorable collisions.

Collision theory states that all reactions are the result of collisions between molecules. For a gaseous reactions, the number of collisions in a liter each second is the collision number, denoted Z_{AB}.

$$Z_{AB} = \text{constant } [A]\,[B]\, k_{AB}\, T^{1/2} \qquad [20]$$

k_{AB} is a constant that depends on the properties of the A and B molecules. The factors are about the same for reactions in aqueous solution. Collision theory also requires that the collisions be good ones. The molecules must be oriented correctly in order to react. Thus, not all collisions leads to reactions.

10. Explain the concept of activation energy.

In addition to being oriented correctly, each collision must possess enough energy for the molecules to react. Molecules repel each other because of the electron clouds that surround them. The collision energy must be large enough to overcome these repulsive forces. The needed energy is called the *activation energy*. If a correctly oriented collision has at least this much energy, a reaction occurs between the colliding molecules. Activation energy can be though of as an energy barrier that must be overcome before reaction can occur. Think for a moment of an automobile on a mountain road. The car easily can be driven off the road and down the mountainside. If a guard rail is present, however, the car must have a high energy (similar to an activation energy) in order to break through the barrier and roll down the mountainside.

11. Explain how transition-state theory extends the theoretical explanation of chemical kinetics.

Collision theory fails to explain many of the details of even simple reactions. Transition-state theory explains the rate of a reaction by considering the intermediate or activated complex that forms as the reactants are transformed into products. Transition-state theory considers the bonding in the activated complex, the orientation of reactants needed for it to form, the energy released when bonds are formed and required to break bonds, and the products that result. A reaction profile depicts the path from reactants through activated complex to products. The reaction profile for the decomposition of $H_2O_2(aq)$, both catalyzed and uncatalyzed, is drawn in Figure 15-3. The reaction profile shows that any exothermic reaction would occur instantaneously if there were no activation energy.

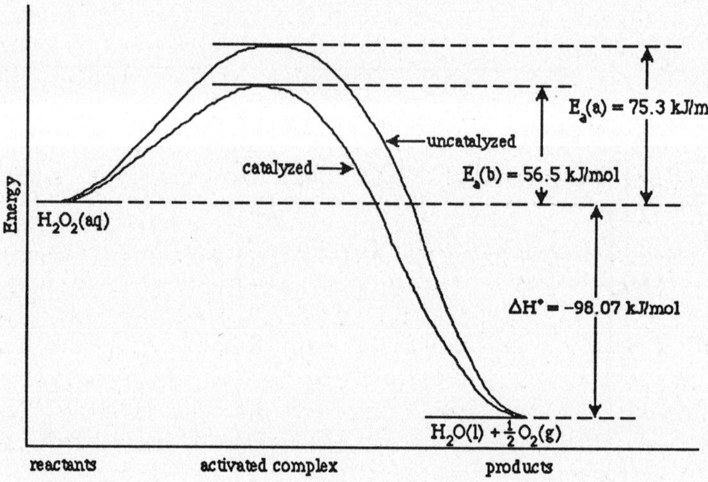

FIGURE 15-3 Reaction Profiles of Catalyzed Reactions.
a. Uncatalyzed: $2\ H_2O_2(aq) \longrightarrow 2\ H_2O(l) + O_2(g)$
b. Catalyzed: $H_2O_2(aq) + I^-(aq) \longrightarrow H_2O(l) + IO^-(aq)$
$IO^-(aq) + H_2O_2(aq) \longrightarrow H_2O(l) + O_2(g) + I^-(aq)$

* 12. Use the Arrhenius equation in calculations involving rate constants, temperatures, and activation energies.

The Arrhenius equation [21] relates the rate constants ($k1$ and k_2) of a chemical reaction at two different kelvin temperatures (T_1 and T_2) to the activation energy, E_a.

$$\ln\frac{k_1}{k_2} = 2.303 \log\frac{k_1}{k_2} = -\frac{E_a}{R}\left(\frac{1}{T_1} - \frac{1}{T_2}\right) \qquad [21]$$

The form of the Arrhenius equation is the same as that of the Clausius-Clapeyron equation, discussed in objective 12-5. If you have forgotten how to use this type of equation, you should review that objective.

13. Describe the role of a catalyst and explain the difference between homogeneous and heterogeneous catalysis.

A *catalyst* is a substance that alters the rate of a chemical reaction while not appearing as a product or a reactant in the overall reaction. Some catalysts speed up reactions (positive catalysts or simply catalysts), and some slow reactions down (negative catalysts or more commonly *inhibitors*). Heterogeneous catalysts are in a different physical state of matter than the reaction mixture (a solid catalyst in contact with a liquid or gas). All of the catalytic activity occurs at the surface of the solid (surface catalysis). Heterogeneous catalysts can be counteracted by poisons that bond strongly to the surface (as lead poisons the catalytic converter in an automobile). In some cases, the catalyst determines the products of the reaction. An example is the reaction between $CO(g)$ and $H_2(g)$.

$$CO(g) + 3\,H_2(g) \xrightarrow{Ni} CH_4(g) + H_2O(g) \qquad [22]$$

$$CO(g) + 2\,H_2(g) \xrightarrow{ZnO/Cr_2O_3} CH_3OH \qquad [23]$$

Homogeneous catalysts are in solution with the reaction mixture. For example, the decomposition of $N_2O(g)$, equation [25], is catalyzed by $Cl_2(g)$, as mechanism [24] shows.

Mechanism:

$$\left.\begin{array}{c} Cl_2(g) \longrightarrow 2\,Cl(g) \\ 2 \times [N_2O(g) + Cl(g) \longrightarrow N_2(g) + ClO(g)] \\ 2\,ClO(g) \longrightarrow Cl_2(g) + O_2(g) \end{array}\right\} \qquad [24]$$

Overall:

$$2\,N_2O(g) \longrightarrow 2\,N_2(g) + O_2(g) \qquad [25]$$

The decomposition of ozone is catalyzed by $N_2O_5(g)$.

Mechanism:

$$2\,N_2O_5(g) \longrightarrow 2\,N_2O_4(g) + O_2(g)$$
$$2 \times [O_3(g) + N_2O_4(g) \longrightarrow O_2(g) + N_2O_5(g)]$$

Overall:

$$2\,O_3(g) \longrightarrow 3\,O_2(g)$$

Catalysts enhance the reaction rate by lowering the activation energy of the reaction.

14. Describe a reaction mechanism, and distinguish between elementary processes and a net chemical reaction.

A reaction mechanism is a series of elementary steps that result in the overall reaction. For example, the catalyzed decomposition of $N_2O(g)$, [25], has the three reactions in [24] as its mechanism. The mechanism of a reaction must meet several criteria. First, each of the elementary steps must involve at most three molecules as reactants. That is, the steps may be unimolecular, bimolecular, or rarely, termolecular. (This is because there is a very small chance that three molecules will collide at the same time.) Second, the elementary steps must sum to the overall reaction. Finally, one must be able to derive the rate law from the mechanism. The chances that the mechanism is correct are greatly enhanced if one can detect some of the reactive intermediates of the mechanism. For example, mechanism [24] is more plausible if one can detect Cl and ClO in the reaction mixture.

*** 15. Derive the rate law from a simple mechanism with the concepts of steady-state condition and rate-determining step.**

An example of a reaction with two mechanisms is equation [26], where $X_2 = H_2$, O_2, Cl_2, or Br_2. (We shall use H_2 as a specific example in this objective.)

$$2\,NO(g) + X_2(g) \longrightarrow N_2O(g) + X_2O(g) \qquad [26]$$

One mechanism is equations [27].

$$\left.\begin{array}{ll} 2\,NO(g) \underset{k_2}{\overset{k_1}{\rightleftharpoons}} N_2O_2(g) & \text{(rapid)} \\ N_2O_2(g) + H_2(g) \xrightarrow{k_3} N_2O(g) + H_2O(g) & \text{(slow)} \end{array}\right\} \qquad [27]$$

The rate of reaction is found from the slow step, the rate-determining step.

$$\text{rate} = \frac{\Delta[N_2O]}{\Delta t} = -\frac{\Delta[N_2O_2]}{\Delta t} = k_3[N_2O_2][H_2]$$

Since N_2O_2 does not appear in the overall equation, we are unable to determine $[N_2O_2]$ experimentally. Thus, we replace $[N_2O_2]$ in the rate law by assuming that the rapid first step is proceeding as quickly forward as reverse. Therefore, $[N_2O_2]$ remains virtually constant during the course of the reaction, a condition known as the "steady-state condition."

forward rate = reverse rate

$k_1[NO]^2 = k_2[N_2O_2]$ *or* $[N_2O_2] = k_1[NO]^2/k_2$

$$\text{rate} = k_3\frac{k_1[NO]^2}{k_2}[H_2] = \frac{k_3k_1}{k_2}[NO]^2[H_2] \qquad [28]$$

This rate law, determined from the mechanism, is in agreement with the experimentally-determined rate law. The other mechanism of reaction [26] is equations [29]. This mechanism gives the same rate law as mechanism [27]. (One way to distinguish between these mechanisms is to detect N_2O_2 or NOH_2 in the reaction mixture.)

$$\left.\begin{array}{lll} NO(g) + H_2(g) \underset{k_2}{\overset{k_1}{\rightleftharpoons}} NOH_2(g) & & \text{(rapid)} \\ NOH_2(g) + NO(g) \xrightarrow{k_3} N_2O(g) + H_2O(g) & & \text{(slow)} \end{array}\right\} \qquad [29]$$

DRILL PROBLEMS

1. Express the rate of each reaction in the drill problems of objective 15-5 in terms of the rates of change of the concentrations of each of the reactants and each of the products in turn.
3. Concentration *vs.* time curves for reactions [5] and [30] are drawn in Figures 15-1 and 15-4.

$$2\ HI(g) \longrightarrow H_2(g) + I_2(g) \text{ at 600 K} \qquad [5]$$

$$C_2H_5Cl \longrightarrow C_2H_4 + HCl \text{ at 700 K} \qquad [30]$$

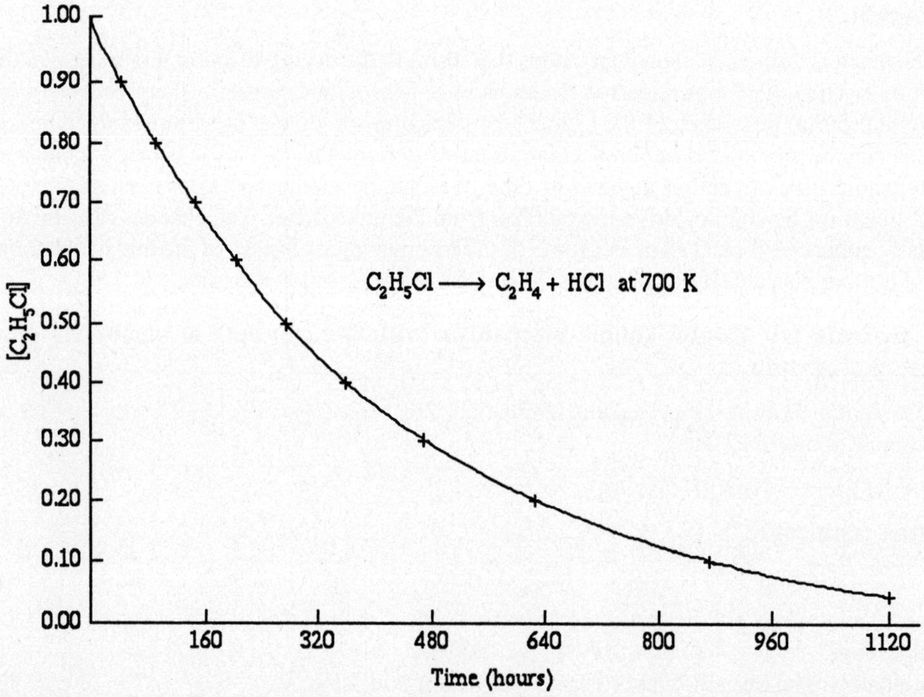

FIGURE 15-4 Concentration vs. Time Data for $C_2H_5Cl \longrightarrow C_2H_4 + HCl$ at 700 K.

(1) Determine the exact rate of the indicated reaction at each of the following times.
For reaction [30]: A. 0 min B. 400 min C. 800 min D. 80 min E. 500 min F. 200 min G. 1000 min
For reaction [6]: H. 5 h I. 10 h J. 20 h K. 60 h L. 30 h M. 40 h N. 35 h

(2) Determine the time at which the indicated reactant has each of the following concentrations and determine the exact rate at this time.
For [C_2H_5Cl] (reaction[30]): O. 0.90 M P. 0.45 M Q. 0.30 M R. 0.18 M
For [HI] (reaction[6]): S. 3.60 M T. 1.80 M U. 0.90 M V. 1.20 M

5. The initial rate of each of the following reactions at various concentrations of the reactant(s) is given on the following lines. Determine the rate law for each reaction (but do not determine a value for k) and state the overall order of the reaction.

A. $2\,N_2O_5(g) \longrightarrow 4\,NO_2(g) + O_2(g)$

$[N_2O_5]$, M	0.170	0.340	0.680
$-\Delta[N_2O_5]/\Delta t$, mol L^{-1} h^{-1}	0.050	0.100	0.200

B. $CH_3CHO(g) \longrightarrow CH_4(g) + CO(g)$

$[CH_3CHO]$, M	0.050	0.100	0.200
$-\Delta[CH_3CHO]/\Delta t$, mol L^{-1} s^{-1}	2.25×10^{-7}	9.00×10^{-7}	3.60×10^{-7}

C. $CO(g) + NO_2(g) \longrightarrow CO_2(g) + NO(g)$

[CO], M	0.10	0.10	0.20
$[NO_2]$, M	0.10	0.20	0.10
$\Delta[CO_2]/\Delta t$, mol L^{-1} s^{-1}	0.012	0.024	0.024

D. $2\,NO(g) + 2\,H_2(g) \longrightarrow N_2(g) + 2\,H_2O(l)$

$[H_2]$, M	0.37	0.37	0.74
[NO], M	0.60	1.20	1.20
$\Delta[N_2]/\Delta t$, mol L^{-1} h^{-1}	0.18	0.72	1.44

E. $NOCl(g) \longrightarrow 2\,NO(g) + Cl_2(g)$

[NOCl], M	0.30	0.60	0.90
$\Delta[Cl_2]/\Delta t$, mol L^{-1} s^{-1}	3.60×10^{-9}	1.44×10^{-8}	3.24×10^{-8}

F. $F_2(g) + 2\,ClO_2(g) \longrightarrow 2\,FClO_2(g)$

$[F_2]$, M	0.10	0.10	0.20
$[ClO_2]$, M	0.010	0.020	0.010
$\Delta[FClO]/\Delta t$, mol L^{-1} s^{-1}	0.0024	0.0048	0.0048

G. $F^- + CH_3Cl \longrightarrow CH_3F + Cl^-$

$[CH_3Cl]$, M	1.2×10^{-6}	2.4×10^{-6}	1.2×10^{-6}	2.4×10^{-6}
$[F^-]$, M	1.2×10^{-6}	1.2×10^{-6}	2.4×10^{-6}	2.4×10^{-6}
$-\Delta[F^-]/\Delta t$, mol L^{-1} s^{-1}	0.69	1.38	1.38	2.7

H. $BrO_3^-(aq) + 5\,Br^-(aq) + 6\,H^+(aq) \longrightarrow 3\,Br_2(l) + 3\,H_2O(l)$

$[BrO_3^-]$, M	0.0333	0.0333	0.0667	0.0333
$[Br^-]$, M	0.0667	0.133	0.0667	0.0667
$[H^+]$, M	0.100	0.100	0.100	0.200
t, s	42.6	21.5	21.1	10.7

(t = time for 0.0020 M BrO_3^- to be consumed. Solution volume is 1.00 L)

I. $NH_4^+(aq) + NO_2^-(aq) \longrightarrow N_2(g) + 2\,H_2O(l)$

$[NH_4^+]$, M	0.0092	0.0092	0.0488	0.0249
$[NO_2^-]$, M	0.098	0.049	0.196	0.196
$\Delta[N_2]/\Delta t$, mol L^{-1} s^{-1}	3.49×10^{-7}	1.75×10^{-7}	3.70×10^{-6}	1.88×10^{-6}

J. $H_2PO_2^-(aq) + OH^-(aq) \longrightarrow HPO_3^{2-}(aq) + H_2(g)$

$[H_2PO_2^-]$, M	0.50	0.25	0.25
$[OH^-]$, M	1.28	1.28	3.84
V_{H2}, cm^3 produced at STP	19.5	25.0	135.0
t, s	10	50	30

(Assume that the solution volume is 1.00 L.)

K. $2\,NO(g) + H_2(g) \longrightarrow N_2O(g) + H_2O(g)$ (at 1000 K)

P_i(NO), atm	0.150	0.075	0.150
$P_i(H_2)$, atm	0.400	0.400	0.200
$-\Delta P_{total}/\Delta t$, atm/min	0.020	0.005	0.010

6. For each of the reactions of the drill problems of objective 15-5, first determine the specific rate constant, making sure to express it with the correct units. Then determine the rate(s) of the reaction when the concentrations are those given below.
   A. $[N_2O_5] = 0.200$ M; $[N_2O_5] = 1.00$ M.
   B. $[CH_3CHO] = 0.150$ M; $[CH_3CHO] = 0.337$ M.
   C. $[CO] = 0.50$ M and $[NO_2] = 0.50$ M; $[CO] = 1.20$ M and $[NO_2] = 0.60$ M.
   D. $[H_2] = 0.12$ M and $[NO] = 0.21$ M; $[H_2] = 0.43$ M and $[NO] = 0.24$ M.
   E. $[NOCl] = 0.50$ M; $[NOCl] = 0.23$ M.
   F. $[F_2] = 0.13$ M and $[ClO_2] = 0.27$ M; $[F_2] = 0.29$ M and $[ClO_2] = 0.11$ M.
   G. $[CH_3Cl] = 9.4 \times 10^{-7}$ M and $[F^-] = 8.7 \times 10^{-7}$ M; $[CH_3Cl] = 4.3 \times 10^{-7}$ M and $[F^-] = 10.2 \times 10^{-7}$ M.
   H. $[BrO_3^-] = 0.100$ M, $[Br^-] = 0.243$ M, and $[H^+] = 0.342$ M;
   $[BrO_3^-] = 0.076$ M, $[Br^-] = 0.111$ M, and $[H^+] = 0.114$ M.
   I. $[NH_4^+] = 0.0100$ M and $[NO_2^-] = 0.100$ M; $[NH_4^+] = 0.0046$ M and $[NO_2^-] = 0.0043$ M.
   J. $[H_2PO_2^-] = 1.04$ M and $[OH^-] = 0.29$ M; $[H_2PO_2^-] = 0.47$ M and $[OH^-] = 0.50$ M.
   K. $P_i(NO) = 0.147$ atm and $P_i(H_2) = 0.089$ atm; $P_i(NO) = 1.06$ atm and $P_i(H_2) = 0.104$ atm.

7. Concentration-time data for various reactions are given below. Determine the order of each reaction for the reactant whose concentration varies. Then determine the value—with the correct units—of the specific rate constant. (mM = 10^{-3} mol/L, μM = 10^{-6} mol/L)

   A. $2\,NH_3(g) \longrightarrow N_2(g) + 3\,H_2(g)$

$[NH_3]$, M	2.000	1.67	1.43	1.11	0.91	0.77	0.67
t, s	0	50.	100.	200.	300.	400.	500.

   B. $CH_3COClN{-}C_6H_5(A) \longrightarrow CH_3COCN{-}C_6H_4{+}Cl^-$

[A], mM	24.5	18.1	13.3	9.7	7.1	5.2
t, min	0	15	30	45	60	75

   C. $2\,AsH_3(g) \longrightarrow 2\,As(s) + 3\,H_2(g)$

$[AsH_3]$, mM	21.6	16.4	15.1	12.6	10.5	8.01	7.32
t, h	0	3.00	4.00	6.00	8.00	11.00	12.00

   D. $NH_4CNO \longrightarrow (NH_2)_2CO$

$[NH_4CNO]$, M	0.400	0.320	0.253	0.207	0.167	0.125
t, min	0	10.0	25.0	40.0	60.0	95.0

   E. $2\,C_4H_6 \longrightarrow C_8H_{12}$

$[C_4H_6]$, M	0.2000	0.1500	0.1200	0.1000	0.0857	0.0750	0.0667	0.0600
t, min	0	500	1000	1500	2000	2500	3000	3500

   F. Cinnamylidene chloride(A) + $C_2H_5OH \longrightarrow$ products (at 23°C, in C_2H_5OH.)

[A], μM	21.1	19.8	17.6	13.3	11.6	9.56	7.43
t, min	0	10	31	67	100	133	178

   G. $(CH_3)_3CBr + H_2O \longrightarrow (CH_3)_3COH + HBr$

$[(CH_3)_3CBr]$, mM	103.9	77.6	63.9	52.9	38.0	27.0	20.7
t, min	0	6.20	10.0	13.5	18.3	26.0	30.8

   H. $CH_3COOCH_3 + H_2O \longrightarrow CH_3COOH + CH_3OH$

$[CH_3OH]$, mM	0	25.16	49.60	73.24	96.21	119.0	140.4
t, min	0	15	30	45	60	75	90

   ($[CH_3COOCH_3] = 0.850$ M and $[H_2O] = 51.0$ M initially.)

   I. $CH_3Br + I^- \longrightarrow CH_3I + Br^-$

$[I^-]$, M	1.00	0.902	0.813	0.662	0.539	0.439	0.357	0.291
t, min	0	10	20	40	60	80	100	120

   J. BzOOPh(aq) + $H_2O \longrightarrow$ BzOOH(aq) + PhOH(aq) $[H_2O] = 55.5$ M

[BzOOPh], μM	33.88	27.83	22.90	18.92	15.38	12.63	10.36
t, min	0	10	20	30	40	50	60

8. Determine the half life of each reaction of objective 15-7. Use the initial concentration of reactant, if necessary.

12. The rate constants—or the rates at the same concentrations of reactants—of various reactions at different temperatures are given in the following lines. Determine the activation energy in kJ/mol of each reaction. Also determine the rate constant, or the rate, at the specified temperature.
   A. $2\,NO_2(g) \longrightarrow 2\,NO(g) + O_2(g)$ $k = 0.775$ s^{-1} at 313 K, $k = 4.02$ s^{-1} at 378 K
   E_a = _______ kJ/mol At 298 K, k = _______ s^{-1}

B. $CO + NO_2 \longrightarrow CO_2 + NO$ $k = 0.220\ M^{-1}\ s^{-1}$ at 650 K, $k = 23.0\ M^{-1}\ s^{-1}$ at 800 K
$E_a =$ ________ kJ/mol At 300 K, $k =$ ________ $M^{-1}\ s^{-1}$

C. $(CH_2COOH)_2CO \longrightarrow (CH_3)_2C{=}O + 2\ CO_2$
$k = 0.0548\ s^{-1}$ at 60°C $k = 2.46 \times 10^{-5}$ at 0°C
$E_a =$ ________ kJ/mol At 100 K, $k =$ ________ s^{-1}

D. $(CH_2)_3 \longrightarrow CH_2CHCH_3$ $E_a =$ ________ kJ/mol At 1000 K, $k =$ ________ s^{-1}

t, °C	470	485	500	510	530
k, s^{-1}	1.10×10^{-4}	2.61×10^{-4}	5.70×10^{-4}	10.21×10^{-4}	2.86×10^{-3}

E. $CH_3I + C_2H_5ONa \longrightarrow CH_3OC_2H_5 + NaI$

t, °C	0	6	12	18	24
k, $M^{-1}\ s^{-1}$	5.60×10^{-5}	1.18×10^{-4}	2.45×10^{-4}	4.88×10^{-4}	1.00×10^{-3}

$E_a =$ ________ kJ/mol At 100 °C, $k =$ ________ $M^{-1}\ s^{-1}$

15. (1) For the reaction $CO + NO_2 \longrightarrow CO_2 + NO$ the rate law under certain conditions is rate = $k[NO_2]^2$. Which of the following mechanisms is the most plausible for this reaction? Find the rate law for each mechanism.

A. $CO + NO_2 \longrightarrow CO_2 + NO$

B. $2\ NO_2 \rightleftharpoons N_2O_4$ (fast)
$N_2O_4 + 2\ CO \longrightarrow 2\ CO_2 + 2\ NO$ (slow)

C. $2\ NO_2 \longrightarrow N_2 + 2\ O_2$ (slow)
$2\ CO + O_2 \longrightarrow 2\ CO_2$ (fast)
$N_2 + O_2 \longrightarrow 2\ NO$ (fast)

D. $2\ NO_2 \longrightarrow NO_3 + NO$ (slow)
$NO_3 + CO \longrightarrow NO_2 + CO_2$ (fast)

(2) For the reaction $CO(g) + Cl_2(g) \longrightarrow COCl_2(g)$ under certain conditions the rate law is rate = $k[CO][Cl_2]^{3/2}$. Which of the following is the most plausible mechanism? Find the rate law for each mechanism.

E. $\frac{1}{2}Cl_2 \rightleftharpoons Cl$ (fast)
$Cl + Cl_2 \rightleftharpoons Cl_3$ (fast)
$Cl_3 + CO \longrightarrow COCl_2 + Cl$ (slow)
$Cl \rightleftharpoons \frac{1}{2}Cl_2$ (fast)

F. $\frac{1}{2}Cl_2 \rightleftharpoons Cl$ (fast)
$Cl + CO \rightleftharpoons COCl$ (fast)
$COCl + Cl_2 \longrightarrow COCl_2 + Cl$ (slow)
$Cl \rightleftharpoons \frac{1}{2}Cl_2$ (fast)

G. $Cl_2 + CO \longrightarrow CCl_2 + O$ (slow)
$O + Cl_2 \longrightarrow Cl_2O$ (fast)
$Cl_2O + CCl_2 \longrightarrow \frac{1}{2}Cl_2$ (fast)

H. $Cl_2 + CO \rightleftharpoons COCl + Cl$ (fast)
$COCl + Cl_2 \longrightarrow COCl_2 + Cl$ (slow)
$Cl + Cl \longrightarrow Cl_2$ (fast)

(3) For the reaction $H_2 + Br_2 \longrightarrow 2\ HBr$ under certain conditions the rate law is rate = $k[H_2][Br_2]^{1/2}$. Which of the following is the most plausible mechanism? Find the rate law for each mechanism.

I. $H_2 + Br_2 \longrightarrow 2\ HBr$

J. $H_2 \rightleftharpoons 2\ H$ (fast)
$H + Br_2 \longrightarrow HBr + Br$ (slow)
$Br + H_2 \longrightarrow HBr + H$ (fast)

K. $Br_2 \rightleftharpoons 2\ Br$ (fast)
$Br + H_2 \longrightarrow HBr + H$ (slow)
$H + Br_2 \longrightarrow HBr + Br$ (fast)

(4) Which of the following mechanisms is the most plausible for $2\ NO(g) + O_2(g) \longrightarrow 2\ NO_2(g)$ if the rate law is rate = $k[NO]^2[O_2]$? Find the rate law for each mechanism.

L. $2\ NO + O_2 \longrightarrow 2\ NO_2$

M. $2\ NO \rightleftharpoons N_2O_2$ (fast)
$N_2O_2 + O_2 \rightleftharpoons 2\ NO_2$ (slow)

N. $2\ NO \rightleftharpoons N_2 + O_2$ (fast)
$N_2 + 2\ O_2 \rightleftharpoons 2\ NO_2$ (slow)

(5) Which mechanism is the most plausible for $C_4H_9Br + OH^- \longrightarrow C_4H_9OH + Br^-$ if the rate law is rate = $k[C_4H_9Br]$? Find the rate law for each mechanism.

O. $C_4H_9Br + OH^- \longrightarrow C_4H_8Br^- + H_2O$
$C_4H_8Br^- \longrightarrow C_4H_8 + Br^-$ (slow)
$C_4H_8 + H_2O \longrightarrow C_4H_9OH$ (fast)

P. $C_4H_9Br \longrightarrow C_4H_9^+ + Br^-$ (slow)
$C_4H_9^+ + OH^- \longrightarrow C_4H_9OH$ (fast)

Q. $C_4H_9Br + OH^- \longrightarrow C_4H_9OH + Br^-$

(6) R. Show that the following three-step mechanism is plausible for $Cl_2 + CHCl_3 \longrightarrow CCl_4 + HCl$. Give the rate law that results

$Cl_2 \rightleftharpoons 2\ Cl$ (slow) $Cl + CHCl_3 \longrightarrow CCl_4 + H$ (slow) $H + Cl \longrightarrow HCl$ (fast)

(7) S. Is the following mechanism plausible for the NO-catalyzed decomposition of O_3?

$$2\ O_3(g) \xrightarrow{NO} 3\ O_2(g)$$

What rate law is obtained from this mechanism?

$NO + O_3 \longrightarrow NO_2 + O_2$ (slow) $NO_2 + O_3 \longrightarrow NO + 2\ O_2$ (fast)

QUIZZES (20 minutes each) Choose the best answer to each question.

Quiz A

1. For $4\,OH + H_2S \longrightarrow SO_2 + 2\,H_2O + H_2$ initial rate data are

[OH], M	13×10^{-12}	39×10^{-12}	39×10^{-12}
$[H_2S]$, M	21×10^{-12}	21×10^{-12}	42×10^{-12}
Rate, M s^{-1}	1.4×10^{-6}	4.2×10^{-6}	8.4×10^{-6}

The rate law is $k[OH]^x[H_2S]^y$ and (a) $x = 2, y = 2$; (b) $x = 2, y = 1$; (c) $x = 1, y = 2$; (d) $x = 1, y = 1$; (e) none of these.

2. The reaction $2\,NO_2Cl \longrightarrow 2\,NO_2 + Cl_2$ is first order. The rate equation is thus: rate = (a) $k[Cl_2]$; (b) $k[NO_2]^2[Cl_2]/[NO_2Cl]^2$; (c) $k[NO_2Cl]^2$; (d) $k[NO_2Cl]$; (e) none of these.
3. For a given reaction, a reactant that has an initial concentration of 0.75 M is completely gone in 30 min. The reaction is thus _____ order with a rate constant of _____. (a) first, 0.0462 min^{-1}; (b) zero, 0.025 M min^{-1}; (c) first, 0.025 min^{-1}; (d) zero, 0.050 M min^{-1}; (e) none of these.
4. What is the first-order rate constant when the half life is 57 seconds? (a) 0.018 s^{-1}; (b) 0.024 s^{-1}; (c) 0.009 s^{-1}; (d) 0.012 s^{-1}; (e) none of these.
5. For a given reaction, rate = $k[A][B]^2$. Which of the following will change the value of k the most? (a) halving [B]; (b) doubling [A]; (c) removing product; (d) increasing both [A] and [B]; (e) none of these.
6. The rate of a specific chemical reaction is independent of the concentrations of the reactants. Thus the reaction is (a) first order; (b) second order; (c) exothermic; (d) catalyzed; (e) none of these.
7. For the reaction $A \longrightarrow$ products time-concentration data are

time, min	0	2	4	6	8
[A], M	1.00	0.80	0.60	0.40	0.20

This reaction is (a) catalyzed; (b) zero order; (c) first order; (d) second order; (e) none of these.

8. In the Haber-Bosch process, nickel metal is used to increase the rate of the reaction $N_2(g) + 3\,H_2(g) \longrightarrow 2\,NH_3(g)$. The nickel is called (a) an inhibitor; (b) a negative catalyst; (c) a homogeneous catalyst; (d) a heterogeneous catalyst; (e) none of these.

Quiz B

1. For the reaction $2\,HgCl_2 + C_2O_4^{2-} \longrightarrow$ products initial rate data are

$[HgCl_2]$, M	0.0836	0.0836	0.0418
$[C_2O_4^{2-}]$, M	0.202	0.404	0.404
Init. rate, M h^{-1}	0.26	1.04	0.52

The rate law is rate = $[HgCl_2]^x[C_2O_4^{2-}]^y$. Thus (a) $x = 1, y = 1$; (b) $x = 2, y = 2$; (c) $x = 2, y = 2$; (d) $x = 1, y = 2$; (e) none of these.

2. For the reaction $2\,NO_2 + F_2 \longrightarrow 2\,NO_2F$ the rate equation is rate = $k[NO_2][F_2]$. The order of the reaction is thus (a) zero; (b) third; (c) first; (d) second order in $[NO_2]$; (e) none of these.
3. A certain reaction is first order in [A] and second order in [B]. When [A] = 3.0 M and [B] = 2.0 M, the rate is 0.144 M s^{-1}. What is the rate constant of this reaction? (a) 0.008; (b) 0.024; (c) 0.004; (d) 0.018; (e) none of these.
4. The rate constant for a first-order reaction is 0.0396 min^{-1}. Its half life is (a) 175 min; (b) 17.5 min; (c) 25.3 min; (d) 35.0 min; (e) none of these.
5. Which of the following does *not* determine the rate of a reaction? (a) value of $\Delta H°$; (b) activation energy; (c) presence of a catalyst; (d) temperature of reactants; (e) none of these.
6. The units of the specific rate constant for a zero-order reaction are (a) s^{-1}; (b) $M^{-1}\,s^{-1}$; (c) $M^{-2}\,s^{-1}$; (d) s; (e) none of these.
7. Which plot is typical of data for a first-order reaction?

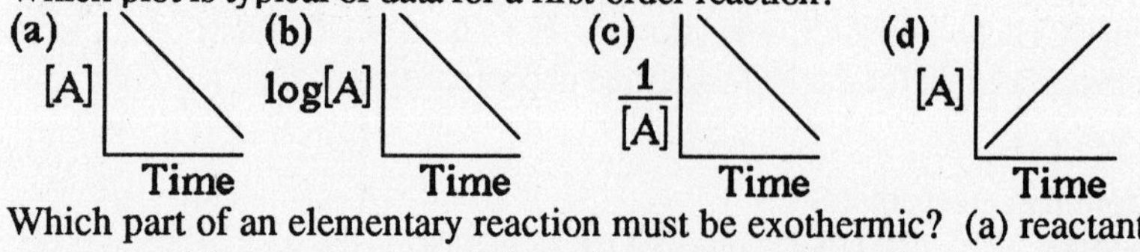

(e) none of these.

8. Which part of an elementary reaction must be exothermic? (a) reactants $\longrightarrow$ products; (b) products $\longrightarrow$ reactants; (c) reactants $\longrightarrow$ activated complex; (d) products $\longrightarrow$ activated complex; (e) activated complex $\longrightarrow$ products.

Quiz C

1. For $2\ NO + O_2 \longrightarrow 2\ NO_2$ initial rate data are

[NO], M	0.010	0.010	0.030
$[O_2]$, M	0.010	0.020	0.020
Rate, M s^{-1}	2.5×10^{-9}	5.0×10^{-9}	45.0×10^{-9}

The rate is rate = $k[NO]^x[O_2]^y$ and (a) $x = 1, y = 1$; (b) $x = 2, y = 1$; (c) $x = 1, y = 2$; (d) $x = 2, y = 2$; (e) none of these.

2. If the half life of a reaction depends on the concentration of the reactant, then the reaction cannot be _____ order. (a) zero; (a) first; (c) second; (d) third; (e) none of these.
3. A reaction has the rate law rate = $k[A][B]^2$. Which of the following will cause the rate to increase the most? (a) tripling [B]; (b) doubling [B]; (c) doubling [A]; (d) quadrupling [A]; (e) lowering temperature.
4. A reaction is first order. If its initial rate is 0.0200 M s^{-1} and 25.0 days later its rate is 6.25×10^{-4} M s^{-1}, then its half life is (a) 25.0 days; (b) 12.5 days; (c) 50.0 days; (d) 5.0 days; (e) none of these.
5. A combination of reactants that exists for an instant at the time of collision between reactants is called a(n) (a) catalyst; (b) activated complex; (c) reaction product; (d) rate-determining step; (e) none of these.
6. The correct units of the specific rate constant for a first-order reaction could be (a) $M^{-2}\ s^{-1}$; (b) s; (c) $M^{-1}\ s^{-1}$; (d) s^{-1}; (e) none of these.
7. For the reaction $A \longrightarrow$ products time-concentration data are

time, min	0	1	3	4
[A], M	1.00	0.50	0.25	0.20

This order of this reaction is (a) zero; (b) first; (c) second; (d) cannot be determined; (e) none of these

8. Which of the following lowers the activation energy of a reaction? (a) adding reactants; (b) lowering the temperature; (c) removing products; (d) adding a catalyst; (e) raising temperature.

Quiz D

1. For $C_2H_4Br_2 + 3\ KI \longrightarrow C_2H_4 + 2\ KBr + KI_3$ initial rate data are

$[C_2H_4Br_2]$, M	0.500	0.500	1.500
[KI], M	1.80	7.20	1.80
Rate, M min^{-1}	0.269	1.08	0.807

The rate law is rate = $k[C_2H_4Br_2]^x[KI]^y$ and (a) $x = 1, y = 2$; (b) $x = 2, y = 1$; (c) $x = 1, y = 1$; (d) $x = 0, y = 0$; (e) none of these.

2. The time required for the initial amount of reactant in a first-order reaction to decrease by a factor of two is known as the (a) reaction time; (b) activation time; (c) half life; (d) reaction rate; none of these.
3. If a reaction has a rate equation of rate = $k[A]^2[B]$, then it is (a) second order in [B]; (b) first order in [A]; (c) second order overall; (d) spontaneous; (e) none of these;
4. If a reaction is first order with a rate constant of $5.48 \times 10^{-2}\ s^{-1}$, how long is required for three-fourths of the initial concentration of reactant to be used up?
5. The effect of a catalyst in a chemical reaction is to change the (a) heat evolved; (b) work done; (c) yield of product; (d) activation energy; (e) none of these.
6. The units of a first-order rate constant could be (a) s^{-1}; (b) M^{-1}; (c) $M^{-1}\ s^{-1}$; (d) M s; (e) none of these.
7. Which plot is typical of data for a second-order reaction?

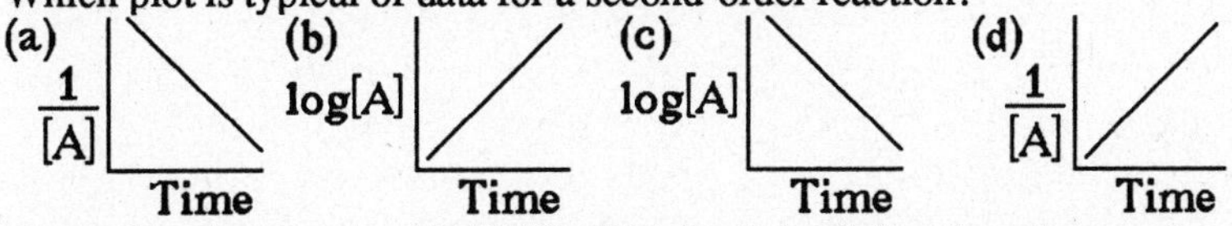

(e) none of these.

8. A reaction has rate = $k[A][B]^2$. Which of the following change the rate constant, k, the least? (a) adding homogeneous catalyst; (b) adding inhibitor; (c) raising temperature 10°C; (d) raising temperature 20°C; (e) increasing [A].

SAMPLE TEST

1. A chemical reaction has a rate constant of 7.00×10^{-9} h^{-1} at 400 K and 1.00×10^{-4} h^{-1} at 500 K. What is the activation energy in kJ/mol for this reaction?
2. Benzene diazonium chloride, $C_6H_5N{\equiv}N^+Cl^-$, decomposes to nitrogen gas and other products. The concentration of benzene diazonium chloride (B) in isoamyl alcohol is measured at various times; the results follow. What is the order of the decomposition reaction and what is the value of the specific rate constant?

[B], mM	3.116	2.822	2.542	2.291	2.064	1.859	1.674
t, min	0	40	80	120	160	200	240
[B], mM	1.505	1.358	1.223				
t, min	280	320	360				

What is the rate of this reaction when [B] = 0.00504 M?

16 Principles of Chemical Equilibrium

CHAPTER OBJECTIVES

1. Describe the condition of equilibrium in a reversible reaction.

Most chemical reactions do not "go to completion," as we assumed in Chapter 4 and in our studies of reactions since then. Rather the final reaction mixture contains some unreacted product. In this way, chemical reactions are similar to many common processes. Consider the "reaction"

woman + man $\longrightarrow$ married couple [1]

Even though equation [1] is written as a one-way process, we know that it occurs both ways. So do chemical reactions. In addition, we can envision reaching a final state from different starting conditions. For example, if everyone in a large city were to marry on January 1, we would expect some to be unmarried (through divorce or death, for example) at the end of the year. In similar manner, if all were single at the beginning of the year, we would expect many to be married by December 31. In fact, we expect the percent of married persons in each of these cases to be essentially the same. The final situation is known as a dynamic equilibrium, dynamic because both the forward (marriage) and the reverse ("unmarriage") reactions are occurring constantly.

Much the same thing occurs for a chemical reaction. Both the forward and the reverse reactions occur continuously, producing a dynamic equilibrium. We can start with either pure reactants or with pure products (or at some intermediate state) and reach the same equilibrium condition. Data for three experiments for the reaction

$2\,NO_2(g) \rightleftharpoons N_2O_4(g)$ at 352.6 K [2]

are given in Table 16-1. The reaction is carried out in a 50.0-L container, except for Trial 3, carried out in a 20.0-L container.

TABLE 16-1 Equilibrium Data for $2\,NO_2(g) \rightleftharpoons N_2O_4(g)$ at 352.6 K

		Amount, mol		*Concentration, M*	
Trial	*Substance*	*Initial*	*Final*	*Initial*	*Final*
1	NO_2	1.000	0.8175	0.0200	0.01635
	N_2O_4	0.000	0.0912	0.000	0.00182
2	NO_2	0.000	0.4457	0.000	0.00891
	N_2O_4	0.250	0.0271	0.00500	0.000542
3	NO_2	0.160	0.210	0.00800	0.01050
	N_2O_4	0.040	0.0150	0.00200	0.000750

* 2. Describe how equilibrium concentrations are established experimentally.

Often an equilibrium mixture can be "frozen" by quickly cooling it or by adding an excess of a reagent that reacts completely with a reactant or a product. After the mixture has been frozen, its composition can be determined by separating the components and identifying them. This procedure can be followed with the mixture resulting from reaction [3].

$2\,HI(g) \rightleftharpoons H_2(g) + I_2(g)$ [3]

Reaction [2] cannot be "frozen" readily because it is very rapid. Thus, this reaction must be studied under the conditions of a *dynamic* equilibrium, 352.6 K and a 50.0-L vessel. $NO_2(g)$ is red-brown, while $N_2O_4(g)$ is colorless. The intensity of the color and thus $[NO_2]$ can be determined visually or with a colorimeter. If $[NO_2]$ is combined with the known initial amounts of NO_2 and N_2O_4, we can determine the final amounts in moles and the concentrations of both species.

EXAMPLE 16-1 0.160 mol $NO_2(g)$ and 0.040 mol $N_2O_4(g)$ are placed in a 20.0-L container and come to equilibrium at 352.6 K. At equilibrium for reaction [2], $[NO_2] = 0.01050$ M. What are the equilibrium amounts and concentrations of all substances?

Since the container has a volume of 20.0 L, the equilibrium amount of NO_2 is

$n_{NO2} = 0.01050 \text{ M} \times 20.0 \text{ L} = 0.210 \text{ mol } NO_2$

0.210 mol − 0.160 mol = 0.050 mol NO_2 is produced, and then we have

Reaction:	$2\ NO_2(g)$	$\rightleftharpoons$	$N_2O_4(g)$
Initial:	0.160 mol		0.040 mol
Changes	+0.050 mol		−0.025 mol
Equilibrium:	0.210 mol		0.015 mol

* 3. **Write the equilibrium constant expression in terms of concentrations, K_c, for a reaction and use the value of K_c and the concentrations of all species but one to determine the equilibrium concentration of that species.**

The equilibrium constant is written as the ratio of the concentrations—product concentrations over those of reactants—each raised to the power of its stoichiometric coefficient in the balanced chemical equation. The equilibrium constant expressions for all reactions are written in the same way (but see objectives 16-4 and 16-5). Finding one equilibrium concentration given all the others and the value of K_c is a matter of rearranging the expression to solve for the necessary concentration.

EXAMPLE 16-2 What is $[N_2O_4]$ at equilibrium if $[NO_2] = 0.100$ M?

$2\ NO_2(g) \rightleftharpoons N_2O_4(g) \quad K_c = 6.81$ at 352.6 K

$$K_c = \frac{[N_2O_4]}{[NO_2]^2} \quad \textit{or} \quad [N_2O_4] = K_c[NO_2]^2$$

Thus, $[N_2O_4] = (6.81)(0.100)^2 = 0.681$ M

* 4. **Derive K values for situations where chemical equations are reversed, multiplied through by constant coefficients, or added together.**

Reaction [4] is the *reverse* of reaction [2].

$N_2O_4(g) \rightleftharpoons 2\ NO_2(g)$ [4]

For reaction [4], the equilibrium constant expression, $K_4 = [NO_2]^2/[N_2O_4]$. K_4 for reaction [4] is merely the inverse of K_2 for reaction [2]. Thus $K_4 = 1/K_2 = 1/6.81 = 0.147$. In similar fashion, reaction [4] can be *multiplied by one-half* to obtain reaction [5].

$\frac{1}{2} N_2O_4(g) \rightleftharpoons NO_2(g)$ [5]

The equilibrium constant expression for reaction [5] is $K_5 = [NO_2]/[N_2O_4]^{1/2} = K_4^{1/2} = (0.147)^{1/2} = 0.383$. Finally, let us consider reaction [6].

$2\ NO_2(g) \rightleftharpoons 2\ NO(g) + O_2(g)$ [6]

For this reaction, $K_6 = [NO]^2[O_2]/[NO_2]^2 = 5.01 \times 10^5$. If we *add reactions* [4] and [6], we obtain reaction [7].

$N_2O_4(g) \rightleftharpoons 2\ NO(g) + O_2(g)$ [7]

The equilibrium constant expression for reaction [7], $K_7 = [NO_2]^2[O_2]/[N_2O_4]$ is the product of the equilibrium constant expressions for reactions [4] and [6].

$$K_4 \times K_6 = \frac{[NO_2]^2}{[N_2O_4]} \times \frac{[NO]^2[O_2]}{[NO_2]^2} = 0.147 \times 5.01 \times 10^5 = 7.36 \times 10^6 = K_7.$$

Thus, when we modify reactions we can follow the following three rules.

1. When a reaction is reversed, the new equilibrium constant is the inverse of the old equilibrium constant.

2. When a reaction is multiplied through by a constant coefficient, the new equilibrium constant is the old K raised to the power of the coefficient.
3. When two reactions are added together, the equilibrium constant of the new reaction is the product of the equilibrium constants of the original two reactions.

5. Assess the relative importance of the forward and reverse reactions from the magnitude of an equilibrium constant.

A large equilibrium constant (≈> 1000) indicates that the products predominate over the reactants in the equilibrium mixture. Thus, the forward reaction occurs more readily than does the reverse reaction. A small equilibrium constant (≈< 0.001), on the other hand, indicates that products do not readily form from the reactants. In fact, the opposite is true: the reverse reaction occurs more readily than the forward reaction.

6. Write an equilibrium constant expression in terms of partial pressures of gases, K_p; and relate a value of K_p to the corresponding value of K_c.

K_p has the same form as does K_c, except partial pressures are used rather than concentrations. K_p values are used in reactions that involve gases only. Although K_p and K_c may not be equal, they both are constant, as long as the temperature is fixed. K_c and K_p are related through the ideal gas equation $PV = nRT$ or $P = (n/V)RT$. Notice that (n/V) is a molar concentration. Thus, for reaction [2]

$$K_p = \frac{P_{N_2O_4}}{(P_{NO_2})^2} = \frac{[N_2O_4]RT}{([NO_2]RT)^2} = \frac{[N_2O_4]}{[NO_2]^2} \times \frac{1}{RT} = \frac{K_c}{RT}$$

(Note that a different expression of K_p will result from another reaction, specifically one in which the stoichiometric coefficients for gases differ.) The value of R is 0.08206 L atm mol^{-1} K^{-1} and the pressure is in atmospheres.

EXAMPLE 16-3 What is K_p for the following reaction?

$$2\,NO_2(g) \rightleftharpoons N_2O_4(g) \quad K_c = 6.81 \text{ at } 352.6 \text{ K}$$

$$K_p = \frac{K_c}{RT} = \frac{6.81}{(0.08206)(352.6)} = 0.235$$

Just as K_c has no units, neither does K_p. By convention, partial pressures always are expressed in atmospheres. It is important to recognize that K_p and K_c are valid only at a fixed temperatures. Merely multiplying or dividing by RT will not produce the value of the equilibrium constant at a different temperature.

*** 7. Know that the concentrations of pure liquids and solids are omitted from equilibrium constant expressions.**

Many experiments have investigated both the rates of chemical reactions and the position of equilibrium. In all of them, it has been found that the amount of pure liquid or pure solid present has no effect as long as *some* of the liquid or solid is present. Thus, the concentrations of pure liquids and solids are omitted from both rate laws and equilibrium constant expressions. Usually this means that the solvent in a liquid solution also is omitted. But in concentrated solutions, when the solvent is a reactant or a product, its concentration must be included. Consider the hydrolysis of ethyl acetate, reaction [8].

$$CH_3COOC_2H_5 + H_2O \rightleftharpoons CH_3COOH + C_2H_5OH \qquad [8]$$

In dilute solution, the solvent—H_2O in this case—acts as if it were pure and $K_c = [CH_3COOH][C_2H_5OH]/[CH_3COOC_2H_5]$. However, in concentrated solution, the $[H_2O]$ varies significantly during the course of the reaction and $K_c = [CH_3COOH][C_2H_5OH]/[CH_3COOC_2H_5][H_2O]$

*** 8. Calculate a numerical value of an equilibrium constant if equilibrium conditions are given.**

We now can determine the value of an equilibrium constant. Often it is helpful to write the concentrations of reactants and products beneath their formulas in the balanced chemical equation. It is very useful to write three lines: initial concentrations, changes of concentrations, and equilibrium concentrations.

EXAMPLE 16-4 The reaction $H_2(g) + CO_2(g) \rightleftharpoons H_2O(g) + CO(g)$ reaches equilibrium at 500 K. Initially, $[H_2]_i = 0.0420$ M and $[CO_2]_i = 0.0160$ M. At equilibrium, there is 0.194 mol CO_2 in the 14.0 L-container. What is the value of K_c?

The equilibrium concentration of CO_2 is $[CO_2]_e = 0.194 \text{ mol}/14.0 \text{ L} = 0.0139$ M. Thus, the concentration of CO_2 that has reacted is $[CO_2]_r = [CO_2]_i - [CO_2]_e = 0.0160 \text{ M} - 0.0139 \text{ M} = 0.0021$ M.

Reaction:	$H_2(g)$ +	$CO_2(g)$	$\rightleftharpoons$	$H_2O(g)$ +	$CO(g)$
Initial:	0.0420 M	0.0160 M		0.0000 M	0.0000 M
Change:		−0.0021 M			
Equilibrium:		0.0139 M			

Since one mole of $H_2(g)$ *reacts* for every mole of $CO_2(g)$ reacting and since one mole each of $H_2O(g)$ and $CO(g)$ are *produced* for each mole of $CO_2(g)$ that reacts, we have

$[H_2]_r = [CO_2]_r = 0.0021$ M *or* $[H_2]_p = [CO_2]_p = -0.0021$ M

and $[H_2O]_p = [CO]_p = [CO_2]_r = 0.0021$ M

Reaction:	$H_2(g)$ +	$CO_2(g)$	$\rightleftharpoons$	$H_2O(g)$ +	$CO(g)$
Initial:	0.0420 M	0.0160 M		0.0000 M	0.0000 M
Change:	−0.0021 M	−0.0021 M		+0.0021 M	+0.0021 M
Equilibrium:	0.0399 M	0.0139 M		0.0021 M	0.0021 M

Notice in every case $[\ ]_e = [\ ]_i + [\ ]_r$. Finally, notice that we can add and subtract concentrations in the same way that we would combine amounts in moles *as long as the species involved are in the same solution* (which they should be in an equilibrium system).

$$K_c = \frac{[CO][H_2O]}{[CO_2][H_2]} = \frac{(0.0021)(0.0021)}{(0.0139)(0.0399)} = 8.0 \times 10^{-3}$$

* **9. Predict the direction in which a reaction proceeds toward equilibrium by comparing the reaction quotient, Q, to K_c.**

The reaction quotient expression is written in the same way as the equilibrium constant expression. But, in the reaction quotient the values of the concentrations and partial pressures are not necessarily those at equilibrium. The reaction quotient is valuable in predicting the direction in which the reaction will shift or move in reaching equilibrium. There are three possibilities.

If $Q = K$, the reaction is at equilibrium

If $Q > K$, the reaction will shift left [9]

If $Q < K$, the reaction will shift right

Shift right means that more products will be produced and reactants will be consumed. This is the case in Trial 1 of Table 16-1.

$$Q_{c,1} = \frac{[N_2O_4]_{1i}}{[NO_2]^2_{1i}} = \frac{0.000}{(0.0200)^2} = 0.00 < 6.81 = K_c$$

Shift left means that reactants are produced at the expense of products. This is the case for Trial 2 of Table 16-1.

$$Q_{c,2} = \frac{[N_2O_4]_{2i}}{[NO_2]^2_{2i}} = \frac{0.00500}{(0.000)^2} = \infty > 6.81 = K_c$$

The same procedure is used in evaluating Q_p and comparing it with K_p, except partial pressures in atmospheres are used rather than molar concentrations.

* **10. Use the ideal gas law and Dalton's law of partial pressures in working with K_p expressions.**

We can add and subtract partial pressures as we did molar concentrations. One advantage is that the total pressure is readily measured and Dalton's law of partial pressures can be used.

EXAMPLE 16-5 A container is filled with $CH_3OH(g)$ at 350 K and equilibrium is achieved for the reaction $CO(g) + 2\ H_2(g) \rightleftharpoons CH_3OH(g)$. At equilibrium, the partial pressure of $CH_3OH(g)$ is 1.000 atm and the total pressure is 1.420 atm. What is the value of K_p?

Because of the stoichiometry of the reaction $P_{H2} = 2P_{CO}$. That is, twice as much H_2 is produced as CO when CH_3OH decomposes.

$P_{total} = P_{CH_3OH} + P_{H_2} + P_{CO}$

$1.420 \text{ atm} = 1.000 \text{ atm} + 2\,P_{CO} + P_{CO} = 1.000 \text{ atm} + 3\,P_{CO}$ *or*

$P_{CO} = (1.420 - 1.000)/3 = 0.140 \text{ atm}$

Thus, $P_{H_2} = 2 \times 0.140 \text{ atm} = 0.280 \text{ atm}$ *and*

$$K_p = \frac{P_{CH_3OH}}{P_{CO}P_{H_2}^2} = \frac{1.000}{(0.140)(0.280)^2} = 91.1$$

* **11. Make qualitative predictions of how equilibrium conditions change when an equilibrium mixture is disturbed.**

LeChâtelier's principle states that, when a stress is applied to an equilibrium system, the system will react ("shift") in a direction that will relieve the stress. One way to account for the change in the equilibrium position brought about by varying concentration is to consider the effect that such variation has on the rates of the forward and reverse reactions.

Chemical equilibria proceed toward products ("shift to the right") when the forward reaction is speeded up relative to the reverse reaction. For example, if the reactant concentrations are increased, the forward reaction speeds up. Or, if the concentrations of the products are decreased, the reverse reaction slows down. In each case the system eventually reaches a new equilibrium. In similar fashion, conditions that cause the reverse reaction's rate to increase relative to the forward rate makes the reaction proceed toward reactants ("shift to the left").

In the special case of reactions involving gaseous species, concentrations can be changed by varying the partial pressures within the system. These pressures can be increased by squeezing the system into a smaller volume. If the pressure is increased on a system at equilibrium, the system tries to make its volume as small as possible. Thus it proceeds toward the side where the sum of the stoichiometric coefficients of gaseous species is the smaller.

Changing the temperature of a system at equilibrium causes the reaction to shift right or left depending on whether the reaction is endothermic or exothermic. An easy way to remember the direction of the shift is to think of heat as a product (in an exothermic reaction) or as a reactant (in an endothermic reaction) of the reaction.

Exothermic: reactants $\longrightarrow$ products + heat [10]

Endothermic: reactants + heat $\longrightarrow$ products

Raising the temperature "increases the concentration of heat" and the exothermic reaction proceeds toward reactants. Of course, heat is not a substance but a form of energy, and relations [10] are misleading. However, they are a convenient memory aid. All of these changes in conditions and their effects are summarized in Table 16-2.

TABLE 16-2 Effects of Changing Conditions on Equilibrium Systems

Changed condition	*Type of reaction*	*Shift direction*
Increase in reactant concentration	General	Right
Increase in product concentration	General	Left
Increase pressure *or* decrease volume	$n_{g,R} < n_{g,P}$	Left
	$n_{g,R} > n_{g,P}$	Right
Decrease pressure *or* increase volume	$n_{g,R} < n_{g,P}$	Right
	$n_{g,R} > n_{g,P}$	Left
Increase temperature	Exothermic	Left
	Endothermic	Right
Decrease temperature	Exothermic	Right
	Endothermic	Left

$n_{g,R}$ = sum of stoichiometric coefficients of gaseous reactants
$n_{g,P}$ = sum of stoichiometric coefficients of gaseous products

* **12. Calculate the final equilibrium condition in a reversible reaction from a given set of initial conditions.**

Once the value of the equilibrium constant is known, it can be used to predict the final concentrations of reactants and products. An easy technique to help keep track of the calculation is that used in objective 16-8. The equation for the reaction is written out. Below the formula of each species we write the initial concentration or partial pressure, with the changes in these values immediately below, and the equilibrium values in turn below them. In this type of problem, one or several concentrations are not known. If possible, we need to express all concentrations or partial pressures in terms of one unknown.

EXAMPLE 16-6 For $H_2(g) + CO_2(g) \rightleftharpoons H_2O(g) + CO(g)$, $K_c = 8.0 \times 10^{-3}$ at 500 K. What are the equilibrium concentrations of all reagents if initial concentrations are $[H_2] = 0.0200$ M, $[CO_2] = 0.0144$ M, and $[CO] = 0.100$ M?

Some $H_2O(g)$ must be produced to attain equilibrium. Let $[H_2O]_{eq} = x$.

Reaction:	$H_2(g)$	+	$CO_2(g)$	$\rightleftharpoons$	$H_2O(g)$ +	$CO(g)$
Initial:	0.0200 M		0.0144 M		0.0000 M	0.0100 M
Changes:	$-x$ M		$-x$ M		$+x$ M	$+x$ M
Equil:	$(0.0200 - x)$ M		$(0.0144 - x)$ M		x M	$(0.0100 + x)$M

$$K_c = \frac{[H_2O]_{eq}[CO]_{eq}}{[H_2]_{eq}[CO_2]_{eq}} = \frac{(x)(0.0100 + x)}{(0.0200 - x)(0.0144 - x)} = 8.0 \times 10^{-3}$$

$$= \frac{0.0100\,x + x^2}{2.88 \times 10^{-4} - 0.0344\,x + x^2}$$

or $\quad 0.992\,x^2 + 0.01028\,x - 2.304 \times 10^{-6} = 0$

We use the quadratic equation to find the solution.

$$x = \frac{-b \pm \sqrt{b^2 - 4ac}}{2a} = \frac{-0.01028 \pm \sqrt{(0.01028)^2 - (4)(0.992)(-2.304 \times 10^{-6})}}{2(0.992)}$$

$$= -5.181 \times 10^{-3} \pm 5.401 \times 10^{-3} = 2.20 \times 10^{-4}\text{ M}$$

We choose the positive root rather than the negative root, because a negative concentration (–0.0010582 M) has no physical meaning. The equilibrium concentrations follow.

$[H_2] = 0.0200 - x = 0.0200 - 0.00022 = 0.0198$ M

$[CO_2] = 0.0144 - x = 0.0144 - 0.00022 = 0.0142$ M

$[H_2O] = x = 0.00022$ M

$[CO] = 0.0100 + x = 0.0100 + 0.00022 = 0.0102$ M

To check our results, we substitute these equilibrium values into K_c.

$$K_c = \frac{[H_2O][CO]}{[H_2][CO_2]} = \frac{(0.00022)(0.0102)}{(0.0198)(0.0142)} = 8.0 \times 10^{-3}$$

EXAMPLE 16-7 For $Br_2(g) + Cl_2(g) \rightleftharpoons 2\,BrCl(g)$, $K_p = 7.21$ at 298 K. What is the equilibrium pressure of BrCl(g) if the initial pressure of $Br_2(g)$ is 1.000 atm and that of Cl_2 is 2.000 atm?

Let x be the decrease in the pressure of $Br_2(g)$. Then x also is the decrease in the pressure of $Cl_2(g)$ and $2x$ is the increase in the pressure of BrCl(g).

Reaction:	$Br_2(g)$	+	$Cl_2(g)$	$\rightleftharpoons$	2 BrCl(g)
Initial:	1.000 atm		2.000 atm		0.000 atm
Changes:	$-x$ atm		$-x$ atm		$+2x$ atm
Equil:	$(1.000 - x)$ atm		$(2.000 - x)$ atm		$2x$ atm

$$K_p = \frac{P^2_{BrCl}}{P_{Br2}P_{Cl2}} = \frac{(2x)^2}{(1.000 - x)(2.000 - x)} = 7.21 \quad or \quad 3.21\,x^2 - 21.630\,x + 14.420 = 0$$

$$x = \frac{21.630 \pm \sqrt{(21.630)^2 - 4(3.21)(14.420)}}{2(3.21)} = 3.369 \pm 2.619 = 0.750\text{ atm}$$

Thus, $P_{Br2} = 1.000 - x = 1.000 - 0.750 = 0.250$ atm

$P_{Cl2} = 2.000 - x = 2.000 - 0.750 = 1.250$ atm

$P_{BrCl} = 2x = 2 \times 0.750 = 1.500$ atm

$$K_p = \frac{P^2_{BrCl}}{P_{Br2}P_{Cl2}} = \frac{(1.500)^2}{(0.250)(1.250)} = 7.20$$

In this case, the positive root is not used because it would produce negative pressures of the reactants. Notice also that the line labeled "Change:" has x multiplied by the stoichiometric coefficients of each species, and that the signs on this line are different for products and reactants.

13. Calculate the new equilibrium concentrations or partial pressures after an equilibrium system has adapted to changed conditions.

Changes in concentrations or pressures cause no change in the value of the equilibrium constant.

EXAMPLE 16-8 The volume of Trial 3 in Table 16-1 is expanded from 20.0 L to 50.0 L. In which direction will the reaction proceed and what will be the final concentrations once equilibrium is reestablished?

The "initial" concentrations after the expansion are determined by multiplying the 20.0-L concentrations by 20.0 L/50.00 L = 0.400. The decrease in pressure that occurs causes the reaction to proceed toward reactants ("shift left").

Reaction:	$2\ NO_2(g)$	$\rightleftharpoons$	$N_2O_4(g)$
Initial equil:	0.01050 M		0.000750 M
Volume to 50.0 L:	0.00420 M		0.000300 M
Change:	$+2x$ M		$-x$ M
Final equil:	$(0.00420 + 2x)$ M		$(0.000300 - x)$ M

The line labeled "Volume to 50.0 L" gives the concentrations at the instant that the volume is increased to 50.0 L, but before the final equilibrium is achieved.

$$K_c = \frac{[N_2O_4]}{[NO_2]^2} = \frac{(0.000300 - x)}{(0.00420 + 2x)^2} = 6.81$$

The quadratic equation gives

$$x = -0.02046 \pm 0.02062 = 0.00016 \text{ M}$$

The positive root is chosen since the negative root gives $x = -0.04104$ M. This produces $[NO_2] = -0.07788$ M. This root is discarded because negative concentrations have no physical meaning. Thus, $[NO_2] = 0.00452$ M and $[N_2O_4] = 0.000140$ M.

Notice that $[NO_2]$ has increased and $[N_2O_4]$ has decreased, as predicted. But have we decided the outcome of the calculation by *assuming* that $[NO_2]$ would increase? Suppose we had assumed that $[NO_2]$ decreased by $2x$ and $[N_2O_4]$ increased by x. Then we would have

$$[NO_2]_e = 0.00420 - 2x \quad \textit{and} \quad [N_2O_4]_e = 0.000300 + x$$

$$K_c = \frac{[N_2O_4]}{[NO_2]^2} = \frac{(0.000300 + x)}{(0.00420 - 2x)^2} = 6.81 \quad \text{or} \quad x = -0.00016 \text{ M}$$

(The positive root gives the physically meaningless result of a negative $[NO_2]_e$.) We obtain the same result as before.

$$[NO_2]_e = 0.00420 - 2(-0.00016) = 0.00452 \text{ M}$$

$$[N_2O_4]_e = 0.000300 + (-0.00016) = 0.00014 \text{ M}$$

When a temperature change occurs, the value of the equilibrium constant changes. Its new value is determined by the van't Hoff equation, as outlined in objective 20-13.

14. Appreciate that the equilibrium constant for a reaction is related to the rate laws of the forward and reverse reactions but that one must have detailed knowledge of the mechanism to establish this relationship.

For some time it was though that the rate law of a chemical reaction could easily be obtained from the balanced chemical equations. The belief was that reaction orders and stoichiometric coefficients were equal, which is not necessarily true. However, one can obtain equilibrium constant expressions from mechanisms. This is one of the various steps that can be used to establish the plausibility of a mechanism.

DRILL PROBLEMS

2. The initial amount of each substance in each of the following reactions is given below the chemical formula of that substance. The equilibrium amount of one substance also is given, along with the volume of the container. Determine the equilibrium concentrations of all substances.

	$CH_4(g)$ +	$CCl_4(g)$ $\rightleftharpoons$	$2\ CH_2Cl_2(g)$	at 298 K	
A.	0.100 mol	0.640 mol	0.220 mol	$n_{CH4} = 0.1502$ mol,	$V = 5.000$ L
B.	0.200 mol	0.200 mol	0.320 mol	$n_{CCl4} = 0.3118$ mol,	$V = 8.00$ L
C.	0.650 mol	0.320 mol	0.320 mol	$n_{CH2Cl2} = 0.1569$ mol,	$V = 4.00$ L
D.	0.3400 mol	0.3400 mol	0.5100 mol	$n_{CH2Cl2} = 0.1593$ mol,	$V = 8.50$ L

	$PCl_5(g)$ $\rightleftharpoons$	$PCl_3(g)$ +	$Cl_2(g)$	at 600 K	
E.	0.100 mol	0.100 mol	0.100 mol	$n_{Cl2} = 0.1848$ mol,	$V = 4.00$ L
F.	0.300 mol	0.204 mol	0.153 mol	$n_{PCl5} = 0.0897$ mol,	$V = 3.00$ L
G.	0.144 mol	0.276 mol	0.072 mol	$n_{PCl3} = 0.408$ mol,	$V = 12.0$ L

	$HCHO(g)$ $\rightleftharpoons$	$H_2(g)$ +	$CO(g)$	at 773 K	
H.	2.50 mol	0.00 mol	0.00 mol	$n_{H2} = 1.00$ mol,	$V = 2.50$ L
I.	0.00 mol	4.10 mol	2.00 mol	$n_{CO} = 0.500$ mol,	$V = 2.00$ L
J.	0.900 mol	0.000 mol	0.600 mol	$n_{HCHO} = 0.500$ mol,	$V = 3.00$ L

	$2\ NO(g)$ +	$Br_2(g)$ $\rightleftharpoons$	$2\ NOBr(g)$	at 350 K	
K.	0.000 mol	0.400 mol	1.000 mol	$n_{NO} = 0.200$ mol,	$V = 0.200$ L
L.	1.50 mol	4.60 mol	0.00 mol	$n_{NOBr} = 1.20$ mol,	$V = 1.60$ L
M.	0.00 mol	8.75 mol	2.00 mol	$n_{NO} = 0.500$ mol,	$V = 6.40$ L

	$3\ H_2(g)$ +	$N_2(g)$ $\rightleftharpoons$	$2\ NH_3(g)$	at 473 K	
N.	0.5800 mol	0.1930 mol	0.0000 mol	$n_{N2} = 0.0097$ mol,	$V = 0.525$ L
O.	0.0000 mol	0.0117 mol	0.5134 mol	$n_{H2} = 0.0200$ mol,	$V = 0.287$ L

3. (1) Write the equilibrium constant expression in terms of concentrations for each of the following reaction. Some numerical values of K_c are given for future problems.

A. $2\ NOBr(g) \rightleftharpoons 2\ NO(g) + Br_2(g)$
B. $4\ HCl(g) + O_2(g) \rightleftharpoons 2\ H_2O(g) + 2\ Cl_2(g)$ $K_c = 1.49 \times 10^3$ at 320 K
C. $N_2(g) + O_2(g) \rightleftharpoons 2\ NO(g)$ $K_c = 6.37 \times 10^{-16}$ at 298 K
D. $CH_4(g) + CCl_4(g) \rightleftharpoons 2\ CH_2Cl_2(g)$
E. $PCl_5(g) \rightleftharpoons PCl_3(g) + Cl_2(g)$
F. $HCHO(g) \rightleftharpoons H_2(g) + CO(g)$
G. $N_2(g) + 3\ H_2(g) \rightleftharpoons 2\ NH_3(g)$
H. $2\ H_2S(g) + 3\ O_2(g) \rightleftharpoons 2\ H_2O(g) + 2\ SO_2(g)$ $K_c = 1.74 \times 10^{45}$ at 1000 K
I. $2.NH_3(g) + 6\ F_2(g) \rightleftharpoons 2\ NF_3(g) + 6\ HF(g)$

(2) For each of the following reactions, the value of K_c and the equilibrium concentrations of all reagents but one are given. Determine the value of the omitted concentration.

	$CH_3ONO(g)$ +	$HCl(g)$ $\rightleftharpoons$	$CH_3OH(g)$ +	$NOCl(g)$	$K_c = 0.242$ at 323 K
J.	0.500 M	0.274 M	1.26 M	___ M	
K.	___ M	4.96 M	1.14 M	1.49 M	

	$CH_4(g)$ +	$CCl_4(g)$ $\rightleftharpoons$	$2\ CH_2Cl_2(g)$	$K_c = 0.0956$ at 298 K
L.	1.124 M	1.124 M	___ M	
M.	0.104 M	___ M	0.256 M	

	$2\ NOBr(g)$ $\rightleftharpoons$	$2\ NO(g)$ +	$Br_2(g)$	$K_c = 0.1563$ at 350 K
N.	0.275 M	0.124 M	___ M	
O.	___ M	0.173 M	0.822 M	

	$N_2(g)$ +	$3\ H_2(g)$ $\rightleftharpoons$	$2\ NH_3(g)$	$K_c = 1.40 \times 10^5$ at 473 K
P.	___ M	0.163 M	0.970 M	
Q.	0.00100 M	0.0220 M	___ M	

	$Br_2(g)$ +	$Cl_2(g)$ $\rightleftharpoons$	$2\ BrCl(g)$	$K_c = 6.30$ at 350 K
R.	0.146 M	___ M	0.227 M	
S.	0.082 M	0.077 M	___ M	

	$PCl_3(g)$ +	$Cl_2(g)$	$\rightleftharpoons$ $PCl_5(g)$	$K_c = 1.784$ at 600 K
T.	0.149 M	____ M	0.206 M	
U.	0.200 M	0.111 M	____ M	

4. Based on your answers to, and the data of, the drill problems in objectives 16-2 and 16-3, write the equilibrium constant expression and determine its value for each of the following equations.
   A. $H_2(g) + CO(g) \rightleftharpoons HCHO(g)$
   B. $PCl_3(g) + Cl_2(g) \rightleftharpoons PCl_5(g)$
   C. $2\ NH_3(g) \rightleftharpoons N_2(g) + 3\ H_2(g)$
   D. $2\ BrCl(g) \rightleftharpoons Br_2(g) + Cl_2(g)$
   E. $2\ CH_2Cl_2(g) \rightleftharpoons CH_4(g) + CCl_4(g)$
   F. $CH_2Cl_2(g) \rightleftharpoons \frac{1}{2}CH_4(g) + \frac{1}{2}CCl_4(g)$
   G. $NO(g) + \frac{1}{2}Br_2(g) \rightleftharpoons NOBr(g)$
   H. $\frac{1}{2}N_2(g) + \frac{3}{2}H_2(g) \rightleftharpoons NH_3(g)$
   I. $\frac{1}{2}Br_2(g) + \frac{1}{2}Cl_2(g) \rightleftharpoons BrCl(g)$
   J. $NO(g) \rightleftharpoons \frac{1}{2}N_2(g) + \frac{1}{2}O_2(g)$
   K. $H_2O(g) + Cl_2(g) \rightleftharpoons 2\ HCl(g) + \frac{1}{2}O_2(g)$
   L. $H_2S(g) + \frac{3}{2}O_2(g) \rightleftharpoons H_2O(g) + SO_2(g)$

6. The value of K_p or K_c is given for each of the following reaction. Write the K_p expression for each reaction and find the value of the one of K_p or K_c that is not given.
   A. $PCl_3(g) + Cl_2(g) \rightleftharpoons PCl_5(g)$ — $K_c = 1.784$ at 600 K
   B. $Br_2(g) + Cl_2(g) \rightleftharpoons 2\ BrCl(g)$ — $K_p = 6.30$ at 350 K
   C. $2\ H_2S(g) + 3\ O_2(g) \rightleftharpoons 2\ H_2O(g) + 2\ SO_2(g)$ — $K_c = 1.74 \times 10^{45}$ at 1000 K
   D. $N_2(g) + 3\ H_2(g) \rightleftharpoons 2\ NH_3(g)$ — $K_c = 1.40 \times 10^5$ at 473 K
   E. $N_2O_4(g) \rightleftharpoons 2\ NO_2(g)$ — $K_p = 0.113$ at 298 K
   F. $2\ NOBr(g) \rightleftharpoons 2\ NO(g) + Br_2(g)$ — $K_c = 0.1563$ at 350 K
   G. $CO(g) + 2\ H_2(g) \rightleftharpoons CH_3OH(g)$ — $K_p = 2.05 \times 10^4$ at 298 K
   H. $CH_4(g) + CCl_4(g) \rightleftharpoons 2\ CH_2Cl_2(g)$ — $K_c = 0.0956$ at 298 K
   I. $2\ NOCl(g) \rightleftharpoons 2\ NO(g) + Cl_2(g)$ — $K_p = 0.0658$ at 298 K
   J. $4\ HCl(g) + O_2(g) \rightleftharpoons 2\ H_2O(g) + Cl_2(g)$ — $K_c = 1.49 \times 10^3$ at 320 K
   K. $2\ H_2S(g) \rightleftharpoons 2\ H_2(g) + S_2(g)$ — $K_p = 3.16 \times 10^{-3}$ at 830 K
   L. $N_2(g) + O_2(g) \rightleftharpoons 2\ NO(g)$ — $K_c = 6.37 \times 10^{-16}$ at 298 K

7. Write the equilibrium constant expression in terms of concentrations, K_c, for each of the following equations.
   A. $CO_2(g) + 2\ NH_3(g) \rightleftharpoons CO(NH_2)_2(s) + H_2O(l)$
   B. $S(s) + 2\ CO(g) \rightleftharpoons SO_2(g) + 2\ C(s)$
   C. $NH_4Cl(s) \rightleftharpoons NH_3(g) + HCl(g)$
   D. $NH_3(g) + CH_3COOH(l) \rightleftharpoons NH_2CH_2COOH(s) + H_2(g)$
   E. $SO_2Cl_2(l) \rightleftharpoons SO_2(g) + Cl_2(g)$
   F. hydrogen(g) + oxygen(g) $\rightleftharpoons$ water(l)
   G. ammonium nitrate(s) $\rightleftharpoons$ dinitrogen oxide(g) + water(g)
   H. hydrogen sulfide(aq) + sodium hydroxide(aq) $\rightleftharpoons$ sodium bisulfide(aq) + water(l)
   I. phosphorus trichloride(g) + water(l) $\rightleftharpoons$ phosphorous acid(aq) + hydrochloric acid(aq)
   J. $4\ HNO_3(aq) + 3\ Ag(s) \rightleftharpoons 3\ AgNO_3(aq) + NO(g) + 2\ H_2O(l)$
   K. $S(s) + 6\ HNO_3(aq) \rightleftharpoons H_2SO_4(aq) + 2\ H_2O(l) + 6\ NO_2(g)$
   L. calcium carbonate(s) $\rightleftharpoons$ calcium oxide(s) + carbon dioxide(g)
   M. $Mg_3N_2(s) + 6\ H_2O(l) \rightleftharpoons 3\ Mg(OH)_2(s) + 2\ NH_3(g)$

8. (1) Use the data from the drill problems of objective 16-2 to determine the value of the equilibrium constant, K_c, for each of the following reactions.
   A. $CH_4(g) + CCl_4(g) \rightleftharpoons 2\ CH_2Cl_2(g)$ at 298 K
   B. $PCl_5(g) \rightleftharpoons PCl_3(g) + Cl_2(g)$ at 600 K
   C. $HCHO(g) \rightleftharpoons H_2(g) + CO(g)$ at 773 K
   D. $2\ NO(g) + Br_2(g) \rightleftharpoons 2\ NOBr(g)$ at 350 K
   E. $3\ H_2(g) + N_2(g) \rightleftharpoons 2\ NH_3(g)$ at 473 K

   (2) The equilibrium pressure of each species is written below its formula in the following equations. Evaluate K_p in each case.

F. $2\ HI(g)$ $\rightleftharpoons$	$H_2(g)$ +	$I_2(g)$ at 298 K	
0.400 atm	0.0136 atm	0.0136 atm	
G. $CO(g)$ +	$2\ H_2(g)$ $\rightleftharpoons$	$CH_3OH(g)$ at 298 K	
0.300 atm	0.300 atm	554 atm	
H. $SO_2Cl_2(l)$ $\rightleftharpoons$	$SO_2(g)$ +	$Cl_2(g)$	
	0.00442 atm	0.00442 atm	

	$H_2(g)$ +	$CO_2(g)$ $\rightleftharpoons$	$H_2O(g)$ +	$CO(g)$ at 350 K
I.	1.00 atm	1.00 atm	0.0110 atm	0.0110 atm

9. The initial pressure or concentration is given below the formula of each species in each of the following equations. Evaluate Q_p or Q_c compare the value obtained with that given for K_c or K_p, and predict whether the reaction will shift left, will shift right, or is at equilibrium.

	$4\,HCl(g)$ +	$O_2(g)$ $\rightleftharpoons$	$2\,H_2O(g)$ +	$Cl_2(g)$	$K_c = 1.49 \times 10^3$ at 320 K
A.	0.100 M	0.0200 M	0.122 M	0.200 M	
B.	0.100 M	0.103 M	0.409 M	0.010 M	

	$CH_3ONO(g)$ +	$HCl(g)$ $\rightleftharpoons$	$CH_3OH(g)$ +	$NOCl(g)$	$K_c = 0.242$ at 323 K
C.	0.104 M	0.246 M	0.115 M	0.0941 M	
D.	0.346 M	0.102 M	0.212 M	0.198 M	

	$CH_4(g)$ +	$CCl_4(g)$ $\rightleftharpoons$	$2\,CH_2Cl_2(g)$	$K_c = 0.0956$ at 298 K
E.	0.1060 M	0.02075 M	0.01450 M	
F.	0.207 M	0.0156 M	0.0142 M	

	$2\,NOBr(g)$ $\rightleftharpoons$	$2\,NO(g)$ +	$Br_2(g)$	$K_c = 0.1563$ at 350 K
G.	0.396 M	0.214 M	0.289 M	
H.	0.204 M	0.311 M	0.149 M	

	$3\,H_2(g)$ +	$N_2(g)$ $\rightleftharpoons$	$2\,NH_3(g)$	$K_c = 1.40 \times 10^5$ at 473 K
I.	0.0163 M	0.214 M	0.270 M	
J.	0.163 M	0.0214 M	0.0270 M	

	$PCl_3(g)$ +	$Cl_2(g)$ $\rightleftharpoons$	$PCl_5(g)$	$K_c = 1.784$ at 600 K
K.	0.100 M	0.100 M	0.173 M	
L.	0.204 M	0.306 M	0.104 M	

	$N_2O_4(g)$ $\rightleftharpoons$	$2\,NO_2(g)$	$K_p = 0.113$ atm at 298 K
M.	0.260 atm	0.152 atm	
N.	0.115 atm	0.214 atm	

	$Br_2(g)$ +	$Cl_2(g)$ $\rightleftharpoons$	$2\,BrCl(g)$	$K_p = 6.30$ at 350 K
O.	0.216 atm	0.216 atm	0.304 atm	
P.	0.143 atm	0.157 atm	0.424 atm	

	$PCl_5(g)$ $\rightleftharpoons$	$PCl_3(g)$ +	$Cl_2(g)$	$K_p = 0.675$ at 500 K
Q.	0.165 atm	0.243 atm	0.114 atm	
R.	0.0464 atm	0.302 atm	0.147 atm	

	$H_2(g)$ +	$CO_2(g)$ $\rightleftharpoons$	$H_2O(g)$ +	$CO(g)$	$K_p = 0.00831$ at 500 K
S.	0.206 atm	0.206 atm	0.114 atm	0.114 atm	
T.	0.814 atm	0.514 atm	0.0621 atm	0.0214 atm	

	$SO_2Cl_2(l)$ $\rightleftharpoons$	$SO_2(g)$ +	$Cl_2(g)$	$K_p = 2.69$ at 360 K
U.		1.76 atm	0.423 atm	
V.		1.70 atm	1.70 atm	

10. Evaluate K_p for each of the following reactions. The pressure written below a species is its equilibrium pressure. P is the total equilibrium pressure.

A. $N_2O_4(g) \rightleftharpoons 2\,NO_2(g)$ at 298 K
158 atm; $P = 2.00$ atm

B. $2\,HI(g) \rightleftharpoons H_2(g) + I_2(g)$ at 320 K
0.464 atm; $P = 1.000$ atm and $P_{H2} = P_{I2}$

C. $CO(g) + 2\,H_2(g) \rightleftharpoons CH_3OH(g)$ at 350 K
0.254 atm, 0.254 atm; $P = 2.000$ atm

D. $Br_2(g) + Cl_2(g) \rightleftharpoons 2\,BrCl(g)$ at 298 K
$P_{Br2} = P_{Cl2}$; 0.543 atm; $P = 2.00$ atm

E. $SO_2Cl_2(l) \rightleftharpoons SO_2(g) + Cl_2(g)$ at 360 K
0.277 atm; $P = 10.00$ atm

F. $H_2(g) + CO_2(g) \rightleftharpoons H_2O(g) + CO(g)$ at 500 K
0.916 atm, 0.916 atm; $P_{H2O} = P_{CO}$; $P = 2.000$ atm

G. $PCl_5(g) \rightleftharpoons PCl_3(g) + Cl_2(g)$ at 500 K
0.551 atm, 0.551 atm; $P = 1.552$ atm

11. Each of the following reacting systems initially is at equilibrium. The indicated changes are made individually on this equilibrium system. Predict whether the system will shift right (R), shift left(L), or be unchanged (U) in regaining equilibrium, by writing the appropriate letter in each blank.

$HCHO(g) \rightleftharpoons CO(g) + H_2(g)$ $\Delta H° = +5.36$ kJ/mol

A. increase temperature ____ B. add 1.00 mol CO(g) ____
C. remove 0.50 mol $H_2(g)$ ____ D. decrease volume ____

$NH_3(g) + CH_3COOH(l) \rightleftharpoons NH_2CH_2COOH(s) + H_2(g)$ $\Delta H° = -6.90$ kJ/mol

E. add 1.00 mol $CH_3COOH(l)$ ____ F. increase pressure ____
G. increase temperature ____ H. remove 0.010 mol $H_2(g)$ ____

$2\ NOCl(g) \rightleftharpoons 2\ NO(g) + Cl_2(g)$ $\Delta H° = +71.38$ kJ/mol

I. increase temperature ____ J. add 1.00 mol $Cl_2(g)$ ____
K. increase volume ____ L. increase pressure ____

$S(s) + 2\ CO(g) \rightleftharpoons SO_2(g) + 2\ C(s)$ $\Delta H° = -75.81$ kJ/mol

M. increase pressure ____ N. decrease temperature ____
O. remove 0.0024 mol S(s) ____ P. add 0.0024 mol $SO_2(g)$ ____

12. Use the data from the problems of objective 16-9. Determine the equilibrium concentrations or partial pressures for each part except A, B, G, H, I, and J. In the additional problems that follow, the initial concentration is given below the formula of each species.

	$HCHO(g)$ $\rightleftharpoons$	$H_2(g)$ +	$CO(g)$	$K_c = 0.267$ at 773 K
W.	0.200 M	0.100 M	0.100 M	
X.	0.400 M	0.000 M	0.000 M	

13. For several of the equilibrium mixtures described in the drill problems of objective 16-3, the total volume of the equilibrium system is varied (by expansion or compression) as specified below. Determine the direction in which the reaction shifts and the new equilibrium value of the indicated concentration.

J. V × 2, [HCl] K. V ÷ 4, [NOCl] L. V × 5, [CH_4] M. V ÷ 3, [CH_2Cl_2]
R. V × 3, [Br_2] S. V ÷ 5, [BrCl] T. V × 4, [Cl_2] U. V ÷ 3, [PCl_3]

QUIZZES (15 minutes each) For each quiz, answer the first question. Then choose the best answer for each of the remaining questions.

Quiz A

1. Write the equilibrium constant expression K_c for the reaction sodium sulfite(aq) + chloric acid(aq) $\rightleftharpoons$ sodium chlorite(aq) + sulfur dioxide(g) + water (l).
2. For the reaction $PCl_3(g) + ICl(g) \rightleftharpoons PCl_4I(g)$, initially there are 1.20 mol PCl_3, 0.48 mol ICl, and 1.26 mol PCl_4I. If there is 1.06 mol PCl_3 present at equilibrium, how many moles of PCl_4I are then present? (a) 1.33; (b) 1.54; (c) 1.68; (d) 1.44; (e) none of these.
3. For $I_2(s) + Cl_2(g) \rightleftharpoons 2\ ICl(g)$, $K_c = [ICl]^2/[Cl_2]$. At equilibrium there is 0.300 mol Cl_2, 0.100 mol I_2, and 2.0 mol ICl in a 5.0-L container. What is the value of K_c? (a) 0.075; (b) 2.7; (c) 0.0075; (d) 133; (e) none of these.
4. $CH_4(g) + 4\ Cl_2(g) \rightleftharpoons CCl_4(l) + 4\ HCl(g)$ $\Delta H° = -397$ kJ/mol. The equilibrium represented by this equation is displaced to the *right* if (a) the temperature is raised; (b) the pressure is lowered; (c) some carbon tetrachloride is removed; (d) some hydrogen chloride is removed; (e) none of these.
5. $N_2(g) + O_2(g) \rightleftharpoons 2\ NO(g)$ $\Delta H° = +21$ kJ/mol The equilibrium represented by this equation is displaced to the *right* if (a) the temperature is raised; (b) the pressure is lowered; (c) some nitrogen is removed; (d) some nitrogen monoxide is added; (e) none of these.
6. Which factor influences the value of the equilibrium constant for a reversible reaction? (a) addition of a catalyst; (b) raising the temperature; (c) removing product; (d) removing reactant; (e) raising the pressure.
7. For the reaction $SbCl_5(g) \rightleftharpoons SbCl_3(g) + Cl_2(g)$, K_c equals (a) K_p; (b) RT/K_p; (c) $K_p(RT)$; (d) K_p/RT; (e) none of these.

Quiz B

1. Write the equilibrium constant expression K_p for the reaction $SCl_4(l) + O_2(g) \rightleftharpoons SO_2(g) + 2\ Cl_2(g)$
2. 10.0 mol NO and 26.0 mol O_2 are placed in an empty container. At equilibrium for $2\ NO + O_2 \rightleftharpoons 2\ NO_2$ there is 8.8 mol NO_2 present. How many moles of O_2 are present at equilibrium? (a) 17.2; (b) 8.4; (c) 21.6; (d) 4.4; (e) none of these.
3. For $CO_2(g) + H_2(g) \rightleftharpoons CO(g) + H_2O(g)$, $K_c = [CO][H_2O]/[CO_2][H_2]$. If there are 1.43 mol each of CO and H_2O, 0.572 mol H_2, and 4.572 mol CO_2 in a 4.00-L container at equilibrium, what is the value of K_c? (a) 0.547; (b) 1.27; (c) 0.782; (d) 0.137; (e) none of these.
4. $N_2H_6CO(s) \rightleftharpoons 2\ NH_3(g) + CO_2(g)$ $\Delta H° = +33$ kJ/mol. The equilibrium represented by this equation is displaced to the *left* if (a) the pressure is lowered; (b) the temperature is lowered; (c) some $N_2H_6CO_2$ is removed; (d) some ammonia is removed; (e) none of these.
5. In the equilibrium $2\ O_3(g) \rightleftharpoons 3\ O_2(g)$, one can conclude that (a) high temperature favors ozone formation; (b) low pressure favors ozone formation; (c) pressure does not affect the equilibrium position; (d) ozone decomposes rapidly into dioxygen; (e) none of these.
6. All of the following may shift the position of a reversible reaction at equilibrium except (a) concentration change; (b) pressure change; (c) temperature increase; (d) addition of reactant; (e) addition of a heterogeneous catalyst.
7. For the reaction $CS_2(l) + 3\ O_2(g) \rightleftharpoons CO_2(g) + 2\ SO_2(g)$, K_c equals (a) K_p; (b) RT/K_p; (c) $K_p(RT)$; (d) K_p/RT; (e) none of these

Quiz C

1. Write the equilibrium constant expression K_c for the reaction $3\ Cu(s) + 8\ HNO_3(aq) \rightleftharpoons 3\ Cu(NO_3)_2(aq) + 2\ NO(g) + 4\ H_2O(l)$.
2. 0.300 mol NO, 0.200 mol Cl_2, and 0.500 mol ClNO are placed in a 25.0-L vessel. At equilibrium for $2\ NO(g) + Cl_2(g) \rightleftharpoons 2\ NOCl(g)$, there is 0.600 mol ClNO present. How many moles of Cl_2 are present at equilibrium? (a) 0.200; (b) 0.050; (c) 0.100; (d) 0.150; (e) none of these.
3. For the formation of NO, $K_c = [NO_2]^2/[NO]^2[O_2]$. At equilibrium in a 2.50-L container, there are 3.00 mol NO, 4.00 mol O_2, and 22.0 mol NO_2. The value of K_c is (a) 13.4; (b) 33.6; (c) 5.38; (d) 0.0116; (e) none of these.
4. $CO_2(g) + 2\ HCl(g) \rightleftharpoons Cl_2CO(g) + H_2O(g)$. The position of equilibrium of this exothermic reaction shifts to the *left* if (a) catalyst is removed; (b) pressure is increase; (c) carbon dioxide is removed; (d) temperature is lowered; (e) none of these.
5. All of the following may shift the position of a reaction at equilibrium except (a) concentration increase; (b) pressure increase; (c) temperature increase; (d) addition of product; (e) addition of a homogeneous catalyst.
6. In a reaction at equilibrium involving only gases, a change in pressure of the reaction mixture shifts the position of equilibrium only when (a) heat is absorbed by the reaction proceeding to the right; (b) the gases are impure; (c) the collision rate increases; (d) the reaction is exothermic as written; (e) none of these.
7. For the reaction $2\ NO(g) \rightleftharpoons N_2O_4(g)$, K_p equals (a) K_c; (b) RT/K_c; (c) $K_c(RT)$; (d) K_c/RT; (e) none of these.

Quiz D

1. Write the equilibrium constant, K_p, for the reaction $CaCO_3(s) + SO_2(g) \rightleftharpoons CaSO_3(s) + CO_2(g)$
2. In the reaction $2\ N_2O(g) + N_2H_4(g) \rightleftharpoons 3\ N_2(g) + 2\ H_2O(s)$, 0.100 mol N_2O and 0.250 mol N_2H_4 are placed in a 10.0-L container. If there is 0.060 mol N_2O present at equilibrium, how many moles of N_2 are then present? (a) 0.090; (b) 0.040; (c) 0.060; (d) 0.020; (e) none of these.
3. For the reaction of $PBr_3(g)$ with $Br_2(g)$, $K_c = [PBr_5]/[PBr_3][Br_2]$. At equilibrium in an 8.00-L vessel there are 1.60 mol PBr_3, 0.800 mol Br_2, and 3.20 mol PBr_5. What is the value of K_c? (a) 2.50; (b) 20.0; (c) 0.313; (d) 0.400; (e) none of these.
4. Consider the equilibrium reaction $N_2(g) + O_2(g) \rightleftharpoons 2\ NO(g)$ $\Delta H° = +\ 192$ kJ/mol. At 2000 K the value of the equilibrium constant is 5.0×10^{-4}. At 2500 K, the value of the equilibrium constant (a) is greater than 5.0×10^{-4}; (b) is less than 5.0×10^{-4}; (c) is 5.0×10^{-4}; (d) depends on [NO]; (e) depends on $[O_2]$.
5. $S_2Cl_2(l) + CCl_4(g) \rightleftharpoons CS_2(g) + 3\ Cl_2(g)$ $\Delta H° = +84.31$ kJ/mol. The equilibrium represented by this equation is displaced to the right is (a) some carbon tetrachloride is added; (b) the temperature is lowered;

(c) the pressure is raised; (d) some carbon disulfide is removed; (e) none of these.

6. For the reaction $2\ N_2O(g) + N_2H_4(g) \rightleftharpoons 3\ N_2(g) + 2\ H_2O(g)$, K_p equals (a) K_c; (b) RT/K_c; (c) $K_c(RT)^2$; (d) K_c/RT; none of these.
7. The system $2\ H_2O(g) \rightleftharpoons 2\ H_2(g) + O_2(g)$ is at equilibrium when (a) $[H_2]/[O_2] = 2$; (b) $[H_2O]$ decreases to 50% of its initial value; (c) $[H_2O]/([H_2] + [O_2]) = 2/3$; (d) H_2O molecules no longer decompose; (e) the total pressure remains constant.

SAMPLE TEST (20 minutes)

1. At 85°C, $K_p = 1.19$ for $PCl_5(g) \rightleftharpoons PCl_3(g) + Cl_2(g)$. What is the partial pressure of $PCl_5(g)$ at equilibrium if the initial pressures are 2.00 atm PCl_3, 1.00 atm Cl_2, and 0.00 atm PCl_5?
2. At 773 K, for the reaction $HCHO(g) \rightleftharpoons H_2(g) + CO(g)$, there are 0.900 mol HCHO(g), 0.000 mol $H_2(g)$, and 0.600 mol CO(g) present initially in a 3.00-L container. At equilibrium, there is 0.500 mol HCHO(g) present in this same container. If the container that holds this equilibrium mixture is now compressed to 1.00 L, what will be $[H_2]$ after equilibrium is re-established?
3. Predict whether each of the following actions will cause the equilibrium system to shift left, shift right, or remain unchanged.

$AgCl(s) \rightleftharpoons Ag^+(aq) + Cl^+(aq)$

a. Add solid $AgNO_3$
b. Add solid AgCl

$2\ NO(g) + O_2(g) \rightleftharpoons 2\ NO_2(g)$ $\Delta H°_{1000} = -116.9$ kJ/mol

c. Increase $[O_2]$
d. Add NO(g)
e. Increase total pressure by compression
f. Raise temperature (endothermic)

$CaCO_3(s) \rightleftharpoons CaO(s) + CO_2(g)$

g. Remove $CO_2(g)$
h. Decrease total pressure by expansion
i. Add CaO(s)
j. Add catalyst

17 Acids and Bases

CHAPTER OBJECTIVES

1. Describe the similarities and differences among the Arrhenius, Brønsted-Lowry, and Lewis theories of acids and bases.

Arrhenius acid-base theory defines an acid as a substance that produces hydrogen ions (H^+) when it ionizes in water. A base produces hydroxide ions (OH^-). All *strong* acids and bases obey the Arrhenius definition. (Recall that the common strong acids and bases are listed in Table 5-1.) The ionization of a typical acid is shown in equation [1], a base in equation [2].

$$HCl \longrightarrow H^+(aq) + Cl^-(aq) \quad [1]$$

$$NaOH \longrightarrow Na^+(aq) + OH^-(aq) \quad [2]$$

The Arrhenius theory does not explain the behavior of some common weak bases in aqueous solution or reactions that seem quite like acid-base reactions but occur in other solvents. The Brønsted-Lowry theory explains these systems. In terms of this theory, an acid is a *proton donor* and a base is a *proton acceptor*. Typical reactions are shown in equations [3] and [4].

$$H_2O(l) + HCN(aq) \longrightarrow CN^-(aq) + H_3O^+(aq) \quad [3]$$

$$NH_3(aq) + H_2O(l) \longrightarrow OH^-(aq) + NH_4^+(aq) \quad [4]$$

In each equation, the base (proton acceptor) appears first on each side, and the acid (proton acceptor) appears second. (A proton is meant to indicate a hydrogen ion, H^+.) The term "conjugate" is used to describe species that differ by only a hydrogen ion. Consider equation [3]. $CN^-(aq)$ is the conjugate base of HCN(aq) *and* HCN(aq) is the conjugate acid of $CN^-(aq)$ *and* HCN(aq) and $CN^-(aq)$ are a conjugate acid-base pair. (The term "conjugate" is used in a similar way to the word "twin": Tom is Jim's twin *and* Jim is Tom's twin *and* Tom and Jim are twins.) All Arrhenius acids and bases are also Brønsted-Lowry acids and bases.

There remain some acid-base reactions that do not involve proton transfer. For example, equation [5] is an acid-base reaction but the reaction between the two anhydrides CaO(s) and $SO_3(g)$ (equation [6]) is also.

$$Ca(OH)_2(aq) + H_2SO_4(aq) \longrightarrow CaSO_4(aq) + 2\ H_2O(l) \quad [5]$$

$$CaO(s) + SO_3(g) \longrightarrow CaSO_4(s) \quad [6]$$

Yet equation [6] cannot be interpreted by either the Arrhenius or the Brønsted-Lowry theory, since there are no protons (H^+ ions) present. (Anhydride means literally "no water." SO_3 is the anhydride of H_2SO_4 because H_2SO_4 can be made by adding H_2O to SO_3. Note that "hydroxide" means "hydrated oxide.") These reactions are explained by Lewis acid-base theory, which defines an acid as an electron-pair acceptor (SO_3 in equation [6]) and a base as an electron-pair donor (the oxide ion of CaO in equation [6]). Lewis acid-base reactions result in the formation of a covalent bond between the acid and the base. The donated (and accepted) pair of electrons forms the bond—a coordinate covalent bond. The acid-base reaction is more easily seen once we draw Lewis structures.

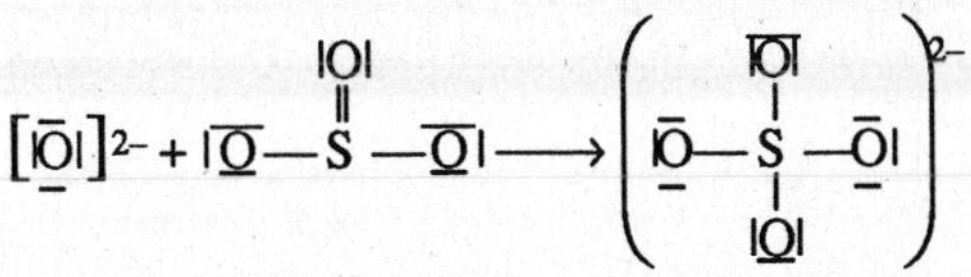

$$[\overset{..}{\underset{..}{O}}|]^{2-} + |\overset{..}{\underset{..}{O}} - \overset{\overset{|O|}{\|}}{S} - \overset{..}{\underset{..}{O}}| \longrightarrow \left(\overset{|O|}{\underset{|O|}{\underset{|}{\overset{|}{O}} - S - \overset{..}{\underset{..}{O}}|}} \right)^{2-} \qquad [7]$$

* **2. Identify Brønsted- Lowry conjugate acids and bases and write equations of acid-base reactions.**

We discussed methods of identifying acids and bases in objective 5-1. With regard to the Brønsted-Lowry theory, there always are four species in a Brønsted-Lowry equation, a base and an acid as reactants and the conjugate base and acid as products. Equation [4] is typical.

$$NH_3(aq) + H_2O(l) \rightleftharpoons OH^-(aq) + NH_4^+(aq) \qquad [4]$$

$NH_3(aq)$ is a base because it accept a proton to become $NH_4^+(aq)$. $H_2O(l)$ is an acid; it donates a proton to $NH_3(aq)$. If the reverse reaction is considered, then $OH^-(aq)$ is a base as it accepts a proton to become $H_2O(l)$. $H_2O(l)$ and $OH^-(aq)$ are a conjugate acid-base pair. In the reverse reaction, $NH_4^+(aq)$ is an acid as it donates a proton to $OH^-(aq)$ and becomes $NH_3(aq)$. $NH_4^+(aq)$ and $NH_3(aq)$ are the other conjugate acid-base pair in this reaction.

* **3. Identify Lewis acid-base reactions and write equations for acid-base reactions that involve them.**

To be able to identify Lewis acids and bases, you must be able to draw Lewis structures. (Review objectives 10-2, 10-3, and 10-6 through 10-10 if necessary.) A Lewis base *must* have at least one lone pair of electrons; this pair will be donated. A Lewis acid will either be an electron-deficient species or one that can have an "expanded octet." When dealing with an ionic compound (such as CaO in equation [6]), consider each ion separately. In the case of CaO, O^{2-} is the Lewis base. Sometime O^{2-} is called the *primary* Lewis base and CaO is called the *secondary* Lewis base.

4. Explain what self-ionization (or autoionization) is and describe the nature of the proton in aqueous solution.

Autoionization is the reaction in which a compound, that normally is thought of as covalent, produces cations and anions. There are many examples.

$$2\, NH_3 \rightleftharpoons NH_4^+ + NH_2^- \qquad [8]$$

$$2\, SO_2 \rightleftharpoons SO^{2+} + SO_3^{2-} \qquad [9]$$

$$2\, HC_2H_3O_2 \rightleftharpoons H_2C_2H_3O_2^+ + C_2H_3O_2^- \qquad [10]$$

$$N_2O_4 \rightleftharpoons NO^+ + NO_3^- \qquad [11]$$

$$COCl_2 \rightleftharpoons COCl^+ + Cl^- \qquad [12]$$

$$2\, H_2O \rightleftharpoons H_3O^+(aq) + OH^-(aq) \qquad [13]$$

The equilibrium constant for reaction [13] is given the symbol K_w and termed the *ion product* of water.

$$K_w = [H_3O^+][OH^-] = 1.00 \times 10^{-14} \text{ at } 25°C \qquad [14]$$

A free proton cannot exists in aqueous solution. It is associated strongly with one water molecules (H_3O^+) and weakly with three others ($H_3O^+\cdot3\, H_2O$ or $H_9O_4^+$). The hydroxide ion (OH^-) is similarly associated with three water molecules ($OH^-\cdot3\, H_2O$ or $H_7O_4^-$). Most chemists write the aqueous hydrogen ion as H_3O^+, often called the hydronium ion, and the aqueous hydroxide ion as OH^-.

* **5. Calculate ionic concentrations in solutions of strong electrolytes, and relate [H_3O^+] and [OH^-] through K_w.**

Strong electrolytes ionize completely in water. This includes strong acids and bases and all salts. When 0.0200 mol NaCl dissolves to form a liter of aqueous solution, that solution contains sodium ions and chloride ions with [Na^+] = 0.0200 M and [Cl^-] = 0.0200 M. Likewise, when 1.5×10^{-4} mol $CaCl_2$ forms a liter of solution, [Ca^{2+}] = 1.5×10^{-4} M and [Cl^-] = 3.0×10^{-4} M. In similar fashion, a 0.0200 M NaOH solution has [Na^+] = 0.0200 M and $[OH^-] = 0.0200$ M. Furthermore, equation [14] relates $[OH^-]$ and $[H_3O^+]$. Since we know $[OH^-] = 0.0200$ M, we can compute $[H_3O^+]$.

$$[H_3O^+](0.0200 \text{ M}) = 1.00 \times 10^{-14} \quad or \quad [H_3O^+] = 5.00 \times 10^{-11} \text{ M}$$

EXAMPLE 17-1 What is $[H_3O^+]$ in a 1.50×10^{-4} M $Ca(OH)_2$ solution?

$$1.50 \times 10^{-4} \text{ M } Ca(OH)_2 \times \frac{2 \text{ mol } OH^-}{1 \text{ mol } Ca(OH)_2} = 3.00 \times 10^{-4} \text{ M} = [OH]^-$$

$$[H_3O^+][OH^-] = [H_3O^+](3.00 \times 10^{-4} \text{ M}) = 1.00 \times 10^{-14}$$

$$or \quad [H_3O^+] = 3.33 \times 10^{-9} \text{ M}$$

* **6. Given a value of any one of $[H_3O^+]$, $[OH^-]$, pH, and pOH, be able to compute values of the other three.**

Four expressions can be used to express the acidity of a solution as $[H_3O^+]$, $[OH^-]$, pH, and pOH. They are: equation [14], which relates $[H_3O^+]$ and $[OH^-]$; the definitions of pH and pOH, equations [15] and [16]; and the relationship between pH and pOH at 25°C, equation [17].

$$pH = -\log [H_3O^+] \qquad [15]$$

$$pOH = -\log [OH^-] \qquad [16]$$

$$pH + pOH = 14.00 \qquad [17]$$

EXAMPLE 17-2 A solution with $[OH^-] = 2.50 \times 10^{-5}$ M has what pH, pOH, and $[H_3O^+]$?

pOH is most readily determined

$$pOH = -\log [OH^-] = -\log(2.50 \times 10^{-5}) = 4.602$$

Then we determine pH.

$$pH + pOH = 14.00 = pH + 4.602 \quad or \quad pH = 9.40$$

Finally, we can find $[H_3O^+]$ in two ways.

$$[H_3O^+] = 10^{-pH} = 10^{-9.40} = 4.0 \times 10^{-10} \text{ M}$$

$$or \quad [H_3O^+][OH^-] = 1.00 \times 10^{-14} = [H_3O^+](2.50 \times 10^{-5} \text{ M}) \qquad [H_3O^+] = 4.00 \times 10^{-10} \text{ M}$$

* **7. Identify a weak acid or base, write a chemical equation to represent its ionization, and set up its ionization constant expression.**

An acid or base that is not strong (Table 5-1) must be weak. It's as simple as that. (Also recall objective 5-1, for identifying weak acids and bases.) The ionization equations for all weak acids in water involve the acid donating a proton to a water molecule, forming H_3O^+ and the anion of the acid.

$$HCN(aq) + H_2O(l) \rightleftharpoons H_3O^+(aq) + CN^-(aq) \qquad [18]$$

$$HF(aq) + H_2O(l) \rightleftharpoons H_3O^+(aq) + F^-(aq) \qquad [19]$$

The ionization constant expression is written in the same way that *all* equilibrium constant expressions are written: concentrations of products in the numerator and those of reactants in the denominator.

$$K_a = \frac{[H_3O^+][CN^-]}{[HCN]} \qquad K_a = \frac{[H_3O^+][F^-]}{[HF]} \qquad [20]$$

The ionization equations of weak bases in water involve the base accepting a proton from a water molecule, producing OH^- and the cation of the base (the protonated original base).

$$NH_3(aq) + H_2O(l) \rightleftharpoons OH^-(aq) + NH_4^+(aq) \qquad [21]$$

$$N_2H_4(aq) + H_2O(l) \rightleftharpoons OH^-(aq) + N_2H_5^+(aq) \qquad [22]$$

The ionization constant expressions are written in the usual manner.

$$K_b = \frac{[NH_4^+][OH^-]}{[NH_3]} \qquad K_b = \frac{[N_2H_5^+][OH^-]}{[N_2H_4]} \qquad [23]$$

Even though the ionization constant of a weak acid usually is symbolized K_a, and called the weak acid ionization constant, sometimes the more general symbol K_i (ionization constant) is used. A similar substitution sometimes is made for K_b.

* **8. Calculate one of K_a, $[H_3O^+]$, or the molarity of a weak acid, given the other two (and perform similar calculations for a weak base). Know how to simplify these calculations by making suitable approximations.**

$$HA(aq) + H_2O(l) \rightleftharpoons H_3O^+(aq) + A^-(aq) \qquad [24]$$

For equation [24] the ionization constant expression is

$$K_a = \frac{[H_3O^+][A^-]}{[HA]} \qquad [25]$$

If HA is the only solute, then it is the only source of $A^-(aq)$ ions.

$$[H_3O^+] = [A^-] \qquad [26]$$

We use these expressions in three types of problems.

(1) *Determining K_a from $[H^+]$ and $[HA]$.* A common way to determine K_a experimentally is to measure the pH of a solution of known initial acid concentration. The initial concentration of acid (or base) also is known as the *formality* since it is based on how the solution is made up or *formulated.*

EXAMPLE 17-3 A 0.10 M solution of HN_3 has pH = 2.860. What is K_a for this acid?

We first write the ionization equation and the ionization constant expression.

$$HN_3(aq) + H_2O(aq) \rightleftharpoons H_3O^+(aq) + N_3^-(aq)$$

$$K_a = \frac{[H_3O^+][N_3^-]}{[HN_3]}$$

Then we determine $[H_3O^+] = 10^{-pH} = 1.38 \times 10^{-3}$ M. Next, we write the concentration of each species below its formula: that before ionization (I = initially), the change in concentration needed to reach equilibrium (C), and that at equilibrium (Eq). (Recall that this is the technique we first employed in objective 16-12 to set up other equilibrium problems.)

Rxn:	$HN_3(aq)$	$+ H_2O(l) \rightleftharpoons$	$H_3O^+(aq)$	$+ N_3^-(aq)$
I:	0.101 M			
C:	−0.00138 M		+0.00138 M	+0.00138 M
Eq:	0.100 M		0.00138 M	0.00138 M

We then use the equilibrium concentrations to compute K_a.

$$K_a = \frac{[H_3O^+][N_3^-]}{[HN_3]} = \frac{(0.00138)(0.00138)}{0.100} = 1.90 \times 10^{-5}$$

(2) *Determining $[H_3O^+]$ from K_a and $[HA]$.* K_a and the initial acid concentration are given and $[H_3O^+]$ is to be computed. The setup is like that of Example 17-3, with $[H_3O^+] = x$.

EXAMPLE 17-4 What is $[H_3O^+]$ in a solution with [HF] = 0.500 M? $K_a = 6.7 \times 10^{-4}$ for HF.

The setup is based on the ionization equation.

Rxn:	HF(aq)	$+ H_2O(l) \rightleftharpoons$	$H_3O^+(aq)$	$+ F^-(aq)$
I:	0.500 M			
C:	$-x$ M		$+x$ M	$+x$ M
Eq:	$(0.500 - x)$ M		x M	x M

Then substitute into the K_a expression.

$$K_a = \frac{[H_3O^+][F^-]}{[HF]} = 6.7 \times 10^{-4} = \frac{(x)(x)}{0.500 - x}$$

There are two ways to solve this equation. The first uses the quadratic equation (objective 15-8). The second uses an approximation. With a small equilibrium constant, we expect very little product. Thus, we assume $x \ll 0.500$ *or* $0.500 - x \approx 0.500$. This gives

$6.7 \times 10^{-4} \approx x^2/0.500$ *or* $x = 1.8 \times 10^{-2}$ M = 0.018 M

0.018 M does not seem small compared to 0.500 M. Let us substitute into the original expression for K_a.

$$K_a = \frac{(0.018)(0.018)}{0.500 - 0.018} = 6.7 \times 10^{-4}$$

We see that $[H_3O^+]$ = 0.018 M does indeed satisfy the original equation.

A convenient rule of thumb is that x is small compared to the original acid concentration, [HA], if x is less than 5% of [HA]. Of course, you can only be sure that x is small enough by substituting into the original equation. The approximation is worth trying if the acid concentration is at least 1000 times larger than K_a. This is a cautious guideline and can be ignored, if you are willing to learn a simple technique, illustrated in Example 17-5.

EXAMPLE 17-5 What is $[H_3O^+]$ in a 1.00 M $HClO_2$ solution? $K_a = 1.2 \times 10^{-2}$ for $HClO_2$.

Rxn:	$HClO_2(aq)$	$+ H_2O(l) \rightleftharpoons$	$H_3O^+(aq)$	$+ ClO_2^-(aq)$
I:	1.00 M			
C:	$-x$ M		$+x$ M	$+x$ M
Eq:	$(1.00 - x)$ M		x M	x M

$$K_a = \frac{[H_3O^+][ClO_2^-]}{[HClO_2]} = \frac{(x)(x)}{1.00 - x} = 1.2 \times 10^{-2} \approx \frac{x^2}{1.00}$$

We have disregarded the guideline cited above and assumed $x << 1.00$.

$x^2 = (1.2 \times 10^{-2})(1.00)$ *or* $x = 0.110$ M

Obviously 0.110 M is not small compared to 1.00 M. However, we can use it as a first guess.

$$\frac{(x)(x)}{1.00 - 0.110} = \frac{x^2}{0.89} \cong 1.2 \times 10^{-2} \quad \text{or} \quad x = 0.103 \text{ M}$$

These last two values (0.103 M and 0.104 M) give a two-significant-figure answer of $[H_3O^+]$ = 0.10 M.

This technique is known as the method of successive approximations. It is quite time saving (especially with an electronic calculator) and much easier than solving the quadratic equation. Note that when two successive approximations agree to three significant figures you should stop, and use the last value. (See the first part of Example 17-8 for another illustration of the method.)

(3) *Determining [HA] from K_a and $[H^+]$.* This last type of calculation determines the concentration of a given weak acid needed to produce a solution of a certain pH.

EXAMPLE 17-6 What $[HNO_2]$ ($K_a = 5.13 \times 10^{-4}$) is needed to produce a solution with pH = 2.200?

We first find $[H_3O^+] = 10^{-pH} = 10^{-2.200} = 6.31 \times 10^{-3}$ M.

Rxn:	$HNO_2(aq)$	$+ H_2O(l) \rightleftharpoons$	$H_3O^+(aq)$	$+ NO_2^-(aq)$
I:	x M			
C:	−0.00631 M		+0.00631 M	+0.00631 M
Eq:	$(x - 0.00631)$ M		0.00631 M	0.00631 M

$$K_a = \frac{(0.00631)^2}{x - 0.00631} = 5.13 \times 10^{-4} \quad or \quad x = 0.0839 \text{ M} = [HNO_2]$$

Often students can solve problems for weak acids but not for weak bases. Weak base problems are easier if you remember $[OH^-]$ is as important for bases as $[H_3O^+]$ is for acids.

EXAMPLE 17-7 What is the pH of a 0.750 M $C_2H_5NH_2$ solution, $K_b = 4.3 \times 10^{-4}$ for $C_2H_5NH_2$.

Rxn:	$C_2H_5NH_2(aq)$	$+ H_2O(l) \rightleftharpoons$	$C_2H_5NH_3^+(aq)$	$+ OH^-(aq)$
I:	0.750 M			
C:	$-x$ M		$+x$ M	$+x$ M
Eq:	$(0.750 - x)$ M		x M	x M

$$K_b = \frac{[C_2H_5NH_3^+][OH^-]}{[C_2H_5NH_2]} = 4.3 \times 10^{-4} = \frac{(x)(x)}{0.750 - x} \approx \frac{x^2}{0.750}$$

We have assumed $x << 0.750$ M. This gives $x = 1.8 \times 10^{-2}$ M = 0.018 M. By substituting,

$$\frac{x^2}{0.750 - x} = \frac{(0.018)^2}{0.750 - 0.018} = 4.4 \times 10^{-4}$$

This agrees with the value of K_a. Thus $x = [OH^-] = 0.018$ M.

pOH = –log (0.018) = 1.74 *and* pH = 14.00 – pOH = 14.00 – 1.74 = 12.26

9. Describe the ionization of a polyprotic acid in aqueous solution and calculate the concentrations of the different species present in such a solution.

Polyprotic acids ionize in several steps, one step for each ionizable hydrogen in the acid.

$H_2C_2O_4(aq) + H_2O(l) \rightleftharpoons H_3O^+(aq) + HC_2O_4^-(aq)$ $K_{a1} = 5.4 \times 10^{-2}$

$HC_2O_4^-(aq) + H_2O(l) \rightleftharpoons H_3O^+(aq) + C_2O_4^{2-}(aq)$ $K_{a2} = 5.4 \times 10^{-5}$

EXAMPLE 17-8 What is the concentration of each species present in a 1.00 M oxalic acid solution?

First we solve the first ionization, that of $H_2C_2O_4(aq)$.

Rxn:	$H_2C_2O_4(aq)$	$+ H_2O(l) \rightleftharpoons$	$H_3O^+(aq)$	$+ HC_2O_4^-(aq)$
I:	1.00 M			
C:	$-x$ M		$+x$ M	$+x$ M
Eq:	$(1.00 - x)$ M		x M	x M

$$K_a = \frac{[H_3O^+][HC_2O_4^-]}{[H_2C_2O_4]} = 5.4 \times 10^{-2} = \frac{(x)(x)}{1.00 - x} \approx \frac{x^2}{1.00}$$

We assumed $x << 1.00$ and found $x = 0.232$ M. (Clearly, our assumption is inaccurate.) Successive approximations yield a final answer.

$x^2/(1.00 - 0.232) = 5.4 \times 10^{-2}$ *or* $x = 0.204$ M

$x^2/(1.00 - 0.204) = 5.4 \times 10^{-2}$ *or* $x = 0.207$ M

$x^2/(1.00 - 0.207) = 5.4 \times 10^{-2}$ *or* $x = 0.207$ M

Then the second ionization, that of $HC_2O_4^-(aq)$, is solved.

Rxn:	$HC_2O_4^-(aq)$	$+ H_2O(l) \rightleftharpoons$	$H_3O^+(aq)$ +	$C_2O_4^{2-}(aq)$
I:	0.207 M		0.207 M	
C:	$-x$ M		$+x$ M	$+x$ M
Eq:	$(0.207 - x)$ M		$(0.207 - x)$ M	x M

$$K_a = \frac{[H_3O^+][C_2O_4^{2-}]}{[HC_2O_4^-]} = 5.4 \times 10^{-5} = \frac{(x)(0.207 + x)}{0.207 - x} \approx \frac{x(0.207)}{0.207}$$

We have assumed $x << 0.207$. This produces $x = 5.4 \times 10^{-5}$ M = $[C_2O_4^{2-}]$. Clearly our assumption is true. Notice that the small $[H_3O^+]$ produced in the second ionization does not significantly alter the $[H_3O^+]$ produced in the first step. Notice in addition that $[C_2O_4^{2-}] = K_{a2}$, a fact that is generally true for solutions of polyprotic acids (but see objective 19-13).

$[C_2O_4^{2-}] = 5.4 \times 10^{-5}$ M

$[H_3O^+] = [HC_2O_4^-] = 0.207$ M

$[H_2C_2O_4] = 0.793$ M

* **10. Predict which ions hydrolyze and whether salt solutions are acidic, basic, or neutral.**

All salts ionize completely when they dissolve in water (objectives 5-1 and 17-5). The anion of a strong acid does not react with water, equation [27]. However, the anion of a weak acid accepts a proton from a water molecule to produce an acid molecule and a hydroxide ion, equation [28].

$$NO_3^-(aq) + H_2O(l) \longrightarrow \text{no reaction} \quad [27]$$

$$NO_2^-(aq) + H_2O(l) \rightleftharpoons HNO_2(aq) + OH^-(aq) \quad [28]$$

This process is known as *hydrolysis*. Notice that NO_2^-, the anion of a weak acid is a base (proton acceptor). In fact, there is little difference in overall form between equations [28] and [29],

$$NH_3(aq) + H_2O(l) \rightleftharpoons NH_4^+(aq) + OH^-(aq) \quad [29]$$

Cations of strong bases do not react with water, equation [30]. Yet, the cation of a weak base hydrolyzes, forming the base and H_3O^+, equation [31]. Thus the cations of weak bases are proton donors (acids).

$$Na^+(aq) + H_2O(l) \longrightarrow \text{no reaction} \quad [30]$$

$$CH_3NH_3^+(aq) + H_2O(l) \rightleftharpoons CH_3NH_2(aq) + H_3O^+(aq) \quad [31]$$

Salts can be of four types, as follows.

1. *Cation of a strong base and anion of a strong acid.* Neither ion hydrolyzes (see reactions [27] and [30]), and the resulting solution is neutral with pH = 7.00.
2. *Cation of a strong base and anion of a weak acid.* The cation does not hydrolyze ([30]), but the anion does (as in [28]). The OH^- thus formed produces a basic solution with pH > 7.00.
3. *Cation of a weak base and anion of a strong acid.* The anion does not hydrolyze ([27]), but the cation does (as in [31]). The resulting solution is acidic with pH < 7.00.
4. *Cation of a weak base and anion of a weak acid.* Both the cation and the anion hydrolyze. The one which does so most extensively determines the acidity of the solution.

The relative strengths of the acid that supplied the anion and the base that supplied the cation determine the acidity of the salt solution. If the base is stronger, the solution is basic. If the acid is stronger, the solution is acidic.

11. Calculate values of K_a for cations and K_b for anions from ionization constants of their conjugates and K_w of water; and calculate the pH values of salt solutions in which hydrolysis occurs.

Hydrolysis problems are set up in the same way as were weak acid and weak base problems in objective 17-8. The sole difference is that the ionization constant (sometimes called the hydrolysis constant, K_h), normally is not tabulated but must be computed. For the anion of a weak acid, where $K_{a,HA}$ is the ionization constant of the weak acid, the ionization constant of the anion (K_b) is obtained as follows.

$$K_w = K_{a,HA}K_b \quad or \quad K_b = K_w / K_{a,HA} \quad [32]$$

For the cation of a weak base, where $K_{b,B}$ is the ionization constant of the weak base, the ionization constant of the cation (K_a) is obtained as follows.

$$K_w = K_{b,B}K_a \quad or \quad K_a = K_w / K_{b,B} \quad [33]$$

Another way to state these two relationships is to realize that

$$K_a = \frac{[H_3O^+][\text{conjugate base}]}{[\text{weak acid}]} \quad and \quad K_b = \frac{[OH^-][\text{conjugate acid}]}{[\text{weak base}]}$$

EXAMPLE 17-9 Determine the pH of a 0.125 M sodium benzoate solution. $K_a = 6.3 \times 10^{-5}$ for benzoic acid.

$$NaC_7H_5O_2 \longrightarrow Na^+(aq) + C_7H_5O_2^-(aq)$$

$Na^+(aq)$ does not hydrolyze but the benzoate anion does.

Rxn:	$C_7H_5O_2^-(aq)$	$+ H_2O(l) \rightleftharpoons$	$OH^-(aq)$	$+ HC_7H_5O_2(aq)$
I:	0.125 M			
C:	$-x$ M		$+x$ M	$+x$ M
Eq:	$(0.125 - x)$ M		x M	x M

$$K_a = \frac{[OH^-][HC_7H_5O_2]}{[C_7H_5O_2^-]} = \frac{(x)(x)}{0.125 - x} = \frac{K_w}{K_a} = \frac{1.00 \times 10^{-14}}{6.3 \times 10^{-5}} \approx \frac{x^2}{0.125}$$

or $x = 4.5 \times 10^{-6} M = [OH^-]$

We have assumed that $x << 0.125$ M. 4.5×10^{-6} M clearly is much smaller than 0.125 M. We obtain pOH = 5.35 *or* pH = 14.00 – pOH = 8.65.

Notice the very strong similarity between this calculation and that of Example 17-7.

*** 12. Use the relative strengths of Brønsted-Lowry acids and bases to predict the direction of acid-base reactions.**

The products of the reaction of an acid with a base are the conjugate acid and the conjugate base.

$$HC_2H_3O_2 + CH_3NH_2 \rightleftharpoons CH_3NH_3^+ + C_2H_3O_2^- \qquad [34]$$

Do products or reactants predominate in this mixture? Remember that the stronger acid is the better proton donor. Yet if the acid is a very good proton donor, there will be very little of it present in solution. Instead, mainly its conjugate base will be present. Since $HC_2H_3O_2$ ($K_a = 1.74 \times 10^{-5}$) is a stronger acid than $CH_3NH_3^+$ ($K_a = K_w/K_b = 2 \times 10^{-9}$), reaction [34] lies to the right. Notice also that CH_3NH_2 ($K_b = 5 \times 10^{-4}$) is a stronger base than $C_2H_3O_2^-$ ($K_b = K_w/K_a = 5.75 \times 10^{-10}$). Hence the equilibrium lies to the side of the weaker acids and the weaker base.

*** 13. Predict whether certain oxides and hydroxo compounds are acidic, basic, or neutral.**

Oxides can be grouped into three types: acidic, basic, and amphoteric. Acidic oxides form an acidic solution when they dissolve in water; basic oxides form an alkaline (a basic) solution. Solutions of amphoteric oxides act as bases when reacted with strong acid, and as acids when reacted with strong base. Acidic oxides are those of the nonmetals: Cl, Br, I, S, Se, N, P, As, C, Si, and B. Basic oxides are those of the metals: Li, Na, K, Rb, Cs, Mg, Ca, Sr, Ba, Tl, and Bi. Amphoteric oxides are those of the metalloids: Be, Al, Ga, Ge, Sn, Pb, and Sb. For transition elements, acidic oxides have the element in a high oxidation state (+6 or higher). The common amphoteric oxides of transition metals are ZnO, CdO, and Cr_2O_3.

The acidic character of hydroxo compounds parallels that of the oxides. Metal hydroxides are bases while nonmetal "hydroxides" are more commonly known as oxoacids. In both cases we are considering the group of atoms: E—O—H. In bases, the E—O bond cleaves more readily than the O—H bond. In acids, the situation is reversed. Breakage of the O—H bond is favored if E is highly electronegative (a nonmetallic atom or an atom of a high oxidation state). Breakage of the E—O bond is favored if E is electropositive.

*** 14. Define what an oxoacid is and predict the relative strengths of oxoacids from their molecular structures.**

An oxoacid is one in which the ionizable proton is bound to oxygen, as in structure [35].

rest of molecule—E—O—H [35]

When the O—H bond breaks readily, the oxoacid is a strong one. The O,H bond will break readily if atom E is highly electronegative. Thus, for the hypohalous acids, HOCl > HOBr > HOI. However, the electronegativities of elements can be altered by the atoms in the "rest of the molecule" bonded to E. In general, the following factors increase the electronegativity of E (and increase the oxoacid strength).

1. Strongly electronegative atoms bonded to E (FCOOH > ClCOOH > HCOOH). Highly electronegative atoms bonded to atoms next to E increase the strength of an acid ($CH_2FCOOH > CH_2ClCOOH > CH_3COOH$), although their effect is less than that of atoms directly bound to E.
2. High positive oxidation state and formal charge of E. Thus, for the chlorine oxoacids, $HClO_4 > HClO_3 > HClO_2 > HClO$, which is the same direction in which chlorine's oxidation state decreases.

DRILL PROBLEMS (Assume throughout that H_2SO_4 is a diprotic strong acid.)

2. (1) Label the Brønsted-Lowry acids and bases in each of the following equations. Make sure to label both conjugate pairs.

A. $H_2S + CO_3^{2-} \rightleftharpoons HCO_3^- + HS^-$
B. $H_3O^+ + HS^- \rightleftharpoons H_2S + H_2O$
C. $HS^- + OH^- \rightleftharpoons H_2O + S^{2-}$
D. $H_2O + NH_2^- \rightleftharpoons NH_3 + OH^-$
E. $HCO_3^- + OH^- \rightleftharpoons H_2O + CO_3^{2-}$
F. $HCN + OH^- \rightleftharpoons H_2O + CN^-$
G. $NH_4^+ + OH^- \rightleftharpoons NH_3 + H_2O$
H. $HSO_4^- + C_2H_3O_2^- \rightleftharpoons HC_2H_3O_2 + SO_4^{2-}$
I. $H_3PO_4 + CN^- \rightleftharpoons HCN + H_2PO_4^-$
J. $HSO_4^- + CN^- \rightleftharpoons HCN + SO_4^{2-}$

(2) Complete each of the following equations. (The acid is first in each.) Label each conjugate acid-base pair.

K. $HCl + NH_3 \rightleftharpoons$ L. $H_3O^+ + H_2PO_4^- \rightleftharpoons$ M. $HC_2H_3O_2 + HS^- \rightleftharpoons$

N. $HN_3 + CH_3NH_2 \rightleftharpoons$ O. $CH_3NH_3^- + OH^- \rightleftharpoons$ P. $HNO_2 + N_3^- \rightleftharpoons$

3. Draw the Lewis structure for each reactant in the following reactions and indicate which is the acid and which is the base.

A. $H_2O + SO_2 \longrightarrow H_2SO_3$ B. $BeF_2 + 2\,F^- \longrightarrow BeF_4^{2-}$

C. $NH_3 + HCl \longrightarrow NH_4Cl$ D. $CO_2 + OH^- \longrightarrow HCO_3^-$

E. $Ag^+ + 2\,NH_3 \longrightarrow Ag(NH_3)_2^+$ F. $O^{2-} + H_2O \longrightarrow 2\,OH^-$

G. $BF_3 + F^- \longrightarrow BF_4^-$ H. $HCl + H_2O \rightleftharpoons H_3O^+ + Cl^-$

I. $H_2O + BF_3 \longrightarrow H_2OBF_3$ J. $S^{2-} + SO_3 \longrightarrow S_2O_3^{2-}$

K. $H^+ + CO_3^{2-} \longrightarrow HCO_3^-$ L. $H_2O + CO_2 \longrightarrow H_2CO_3$

M. $BF_3 + CH_3OH \longrightarrow CH_3OH{\cdot}BF_3$

5. & 6. Fill in the blanks in the table that follows. The first line is an example.

	Solute	*Concentration*		*Acidity*			
	Formula	M	g/L	[H⁺]	[OH⁻]	pH	pOH
	HCl	0.020	0.73	0.020	5.0×10^{-13}	1.70	12.30
A.	NaOH	0.040	____	____	____	____	____
B.	H_2SO_4	____	____	0.00125	____	____	____
C.	$Ca(OH)_2$	____	____	____	0.0333	____	____
D.	RbOH	____	____	____	0.0375	____	____
E.	$HClO_4$	____	3.84	____	____	____	____
F.	$Ba(OH)_2$	____	0.513	____	____	____	____
G.	HNO_3	0.0125	____	____	____	____	____
H.	KOH	____	____	____	0.045	____	____
I.	$Sr(OH)_2$	____	____	____	____	9.13	____
J.	CsOH	____	____	____	____	____	4.65
K.	HBr	____	____	____	____	3.15	____
L.	HI	____	____	____	____	____	12.75

7. For each weak acid or weak base below, write the equation for ionizations in water and the ionization constant expression.

A. HOBr B. $(CH_3)_2NH$ C. $HC_6H_7O_2$ D. N_2H_4 E. $HC_3H_5O_3$

F. $(CH_3)_3N$ G. $HC_3H_5O_2$ H. $HC_2HCl_2O_2$ I. H_3BO_3 J. $HC_2Cl_3O_2$

8. (1) The initial concentration (or formality) of each acid or base is given below, along with an indication of the solution's acidity. Compute the ionization constant of the acid or base in each case.

A. HOBr, 0.250 M, pH = 4.640 B. $(CH_3)_2NH$, 0.400 M, $[OH^-] = 0.0165$ M

C. $HC_6H_7O_2$, 0.300 M, pOH = 11.687 D. N_2H_4, 0.0100 M, pH = 9.996

E. $HC_3H_5O_3$, 1.20 M, $[H^+] = 0.0313$ M F. $(CH_3)_3N$, 0.140 M, pH = 11.503

G. $HC_3H_5O_2$, 0.0200 M, pH = 3.282 H. $HC_2HCl_2O_2$, 2.20 M, $[H^+] = 0.254$ M

I. H_3BO_3, 0.0400 M, pOH = 9.683 J. $HC_2Cl_3O_2$, 1.50 M, $[H^+] = 0.466$ M

(2) Determine the pH of each of the following solutions. Use the data of *Table 17-2 in the text.*

K. 0.250 M $HC_2H_3O_2$ L. 1.20 M $HC_7H_5O_2$ M. 0.820 M $HCHO_2$

N. 0.400 M HNO_2 O. 0.230 M HF P. 1.00 M $HClO_2$

Q. 1.20 M NH_3 R. 2.60 M CH_3NH_3 S. 5.10 M $C_2H_5NH_2$

T. 0.510 M HN_3 U. 0.510 M $HONH_2$ V. 0.150 M $C_6H_5NH_2$

($K_a = 1.90 \times 10^{-5}$ for HN_3)

9. Determine the pH, the final concentration of undissociated acid, and the concentration of all anions in the following solutions. Use the data of this objective, and of *Table 17-3 in the text.*

A. 0.00460 M H_2CO_3 B. 0.0100 M H_2S C. 1.20 M $H_2C_2O_4$

D. 0.750 M H_3PO_4 E. 0.900 M H_3PO_3 F. 1.00 M H_2SO_3

10. For each of the following salts, write the reaction for the salt dissolving in water. Based on the ions present, predict whether the solution is acidic, basic, or neutral. Finally, write any hydrolysis reactions that occur.

A. NaCl B. NH_4Cl C. N_2H_5Br D. KNO_3 E. $C_2H_5NH_3NO_3$

F. Na_2CO_3 G. $C_6H_5NH_3I$ H. $Ca(ClO)_2$ I. $RbClO_2$ J. $(HONH_3)_2SO_4$

K. SrF2 L. $CsC_2H_3O_2$ M. NH_4CN N. $HONH_3F$ O. $CH_3NH_3ClO_2$

11. (1) 1.00 mol of each of the salts in parts A through L of the drill problems of objective 17-10 is used to form 1.00 L of aqueous solution. Determine the pH of each solution.
(2) Determine the pH of each of the following solutions.
M. 3.00 mmol sodium acetate in 15.0 mL solution
N. 5.00 mmol calcium benzoate in 50.0 mL solution
O. 12.5 mmol potassium formate in 12.5 mL solution
P. 3.20 mmol barium nitrite in 16.0 mL solution
Q. 1.79 mmol rubidium fluoride in 20.0 mL solution
R. 12.1 mmol cesium chlorite in 10.0 mL solution
S. 6.10 mmol ammonium sulfate in 25.0 mL solution
T. 5.37 mmol methylammonium sulfate in 20.0 mL solution
U. 26.4 mmol ethylammonium nitrate in 60.0 mL solution
V. 177.44 mmol hydrazine chloride (N_2H_5Cl) in 180.0 mL solution ($K_b = 9.8 \times 10^{-7}$)
W. 2.28 mmol aniline bromide in 250.0 mL solution
X. 11.0 mmol ammonium iodide in 80.0 mL solution

12. Equilibrium in each reaction in the two groups of drill problems of objective 17-2 lies to the right. Rank all the conjugate acid-base pairs in order of decreasing acid strength (strongest to weakest). Use the data of *Table 17-2 in the text* if necessary. Then complete the following equations and use your ranking to predict whether each equilibrium lies to the left or to the right.
A. $H_2O + N_3^- \rightleftharpoons$ B. $HNO_2 + OH^- \rightleftharpoons$ C. $H_3O^+ + CN^- \rightleftharpoons$
D. $NH_3 + CN^- \rightleftharpoons$ E. $HCN + H_2PO_4^- \rightleftharpoons$ F. $H_3PO_4 + NH_2^- \rightleftharpoons$
G. $H_3PO_4 + OH^- \rightleftharpoons$ H. $H_3O^+ + NH_2^- \rightleftharpoons$ I. $HCO_3^- + C_2H_3O_2^- \rightleftharpoons$
J. $HSO_4^- + HS^- \rightleftharpoons$ K. $H_2S + OH^- \rightleftharpoons$ L. $H_2O + C_2H_3O_2^- \rightleftharpoons$
M. $HCO_3^- + SO_4^{2-} \rightleftharpoons$ N. $HSO_4^- + OH^- \rightleftharpoons$

13. (1) Classify each of the following as acidic, basic, or amphoteric oxides.
A. CaO B. CO_2 C. PbO D. SiO_2 E. SO_2
F. ClO_2 G. BeO H. NO_2 I. SnO J. BaO
K. Al_2O_3 L. Tl_2O M. Na_2O N. P_2O_5 O. MgO
(2) Complete and balance each of the following reactions. ("No reaction" may be the correct answer.)
P. $CaO + SiO_2 \longrightarrow$ Q. $Al_2O_3 + SO_2 \longrightarrow$ R. $NO + CO_2 \longrightarrow$
S. $NaO + CaO \longrightarrow$ T. $BaO + SO_2 \longrightarrow$ U. $PbO + P_2O_5 \longrightarrow$
V. $Na_2O + Al_2O_3 \longrightarrow$

14. Predict which acid of each of the following pairs is the stronger and briefly explain why.
A. H_3BO_3 *or* H_2CO_3 B. H_2SO_4 *or* H_2SO_3 C. H_3PO_4 *or* H_3AsO_4
D. $HClO_3$ *or* $HClO_4$ E. $HBrO$ *or* $HClO$ F. HNO_2 *or* HNO_3
G. H_2SO_4 *or* H_2SeO_4 H. CF_3OH *or* CH_3OH I. FC_6H_4OH *or* ClC_6H_4OH
J. $HBrO_3$ *or* H_2SeO_3 K. H_3PO_4 *or* $HClO_4$ L. H_2CO_3 *or* H_4SiO_4

QUIZZES (20 minutes each) Choose the best answer for each question.

Quiz A

1. In the reaction $BF_3 + NH_3 \longrightarrow F_3B{:}NH_3$, BF_3 accepts an electron pair and acts as (a) an Arrhenius base; (b) a Brønsted acid; (c) a Lewis acid; (d) a Lewis base; (e) none of these.
2. Which of the following five statements about Brønsted-Lowry acids and bases is true? (a) An acid and its conjugate base react to form a salt and water. (b) The acid H_2O is its own conjugate base. (c) The conjugate base of a strong acid is a strong base. (d) The conjugate base of a weak acid is a strong base. (e) A base and its conjugate acid react to form a neutral solution.
3. Nitrous acid, HNO_2, is 0.37% ionized in 3.0 M solution. What is the value of K_a for HNO_2? (a) $(3.0)(0.37)^2/99.63$; (b) $(3.0)(0.0037)^2/0.9963$; (c) $(0.0037)^2/0.9963$; (d) $[0.9963/(0.0037)^2]/3.0$; (e) none of these.
4. A 2.4 M solution of a weak base has $[OH^-] = 0.0034$ M. What is K_b for this base? (a) 0.0034; (b) 0.0034/2.4; (c) $(2.4)^2/0.0034$; (d) $(0.0034/2.4)^2$; (e) none of these.
5. 7200. mL of solution contains 0.216 mmol of potassium hydroxide. What is the pH of this solution? (a) 10.143; (b) 4.523; (c) 3.857; (d) 9.477; (e) none of these.

6. pOH = 3.14 is equivalent to (a) pH = 11.86; (b) $[H^+] = 1.4 \times 10^{-10}$; (c) $[OH^-] = 7.3 \times 10^{-4}$; (d) acidic solution; (e) none of these.
7. $CO_2(g)$ acts as an acid in the reaction $CaO(s) + CO_2(g) \longrightarrow CaCO_3(s)$ because it (a) turns blue litmus red; (b) reacts with a metal; (c) is a proton donor; (d) is an electron-pair acceptor; (e) none of these.
8. In order of decreasing strength: $OH^- > ClO^- > NO_2^-$. Which of the following reactions lies to the right? (a) $NO_2^- + HClO \rightleftharpoons HNO_2 + ClO^-$; (b) $H_2O + HClO \rightleftharpoons H_3O^+ + ClO^-$; (c) $OH^- + HNO_2 \rightleftharpoons H_2O + NO_2^-$; (d) $H_2O + ClO^- \rightleftharpoons OH^- + HClO$; (e) none of these.
9. Which of the following is the strongest acid? (a) H_2SO_4; (b) H_2SO_3; (c) H_2SeO_4; (d) $HONH_2$; (e) none of these.

Quiz B

1. In the reversible reaction $HCO_3^-(aq) + OH^-(aq) \rightleftharpoons CO_3^{2-}(aq) + H_2O$ the Brønsted acids are (a) HCO_3^- and CO_3^{2-}; (b) HCO_3^- and H_2O; (c) OH^- and H_2O; (d) OH^- and CO_3^{2-}; (e) none of these.
2. *Proton donor* is an abbreviated definition of a (a) Brønsted-Lowry acid; (b) Brønsted-Lowry base; (c) Lewis acid; (d) Lewis base; (e) none of these.
3. Propionic acid, $HC_3H_5O_2$, is 0.42% ionized in 0.80 M solution. What is K_a for this acid? (a) $0.80(0.0042)^2/0.9958$; (b) $(0.80)(0.42)^2/99.58$; (c) $(0.0042)^2/0.9958$; (d) $(0.9958/0.80)/(0.0042)^2$; (e) none of these.
4. In a 2.00 M solution of HNO_2, $[H^+] = 0.030$ M. What is K_a for this acid? (a) 2.0/0.030; (b) 0.30/2.0; (c) $(0.030)^2/2.00$; (d) $2.00/(0.030)^2$; (e) none of these.
5. 250. mL of solution contains 0.132 mmol of perchloric acid. What is the pH of this solution? (a) 3.880; (b) 2.398; (c) 10.120; (d) 3.277; (e) none of these.
6. $[OH^-] = 3.0 \times 10^{-10}$ M is equivalent to (a) pH = 4.522; (b) pOH = 10.478; (c) 3.33×10^{-4} M = $[H_3O^+]$; (d) acidic solution; (e) none of these.
7. The following all are polyprotic acids *except* (a) H_2SO_4; (b) H_3BO_3; (c) $HClO_4$; (d) H_3PO_3; (e) $H_2C_2O_4$.
8. In order of decreasing strength $HF > HC_2H_3O_2 > HCN$. Which of the following equilibria lies to the right? (a) $HCN + C_2H_3O_2^- \rightleftharpoons CN^- + HC_2H_3O_2$; (b) $HF + CN^- \rightleftharpoons HCN + F^-$; (c) $HCN + H_2O \rightleftharpoons H_3O^+ + CN^-$; (d) $HC_2H_3O_2 + F^- \rightleftharpoons C_2H_3O_2^- + HF$; (e) none of these.
9. Which of the following is the strongest acid? (a) $HC_2H_3O_2$; (b) $HC_2HF_2O_2$; (c) H_2O; (d) NH_3; (e) $HC_2H_2FO_2$.

Quiz C

1. In the reaction $HCl(g) + NaOH(s) \longrightarrow NaCl(s) + H_2O(g)$, HCl(g) acts as an acid because it (a) turns blue litmus red; (b) reacts with a metal; (c) is an electron-pair acceptor; (d) is a proton donor; (e) none of these.
2. A Lewis base is defined as a (a) proton donor; (b) proton acceptor; (c) hydrogen ion donor; (d) hydrogen ion acceptor; (e) none of these.
3. Acetic acid, $HC_2H_3O_2$, is 0.11% ionized in 1.5 M solution. What is K_a for this acid? (a) $(0.11)^2/99.89$; (b) $(0.0011)^2/0.9989$; (c) $99.89/(0.11)^2$; (d) $(0.9989/1.5)/(0.0011)^2$; (e) none of these.
4. A 0.30 M weak acid solution has $[H^+] = 6.8 \times 10^{-5}$. The value of K_a for this acid is (a) 1.5×10^{-8}; (b) 6.8×10^{-5}; (c) 4.6×10^{-9}; (d) 2.3×10^{-4}; (e) none of these.
5. 500. mL of solution contains 1.50 mmol of nitric acid. What is the pH of this solution? (a) 2.824; (b) 2.699; (c) 2.523; (d) 11.477; (e) none of these.
6. pH = 4.25 is equivalent to (a) pOH = 9.75; (b) $[H^+] = 5.61 \times 10^{-4}$ M; (c) $[OH^-] = 5.61 \times 10^{-4}$ M; (d) an alkaline solution; (e) none of these.
7. Which pair does not represent and acid and its anhydride? (a) H_3PO_4, P_4O_{10}; (b) H_2CO_3, CO_2; (c) $HClO_2$, Cl_2O; (d) HNO_3, N_2O_5; (e) H_3BO_3, B_2O_3.
8. The equilibria $HClO + CN^- \rightleftharpoons ClO^- + HCN$ *and* $HC_2H_3O_2 + ClO^- \rightleftharpoons C_2H_3O_2^- + HClO$ lie to the right. Which list in order of decreasing strength is correct? (a) $C_2H_3O_2^- > ClO^- > CN^-$; (b) $C_2H_3O_2^- > CN^- > ClO^-$; (c) $HC_2H_3O_2 > HCN > HClO$; (d) $HCN > HClO > HC_2H_3O_2$; (e) none of these.
9. Which of the following is the strongest acid? (a) HBrO; (b) HIO_2; (c) $HBrO_2$; (d) $HClO_2$; (e) $HClO_3$.

Quiz D

1. The conjugate acid of HPO_4^{2-} is (a) PO_4^{3-}; (b) $H_2PO_4^-$; (c) H_3PO_4; (d) H_3O^+; (e) P_2O_5.
2. Proton acceptor is an abbreviated definition of a (a) Brønsted-Lowry base; (b) Brønsted-Lowry acid; (c) Lewis base; (d) Lewis acid; (e) none of these.
3. Formic acid, $HCHO_2$, is 0.915% ionized in 2.5 M solution. What is K_a for this acid?

(a) (0.99085/2.5)/(0.00915)^2; (b) (0.00915)^2/0.99085; (c) (2.5)(0.00915)^2/0.99085; (d) (2.5)(0.915)^2/99.085; (e) none of these.

4. In a 2.00 M solution of weak base, $[OH^-] = 0.040$ M. What is the value of K_b for this base? (a) 0.02; (b) 8×10^{-4}; (c) 4×10^{-4}; (d) 1.6×10^{-3}; (e) none of these.
5. 1200. mL of solution contains 0.0720 mmol of sulfuric acid. What is the pH of this solution? (a) 3.921; (b) 4.301; (c) 9.699; (d) 1.143; (e) none of these.
6. $[H^+] = 2.5 \times 10^{-4}$ M is equivalent to (a) pH = 4.40; (b) pOH = 11.60; (c) $[OH^-] = 4.0 \times 10^{-10}$; (d) acidic solution; (e) none of these.
7. P_4O_6 is the anhydride of (a) HPO_3; (b) $H_4P_2O_7$; (c) H_3PO_4; (d) H_3PO_3; (e) none of these.
8. The equilibria $OH^- + HClO \rightleftharpoons H_2O + ClO^-$ *and* $ClO^- + HNO_2 \rightleftharpoons HClO + NO_2^-$ both lie to the right. Which list in order of decreasing strength is correct? (a) $HClO > HNO_2 > H_2O$; (b) $ClO^- > NO_2^- > OH^-$; (c) $NO_2^- > ClO^- > OH^-$; (d) $HNO_2 > HClO > H_2O$; (e) none of these.
9. Which of the following is the strongest acid? (a) H_3AlO_3; (b) H_2SiO_3; (c) H_3PO_3; (d) H_2SO_3; (e) $HClO_3$.

SAMPLE TEST

1. Benzoic acid (C_6H_5COOH, 122.0 g/mol) has $K_a = 6.3 \times 10^{-5}$. What is the pH of a solution formulated by dissolving 3.050 g of benzoic acid in enough water to make 0.250 L of solution?
2. Classify the following acid-base reactions in terms of the acid-base concept that they represent. In each case, choose the most elementary concept that is applicable. In order of increasing complexity, the acid-base concepts are Arrhenius, Brønsted-Lowry, and Lewis. Indicated which substance(s) is (are) acids and which is (are) bases.
   a. $HF + N_2H_4 \rightleftharpoons F^- + N_2H_5^+$
   b. $O^{2-} + H_2O \rightleftharpoons 2\ OH^-$
   c. $SOI_2 + BaSO_3 \rightleftharpoons Ba^{2+} + 2\ I^- + 2\ SO_2$
   d. $HgCl_3^- + Cl^- \rightleftharpoons HgCl_4^{2-}$
3. All four of the equilibria below lie to the right.
   $H_2SO_3 + F^- \rightleftharpoons HSO_3^- + HF$ $\qquad$ $CH_3NH_3^+ + OH^- \rightleftharpoons CH_3NH_2 + H_2O$
   $HF + N_2H_4 \rightleftharpoons F^- + N_2H_5^-$ $\qquad$ $N_2H_5^+ + CH_3NH_2 \rightleftharpoons N_2H_4 + CH_3NH_3^+$
   a. Rank all of the acids involved in order of decreasing acid strength.
   b. Rank all of the bases involved in order of decreasing base strength.
   c. State whether each of the following two equilibria lies primarily to the right or to the left.
   (i) $HF + OH^- \rightleftharpoons F^- + H_2O$ $\qquad$ (ii) $CH_3NH_3^+ + HSO_3^- \rightleftharpoons CH_3NH_2 + H_2SO_3$
4. 3.00 mol of calcium chlorite is dissolved in enough water to produce 2.50 L of solution. $K_a = 2.95 \times 10^{-8}$ for HClO and $K_a = 1.2 \times 10^{-2}$ for $HClO_2$. Compute the pH of the solution produced.

18 Additional Aspects of Acid-Base Equilibria

CHAPTER OBJECTIVES

* **1. Describe the effect of common ions on the ionization of weak acids and bases, and calculate the concentrations of all species present in solutions of weak acids or bases and their common ions.**

The ionization of a weak acid or base is a dynamic equilibrium, governed by LeChâtelier's principle. If we increase the concentration of one of the products, the equilibrium will shift left. Consider the ionization of a weak acid. We use the techniques of Chapter 17 to compute $[H_3O^+]$ and $[F^-]$ in an aqueous solution of HF.

EXAMPLE 18-1 What are $[F^-]$ and $[H_3O^+]$ in 0.500 M HF? From *Table 17-2 in the text*, $K_a = 6.7 \times 10^{-5}$.

Rxn:	HF(aq)	$+ H_2O(l) \rightleftharpoons$	$H_3O^+(aq)$	$+ F^-(aq)$
I:	0.500 M			
C:	$-x$ M		$+x$ M	$+x$ M
Eq:	$(0.500 - x)$ M		x M	x M

$$K_a = \frac{[H_3O^+][F^-]}{[HF]} = 6.7 \times 10^{-4} = \frac{(x)(x)}{0.500 - x} \approx \frac{x^2}{0.500}$$

We assumed $x << 0.500$ M. Then, $x = 0.018$ M $= [H_3O^+] = [F^-]$. x is small (less than 5%) compared to 0.500 M. The percent ionization of HF is now computed.

$$\% \text{ ionization} = \frac{0.018 \text{ M}}{0.500 \text{ M}} \times 100 = 3.6\%$$

If we add H_3O^+ or F^- to the solution, the equilibrium will shift to the left, and the percent ionization of HF will decrease. Example 18-2 shows what occurs if we add 0.100 mol H_3O^+ (for example, 0.100 mol HCl) to each liter of the solution of Example 18-1.

EXAMPLE 18-2 What are $[H_3O^+]$ and $[F^-]$ in a solution that has $[HF]_i = 0.500$ M and $[HCl]_i = 0.100$ M?

Rxn:	HF(aq)	$+ H_2O(l) \rightleftharpoons$	$H_3O^+(aq)$	+	$F^-(aq)$
I:	0.500 M		0.100 M		
C:	$-x$ M		$+x$ M		$+x$ M
Eq:	$(0.500 - x)$ M		$(0.100 + x)$M		x M

$$K_a = \frac{[H_3O^+][F^-]}{[HF]} = 6.7 \times 10^{-4} = \frac{(0.100 + x)(x)}{0.500 - x} \approx \frac{(0.100 + x)(x)}{0.500}$$

We have assumed $x << 0.100$ M. This gives $x = (0.500)(6.7 \times 10^{-4})/0.100 = 3.4 \times 10^{-3} = [F^-]$. Thus x is small (less than 5%, actually 3.4%) compared to 0.100 .

$$\% \text{ ionization} = \frac{0.0034 \text{ M}}{0.500 \text{ M}} \times 100 = 0.68\% \quad ([H_3O^+] = 0.103 \text{ M})$$

This 0.68% is less than the 3.6% of Example 18-2, as we predicted. The presence of one of the product ions (H_3O^+) has suppressed the ionization of the weak acid, as predicted by LeChâtelier's principle. This effect on a reaction that has ions as products is known as the *common ion effect*. Adding either one of the product ions to the solution suppresses ionization. Let us see what effect the addition of F^- has. If we add less F^- than we did H_3O^+ in Example 18-2, the suppression of ionization should be less. Thus the percent ionization we calculate should be larger than in Example 18-2 but smaller than in Example 18-1.

EXAMPLE 18-3 What are $[H_3O^+]$ and $[F^-]$ in a solution with $[HF]_i = 0.500$ M and $[F^-]_i = 0.0400$ M?

Rxn:	HF(aq)	+ $H_2O(l) \rightleftharpoons$	$H_3O^+(aq)$	+ $F^-(aq)$
I:	0.500 M			0.0400 M
C:	$-x$ M		$+x$ M	$+x$ M
Eq:	$(0.500 - x)$ M		x M	$(0.0400 + x)$ M

$$K_a = \frac{[H_3O^+][F^-]}{[HF]} = 6.7 \times 10^{-4} = \frac{(x)(0.0400 + x)}{0.500 - x} \approx \frac{(0.0400)(x)}{0.500}$$

We assumed $x << 0.0400$. This gives $x = 8.38 \times 10^{-3}$, which is *not* small compared to 0.0400. We use the technique of successive approximations (Example 17-5, Objective 17-8).

$$\frac{(x)(0.0400 + x)}{0.500 - x} \cong \frac{(x)(0.0400 + 0.0083)}{0.500 - 0.0083} = \frac{0.0483\,x}{0.492} \approx 6.7 \times 10^{-4}$$

Then $x = (6.7 \times 10^{-4})(0.492)/0.0483 = 0.00682$ M. We use another cycle of approximation.

$$\frac{(x)(0.0400 + x)}{0.500 - x} \cong \frac{(x)(0.0400 + 0.0068)}{0.500 - 0.0068} = \frac{0.0468\,x}{0.494} \approx 6.7 \times 10^{-4}$$

Then $x = (6.7 \times 10^{-4})(0.494)/0.0468 = 7.07 \times 10^{-3}$ M *or* 0.00707 M. The last two values are in fair agreement. The quadratic equation gives 0.0070 M, as does one more cycle of successive approximations. Thus, $[H_3O^+] = 0.0070$ M and $[F^-] = 0.0400 + 0.0070 = 0.0470$ M.

$$\%\ \text{ionization} = \frac{0.0070\ \text{M}}{0.500\ \text{M}} \times 100 = 1.4\%$$

This is more than the 0.68% of Example 18-2, as we predicted. But it still is less than the 3.6% of Example 18-1, where HF was the only solute. We now work the same types of problems for a solution of a weak base [trimethylamine, $(CH_3)_3N$, $K_b = 7.4 \times 10^{-5}$].

EXAMPLE 18-4 What are $[OH^-]$, $[(CH_3)_3NH^+]$, and the percent ionization in solutions with the following initial concentrations? (a) $[(CH_3)_3N]_i = 0.400$ M; (b) $[(CH_3)_3N]_i = 0.400$ M and $[OH^-]_i = 0.0900$ M; (c) $[(CH_3)_3N]_i = 0.400$ M and $[(CH_3)_3NH^+]_i = 0.0500$ M.

(a) Rxn:	$(CH_3)_3N(aq)$	+ $H_2O(l) \rightleftharpoons$	$(CH_3)_3NH^+(aq)$	+ $OH^-(aq)$
I:	0.400 M			
C:	$-x$ M		$+x$ M	$+x$ M
Eq:	$(0.400 - x)$ M		x M	x M

$$K_b = \frac{[(CH_3)_3NH^+][OH^-]}{[(CH_3)_3N]} = 7.4 \times 10^{-5} = \frac{(x)(x)}{0.400 - x} \approx \frac{x^2}{0.400}$$

We assumed $x << 0.40$ M. Then $x = 5.4 \times 10^{-3}$ M $= [OH^-] = [(CH_3)_3NH^+] << 0.400$ M

$$\%\ \text{ionization} = \frac{5.4 \times 10^{-3}}{0.400} \times 100 = 1.4\%$$

(b) Rxn:	$(CH_3)_3N(aq)$	+ $H_2O(l) \rightleftharpoons$	$(CH_3)_3NH^+(aq)$	+ $OH^-(aq)$
I:	0.400 M			0.0900 M
C:	$-x$ M		$+x$ M	$+x$ M
Eq:	$(0.400 - x)$ M		x M	$(0.0900 + x)$ M

$$K_b = \frac{[(CH_3)_3NH^+][OH^-]}{[(CH_3)_3N]} = 7.4 \times 10^{-5} = \frac{(x)(0.0900 + x)}{0.400 - x} \approx \frac{(0.0900)(x)}{0.400}$$

We assumed $x << 0.0900$ M. Then $x = 3.3 \times 10^{-4}$ M = $[(CH_3)_3NH^+] << 0.0900$ M and $[OH^-] = 0.0900 + 0.0003 = 0.0903$ M

$$\% \text{ ionization} = \frac{3.3 \times 10^{-4}}{0.400} \times 100 = 0.083\%$$

(c)

Rxn:	$(CH_3)_3N(aq)$ + $H_2O(l)$ $\rightleftharpoons$	$(CH_3)_3NH^+(aq)$	+ $OH^-(aq)$
I:	0.400 M	0.0500 M	
C:	$-x$ M	$+x$ M	$+x$ M
Eq:	$(0.400 - x)$ M	$(0.0500 + x)$ M	x M

$$K_b = \frac{[(CH_3)_3NH^+][OH^-]}{[(CH_3)_3N]} = 7.4 \times 10^{-5} = \frac{(0.0500 + x)(x)}{0.400 - x} \approx \frac{(0.0500)(x)}{0.400}$$

We assumed $x << 0.0500$ M. Then $x = 5.9 \times 10^{-4}$ M = $[OH^-] << 0.0500$ M and $[(CH_3)_3NH^+] = 0.0500 + 0.0006 = 0.0506$ M

$$\% \text{ ionization} = \frac{5.9 \times 10^{-4}}{0.400} \times 100 = 0.15\%$$

You should be able to identify the source of the common ion based on the following information. H_3O^+ or OH^- comes from a strong acid or base (See Table 5-1 in objective 5-1). The anion of a weak acid comes from a salt in which the cation is that of a strong base. For example, NaF will add F^- ion to solution. The cation of a weak base comes from a salt in which the anion is that of a strong acid. For instance, $(CH_3)_3NHCl$ will add $(CH_3)_3NH^+$ to solution.

2. Explain why the pH of water changes markedly when a small amount of H_3O^+ or OH^- is added, and why the pH of a buffer does not change very much with a similar addition.

A buffer can be either (a) a solution of a weak acid and about the same amount of the weak acid's anion *or* (b) a solution of a weak base and about the same amount of the weak base's cation. The solution of Example 18-4(c) is a buffer. It contains a weak base, 0.400 M $(CH)_3N$, and its cation, 0.0500 M $(CH_3)_3NH^+$. The solution of Example 18-3 contains a weak acid, 0.500 M HF, and its anion, 0.0400 M F^-. However, most chemists would not think of this latter solution as a buffer because the concentrations of the two species should differ by less than a factor of 10. Thus, for a buffer that contains a weak acid,

$$0.10 \leq \frac{[\text{acid's anion}]}{[\text{acid}]} = \frac{[A^-]}{[HA]} \leq 10.0 \qquad [2]$$

and for a buffer that contains a weak base,

$$0.10 \leq \frac{[\text{base's cation}]}{[\text{base}]} = \frac{[BH^+]}{[B]} \leq 10.0 \qquad [3]$$

Both equations ([2] and [3]) can be combined in a single expression.

$$0.10 \leq \frac{[\text{conjugate acid}]}{[\text{conjugate base}]} \leq 10.0 \qquad [4]$$

In Example 18-3, $[F^-]/[HF] = 0.0400$ M/0.500 M = 0.0800, and we would not consider the solution to be a buffer.

A buffer keeps the pH of a solution constant by reacting with small amounts of added H_3O^+ or OH^- to remove them from solution. Consider first the buffer of a weak acid, a solution with [HF]: = 1.00 M and $[F^-]$ = 0.500 M. A small amount of added H_3O^+ reacts with the weak acid's anion and produces the weak acid.

$$H_3O^+(aq) + F^-(aq) \rightleftharpoons H_2O + HF(aq) \qquad [5]$$

A small amount of added OH^- reacts with the weak acid and produces its anion.

$$OH^-(aq) + HF(aq) \longrightarrow H_2O + F^-(aq) \qquad [6]$$

Notice in both reactions [5] and [6] one of the fluorine-containing species, either acid or anion, is consumed and the other is produced. Therefore, the only effects of adding H_3O^+ or OH^- are to produce H_2O and change the ratio of $[F^-]/[HF]$.

A buffer that contains a weak base operates somewhat differently. Consider a buffer of 0.400 M CH_3NH_2 and 1.00 M $CH_3NH_3^+$. A small amount of added H_3O^+ reacts with the weak base and produces its cation.

$$H_3O^+(aq) + CH_3NH_2(aq) \longrightarrow CH_3NH_3^+(aq) + H_2O \qquad [7]$$

A small amount of added OH^- reacts with the weak base cation to produce the weak base.

$$OH^-(aq) + CH_3NH_3^+(aq) \rightleftharpoons CH_3NH_2(aq) + H_2O \qquad [8]$$

Thus, the sole effects of adding H_3O^+ or OH^- are to produce H_2O and change the $[CH_3NH_3^+]/[CH_3NH_2]$ ratio. It now is clear why pure water cannot act as a buffer. There are no species present to react with added H_3O^+ or OH^-.

Finally, some thoughts on naming buffer solutions. As equation [4] reveals, every buffer solution contains both a weak acid and its conjugate base, a weak base. We will, however, find it useful to distinguish between two types of buffer solutions. The first of these consists of an uncharged weak acid (such as acetic acid, $HC_2H_3O_2$) and its anion ($C_2H_3O_2^-$) as the conjugate base. We shall call this a *weak acid buffer*. The second type of buffer solution consists of an uncharged weak base (such as ammonia, NH_3) and its cation (NH_4^+) as the conjugate acid. We shall call this a *weak base buffer*.

3. Describe how buffer solutions can be prepared.

There are three ways of making a buffer that contains a weak acid.

1. Place nearly equal quantities of a weak acid and the salt of its anion in solution. 0.600 mol $HC_2H_3O_2$ plus 0.400 mol $KC_2H_3O_2$ in solution is an acetic acid-acetate buffer.
2. Partly neutralize the weak acid with a strong base. 1.000 mol $HC_2H_3O_2$ plus 0.400 mol KOH in solution yields the same acetic acid-acetate buffer as in part 1.

	$HC_2H_3O_2$ +	KOH $\longrightarrow$	$C_2H_3O_2^-$ +	K^+ +	H_2O
before rxn:	1.000 mol	0.400 mol			
after rxn:	0.600 mol	0 mol	0.400 mol	0.400 mol	

3. Partly neutralize the salt of a weak acid's anion with strong acid. 1.000 mol $KC_2H_3O_2$ (a source of $C_2H_3O_2^-$) plus 0.600 mol HCl (a source of H_3O^+) in solution also produces a buffer. This buffer has more $K^+(aq)$ and $Cl^-(aq)$ ions (0.600 mol more of each) than did the previous two. However, these two ions are spectator ions and do not influence the pH of the solution. Specifically, neither $K^+(aq)$ nor $Cl^-(aq)$ hydrolyzes.

	$C_2H_3O_2^-$ +	H_3O^+ $\longrightarrow$	$HC_2H_3O_2$ +	H_2O
before rxn:	1.000 mol	0.600 mol		
after rxn:	0.400 mol	0 mol	0.600 mol	

The buffer of a weak base also can be made in three ways.

1. The weak base and about the same amount of its salt (0.700 mol NH_3 + 0.300 mol NH_4Cl).
2. The weak base partly titrated with a strong acid (1.000 mol NH_3 + 0.300 mol HCl).
3. A salt of the weak base's cation partly titrated with a strong base (1.000 mol NH_4Cl + 0.700 mol NaOH).

4. Know the limitations of the basic equations used to determine the pH of buffer solutions (specifically the Henderson-Hasselbalch equation) by understanding their derivations.

EXAMPLE 18-5 What are the $[H_3O^+]$ and pH of a solution with $[HF]_i$ = 1.00 M and $[F^-]_i$ = 0.500 M?

Rxn:	HF(aq)	+ $H_2O(l)$ $\rightleftharpoons$	$H_3O^+(aq)$	+ $F^-(aq)$
I:	1.00 M			0.500 M
C:	$-x$ M		$+x$ M	$+x$ M
Eq:	$(1.00 - x)$ M		x M	$(0.500 + x)$ M

$$K_a = \frac{[H_3O^+][F^-]}{[HF]} = 6.7 \times 10^{-4} = \frac{(x)(0.500 + x)}{1.00 - x} \approx \frac{(0.500)(x)}{1.00}$$

We assumed that $x << 0.500$ M and find $x = 1.4 \times 10^{-3}$ M $<< 0.500$ M, pH = 2.85.

The exact expression for K_a is equation [9]

$$K_a = \frac{x([F^-]_i + x)}{([HF]_i - x)} \qquad [9]$$

where the concentrations of HF and F^- are initial ones before dissociation occurs. The approximation for this situation, that $[F^-]_i >> x << [HF]_i$, produces equation [10]

$$K_a = \frac{x([F^-]_i)}{([HF]_i)} \qquad [10]$$

which is easier to solve that the exact expression. Equation [10] can be used if $[HF]_i$ and $[F^-]_i$ differ by less than a factor of 10. This is the same condition that we imposed on a buffer, expression [2]. Recognizing that $x = [H_3O^+]$, we obtain

$$pK_a = pH - \log \frac{[F^-]_i}{[HF]_i} \quad or \quad pH = pK_a + \log \frac{[F^-]_i}{[HF]_i} \qquad [11]$$

If expression [2] is true, then for any buffer that contains a weak acid (where HA = weak acid and A^- = its anion),

$$pH = pK_a + \log \frac{[A^-]_i}{[HA]_i} \qquad [12]$$

If expression [3] is true, then for any buffer that contains a weak base (where B = weak base and BH^+ = its cation),

$$pOH = pK_b + \log \frac{[BH^+]_i}{[B]_i} \qquad [13]$$

*** 5. Calculate the pH of a buffer solution from concentrations of the buffer components and a value of K_a or K_b, and describe how to prepare a buffer that has a specific pH.**

This objective illustrates the application of equations [12] and [13].

EXAMPLE 18-6 A solution has $[HCOOH]_i = 0.800$ M and $[Ca(HCOO)_2]_i = 0.230$ M. What is its pH?

We first compute a value of $[HCOO^-]_i$, the initial concentration of formate anion.

$$[HCOO^-]_i = \frac{0.230 \text{ mol } Ca(HCOO)_2}{L} \times \frac{2 \text{ mol } HCOO^-}{\text{mol } Ca(HCOO)_2} = 0.460 \text{ M}$$

The ratio $[HCOO^-]/[HCOOH]$ equals (0.460 M/0.800 M) = 0.575 and satisfies expression [2]. From *Table 17-2 in the text,* for formic acid $K_a = 1.8 \times 10^{-4}$ and $pK_a = 3.74$.

$$pH = pK_a + \log \frac{[A^-]_i}{[HA]_i} = 3.74 + \log 0.575 = 3.50$$

EXAMPLE 18-7 What is the pH of a solution in which $[C_2H_5NH_2]_i = 0.750$ M and $[C_2H_5NH_3NO_3]_i = 0.250$ M?

The ratio $[C_2H_5NH_3^+]_i/[C_2H_5NH_2]_i$ is (0.250 M/0.750 M) = 0.333 and satisfies expression [3]. From *Table 17-2 in the text* for ethylenediamine, $K_b = 4.3 \times 10^{-4}$ and $pK_b = 3.37$

$$pOH = pK_b + \log \frac{[BH^+]_i}{[B]_i} = 3.37 + \log 0.333 = 2.89$$

$$pH = 14.00 - pOH = 14.00 - 2.89 = 11.11$$

Buffer preparation involves first choosing an appropriate acid or base. Equation [12] in conjunction with condition [2] (log 10.0 = +1.0 and log 0.10 = −1.0), limit the pK_a of the acid to within one unit of the desired pH for a weak acid buffer. In like fashion, equation [13] in conjunction with condition [3], limits the pK_b of the base to within one unit of the desired pOH for a weak base buffer. After the weak acid or base is selected, obtaining the desired pH or pOH requires that the correct ratio of either $[A^-]_i/[HA]_i$ or $[BH^+]_i/[B]_i$ be obtained by adjusting the concentrations of the species involved.

*** 6. Determine the changes in pH of buffer solutions that result from the addition of acids or bases.**

When a small amount of H_3O^+ is added to a weak acid buffer, some of the weak acid's anion reacts.

$$H_3O^+(aq) + A^-(aq) \longrightarrow H_2O + HA(aq) \qquad [5a]$$

When a small amount of OH^- is added to a weak acid buffer, some of the weak acid reacts.

$$OH^-(aq) + HA(aq) \longrightarrow H_2O + A^-(aq) \qquad [6a]$$

When a small amount of H_3O^+ is added to a weak base buffer, some of the weak base reacts.

$$H_3O^+(aq) + B \longrightarrow H_2O + BH^+(aq) \qquad [7a]$$

When a small amount of OH^- is added to a weak base buffer, some of the weak base's cation reacts.

$$OH^-(aq) + BH^+(aq) \longrightarrow H_2O + B(aq) \qquad [8a]$$

The effects of these small additions are to produce water and to change the ratio $[A^-]/[HA]$ for a weak acid buffer or the ratio $[BH^+]/[B]$ for a weak base buffer. Thus, to calculate the new pH when a small amount of strong acid or base is added, we first compute the concentrations of the components of the buffer present after the appropriate reaction [5a], [6a], [7a], or [8a] occurs. Then we compute the ratio $[A^-]/[HA]$ or $[BH^+]/[B]$, make sure it satisfies condition [2] or [3], and determine pH or pOH from equation [12] or [13].

EXAMPLE 18-8 0.050 mol NaOH is added to 1 L of the buffer of Example 18-7. What is the pH?

Rxn:	$C_2H_5NH_2(aq) + H_2O \rightleftharpoons$	$C_2H_5NH_3^+$	$+ OH^-$
I:	0.750 M	0.250 M	0.050 M
after rxn [7a]:	0.800 M	0.200 M	

$[C_2H_5NH_3^+]/[C_2H_5NH_2]$ = (0.200 M / 0.800 M) = 0.250, which satisfies condition [3]. From equation [13] we determine the pOH.

$$pOH = pK_b + \log \frac{[C_2H_5NH_3^+]}{[C_2H_5NH_2]} = 3.37 + \log 0.250 = 2.77$$

$$pH = 14.00 - pOH = 14.00 - 2.77 = 11.23$$

This is an increase of 0.12 pH unit (from 11.11 to 11.23). If NaOH had been added to pure water, the pH would have gone from pH = 7.00 to pH =12.70, an increase of 5.70 pH units.

* **7. Define and compute values for: "buffer range" and "buffer capacity."**

Buffer range is the range of pH values over which a buffer satisfies expression [2] or [3]. We have

$$\text{pH range of acid buffer} = pK_a \pm 1.00$$

$$\text{pOH range of base buffer} = pK_b \pm 1.00 \qquad [14]$$

The capacity of a buffer is the number of moles of strong acid (H_3O^+) or strong base (OH^-) that the buffer can absorb before condition [2] or [3] becomes invalid. The capacity of a buffer for OH^- may differ from its capacity for H_3O^+. Buffer capacity depends also on the total volume of buffer solution; 10 L of solution has five time the buffer capacity of 2.0 L. Another factor is the concentrations of the two components of the buffer. A buffer with $[HF]_i$ = 1.00 M and $[F^-]_i$ = 0.500 M has twice the capacity of one with $[HF]_i$ = 0.500 M and $[F^-]_i$ = 0.250 M.

EXAMPLE 18-9 What is the capacity toward (a) strong acid and (b) strong base of 2.00 L of buffer with $[HF]_i$ = 0.400 M and $[F^-]_i$ = 0.700 M? (Note that $[HF] + [F^-]$ = 1.100 M throughout this problem since no fluorine-containing species are added.)

(a) The buffer reacts with all of the added strong acid.

Rxn:	$F^-(aq) +$	$H_3O^+(aq) \rightleftharpoons$	$HF(aq) +$	H_2O
I:	0.700 M	x M	0.400 M	
C:	$-x$ M	$-x$ M	$+x$ M	
Eq:	$(0.700 - x)$ M		$(0.400 + x)$ M	

The buffer is exhausted when enough acid is added so that $[F^-]/[HF] = 0.100 = (0.700 - x)(0.400 + x)$. This gives $x = 0.600$. Thus the buffer capacity toward H_3O^+ is

$$0.600 \text{ mol } H_3O^+/L \times 2.00 \text{ L solution} = 1.20 \text{ mol } H_3O^+$$

(b) the buffer reacts with all of the added strong base.

Rxn:	$HF(aq) +$	$OH^-(aq) \rightleftharpoons$	$F^-(aq) +$	H_2O
I:	0.400 M	x M	0.700 M	
C:	$-x$ M	$-x$ M	$+x$ M	
Eq:	$(0.400 - x)$ M		$(0.700 + x)$ M	

The buffer is exhausted when enough base is added so that $[F^-]/[HF] = 10.0 = (0.700 + x)(0.400 - x)$. This gives $x = 0.300$. Thus the buffer capacity toward OH^- is

$$0.300 \text{ mol } OH^-/L \times 2.00 \text{ L soln} = 0.600 \text{ mol } OH^-$$

8. Describe how the blood buffer system works.

Blood contains at least three buffers, the most important being a H_2CO_3—HCO_3^- buffer.

$$H_2CO_3(aq) + H_2O \rightleftharpoons H_3O^+(aq) + HCO_3^-(aq) \qquad [15]$$

To maintain pH = 7.4, the ratio $[HCO_3^-]/[H_2CO_3] \cong 20$. Thus, there is a large amount of $HCO_3^-(aq)$ present to absorb excess acidity. Excess base reacts with H_2CO_3 which is replenished from CO_2 in the lungs.

$$CO_2(g) + H_2O \rightleftharpoons H_2CO_3(aq) \qquad [16]$$

9. Explain how an acid-base indicator works to determine the equivalence point in a titration.

An indicator is a weak acid, HIn, which has a different color than its anion, In^-. The pH color change range of an indicator spans two pH units, from one unit below to one unit above pK_{In} of the indicator. In the titration procedure, the material to be titrated is placed in a flask, indicator is added, and titrant of known concentration is added from a buret until the indicator changes color—at the *endpoint*. The *equivalence point* occurs when chemically equivalent amounts of acid and base are present in solution. If the indicator has been chosen correctly (see objective 18-12), the endpoint is the same as the equivalence point.

* **10. Calculate pH values and plot the titration curve of a strong acid with a strong base.**

The common features of plotting titration curves are most easily seen in the titration of a strong acid with a strong base where the complications of a small ionization constant are absent. Consider the titration of 25.00 mL of 0.112 M NaOH with 0.100 M HCl solution. We use strong acid as titrant to provide an example that is different from that of the text. We have divided the procedure into seven steps.

1. Write the titration reaction as a net ionic equation.

$$NaOH(aq) + HCl(aq) \longrightarrow NaCl(aq) + H_2O \qquad [17]$$

$$OH^-(aq) + H_3O^+(aq) \longrightarrow 2\,H_2O \qquad [18]$$

2. Determine the amount of substance being titrated.

$$25.00 \text{ mL NaOH soln} \times \frac{0.112 \text{ mmol NaOH}}{\text{mL soln}} = 2.80 \text{ mmol NaOH}$$

3. Find the volume of titrant needed to reach the equivalence point.

$$2.80 \text{ mmol NaOH} \times \frac{1 \text{ mmol HCl}}{1 \text{ mmol NaOH}} \times \frac{1.00 \text{ mL HCl soln}}{0.100 \text{ mmol HCl}} = 28.0 \text{ mL HCl soln}$$

This volume is known also as the 100% titration point. Now consider that you want to draw a curve like Figure 18-1 by calculating as few points as possible. A good titration curve can be drawn from the pH at 0%, 10%, 50%, 90%, 100%, and 110% titration. For each point (except 0%) you need (a) the volume of titrant added, V_t, (b) the total volume of solution, V_s, (c) the amount (in moles) of each species in solution, and (d) the concentration of each species.

4. Determine the pH at 0% titration.

$$[OH^-] = 2.80 \text{ mmol}/25.00 \text{ mL} = 0.112 = [Na^+]$$

$$pOH = -\log(0.112) = 0.951 \quad \textit{or} \quad pH = 13.049$$

5. Determine the pH at 100% titration ($V_t = 28.00$ mL and $V_s = 53.00$ mL). we find the number of moles of each species.

	$OH^-(aq)$ +	$H_3O^+(aq)$	$\longrightarrow 2\,H_2O(l)$
before rxn:	2.80 mmol	2.80 mmol	
after rxn:	0 mmol	0 mmol	

Thus, the solution is neutral and pH = 7.000. At the endpoint of a *strong* acid-*strong* base titration, pH = 7.000.

6. Determine the pH at 110% titration ($V_t = 30.8$ mL and $V_s = 55.80$ mL).

	$OH^-(aq)$ +	$H_3O^+(aq)$	$\longrightarrow 2\,H_2O(l)$
before rxn:	2.80 mmol	3.08 mmol	
after rxn:	0 mmol	0.28 mmol	

Since neither Na^+ nor Cl^- hydrolyzes, they do not affect the pH of the solution and we need not calculate their concentrations.

$$[H_3O^+] = (0.28 \text{ mmol}/55.8 \text{ mL}) = 5.0 \times 10^{-3} \quad \textit{or} \quad pH = 2.30$$

The pH after the equivalence point is based on the total solution volume and the excess amount (mol or mmol) of titrant added. This is true of *every* titration in which the titrant is a strong acid or strong base.

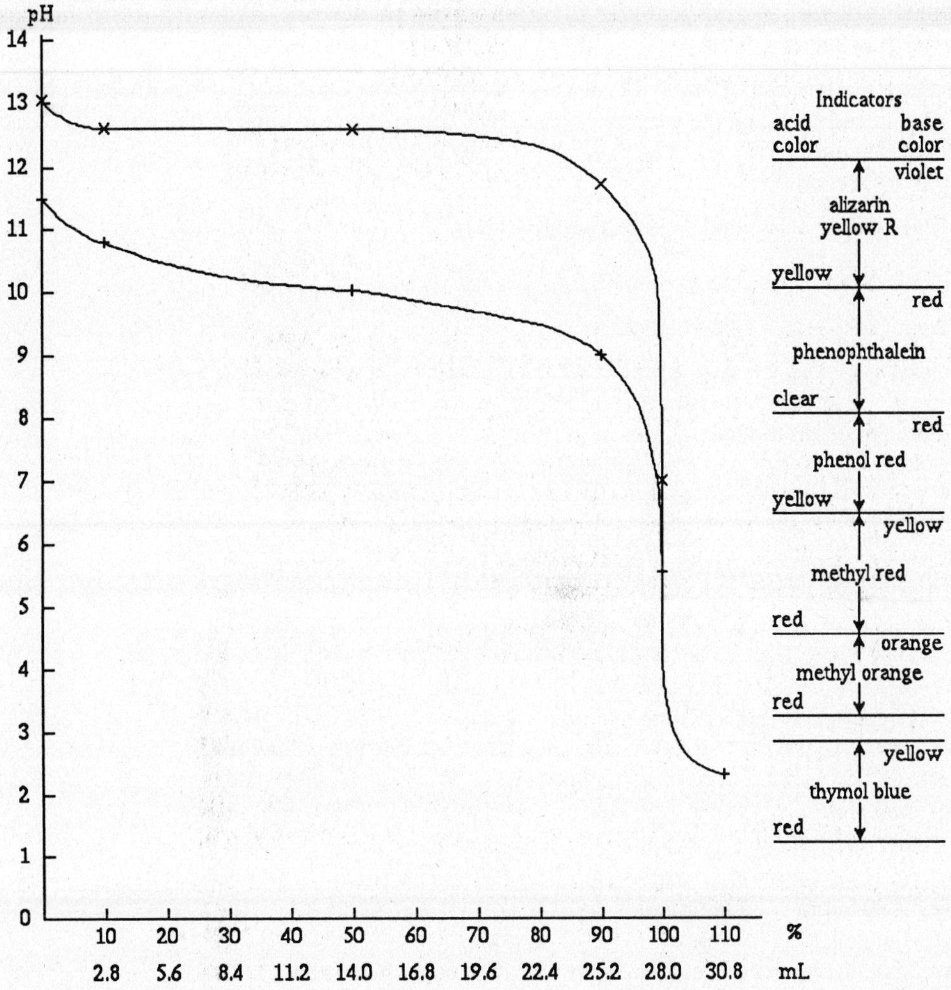

FIGURE 18-1 Titration curves: titrating 25.00 mL of 0.112 M NaOH (X) or $(CH_3)_3N$ (+) with 0.100 M HCl.

7. Determine the pH between 0% and 100% titration.

50% titration (V_t = 14.00 mL and V_s = 39.00 mL).

$$OH^-(aq) + H_3O^+(aq) \rightarrow 2\,H_2O(l) \qquad [OH^-] = \frac{1.40 \text{ mmol}}{39.0 \text{ mL}} = 0.0359 \text{ M}$$

before: 2.80 mmol 3.08 mmol
after: 0 mmol 0.28 mmol pOH = 1.444 and pH = 12.556

10% titration (V_t = 2.80 mL and V_s = 27.80 mL).

$$OH^-(aq) + H_3O^+(aq) \rightarrow 2\,H_2O(l) \qquad [OH^-] = \frac{2.52 \text{ mmol}}{27.8 \text{ mL}} = 0.0906 \text{ M}$$

before: 2.80 mmol 0.28 mmol
after: 2.52 mmol 0. mmol pOH = 1.444 and pH = 12.556

90% titration (V_t = 25.20 mL and V_s = 50.20 mL).

$$OH^-(aq) + H_3O^+(aq) \rightarrow 2\,H_2O(l) \qquad [OH^-] = \frac{0.28 \text{ mmol}}{50.2 \text{ mL}} = 0.00558 \text{ M}$$

before: 2.80 mmol 2.52 mmol
after: 0.28 mmol 0 mmol pOH = 2.253 and pH = 11.747

* **11. Calculate pH values and plot the titration curve of a weak acid with a strong base or of a weak base with a strong acid.**

We now consider the titration of 25.00 mL of 0.112 M $(CH_3)_3N$ ($K_b = 7.4 \times 10^{-5}$) with 0.100 M HCl. The concentrations of acid and base and the volume of base to be titrated are the same as objective 18-10, and we follow the procedure of that objective.

1. The titration reaction:

$$(CH_3)_3N(aq) + HCl(aq) \longrightarrow (CH_3)_3NHCl(aq)$$

$$(CH_3)_3N(aq) + H_3O^+(aq) \longrightarrow (CH_3)_3NH^+(aq)$$

2. The amount of $(CH_3)_3N$ being titrated:

$$25.00 \text{ mL } (CH_3)_3N \text{ soln} \times \frac{0.112 \text{ mmol } (CH_3)_3N}{\text{mL soln}} = 2.80 \text{ mmol } (CH_3)_3N$$

3. The volume of titrant needed:

$$2.80 \text{ mmol } (CH_3)_3N \times \frac{\text{mmol HCl}}{\text{mmol } (CH_3)_3N} \times \frac{1.00 \text{ mL HCl soln}}{0.100 \text{ mmol HCl}} = 28.0 \text{ mL HCl soln}$$

4. pH at 0% titration. This is a weak base problem (see objective 17-8 if necessary).

Rxn:	$(CH_3)_3N(aq)$ + $H_2O(l)$ $\rightleftharpoons$	$(CH_3)_3NH^+(aq)$		+ $OH^-(aq)$
I:	0.112 M			
C:	$-x$ M	$+x$ M		$+x$ M
Eq:	$(0.112 - x)$ M	x M		x M

$$K_b = \frac{[(CH_3)_3NH^+][OH^-]}{[(CH_3)_3N]} = 7.4 \times 10^{-5} = \frac{(x)(x)}{0.112 - x} \approx \frac{x^2}{0.112} \text{ (assuming } x << 0.112)$$

$$x = 2.9 \times 10^{-3} \text{ M} = [OH^-] << 0.122 \quad pOH = 2.54 \quad \textit{and} \quad pH = 11.46$$

5. pH at 100% titration ($V_t = 28.0$ mL and $V_s = 53.0$ mL).

	$(CH_3)_3N(aq)$ +	$H_3O^+(aq) \longrightarrow$	$(CH_3)_3NH^+(aq)$	+ H_2O
before:	2.80 mmol	2.80 mmol		
after:	0 mmol	0 mmol	2.80 mmol	

Now we have the hydrolysis of a weak base's cation (see objective 17-11).

$$[(CH_3)_3NH^+]_i = 2.80 \text{ mmol}/53.0 \text{ mL} = 0.0528 \text{ M}$$

Rxn:	$(CH_3)_3NH^+(aq) + H_2O(l) \rightleftharpoons$	$(CH_3)_3N(aq)$	+ $H_3O^+(aq)$
I:	0.0528 M		
C:	$-x$ M	$+x$ M	$+x$ M
Eq:	$(0.0528 - x)$ M	x M	x M

$$K_a = \frac{[(CH_3)_3N][H_3O^+]}{[(CH_3)_3NH^+]} = \frac{1.00 \times 10^{-14}}{7.4 \times 10^{-5}} = 1.35 \times 10^{-10} = \frac{(x)(x)}{0.0528 - x} - \frac{x^2}{0.0528}$$

(assuming $x << 0.0528$)

$$x = 2.76 \times 10^{-6} \text{ M} = [H_3O^+] << 0.0528 \quad \textit{and} \quad pH = 5.57$$

pH = 7.000 at the equivalence point *only* for strong acid-strong base titrations.

6. pH at 110% titration: This is the same calculation with the same result as step (6) of objective 18-10. pH = 2.30.

7. pH at points between 0% and 100% titration:
50% titration ($V_t = 14.00$ mL and $V_s = 39.00$ mL)

	$(CH_3)_3N(aq)$ +	$H_3O^+(aq) \longrightarrow$	$(CH_3)_3NH^+(aq)$	+ H_2O
before:	2.80 mmol	1.40 mmol		
after:	1.40 mmol	0 mmol	1.40 mmol	

$$[(CH_3)_3N] = 1.40 \text{ mmol}/39.00 \text{ mL} = [(CH_3)_3NH^+]$$

This is a buffer solution and we can use equation [13].

$$pOH = pK_b + \log \frac{[(CH_3)_3NH^+]_i}{[(CH_3)_3N]_i} = 4.13 + \log \frac{1.40 \text{ mmol}/39.0 \text{ mL}}{1.40 \text{ mmol}/39.0 \text{ mL}}$$

Notice that the two volumes are the same and thus they cancel. This is particularly helpful in titration calculations. Note also that the number of millimoles is the same at the 50% point.

$$pOH = pK_b + \log 1.00 = pK_b = 4.13 \quad and \quad pH = 9.87$$

The 50% titration point is the *half-equivalence point*. When a weak acid is titrated with a strong base, pH = pK_a at the 50% titration point.

10% titration ($V_t = 2.80$ mL and $V_s = 27.80$ mL)

$$(CH_3)_3N(aq) + H_3O^+(aq) \longrightarrow (CH_3)_3NH^+(aq) + H_2O$$

	$(CH_3)_3N$	H_3O^+	$(CH_3)_3NH^+$
before:	2.80 mmol	0.28 mmol	
after:	2.52 mmol	0 mmol	0.28 mmol

Again we have a buffer and thus we use equation [12].

$$pOH = 4.13 + \log\frac{0.28 \text{ mmol}}{25.2 \text{ mmol}} = 4.13 - 0.95 = 3.18 \quad and \quad pH = 10.82$$

We used the amounts in mmol of weak base and its cation in the ratio, since they are in the same volume. A further simplification is possible. Consider that 10% titration means that the fraction titrated = 0.10 and the fraction untitrated = 0.90. Notice that in this case 0.28 mmol = (0.10)(2.80 mmol) and 2.52 mmol = (0.90)(2.80 mmol), where 2.80 mmol is the amount of $(CH_3)_3N$ being titrated (step 2). Thus,

$$pOH = 4.13 + \log\frac{(0.10)(2.80 \text{ mmol})}{(0.90)(2.80 \text{ mmol})} = 4.13 + \log\frac{0.10}{0.90}$$

In general, for the titration of a weak base by a strong acid,

$$pOH = pK_b + \log\frac{\text{fraction of base titrated}}{\text{fraction of base untitrated}} \quad [19]$$

For the titration of a weak acid by a strong base,

$$pH = pK_a + \log\frac{\text{fraction of acid titrated}}{\text{fraction of acid untitrated}} \quad [20]$$

Equations [19] and [20] are based on equations [14] and [13], which in turn depend on conditions [3] and [2]. Thus, equations [19] and [20] are restricted to the region from 9% titrated to 91% titrated—the *buffer region.*

90% titrated: We use equation [19] directly.

$$pOH = pK_b + \log(0.90/0.10) = 4.13 + 0.95 = 5.08 \quad and \quad pH = 8.92$$

* **12. Plot titration curves and use those curves to determine the initial pH, buffer region, and the pH of the equivalence point, and to select an appropriate indicator.**

The results of objectives 18-10 and 18-11 are collected in Table 18-1 and plotted in Figure 18-1, with the NaOH results as x's and the $(CH_3)_3N$ results as +'s. The abscissa (the horizontal axis) can be labeled either "percent titration" or "volume of titrant." Notice the gradual change in pH for the weak base in the buffer region.

TABLE 18-1 Titration of a Strong and a Weak Base with a Strong Acid (Results of Objectives 18-10 and 18-11).

Percent titration	Volume of titrant, mL	pH for the titration of NaOH	pH for the titration of $(CH_3)_3N$
0	0.00	13.049	11.46
10	2.80	12.597	10.82
50	14.00	12.556	9.87
90	25.20	11.747	8.92
100	28.00	7.000	5.57
110	30.80	2.30	2.30

To choose an indicator, we consider the equivalence point. (Note the indicators given at the right of Figure 18-1.) For the NaOH-HCl titration we could choose phenophthalein, which is red at pH = 10 and colorless at pH = 8, but the color change from very faint red to just colorless is a bit hard to see. A better indicator is phenol red: red at pH = 8 and yellow at pH = 6.4. Phenol red would not work for the $(CH_3)_3N$-HCl titration because it would be starting to change color at about 96% titration. A better indicator would be methyl red: yellow at pH = 6.2 and red at pH =

4.5. It might seem that this indicator would be suitable for the NaOH-HCl titration. But we prefer that the equivalence point fall within the pH color change range of the indicator.

* 13. Calculate the pH of certain solutions of salts of polyprotic acids.

Polyprotic acids are related to three distinct kinds of chemical species in aqueous solution: (1) the acid itself [such as $H_2CO_3(aq)$], (2) the anion of the partly neutralized acid [such as $HCO_3^-(aq)$], and (3) the anion of the totally neutralized acid [such as $CO_3^{2-}(aq)$]. Calculation of the pH of a solution of the original acid was discussed in objective 17-9. The pH of a solution of the anion of the totally neutralized acid is a hydrolysis problem (objective 17-11). Calculation of the pH of a solution of the anion of the partly neutralized acid [$HCO_3^-(aq)$] is not a straightforward task, because two reactions are occurring: ionization and hydrolysis.

$$HCO_3^-(aq) + H_2O \rightleftharpoons CO_3^{2-}(aq) + H_3O^+(aq) \qquad [21]$$

$$HCO_3^-(aq) + H_2O \rightleftharpoons H_2CO_3(aq) + OH^-(aq) \qquad [22]$$

For equation [21], $K_a = K_{a2}$, and for reaction [22], $K_b = K_w/K_{a1}$. The pH of a solution that contains significant carbonate ion (at a concentration of $[HCO_3^-]_i > 0.10$ M) is the average of pK_{a1} and pK_{a2}.

$$pH = \frac{1}{2}(pK_{a1} + pK_{a2}) \qquad [23]$$

Note that the ionization constants used in equation [23] are those of two species: the anion placed in solution and the species that results from hydrolysis of that anion. Solutions of these anions (of partly neutralized polyprotic acids) are important for a reason other than the uniqueness of the calculation of their pH's. These ions are both anions of weak acids and "cations" of weak bases. Thus, their solutions can act as buffers, reacting with both strong acid and with strong base, equations [24] and [25].

$$H_3O^+(aq) + HCO_3^-(aq) \longrightarrow H_2CO_3(aq) + H_2O \qquad [24]$$

$$OH^-(aq) + HCO_3^-(aq) \longrightarrow CO_3^{2-}(aq) + H_2O \qquad [25]$$

14. Define and compute values for equivalent weight, equivalent, and solution normality.

An equivalent of an acid or base is the quantity of the substance that produces or consumes one mole of hydrogen ions. It is convenient to define n as follows.

$$n = \text{number of equivalents/mole of acid or base} \qquad [26]$$

n will always be a positive whole number (n = 1, 2, 3, and so forth). For many acids, n equals the number of hydrogens in the formula of the acid. This is the case for simple acids such as HF, HCl, HBr, HI, $HClO_4$, HNO_3, and HCN. For organic acids, n is the number of —COOH groups in the formula of the acid. Thus $n = 1$ for HCOOH (formic acid) and CH_3COOH (acetic acid).

For many bases, n is the number of OH groups in the formula of the base. Thus, $n = 1$ for NaOH, TlOH, and CsOH; $n = 2$ for $Mg(OH)_2$, $Fe(OH)_2$, and $Ba(OH)_2$; and $n = 3$ for bases such as $Al(OH)_3$. For organic bases, n is the number of $—NH_2$ groups in the formula of the base. Thus $n = 1$ for CH_3NH_2 (methylamine) and $C_2H_5NH_2$ (ethylamine) as well as for NH_3 (ammonia). For polyprotic acids, the maximum value of n, n_{max} is the number of hydrogens (or —COOH groups) in the formula. Thus $n_{max} = 2$ for H_2SO_4, H_2SO_3, and $(COOH)_2$ (oxalic acid), and $n_{max} = 3$ for H_3PO_4. For polyprotic acids, the value of n may be different for the same acid in different reactions.

$$\text{equivalent weight (g/eq)} = \text{mole weight (g/mol)} \div n \text{ (eq/mol)} \qquad [26]$$

$$\text{normality} = \text{eq solute / L solution} = \text{molarity (mol/L)} \times n \text{ (eq/mol)} \qquad [27]$$

EXAMPLE 18-11 What is the equivalent weight of $Ba(OH)_2$, and what is the normality of a 0.102 M $Ba(OH)_2$ solution?

Since there are two hydroxide groups in the formula of $Ba(OH)_2$, $n = 2$.

equivalent weight = [171.35 g $Ba(OH)_2$/mol]÷[2 eq/mol]= 85.67 g $Ba(OH)_2$/eq

Normality = 0.102 mol $Ba(OH)_2$/L × 2 eq $Ba(OH)_2$/mol = 0.204 N

* 15. Explain why a substance may have more than one equivalent weight and why one equivalent of acid always reacts with one equivalent of base.

Consider reactions [29] and [30] between NaOH and H_2SO_4.

$$NaOH(aq) + H_2SO_4(aq) \longrightarrow NaHSO_4(aq) + H_2O \quad [29]$$

$$2\ NaOH(aq) + H_2SO_4(aq) \longrightarrow Na_2SO_4(aq) + 2\ H_2O \quad [30]$$

In equation [29] one mole of NaOH reacts with each mole of H_2SO_4, and in equation [30] two moles of NaOH reacts with each mole of H_2SO_4. Yet in both equations one equivalent of NaOH reacts with one equivalent of H_2SO_4! In equation [29] one mole of H_2SO_4 produces one mole of hydrogen ions. Thus one mole of H_2SO_4 equals one equivalent in equation [29] or $n = 1$. The equivalent weight of sulfuric acid is 98 g/mol ÷ 1 eq/mol = 98 g/eq. In equation [30], on the other hand, $n = 2$ and thus the equivalent weight of H_2SO_4 is 98 g/mol ÷ 2 eq/mol = 49 g/eq.

*** 16. Apply the concepts of equivalent weight and normality to solve solution stoichiometry problems.**

Volume of acid × normality of acid = volume of base × normality of base.

$$V_A \times N_A = V_B \times N_B \quad [31]$$

EXAMPLE 18-12 17.4 mL of 0.100 N NaOH titrates 12.3 mL of sulfuric acid solution. What is the normality of the acid?

$$12.3\ \text{mL} \times N_A = 17.4\ \text{mL} \times 0.100\ \text{N} \quad \textit{or} \quad N_A = 0.141\ \text{N}$$

Thus, we label the acid of Example 18-12: "0.141 N H_2SO_4." This can cause problems. Notice the difference if the normality is defined according to equation [29] or according to equation [30]. According to equation [29] the mass of H_2SO_4 per liter of solution is

$$0.141\ \text{eq}\ H_2SO_4/\text{L} \times 98\ \text{g}\ H_2SO_4/\text{eq} = 14\ \text{g}\ H_2SO_4/\text{L}$$

According to equation [30], the mass of H_2SO_4 per liter is

$$0.141\ \text{eq}\ H_2SO_4/\text{L} \times 49\ \text{g}\ H_2SO_4/\text{eq} = 6.9\ \text{g}\ H_2SO_4/\text{L}$$

We cannot know which mass is correct unless we know the reaction that occurred during the titration.

DRILL PROBLEMS

1. Determine the pH of each of the following solutions. First, when only the uncharged weak acid or weak base is present. And then when both substances are present. Initial concentrations are given in each case.
   A. $[HC_2H_3O_2]$ = 0.250 M, $[NaC_2H_3O_2]$ = 0.120 M
   B. $[HC_2H_3O_2]$ = 0.250 M, [HCl] = 0.140 M
   C. $[HC_7H_5O_2]$ = 1.20 M, $[Ca(C_7H_5O_2)_2]$ = 0.130 M
   D. $[HCHO_2]$ = 0.820 M, $[HNO_3]$ = 0.240 M
   E. $[HNO_2]$ = 0.400 M, $[Ba(NO_2)_2]$ = 0.0480 M
   F. $[HCHO_2]$ = 0.620 M, $[KCHO_2]$ = 0.340 M
   G. $[HClO_2]$ = 1.00 M, $[CsClO_2]$ = 0.138 M
   H. $[NH_3]$ = 1.20 M, [KOH] = 0.320 M
   I. $[CH_3NH_2]$ = 0.260 M, $[CH_3NH_3Cl]$ = 0.130 M
   J. $[C_2H_5NH_2]$ = 0.150 M, [NaOH] = 0.216 M
   K. $[(CH_3)_3N]$ = 0.460 M, $[(CH_3)_3NHI]$ = 0.819 M ($K_b = 7.4 \times 10^{-5}$)
   L. $[NH_3]$ = 0.510 M, $[(NH_4)_2SO_4]$ = 0.774 M
   M. $[CH_3NH_2]$ = 0.104 M, $[Ca(OH)_2]$ = 0.0102 M
   N. $[C_2H_5NH_2]$ = 0.630 M, $[C_2H_5NH_3NO_3]$ = 1.89 M

5. The buffer solutions in the following lines are made up as described in objective 18-3. Determine the initial concentrations of the weak acid and its anion, or of the weak base and its cation. Use these concentrations to determine the pH of each buffer. (The concentrations given below are formalities, that is, how the solution was made up, not the concentrations at equilibrium.)
   A. $[HC_2H_3O_2]$ = 1.450 M, $[NaC_2H_3O_2]$ = 0.270 M
   B. $[HC_7H_5O_2]$ = 1.300 M, [KOH] = 0.640 M
   C. $[NaCHO_2]$ = 0.815 M, [HBr] = 0.015 M
   D. $[NH_3]$ = 0.750 M, $[NH_4I]$ = 0.250 M

E. $[C_2H_5NH_2] = 2.31$ M, $[HNO_3] = 1.02$ M
F. $[CH_3NH_3Cl] = 0.555$ M, $[CsOH] = 0.114$ M
G. $[HF] = 1.296$ M, $[NaF] = 1.045$ M
H. $[HC_2H_2ClO_2] = 2.93$ M, $[Mg(OH)_2] = 1.07$ M
I. $[Ca(NO_2)_2] = 0.926$ M, $[HClO_4] = 1.302$ M
J. $[CH_3NH_2] = 0.990$ M, $[CH_3NH_3ClO_4] = 0.443$ M
K. $[(CH_3)_3N] = 1.236$ M, $[HI] = 0.341$ M ($K_b = 7.4 \times 10^{-5}$)
L. $[(NH_4)_2SO_4] = 0.1140$ M, $[KOH] = 0.0731$ M
M. $[HCHO_2] = 1.946$ M, $[Mg(CHO_2)_2] = 0.104$ M
N. $[HC_2H_3O_2] = 0.0561$ M, $[Ca(OH)_2] = 0.0112$ M
O. $[NaC_7H_5O_2] = 0.743$ M, $[H_2SO_4] = 0.256$ M
P. $[(CH_3)_3NHBr] = 0.831$ M, $[KOH] = 0.218$ M ($K_b = 7.4 \times 10^{-5}$)
Q. $[CH_3NH_2] = 1.214$ M, $[CH_3NH_3NO_3] = 0.885$ M
R. $[C_2H_5NH_2] = 1.149$ M, $[HClO_4] = 1.001$ M

6. The amount of strong acid or strong base given on each line below is added to the given volume of the corresponding buffer in the drill problems of objective 18-5. Determine the final pH in each case (addition of strong acid, and addition of strong base) for each solution.

A. 0.0520 mol to 0.750 L B. 0.0130 mol to 0.800 L C. 0.00780 mol to 1.00 L
D. 0.0110 mol to 121 mL E. 0.544 mol to 2.04 L F. 0.123 mol to 5.91 L
G. 0.247 mol to 1.20 L H. 1.32 mol to 11.4 L I. 0.963 mol to 8.71 L
J. 0.0231 mol to 0.604 L K. 0.150 mol to 0.890 L L. 1.09 mol to 75.9 L
M. 0.0231 mol to 0.561 L N. 0.196 mol to 103 L O. 0.0931 mol to 1.52 L
P. 0.161 mol to 3.75 L Q. 7.14 mol to 10.5 L R. 0.100 mol to 1.00 L

7. For each of the buffers in the drill problems of objective 18-5, determine the buffer range. Then, using the volume of each buffer given in the problems of objective 18-6, determine the capacity of each buffer for both strong acid and strong base. Express each capacity in moles of strong acid, and moles of strong base.

10. & 11. Determine the pH at 0%, 10%, 50%, 90%, 100%, and 110% titration for each of the following titrations. (objective 18-10: A through G; objective 18-11: H through N.) Volumes are given in mL.

	Substance being titrated			*Titrant*	
	Volume	*Formula*	*Molarity*	*Formula*	*Molarity*
A.	25.00	HCl	0.245	KOH	0.200
B.	15.00	NaOH	0.250	HNO_3	0.100
C.	30.00	CsOH	2.50	HI	1.00
D.	10.00	$HClO_4$	0.120	RbOH	0.100
E.	20.00	HBr	0.250	KOH	0.100
F.	16.00	NaOH	0.624	H_2SO_4	0.100
G.	40.00	$Ca(OH)_2$	0.0100	HCl	0.0250
H.	25.00	HNO_2	0.245	KOH	0.200
I.	15.00	CH_3NH_2	0.250	HNO_3	0.100
J.	30.00	$C_2H_5NH_2$	2.50	HI	1.00
K.	10.00	HF	0.120	RbOH	0.100
L.	20.00	$HCHO_2$	0.250	KOH	0.100
M.	16.00	CH_3NH_2	0.624	H_2SO_4	0.100
N.	40.00	C_5H_5N	0.0200	HCl	0.0250

12. For each of the titrations in the drill problems of objective 18-10 and 18-11, plot the titration curve. Indicate the equivalence point and, for the titration of a weak acid or weak base, the buffer region. Finally, choose a suitable indicator from those given in Figure 18-1.

13. Determine the pH of a solution of each of the following compounds. Use the data of *Table 17-3 in the text.* (The numbers in parentheses are the pK_a values for successive ionizations of the parent acid.)

A. $KHCO_3$ B. $Ca(HS)_2$ C. NaH_2PO_3 D. KHC_2O_4
E. NaH_2AsO_3 (2.22, 6.98, 11.50) F. $Ca(HCrO_4)_2$ (−0.98, 6.50)
G. $Ca(HAsO_4)$ (2.22, 6.98, 11.50) H. $KHC_8H_4O_4$ (2.95, 5.41)
I. $NaHN_2O_2$ (7.05, 11.4) J. $CaHVO_4$ (3.78, 7.8, 13.0)
K. $NaHGeO_3$ (9.0, 12.4) L. $CaHC_9H_3O_6$ (2.52, 3.84, 5.20)

15. Compute the equivalent weight and the normality of each reactant in the equations that follow. The molarity of each reactant is given in parentheses after its formula.
A. 2 HCl (0.104 M) + $Ca(OH)_2$ (0.0521 M) ⟶ $CaCl_2 + H_2O$
B. H_2SO_4 (0.704 M) + KOH (0.197 M) ⟶ $KHSO_4 + H_2O$
C. HNO_3 (0.146 M) + LiOH (0.412 M) ⟶ $LiNO_3 + H_2O$
D. 3 HBr (0.904 M) + $Al(OH)_3$ (0.222 M) ⟶ $AlBr_3 + 3\ H_2O$
E. H_2SO_4 (0.818 M) + $Ca(OH)_2$ (0.525 M) ⟶ $CaSO_4 + 2\ H_2O$
F. H_3PO_4 (0.200 M) + $Al(OH)_3$ (0.332 M) ⟶ $AlPO_4 + 3\ H_2O$
G. H_3PO_4 (0.165 M) + $Sr(OH)_2$ (0.227 M) ⟶ $SrHPO_4 + 2\ H_2O$
H. $H_2C_2O_4$ (0.774 M) + NaOH (0.146 M) ⟶ $NaHC_2O_4 + H_2O$

16. Fill in the blanks in the table that follows, assuming that the two solutions react completely with each other in a titration. Volumes are given in mL.

	Acid Solution		*Base solution*	
	Volume	*Normality*	*Volume*	*Normality*
A.	_____	0.2431	27.46	0.1000
B.	25.00	_____	19.46	0.1000
C.	7.265	0.8417	_____	0.1746
D.	27.84	0.2500	20.00	_____
E.	_____	0.2000	21.04	0.1946
F.	40.00	_____	37.43	0.1000
G.	33.14	0.7142	_____	0.2000
H.	49.24	0.1500	30.00	_____

QUIZZES (20 minutes each) Choose the best answer for each question.

Quiz A

1. What volume in mL of 2.0 M H_2SO_4 is needed to neutralize 7.8 g of $Al(OH)_3$ (78.0 g/mol)? $Al_2(SO_4)_3$ and water are the titration products. (a) 150.; (b) 75; (c) 300.; (d) 500.; (e) none of these.
2. A weak base, B, has $K_b = 4.0 \times 10^{-5}$. If $[OH^-] = 2.0 \times 10^{-5}$ M, what must $[B^+]/[B]$ equal? (a) 1.0; (b) 0.50; (c) 1.5; (d) 8.0; (e) none of these.
3. Which of the following substances, when added to 1.00 L of water, will produce a buffer solution? (a) 1 mol $HC_2H_3O_2$ and 0.5 mol HCl; (b) 1 mol $NH_3(aq)$ and 0.5 mol NaOH; (c) 1 mol NH_4Cl and 0.5 mol HCl; (d) 1 mol $HC_2H_3O_2$ and 0.5 mol NaOH; (e) none of these.
4. 20.0 mL of 0.200 M NaOH is added to 50.0 mL of 0.100 M $HC_2H_3O_2$ ($K_a = 1.74 \times 10^{-5}$). What is the pH of this solution? (a) 4.23; (b) 9.77; (c) 5.35; (d) 9.26; (e) 3.00.
5. The solution that is added from the buret during a titration is the (a) buffer; (b) titrant; (c) indicator; (d) base; (e) none of these.
6. A weak acid and the sodium salt of its anion in solution in equimolar amounts has (a) pH > 7; (b) pH < 7; (c) pH = 7; (d) depends only on concentration; (e) none of these.
7. An indicator, HIn, is purple, and its anion, In^-, is yellow. The indicator changes color in the range pH 8-10. This indicator will be yellow in which of the following? (a) 0.010 M HCl(aq); (b) 0.10 M $NaC_2H_3O_2$ ($K_a = 1.8 \times 10^{-5}$ for $HC_2H_3O_2$); (c) pure H_2O; (d) 0.010 M NaOH; (e) none of these.

Quiz B

1. What volume in mL of 1.50 M NaOH is needed to neutralize 45.0 mL of 0.500 M acetic acid? (a) 15.0; (b) 22.5; (c) 33.8; (d) 135; (e) none of these.
2. A weak acid has $K_a = 4.2 \times 10^{-3}$. If $[A^-] = 2.0$ M, what must be [HA] so that $[H^+] = 2.1 \times 10^{-3}$ M? (a) 0.500 M; (b) 2.0 M; (c) 1.0 M; (d) 1.5 M; (e) none of these.
3. The addition of NH_4Cl to a solution of NH_3 would (a) decrease total $[NH_4^+]$; (b) decrease $[NH_3]$; (c) decrease the K_i of NH_3; (d) increase $[OH^-]$ in the solution; (e) increase the pH of the solution.
4. A strong base and the chlorine salt of its cation in water has (a) pH >7; (b) pH < 7; (c) pH = 7; (d) pH that depends on $[Cl^-]$; (e) none of these.
5. Phenophthalein (color change range of 8-10) may be used for the titration of (a) a weak acid with a weak base; (b) a weak base with a strong acid; (c) a weak acid with a strong base; (d) any acid with any base;

(e) any base with any acid.

6. 10.0 mL of 0.100 M NH_3 (K_b = 1.74 × 10^{-5}) is mixed with 5.00 mL of 0.200 M NH_4Cl. The resulting solution has (a) pH = 4.76; (b) [NH_4^+] larger than in 01.00 M NH_3; (c) [OH^-] = 3.6 × 10^{-5} M; (d) [H^+] of about 10^{-3} M; (e) none of these.
7. In the neutralization of 50.0 mL of 0.100 M B (a weak base with K_b = 1.6 × 10^{-7}) with 0.100 M H_2SO_4, the most correct description of the solution at the midpoint of the titration (that is, half-neutralized) is a solution (a) whose volume is 75.0 mL and which contains significant concentrations of undissociated B molecules and some B^+, OH^-, and HSO_4^-; (b) whose volume is 62.5 mL and which contains the same species listed in (a); (c) in which [BH^+] is about the same as [B]; (d) in which [BH^+] is about 4 × 10^{-4} M; (e) none of these.

Quiz C

1. What volume in mL of 1.7 M H_2SO_4 is needed to neutralize 68 mL of 2.5 M NaOH? Na_2SO_4 is produced. (a) 100.; (b) 200.; (c) 50.; (d) 46; (e) none of these.
2. A weak acid has K_a = 1.00 × 10^{-3}. If [HA] = 1.00 M, what must be [A^-] for the pH to be 2.7? (a) 2.7 M; (b) 2.0 M; (c) 0.50 M; (d) 0.37 M; (e) none of these.
3. Which of the following would *not* be a good buffer pair? (a) potassium carbonate and potassium bicarbonate; (b) ammonium chloride and ammonia; (c) boric acid and sodium borate; (d) sodium chloride and sodium hydroxide; (e) all would be good.
4. A substance that displays little change in pH when small amounts of acid or base are added is called a(n) (a) buffer; (b) indicator; (c) salt; (d) titrant; (e) none of these.
5. Which condition characterizes the equivalence point of the titration of a weak acid with a strong base? (a) a slight excess of titrant is present; (b) pH = 7.00; (c) the indicator must change color; (d) pH = pK_a; (e) none of these.
6. Which factor governs the selection of an indicator? (a) volume of titrant; (b) pH at the equivalence point; (c) concentration of the titrant; (d) final solution volume; (e) none of these.
7. 25.0 mL of 0.200 M NaOH is added to 50.0 mL of 0.100 M $HC_2H_3O_2$ (K_a = 1.74 × 10^{-5}). What is the pH of the final solution? (a) 4.74; (b) 9.26; (c) 9.74; (d) 12.78; (e) none of these.

Quiz D

1. What volume in mL of 2.50 M H_3PO_4 is needed to neutralize 150. mL of 0.500 M KOH. Potassium phosphate is produced. (a) 30.0; (b) 10.0; (c) 90.0; (d) 750.; (e) none of these.
2. A weak acid HA has K_a = 1.4 × 10^{-3}. in order to produce a solution with [H^+] = 5.6 × 10^{-3}, the ratio [A^-]/[HA] must equal (a) 4.0; (b) 0.25; (c) 2.0; (d) 1.0; (e) none of these.
3. $HC_2H_3O_2$ will be ionized to the greatest degree in which of the following solutions? (a) 0.010 M $HC_2H_3O_2$ and 0.10 M HCl; (b) 0.10 M $HC_2H_3O_2$ and 0.10 M $KC_2H_3O_2$; (c) 0.0010 M $HC_2H_3O_2$ and 0.10 M $NaC_2H_3O_2$; (d) 0.010 M $HC_2H_3O_2$; (e) 0.0010 M $HC_2H_3O_2$.
4. The region of pH over which an indicator changes color is the (a) buffer region; (b) equivalence point; (c) overtitration region; (d) pH color change range; (e) none of these.
5. At the equivalence point of a titration (a) the indicator must change color; (b) pH = pK_a; (c) pH = 7.00; (d) equivalents of acid = equivalents of base; (e) none of these.
6. The indicator HIn is yellow and its anion In^- is green. The indicator changes color from pH = 5.1 to pH = 7.1. This indicator will be yellow in which of the following? (a) 0.01 M HCl(aq); (b) 0.10 M NH_4Cl (K_b = 1.8 × 10^{-5} for NH_3); (c) pure H_2O; (d) 0.10 M NaOH; (e) none of these.
7. 50.0 mL of 0.100 M NH_3, 10.0 mL of 0.100 M NH_4Cl, and 40.0 mL of 0.050 M HCl are mixed. The resulting solution (a) contains 5.0 mmol NH_3, 2.0 mmol HCl, and 0.1 mmol NH_4Cl; (b) is a buffer with pH = pK_b for ammonia; (c) has [Cl^-] ≅ 0.30 M; (d) has [NH_3] ≅ 0.30 M; (e) none of these.

SAMPLE TEST (15 minutes)

1. 7.500 g of weak acid HA is added to sufficient distilled water to produce 500.0 mL of solution with pH = 2.716. This solution is titrated with NaOH(aq) solution. At the half-equivalence point, pH = 4.602. What is the equivalent weight of the acid?
2. The titration curve at right was obtained in a general chemistry laboratory with a pH meter. The titrant was either 0.100 M NaOH or 0.100 M HCl. The unknown being titrated weighed 0.400 g.
   a. Was the unknown a weak acid or a weak base?
   b. What is the mol weight of the unknown?
   c. What is the value of the ionization constant of the unknown?
   d. What would be the pH color change range of a suitable indicator for this titration? Choose a suitable indicator from those given in Figure 18-1.

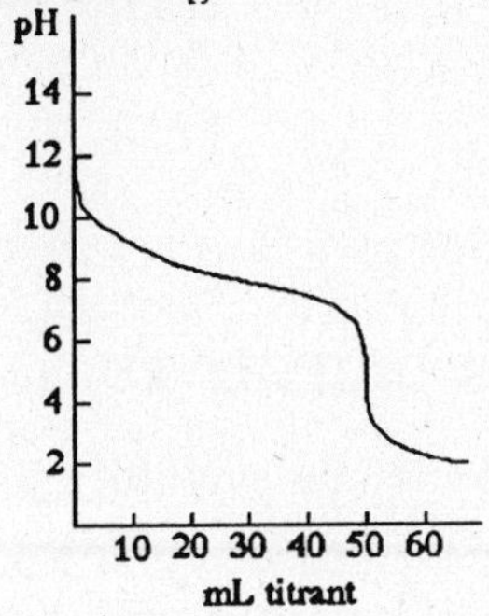

19 Solubility and Complex Ion Equilibria

CHAPTER OBJECTIVES

* 1. **Write the solubility product expression, K_{sp}, for a slightly soluble ionic compound.**

The K_{sp} expression is written in the same way as an equilibrium constant expression. The solubility equilibrium has a solid ionic compound as reactant and its ions in solution as products. The concentration of solid does not appear in the equilibrium constant expression. Thus, the K_{sp} expression simply is a product of two ionic concentrations, raised to appropriate powers. (Sometimes the solubility products expression is called the ion product constant.) There are twelve different types of ionic compounds (objective 3-11, Table 3-4). These produce six different forms of solubility product expressions, given in Table 19-1.

TABLE 19-1 Types of Solubility Equilibria

Salt formula	Solubility equilibrium	Solubility constant expression: General	Solubility constant expression: In terms of S
MX	$MX(s) \rightleftharpoons M^{+}(aq) + X^{-}(aq)$	$[M^{+}][X^{-}]$	S^2
	or $M^{2+}(aq) + X^{2-}(aq)$	$[M^{2+}][X^{2-}]$	
	or $M^{3+}(aq) + X^{3-}(aq)$	$[M^{3+}][X^{3-}]$	
MX_2	$MX_2(s) \rightleftharpoons M^{2+}(aq) + 2\,X^{-}(aq)$	$[M^{2+}][X^{-}]^2$	$4S^3$
	or $M^{4+}(aq) + 2\,X^{2-}(aq)$	$[M^{4+}][X^{2-}]^2$	
M_2X	$M_2X(s) \rightleftharpoons 2\,M^{+}(aq) + X^{2-}(aq)$	$[M^{+}]^2[X^{2-}]$	
MX_3	$MX_3(s) \rightleftharpoons M^{3+}(aq) + 3\,X^{-}(aq)$	$[M^{3+}][X^{-}]^3$	$27S^4$
M_3X	$M_3X(s) \rightleftharpoons 3\,M^{+}(aq) + X^{3-}(aq)$	$[M^{+}]^3[X^{3-}]$	
M_2X_3	$M_2X_3(s) \rightleftharpoons 2\,M^{3+}(aq) + 3\,X^{2-}(aq)$	$[M^{3+}]^2[X^{2-}]^3$	$108S^5$
M_3X_2	$M_3X_2(s) \rightleftharpoons 3\,M^{2+}(aq) + 2\,X^{3-}(aq)$	$[M^{2+}]^3[X^{3-}]^2$	
MX_4	$MX_4(s) \rightleftharpoons M^{4+}(aq) + 4\,X^{-}(aq)$	$[M^{4+}][X^{-}]^4$	$256S^5$
M_3X_4	$M_3X_4(s) \rightleftharpoons 3\,M^{4+}(aq) + 4\,X^{3-}(aq)$	$[M^{4+}]^3[X^{3-}]^4$	$6912S^7$

K_{sp} values are not easy to determine because they are based on a very small mass of compound in solution. Even a small error in this mass is magnified when the concentration is squared or cubed. For example, for AgCl the solubility is 1.9 mg/L. This gives $K_{sp} = 1.8 \times 10^{-10}$. If the solubility were determined as 1.8 mg/L then $K_{sp} = 1.6 \times 10^{-10}$. This difficulty in measurement is reflected by the various values of K_{sp} cited by different sources. For AgCl, five sources give: 1.56×10^{-10}, 1.7×10^{-10}, 1.78×10^{-10}, 1.79×10^{-10}, and 1.8×10^{-10}.

* 2. **Calculate K_{sp} from the solubility of an ionic compound or solubility from the value of K_{sp}.**

The right-hand column of Table 19-1 gives the expression for the solubility product constant of an ionic compound (a salt). In each expression, S is the solubility of the salt in moles per liter. These expressions can be used when the salt is the only source of its ions in solution. They cannot be used when one of the ions of the salt already is present in solution, that is, in common ion problems.

EXAMPLE 19-1 What is the molar solubility of aluminum hydroxide?

From *Table 19-1 in the text*, $K_{sp} = 1.3 \times 10^{-33}$ for $Al(OH)_3$. The solubility equilibrium is

$$Al(OH)_3(s) \rightleftharpoons Al^{3+}(aq) + 3\,OH^{-}(aq) \qquad [1]$$

The solubility product constant expression for $Al(OH)_3$ is then

$K_{sp} = [Al^{3+}][OH^-]^3$ [2]

Let S be the molar solubility of $Al(OH)_3$. Then $[Al^{3+}] = S$ and $[OH^-] = 3S$ since three moles of OH^- are in solution for each mole of $Al(OH)_3$ that dissolves.

$[Al^{3+}] = S$ *and* $[OH^-] = 3S$ [3]

Substitution of expressions [3] into equation [2] produces equation [4].

$K_{sp} = (S)(3S)^3 = 27S^4$ [4]

Notice that the forms of the solubility equilibrium [1] and of the solubility product expression ([2] and [4]) are the same as those in Table 19-1 for a salt with the general formula MX_3.

$K_{sp} = 1.3 \times 10^{-33} = 27S^4$ *or* $S^4 = (1.3 \times 10^{-33})/27$

This gives $S = 2.6 \times 10^{-9}$ M.

EXAMPLE 19-2 The solubility of $Ca_3(PO_4)_2$ is 0.002 g/100 mL of solution at 25°C. What is the value of K_{sp} at this temperature?

The solubility equilibrium for $Ca_3(PO_4)_2$ is

$Ca_3(PO_4)_3(s) \rightleftharpoons 3\ Ca^{2+}(aq) + 2\ PO_4^{3-}(aq)$ [5]

The K_{sp} expression is

$[Ca^{2+}]^3[PO_4^{3-}]^2$ [6]

Let S is the molar solubility of $Ca_3(PO_4)_2$. Remember that for every mole of $Ca_3(PO_4)_2$ that dissolves, there are in solution three moles of Ca^{2+} and two moles of PO_4^{3-}. Thus,

$[Ca^{2+}] = 3S$ *and* $[PO_4^{3-}] = 2S$ [7]

Substitution of expressions [7] into equation [6] produces equation [8].

$K_{sp} = (3S)^3(2S)^2 = 27S^3\,4S^2 = 108S^5$ [8]

Notice that equations [5], [6], and [8] have the same form as that given for a salt of general formula M_3X_2 in Table 19-1. The molar solubility of $Ca_3(PO_4)_2$ is computed.

$$S = \frac{0.002\ \text{g}\ Ca_3(PO_4)_2}{100\ \text{mL soln}} \times \frac{1000\ \text{mL}}{\text{L}} \times \frac{1\ \text{mol}\ Ca_3(PO_4)_2}{310.3\ \text{g}}$$

$= 6.4 \times 10^{-5}$ M $= 6 \times 10^{-5}$ M (significant figures)

With this value of S in equation [8] we obtain a value of

$K_{sp} = 108(6.4 \times 10^{-5})^5 = 1 \times 10^{-19}$

We used two significant figures throughout the calculation of Example 19-2 and then rounded to one significant figure in the final result. This is an acceptable and generally followed practice. Also notice that we used the term *salt* to mean ionic compound. Strictly a chemist would consider only a soluble ionic compound to be a salt. However, chemists freely use the phrase "sparingly soluble salt."

* 3. **Calculate the effect of common ions on the aqueous solubilities of sparingly soluble salts.**

Objective 19-2 deals with the solubility of a salt in solutions where it is the only solute. Both ions come from the salt. If there is an ion in solution that is the same as one of the ions of the salt, then less salt will dissolve than in pure water. Because the solution and the salt have one ion in common, this lower solubility is called the common ion effect. It can be explained with LeChâtelier's principle. Since some of the product (one of the ions) of the solubility equilibrium is present in solution, the equilibrium will lie to the left (toward undissolved solute) compared to where it would be if no common ion were present. The expressions in the right-hand column of Table 19-1 *cannot* be used in common ion problems.

EXAMPLE 19-3 What is the molar solubility of $Al(OH)_3$ in a solution in which $[Al(NO_3)_3] = 0.0200$ M?

The solubility equilibrium follows, with the equilibrium concentration written below the formula of each ion. s is the solubility of $Al(OH)_3$.

Rxn:	$Al(OH)_3(s) \rightleftharpoons$	$Al^{3+}(aq)$	+	$3\ OH^-(aq)$
Initial		0.0200 M		
Change		$+S$		$+3S$
Equil.		$(0.0200 + S)$ M		$3S$

$K_{sp} = [Al^{3+}][OH^-]^3 = (0.0200\ \text{M} + S)(3S)^3 = 1.3 \times 10^{-33}$

Let us assume that S is small compared to 0.200 M. Then we have

$$(0.0200 \text{ M})(3S)^3 = 1.3 \times 10^{-33} \quad and \quad S = 1.3 \times 10^{-11} \text{ M}$$

We see that s *is* small with respect to 0.0200 M. The presence of the common ion, Al^{3+}, greatly reduces the solubility of $Al(OH)_3$ from 2.6×10^{-9} M (Example 19-1) to 1.3×10^{-11} M.

4. Describe how the presence of "uncommon" ions in solution or the formation of ion pairs in concentrated solutions increases the solubility of a sparingly soluble salt.

In a dilute solution, each ion is completely surrounded by water molecules and does not feel the presence of the other ions because they are too far away. When the solution becomes more concentrated, the ions are closer together. Ions of opposite charge attract and those of like charge repel. Hence ions clump into ion pairs. An ion pair acts more like a molecule than like two ions. Therefore more of the solute dissolves because more ions can be present in solution. The effective concentration or the activity becomes less than the formal concentration, the amount of solute that dissolved. One indication of this difference between activity (a) and formal concentration (c) is that the activity coefficient (γ) becomes steadily smaller as the solution becomes more concentrated.

$$\gamma = a/c \qquad [9]$$

Another indication of this difference is the decrease in the van't Hoff factor as the solution becomes concentrated, as shown by the data in Table 13-2 (objective 13-10). The charge, not the identity, of ions influences the activity coefficient. Hence, a sodium ion (Na^+) and a silver ion (Ag^+) should have about the same effect.

* **5 Determine whether a salt will precipitate from solution based on the concentrations of its ions.**

The ion product expression has the same form as the solubility product expression but the ionic concentrations are not necessarily those at equilibrium. The ion product, Q, is related to the solubility product, K_{sp}, as shown in expression [10] (recall objective 16-9).

If $Q < K_{sp}$, more solute can dissolve

If $Q > K_{sp}$, precipitate forms [10]

If $Q = K_{sp}$, the solution is saturated.

EXAMPLE 19-3 150 mL of solution with $[Ca^{2+}] = 1.2 \times 10^{-3}$ M is mixed with 100 mL of solution with $[PO_4^{3-}] = 4.0 \times 10^{-3}$ M. $K_{sp} = 1.3 \times 10^{-32}$ for $Ca_3(PO_4)_2$. Should $Ca_3(PO_4)_2$ precipitate when the two solutions are mixed?

We first find the amount of each ion.

Amount of Ca^{2+} = 150 mL $\times$ 1.2×10^{-3} M = 0.18 mmol Ca^{2+}

Amount of PO_4^{3-} = 100 mL $\times$ 4.0×10^{-4} M = 0.040 mol PO_4^{3-}

Solution volume = 150 mL + 100 mL = 250 mL

$[Ca^{2+}]$ = 0.18 mmol/250 mL = 7.2×10^{-4} M

$[PO_4^{3-}]$ = 0.040 mmol/250 mL = 1.6×10^{-3} M

Q is computed and compared to K_{sp}.

$$Q = (7.2 \times 10^{-4})^3(1.6 \times 10^{-3})^2 = 9.56 \times 10^{-16} > 1.3 \times 10^{-32} = K_{sp}$$

Thus precipitation should occur.

* **6. Determine the concentration of ions remaining in solution after precipitation and predict whether precipitation will be complete.**

Determining whether precipitation is complete when the concentration of one ion remains constant involves using the constant concentration in the solubility product constant expression. This gives the maximum possible concentration of the other ion. Most chemists agree that 99.9% of the original amount of an ion present in a solution must precipitate in order for precipitation to be considered complete. That is, when C is the molar concentration of the ion being considered, precipitation is complete if

$$C_{final} <= 0.001\ C_{initial} \qquad [11]$$

EXAMPLE 19-5 A solution has $[I^-] = 1.3 \times 10^{-3}$ M. If $[Pb^{2+}]$ is kept equal to 2.0×10^{-2} M, will precipitation of I^- be complete? For PbI_2, $K_{sp} = 7.1 \times 10^{-9}$.

The maximum $[I^-]$ that can exist in this solution is obtained first.

$$K_{sp} = [Pb^{2+}][I^-]^2 = 7.1 \times 10^{-9} = (2.0 \times 10^{-2})[I^-]^2$$

$$[I^-] = \sqrt{\frac{7.1 \times 10^{-9}}{2.0 \times 10^{-2}}} = 5.9 \times 10^{-5}\text{ M}$$

This is 4.5% of the original $[I^-]$. Thus, according to condition [11], precipitation of I^- is not complete.

When the concentrations of both ions decreases, the calculation is somewhat more difficult. The easiest method is to assume that solute precipitates until all of one ion is consumed (its concentration reaches zero), and then solve a common ion problem.

EXAMPLE 19-6 What will be the final concentration of each ion in the solution of Example 19-4?

Rxn:	$Ca_3(PO_4)_2(s) \rightleftharpoons$	$3\ Ca^{2+}(aq)$	+	$2\ PO_4^{3-}(aq)$
Init:		7.2×10^{-4} M		1.6×10^{-3} M
Max. pptn:		0 M		$(1.6 \times 10^{-3} - 4.8 \times 10^{-4})$ M
Changes:		$+3S$		$+2S$
Equil:		$3S$		$(1.12 \times 10^{-3} + S)$ M

The 4.8×10^{-4} M decrease in $[PO_4^{3-}]$ was computed as follows.

$$\Delta[PO_4^{3-}] = \frac{7.2 \times 10^{-4}\text{ mol }Ca^{2+}}{\text{L}} \times \frac{2\text{ mol }PO_4^{3-}}{3\text{ mol }Ca^{2+}} = 4.8 \times 10^{-4}\text{ M}$$

The equilibrium expressions for ionic concentrations then are substituted into the K_{sp} expression. Remember, $K_{sp} = 1.3 \times 10^{-32}$ for $Ca_3(PO_4)_2$.

$$[Ca^{2+}]^3[PO_4^{3-}]^2 = (3S)^3(1.12 \times 10^{-3} + 2S)^2 = 1.3 \times 10^{-32}$$

Let us assume $2S << 1.12 \times 10^{-3}$ M.

$$(3S)^3(1.12 \times 10^{-3})^2 = 1.3 \times 10^{-32} = (27S^3)(1.25 \times 10^{-6}) \quad or \quad S = 7.3 \times 10^{-10}\text{ M}$$

We now need to evaluate our assumption that $2S << 1.12 \times 10^{-3}$. We determine whether the following equation is valid by substituting $S = 7.3 \times 10^{-10}$ M and evaluating both sides.

$$(3S)^3(1.12 \times 10^{-3})^2 = (3S)^3(1.12 \times 10^{-3} + 2s)^2$$

$$(3 \times 7.3 \times 10^{-10})^3(1.12 \times 10^{-3})^2 = (3 \times 7.3 \times 10^{-10})^3(1.12 \times 10^{-3} + 2 \times 7.3 \times 10^{-10})^2$$

$$1.3 \times 10^{-32} = 1.3 \times 10^{-32}$$

Thus, the assumption is valid. An easier way to evaluate the assumption is to see of the omitted quantity ($2S$ in this case) is more than 5% of the value to which it is added (1.12×10^{-3} M). In this problem, $2S = 1.5 \times 10^{-9}$ M = 0.00013% of 1.12×10^{-3} M and thus the assumption is valid. The final concentrations now are computed.

$$[PO_4^{3-}] = 1.12 \times 10^{-3}\text{ M}$$

$$[Ca^{2+}] = 3S = 21.9 \times 10^{-9}\text{ M}$$

Notice that the concentration of calcium ion has decreased to less than 0.1% (actually to 0.0003%) of its original value. Hence its precipitation is complete.

* 7. Explain how fractional precipitation works and when it can be used.

Fractional precipitation is the technique of separating one or more ions from solution in a precipitate while leaving others behind in solution. To separate two cations, an anion is added to the solution. As the concentration of the anion increases, eventually the value of the ion product, Q, for one of the salts exceeds the value of its solubility product constant, K_{sp}. The anion concentration is increased until the first cation has completely precipitated (99.9% of the original amount of cation has precipitated). It is hoped that other cations will not have begun to precipitate at this point. If they have, fractional precipitation is not possible, at least not using this anion as a precipitating agent.

EXAMPLE 19-7 A solution has $[Ba(NO_3)_2] = 0.50$ M, $[AgNO_3] = 3.0 \times 10^{-5}$ M, and $[Zn(NO_3)_2] = 2.0 \times 10^{-5}$ M. Solid sodium oxalate is added gradually. What is the $[C_2O_4^{2-}]$ when each cation (Ba^{2+}, Ag^+, Zn^{2+}): (a) just begins to precipitate? (b) is completely precipitated? Which of the ions is completely separated from the others by fractional precipitation and which is not? Values of K_{sp} at 25°C are 1.5×10^{-8} for BaC_2O_4, 1.35×10^{-9} for ZnC_2O_4, and 1.1×10^{-11} for $Ag_2C_2O_4$.

We first compute the $[C_2O_4^{2-}]$ needed to just start the precipitation of each cation. At this point $Q = K_{sp}$.

BaC_2O_4: $K_{sp} = [Ba^{2+}][C_2O_4^{2-}]$

$1.5 \times 10^{-8} = (0.50)[C_2O_4^{2-}]$ *or* $[C_2O_4^{2-}] = 3.0 \times 10^{-8}$ M

ZnC_2O_4: $K_{sp} = [Zn^{2+}][C_2O_4^{2-}]$

$1.35 \times 10^{-9} = (2.0 \times 10^{-9})[C_2O_4^{2-}]$ *or* $[C_2O_4^{2-}] = 6.8 \times 10^{-5}$

$Ag_2C_2O_4$: $K_{sp} = [Ag^+]^2[C_2O_4^{2-}]$

$1.11 \times 10^{-11} = (3.0 \times 10^{-8})^2[C_2O_4^{2-}]$ *or* $[C_2O_4^{2-}] = 1.2 \times 10^{-2}$ M = 0.012 M

Next we determine the $[C_2O_4^{2-}]$ when each of the ions has completely precipitated, by substituting 0.001 of each original ion concentration into the K_{sp} expression.

BaC_2O_4: $1.5 \times 10^{-8} = 5.0 \times 10^{-4}[C_2O_4^{2-}]$ *or* $[C_2O_4^{2-}] = 3.0 \times 10^{-5}$ M

ZnC_2O_4: $1.35 \times 10^{-9} = 2.0 \times 10^{-8}[C_2O_4^{2-}]$ *or* $[C_2O_4^{2-}] = 6.8 \times 10^{-2}$ M = 0.068 M

$Ag_2C_2O_4$: $1.1 \times 10^{-11} = (3.0 \times 10^{-8})^2[C_2O_4^{2-}]$ *or* $[C_2O_4^{2-}] = 1.2 \times 10^{4}$ M

Thus, as $[C_2O_4^{2-}]$ increases Ba^{2+} is the first ion to precipitate at $[C_2O_4^{2-}] = 3.0 \times 10^{-8}$ M. The precipitation of Ba^{2+} is complete when $[C_2O_4^{2-}] = 3.0 \times 10^{-5}$ M. When $[C_2O_4^{2-}]$ rises to 6.8×10^{-5} M, Zn^{2+} begins to precipitate. Thus Ba^{2+} is completely separated. When $[C_2O_4^{2-}] = 0.068$ M, Zn^{2+} is completely precipitated, but Ag^+ begins to precipitate before this, when $[C_2O_4^{2-}] = 0.012$ M. Thus Ag^+ and Zn^{2+} are not separated by fractional precipitation with oxalate ion. Also, we cannot completely precipitate Ag^+; 1.2×10^4 M is an impossible concentration to reach.

* **8. Describe, through net ionic equations and calculations, the effect of pH on the precipitation and dissolving of certain substances.**

Many metal hydroxides are insoluble. (Recall the solubility rules in objective 5-3.) $Fe(OH)_3$ is an example.

$$Fe(OH)_3(s) \rightleftharpoons Fe^{3+}(aq) + 3\ OH^-(aq) \quad K_{sp} = 4.0 \times 10^{-38} \qquad [12]$$

If we add strong acid to a mixture of $Fe(OH)_3(s)$ and its saturated solution, $H_3O^+(aq)$ from the acid reacts with $OH^-(aq)$ from the $Fe(OH)_3$. The equilibrium of equation [12] shifts right. $Fe(OH)_3$ is soluble in a strongly acidic solution. This is an acid-base neutralization.

$$Fe(OH)_3(s) + 3\ H_3O^+(aq) \rightleftharpoons Fe^{3+}(aq) + 6\ H_2O(aq) \qquad [13]$$

All acids will enhance the solubility of $Fe(OH)_3$, *including* the cation of a weak base.

$$Fe(OH)_3(s) + 3\ NH_4^+(aq) \rightleftharpoons Fe^{3+}(aq) + 3\ NH_3(aq) + 3\ H_2O \qquad [14]$$

An insoluble hydroxide can precipitate from a neutral solution because of hydrolysis of the cation.

$$Fe^{3+}(aq) + 6\ H_2O \rightleftharpoons Fe(OH)_3(s) + 3\ H_3O^+(aq) \qquad [15]$$

EXAMPLE 19-8 What is the solubility of $Fe(OH)_3$, $K_{sp} = 4.0 \times 10^{-38}$, in (a) pure water and (b) a solution maintained at pH = 2.00?

a. $K_{sp} = [Fe^{3}+][OH^-]^3 = (S)(3S)^3 = 4.0 \times 10^{-38}$ *or* $S = 2 \times 10^{-10}$ M

b. pH = 2.00 is pOH = 12.00; thus $[OH^-] = 1.0 \times 10^{-12}$ M.

$4.0 \times 10^{-38} = [Fe^{3+}](1.0 \times 10^{-12})^3$ or $[Fe^{3+}] = 4.0 \times 10^{-2}$ M

Another group of compounds whose solubilities are affected by pH are those in which the anion is the anion of a weak acid.

$$SrF_2(s) \rightleftharpoons Sr^{2+}(aq) + 2\ F^-(aq) \qquad [16]$$

The addition of $H_3O^+(aq)$ removes $F^-(aq)$ from solution, and equation [16] shifts to the right.

$$H_3O^+(aq) + F^-(aq) \rightleftharpoons HF(aq) + H_2O \qquad [17]$$

* **9. Write equations showing the effect of complex ion formation on other equilibrium processes such as solubility equilibria.**

In the analysis of the ammonium sulfide group in qualitative analysis, Fe^{3+} is precipitated as the hydroxide (see equation [12]). If NaCN(aq) is added, free $Fe^{3+}(aq)$ forms a complex ion.

$$Fe^{3+}(aq) + 6\,CN^-(aq) \rightleftharpoons [Fe(CN)_6]^{3-}(aq) \qquad K_f = 1 \times 10^{12} \qquad [18]$$

Equation [12] is displaced to the left, and some of the $Fe(OH)_3(s)$ dissolves. Notice that the stronger Lewis base (CN^-) displaces the weaker one (OH^-) from the Lewis acid (Fe^{3+}). The relative strengths of Lewis bases is given by the spectrochemical series, expression [2] in Chapter 25. Notice, in that series, that halide ligands generally are weak, nitrogen-containing ligands generally are strong, while ligands that contain oxygen are intermediate in strength.

* **10. Use complex ion formation constants, K_f (from *Table 19-2 in the text*), to compute the concentrations in solution of: free ions, ligands, and complex ions.**

Although Pb^{2+} will precipitate as $PbCl_2(s)$, the addition of excess Cl^- will redissolve the solid because of the formation of $[PbCl_3^-](aq)$.

$$Pb^{2+}(aq) + 3\,Cl^-(aq) \rightleftharpoons [PbCl_3^-](aq) \qquad K_f = 24 \qquad [19]$$

EXAMPLE 19-9 What is the $[Pb^{2+}]$ at equilibrium if a 0.0100 M Pb^{2+} solution is made 0.400 M in Cl^-?

Line Right: in the set-up below gives the concentrations when all the $Pb^{2+}(aq)$ reacts.

Rxn:	$Pb^{2+}(aq)$ +	$3\,Cl^-(aq)$	$\rightleftharpoons$	$[PbCl_3]^-(aq)$
Init:	0.0100 M	0.400 M		
Right:	0 M	0.370 M		0.0100 M
Change:	$+x$ M	$+ 3x$ M		$-x$ M
Eq:	x M	$(0.370 + 3x)$ M		$(0.0100 - x)$ M

$$K_f = \frac{0.0100 - x}{x(0.370 - 3x)^3} \approx \frac{0.0100 - x}{x(0.370)^3} = 24 \quad \text{or} \quad x = 0.00451$$

We have assumed $3x << 0.370$ M. This is not quite the case, so we go through two cycles of approximation to find $x = 0.0026$ M. $[Pb^{2+}] = 0.00425$ M, the concentration of free $Pb^{2+}(aq)$.

We used the method of successive approximations before (objective 17-8, Example 17-5). This method is essential here since the exact equation is fourth order and has no simple solution.

* **11. Use K_f values along with K_{sp} values to determine the solubilities of slightly soluble solutes in the presence of complexing ligands.**

The presence of a complexing ligand reduces the concentration of the free metal ion. This reduction may be enough that no precipitate forms even in the presence of a precipitating reagent. For example, CdS will precipitate when $Cd^{2+}(aq)$ and $S^{2-}(aq)$ are present even at low concentrations, equation [20]. But the addition of $CN^-(aq)$ lowers the concentration of free $Cd^{2+}(aq)$, equation [21], so that precipitation may not occur.

$$CdS(s) \rightleftharpoons Cd^{2+}(aq) + S^{2-}(aq) \qquad K_{sp} = 8.0 \times 10^{-27} \qquad [20]$$

$$Cd^{2+}(aq) + 4\,CN^-(aq) \rightleftharpoons [Cd(CN)_4]^{2-} \qquad K_f = \frac{[Cd(CN)_4^{2-}]}{[Cd^{2+}][CN^-]^4} = 7.1 \times 10^{18} \qquad [21]$$

EXAMPLE 19-10 One liter of solution has $[S^{2-}] = 3.2 \times 10^{-8}$ M. (a) What is the maximum amount of $Ca(NO_3)_2$ that may be added without a precipitate forming? (b) If the solution also has $[CN^-] = 0.100$ M, what is the maximum amount of $Cd(NO_3)_2$ that may be added?

a. Substitution into the K_{sp} expression for reaction [20] produces the maximum $[Cd^{2+}]$.

$$[Cd^{2+}][S^{2-}] = [Cd^{2+}](3.2 \times 10^{-8}\text{ M}) = 8.0 \times 10^{-27}$$

$$[Cd^{2+}] = 2.5 \times 10^{-19}\text{ M}$$

Thus, we could add only 2.5×10^{-19} mol $Cd(NO_3)_2$.

b. We know $[Cd^{2+}] = 2.5 \times 10^{-19}$ M at equilibrium before $CN^-(aq)$ is added. This value will remain fixed because the $[S^{2-}]$ does not change.

Rxn:	$Cd^{2+}(aq)$ +	$4\ CN^-(aq)$	$\rightleftharpoons$	$[Cd(CN)_4]^{2-}(aq)$
Init:	2.5×10^{-19} M	0.100 M		
Change:		$-4x$ M		$+x$ M
Equil:	2.5×10^{-19} M	$(0.100 - 4x)$ M		x M

$$K_f = \frac{x}{(2.5 \times 10^{-19})(0.100 - 4x)^4} = 7.1 \times 10^{18} \approx \frac{x}{(2.5 \times 10^{-19})(0.100)^4}$$

We assumed $4x << 0.100$ M. This gives $x = 1.78 \times 10^{-4}$ M = $[Cd(CN)_4^{2-}]$. Practically all of the cadmium ion is present in the complex, and thus we can add 1.78×10^{-4} mol $Cd(NO_3)_2$ to 1 L of solution, 7.12×10^{14} times more than in part (a).

12. Calculate the solubilities of certain solutes in the presence of complexing ligands.

This is similar to the previous objective except that the complexing reagent is added to the already formed precipitate. The problem is solved by combining the equilibrium constants of the precipitation reaction (equation [22]) and the complex ion formation reaction (equation [23]) to obtain an equilibrium constant for the overall reaction (equation [24]). (Objective 16-4 discusses how equilibrium constants are calculated when two reactions are combined.)

$$ZnS(s) \rightleftharpoons Zn^{2+}(aq) + S^{2-}(aq) \qquad K_{sp} = 1.0 \times 10^{-21} \qquad [22]$$

$$Zn^{2+}(aq) + 4\ NH_3(aq) \rightleftharpoons [Zn(NH_3)_4]^{2+}(aq) \qquad K_f = 4.1 \times 10^8 \qquad [23]$$

$$ZnS(s) + 4\ NH_3(aq) \rightleftharpoons [Zn(NH_3)_4]^{2+}(aq) + S^{2-}(aq) \qquad K = K_{sp} \times K_f = 4.1 \times 10^{-13} \qquad [24]$$

EXAMPLE 19-11 What is the molar solubility of ZnS in (a) pure water and (b) 0.800 M $NH_3(aq)$?

a. The K_{sp} expression is solved to determine the molar solubility.
$[Zn^{2+}][S^{2-}] = 1.0 \times 10^{-21} = s^2$ *or* $s = 3.2 \times 10^{-11}$ M

b. We solve this problem with the equilibrium constant of equation [24].

Rxn:	$ZnS(s)$ +	$4\ NH_3(aq)$	$\rightleftharpoons$	$[Zn(NH_3)_4^{2+}](aq)$ +	$S^{2-}(aq)$
Init:		0.800 M			
Change:		$-4x$ M		$+x$ M	$+x$ M
Equil:		$(0.800 - 4x)$M		x M	x M

$$K = \frac{[[Zn(NH_3)_4]^{2+}][S^{2-}]}{[NH_3]^4} = 4.1 \times 10^{-13} = \frac{x \cdot x}{(0.800 - 4x)^4}$$

or $x/(0.800 - 4x)^2 = 6.4 \times 10^{-7}$

or $x = 4.1 \times 10^{-7}$ M = $[[Zn(NH_3)_4^{2+}]]$

We have assumed $4x << 0.800$ M, which we now see is true. Practically all of the zinc ion in solution is present in the complex ion.

13. Relate $[H_3O^+]$ and $[S^{2-}]$ in $H_2S(aq)$ solutions through the combined equilibrium constant expression, $K_{a1} \times K_{a2}$.

For a diprotic acid, H_2A, $[A^{2-}] = K_{a2}$ (objective 17-9). There are two assumptions: the two ionization constants must differ by at least a factor of 1000, and *all* of the A^{2-} and H_3O^+ must come from the diprotic acid. When a common ion is present, we can relate $[H_3O^+]$ and $[A^{2-}]$ through a combined equilibrium constant, $K_{a1} \times K_{a2}$. For H_2S,

$$H_2S(aq) + H_2O \rightleftharpoons H_3O^+(aq) + HS^-(aq) \qquad K_{a1} = 1.1 \times 10^{-7}$$

$$\underline{HS^-(aq) + H_2O \rightleftharpoons H_3O^+(aq) + S^{2-}(aq)} \qquad K_{a2} = 1.0 \times 10^{-14}$$

$$H_2S(aq) + 2\ H_2O \rightleftharpoons 2\ H_3O^+(aq) + S^{2-}(aq) \qquad K = K_{a1}\ K_{a2} = 1.1 \times 10^{-21} \qquad [25]$$

$$K = \frac{[H_3O^+]^2[S^{2-}]}{[H_2S]} = 1.1 \times 10^{-21} \qquad [26]$$

H_2S is considered here because very often we wish to precipitate metal ions as their sulfides. Precipitation is normally done by saturating the solution with H_2S. At 25°C, the concentration of a saturated solution is $[H_2S] = 0.10$ M. Thus equation [26] becomes

$$K = [H_3O^+]^2[S^{2-}] = 1.1 \times 10^{-22} \quad \text{(in a saturated } H_2S \text{ solution)} \qquad [27]$$

*** 14. Predict whether metal sulfides will precipitate from saturated $H_2S(aq)$ solutions of known pH.**

Equation [27] can be used to find $[S^{2-}]$ in a solution of known pH. This $[S^{2-}]$ then can be used in the solubility product constant expression for a metal sulfide.

EXAMPLE 19-12 A solution is maintained at pH = 4.75 by an $HC_2H_3O_2$-$NaC_2H_3O_2$ buffer. The solution contains Pb^{2+} and Mn^{2+}, each at a concentration of 0.100 M. The solution is saturated with H_2S. Will either of the metal sulfides precipitate? $K_{sp}(PbS) = 8.0 \times 10^{-28}$ and $K_{sp}(MnS) = 2.5 \times 10^{-13}$.

pH = 4.75 means $[H_3O^+] = 1.78 \times 10^{-5}$ M. From equation [27]

$$[H_3O^+]^2[S^{2-}] = (1.78 \times 10^{-5})^2[S^{2-}] = 1.1 \times 10^{-22} \quad or \quad [S^{2-}] = 3.5 \times 10^{-13}$$

From this, we compute the maximum concentration of each ion that can be present in solution based on the value of K_{sp}.

$$[Pb^{2+}][S^{2-}] = [Pb^{2+}](3.5 \times 10^{-13}) = 8.0 \times 10^{-28} \quad or \quad [Pb^{2+}] = 2.3 \times 10^{-15}\ M$$

$$[Mn^{2+}][S^{2-}] = [Mn^{2+}](3.5 \times 10^{-13}) = 2.5 \times 10^{-15} \quad or \quad [Mn^{2+}] = 7.1 \times 10^{2}\ M$$

PbS will precipitate, while MnS will not. Of course, 710 mol Mn^{2+} will not dissolve in a liter of solution; no manganese(II) compound is *that* soluble.

DRILL PROBLEMS

1. Write the solubility equilibrium and the solubility product expression for each of the following salts. The first is done as an example.

Ag_2SO_4: $Ag_2SO_4(s) \rightleftharpoons 2\ Ag^+(aq) + SO_4^{2-}(aq)$ $\quad K_{sp} = [Ag^+]^2[SO_4^{2-}]$

A. $AgCN$ B. SrF_2 C. $TlBr$ D. $Zn_3(AsO_4)_3$
E. Hg_2Br_2 F. Ag_3AsO_4 G. Li_2CO_3 H. PbC_2O_4
I. $La(IO_3)_3$ J. $Ce_2(C_2O_4)_2$ K. $CdSO_3$ L. $AlPO_4$
M. Al_2S_3 N. $Ca(IO_3)_2$ O. CoS P. $FeAsO_4$

2. Fill in the blanks in the following lines. The first line is completed as an example.

Compound	K_{sp}	*Molar solubility of* cation, M	*Molar solubility of* anion, M	*Cmpd solubility, g/100 mL soln*
Ag_2SO_4	1.4×10^{-5}	0.030	0.015	0.47
A. $AgCNS$	____	1.08×10^{-6}	____	____
B. $SrCrO_4$	____	____	4.7×10^{-3}	____
C. $TlIO_3$	____	____	____	6.7
D. Hg_2CrO_4	2.0×10^{-9}	____	____	____
E. $AlPO_4$	____	7.9×10^{-10}	____	____
F. $Ba(BrO_3)_2$	____	____	1.9×10^{-2}	____
G. $Be(OH)_2$	____	____	____	1.5×10^{-5}
H. $Ca(IO_3)_2$	7.1×10^{-7}	____	____	____
I. Cu_2S	____	1.7×10^{-16}	____	____
J. Tl_2CrO_4	____	____	6.3×10^{-5}	____
K. $Ag_2C_2O_4$	____	____	____	0.62
L. BiI_3	8.1×10^{-19}	____	____	____
M. Ag_3PO_4	____	1.4×10^{-4}	____	____
N. $La(IO_3)_3$	____	____	2.1×10^{-3}	____
O. $Ca_3(PO_4)_2$	____	____	____	2.2×10^{-3}
P. Al_2S_3	2.0×10^{-7}	____	____	____

3. Use the data of *Table 19-1 in the text* to predict the molar solubility of each of the following salts in a solution that contains the given concentration of one of its ions.

A. $BaSO_4$; $[SO_4^{2-}] = 4.3 \times 10^{-5}$ M B. AgI; $[I^-] = 7.2 \times 10^{-6}$ M
C. SnS; $[S^{2-}] = 3.1 \times 10^{-2}$ M D. $MgCO_3$; $[Mg^{2+}] = 1.7 \times 10^{-4}$ M
E. Ag_2CrO_4; $[Ag^+] = 1.3 \times 10^{-3}$ M F. Hg_2Cl_2; $[Hg_2^{2+}] = 0.85$ M
G. $Fe(OH)_3$; $[Fe^{3+}] = 1.6 \times 10^{-2}$ M H. PbI_2; $[I^-] = 3.4$ M
I. Li_3PO_4; $[PO_4^{3-}] = 0.29$ M J. $Mg_3(PO_4)_2$; $[PO_4^{3-}] = 0.87$ M

5. (1) In each part, determine the concentration of the second compound that is just large enough to cause precipitation from a solution of the given concentration. Use the data of *Table 19-1 in the text.*
A. $[Cu(NO_3)_2] = 1.45 \times 10^{-13}$ M; Na_2S
B. $[AgNO_3] = 1.2 \times 10^{-6}$ M; NaBr
C. $[Sr(NO_3)_2] = 1.4 \times 10^{-3}$ M; Na_2CO_3
D. $[NiSO_4] = 4.7 \times 10^{-16}$ M; Na_2S
E. $[Pb(NO_3)_2] = 8.9 \times 10^{-3}$ M; $MgSO_4$
F. $[MgCl_2] = 3.4 \times 10^{-5}$ M; NaF
G. $[AgNO_3] = 5.0 \times 10^{-3}$ M; Na_2SO_4
H. $[Pb(NO_3)_2] = 1.6 \times 10^{-3}$ M; MgI_2
I. $[Al(NO_3)_3] = 1.7 \times 10^{-14}$ M; NaOH
J. $[MgSO_4] = 2.4 \times 10^{-8}$ M; Na_3PO_4
(2) Determine if precipitation will occur in each of the following solutions. Use the data of *Table 19-1 in the text.* All concentrations are given in moles per liter of solution.
K. $[Co^{2+}] = 3.2 \times 10^{-5}$; $[S^{2-}] = 1.6 \times 10^{-12}$
L. $[Pb^{2+}] = 0.0047$; $[SO_4^{2-}] = 7.0 \times 10^{-6}$
M. $[Ag^+] = 1.2 \times 10^{-5}$; $[I^-] = 9.2 \times 10^{-8}$
N. $[Ba^{2+}] = 0.064$; $[SO_4^{2-}] = 1.6 \times 10^{-5}$
O. $[Sr^{2+}] = 3.2 \times 10^{-5}$; $[CO_3^{2-}] = 7.2 \times 10^{-7}$
P. $[Pb^{2+}] = 8.6 \times 10^{-6}$; $[CrO_4^{2-}] = 0.0017$
Q. $[Ag^+] = 0.017$; $[CrO_4^{2-}] = 4.6 \times 10^{-5}$
R. $[Mg^{2+}] = 3.7 \times 10^{-6}$; $[F^-] = 0.0088$
S. $[Li^+] = 1.6$; $[PO_4^{3-}] = 0.0042$
T. $[Mg^{2+}] = 1.4 \times 10^{-5}$; $[PO_4^{3-}] = 7.9 \times 10^{-6}$

6. (1) For each of that compounds that precipitates in the group (2) drill problems of objective 19-5, determine the final concentration of each ion after precipitation and then determine if precipitation of either the anion or the cation is complete.
(2) Do the same for each of the following solutions. In each of these cases, precipitation *does* occur. Again, all concentrations are given in moles per liter of solution.
A. $[Zn^{2+}] = 1.1 \times 10^{-5}$; $[S^{2-}] = 9.5 \times 10^{-10}$
B. $[Hg_2^{2+}] = 3.4 \times 10^{-14}$; $[Cl^-] = 1.2 \times 10^{-7}$
C. $[Al^{3+}] = 1.6 \times 10^{-12}$; $[OH^-] = 2.7 \times 10^{-4}$
D. $[Pb^{2+}] = 1.0$; $[Cl^-] = 0.015$
E. $[Ag^+] = 1.7 \times 10^{-6}$; $[Br^-] = 3.3 \times 10^{-3}$
F. $[Bi^{3+}] = 2.7 \times 10^{-6}$; $[S^{2-}] = 4.6 \times 10^{-15}$

7. Use the solubility rules in objective 5-3 and the data of *Table 19-1 in the text* to determine what anion or cation will produce a precipitate with each member of the following pairs of ions. Determine the concentration of precipitating agent that must be present to just cause precipitation of each ion, and the concentration needed to completely precipitate each ion. Finally, determine if the two ions can be separated by fractional precipitation or not.
A. $[Cl^-] = 1.2 \times 10^{-6}$ M; $[Br^-] = 8.9 \times 10^{-4}$ M
B. $[Ba^{2+}] = 1.6 \times 10^{-5}$ M; $[Pb^{2+}] = 1.3 \times 10^{-5}$ M
C. $[Pb^{2+}] = 3.9 \times 10^{-4}$ M; $[Hg_2^{2+}] = 3.9 \times 10^{-4}$ M
D. $[Sr^{2+}] = 1.6 \times 10^{-8}$ M; $[Pb^{2+}] = 3.4 \times 10^{-5}$ M (use SO_4^{2-})
E. $[OH^-] = 9.6 \times 10^{-4}$ M; $[F^-] = 0.13$ M (use Mg^{2+})
F. $[Ag^+] = 4.6 \times 10^{-3}$ M; $[Pb^{2+}] = 0.023$ M (use CrO_4^{2-})
G. $[Zn^{2+}] = 7.9 \times 10^{-14}$ M; $[Sn^{2+}] = 2.9 \times 10^{-12}$ M
H. $[Li^+] = 0.60$ M; $[Mg^{2+}] = 7.4 \times 10^{-4}$ M

8. Determine the solubility of each of the following compounds in a solution maintained at pH = 2.00, 7.00, and 10.00. Use the data of *Table 19-1 in the text* where needed.
A. $Al(OH)_3$ B. $Cr(OH)_3$ C. $Mg(OH)_2$ D. $Be(OH)_2$ $K_{sp} = 1.6 \times 10^{-27}$
E. $Ba(OH)_2$ F. $Fe(OH)_3$ G. $Ca(OH)_2$ H. AgOH $K_{sp} = 2.0 \times 10^{-8}$

9. Write equations for the equilibrium reactions that occur in aqueous solutions that contain the following species. Does the formation of a complex ion increase or decrease the amount of precipitate?
A. Ag^+, $S_2O_3^{2-}$, Cl^-, Na^+
B. Hg^{2+}, I^-, Ag^+, NO_3^-
C. Zn^{2+}, CN^-, S^{2-}, Na^+
D. Fe^{2+}, CN^-, OH^-, K^+
E. Al^{3+}, F^-, Ca^{2+}, Li^+
F. Cu^+, CN^-, Ag^+, Cl^-
G. Al^{3+}, F^-, OH^-, Na^+
H. Co^{3+}, S^{2-}, NH_3, Cl^-
I. Cu^{2+}, NH_3, OH^-, Na^+

10. Determine the concentrations of free metal ion and complex ion in each solution in which the initial metal ion and ligand concentrations (all in moles per liter) are given below. Neglect the acid and base reactions of the ligands and cations. Use the data of *Table 19-2 in the text* as needed.
A. $[F^-] = 0.100$; $[Al^{3+}] = 0.002$
B. $[Cl^-] = 1.00$, $[Pb^{2+}] = 8.2 \times 10^{-5}$
C. $[S_2O_3^{2-}] = 0.310$; $[Ag^+] = 0.0067$
D. $[CN^-] = 0.850$; $[Zn^{2+}] = 5.0 \times 10^{-4}$
E. $[CN^-] = 0.850$; $[Cu^{2+}] = 5.0 \times 10^{-4}$
F. $[CN^-] = 0.850$; $[Fe^{2+}] = 5.0 \times 10^{-4}$
G. $[CN^-] = 0.850$; $[Fe^{3+}] = 5.0 \times 10^{-4}$
H. $[NH_3] = 1.52$; $[Co^{3+}] = 1.4 \times 10^{-5}$
I. $[NH_3] = 1.52$; $[Ag^+] = 1.4 \times 10^{-5}$
J. $[I^-] = 1.00$; $[Hg^{2+}] = 2.4 \times 10^{-6}$
K. $[Cl^-] = 1.00$; $[Hg^{2+}] = 2.4 \times 10^{-6}$

11. Determine the concentration of free metal ion (after a complex forms, but before precipitation occurs) and state whether a precipitate will form or not. The initial concentration of each species is given. You may neglect the acid and base reactions between water and the ligand or the cation.
A. $[Ag^+] = 0.0100$ M, $[I^-] = 0.0100$ M, $[NH_3] = 5.00$ M
B. $[Ag^+] = 0.0100$ M, $[Cl^-] = 0.0100$ M, $[NH_3] = 5.00$ M
C. $[Zn^{2+}] = [S^{2-}] = 0.0100$ M. $[EDTA^{4-}] = 0.100$ M $K_f = 3.9 \times 10^{16}$ for $[Zn(EDTA)]^{2-}$
D. $[Ag^+] = 0.0200$ M, $[Br^-] = 0.0100$ M, $[S_2O_3^{2-}] = 0.250$ M
E. $[Hg^{2+}] = 0.0100$ M, $[SO_4^{2-}] = 0.0200$ M, $[I^-] = 0.400$ M $K_{sp} = 7.4 \times 10^{-7}$ for $HgSO_4$
F. $[Cu^{2+}] = 0.0400$ M, $[S^{2-}] = 0.0200$ M, $[NH_3] = 2.00$ M
G. $[Ag^+] = 0.0100$ M, $[S^{2-}] = 0.00500$ M, $[CN^-] = 1.00$ M

14. Use the data of *Table 19-1 in the text* and determine the pH at which precipitation of each metal sulfide will just begin to occur if the metal ion concentration is as given below and the solution is saturated with H_2S.
A. $[Co^{2+}] = 1.40$ M B. $[Fe^{2+}] = 0.140$ M C. $[Ni^{2+}] = 0.210$ M
D. $[Bi^{3+}] = 3.0 \times 10^{-20}$ M E. $[Cu^{2+}] = 3.0 \times 10^{-20}$ M F. $[Hg^{2+}] = 1.0 \times 10^{-10}$ M
G. $[Ag^+] = 5.0 \times 10^{-5}$ M

QUIZZES (20 minutes each) Choose the best answer for each question.

Quiz A

1. A saturated solution of copper(I) bromide has a concentration of 7.0×10^{-5} M. What is the value of K_{sp} of this compound? (a) 4.9×10^{-9}; (b) 4.9×10^{-10}; (c) 7.0×10^{-5}; (d) 14.0×10^{-5}; (e) none of these.
2. The value of the solubility product constant is 2.8×10^{-9}. What is the molar solubility of this compound? (a) 2.8×10^{-9} M; (b) $\sqrt{28} \times 10^{-5}$ M; (c) $(2.8)^{1/3} \times 10^{-3}$ M; (d) $(2.8/4)^{1/3} \times 10^{-3}$ M; (e) none of these.
3. The value of the solubility product constant of $Mg(OH)_2$ is 9.0×10^{-12}. If a solution has $[Mg^{2+}] = 0.010$ M, $[OH^-]$ required to just start the precipitation of $Mg(OH)_2$ is (a) 9.0×10^{-10} M; (b) 1.5×10^{-7} M; (c) 3.0×10^{-7} M; (d) 1.5×10^{-5} M; (e) 3.0×10^{-5} M.
4. The value of K_{sp} for AgCl is 1.10×10^{-10}. AgCl(s) will precipitate from solution whenever (a) $[Ag^+]$ exceeds 1.10×10^{-10}; (b) a solution containing Ag^+ is added to a solution containing Cl^-; (c) $[Ag^+] + [Cl^-]$ exceeds 1.10×10^{-10}; (d) $[Ag^+][Cl^-]$ exceeds 1.10×10^{-10}; (e) $\sqrt{[Ag^+][Cl^-]}$ exceeds 1.10×10^{-10}
5. Which of the following has the largest molar solubility? (a) CuS, $K_{sp} = 8 \times 10^{-37}$; (b) Bi_2S_3, $K_{sp} = 1 \times 10^{-70}$; (c) Ag_2S, $K_{sp} = 6 \times 10^{-51}$ (d) MnS, $K_{sp} = 7 \times 10^{-16}$
6. Saturated solutions of cesium iodide, mercury(I) acetate, and calcium chlorate are mixed together. The precipitate that forms is (a) calcium acetate; (b) mercury(I) chlorate; (c) mercury(I) iodide; (d) cesium acetate; (e) nothing precipitates.
7. AgI has the highest solubility in which of the following solutions? (a) 0.100 M $Mg(NO_3)_2$; (b) 0.100 M $AgNO_3$; (c) 0.100 M NaI; (d) 0.100 M MgI_2; (e) 0.100 M NaCN.
8. Which would be the least effective in dissolving $CuSO_3$? (a) add SO_2(aq); (b) add $Na_2S_2O_3$(aq); (c) add $HC_2H_3O_2$(aq); (d) add H_2O; (e) all would be equally effective.

Quiz B

1. A saturated solution of lithium carbonate has a concentration of 1.62×10^{-2} M. What is the value of K_{sp} of this compound? (a) $4(1.62 \times 10^{-2})^3$; (b) $(1.62 \times 10^{-2})^3$; (c) $(1.62 \times 10^{-2})^2$; (d) 1.62×10^{-2}; (e) none of these.
2. The value of the solubility product constant of copper(II) sulfide is 9.0×10^{-36}. What is the molar solubility of this compound? (a) 1.7×10^{-18} M; (b) 1.5×10^{-12} M; (c) 1.2×10^{-18} M; (d) 3.0×10^{-11} M; (e) none of these.
3. Saturated solutions of sodium sulfate, ammonium carbonate, and nickel(II) nitrate are mixed together. The precipitate that forms is (a) nickel(II) carbonate; (b) sodium carbonate; (c) nickel(II) sulfate; (d) ammonium sulfate; (e) nothing precipitates.
4. Which of the following has the largest molar solubility? (a) $BaSO_4$, $K_{sp} = 1.1 \times 10^{-10}$; (b) AgCl, $K_{sp} = 1.6 \times 10^{-10}$; (c) $Mg(OH)_2$, $K_{sp} = 2 \times 10^{-11}$; (d) $Cr(OH)_3$, $K_{sp} = 6.3 \times 10^{-11}$.
5. The value of K_{sp} for AgCl is a.1.6×10^{-10}. What is the maximum $[MnCl_2]$ that can exist in solution before precipitation occurs if the solution has $[AgNO_3] = 3.4 \times 10^{-4}$ M? (a) 4.7×10^{-5} M; (b) 2.4×10^{-5} M;

(c) 4.7×10^{-7} M; (d) 2.4×10^{-7} M; (e) none of these.

6. A substance is said to be completely precipitated when it (a) has a concentration of 0.00 M; (b) has 0.1% of its initial concentration; (c) cannot be precipitated by any known method; (d) has a concentration of less than $1/(6.022 \times 10^{23})$ M; (e) none of these.
7. Which would be the least effective in dissolving $MnCO_3$? (a) add $NaC_2O_4(aq)$; (b) add $NaCN(aq)$; (c) add $CO_2(aq)$; (d) add $H_2O(l)$; (e) all would be equally effective.
8. $CuCO_3$ is least soluble in which of the following solutions? (a) 0.100 M HCl; (b) 1.00 M $NH_3(aq)$; (c) 0.100 M $CuCl_2$; (d) 0.100 M NaCl; (e) 0.100 M $MnSO_4$.

Quiz C

1. A saturated solution of barium chromate has a concentration of 1.55×10^{-5} M. What is the value of K_{sp} of this compound? (a) $.55 \times 10^{-5}$; (b) $(2 \times 1.55 \times 10^{-5})^2$; (c) $2(1.55 \times 10^{-5})$; (d) $1.55^2 \times 10^{-10}$; (e) none of these.
2. The value of the solubility product constant of silver carbonate is 8.2×10^{-12}. What is the molar solubility of this compound? (a) 1.9×10^{-6}; (b) 1.4×10^{-4}; (c) 8.2×10^{-12}; (d) 9.1×10^{-7}; (e) none of these.
3. Saturated solutions of sodium phosphate, copper(II) chloride, and ammonium acetate are mixed together. The precipitate is (a) copper(II) acetate; (b) copper(II) phosphate; (c) sodium chloride; (d) ammonium phosphate; (e) nothing precipitates.
4. Which of the following has the largest molar solubility? (a) MgF_2, $K_{sp} = 3.7 \times 10^{-8}$; (b) $MgCO_3$, $K_{sp} = 3.5 \times 10^{-8}$; (c) $Mg_3(PO_4)_2$, $K_{sp} = 1 \times 10^{-25}$; (d) Li_3PO_4, $K_{sp} = 3.2 \times 10^{-9}$.
5. A solution has $[F^-] = 1.4 \times 10^{-3}$ M. What is the maximum $[Ca^{2+}]$ that can be present before CaF_2 ($K_{sp} = 2.7 \times 10^{-11}$) will precipitate? (a) 1.4×10^{-5}; (b) 7.0×10^{-6}; (c) 3.5×10^{-6}; (d) 5.6×10^{-5}; (e) none of these.
6. In which solution will $AgNO_3$ be the most soluble? (a) 0.01 M HCl; (b) pure water; (c) 0.01 M HNO_3; (d) cannot be determined; (e) equally soluble in the two solutions.
7. Which of the following would be the least effective in producing a solution from $CuSO_4$? (a) add H_2O; (b) add $H_2SO_4(aq)$; (c) add $CO_2(aq)$; (d) add $CuCl_2(aq)$; (e) all are equally effective.
8. In which of the following solutions is $PbCl_2$ the most soluble? (a) 0.100 M NaCl; (b) 0.100 M $Na_2S_2O_3$; (c) 0.100 M $Pb(NO_3)_2$; (d) 0.100 M $NaNO_3$; (e) 0.100 $MnSO_4$.

Quiz D

1. A saturated solution of magnesium fluoride has a concentration of 1.17×10^{-3} M. What is the value of K_{sp} of this compound? (a) 1.17×10^{-3}; (b) $(1.17 \times 10^{-3})^2$; (c) $4(1.17 \times 10^{-3})^3$; (d) $(1.17 \times 10^{-3})^3$; (e) none of these.
2. The value of the solubility product constant of silver chloride is 1.6×10^{-10}. What is the molar solubility of this compound? (a) 7.5×10^{-4}; (b) 9.2×10^{-5}; (c) 1.3×10^{-5}; (d) 1.7×10^{-10}; (e) none of these.
3. Which of the following has the largest molar solubility? (a) $Al(OH)_3$, $K_{sp} = 1.3 \times 10^{-33}$; (b) CoS, $K_{sp} = 4.0 \times 10^{-21}$; (c) CuS, $K_{sp} = 6.3 \times 10^{-36}$; (d) Ag_2S, $K_{sp} = 6.3 \times 10^{-50}$.
4. Saturated solutions of barium chloride, calcium nitrate, and cobalt(III) iodide are mixed together. The precipitate is (a) calcium iodide; (b) calcium chloride; (c) barium iodide; (d) cobalt(III) chloride; (e) none of these.
5. In which solution will $BaSO_4$ be the most soluble? (a) $BaSO_4$ is insoluble in everything; (b) pure water; (c) 0.010 M $NaNO_3$; (d) 0.010 M $Ba(NO_3)_2$; (e) cannot be determined.
6. A solution has $[Ag^+] = 1.3 \times 10^{-4}$ M. What is the maximum $[CO_3^{2-}]$ that can be present before Ag_2CO_3 ($K_{sp} = 8.1 \times 10^{-12}$) will precipitate? (a) 1.2×10^{-4} M; (b) 9.6×10^{-4} M; (c) 4.8×10^{-4} M; (d) 3.1×10^{-8} M; (e) none of these.
7. Which constant is numerically equal to the inverse of the formation constant of $[AgCl_2]^-$? (a) step-wise dissociation constant; (b) lissociation constant; (c) solubility product constant; (d) step-wise formation constant; (e) ionization constant.
8. In which of the following solutions is ZnS the least soluble? (a) 1.00 M $NH_3(aq)$; (b) 1.00 M $Na_2S_2O_3$; (c) 1.00 M NaCl; (d) 1.00 M $Mg(NO_3)_2$; (e) 1.00 M Na_2CO_3.

SAMPLE TEST (20 minutes)

1. A holding pond has $[PO_4^{3-}] = 8.06 \times 10^{-4}$ M. It measures 100. m $\times$ 200. m $\times$ 8.00 m. What mass, in tons, of $Ca(NO_3)_2$ must be added to this holding pond to reduce $[PO_4^{3-}]$ to 1.00×10^{-12} M? For $Ca_3(PO_4)_2$, $K_{sp} = 1.30 \times 10^{-32}$.
2. An aqueous solution has $[CrO_4^{2-}] = 2.0 \times 10^{-3}$ M and $[Cl^-] = 1.0 \times 10^{-5}$ M. Concentrated $AgNO_3$ solution is added. Will AgCl ($K_{sp} = 1.6 \times 10^{-10}$) or Ag_2CrO_4 ($K_{sp} = 2.4 \times 10^{-12}$) precipitate first? Eventually $[Ag^+]$ should increase enough to cause precipitation of the second compound. What is that remaining concentration of the first ion that precipitates when the second ion just begins to precipitate?
3. 350.0 mL of 0.200 M $AgNO_3$(aq) is added to 250.0 mL of 0.240 M Na_2SO_4(aq) and Ag_2SO_4 ($K_{sp} = 1.4 \times 10^{-5}$) precipitates. 400.0 mL of 0.500 M $Na_2S_2O_3$(aq) is added to this mixture. What mass of Ag_2SO_4(s) remains? $K_f = 1.7 \times 10^{13}$ for $[Ag(S_2O_3)_2]^{3-}$.

20 Spontaneous Change: Entropy and Free Energy

CHAPTER OBJECTIVES

1. State the first law of thermodynamics and the sign conventions used for heat, *q*, and work, *w*.

* **2. Calculate the value of one of the following from known values of the other two: ΔE, *q*, and *w*.**

The first law of thermodynamics also is known as the law of conservation of energy. A convenient statement is that energy is neither created nor destroyed, merely changed from one form into another.

$$\Delta E = q + w \qquad [1]$$

ΔE is the change in the internal energy of the system. *q* is defined as the heat absorbed by the system. If the system actually gives off heat during a process, that heat has a negative sign by universal convention. *w* is defined as the work done on the system by its surroundings. If the system does work on the surroundings, that work has a negative sign by convention. (Although the convention for heat is universal, that for work is not. Many writers use exactly the opposite convention: positive work is that done *by* the system, and if the surroundings do work on the system, that work is negative. Thus, the change in internal energy becomes $\Delta E = q - w$. In addition, the SI recommendation is to symbolize internal energy by *U*, rather than by *E*.)

When work is done on a system or when heat is absorbed by the system, energy is flowing into the system and ΔE must be positive; the internal energy of the system must increase. When the system is giving off heat or doing work, energy is leaving the system and ΔE must be negative; the internal energy of the system must decrease. Remember that we can determine only changes in internal energy, never its absolute value.

3. Explain the purpose served by the thermodynamic property of enthalpy (*H*), describe how ΔH is related to ΔE, and calculate one from the other for reactions involving gases.

The heat given off at constant volume equals the change in internal energy (ΔE). However, most reactions run at constant pressure. The heat given off at constant pressure equals the change in enthalpy (ΔH).

$$\Delta E = q_V \qquad [2]$$

$$\Delta H = q_P \qquad [3]$$

Enthalpy is a state function related to internal energy.

$$\Delta H = \Delta E + P\Delta V \qquad [4]$$

$$\Delta H = \Delta E + \Delta n_g RT \qquad [5]$$

Equation [5] is the more convenient to use. Δn_g is the difference between the stoichiometric coefficients of the gaseous products and those of the gaseous reactants in the balanced equation for the chemical reaction. Values of Δn_g for five reactions follow.

$N_2(g) + 3\,H_2(g) \longrightarrow 2\,NH_3(g) \quad \Delta n_g = 2 - 4 = -2$ mol gas/mol rxn

$Mg_3N_2(s) + 6\,H_2O(l) \longrightarrow 3\,Mg(OH)_2(s) + 2\,NH_3(g) \quad \Delta n_g = 2 - 0 = +2$ mol gas/mol rxn

$C_3H_7OH(l) + \frac{9}{2} O_2(g) \longrightarrow 3 CO_2(g) + 4 H_2O(l)$ $\quad \Delta n_g = 3 - 4.5 = -1.5$ mol gas/mol rxn

$P_4(s) + 5 O_2(g) \longrightarrow P_4O_{10}(s)$ $\quad \Delta n_g = 0 - 5 = 5$ mol gas/mol rxn

$4 NH_3(g) + 3 O_2(g) \longrightarrow 2 N_2(g) + 6 H_2O(l)$ $\quad \Delta n_g = 2 - 7 = -5$ mols gas/mol rxn

Equations [4] and [5] depend strongly on the concept of a state function. Equations [2] and [3] tell us how to measure the state functions ΔE and ΔH.We determine the heat given off by the system at constant volume or constant pressure. But once ΔE and ΔH are measured, they can be combined, disregarding the different conditions of measurement. This works because ΔE and ΔH are state functions. We use state functions in many common situations. For example, the difference in your body weight from week to week depends on how much food you eat and how much you exercise. Food and exercise are converted into common units, nutritional Calories.

Notice that both ΔH and ΔE in equation [5] are molar quantities, expressed in units of energy per mole (J/mol). "Per mole of what?" you might ask. In the first three chemical equations above, the answer is easier: per mole of $N_2(g)$, $Mg_3N_2(s)$, or $C_3H_7OH(l)$, respectively. In the last two equations, per mole means: for 5 moles of $O_2(g)$, 1 mole of $P_4O_{10}(s)$; for 4 moles of $NH_3(g)$, 6 moles of $H_2O(l)$, and so on. Thus, if the energy of a reaction is expressed in kJ/mol, as usually is the case, the stoichiometric coefficients in the balanced equation indicate the number of moles of reactants consumed and products produced when that much energy is evolved. A concise way of stating these ideas is the enthalpy change *per mole of reaction*. Note that ΔH and ΔE usually are given in kJ/mol whereas $R = 8.314$ joules mol^{-1} K^{-1}.Thus $\Delta n_g RT$ must be converted to kJ/mol before it is combined with ΔH~ and ΔE~.

EXAMPLE 20-1 0.1640 g $NH_3(g)$ reacts at 293 K via $4 NH_3(g) + 3 O_2(g) \longrightarrow 2 N_2(g) + 6 H_2O(l)$ and $q_V = -3.651$ kJ. Compute the value of ΔH for this reaction.

$$\Delta E = \frac{-3.651 \text{ kJ}}{0.1640 \text{ g NH}_3} \times \frac{17.02 \text{ g NH}_3}{\text{mol NH}_3} \times \frac{4 \text{ mol NH}_3}{\text{mol rxn}} = -1517 \text{ kJ/mol}$$

$$\Delta H = \Delta E + \Delta n_g RT$$

$$= -1517 \frac{\text{kJ}}{\text{mol}} - \frac{5 \text{ mol gas}}{\text{mol rxn}} \times 8.314 \frac{\text{J}}{\text{mol K}} \times 293 \text{ K} \times \frac{\text{kJ}}{1000 \text{ J}}$$

$$= -1517 \text{ kJ/mol} - 12.2 \text{ kJ/mol} = -1529 \text{ kJ/mol (of reaction)}$$

4. Explain the meaning of "spontaneous change" as it applies to chemical reactions.

A spontaneous reaction is one that eventually will occur when left to itself. A nonspontaneous reaction will not eventually occur. The distinction between spontaneity and speed is important. Simply because a reaction is spontaneous does not mean that it is rapid. Silver tarnishes spontaneously and diamond spontaneously changes to graphite, but neither process is particularly rapid.

5. Explain why entropy is important and how it is related to the disorder of the system.

The evolution of heat was first thought to be a sign of a spontaneous change. But not all spontaneous reactions are exothermic. Vaporizing water is an example of a spontaneous endothermic process. (Water spontaneously vaporizes from your skin to cool you on a warm day.)

To arrive at a criterion for spontaneous change, let us consider everyday changes: diving into a pool, spilling milk, and an automobile collision. The reverse of each of these spontaneous changes is nonspontaneous. In the reverse of the automobile collision, the scattered parts fly off the roadway and reassemble into two undamaged cars that back away from each other. Nonspontaneous changes "look funny" because in them energy is not dissipated but is assembled. When an automobile part strikes the ground, its energy of motion is scattered: some causing the ground to shake, some producing sound, and some scattering soil and pebbles. Reassembling all of this energy is extremely unlikely.

Thus, a nonspontaneous change does not occur because in the reverse change—the spontaneous one—energy has been scattered. Another way to say this is that entropy increases in a spontaneous process. Entropy is related to the amount of disorder present. The more disorder, the higher is the entropy.

* **6. Predict whether entropy increases or decreases for certain processes.**

Determining entropy change is easier if we think of entropy as a measure of disorder. When a solid melts, its orderly crystalline structure is broken apart, resulting in a liquid that is more disordered. Likewise, a gas is more disordered than the liquid because the gas molecules are not confined to the small volume of the liquid.

Entropy also increases when two substances are mixed together since the molecules or atoms now are mixed up. Placing a solid solute in a liquid solution is accompanied by a large entropy increase because the regular crystal structure of the solid is broken apart and the ions or molecules of the solid are scattered throughout the solution. This entropy effect is large enough to cause ionic solids to dissolve even when a great deal of energy must be supplied (the solution process is very endothermic). On the other hand, entropy *decreases* when a gas dissolves in a liquid solution. Gases are very disordered and confining them to the small volume of a liquid increases their order. Whenever the balanced chemical equation shows more moles of gaseous products than of gaseous reactants, there will be an entropy increase. Thus, when $\Delta n_g > 0$, $\Delta S > 0$.

6a. Explain why entropy alone is not used to predict a spontaneous change and why free energy is needed.

It seems very easy to determine whether the entropy of the system increases ($\Delta S > 0$) or decreases ($\Delta S < 0$) for a given process. However, it is the *total* entropy change, that of the system *plus* that of the surroundings, that determines whether or not a process is spontaneous. Only if this total entropy change (the entropy change of the universe) is positive is the process spontaneous. For a spontaneous change,

$$\Delta S_{univ} = \Delta S_{syst} + \Delta S_{surr} > 0 \qquad [6]$$

ΔS_{surr} often is hard to compute. In addition, other measured properties (ΔH, T, P, volume, density, and so on) are properties of the *system*. We wish the criterion of spontaneous change to be a property of the system, and only the system, also. The free energy change is defined by

$$\Delta G = \Delta H - T\Delta S \qquad [7]$$

Free energy is *negative* for a spontaneous change: $\Delta G < 0$. In fact, we can show that $\Delta G = -T\Delta S_{univ}$, as is done in the material preceding *equation (20.8) in the text.*

7. Write the equations that define free energy and free energy change and know that free energy decreases ($\Delta G < 0$) for a spontaneous change.

The defining equations are [7] and [8]

$$G = H - TS \qquad [8]$$

Equation [7] is produced from equation [8] only when the temperature is constant or when the initial and final temperatures of a process are the same. Recall that ΔG decreases whenever ΔS_{univ} increases. Another way to regard equation [7] is as a balance between two trends. First, the system tries to become disordered, making ΔS positive and thus $-T\Delta S$ negative. The second trend is that the system tries to reach the lowest possible energy, making ΔH negative.

* **7a. Qualitatively predict whether reactions are spontaneous or nonspontaneous.**

There are four possibilities in applying equation [7].

1. An exothermic reaction ($\Delta H < 0$) which decreases the order of the system ($\Delta S > 0$ or $-T\Delta S < 0$) must be spontaneous.
2. An exothermic reaction ($\Delta H < 0$) which increases the order of the system ($\Delta S < 0$ or $-T\Delta S > 0$) is spontaneous at low temperatures where the $-T\Delta S$ term is small and ΔH determines the sign of ΔG. As the temperature rises, $-T\Delta S$ gets larger and the reaction becomes nonspontaneous. An example of such a process is the condensation of a gas to form a liquid.
3. An endothermic reaction ($\Delta H > 0$) which decreases the order of the system ($\Delta S > 0$ or $-T\Delta S < 0$) is nonspontaneous at low temperatures where $-T\Delta S$ has a small effect. At high temperatures, the $T\Delta S$ term increases and eventually the reaction becomes spontaneous. The dissolving of most ionic salts and the dissociation of any substance into its atoms both fall into this category.
4. An endothermic reaction ($\Delta H > 0$) that increases the order of the system ($\Delta S < 0$ or $-T\Delta S > 0$) is nonspontaneous at all temperatures.

* **7b. Determine $\Delta G°$ from tabulated data, both tables of $\Delta G_f°$ and those of $\Delta H_f°$ and S.**

$\Delta G°$ is the standard molar free energy change of a reaction. If $\Delta G°$ is negative, the reaction is spontaneous and the products in their standard states will be spontaneously produced from the reactants in their standard states. Consider reaction [9].

$$2\,NO(g) + O_2(g) \longrightarrow 2\,NO_2(g) \quad \Delta G_f^\circ = -69.71 \text{ kJ/mol} \qquad [9]$$

69.71 kJ of free energy is produced when 2 mol $NO_2(g)$ at a total pressure of 1.00 atm is produced from 2 mol NO(g) and 1 mol $O_2(g)$, with each reactant having a partial pressure of 1.00 atm. ΔG_f° is the standard molar free energy of formation. ΔG_f° is thus ΔG of a reaction in which one mole of product is formed and in which the reactants are elements in their most stable forms. Products and reactants are in their standard states, as the superscript ° indicates. For example, equation [10] is the formation reaction of $NO_2(g)$.

$$\tfrac{1}{2} N_2(g) + O_2(g) \longrightarrow NO_2(g) \qquad \Delta G_f^\circ = 33.85 \text{ kJ/mol} \qquad [10]$$

ΔG° is determined from values of ΔG_f° in the same way that ΔH° is determined from values of ΔH_f° (see objective 7-10).

$$\Delta G^\circ_{rxn} = \sum_{products} \nu_{prod}\, \Delta G^\circ_{f,prod} - \sum_{reactants} \nu_{reac}\, \Delta G^\circ_{f,reac} \qquad [11]$$

where the v's are the stoichiometric coefficients in the balanced chemical equation. For example, for the reaction

$$4\, NH_3(g) + 5\, O_2(g) \longrightarrow 4\, NO(g) + 6\, H_2O(l) \qquad [12]$$

ΔG° is computed as follows.

$$\Delta G^\circ_{rxn} = 4\, \Delta G_f^\circ[NO(g)] + 6\, \Delta G_f^\circ[H_2O(l)] - 4\, \Delta G_f^\circ[NH_3(g)] - 5\, \Delta G_f^\circ[O_2(g)]$$

$$= 4(86.69) + 6(-237.19) - 4\,(-16.64) - 5\,(0.00) \text{ kJ/mol} = -1009.82 \text{ kJ/mol}$$

Tabulated values of ΔG_f° are valid for only one temperature, often 298 K. One also can use the Gibbs-Helmholtz equation [14] to determine ΔG for a reaction.

$$\Delta G^\circ = \Delta H^\circ - T\Delta S^\circ \qquad [14]$$

ΔH° is computed from ΔH_f° data as described in objective 7-10. ΔS° is computed from S° data in a similar manner.

$$\Delta S^\circ_{rxn} = \sum_{products} \nu_{prod}\, S^\circ_{prod} - \sum_{reactants} \nu_{reac}\, S_{reac} \qquad [15]$$

For reaction [12], ΔS) is computed as follows.

$$\Delta S^\circ_{rxn} = 4\, S^\circ[NO(g)] + 6\, S^\circ[H_2O(l)] - 4\, S^\circ[NH_3(g)] - 5\, S^\circ[O_2(g)]$$

$$= 4(210.62) + 6(69.96) - 4\,(192.51) - 5\,(205.03) \text{ kJ/mol} = -532.97 \text{ J mol}^{-1}\text{ K}^{-1}$$

We computed ΔH° for this reaction in Chapter 7 and obtained $\Delta H^\circ = -1168.86$ kJ/mol. Thus, from the Gibbs-Helmholtz equation at $T = 298.15$ K,

$$\Delta G^\circ = -1168.86 \text{ kJ/mol} - 298.15 \text{ K} \times (-532.97 \text{ J mol}^{-1}\text{ K}^{-1}) \times \text{kJ/1000 J}$$

$$= -1168.86 \text{ kJ/mol} + 158.91 \text{ kJ/mol} = -1009.95 \text{ kJ/mol}$$

* **8. Use $\Delta G^\circ = \Delta H^\circ - T\Delta S^\circ$, the Gibbs-Helmholtz equation, to determine ΔG° at various temperatures.**

In objective 20-7b we learned how to determine ΔG° from values of ΔG_f° and from the Gibbs-Helmholtz equation [14]. The Gibbs-Helmholtz equation can be used with ΔH° and ΔS° values for 298 K only when ΔH° and ΔS° do not change with temperature, which is a good assumption over small temperature ranges (of one hundred degrees or less).

EXAMPLE 20-2 For the reaction $4\, NH_3(g) + 5\, O_2(g) \rightleftharpoons 4\, NO(g) + 6\, H_2O(l)$, $\Delta H^\circ = -1168.86$ kJ/mol and $\Delta S^\circ = -532.97$ J mol^{-1} K^{-1}. Determine the value of ΔG° at 350 K.

$$\Delta G^\circ = -1168.86 \text{ kJ/mol} - (350 \text{ K})(-532.97 \text{ J mol}^{-1}\text{ K}^{-1})(\text{kJ/1000 J})$$

$$= -982.32 \text{ kJ/mol}$$

* **9. Know that $\Delta G_{tr} = 0$ at equilibrium. For phase changes, use $\Delta S_{tr} = \Delta H_{tr}/T_{tr}$ = the molar entropy of transition.**

$\Delta G = 0$ at equilibrium, the point where the reaction changes from being spontaneous in the forward direction to being spontaneous in the reverse direction. We can reach equilibrium by varying temperature (considered here), or by varying concentration. Recall the Gibbs-Helmholtz equation ($\Delta G = \Delta H - T\Delta S$). If ΔH and ΔS have the same sign (either both positive or both negative), the temperature can be varied so that ΔG changes sign. At the temperature where the sign change occurs, $\Delta G = 0$. For a phase change such as freezing or boiling, this temperature is the freezing point or the boiling point. For a general transition, it is the transition temperature, T_{tr}. At the normal transition point (where the external pressure is 1.000 atm),

$$\Delta G°_{tr} = 0 = \Delta H°_{tr} - T\Delta S°_{tr} \quad or \quad \Delta S°_{tr} = \Delta H°_{tr}/T_{tr} \qquad [16]$$

This provides a convenient way of determining $\Delta S°_{tr}$, since $\Delta H°_{tr}$ can be measured readily.

EXAMPLE 20-2 Determine T_{tr} for the change S(monoclinic) $\rightleftharpoons$ S(rhombic) given the data that follow.

S(mono): $\Delta H^\circ_f = 0.297$ kJ/mol $S° = 32.55$ J mol^{-1} K^{-1}

S(rhom): $\Delta H^\circ_f = 0.000$ kJ/mol $S° = 31.88$ J mol^{-1} K^{-1}

For the transition reaction,

$$\Delta H°_{tr} = 0.000 - 0.297 \text{ kJ/mol} = -0.297 \text{ kJ/mol}$$

$$\Delta S°_{tr} = 31.88 - 32.55 \text{ J mol}^{-1} \text{ K}^{-1} = -0.67 \text{ J mol}^{-1} \text{ K}^{-1}$$

T_{tr} is computed from equation [16]

$$-0.67 \text{ J mol}^{-1} \text{ K}^{-1} = \frac{-0.297 \text{ kJ/mol} \times 1000 \text{ J/kJ}}{T_{tr}} \quad or \quad T_{tr} = (0.297 \times 1000/0.76) \text{ K} = 433 \text{ K}$$

*** 10. Write thermodynamic equilibrium constant expressions K_{eq} for reactions and relate these to K_p and K_c.**

*** 11. Compute values of K_{eq} from tabulated data and $\Delta G° = -RT \ln K_{eq}$**

$$\Delta G = \Delta G° + RT \ln Q \qquad [17]$$

relates ΔG and $\Delta G°$. Q is the reaction quotient but it is somewhat more restricted than shown in objectives 15-3 and 15-6. If the reactants or products are gases, we *must* use partial pressures, *not* concentrations, because Q expresses a change from standard conditions.

Standard conditions:

Gases:	Ideal gas at 1.000 atm pressure
Liquids:	Pure liquid under 1.000 atm total pressure
Solids:	Pure solid under 1.000 atm total pressure
Solutes:	Ideal solution at 1.000 M concentration.

$$2\,NO(g) + O_2(g) \rightleftharpoons 2\,NO_2(g) \qquad [9]$$

$$Q = \frac{P(NO_2)^2}{P(NO)^2 P(O_2)} \qquad [18]$$

The free energy of pure liquids and solids does not change very much in the normal range of pressures that can be produced in the laboratory (0 to 5 atm). Pure liquids and solids are not included in the reaction quotient, Q. The free energy of a gas does depend on pressure and thus we use the partial pressure in the reaction quotient. The free energy of solutes depends on their concentrations, which are used in the reaction quotient. Equation [17] is quite important at equilibrium.

$$\Delta G = 0 = \Delta G° + RT \ln Q_{eq} \quad \text{or} \quad -RT \ln Q_{eq} = \Delta G° = -RT \ln K_{eq} \qquad [19]$$

We use K_{eq}, the thermodynamic equilibrium constant, to replace Q_{eq}. K_{eq} has the same restrictions as does Q_{eq}. For example, the partial pressures of gases must be used. For equilibrium [20], Q_{eq} is given in [21].

$$CO_2(aq) \rightleftharpoons CO_2(g) \qquad [20]$$

$$K_{eq} = \frac{P(CO_2)}{[CO_2]} \qquad [21]$$

EXAMPLE 20-3 At 298 K, standard molar free energies of formation are $CO_2(g) = -394.38$ kJ/mol and $CO_2(aq) = -386.23$ kJ/mol. What is $[CO_2]$ if the pressure of $CO_2(g)$ above the solution is 2.000 atm?

$$\Delta G° = -394.38 - (-386.23) \text{ kJ/mol} = -8.15 \text{ kJ/mol} = -8150 \text{ J/mol}$$

$$= -(8.314 \text{ J mol}^{-1} \text{ K}^{-1})(298 \text{ K}) \ln K$$

$$\ln K = 3.290 \quad or \quad K = 26.8$$

$$26.8 = K_{eq} = \frac{P(CO_2)}{[CO_2]} = \frac{2.000 \text{ atm}}{[CO_2(aq)]} \quad or \quad [CO_2(aq)] = 0.0745 \text{ M}$$

12. Explain how absolute entropies of substances can be determined with the third law of thermodynamics.

The third law states that the entropy of a pure perfect crystalline solid is zero at 0 K. All that is necessary to determine $S°$ at 298 K is to find the entropy change in heating from 0 K to 298 K. This has been done for several hundred substances. With equation [15], we can determine $S°$ for one substance in a reaction if we know $\Delta S°_{rxn}$ and have values of $S°$ for all the other reactants and products. In this way we know that absolute entropies of most substances. We have no such absolute values of enthalpy (and thus not of free energy either). All enthalpies and free energies are expressed relative to the stable elements rather than as absolute values.

* **13. Relate the equilibrium constant to the standard molar enthalpy of reaction, $\Delta H°$, and to kelvin temperature, both graphically and algebraically.**

Note that there are two ways to compute the variation of the equilibrium constant with temperature. The first uses the van't Hoff equation, the subject of this objective. The second method uses the Gibbs-Helmholtz equation ($\Delta G° = \Delta H° - T\Delta S°$) followed by equation [19] ($\Delta G° = -RT \ln K_{eq}$). In both cases $\Delta H°$ and $\Delta S°$ are assumed not to vary with temperature. The van't Hoff equation, [22],

$$\ln\frac{K_1}{K_2} = \frac{-\Delta H°}{R}\left(\frac{1}{T_1} - \frac{1}{T_2}\right) \qquad [22]$$

is similar in form to both the Clausius-Clapeyron equation and the Arrhenius equation. The method used in solving these equations is described in objective 12-5, which you should review if necessary. One way to evaluate $\Delta H°$ is graphically, by plotting ln K against $1/T$. The slope of this graph equals $-\Delta H°/R$.

EXAMPLE 20-4 Consider the decomposition: $PCl_5(g) \rightleftharpoons PCl_3(g) + Cl_2(g)$. At 500 K, equilibrium partial pressures are $PCl_5(g)$ = 1.285 atm, $PCl_3(g)$ = 0.715 atm, and $Cl_2(g)$ = 1.215 atm; $\Delta H°$ = 92.59 kJ/mol. If the temperature is raised to 550 K, what is the partial pressure of each species present once equilibrium is reestablished?

The first step is to use the van't Hoff equation to compute the equilibrium constant, K_2, at 550 K.

$$\ln\frac{K_1}{K_2} = \frac{-\Delta H°}{R}\left(\frac{1}{T_1} - \frac{1}{T_2}\right) = \frac{-92590 \text{ J/mol}}{8.314 \text{ J mol}^{-1} \text{ K}^{-1}}\left(\frac{1}{500} - \frac{1}{550}\right) = -2.0248$$

Thus $\frac{K_1}{K_2} = e^{-2.0248} = 0.1320$ (where e = 2.718)

Now $K_1 = \frac{P_{PCl_3} P_{Cl_2}}{P_{PCl_5}} = \frac{(0.715)(1.215)}{1.285} = 0.676$

And $\frac{K_1}{K_2} = \frac{0.676}{K_2} = 0.1320$ *or* $K_2 = \frac{0.676}{0.1320} = 5.12$

Reaction:	$PCl_5(g)$	$\rightleftharpoons$	$PCl_3(g)$	+	$Cl_2(g)$
Initial:	1.285 atm		0.715 atm		1.215 atm
Change:	$-x$ atm		$+x$ atm		$+x$ atm
Equil:	$(1.285 - x)$ atm		$(0.715 \text{ atm} + x)$ atm		$(1.215 + x)$ atm

$$K_p = \frac{(0.715 + x)(1.215 + x)}{1.285 - x} = 5.12 \qquad x = -3.525 \pm 4.259 \text{ atm} = 0.734 \text{ atm}$$

Partial pressures: $PCl_5(g)$ = 0.551 atm, $PCl_3(g)$ = 1.449 atm, $Cl_2(g)$ = 1.949 atm.

14. Describe how a heat engine functions and the factors that limit its efficiency.

A heat engine takes a quantity of heat, q_h, from a high temperature (T_h) reservoir. Some of this heat is converted into work (w) and the rest of the heat, q_l, is emitted at a low temperature, T_l. The fraction of the heat absorbed that is turned into work is called the efficiency of the engine.

$$\text{efficiency} = \frac{w}{q_h} = \frac{T_h - T_l}{T_h} = \frac{q_h - q_l}{q_h} \qquad [23]$$

Air conditioners and heat pumps operate on the same principles as heat engines except they do it "backwards." The efficiency of the heat pump also is given by equation [23]. Thus, if the temperature of the outside air is –10°C (14°F) and the temperature of a house is to be 20°C (68°F), the efficiency of the heat pump is

$$\text{efficiency} = \frac{T_h - T_l}{T_h} = \frac{293\text{ K} - 263\text{ K}}{293\text{ K}} = 0.102$$

Thus, the work needed to "pump" 400 kJ of heat into the house is given by

$$\text{efficiency} = \frac{w}{q_h} = \frac{w}{400\text{ kJ}} = 0.0102 \quad or \quad w = 40.8\text{ kJ}$$

A large temperature difference means that a great deal of work must be used to "pump" heat into a house. However, you also should notice that by using a heat pump, more heat can be delivered into the house than if the energy used to run the pump (usually electricity) were converted directly into heat.

15. Describe the relationship of thermal pollution to the use of heat engines.

Notice from equation [23] that every time a heat engine produces work, it also must discard some heat (q_l). A large temperature difference means that a large quantity of work can be produced (and relatively little heat discarded) for a given quantity of heat absorbed. But there are practical limitations. The boiler (high temperature reservoir) cannot be infinitely hot or the materials of which it is made would melt. And it is impractically expensive to operate the condenser (at T_l) at a temperature much lower than that of the surroundings. As a consequence most heat engines (and this includes all power plants except hydroelectric and wind driven ones) operate at efficiencies below 50%, discarding half of the heat they absorb. This discarded heat is known as *thermal pollution.* If excessive heat is discarded into a river, it can be harmful or deadly to fish. Thus, many power plants now dispose of the unwanted heat into the atmosphere through cooling towers.

DRILL PROBLEMS

2. Determine q, w, and ΔE in each of the following cases.

	Heat (of system)	*Work done*		*Heat (of system)*	*Work done*
A.	196 J evolved	123 J by system	B.	189 J evolved	247 J on system
C.	647 J absorbed	92.4 J on surr.	D.	1362 J absorbed	321 J by system
E.	142 J evolved	254 J by surr.	F.	802 J absorbed	897 J by surr.
G.	–1432 J absorbed	937 J by surr.	H.	972 J evolved	432 J by system

3. Compute ΔH and ΔE in kJ/mol for each reaction below. q_p or q_v is given in kJ/g of the first reactant. ($T = 298$ K)
   A. $C_4H_9OH(l) + 6\,O_2(g) \longrightarrow 4\,CO_2(g) + 5\,H_2O(l)$ $q_p = -36.11$ kJ/g
   B. $C_6H_5OH(s) + 7\,O_2(g) \longrightarrow 6\,CO_2(g) + 3\,H_2O(l)$ $q_v = -32.56$ kJ/g
   C. $PCl_5(g) + H_2O(l) \longrightarrow POCl_3(g) + 2\,HCl(g)$ $q_p = -0.4410$ kJ/g
   D. $2\,Cl_2(g) + 7\,O_2(g) \longrightarrow 2\,Cl_2O_7(g)$ $q_v = -3.864$ kJ/g
   E. $H_2(g) + Br_2(g) \longrightarrow 2\,HBr(g)$ $q_p = -35.95$ kJ/g
   F. $2\,B_5H_9(g) + 12\,O_2(g) \longrightarrow 5\,B_2O_3(s) + 9\,H_2O(l)$ $q_v = -71.137$ kJ/g
   G. $SOCl_2(l) + H_2O(g) \longrightarrow SO_2(g) + 2\,HCl(g)$ $q_p = -2.842$ kJ/g
   H. $CCl_4(g) + 2\,H_2O(g) \longrightarrow CO_2(g) + 4\,HCl(g)$ $q_v = 1.1526$ kJ/g

6. Predict whether ΔS is positive or negative for each of the following processes. (Do not *calculate* a value of ΔS.) In which case(s) is a prediction unclear? All substances are gases unless otherwise indicated.
   A. $NH_3(g) + HCl(g) \longrightarrow NH_4Cl(s)$
   B. $Li_2CO_3(s) \longrightarrow Li_2O(s) + CO_2(g)$
   C. $NH_4HCO_3(s) \rightarrow NH_3 + H_2O(l) + CO_2$
   D. $N_2O_4(g) \longrightarrow 2\,NO_2(g)$
   E. $CaCO_3(s) \longrightarrow CaO(s) + CO_2(g)$
   F. $CaCO_3(s) + SO_2 \longrightarrow CaSO_3(s) + CO_2$
   G. $PCl_3(g) + Cl_2(g) \longrightarrow PCl_5(g)$
   H. $2\,SO_2(g) + O_2(g) \longrightarrow 2\,SO_3(g)$
   I. $Br_2(g) + Cl_2(g) \longrightarrow 2\,BrCl(g)$
   J. $2\,H_2S + 3\,O_2 \longrightarrow 2\,H_2O(l) + 2\,SO_2(g)$
   K. $N_2(g) + 3\,H_2(g) \longrightarrow 2\,NH_3(g)$
   L. $CO(g) + 2\,H_2(g) \longrightarrow CH_3OH(g)$
   M. $CH_4(g) + CCl_4(g) \longrightarrow 2\,CH_2Cl_2$
   N. $2\,NOCl(g) \longrightarrow 2\,NO(g) + Cl_2(g)$
   O. $N_2(g) + O_2(g) \longrightarrow 2\,NO(g)$
   P. $CO_2 + 2\,NH_3 \rightarrow CO(NH_2)_2(l) + H_2O(l)$
   Q. $S(s) + 2\,CO(g) \longrightarrow SO_2(g) + 2\,C(gr)$
   R. $SO_2Cl_2(l) \longrightarrow SO_2(g) + Cl_2(g)$
   S. $CO_2(g) \longrightarrow CO_2(aq)$
   T. $2\,O_3(g) \longrightarrow 3\,O_2(g)$
   U. $4\,HCl + O_2 \longrightarrow 2\,H_2O + 2\,Cl_2$
   V. $H_2 + CO_2 \longrightarrow H_2O + CO$

W. $2\ HI(g) \longrightarrow H_2(g) + I_2(g)$ X. $HCHO(g) \longrightarrow H_2(g) + CO(g)$

7a. & 7b. For each of the reactions in the drill problems of objective 20-6 compute $\Delta G_f°$, $\Delta H_f°$, and $\Delta S°$. Use the data of Table 20-1 (below). State whether the reaction is spontaneous (S) or nonspontaneous (N), and to which one of the categories (1, 2, 3, or 4) described in objective 20-7a it belongs.

TABLE 20-1 Thermodynamic Properties of Selected Substances $\Delta G_f°$ and $\Delta H_f°$ in kJ/mol; $\Delta S°$ in J mol^{-1} K^{-1}

Substance	$\Delta G_f°$	$\Delta H_f°$	$\Delta S°$.	Substance	$\Delta G_f°$	$\Delta H_f°$	$\Delta S°$
$NH_3(g)$	–46.19	–16.64	192.5	$I_2(g)$	62.26	19.37	260.6
$NH_4Cl(s)$	–315.38	–203.89	94.6	$HI(g)$	25.9	1.3	206.3
$NH_4HCO_3(s)$	–852.3	–660.1	120.9	$Li_2CO_3(s)$	–1215.6	–1132.4	90.37
$BaCO_3(s)$	–1218.8	–1138.9	112.1	$LiCl(s)$	–408.8	–243	213.4
$BaCO_3(aq)$	–1214.62	–1088.7	–41.8	$LiOH(s)$	–487.23	–443.9	50
$Br_2(g)$	30.7	3.14	245.3	$Li_2O(s)$	–595.8	–560.49	37.9
$BrCl(g)$	14.7	–0.88	239.9	$LiNO_3(s)$	–482.33	–390	105
$CaCO_3(s)$	–1206.87	–1128.76	92.9	$N_2(g)$	0.00	0.00	191.5
$CaCl_2(s)$	–795.0	–750.2	114	$NO(g)$	90.37	86.69	210.6
$Ca(OH)_2(s)$	–986.59	–896.76	76.1	$NO_2(g)$	33.85	51.84	240.5
$CaO(s)$	–635.5	–478.6	40	$N_2O_4(g)$	9.67	98.28	304.3
$CaSO_4(s)$	–1432.7	–1320.3	106.7	$NOCl(g)$	52.59	66.36	264
C(graphite)	0.00	0.00	5.69	$HNO_3(l)$	–173.2	–79.91	155.6
$CO(g)$	–110.5	–137.3	197.9	$O_2(g)$	0.00	0.00	205.03
$CO_2(g)$	–393.5	–394.4	213.6	$O_3(g)$	142	163.4	238
$CO_2(aq)$	–412.9	–386.2	121.3	$PCl_3(g)$	–306.4	–286.3	311.6
$CH_4(g)$	–74.848	–50.794	186.2	$PCl_5(g)$	–398.9	–324.6	353
$CH_3OH(g)$	–201.2	–161.9	238	$AgCl(s)$	–127.03	109.70	96.11
$CH_2Cl_2(g)$	–82.0	–58.6	234.2	$AgNO_3(aq)$	–100.67	–33.4	220.4
$CCl_4(g)$	–106.7	–64.0	309.4	$NaCl(aq)$	–407.11	–393.0	115.5
$HCHO(g)$	–115.9	–110.0	218.7	$NaNO_3(aq)$	–446.22	–372.4	206.7
$CO(NH_2)_2(s)$	–333.19	–197.15	104.6	S(s, rhombic)	0.00	0.00	31.9
$Cl_2(g)$	0.00	0.00	223.0	$H_2S(g)$	–20.15	–33.02	205.6
$HCl(g)$	–92.30	–95.27	186.7	$SO_2(g)$	–269.6	–300.4	248.5
$H_2(g)$	0.00	0.00	130.6	$SO_3(g)$	–395.2	–370.4	256.2
$H_2O(g)$	–241.8	–228.6	118.7	$SO_2Cl_2(l)$	–389	–314	207
$H_2O(l)$	–285.9	–237.2	69.96				

8. Determine the value of ΔG° and K for each of the reactions in the drill problems of objective 20-6 at 200 K and 400 K. Assume that no phase changes occur.

9. Fill in the blanks in the following table. All values of $\Delta H_f°$ and $\Delta H_{tr}°$ are in kJ/mol, all values of $\Delta S°$ and $\Delta S_{tr}°$ are in J mol^{-1} K^{-1}, and all temperatures are in kelvins. Each line represents one transition, from substance *A* to substance *B*.

	Substance A	$\Delta H_f°$	$S°$	*Substance B*	$\Delta H_f°$	$S°$	$\Delta H_{tr}°$	$\Delta S_{tr}°$	T_{tr}
A.	$SbCl_3(g)$	–315	338	$SbCl_3(s)$	–382.2	186	____	________	
B.	$AsF_3(g)$	–314	327	$AsF_3(l)$	–336	____	____	____	228
C.	BeO(c)	–610.9	14.1	BeO(g)	49.4	197.4	____	____	____
D.	$BCl_3(l)$	–418.4	209	$BCl_3(g)$	–395	289.9	____	____	____
E.	$Br_2(g)$	____	245.3	$Br_2(l)$	0.00	152	____	____	266
F.	$Ca_3(PO_4)_2(\alpha)$	–4126.25	241	$Ca_3(PO_4)_2(\beta)$	–4138	236.9	____	____	____
G.	$I_2(g)$	62.25	260.6	$I_2(s)$	0.00	116.7	____	____	____
H.	Mn(a)	0.00	31.8	Mn(g)	1.55	32.3	____	____	____
I.	HgO(red)	–90.71	____	HgO(yellow)	–90.21	73.2	____	____	400

11. (1) For each of the reactions in the drill problems of objective 20-6, write the expression for the thermodynamic equilibrium constant, and evaluate the constant at 298 K.

(2) Determine the value of $\Delta G°$ that corresponds to each of the following sets of equilibrium constant (K) and temperature (T).

A. $K = 1, T = 298$ K B. $K = 1, T = 500.$ K C. $K = 10^{-3}, T = 298$ K

D. $K = 10^{-3}, T = 200.$ K E. $K = 10^{-3}, T = 500.$ K F. $K = 10^3, T = 200.$ K

G. $K = 10^3, T = 298$ K H. $K = 10^3, T = 500.$ K

13. The value of $\Delta H°$ and/or the equilibrium constant at two different temperatures is given or requested for each of the following equations. Fill in each blank.

A. $CO(g) + 2 H_2(g) \rightleftharpoons CH_3OH(g)$ $\Delta H° =$ ____ kJ/mol
$K_p = 91.4$ at 350 K $K_p = 2.05 \times 10^{-4}$ at 298 K

B. $HCHO(g) \rightleftharpoons H_2(g) + CO(g)$ $\Delta H° = 5.36$ kJ/mol
$K_c =$ ____ at 450 K $K_c = 0.267$ at 773 K

C. $4 HCl(g) + O_2(g) \rightleftharpoons 2 H_2O(g) + 2 Cl_2(g)$ $\Delta H° = -114.47$ kJ/mol
$K_c =$ ____ at 298 K $K_c = 1.49 \times 10^3$ at 320 K

D. $2 P(white) + 3 H_2(g) \rightleftharpoons 2 PH_3(g)$ $\Delta H° =$ ____ kJ/mol
$K_p = 3.10 \times 10^6$ at 270 K $K_p = 7.62 \times 10^5$ at 330 K

E. $SO_2Cl_2(l) \rightleftharpoons SO_2(g) + Cl_2(g)$ $\Delta H° = 29.37$ kJ/mol
$K_p = 6.30$ at 350 K $K_p =$ ____ at 400 K $K_p = 1.03$ at ____ K

F. $PCl_5(g) \rightleftharpoons PCl_3(g) + Cl_2(g)$ $\Delta H° = 92.59$ kJ/mol
$K_p = 0.675$ at 500 K $K_p =$ ____ at 60 K $K_p = 1.000$ at ____ K

QUIZZES (20 minutes each) Choose the best answer for each question. Accept a numerical answer if it is within 5% of the correct value.

Quiz A

1. $2 NH_3(g) + \frac{3}{2} O_2(g) \longrightarrow N_2(g) + 3 H_2O(g)$
$S°$, J mol^{-1} K^{-1} 192.5 205.0 191.5 188.7
What is the value of $\Delta S°$ in J mol^{-1} K^{-1} for the above reaction? (a) +65.1; (b) –65.1; (c) –17.3; (d) +17.3; (e) none of these

2. $Cl_2O(g) + \frac{3}{2} O_2(g) \longrightarrow 2 ClO(g)$
$\Delta G_f°$, kJ/mol 97.9 0.00 123.4
What is the value of $\Delta G°$ in kJ/mol for the above reaction? (a) +25.5; (b) +148.9; (c) +221.3; (d) –25.5; (e) none of these.

3. For the reaction $SO_3(g) \longrightarrow SO_2(g) + \frac{1}{2}O_2(g)$, $\Delta H° = 93.8$ kJ/mol and $\Delta S° = 326$ J mol^{-1} K^{-1}. $\Delta G°$ for this reaction at 27°C in kJ/mol equals (a) 89.5; (b) 196; (c) 0.42; (d) 107.1; (e) none of these.

4. For $Cl_2(g) + F_2(g) \longrightarrow 2 ClF(g)$, $\Delta H° = -11.3$ kJ/mol and $\Delta S° = 9.25$ J mol^{-1} K^{-1}. As temperature decreases for this reaction, (a) it eventually becomes nonspontaneous; (b) $\Delta G°$ decreases; (c) it reaches equilibrium; (d) it shifts left; (e) none of these.

5. The change in free energy of a reaction (a) = $\Delta H + T\Delta S$; (b) = work; (c) predicts speed; (d) depends on the standard state chosen; (e) none of these.

6. An endothermic reaction is one for which (a) $\Delta H < 0$; (b) $\Delta H > 0$; (c) $\Delta G > 0$; (d) $\Delta G < 0$; (e) none of these.

7. Which of the following phrases is *not* used in the definition of the standard enthalpy of formation? (a) most stable form; (b) heat evolved at constant pressure; (c) elements in standard states; (d) work at constant volume; (e) none of these.

8. Which of the following best expresses the increased degree of randomness associated with melting and sublimation, respectively? (a) $\Delta S_{fus} = 4$ J mol^{-1} K^{-1}, $\Delta S_{sub} = 120$ J mol^{-1} K^{-1}; (b) $\Delta S_{fus} = 120$ J mol^{-1} K^{-1}, $\Delta S_{sub} = 4$ J mol^{-1} K^{-1}; (c) $\Delta S_{fus} = \Delta S_{sub} = 0$; (d) $\Delta S_{fus} = 4$ J mol^{-1} K^{-1}, $\Delta S_{sub} = -120$ J mol^{-1} K^{-1}; (e) $\Delta S_{fus} - \Delta S_{sub} = 0$.

Quiz B

1. $C(graph) + 2 H_2(g) \longrightarrow CH_4(g)$
$S°$, J mol^{-1} K^{-1} 5.69 130.58 186.19
What is the value of $\Delta S°$ in J mol^{-1} K^{-1} for the above reaction? (a) +80.67; (b) +49.92; (c) –49.92; (d) +115.23; (e) none of these

2. $B_2H_6(g) + 3 Cl_2(g) \longrightarrow 2 BCl_3(g) + 6 HCl(g)$
$\Delta G_f°$, kJ/mol 86.6 0.00 –388.7 22.8
What is the value of $\Delta G°$ in kJ/mol for the above reaction? (a) –291.6; (b) –279.3; (c) –727.2; (d) –452.5; (e) none of these.

3. For the reaction $Cl_2O(g) + \frac{3}{2} O_2(g) \longrightarrow 2\ ClO(g)$, $\Delta H° = +126.4$ kJ/mol and $\Delta S° = -74.9$ J mol^{-1} K^{-1}. $\Delta G°$ for this reaction at 377°C in kJ/mol equals (a) +98.3; (b) +77.8; (c) +175.1; (d) +51.5; (e) none of these.

4. For $ICl_3(s) \longrightarrow ICl(g) + Cl_2(g)$, $\Delta H° = +105.9$ kJ/mol and $\Delta S° = +298.4$ J mol^{-1} K^{-1}. As temperature decreases for this reaction, (a) it eventually becomes spontaneous; (b) $\Delta G°$ decreases; (c) it shifts right; (d) equilibrium is eventually reached; (e) none of these.

5. In a spontaneous process the entropy of the universe increase, is a statement of the (a) zeroth law; (b) third law; (c) first law; (d) second law; (e) none of these.

6. The free energy change for a particular process is +136 kJ. Thus the process is (a) exothermic; (b) endothermic; (c) nonspontaneous; (d) spontaneous; (e) none of these.

7. Which of the following is *not* a state property? (a) heat; (b) enthalpy; (c) entropy; (d) free energy; (e) all are state properties.

8. The enthalpy of a reaction could be determined, at least approximately, from complete tables of each of the following kinds of data except (a) bond enthalpies; (b) heats of combustion; (c) ionization energies; (d) heats of formations; (e) all could be used.

Quiz C

1.
$$CO_2(g) + 2\ HCl(g) \longrightarrow COCl_2(g) + H_2O(g)$$
$\Delta G_f°$, kJ/mol   −394.4   −95.3   −210.5   −228.6
What is the value of $\Delta G°$ in kJ/mol for the above reaction? (a) −145.8; (b) −439.1; (c) +50.6; (d) +146.0; (e) none of these.

2.
$$C_3H_4(g) + 2\ H_2(g) \longrightarrow C_3H_8(g)$$
$S°$, J mol^{-1} K^{-1} 266.9   130.6   269.9
What is the value of $\Delta S°$ in J mol^{-1} K^{-1} for the above reaction? (a) −127.6; (b) −258.2; (c) +3.0; (d) +127.0; (e) none of these

3. For the reaction $B_2H_6(g) + 3\ Cl_2(g) \longrightarrow 2\ BCl_3(g) + 6\ HCl(g)$, $\Delta H° = -1396.9$ kJ/mol and $\Delta S° = 800.0$ J mol^{-1} K^{-1}. $\Delta G°$ for this reaction at 227°C in kJ/mol equals (a) −996.8; (b) −2196.9; (c) −400.0; (d) −1578.5; (e) none of these.

4. For $2\ NO(g) + Cl_2(g) \longrightarrow 2\ NOCl(g)$, $\Delta H° = -37.78$ kJ/mol and $\Delta S° = -117.03$ J mol^{-1} K^{-1}. As temperature decreases for this reaction, (a) $\Delta G°$ increases; (b) it becomes more spontaneous; (c) it shifts right; (d) $\Delta S°$ increases; (e) none of these.

5. If $\Delta G = -10$ kJ for a reaction, then the reaction is said to be (a) spontaneous; (b) reversible; (c) endothermic; (d) exothermic; (e) none of these.

6. Thermodynamics *cannot* predict for a reaction that (a) speed; (b) position of equilibrium; (c) maximum amount of work available; (d) feasibility of the reaction; (e) it can predict them all.

7. $C(graphite) + O_2(g) \longrightarrow CO_2(g)$ $\Delta H° = -393.5$ kJ/mol is not a (a) combustion reaction; (b) bond enthalpy; (c) formation reaction; (d) thermochemical equation; (e) it is all of these.

8. The energy available for useful work is (a) constant; (b) the free energy; (c) the entropy; (d) the internal energy; (e) the enthalpy.

Quiz D

1.
$$2\ KClO_4(s) \longrightarrow 2\ KClO_3(s) + O_2(g)$$
$S°$, J mol^{-1} K^{-1} 151.0   143.0   205.0
What is the value of $\Delta S°$ in J mol^{-1} K^{-1} for the above reaction? (a) +196.9; (b) −196.9; (c) +188.9; (d) −188.9; (e) none of these

2.
$$4\ NH_3(g) + 7\ O_2(g) \longrightarrow 4\ NO_2(g) + 6\ H_2O(l)$$
$\Delta G_f°$, kJ/mol   16.65   0.00   51.84   −237.19
What is the value of $\Delta G°$ in kJ/mol for the above reaction? (a) −1282.40; (b) +168.70 (c) −168.70; (d) −210.97; (e) none of these.

3. For the reaction $NH_3(g) + HCl(g) \longrightarrow NH_4Cl(s)$, $\Delta H° = -176.90$ kJ/mol and $\Delta S° = -284.6$ J mol^{-1} K^{-1}. $\Delta G°$ for this reaction at 400. K in kJ/mol equals (a) +63.05; (b) −368.4; (c) +290.7; (d) −63.05; (e) none of these.

4. For $CaCO_3(s) \longrightarrow CaO(s) + CO_2(g)$, $\Delta H° = +178$ kJ/mol and $\Delta S° = +160.5$ J mol^{-1} K^{-1}. As temperature decreases for this reaction, (a) it eventually becomes spontaneous; (b) $\Delta G°$ increases; (c) $\Delta S°$ increases; (d) it becomes nonspontaneous; (e) nothing happens.

5. Which material has the largest entropy? (a) salt; (b) powdered sugar; (c) pure water; (d) salt water;

(e) cannot be determined.

6. A reaction is spontaneous if (a) $\Delta H < 0$; (b) $\Delta S < 0$; (c) $\Delta G < 0$; (d) $\Delta S > 0$; (e) none of these.
7. The state function most closely associated with the second law is (a) enthalpy; (b) entropy; (c) internal energy; (d) temperature; (e) none of these.
8. What must be true of an isothermal system? (a) constant temperature; (b) no heat exchanged; (c) no work done; (d) closed to mass transfer; (e) none of these.

SAMPLE TEST (20 minutes)

1. Write the thermodynamic equilibrium constant for each of the following reactions.
   a. $Mg_3N_2(s) + 6\,H_2O(l) \longrightarrow 3\,Mg(OH)_2(aq) + 2\,NH_3(aq)$
   b. $SO_2(g) + Cl_2(g) \longrightarrow SO_2Cl_2(g)$
   c. $CO(g) + 2\,H_2(g) \longrightarrow CH_3OH(l)$
   d. $Ag_2O(s) + H_2SO_4(aq) \longrightarrow H_2O(l) + Ag_2SO_4(aq)$
2. For reaction $PCl_5(g) \rightleftharpoons PCl_3(g) + Cl_2(g)$ at 298 K, $K_{eq} = 1.87 \times 10^{-7}$, $\Delta S° = 181.92$ J mol^{-1} K^{-1}.
   a. Is this reaction spontaneous at 298 K? Explain why or why not.
   b. Compute the value of ΔH° for this reaction.
   c. Without performing any calculations, predict whether K_{eq} will increase or decrease when the temperature changes from 298 K to 200. K. Explain your prediction.
   d. Compute ΔG° for this reaction at 46°C. Carefully state any assumptions that you make during this calculation.
3. Consider the following four compounds.

Compound	$CaO(s)$	$CO(g)$	$CO_2(g)$	$COCl_2(g)$
$S°$, J mol^{-1} K^{-1}	40	198	214	289

   Explain the increase in standard molar entropies in terms of disorder.

4. What is $\Delta H°$ for a reaction that has $K_p = 1.456$ at 273 K and $K_p = 14.2$ at 298 K?

21 Electrochemistry

CHAPTER OBJECTIVES

* **1. Describe how a voltaic cell operates, with the concepts of electrodes, salt bridges, half-cell reactions, net cell reaction, and cell diagram.**

An oxidation-reduction reaction can be run in an electrochemical cell. If the reaction is spontaneous, a chemical change produces an electrical current. This is a *galvanic* or a *voltaic* cell. If the cell reaction is not spontaneous, electricity produces a chemical change. This is an *electrolytic* cell (see objective 21-8). To understand the working of a galvanic cell, consider the following spontaneous reaction.

$$CuSO_4(aq) + Fe(s) \longrightarrow Cu(s) + FeSO_4(aq)$$

This reaction produces two half-equations.

$$Cu^{2+}(aq) + 2\ e^- \longrightarrow Cu(s) \qquad [1]$$

$$Fe(s) \longrightarrow Fe^{2+}(aq) + 2\ e^- \qquad [2]$$

If we physically separate these half-reactions, the electrons from Fe(s) have to travel to combine with the $Cu^{2+}(aq)$ ions. This flow of electrons is an electrical current.

The first step in physically separating the half-reactions is to create two half-cells: (1) an iron bar immersed in a 1.00 M solution of $FeSO_4(aq)$ and (2) a copper bar immersed in a 1.00 M $CuSO_4(aq)$ solution. If we connect the two electrodes (the iron and the copper bars) with a wire, as in Figure 20-1(a), no current flows. This is because when electrons come from the iron bar, Fe^{2+} ions go into solution. Thus, the $FeSO_4$ solution becomes positively charged. This positive charge attracts electrons and prevents current from flowing.

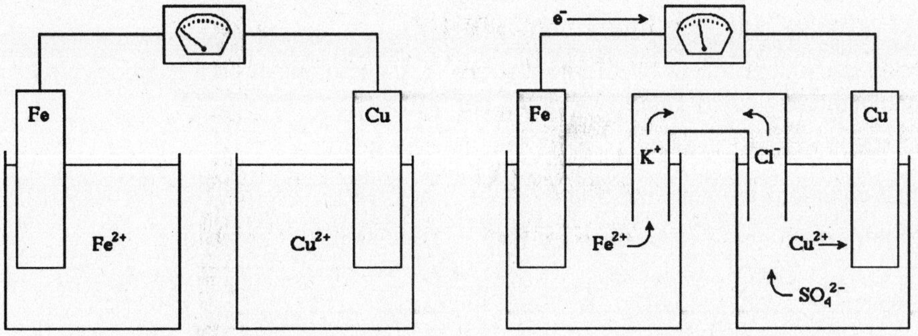

(a) Two Half-Cells (b) Two Half-Cells with a Salt Bridge

FIGURE 21-1 Construction of a Galvanic or Voltaic Cell

One way to make sure the solutions are uncharged is to allow them to mix together. But then the Cu^{2+} ions will move into the iron half-cell and react directly with the iron bar. If we place another solution between the $CuSO_4(aq)$ and $FeSO_4(aq)$ solutions (Figure 21-1b), the ions can move through that solution to keep both solutions uncharged. Yet the Cu^{2+} ions will not reach the iron nail. The solution in the middle is called a salt bridge. It is an aqueous solution of a strong electrolyte.

The *electrode* is the solid metal bar that is placed in the solution, although sometimes the entire half-cell is called the electrode. Oxidation occurs at the anode. Reduction occurs at the cathode. (To help you remember, note that both anode and oxidation begin with a vowel and both cathode and reduction begin with a consonant. Note also that anions move toward the anode and cations move toward the cathode.) The electrolyte is either (1) the solution containing ions or (2) the solute of this solution. For instance, either 1.00 M $CuSO_4$(aq), or $CuSO_4$ is the electrolyte.

A cell diagram is a more compact way of representing an electrochemical cell than is a sketch like Figure 21-1. In cell diagrams, the anode always is on the left, the cathode is on the right. The electrolytes are in the center of the diagram and the electrodes are on the edges. A solid bar (|) indicates a boundary between two phases. The cell diagram for Figure 20-1(b) is Fe|$FeSO_4$(1.00 M)||$CuSO_4$(1.00 M)|Cu. The double bar represents the salt bridge. Spectator ions often are not included: Fe|Fe^{2+}(1.00 M)||Cu^{2+}(1.00 M)|Cu. Perhaps the easiest way to think of cell diagrams is to trace the path of the positive charge from the anode to the cathode through the cell. Simply write down every species you encounter.

Sometimes neither the oxidized nor the reduced species in a half-cell can be made into a metal bar. Consider the reaction

$$Zn(s) + Cl_2(g) \longrightarrow ZnCl_2(aq)$$

The half-equations for this reaction are

$$\text{anode:}\quad Zn(s) \longrightarrow Zn^{2+}(aq) + 2\,e^-$$

$$\text{cathode:}\quad Cl_2(g) + 2\,e^- \longrightarrow 2\,Cl^-(aq)$$

The best way to make the cathode is to pass a stream of chlorine gas over a surface of finely divided platinum, called platinum black. The cell diagram is Zn|Zn^{2+}(aq)||Cl^-(aq)|Cl_2(g)|Pt.

2. Describe the standard hydrogen electrode and explain how other standard electrode potentials are related to it.

The standard hydrogen electrode (S.H.E.) is similar to the chlorine electrode described above. A platinum black electrode is immersed in a solution with $[H^+]$ = 1.00 M, and H_2(g) at a pressure of 1.00 atm is bubbled past the electrode. The potential of this electrode (or half-cell) is set equal to 1.000 V. The potential of any other electrode is the voltage of a cell in which the other electrode is paired with the S.H.E. The sign of that potential is positive if the S.H.E is the anode, and negative if the S.H.E is the cathode.

* **3. Use tabulated standard potentials, $E°$, to determine $E°_{cell}$ for an oxidation-reduction reaction and predict whether the reaction is spontaneous.**

An oxidation-reduction reaction produces two half-equations. The potentials of both reactions are given in a table of standard reduction potentials such as *Table 21-1 in the text.*

$$5\,H_2O_2(aq) + 2\,KMnO_4(aq) + 6\,HCl(aq) \longrightarrow 2\,MnCl_2(aq) + 8\,H_2O + 5\,O_2(g) \qquad [3]$$

The two reduction half-equations for equation [3] are

$$MnO_4^-(aq) + 8\,H^+(aq) + 5\,e^- \longrightarrow Mn^{2+}(aq) + 4\,H_2O \qquad E° = +1.51\text{ V}$$

$$O_2(g) + 2\,H^+(aq) + 2\,e^- \longrightarrow H_2O_2(aq) \qquad E° = +0.682\text{ V}$$

The balanced net ionic equation [3] is twice half-equation [4] minus five times half-equation [5]. (Recall that balancing oxidation-reduction equations, by combining half-equations, was discussed in objective 5-8.) The half-cell potentials are added in the same way as the half-equations *except* that the potentials are *not* multiplied by integers.

$$E° = +1.51 - (+0.682) = +0.83\text{ V}$$

When the resulting potential of the cell is positive, as it is for reaction [3], the reaction is spontaneous. A negative cell potential indicates a nonspontaneous reaction.

* **4. Quantitatively and qualitatively predict the effect of varying conditions (concentrations and gas pressures) on values of E_{cell}.**

The Nernst equation [6] describes how cell potential depends on concentration and gas pressure.

$$E_{cell} = E°_{cell} - \frac{RT}{n\mathcal{F}} \ln Q = E°_{cell} - \frac{0.05915}{n} \log Q \quad \text{at } 25°C \qquad [6]$$

Increasing the concentration or gas pressure of the products decreases the cell voltage, making the reaction less spontaneous. Increasing the concentration or the gas pressure of the reactants increases the cell voltage. *n* is the number of electrons transferred in the balanced oxidation-reduction equation.

EXAMPLE 21-1 What is the potential of the cell $Cu|Cu(NO_3)_2(2.00\ M)||AgNO_3(0.100\ M)|Ag$?

$[Ag^+(aq) + e^- \longrightarrow Ag(s)] \times 2 \quad E° = 0.799\ V$

$Cu(s) \longrightarrow Cu^{2+}(aq) + 2\ e^- \quad -E° = -(0.337\ V)$

$2\ Ag^+ + Cu \longrightarrow 2\ Ag + Cu^{2+} \quad E° = +0.462\ V$

$n = 2$ for this reaction. The Nernst equation gives

$$E = +0.462\ V - (0.05915/2) \log \frac{[Cu^{2+}]}{[Ag^+]^2} = 0.462\ V - 0.02958 \log \frac{2.00}{(0.100)^2}$$

$$= 0.462\ V - 0.068\ V = 0.394\ V$$

* **5. Know and be able to use the equations that relate $\Delta G°$, $E°_{cell}$, and K.**

A negative ΔG and a positive E_{cell} indicate a spontaneous reaction. However, ΔG is an extensive property, whereas E is intensive. The product of E and n, the number of moles of electrons transferred, is extensive.

$$\Delta G = -n\mathcal{F}E \qquad [7]$$

$\mathcal{F}$ is the number of coulombs of charge per mole of electrons. $\mathcal{F}$ = 96,500 C/mol = Faraday's constant. Now, if the reactants and products are in their standard states, we can compute the equilibrium constant of a reaction from its value of $E°$.

$$\Delta G° = -n\mathcal{F}E° = -RT \ln K_{eq} \qquad [8]$$

$$n\mathcal{F}E° = RT \ln K_{eq} \qquad [9]$$

EXAMPLE 21-2 What are $\Delta G°$ and K_{eq} for $2\ AgNO_3(aq) + Cu(s) \longrightarrow Cu(NO_3)_2(aq) + 2\ Ag(s)$?

From Example 21-1, $E° = 0.462$ V. (Note also that 1 joule = 1 volt-coulomb.)

$\Delta G° = -n\mathcal{F}E° = -(2\ mol\ e^-)(96{,}500\ C/mol)(0.462\ V) = -89.2 \times 10^3\ J/mol = -RT \ln K_{eq}$

$-89.2 \times 10^3\ J/mol = -(8.314\ J\ mol^{-1}\ K^{-1})(298\ K) \ln K_{eq}$

$\ln K_{eq} = 36.0 \quad or \quad K_{eq} = 4.3 \times 10^{15}$

EXAMPLE 21-3 What is ΔG for $2\ AgNO_3(aq) + Cu(s) \longrightarrow Cu(NO_3)_2(aq) + 2\ Ag(s)$ when $[AgNO_3]$ = 0.100 M and $[Cu(NO_3)_2]$ = 2.00 M?

In Example 21-1 we found that $E = 0.394$ V with these same concentrations.

$\Delta G = -n\mathcal{F}E = -(2\ mol\ e^-)(96{,}500\ C/mol\ e^-)(0.394\ V) = -76.0 \times 10\ J/mol$

6. Describe some common voltaic cells: the flashlight or dry cell, the lead storage battery, the silver-zinc cell, and the fuel cell.

A battery is a device in which several cells are linked together. For the dry cell, the cells of the lead storage battery, and the silver-zinc cell, you should know the answers to the following questions.

1. What are the materials of the anode and of the cathode?
2. What are the cell equation and the anode and cathode half-cell equations?
3. What does the cell look like? Sketch the cell.
4. What is the cell diagram?
5. What are the standard cell potentials of each half-cell and of the full cell?

Fuel cells often are based on combustion reactions that are run so that an electrical current is generated. The standard cell potential is best calculated from ΔG by means of equation [8].

7. Explain the corrosion of metals in electrochemical terms and describe methods of corrosion protection.

The corrosion of a metal is simply its oxidation in the presence of oxygen. Metals often are protected from corrosion by coating: either with another metal like zinc or chromium or with paint. Another means of protection is a sacrificial anode. An active metal, such as magnesium, is connected to the metal to be protected. The active metal oxidizes first and only when it is consumed will the protected metal corrode. Some metals protect themselves with an insoluble and firmly bound compound on their surface. Aluminum forms Al_2O_3 which gives the metal a dull appearance, and copper forms a blue-green mixture of $CuCO_3$ and $Cu(OH)_2$. If the insoluble compound is removed, the fresh surface soon corrodes.

8. Describe an electrolytic cell and how it differs from a voltaic cell.

In electrolytic cells, electricity causes a chemical change. The cell reaction often is a decomposition. (Electrolytic literally means breaking with electricity.) The cell reaction usually is nonspontaneous and electricity overcomes the nonspontaneity. The principal problem is keeping the products from reacting. For example, an important industrial process is the electrolysis of brine (NaCl solutions).

$$2\,NaCl(aq) + 2\,H_2O \longrightarrow 2\,NaOH(aq) + Cl_2(g) + H_2(g) \qquad [10]$$

$H_2(g)$ and $Cl_2(g)$ explode when mixed and exposed to light. Thus, the electrolysis cell is designed to keep these gases separate. Reaction [10] is nonspontaneous.

$$\begin{array}{lll} 2\,Cl^-(aq) & \longrightarrow Cl_2(g) + 2\,e^- & E° = -1.360\text{ V} \\ 2\,H_2O + 2\,e^- & \longrightarrow H_2(g) + 2\,OH^-(aq) & -E° = -(0.828\text{ V}) \\ \hline 2\,H_2O + 2\,Cl^- & \longrightarrow Cl_2(g) + H_2(g) + 2\,OH^- & E° = -2.188\text{ V} \end{array}$$

The applied voltage must be larger than +2.188 V to cause reaction [10] to occur.

* **9. Identify the possible half-reactions that might occur in an electrolysis and choose the pair that will occur based on the highest (least negative) cell potential.**

First write all the half-reactions that might occur, along with their standard potentials by referring to a table of standard electrode potentials like *Table 21-1 in the text.* Then select the one oxidation and the one reduction that sum to the largest (least negative) cell potential.

EXAMPLE 21-4 A solution with $[Cu(NO_3)_2] = 0.500$ M, $[ZnCl_2] = 0.500$ M, and $[FeBr_2] = 0.500$ M is electrolyzed between inert electrodes. What are the initial products at the anode and at the cathode, and what minimum voltage must be used?

There are eight possible half-equations, including the electrolysis of water.

$$\begin{array}{ll} Cu^{2+}(aq) + 2\,e^- \longrightarrow Cu(s) & E° = +0.337\text{ V} \\ Zn^{2+}(aq) + 2\,e^- \longrightarrow Zn(s) & E° = -0.763\text{ V} \\ Fe^{2+}(aq) + 2\,e^- \longrightarrow Fe(s) & E° = -0.440\text{ V} \\ 2\,H^+(aq) + 2\,e^- \longrightarrow H_2(g) & E° = \ \ 0.000\text{ V} \\ NO_3^-(aq) + 4\,H^+ + 3\,e^- \longrightarrow NO(g) + 2\,H_2O & E° = +0.96\text{ V} \\ 2\,Br^-(aq) \longrightarrow Br_2(l) + 2\,e^- & E° = -1.065\text{ V} \\ 2\,Cl^-(aq) \longrightarrow Cl_2(g) + 2\,e^- & E° = -1.360\text{ V} \\ 2\,H_2O \longrightarrow O_2(g) + 4\,H^+(aq) + 4\,e^- & E° = -1.229\text{ V} \end{array}$$

We are tempted to pick the NO_3^- reduction until we realize that anions move away from the cathode where reduction occurs. The cation with the highest half-cell potential is $Cu^{2+}(aq)$. $Br^-(aq)$ has the highest oxidation potential.

$$\begin{array}{ll} Cu^{2+}(aq) + 2\,e^- \longrightarrow Cu(s) & E° = +0.337\text{ V} \\ 2\,Br^-(aq) \longrightarrow Br_2(l) + 2\,e^- & -E° = -(+1.065\text{ V}) \\ \hline Cu^{2+} + 2\,Br^- \longrightarrow Cu + Br_2 & E° = -0.728\text{ V} \end{array}$$

The Nernst equation now is used to compute E.

$$E = E° - \frac{0.05915}{n}\log\frac{1}{[Cu^{2+}][Br^-]^2} = -0.728\text{ V} - 0.02958\log\frac{1}{0.500(1.00)^2}$$

$$= -0.737\text{ V}$$

*** 10. Use Faraday's laws to relate the quantity of chemical change produced by a given amount of charge.**

Voltage is the driving force of electrochemical change. The quantity of chemical change is measured by the quantity of charge produced or consumed. The rate of chemical change is measured by the current. If current and voltage seem similar, an analogy with water may help. Voltage is similar to water pressure and current is similar to rate of flow (perhaps measured in gallons/minute).

Current is measured in amperes where 1 ampere (A) = 1 coulomb/second = 1C/s. Charge is measured in coulombs (C) or in Faradays ($\mathcal{F}$), where 1 Faraday = 96,500 C = 1 mole of electrons. The conversion factor method uses these relationships.

EXAMPLE 21-5 A 20.0-A current is used to plate copper out of a $CuSO_4(aq)$ solution. How long must the current be applied in order to plate out 1.00 lb of Cu?

The half-equation is that of the reduction of $Cu^{2+}(aq)$.

$$Cu^{2+}(aq) + 2\ e^- \longrightarrow Cu(s)$$

Then we determine the time required.

$$1.00\text{ lb Cu} \times \frac{454\text{ g Cu}}{\text{lb Cu}} \times \frac{\text{mol Cu}}{63.5\text{ g Cu}} \times \frac{2\text{ mol e}^-}{\text{mol Cu}} \times \frac{\text{s}}{20.0\text{ C}} \times \frac{\text{hr}}{3600\text{ s}} = 19.2\text{ h}$$

EXAMPLE 21-6 A 12.4 A current is used to electrolyze 5.20 L of a 3.00 M $AgNO_3(aq)$ solution. What is $[Ag^+]$ after 1.50 h?

We first compute the amount of Ag plated out.

$$1.50\text{ hr} \times \frac{3600\text{ s}}{\text{hr}} \times \frac{12.4\text{ C}}{\text{s}} \times \frac{\text{mol e}^-}{96{,}500\text{ C}} \times \frac{\text{mol Ag}^+}{\text{mol e}^-} = 0.964\text{ mol Ag}^+$$

The amount of Ag originally present is found.

$$5.20\text{ L} \times \frac{3.00\text{ mol AgNO}_3}{\text{L}} \times \frac{\text{mol Ag}^+}{\text{mol AgNO}_3} = 15.6\text{ mol Ag}^+$$

Then we find the amount of Ag^+ left.

$$15.750\text{ mol Ag}^+ - 0.694\text{ mol Ag}^+ = 15.056\text{ mol Ag}^+$$

Finally we compute the final [Ag+].

$$[Ag^+] = 15.056\text{ mol Ag}^+/5.520\text{ L} = 2.868\text{ M}$$

*** 11. Use the concepts of equivalent weight and normality to perform stoichiometric calculations based on oxidation-reduction reactions.**

In an oxidation-reduction reaction, one equivalent is the amount of oxidizing or reducing agent that produces or consumes one mole of electrons. In equation [5], 1 mol $O_2(g)$ consumes 2 mol electrons. Thus, one mol $O_2(g)$ is two equivalents (2 eq). In equation [4], 1 mol $MnO_4^-(aq)$ consumes 5 mol e^- and equals 5 eq. The equivalent weight is the mole weight divided by the number of equivalents per mole.

EXAMPLE 21-7 What is the equivalent weight of $KMnO_4$ in equation [3]?

According to equation [4], 1 mol MnO_4^- or 1 mol $KMnO_4$ consumes 5 mol e^- and equals 5 eq.

$$\frac{158.0\text{ g KMnO}_4}{\text{mol KMnO}_4} \times \frac{\text{mol KMnO}_4}{5\text{ eq KMnO}_4} = 52.67\text{ g KMnO}_4\text{/eq}$$

One equivalent of oxidizing agent reacts with one equivalent of reducing agent. Thus, in equation [11] one equivalent of $KMnO_4$ reacts completely with one equivalent of H_2S. *Normality* is defined as the number of equivalents of solute per liter of solution. The normality of a solution is its molarity multiplied by the number of equivalents per mole.

$$2\ KMnO_4(aq) + 3\ H_2S(s) \longrightarrow 2\ MnO_2(s) + 2\ KOH(aq) + 3\ S(s) + 2\ H_2O \qquad [11]$$

EXAMPLE 21-8 0.0250 eq H_2S reacts completely with 74.2 mL of $KMnO_4$ solution, as in equation [11]. What is the normality and the molarity of the solution?

Since 1 eq H_2S = 1 eq $KMnO_4$

$$\text{Normality} = \frac{0.0250 \text{ eq } H_2S}{72.4 \text{ mL soln}} \times \frac{1 \text{ eq } KMnO_4}{1 \text{ eq } H_2S} \times \frac{1000 \text{ mL}}{L} = 0.337 \text{ N}$$

The molarity of this solution is then

$$[KMnO_4] = \frac{0.337 \text{ eq } KMnO_4}{L \text{ soln}} \times \frac{\text{mol } KMnO_4}{3 \text{ eq } KMnO_4} = 0.112 \text{ M}$$

DRILL PROBLEMS

1. Draw the voltaic cells for each of the equations given below. Label the anode and the cathode. Indicate the direction of movement of all ions, including those of the salt bridge, $K^+(aq)$ and $Cl^-(aq)$. Draw the cell diagram. You may use either platinum or graphite as inert electrodes.
   A. $CuSO_4(aq) + H_2O_2(aq) \longrightarrow Cu(s) + O_2(g) + H_2SO_4(aq)$
   B. $Mg(s) + AgNO_3(aq) \longrightarrow Ag(s) + Mg(NO_3)_2(aq)$
   C. $Al(s) + H_2SO_4(aq) \longrightarrow Al_2(SO_4)_3(aq) + H_2(g)$
   D. $Fe_2(SO_4)_3(aq) + Pb(s) \longrightarrow PbSO_4(s) + Fe(s)$
   E. $NaI(aq) + F_2(g) \longrightarrow I_2(s) + NaF(aq)$
   F. $H_2(g) + KClO(aq) \longrightarrow KCl(aq) + H_2O(l)$, in alkaline solution
   G. $MnO_2(s) + HCl(aq) \longrightarrow MnCl_2(aq) + H_2O(l) + Cl_2(g)$
   H. $I_2(s) + H_2O(l) + Br_2(l) \longrightarrow HBr(aq) + HIO_3(aq)$
   I. $Ag(s) + HNO_3(aq) \longrightarrow AgNO_3(aq) + NO(g) + H_2O(l)$
   J. $S(s) + H_2O(l) + Pb(NO_3)_2(aq) \longrightarrow Pb(s) + H_2SO_3(aq) + HNO_3(aq)$

3. For each reaction in the drill problems of objective 21-1, along with each of the reactions listed below, use the data of *Table 21-1 in the text* to determine E°_{cell} and state whether the standard cell reaction is spontaneous or nonspontaneous.
   K. $O_2(g) + H_2O(l) \longrightarrow O_3(g) + H_2O_2(aq)$
   L. $H_2S(g) + Br_2(l) \longrightarrow S(s) + HBr(aq)$
   M. $SnSO_4(aq) + FeSO_4(aq) \longrightarrow Sn(s) + Fe_2(SO_4)_3(aq)$
   N. $H_2SO_4(aq) + S(s) + H_2O(l) \longrightarrow H_2SO_3(aq)$
   O. $I_2(s) + H_2O(l) \longrightarrow HI(aq) + HIO_3(aq)$

4. Determine the cell potential of each of the ten cells in the drill problems of objective 21-1, if all concentrations are 1.00 M and all gas pressures are 1.00 atm, except the following.
   A. $[CuSO_4] = 4.00$ M, $P(O_2) = 0.210$ atm   B. $[AgNO_3] = 0.150$ M, $[Mg(NO_3)_2] = 2.75$ M
   C. $[H_2SO_4] = 6.00$ M, $P(H_2) = 0.100$ atm   D. $[PbSO_4] = 1.30 \times 10^{-5}$ M
   E. $[NaF] = 2.50$ M, $[NaI] = 0.400$ M   F. $P(H_2) = 6.40$ atm, $[KCl] = 1.50$ M
   G. $[HCl] = 6.00$ M, $P(Cl_2) = 0.0120$ atm   H. $[HBr] = 0.100$ M, $[HIO_3] = 0.100$ M
   I. $[HNO_3] = 3.00$ M, $P(NO) = 4.00$ atm   J. $[HNO_3] = 4.00$ M, $[H_2SO_3] = 0.250$ M

5. Determine values of ΔG° and K_{eq} for the first ten cells of the drill problems of objective 21-1. Determine the value of ΔG for each of the cells in the drill problems of objective 21-4.

9. For each of the following solutions, use the data of Table 21-1 in the text to predict the products of electrolysis at the anode and at the cathode. Write the net ionic equation for the electrolysis reaction. Then use the Nernst equation to predict the minimum voltage that must be imposed for electrolysis to occur. Disregard overpotentials. Assume that any products formed are in their standard states.
   A. $[Na_2SO_4] = 1.00$ M, $[MgCl_2] = 1.50$ M   B. $[CuCl_2] = 1.00$ M, $[NaNO_3] = 0.500$ M
   C. $[HCl] = 6.20$ M, $[AlBr_3] = 2.00$ M   D. $[KOH] = 1.25$ M, $[NaNO_3] = 2.00$ M
   E. $[Cu(NO_3)] = 0.13$ M, $[SnCl_2] = 3.15$ M   F. $[FeSO_4] = 1.25$ M, $[AlI_3] = 0.0240$ M

10. In the following problems, assume that the ions present in solution are not depleted during the course of the electrolysis.
   A. A constant current of 10.0 A is passed through an electrolytic cell for 90. min. How many Faradays of charge are passed through the cell?
   B. A solution of $Sn(NO_3)_2$ is electrolyzed by passing 85.0 A of current for 72.0 min. How many grams of tin plate out?

C. Two cells containing solutions of $AgNO_3$ and $CuSO_4$ are connected in series (that is, the current goes in the cathode of one and out its anode, and then passes into the cathode of the other cell and out its anode) and electrolyzed. The cathode in the $AgNO_3$ cell gains 1.078 g. What mass has the cathode in the other cell gained?
D. A solution of $Cu_3(PO_4)_2$ is electrolyzed by passing 17.2 A of current for 145 min. How many grams of Cu plate out?
E. A solution of $Hg_2(NO_3)_2$ is electrolyzed by passing 108 A of current for 95.0 min. How many grams of elemental mercury are produced?
F. A solution of copper(II) sulfate is electrolyzed. Assume that no gas evolution occurs at the cathode. For how long must a current of 1.93 A flow to deposit 64.0 g of copper?
G. A solution of $Au_2(SO_4)_3$ is electrolyzed by passing 104 A of current for 165 min. What mass of gold plates out?
H. What volume of oxygen (at STP) can be liberated during the passage of 5.00 $\mathcal{F}$ of electricity?
I. A solution of $Al(NO_3)_3$ is electrolyzed by passing 144 A of current for 102 min. What mass of aluminum plates out?
J. A solution of $Cr(NO_3)_3$ is electrolyzed with a current of 15.0 A of current for 36.2 min. What mass of chromium plates out?
K. A solution of $AuCl_3$ is electrolyzed by passing 85.0 A of current for 17.6 min. What mass of gold plates out?
L. What mass of silver can be deposited by the passage of a constant current of 5.00 A through a $AgNO_3$ solution?

11. Determine the equivalent weights of the oxidizing agent and the reducing agent in each equation of the drill problems of objectives 21-1 and 21-3. The use each of the following sets of titration data to determine the normality of the titrant. (There is no titration data for equations H, J, K, L, N, and O.)
A. 25.00 mL $CuSO_4(aq)$ titrates 16.00 mL of 0.250 M H_2SO_4.
B. 14.70 mL $AgNO_3(aq)$ titrates 27.2 mg Mg(s).
C. 52.15 mL $H_2SO_4(aq)$ titrates 19.2 mg Al(s).
D. 85.17 mL $Fe_2(SO_4)_3(aq)$ titrates 41.7 mg Pb(s).
E. 24.25 mL NaI(aq) titrates 17.4 mg $F_2(g)$.
F. 19.71 mL KClO(aq) titrates 1.25 mg $H_2(g)$.
G. 25.41 mL HCl(aq) titrates 57.6 mg $MnO_2(s)$.
I. 19.75 mL $HNO_3(aq)$ titrates 174 mg Ag(s).
M. 27.21 mL $SnSO_4(aq)$ titrates 19.25 mL 0.250 M $FeSO_4(aq)$.

QUIZZES (20 minutes each) Choose the best answer for each question. Since balancing oxidation-reduction equations is an important part of electrochemistry, some equation-balancing problems are included for review. (Balancing is discussed in objectives 5-7 and 5-8.)

Quiz A

1. When the equation $HClO_3 + H_2O + I_2 \longrightarrow HIO_3 + HCl$ is balanced with the smallest integer coefficients, the coefficient of chloric acid is (a) 10; (b) 12; (c) 6; (d) 8; (e) none of these.
2. $Cr_2O_7^{2-} + I^- + H^+ \longrightarrow Cr^{3+} + I_2 + H_2O$ The correct coefficient for I^- in the balanced equation is (a) 2; (b) 3; (c) 6; (d) 14; (e) none of these.
3. When the equation $Br_2 + KOH + MnBr_2 \longrightarrow MnO_2 + KBr + H_2O$ is balanced with the smallest integer coefficients, the coefficient of H_2O is (a) 3; (b) 4; (c) 1; (d) 2; (e) none of these.
4. In the reaction $Br_2 + 2\,H_2O + SO_2 \longrightarrow 2\,HBr + H_2SO_4$ the equivalent weight of SO_2 is (a) 64.1 g; (b) 128.2 g; (c) 21.4 g; (d) 32.1 g; (e) none of these.
5. The cell reaction and $E°$ of the spontaneous cell made from the Zn/Zn^{2+} (0.76 V) and $2\,Br^-/Br_2$ (−1.07 V) half-cells are (a) $Zn^{2+} + 2\,Br^- \longrightarrow Br_2 + Zn$, +1.83 V; (b) $Br_2 + Zn \longrightarrow Zn^{2+} + 2\,Br^-$, +1.83 V; (c) $Br_2 + Zn \longrightarrow Zn^{2+} + 2\,Br^-$, +0.31 V; (d) $Zn^{2+} + 2\,Br^- \longrightarrow Br_2 + Zn$, +1.30 V; (e) none of these.
6. Consider the half-cell of Pb metal in contact with 1 M $Pb(NO_3)_2$, connected to the half-cell of Ag metal in contact with 1.0 M $AgNO_3$ in a complete circuit connected with an exterior wire. (standard reduction potentials: Pb −0.13 V, Ag +0.80 V) If NaCl is added to the half-cell containing $AgNO_3$, and AgCl precipitates, (a) the voltage of the cell would become more positive; (b) the voltage of the cell would become less positive;

(c) the voltage of the cell would not change; (d) the equilibrium constant of the cell reaction would change; (e) the ΔG of the cell reaction would remain the same.

7. The law that relates the amount of material oxidized or reduced to the amount of charge passed is due to (a) Volta; (b) Faraday; (c) Nernst; (d) Mendeleev; (e) Avogadro.
8. In the cell $Zn|Zn^{2+}||Ni^{2+}|Ni$ the zinc is called the (a) electrolyte; (b) cathode; (c) anode; (d) oxidizing agent; (e) none of these.
9. What volume of H_2 (measured at STP) is produced by the electrolysis of a 2.0 M H_2SO_4 solution? A current of 0.50 A is passed for 30.0 min. (a) 418 mL; (b) 52 mL; (c) 209 mL; (d) 104 mL; (e) none of these.
10. Suppose that the cell $Sn^{2+} + Zn° \longrightarrow Sn° + Zn^{2+}$, $E° = 0.537$ V will run until one of the reactants is completely gone. If the cell consists of 30. mL of 0.50 M Sn^{2+}, 30. mL of 0.50 M Zn^{2+}, 10. g Zn and 18.2 g Sn, which one will run out first? (a) Sn^{2+} solution; (b) Zn^{2+} solution; (c) Sn strip; (d) Zn strip; (e) none of these, two reactants will run out simultaneously.

Quiz B

1. When the half-equation $NO_3^- + H_2O \longrightarrow NO + OH^-$ is balanced with the smallest integer coefficients, the coefficient of OH^- is (a) 3; (b) 2; (c) 4; (d) 1; (e) none of these.
2. When the equation $KMnO_4 + KNO_2 + H_2O \longrightarrow MnO_2 + KNO_3 + KOH$ is balanced with the smallest integer coefficients, the coefficient of potassium nitrate is (a) 1; (b) 2; (c) 3; (d) 4; (e) none of these.
3. When the equation $HI + HNO_3 \longrightarrow H_2O + NO + I_2$ is balanced with the smallest integer coefficients, the sum of all the coefficients is (a) 5; (b) 7; (c) 6; (d) 8; (e) none of these.
4. In the reaction $2\,H_2O + 3\,KCN + 2\,KMnO_4 \longrightarrow 3\,KCNO + 2\,MnO_2 + 2\,KOH$ the equivalent weight of $KMnO_4$ is (a) 158 g; (b) 52.7 g; (c) 79.0 g; (d) 316 g; (e) none of these.
5. The cell reaction and $E°$ of the spontaneous cell made from the Ag/Ag (–0.80 V) and Cl^-/Cl_2 half-cells are (a) $Cl_2 + 2\,Ag \longrightarrow 2\,Cl^- + 2\,Ag^+$, 0.56 V; (b) $Cl_2 + 2\,Ag \longrightarrow 2\,Cl^- + 2\,Ag^+$, 1.12 V; (c) $Cl_2 + 2\,Ag \longrightarrow 2\,Cl^- + 2\,Ag^+$, 2.16 V; (d) $2\,Ag^+ + 2\,Cl^- \longrightarrow 2\,Ag + Cl_2$, 1.08 V; (e) none of these.
6. In the cell $Fe^{2+}|Fe^{3+}||Cu^{2+}|Cu$ which will increase the cell voltage the *most*? (a) halve $[Cu^{2+}]$; (b) halve $[Fe^{2+}]$; (c) double $[Cu^{2+}]$; (d) double $[Fe^{2+}]$; (e) cut Cu electrode in half.
7. If the voltage of an electrochemical sell is negative then the cell reaction is (a) nonspontaneous; (b) slow; (c) exothermic; (d) spontaneous; (e) none of these.
8. The equation that relates ion concentrations and gas pressures to cell voltages is called (a) Raoult's law; (b) Le Châtelier's principle; (c) Volta's theory; (d) Faraday's law; (e) none of these.
9. What mass of copper (63.54 g/mol) can be deposited by the passage of 9650 C of electricity through a $CuSO_4$ solution? (a) 12.7 g; (b) 6.35 g; (c) 3.18 g; (d) 7.98 g; (e) none of these.
10. A copper electrode weighs 35.42 g before the electrolysis of a copper(II) sulfate solution and 36.69 g after the electrolysis has run for 20.0 s. The equivalent weight of copper for this electrolysis is 31.77 g. Compute the amperage of the current that is used. (a) 3860 A; (b) 157 A; (c) 197 A; (d) 259 A; (e) none of these.

Quiz C

1. When the half-reaction $ClO_3^- + H^+ \longrightarrow Cl^- + H_2O$ is balanced with the smallest integer coefficients, the coefficient of H_2O is (a) 6; (b) 5; (c) 2; (d) 3; (e) none of these.
2. When the equation $Cs_2MnO_4 + H_2O \longrightarrow CsMnO_4 + CsOH + MnO_2$ is balanced with the smallest integer coefficients, the coefficient of MnO_2 is (a) 1; (b) 2; (c) 3; (d) 4; (e) none of these.
3. When the equation $HCl + H_2C_2O_4 + MnO_2 \longrightarrow MnCl_2 + CO_2 + H_2O$ is balanced with the smallest integer coefficients, the coefficient of CO_2 is (a) 1; (b); (c) 3; (d) 4; (e) none of these.
4. In the reaction $HMnO_4 + 5\,AsH_3 + 8\,H_2SO_4 \longrightarrow 5\,H_3AsO_4 + 8\,MnSO_4 + 12\,H_2O$ the equivalent weight of AsH_3 is (a) 77.9 g; (b) 26.0 g; (c) 15.6 g; (d) 9.74 g; (e) none of these.
5. The cell reaction and $E°$ of the spontaneous cell made from Fe/Fe^{2+} (0.44 V) and Sn/Sn^{2+} (0.14 V) half-cells are (a) $Fe^{2+} + Sn \longrightarrow Sn^{2+} + Fe$, –0.30 V; (b) $Fe^{2+} + Sn \longrightarrow Sn^{2+} + Fe$, +0.30 V; (c) $Fe° + Sn^{2+} \longrightarrow Sn° + Fe^{2+}$, –0.30 V; (d) $Fe° + Sn^{2+} \longrightarrow Sn° + Fe^{2+}$, +0.58 V; (e) none of these.
6. For the cell $Cu|Cu^+||Al^{3+}|Al$ which will increase the cell voltage the most? (a) double $[Al^{3+}]$; (b) halve $[Cu^+]$; (c) double $[Cu^+]$; (d) halve $[Al^{3+}]$; (e) cut the Al electrode in half.
7. One mole of electrons has a charge of (a) 96,500 A; (b) 1.60×10^{-19} C; (c) 6.02×10^{23} A; (d) 96,500 C; (e) none of these.
8. In a galvanic cell, oxidation occurs at the (a) anode; (b) cathode; (c) salt bridge; (d) electrolyte; (e) none of these.

9. 1.00 L of a 1.500 M $Al_2(SO_4)_3$ solution is electrolyzed between platinum electrodes for 1.20 h with a current of 62.0 A. What is $[Al^{3+}]$ at the end of the electrolysis? (a) 0.57 M; (b) 0.93 M; (c) 2.07 M; (d) not enough Al^{3+} present; (e) none of these.
10. A solution of CuBr is electrolyzed by passing 35.0 A of current for 17.0 min. What mass in grams of Cu (63.54 g/mol) plates out? (a) 26.5; (b) 53.1; (c) 11.8; (d) 23.5; (e) none of these.

Quiz D

1. After balancing the ionic equation $MnO_4^- + Cl^- + H^+ \longrightarrow Mn^{2+} + Cl_2 + H_2O$ the sum of the simplest set of integer coefficients of the six substances would be (a) 21; (b) 24; (c) 38; (d) 43; (e) 51.
2. When the equation $ClO_2 + H_2O_2 + KOH \longrightarrow H_2O + KClO_2 + O_2$ is balanced with the smallest integer coefficients, the coefficient of H_2O is (a) 3; (b) 5; (c) 2; (d) 4; (e) none of these.
3. When the equation $HNO_2 \longrightarrow HNO_3 + NO + H_2O$ is balanced with the smallest integer coefficients, the sum of all four of the coefficients is (a) 7; (b) 6; (c) 9; (d) 10; (e) none of these.
4. In the reaction $2\ Bi(OH)_3 + 3\ Na_2SnO_2 \longrightarrow 2\ Bi + 3\ H_2O + 3\ Na_2SnO_3$ the equivalent weight of $Bi(OH)_3$ is (a) 86.7 g; (b) 260. g; (c) 130. g; (d) 173 g; (e) none of these.
5. The cell reaction and $E°$ of the spontaneous cell made from the $Sn^{2+}/Sn+4+$ (–0.15 V) and $H_2S/S°$ (–0.14 V) half-cells are (a) $Sn^{4+} + H_2S \longrightarrow 2\ H^+ + S° + Sn^{2+}$, 0.29 V; (b) $2\ H^+ + S° + Sn^{2+} \longrightarrow Sn^{4+} + H_2S$, –0.29 V; (c) $Sn^{4+} + H_2S \longrightarrow 2\ H^+ + S° + Sn^{2+}$, 0.01 V; (d) $2\ H^+ + S° + Sn^{2+} \longrightarrow Sn^{4+} + H_2S$, –0.01 V; (e) none of these.
6. For the cell $Ag|Ag^+||Br^-|Br_2$ which of the following will increase the cell voltage the *most*? (a) double $[Ag^+]$; (b) triple $[Br^-]$; (c) cut the Ag electrode in half; (d) remove half of the $Br_2(l)$; (e) halve $[Ag^+]$.
7. Consider the half-cell of Pb metal in contact with 1.00 M $Pb(NO_3)_2$, connected to the half-cell of Ag metal in contact with 1.00 M $AgNO_3$ in a complete circuit connected with an external wire. (standard reduction potentials: Pb = –0.13 V, Ag = +0.80 V) As this cell operates, (a) the Pb electrode increases in mass; (b) electrons flow from the Pb electrode to the Ag electrode; (c) the Ag electrode is the anode of the cell; (d) NO_3^- ions migrate from the lead half-cell to the silver half-cell; (e) the Ag electrode dissolves slowly.
8. A solution of $CuCl_2$ is electrolyzed by passing 35.0 A for 17.0 min. What mass in grams of Cu (63.54 g/mol) plates out? (a) 24.9; (b) 49.8; (c) 47.0; (d) 23.5; (e) none of these.
9. The amount of substance that produces a mole of electrons is known as a(n) (a) Lewis base; (b) equivalent; (c) reducing agent; (d) oxidizer; (e) none of these.
10. For how many hours must a current of 14.0 A flow to produce 5.02 g Al (27.0 g/mol) from a solution of $Al(NO_3)_3$ (213 g/mol)? (a) 1.07; (b) 0.135; (c) 0.356; (d) 0.405; (e) none of these.

SAMPLE TEST (20 minutes)

1. The reaction $3\ Pb(NO_3)_2 + 2\ Al \longrightarrow 2\ Al(NO_3)_3 + 3\ Pb$ is used in a galvanic cell.
   a. Write the cell diagram, with a KCl salt bridge
   b. From the following list of reduction couples, determine $E°$ of the cell.
      Pb^{2+}/Pb, $E° = -0.126$ V; Al^{3+}/Al, –1.66 V; Pb^{4+}/Pb^{2+} 1.455 V, NO_3^-/NO, 0.96 V
   c. Determine the value of E when $[Pb(NO_3)_2] = 4.00$ M and $[Al(NO_3)_3] = 0.0200$ M.
2. 50.0 mL of 0.100 M $CuSO_4(aq)$ is electrolyzed with a 0.965-A current for 1.00 min. When electrolysis is complete what is
   a. $[Cu^{2+}]$? b. pH, assuming pH = 7.00 initially?
3. Balance the following oxidation-reduction equations.
   a. $KMnO_4(aq) + H_2S(g) \longrightarrow MnO_2(s) + S(s) + KOH(aq) + H_2O(l)$
   b. $NiS(s) + KClO_3(aq) + HCl(aq) \longrightarrow NiCl_2(aq) + KCl(aq) + S(s) + H_2O(l)$

22 Chemistry of the Representative (Main-Group) Elements I: Metals

CHAPTER OBJECTIVES

1. Describe how standard electrode potentials are related to enthalpies of sublimation, ionization and hydration; and how lattice energies and enthalpies of hydration affect the solubilities of ionic compounds.

We can relate the standard cell potential, $E°_{cell}$, to the standard enthalpy change of a reaction, $\Delta H°_{rxn}$.

$$\Delta G°_{rxn} = \Delta H°_{rxn} - T\Delta S°_{rxn} = -n\mathcal{F}E°\text{cell} \quad [1]$$

We first assume that $T\Delta S°_{rxn}$ is so small that $\Delta H°_{rxn}$ determines the value of $\Delta G°_{rxn}$ (and that of $E°$cell), and thus the spontaneity of the reaction. Since the standard half-cell potential for hydrogen equals zero ($E°_H = 0$) the standard cell potential for the reaction in which a metal displaces H^+ ion from aqueous solution equals the oxidation potential of the metal ($E°_{cell} = E°_{ox}$). With these two conclusions, we can relate $\Delta H°_{ox}$ to $E°_{ox}$.

$$M(s) \longrightarrow M^+(aq) + e^- \qquad E°_{ox} = \Delta H°_{ox} \quad [2]$$

$$H^+ + e^- \longrightarrow \tfrac{1}{2} H_2 \qquad E°_{red} = 0.000 \text{ V} \qquad \Delta H°_H = -449 \text{ kJ/mol} \quad [3]$$

$$M + H^+ \longrightarrow M^+ + \tfrac{1}{2} H_2 \qquad E°\text{cell} = E°_{ox} \qquad \Delta H°_{rxn} = (\Delta H°_{ox} - 449) \text{ kJ/mol} \quad [4]$$

Thus, the smaller the value of $\Delta H°_{ox}$, the more spontaneous is the reaction. We now consider the contributions to $\Delta H°_{ox}$. It is composed of three parts.

$$M(s) \longrightarrow M(g) \qquad \Delta H°_{subl} (> 0) \quad [5]$$

$$M(g) \longrightarrow M^+(g) + e^- \qquad \Delta H°_{ioniz} (> 0) \quad [6]$$

$$\underline{M^+(g) \longrightarrow M^+(aq) \qquad \Delta H°_{hydr} (< 0)} \quad [7]$$

$$M(s) \longrightarrow M^+(aq) + e^- \qquad \Delta H°_{ox} \quad \text{where} \quad \Delta H°_{ox} = \Delta H°_{subl} + \Delta H°_{ioniz} + \Delta H°_{hyd} \quad [8]$$

Within the alkali metals, both $\Delta H°_{subl}$ and $\Delta H°_{ioniz}$ become smaller as atomic number increases. These two trends are offset by the large negative $\Delta H°_{hyd}$ of Li^+. Thus, lithium has the largest $E°_{ox}$ and is the best reducing agent in the alkali metal family. In other families of the periodic table, the $\Delta H°_{hyd}$ is not sufficiently negative to overcome the trends in $\Delta H°_{subl}$ and $\Delta H°_{ioniz}$ and thus reducing power increases with atomic number.

$$M^+(g) + X^-(g) \longrightarrow MX(s) \qquad L.E. \quad [9]$$

Lattice energy, defined for an alkali metal halide in equation [9], is negative (an exothermic process) and has terms such as $Z_A Z_C/(r_A + r_C)$ that depend on the size (r) and charge (z) of the ions involved. Thus, small ions of high charge (those with high charge density: z/r, with r in Å) produce compounds with a very negative lattice energy—compounds which are quite difficult to break apart into ions. On the other hand, ions with high charge density also have very negative values of $\Delta H°_{hyd}$. The first effect predicts that compounds composed of ions with high charge density will not readily dissolve whereas the second effect predicts that they will. Experimentally, we find that if both the cation and the anion have high charge density (ions such as Li^+, 1.47; Mg^{2+}, 3.03; Al^{3+}, 5.88; F^-, 0.75; OH^-, 0.83; CO_3^{2-}, 1.17; and PO_4^{3-}, 1.34), the compound is insoluble. On the other hand, if only the cation has high charge density, the resulting compound is often quite soluble, so much so that it may be deliquescent (it will absorb water from the atmosphere and form a saturated solution).

*** 2. Outline the methods that are used to produce Na, Mg, and Al.**

The raw material for the production of *sodium* is NaCl which is either mined from underground deposits of rock salt or obtained from brine—NaCl(aq)—by evaporation. Sodium is produced by electrolysis of molten NaCl (mixed with $CaCl_2$ or Na_2CO_3 to lower the fusion point of the mixture below the boiling point of Na), equation [10]. This reaction occurs in the Downs cell, specially designed to keep Na(l) separated from Cl_2(g).

$$2\ NaCl(l) \xrightarrow{\text{electrolysis}} 2\ Na(l) + Cl_2(g) \qquad [10]$$

The source of *magnesium* is seawater, from which it is precipitated as $Mg(OH)_2$ by the addition of $Ca(OH)_2$, equation [13]. (The $Ca(OH)_2$ is produced by first calcining limestone, equation [11], and slaking the resulting lime, equation [12]). Treatment of $Mg(OH)_2$(s) with HCl(aq) produces an aqueous solution of $MgCl_2$, equation [14]. The water is evaporated and the resulting $MgCl_2$ is electrolyzed to produce the metal, equation [15].

$$CaCO_3(s) \xrightarrow{\Delta} CaO(s) + CO_2(g) \qquad [11]$$

$$CaO(s) + H_2O \longrightarrow Ca(OH)_2(aq) \qquad [12]$$

$$Mg^{2+}(aq) + Ca(OH)_2(aq) \longrightarrow Mg(OH)_2(s) + Ca^{2+}(aq) \qquad [13]$$

$$Mg(OH)_2(s) + 2\ HCl(aq) \longrightarrow MgCl_2(aq) + 2\ H_2O \qquad [14]$$

$$MgCl_2(l) \xrightarrow{\text{electrolysis}} Mg(l) + Cl_2(g) \qquad [15]$$

The principal ore of *aluminum* is bauxite, which contains large quantities of SiO_2 and Fe_2O_3 as impurities. Treatment of bauxite with concentrated NaOH(aq) dissolves Al_2O_3 in the ore, equation [16], and leaves Fe_2O_3 and SiO_2 behind. The solution is filtered and either diluted or treated with acid to precipitate $Al(OH)_3$, equation [17]. Al_2O_3(s) is formed by heating the $Al(OH)_3$(s), equation [18]. Finally, Al_2O_3 is mixed with cryolite (Na_3AlF_6) to lower the melting temperature and electrolyzed between carbon electrodes, equation [19].

$$Al_2O_3(s) + 2\ OH^-(aq) + 3\ H_2O \longrightarrow 2\ [Al(OH)_4]^-(aq) \qquad [16]$$

$$Al(OH)_4^-(aq) + H_3O^+(aq) \longrightarrow Al(OH)_3(s) + 2\ H_2O \qquad [17]$$

$$2\ Al(OH)_3(s) \xrightarrow{\Delta} Al_2O_3(s) + 3\ H_2O(g) \qquad [18]$$

$$2\ Al_2O_3(l) + 3\ C(s) \xrightarrow{\text{electrolysis}} 4\ Al(l) + 3\ CO_2(g) \qquad [19]$$

3. Outline the Solvay process for the production of $NaHCO_3$, showing how substances are recycled in the process.

The raw materials for Na_2CO_3 in the Solvay process are NaCl (from rock salt or brine) and $CaCO_3$ (limestone). The limestone is calcined, equation [11], to obtain CO_2(g). The CO_2(g) and NH_3(g) are passed through an aqueous solution of NaCl, equation [20]. The $NaHCO_3$(s) that precipitates is heated to produce Na_2CO_3(s), equation [21]. (The CO_2 produced is recycled to yield more $NaHCO_3$.) NH_3 is recycled by treating the NH_4Cl(aq) with strong base (equation [22]), produced by slaking CaO(s), equation [12]. The net reaction for the process is equation [23] and the only by-product is $CaCl_2$.

$$NaCl(aq) + NH_3(g) + CO_2(g) + H_2O \longrightarrow NaHCO_3(aq) + NH_4Cl(aq) \qquad [20]$$

$$2\ NaHCO_3(s) \xrightarrow{\Delta} Na_2CO_3(s) + CO_2(g) + H_2O(g) \qquad [21]$$

$$2\ NH_4Cl(aq) + Ca(OH)_2(aq) \longrightarrow CaCl_2(aq) + 2\ NH_3(g) + 2\ H_2O \qquad [22]$$

$$2\ NaCl(aq) + CaCO_3(s) \longrightarrow Na_2CO_3(s) + CaCl_2(s) \qquad [23]$$

4. Outline the series of reactions whereby rain water acquires temporary hardness and how this hardness is removed by heating; and write chemical equations for reactions that can be used to soften temporary and permanent hard water.

When rain falls through the atmosphere, it absorbs $CO_2(g)$, forming a weakly acidic solution, equation [24]. As this solution flows through underground limestone beds, some of the limestone dissolves, equation [25]. Heating this solution drives off $CO_2(g)$, reversing equilibrium [24]. This in turn reverses equilibrium [25] and $CaCO_3(s)$ precipitates. Although the water now is soft, the $CaCO_3$ precipitate is undesirable if it occurs in the wrong place. For example, as $CaCO_3(s)$ builds up on the inside of a boiler it acts as an insulator, and more heat then is required to heat the water in the boiler.

$$CO_2(g) + H_2O \rightleftharpoons H_2CO_3(aq) \qquad [24]$$

$$CaCO_3(s) + H_2CO_3(aq) \rightleftharpoons Ca(HCO_3)_2(aq) \qquad [25]$$

Note that both a divalent (+2) cation ($M^{2+} = Ca^{2+}$, Mg^{2+}, or Fe^{2+}) and a divalent (–2) anion (CO_3^{2-}, SO_4^{2-}) must be present in solution in order for water to be considered hard. Permanent hard water differs from the soft variety in that the hardness cannot be removed by heating; the anion usually is SO_4^{2-}. Chemical methods of removing hardness generally work by removing the cation. The first of these is to precipitate $M(OH)_2$ by adding a strong base, equation [26]. Slaked lime, $Ca(OH)_2$, often is used as a source of OH^- to treat municipal water supplies. Na_2CO_3 often is used in cleaning products for the same purpose, equation [27]. In the second chemical method, a chelating agent, such as a polyphosphate ($P_2O_7^{4-}$) compound, is added to the water to sequester the hard water cation as a soluble complex ion, equation [28]. Finally, zeolites (naturally occurring silicate minerals) remove hard water cations by the process of ion exchange. When water containing the divalent cations is passed through the channels in the zeolite (—Z—), the dipositive cations are attracted to its silicate structure and held there. The Na+ ions originally present in the zeolite go into solution, equation [29]. Thus, temporary hard water becomes a solution of $NaHCO_3$ (baking soda) on passage through a zeolite. The zeolite is "recharged" by passing a concentrated NaCl solution through it. The very high $[Na^+]$ forces the dipositive cations out of the zeolite, by reversing equilibrium [29].

$$M^{2+}(aq) + 2\,OH^-(aq) \longrightarrow M(OH)_2(s) \quad M = Ca, Mg, \text{ or } Fe \qquad [26]$$

$$Na_2CO_3 + H_2O + Ca(HCO_3)_2(aq) \longrightarrow 2\,Na^+(aq) + 2\,HCO_3^-(aq) + CaCO_3(s) \qquad [27]$$

$$M^{2+}(aq) + 2\,P_2O_7^{4-}(aq) \longrightarrow M(P_2O_7)^{6-}(aq) \qquad [28]$$

$$—Z(—Na^+)_2 + Ca^{2+}(aq) \rightleftharpoons —Z—Ca^{2+} + 2\,Na^+(aq) \qquad [29]$$

* **5. Write formulas for some typical alums.**

Alums are composed of three ions: $M(H_2O)_6^+$, $M'(H_2O)_6^{3+}$, and SO_4^{2-} in the ratio 1:1:2, that is $MM'(SO_4)_2{\cdot}12\,H_2O$. M(I) is any unipositive (+1) cation except Li+, with K^+ and NH_4^+ being the most common. M'(III) is any of the ions Al^{3+}, Ti^{3+}, V^{3+}, Cr^{3+}, Mn^{3+}, Fe^{3+}, Co^{3+}, Ga^{3+}, In^{3+}, Re^{3+}, or In^{3+}, with Al^{3+} being the most common. $KAl(SO_4)_2{\cdot}12\,H_2O$ is *alum*, and other alums are often named as derivatives of this compound. Thus, $NH_4Al(SO_4)_2{\cdot}12\,H_2O$ is ammonium alum (or ammonium aluminum sulfate duodecahydrate) and $KCr(SO_4)_2{\cdot}12\,H_2O$ is chromium(III) alum [or potassium chromium(III) sulfate duodecahydrate].

* **6. Describe the four fundamental processes of traditional extractive metallurgy—concentration, roasting, reduction, and refining—with specific reference to metals such as Sn, Pb, Zn, Cd, and Hg.**

The first step in processing metallic ores is *concentration* or removal of the unwanted material, the gangue. This gangue usually is silicate rock of low density (that of SiO_2 is 2.66 g/cm^3). One method of concentration is washing—allowing a stream of rapidly moving water to carry the lighter gangue away from the metal-containing mineral. (Some mineral densities are: SnO_2, 6.95 g/cm^3; PbS, 7.5 g/cm^3; ZnS, 4.09 g/cm^3; HgS, 8.10 g/cm^3; Cu_2S, 5.6 g/cm^3; and CuS, 4.6 g/cm^3). A second method of concentration is flotation. The ore is mixed with water, a wetting agent, and a foaming agent. The mixture is agitated and air is blown through. The particles of the mineral stick to the foam on the surface which is skimmed off. Concentration normally removes more than 99% of the gangue.

During *roasting* the concentrated ore is heated strongly in the presence of air. This calcines carbonates (equation [30]), transforms sulfides to oxides (equations [31] and [32]), or oxidizes metallic impurities and converts nonmetallic ones to volatile oxides. $SO_2(g)$ is not vented to the atmosphere; it can be reduced to S(s) by reaction with red hot C, equation [33].

$$ZnCO_3 \xrightarrow{\Delta} ZnO(s) + CO_2(g) \qquad [30]$$

$$2\ MS(s) + 3\ O_2(g) \xrightarrow{\Delta} 2\ MO(s) + 2\ SO_2(g) \qquad M = Pb, Zn, Cd \qquad [31]$$

$$HgS(s) + O_2(g) \xrightarrow{\Delta} Hg(l) + SO_2(g) \qquad [32]$$

$$SO_2(g) + 2\ C(s) \xrightarrow{\Delta} S(l) + 2\ CO(g) \qquad [33]$$

The oxide is transformed to the metal during *reduction*. Typical reducing agents are C(s) obtained from coking coal (heating it in the absence of air) and CO(g) obtained from coke (as in equation [34]). The reductions of SnO_2 and PbO are typical reactions; the reductions of ZnO and CdO are similar to that of PbO, except that the metal is produced as a vapor in each case. Note that HgS is reduced when it is roasted, equation [32].

$$2\ C(s) + O_2(g) \rightleftharpoons 2\ CO(g) \qquad [34]$$

$$SnO_2(s) + 2\ C(s) \xrightarrow{\Delta} Sn(l) + 2\ CO(g) \qquad [35]$$

$$PbO(s) + C(s) \xrightarrow{\Delta} Pb(l) + CO(g) \qquad [36]$$

$$PbO(s) + CO(g) \xrightarrow{\Delta} Pb(l) + CO_2(g) \qquad [37]$$

Refining removes the impurities from the metal. Tin is refined by remelting. Impurities are oxidized; the oxides float on the surface and are skimmed off. Metallic impurities remain unmelted and the molten pure tin is poured off. Pb also is refined by remelting; copper crystallizes out and remains behind when the molten lead is poured off. When air is blown through the melt, the nonmetallic impurities form lead oxoanion compounds that float on the surface and are skimmed off. Finally, 1-2% Zn is added to the melt. Ag impurity dissolves in the Zn(l) and floats on the surface; it is skimmed off and the Ag is recovered by distilling off the Zn. Zinc is refined either by fractional distillation of the molten impure metal or electrolytically. The latter process combines reduction and refining. ZnO(s) is dissolved in $H_2SO_4(aq)$, equation [38]. Zn dust is added to the solution to displace Cd, and the solution is electrolyzed to produce pure Zn(s), equation [39]. Cadmium is refined in a similar fashion. Mercury is refined by first adding dilute HNO_3 to oxidize the impurities; the oxides float on the surface and are skimmed off. The mercury is further refined by distillation.

$$ZnO(s) + H_2SO_4(aq) \longrightarrow Zn^{2+}(aq) + SO_4^{2-}(aq) + H_2O \qquad [38]$$

$$Zn^{2+}(aq) + SO_4^{2-}(aq) + H_2O \xrightarrow{\text{electrolysis}} Zn(s) + H_2SO_4(aq) + {}^1\!/_2\ O_2(g) \qquad [39]$$

* **7. Write equations to show the effect of H_2O, of dilute and concentrated HCl(aq), $H_2SO_4(aq)$, and $HNO_3(aq)$, and of NaOH(aq) on metals, such as those of groups 1A and 2A, Al, Sn, Pb, Zn, Cd, and Hg.**

All of the alkali metals (group 1A) and alkaline earth metals (2A) except for Be readily react with water. The reactions of Na (equation [40]) and Sr (equation [41]) are typical. Mg requires steam for reaction, equation [42]. Zn also reacts with steam in the same manner as Mg or, if the protective coating of basic zinc carbonate is removed, in the same manner as Ca. If its protective oxide coating is removed, Al also reacts with water (equation [43]); otherwise Al is unaffected. Cd, Hg, Bi, Sn, and Pb do not react with pure water. Pb dissolves slowly if $O_2(aq)$ is present.

$$2\ Na(s) + 2\ H_2O \longrightarrow 2\ NaOH(aq) + H_2(g) \qquad [40]$$

$$Ca(s) + 2\ H_2O \longrightarrow Ca(OH)_2(aq) + H_2(g) \qquad [41]$$

$$Mg(s) + H_2O \xrightarrow{\Delta} MgO(s) \qquad [42]$$

$$2\ Al(s) + 6\ H_2O \longrightarrow 2\ Al(OH)3(s) + 3\ H_2(g) \qquad [43]$$

All of the alkali metals and the alkaline earth metals, along with Zn, Cd, Sn, and Al react with dilute HCl(aq) and $H_2SO_4(aq)$ as in equations [44] through [46]. The reactions of the alkali metals are quite violent. Pb reacts initially, but surface coatings of $PbCl_2(s)$ and $PbSO_4(s)$ prevent further reaction. Neither Hg nor Bi reacts with dilute HCl(aq) or H_2SO_4(aq). The reactions of the alkali metals and the alkaline earth metals (except Be) with concentrated HCl(aq) and H_2SO_4(aq), with HNO_3(aq), and with NaOH(aq) are the same as their reactions with water.

$$Be(s) + H_2SO_4(aq) \longrightarrow BeSO_4(aq) + H_2(g) \quad [44]$$

$$Cd(s) + 2\ HCl(aq) \longrightarrow CdCl_2(aq) + H_2(g) \quad [45]$$

$$2\ Al(s) + 6\ HCl(aq) \longrightarrow 2\ AlCl_3(aq) + 3\ H_2(g) \quad [46]$$

In their reactions with concentrated HCl(aq), the metals Be, Zn, Cd, Sn, Pb, and Al all form chloro complexes. The reactions of Be, Zn, Cd, and Sn are similar to that of Pb, equation [47]. The reaction of Al is given in equation [48]. Hg and Bi do not react with concentrated HCl(aq).

$$Pb(s) + 2\ H^+ + 4\ Cl^- \longrightarrow [PbCl_4]^{2-} + H_2(g) \quad [47]$$

$$2\ Al(s) + 6\ H^+ + 12\ Cl^- \longrightarrow 2\ [AlCl_6]^{3-} + 3\ H_2(g) \quad [48]$$

The reaction of Sn with concentrated H_2SO_4(aq) is similar to its reaction with dilute H_2SO_4(aq). Be, Zn, Cd, Hg, and Pb all react with concentrated H_2SO_4(aq) to yield SO_2(g) and the divalent cation in aqueous solution, as in equation [49] for Hg. Bi does not react with concentrated H_2SO_4(aq); Al does not either because of the formation of a protective coating of Al_2O_3.

$$Hg(l) + 2\ H_2SO_4(conc.) \longrightarrow HgSO_4(aq) + SO_2(g) + 2\ H_2O \quad [49]$$

In reactions with dilute HNO_3(aq), moderately active metals like Zn and Be yield N_2O(g) and the divalent cation in aqueous solution, equation [50]. Less active metals yield NO(g), equation [51], with Hg giving Hg_2^{2+}, equation [52]. With concentrated HNO_3(aq), moderately active metals give N_2 (equation [53]), less active metals give NO_2(g) (equation [54]), and Hg yields Hg^{2+}. Aluminum does not react with dilute or concentrated HNO_3(aq) because of the formation of an oxide coating.

$$4\ Zn(s) + 10\ HNO_3(aq) \longrightarrow 4\ Zn(NO_3)_2(aq) + N_2O(g) + 5\ H_2O \quad [50]$$

$$3\ Pb(s) + 8\ HNO_3(aq) \longrightarrow 3\ Pb(NO_3)_2(aq) + 2\ NO(g) + 4\ H_2O \quad [51]$$

$$6\ Hg(l) + 8\ HNO_3(aq) \longrightarrow 3\ Hg_2(NO_3)_2(aq) + 2\ NO(g) + 4\ H_2O \quad [52]$$

$$5\ Zn(s) + 12\ HNO_3(conc.) \longrightarrow 5\ Zn(NO_3)_2(aq) + N_2(g) + 6\ H_2O \quad [53]$$

$$Cd(s) + 4\ HNO_3(conc.) \longrightarrow Cd(NO_3)_2(aq) + 2\ NO_2(g) + 2\ H_2O \quad [54]$$

Be, Zn, Sn, Pb, and Al all react with NaOH(aq) by forming hydroxo complexes. The reactions of Be and Sn are similar to that of Zn, equation [55]. Al forms $Al(OH)_4^-$ and Pb forms $Pb(OH)_6^{4-}$; other than that their reactions are similar.

$$Zn(s) + 2\ OH^-(aq) + 2\ H_2O \longrightarrow [Zn(OH)_4]^{2-}(aq) + H_2(g) \quad [55]$$

* **8. Write equations to represent the amphoteric nature of the oxides and hydroxides of Al, Sn, and Zn.**

By amphoteric, we mean that the substance reacts as an acid when treated with a strong base and as a base when treated with a strong acid. Consider the oxide and hydroxide of aluminum as examples. When treated with strong acid, these substances react as bases.

$$2\ Al(OH)_3(s) + 3\ H_2SO_4(aq) \longrightarrow Al_2(SO_4)_3(aq) + 6\ H_2O \quad [56]$$

$$Al_2O_3(s) + 3\ H_2SO_4(aq) \longrightarrow Al_2(SO_4)_3(aq) + 3\ H_2O \quad [57]$$

When treated with strong bases, $Al(OH)_3$ and Al_2O_3 react as acids.

$$Al(OH)_3(s) + 3\ NaOH(aq) \longrightarrow Na_3AlO_3(aq) + 3\ H_2O \quad [58]$$

$$Al_2O_3(s) + 6\ NaOH(aq) \longrightarrow 2\ Na_3AlO_3(aq) + 3\ H_2O \quad [59]$$

The anion produced depends on the oxide that reacts with the strong base. ZnO yields $Zn(OH)_4^{2-}$, SnO yields $Sn(OH)_3^-$, and SnO_2 yields $Sn(OH)_6^{2-}$. The oxides of Pb yield hydroxo complexes of the same formula as the oxides of Sn.

9. Explain ways in which Li differs from other members of group 1A; Tl differs from other members of 3A; Pb differs from Sn; and Hg differs from Zn and Cd.

Lithium differs from the other alkali metals in that it forms a nitride; it forms a normal oxide (Li_2O) on reaction with $O_2(g)$ rather than a superoxide or a peroxide; both its carbonate and its hydroxide can be thermally decomposed to the oxide; and its carbonate, hydroxide, fluoride, and phosphate have low solubility in water. *Thallium* differs from the other metals of group 3A in that its monovalent cation, Tl^+, is more stable than its trivalent cation, Tl^{3+}; $Tl^{3+}(aq)$ is a strong oxidizing agent; Tl^+ behaves much like K^+ in aqueous solution; and TlOH is a strong base. *Lead* differs from Sn in that Pb^{2+} is more stable than Pb^{4+} while Sn^{2+} and Sn^{4+} are almost equally stable; Pb(IV) compounds are good oxidizing agents while Sn(II) compounds are reducing agents; and Sn exists in three allotropic forms of which one (gray tin) is distinctly nonmetallic with a crystal structure similar to diamond while Pb has only the metallic allotrope. *Mercury* differs from Zn and Cd in that Hg has little tendency to combine with oxygen; it does not dissolve in nonoxidizing acids; it forms few water soluble compounds; it forms many covalent compounds including a wide variety of organic mercury compounds; mercury halides are only slightly ionized in aqueous solution; and it forms Hg_2^{2+}, the only common diatomic metal cation.

* **10. State some of the principal uses of the representative metals; and describe some of the more important reactions and uses of compounds of the representative elements.**

Many of the important uses of the representative metals and their compounds are given in Table 14-3. Additional metals and their compounds have been introduced in this chapter along with additional uses for the compounds discussed in Chapter 14. These additions, arranged by groups in the periodic table, are given in Table 22-1. That material should be reviewed carefully and committed to memory.

TABLE 22-1 Uses of Representative Metals and Some of Their Compounds

Substance	*Important or unique uses*
	Group 1A
KOH	manufacture of soft soap
K_2CO_3	(potash, which gave the element its name) in hard glass; to make textile dyes and bleaches and soft soap.
KCl	fertilizer production (K^+ is a plant nutrient)
KNO_3	oxidizing agent in gun powder; dye production; fertilizer
$KMnO_4$	common laboratory and small scale industrial oxidizing agent
$KClO_3$	strong oxidizing agent in fireworks, matches, and explosives
Rb metal	to degas vacuum tubes; in photoelectric cells
Cs metal	in photoelectric cells
Cs_2SO_4	with V_2O_5 to catalyze oxidation of SO_2 to SO_3
	Group 2A
Ca metal	to remove O_2 and N_2 from molten steel; hardener for Pb alloys in storage batteries; reducing agent in Cr metallurgy
CaO	(lime) in plaster and mortar; in agriculture; as a cheap base in the chemical industry; basic furnace lining
$CaCO_3$	(limestone, marble, chalk) glass manufacture; building material; flux to remove SiO_2 in metallurgy: $CaCO_3 + SiO_2 \longrightarrow CaSiO_3 + CO_2(g)$; paper whitener
$CaCl_2$	by-product of Na_2CO_3 manufacture; to melt ice on roads; to absorb water to settle dust on roads
$CaSO_4 \cdot 2\,H_2O$	(gypsum) reversibly dehydrates to plaster of Paris
$Ca(NO_3)_2 \cdot 2\,H_2O$	fertilizer
CaC_2	portable source of acetylene
$Ca_3(PO_4)_2$	fertilizer
$SrCl_2$, $Sr(NO_3)_2$	red color in flares and fireworks
Ba metal	in alloys for cathodes in vacuum tubes and spark plugs
$BaSO_4$	with ZnS in the white pigment lithopone; whitener and filler in paper, rubber, and flooring tile

$BaCl_2$	laboratory reagent
$Ba(NO_3)_2$, $Ba(ClO_3)_2$	green color in fireworks
$BaSiF_6$	rat poison
$BaCO_3$	insecticide
$BaCrO_4$	yellow porcelain pigment
$BaMnO_4$	green porcelain pigment
BaU_2O_7	yellow or orange porcelain pigment

Group 2B

Zn metal	in Pb and Zn metallurgy to recover Ag and Cd, respectively; to galvanize Fe; in brass; manufacture of dry cells; roofing materials
ZnO	reinforcing agent and white pigment in rubber; in burn ointments; in cosmetics; dietary supplement; photoconductors in copying machines
ZnS	phosphors in x-ray and television screens; with $BaSO_4$ in the white pigment lithopone; in luminous paints
$ZnSO_4$	rayon manufacture; in animal feeds; wood preservative
ZnCrO4	yellow corrosion resistant paint pigment
$ZnCl_2$	fireproofing agent; solutions dissolve cellulose to produce fiberboard when dried
$ZnHPO_4$	dental cement
Cd metal	in bearing alloys, low melting solders, copper alloys; in voltaic cells (Ni-Cd cell); low cost plating to replace Cr; control rods and shielding in nuclear reactors
CdO	in electroplating; in batteries; as catalyst; nematocide
CdS	yellow pigment in glass, textiles, paper, rubber, ceramics, and soap; in solar cells; photoconductor in xerography; phosphors
$CdSO_4$	electroplating of Cd, Cu, Ni; standard voltaic cells (Weston cell).
Hg metal	thermometers; barometers; electrical relays; electrodes in electrochemical cells; fluorescent tubes; dental amalgams
HgO	polishing compounds; in dry cells; antifouling paint for ships' bottoms; fungicide; red pigment
$HgCl_2$	manufacture of Hg compounds; disinfectant; fungicide; insecticide; wood preservative; formerly used to make felt for hats
Hg_2Cl_2	in electrodes; pharmaceuticals (cathartic and diuretic); fungicide
$Hg(CNO)_2$	highly sensitive explosive for detonators and percussion caps
$C_6H_5HgC_2H_3O_2$, C_2H_5HgCl	fungicide for treating seeds stored for planting

Group 3A

$Al_2(SO_4)_3$	in fire extinguishers; paper making; food additives; treatment of industrial wastes and municipal waste water
Na_3AlF_6	flux for Al electrolytic reduction
Ga metal	in high temperature thermometers
GaAs, GaP	in semiconductors
In metal	in bearing alloys; to "silver" mirrors; in nuclear control rods
InSb, InAs	in semiconductors
Tl metal	in semiconductors
Tl_2SO_4	rat poison

Group 4A

Sn metal	tin plating; in solders, pewter, and bronze
$SnCl_2$	reducing agent
SnF_2	anticavity additive in toothpaste
SnO_2	jewelry abrasive
SnS_2	bronze pigment; for imitation gilding
$SnCl_4$	mordant; durable ceramic glaze
$Sn(CrO_4)_2$	rose and violet ceramic paint
Pb metal	in storage batteries, solder, and type metal; plumbing for chemical laboratories; radiation shields
PbO	in cement; pottery glaze; additive to glass to increase brilliance
Pb_3O_4	corrosion-preventing paint pigment for structural steel

$Pb(C_2H_3O_2)_4$	oxidizing agent in organic synthesis
PbO_2	in explosives and matches
$PbCrO_4$	yellow paint pigment
$Pb(C_2H_5)_4$	to increase the octane number of gasoline
$2PbCO_3 \cdot Pb(OH)_2$	white ceramic glaze and former paint pigment
	Group 5A
Bi metal	with Sn, Pb, and Cd in low melting alloys; in casting alloys; to "silver" mirrors.
$BiONO_3$	to treat gastrointestinal infections and ulcers, and skin infections such as eczema
$BiOCl$	white pigment; in face powders
Bi_2O_3	disinfectant; to fireproof fabrics and plastics
$(BiO)_2CO_3$	in white opalescent ceramic glazes; also pharmaceutically as for $BiONO_3$

Most of the reactions of the representative metals and their compounds have been presented in the preceding objectives of this chapter. Many of the others were presented in Chapter 14. The remaining important reactions—those for metals and compounds of groups 1A and 2A— are presented in Table 22-2.

TABLE 22-2 Some Reactions of the Representative Metals and Their Compounds

Group 1A metals and compounds	
$4\,Li(s) + O_2(g) \longrightarrow 2\,Li_2O(s)$	[60]
$2\,Na(s) + O_2(g) \longrightarrow Na_2O_2(s)$	[61]
$K(s) + O_2(g) \longrightarrow KO_2(s)$ (Rb and Cs also)	[62]
$2\,M(s) + H_2(g) \longrightarrow 2\,MH(s)$	[63]
$M_2O(s) + H_2O \longrightarrow M_2CO_3(s)$	[64]
$6\,Li(s) + N_2(g) \longrightarrow 2\,Li_3N(s)$	[65]
$2\,LiOH(s) \xrightarrow{\Delta} Li_2O(s) + H_2O(g)$	[66]
$Li_2CO_3(s) \xrightarrow{\Delta} Li_2O(s) + CO_2(g)$	[67]
$2\,MHCO_3(s) \xrightarrow{\Delta} M_2CO_3(s) + CO_2(g) + H_2O(g)$	[68]
$2\,MNO_3(s) \xrightarrow{\Delta} 2\,MNO_2(s) + O_2(g)$ (except for $LiNO_3$)	[69]
$4\,LiNO_3(s) \xrightarrow{\Delta} 2\,Li_2O(s) + 4\,NO_2(g) + O_2(g)$	[70]
$2\,MOH(aq) + SO_2(aq) \longrightarrow M_2SO_3(aq) + H_2O$	[71]
$Na_2SO_3(aq) + S \xrightarrow{\Delta} Na_2S_2O_3(aq)$	[72]
Group 2A elements and compounds	
$MCO_3(s) \xrightarrow{\Delta} MO(s) + CO_2(g)$	[73]
$M(OH)_2(s) \xrightarrow{\Delta} MO(s) + H_2O(g)$	[74]

11. Describe how Pb, Hg, and Cd function as poisons in the human body, some of their toxic effects, and some of the environmental sources of these poisons.

The serious nature of poisoning by lead, mercury, and cadmium is enhanced by their tenacity. Once these metals are absorbed by the body, they are excreted slowly if at all. Both Pb and Hg are eliminated over a period of months. There is no evidence that Cd is eliminated in any way.

Lead poisoning seems to be due to interference with the synthesis of the heme group in hemoglobin. Initial symptoms of lead poisoning include listlessness, vomiting, and convulsions. Long term symptoms include anemia, weakness, weight loss and permanent brain damage. Sources of lead in the environment include old paint and automobile exhaust. $Pb(OH)_2 \cdot 2PbCO_3$ formerly was used as a white pigment; it now is banned. $Pb(C_2H_5)_4$ is present in leaded gasoline to improve its octane rating; blood levels of Pb have dropped since unleaded gasoline has been required for new automobiles.

Mercury functions as a poison by interfering with the action of sulfur containing enzymes. The symptoms of mercury poisoning include personality changes and lack of coordination. Mercury attacks the central nervous system and causes brain damage, paralysis, and blindness. Because of the chemical inertness of Hg, it originally was thought that elemental Hg could safely be dumped in the environment. It since has been found that microorganisms produce $Hg(CH_3)_2$ from elemental mercury. Hg discarded in the past is being cleaned up because of the hazard it poses. Hg compounds also are being phased out as fungicides. One further source remains: discarded mercury batteries, such as those for hearing aids.

Cadmium functions biochemically as a poison by replacing Zn in enzymes. Symptoms of Cd poisoning include liver and kidney damage, lung disease, high blood pressure, and skeletal disorders. Environmental sources include metal plating, mining, cigarette smoke, and zinc metal and Zn compounds in which Cd is a contaminant. A major source of Cd is the dust produced from the wear of automobile tires in which ZnO (and thus CdO) is used as a reinforcing agent and a white pigment.

DRILL PROBLEMS

2. Write equations for the following processes.
   A. production of Mg by electrolysis
   B. production of Na by electrolysis
   C. production of Al by electrolysis
   D. concentration of aluminum ore
   E. concn. of Mg from its natural source
   F. calcining of limestone
   G. production of a cheap strong base from limestone

5. Give the formulas and names of the sixteen alums that can be obtained from the cations: Na^+, K^+, NH_4^+, Tl^+, Al^{3+}, V^{3+}, Cr^{3+}, and Fe^{3+}

6. Write equations for the following processes.
   A. roasting of $ZnCO_3$
   B. roasting of ZnS
   C. reduction of NiO with H_2
   D. roasting of HgS
   E. reduction of Fe_2O_3 with H_2
   F. recovery of SO_2
   G. reduction of SnO_2 with C
   H. reduction of PbO with CO
   I. roasting of PbS
   J. reduction of CuO with H_2
   K. roasting of CdS

7. Complete and balance the following reactions. "No reaction" may be the correct answer.
   A. $Be + H_2O \longrightarrow$
   B. $K + HCl(aq) \longrightarrow$
   C. $Zn + HCl(aq) \longrightarrow$
   D. $Pb + HCl(dil) \longrightarrow$
   E. $Al + NaOH(aq) \longrightarrow$
   F. $Sn + HCl(conc.) \longrightarrow$
   G. $Pb + H_2SO_4(conc.) \longrightarrow$
   H. $Al + HCl(conc.) \longrightarrow$
   I. $Cd + H_2SO_4(aq) \longrightarrow$
   J. $Hg + HNO_3(dil.) \longrightarrow$
   K. $Pb + HNO_3(conc.) \longrightarrow$
   L. $Be + HNO_3(dil.) \longrightarrow$
   M. $Be + NaOH(aq) \longrightarrow$
   N. $Bi + H_2SO_4(conc.) \longrightarrow$

8. Complete and balance the following reactions.
   A. $ZnO + HCl(aq) \longrightarrow$
   B. $SnO_2 + H_2SO_4 \longrightarrow$
   C. $ZnO + NaOH(aq) \longrightarrow$
   D. $SnO + HCl(aq) \longrightarrow$
   E. $Zn(OH)_2 + H_2SO_4(aq) \longrightarrow$
   F. $Sn(OH)_2 + NaOH(aq) \longrightarrow$
   G. $Al_2O_3 + CaO(s) + H_2O \longrightarrow$
   H. $Sn(OH)_4 + H_2SO_4(aq) \longrightarrow$
   I. $Sn(OH)_4 + CaO(s) + H_2O \longrightarrow$
   J. $SnO_2 + NaOH(aq) \longrightarrow$

10. Give at least two substances with each of the following uses. Use the data in Table 22-1.
   A. in glass
   B. in nuclear reactors
   C. in soap
   D. fertilizer
   E. in fireworks
   F. in vacuum tubes
   G. to protect metals
   H. in metallurgy
   I. in photoelectricity
   J. in explosives
   K. in medications
   L. fireproofing

M. in wood manufacture
N. phosphors
O. in paper manufacture
P. reducing agents
Q. oxidizing agent
R. in batteries
S. catalyst
T. in textile manufacture
U. in cosmetics

QUIZZES (15-20 minutes each) Choose the best answer for each question.

Quiz A

1. What method is *not* used to treat hard water to make it soft? (a) add CaO; (b) add Na_2CO_3; (c) boil; (d) add $(NaPO_3)_n$; (e) pass through charcoal.
2. Sea water is the source of which metal? (a) Ca; (b) Na; (c) Au; (d) I2; (e) Mg.
3. Which of the following is not a heavy metal poison? (a) Pb; (b) Cd; (c) Zn; (d) Hg; (e) all are poisons.
4. In the Solvay process, $NH_3(g)$ is recovered by adding which of these materials? (a) NaCl(aq); (b) $CaCO_3(s)$; (c) $CO_2(g)$; (d) $Ca(OH)_2(aq)$; (e) none of these.
5. In alkaline solution, the ionic form of tin(IV) is (a) $Sn^{4+}(aq)$; (b) $Sn(OH)_4(aq)$; (c) $Sn(OH)_6^{2-}(aq)$; (d) $Sn(OH)_5^-$; (e) none of these.
6. Which of the following is the formula of a dehydrated alum? (a) NH_4KSO_4; (b) $KCr(SO_4)_2$; (c) $FeCr(SO_4)_2$; (d) $LiAl(SO_4)_2$; (e) none of these.
7. One of the unique characteristics of Li among the alkali metals is its (a) basic hydroxide; (b) thermally stable carbonate; (c) nitride formation; (d) soluble chloride; (e) none of these.
8. Which of the following is used in nuclear reactors? (a) Sn; (b) ZnO; (c) Cd; (d) $PbCrO_4$; (e) none of these.

Quiz B

1. Which of the following does not dissolve in concentrated $HNO_3(aq)$? (a) Zn; (b) Hg; (c) Cd; (d) Al; (e) Pb.
2. Which ion must be present in temporary soft water? (a) Ca^{2+}; (b) Mg^{2+}; (c) Fe^{2+}; (d) SO_4^{2-}; (e) CO_3^{2-}.
3. In the metallurgy of which metal do roasting and reduction occur in one step? (a) Cd; (b) Hg; (c) Sn; (d) Pb; (e) Mg.
4. Pb differs from Sn in which of the following ways? (a) Pb(IV) compounds exist; (b) Pb(II) compounds are reducing agents; (c) Pb has two allotropes; (d) Sn(II) compounds are reducing agents; (e) none of these.
5. Which of the following is not the natural source of an element? (a) NaCl; (b) HgS; (c) $ZnCO_3$; (d) $PbSO_4$ (e) Al_2O_3.
6. Which of the following is insoluble in NaOH(aq)? (a) ZnO; (b) CdO; (c) PbO; (d) Al_2O_3; (e) none of these.
7. Which of the following is not used in explosives? (a) $Hg(CNO)_2$; (b) PbO_2; (c) $KClO_3$; (d) KNO_3; (e) none of these.
8. Which element readily forms a peroxide with $O_2(g)$? (a) Li; (b) Na; (c) Rb; (d) Mg; (e) Ba.

Quiz C

1. Which of the following dissolves in dilute $H_2SO_4(aq)$? (a) Hg; (b) Pb; (c) Bi; (d) Cd; (e) none of these.
2. Permanent hard water: (a) softens when boiled; (b) contains CO_3^{2-}; (c) forms from limestone; (d) forms boiler scale; (e) is softened by zeolites.
3. "Red mud" is the by-product of the metallurgy of (a) Sn; (b) Zn; (c) Hg; (d) Al; (e) none of these.
4. Thallium differs from other metals of group 3A in what way? (a) Tl_2O_3 is amphoteric; (b) Tl has only one allotrope; (c) Tl_2O is a strong reducing agent; (d) TlOH is a strong base; (e) none of these.
5. The ore of which element is concentrated chemically? (a) Na; (b) Sn; (c) Mg; (d) Pb; (e) Zn.
6. Which of the following is anhydrous chromium(III) alum? (a) $NaCr(SO_4)_2$; (b) $CrAl(SO_4)_2$; (c) $MgCr(SO_4)_2$; (d) $KCr(SO_4)_2$; (e) none of these.
7. Which of the following is used in paper production? (a) SnO_2; (b) $Al_2(SO_4)_3$; (b) CaO; (d) $CaCl_2$; (e) none of these.
8. Which element dissolves in both dilute and concentrated acids and bases but not in water? (a) Ca; (b) Cd; (c) Zn; (d) Pb; (e) Hg.

Quiz D

1. Which of the following does not dissolve in NaOH(aq)? (a) Cd; (b) Zn; (c) Be; (d) Sn; (e) Al.
2. After hard water is passed through a zeolite, the effluent has (a) high $[CO_3^{2-}]$; (b) high $[Na^+]$; (c) high $[H^+]$; (d) low $[CO_3^{2-}]$; (e) high $[Cl^-]$.

3. Which of the following is not produced or refined electrically? (a) Na; (b) Mg; (c) Al; (d) Pb; (e) Zn.
4. In which of the following ways does Hg differ from Zn and Cd? (a) Hg_2^{2+} has a covalent bond; (b) Hg is poisonous; (c) Hg has a low melting point; (d) Hg forms several useful alloys; (e) none of these.
5. Which element is found in nature as the chloride? (a) Ca; (b) Na; (c) Zn; (d) Mg; (e) Pb.
6. Which of the following is insoluble in dilute HCl(aq)? (a) Na; (b) Mg; (c) Al; (d) Zn; (e) none of these.
7. Which of the following is not used in glass? (a) PbO; (b) CdS; (c) $CaCO_3$; (d) K_2CO_3; (e) none of these.
8. Which element is the most inert chemically? (a) Al; (b) Cd; (c) Sn; (d) Be; (e) Pb.

SAMPLE TEST (40 minutes)

1. Throughout the temperature range 0°C to 2000°C, $\Delta S° = 179.5$ J mol^{-1} K^{-1} and $\Delta H° = -221$ kJ/mol for $2\ C(s) + O_2(g) \longrightarrow 2\ CO(g)$; and $\Delta S° = -138.1$ J mol^{-1} K^{-1} and $\Delta H° = -554$ kJ/mol for $2\ Fe(s) + O_2(g) \longrightarrow 2\ FeO(s)$. What is the minimum kelvin temperature at which CO(g) spontaneously will reduce FeO(s) to Fe(s)?
2. Write equations for the synthesis of the following substances. Use the naturally occurring compounds (those in the ores) as sources of raw materials. In addition, you may use water, $O_2(g)$, and aqueous solutions (concentrated or dilute) of HCl, HNO_3, H_2SO_4, and NaOH. Use laboratory conditions (moderate ones) rather than industrial conditions, when possible.

a. $PbCl_2(s)$
b. $Hg_2SO_4(s)$
c. $Ca_3(AlO_3)_2(s)$
d. $CdCO_3(s)$
e. $MgSO_4(s)$
f. $Na_2SnO_2(s)$
g. ZnO(s) without using $O_2(g)$
h. $BaSO_4(s)$ from $BaCO_3$ ore
i. $KNO_3(s)$ from KCl ore

23 Chemistry of the Representative (Main-Group) Elements II: Nonmetals

CHAPTER OBJECTIVES

* **1. Use electrode (reduction) potential diagrams as a means of writing half equations and predicting oxidation-reduction equations.**

In electrode potential diagrams, the oxidized and reduced forms of an element are connected with a line. The standard reduction potential is written above each line. The two forms of the element are called a couple. Thus, $SO_4^{2-}|SO_3^{2-}$ is a couple. The couple can be expanded into a complete half equation in the same way we balanced half equations (see objective 5-8).

in base: $SO_4^{2-} + H_2O + 2\,e^- \longrightarrow SO_3^{2-} + 2\,OH^- \quad E° = -0.93\text{ V}$

When we wish to combine two couples (as $SO_4^{2-}|SO_3^{2-}$ and $SO_3^{2-}|S$) to obtain the potential for a third couple ($SO_4^{2-}|S$) we cannot simply add potentials, since the result is a half equation. But we *can* add values of $\Delta G° = -n\mathcal{F}E°$.

$$SO_4^{2-} + H_2O + 2\,e^- \longrightarrow SO_3^{2-} + 2\,OH^- \quad E°_1 = -0.93\text{ V};\ \Delta G°_1 = -2(-0.93)\mathcal{F}$$

$$SO_3^{2-} + 3\,H_2O + 4\,e^- \longrightarrow S + 6\,OH^- \quad E°_2 = -0.67\text{ V};\ \Delta G°_2 = -4(-0.67)\mathcal{F}$$

$$SO_4^{2-} + 4\,H_2O + 6\,e^- \longrightarrow S + 8\,OH^- \quad \Delta G° = 2(0.93)\mathcal{F} + 4(0.67)\mathcal{F} = -6\mathcal{F}E°_3$$

Thus, $-6E°_3 = 4.54$ V *or* $E°_3 = -4.54\text{ V}/6 = -0.76$ V. Electrode potential diagrams of several elements are in *Chapter 23 of the text:* that for chlorine is *Figure 23-1*, that that for oxygen is *Figure 23-6*, that for sulfur is *Figure 23-7*, and that for nitrogen is *Figure 23-10*.

* **2. Outline the principal methods of preparing the halogen elements, the hydrogen halides, and the oxoacids and oxoanions of the halogens.**

Methods of preparing F_2 and Cl_2 are given in equations [12] and [2], respectively, of chapter 14. *Bromine* is extracted from Br^- in seawater with Cl_2 as an oxidizing agent, equation [1]. An abundant natural source of *iodine* is $NaIO_3$, from which it is extracted by a reducing agent, equations [2] and [3].

$$Cl_2(g) + 2\,Br^-(aq) \longrightarrow Br_2(l) + 2\,Cl^-(aq) \quad [1]$$

$$IO_3^- + 3\,HSO_3^- \longrightarrow I^- + 3\,SO_4^{2-} + 3\,H^+ \quad [2]$$

$$5\,I^- + IO_3^- + 6\,H^+ \longrightarrow 3\,I_2 + 3\,H_2O \quad [3]$$

The *hydrogen halides* can be prepared in several ways. Direct union of hydrogen with the halogen, equation [4], occurs explosively with F_2, explosively with Cl_2 in the presence of light, at a moderate rate with Br_2, and slowly with I_2. Combination of the halide ion with hydrogen ion is another method. For HF, almost any acid is suitable, equation [5]. (This is why acids should not be stored near fluoride salts; HF(g) is exceedingly toxic.) For the other hydrogen halides, a rich source of protons (a concentrated strong acid) is needed. H_2SO_4(conc.) is suitable for HCl, equation [6], but not for HBr and HI as H_2SO_4 will oxidize the halide ion to the element, equation [7]. Thus, H_3PO_4(conc.) is used to produce HBr and HI from bromide and iodide salts, equation [8]. Finally, many hydrogen halides can be produced by the hydrolysis of nonmetal halides, equations [9] and [10].

$H_2(g) + X_2(g) \longrightarrow 2\ HX(g)$ [4]

$NaF(s) + HC_2H_3O_2(aq) \longrightarrow NaC_2H_3O_2(aq) + HF(g)$ [5]

$NaCl(s) + H_2SO_4(conc.) \longrightarrow NaHSO_4(aq) + HCl(g)$ [6]

$2\ HI + H_2SO_4 \longrightarrow 2\ H_2O + I_2(g) + SO_2(g)$ [7]

$NaI(s) + H_3PO_4(conc.) \longrightarrow NaH_2PO_4(aq) + HI(g)$ [8]

$PBr_5(g) + 4\ H_2O \longrightarrow H_3PO_4(aq) + 5\ HBr(g)$ [9]

$SiCl_4(g) + 2\ H_2O \longrightarrow SiO_2(s) + 4\ HCl(g)$ [10]

The preparation of the *oxoacids and oxoanions* of chlorine are typical of preparations of the other halogens. The hypohalous acids (HOX) are prepared by adding the element to water, equation [11]. The equilibrium is shifted to the right and the hypohalite ion (OX^-) in aqueous solution is prepared, if dilute cold base is added to the solution, equation [12]. All of the hypohalites are thermally unstable. This instability is used to advantage in preparing the halates (XO_3^-); the element is added to a hot concentrated solution of strong base, equation [13]. All of the halic acids can be prepared as is chloric acid, by reaction of $Ba(ClO_3)_2$ with H_2SO_4, equation [14].

$X_2 + H_2O \rightleftharpoons HOX(aq) + HX(aq)$ [11]

$X_2 + OH^-(aq) \longrightarrow XO^-(aq) + X^-(aq) + H_2O$ [12]

$3\ X_2 + 6\ OH^-(aq) \longrightarrow XO_3^-(aq) + 5\ X^-(aq) + 3\ H_2O$ [13]

$Ba(ClO_3)_2(aq) + H_2SO_4 \longrightarrow 2\ HClO_3(aq) + BaSO_4(s)$ [14]

Chlorous acid can be prepared from a suspension of $Ba(ClO_2)_2$ by reaction with H_2SO_4, equation [15]. Chlorites are formed by the disproportionation of $ClO_2(g)$ in water, equation [16], and $ClO_2(g)$ in turn is prepared by the reaction of $HClO_3$ with oxalic acid, equation [17]. Finally, perchloric acid is produced by the electrolysis of perchlorate salts, half equation [18], followed by distilling the resulting mixture in the presence of H_2SO_4 under reduced pressure (since $HClO_4$ explodes above 92°C), equation [19].

$Ba(ClO_2)_2(aq,susp.) + H_2SO_4(aq) \longrightarrow BaSO_4(s) + 2\ HClO_2(aq)$ [15]

$2\ ClO_2 + 2\ OH^- \longrightarrow ClO_2^- + ClO_3^- + H_2O$ [16]

$2\ HClO_3(aq) + H_2C_2O_4(s) \longrightarrow 2\ ClO_2(g) + 2\ CO_2(g) + 2\ H_2O$ [17]

$ClO_3^-(aq) + H_2O \longrightarrow ClO_4^-(aq) + 2\ H^+(aq) + 2\ e^- \quad E° = -0.36\ V$ [18]

$ClO_4^-(aq) + H_2SO_4(conc.) \longrightarrow HSO_4^-(aq) + HClO_4(g)$ [19]

* **3. Describe bonding and structures of the oxoanions of the halogens, sulfur, and nitrogen; and predict the shapes of interhalogen compounds, polyhalide ions, and noble gas compounds.**

To predict the shapes of oxoanions, interhalogens, polyhalide ions and noble gas compounds, one uses the VSEPR theory (recall objective 10-12). This requires first drawing a plausible Lewis structure (recall objectives 10-6 through 10-10). The total number of electron pairs (= ligands + lone pairs) determines the electron pair geometry of the molecule and the hybridization of the central atom (see Table 11-1). The number of ligands and the number of lone pairs determines the molecular shape (see Table 10-3 and Figure 10-3). You should be aware of one additional hybridization: sp^3d^3 in IF_7 with a pentagonal bipyramid electron pair geometry and molecular shape.

4. Cite similarities and differences between the first and second members of a group in the periodic table.

Here we are comparing the second period elements F, O, N, C, and B with the third period elements Cl, S, P, Si, and Al. The first evident difference is that the second period elements are the more nonmetallic; they have the more positive reduction potentials and thus are the better oxidizing agents. Second, the third period elements are able to form ions and compounds that have more ligands, largely because of the availability of *d* orbitals in those elements. For example, Cl_2, ClF, ClF_3, ClF_5, ClO^-, ClO_2^-, ClO_3^-, and ClO_4^- are all well characterized while the only analogous fluorine species are F_2 and FO^- (the latter having been prepared only during the 1970's). Third, the hydrides of the second period nonmetals (with the exceptions of those of C and B) form strong hydrogen bonds while those of the third period elements do not. Finally, the oxoanions of the third period elements readily join together ($P_2O_7^{4-}$, $S_2O_7^{2-}$, and silicates); this is not so common with second period elements.

These differences are not as important as the similarities of elements in the same family. Cl_2 behaves more like F_2 than it does like O_2 despite the diagonal relationship between O_2 and Cl_2. These similarities are most evident in the organization of how we study the elements. Both Chapters 14, 22, and 23 in the text are organized by families in the periodic table.

5. List several methods of preparing oxygen.

Oxygen is prepared commercially by the fractional distillation of liquid air. This method requires a large capital investment and in general is not practical for laboratory work. Several methods that are suitable for laboratory preparations include: (1) heating oxides of metals of low reactivity, equations [20] and [21]; (2) heating certain oxygen-containing salts, equation [22]; (3) decomposition of water by a thermochemical cycle or by electrolysis, equations [23] and [24]; (4) reaction of an ionic peroxide or a superoxide with water, equations [25] and [26]; (5) decomposition or oxidation of hydrogen peroxide, equations [27] and [28].

$$2\ HgO(s) \xrightarrow{\Delta} 2\ Hg(l) + O_2(g) \qquad [20]$$

$$2\ Ag_2O(s) \xrightarrow{\Delta} 4\ Ag(l) + O_2(g) \qquad [21]$$

$$2\ KClO_3(s) \xrightarrow{\Delta} 2\ KCl(s) + 3\ O_2(g) \qquad [22]$$

$$2\ H_2O \xrightarrow{\Delta} 2\ H_2(g) + O_2(g) \qquad [23]$$

$$2\ H_2O \xrightarrow{\text{electrolysis}} 2\ H_2(g) + O_2(g) \qquad [24]$$

$$2\ O_2^{2-} + 2\ H_2O \longrightarrow 4\ OH^-(aq) + O_2(g) \qquad [25]$$

$$4\ O_2^- + 2\ H_2O \longrightarrow 4\ OH^-(aq) + 3\ O_2(g) \qquad [26]$$

$$2\ H_2O_2(aq) \longrightarrow 2\ H_2(g) + O_2(g) \qquad [27]$$

$$5\ H_2O_2 + 2\ MnO_4^- + 6\ H^+ \longrightarrow 2\ Mn^{2+} + 8\ H_2O + 5\ O_2(g) \qquad [28]$$

6. Describe a number of thio compounds by name, formula, and structure.

The common thio anions are shown in Figure 23-1. The bond angles are approximately tetrahedral in all of these structures. Both dithionite and thiosulfate are good reducing agents. Sodium dithionite is used as a bleach for dyes.

$$S_2O_4^{2-} + 4\ OH^- \longrightarrow 2\ SO_3^{2-} + 2\ H_2O + 2\ e^- \qquad E° = +1.12\ V$$

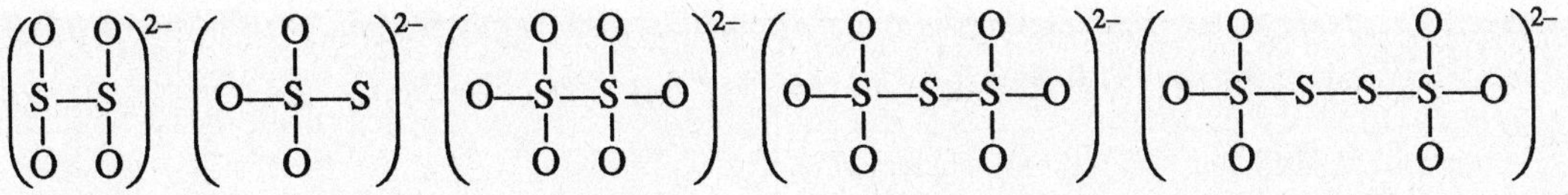

dithionite thiosulfate dithionate trithionate tetrathionate

FIGURE 23-1 Common Thio Anions

* 7. Predict whether a metal sulfide will dissolve in water, acids, bases, $HNO_3(aq)$, or aqua regia.

This is a 3 × 5 card project. Use the data of *Table 23-7 in the text.* Write the conditions on one side of each card and the formulas of the sulfides that dissolve on the other. Note that the water-soluble sulfides also are soluble in 0.3 M HCl, 3 M HNO_3, and aqua regia. Those soluble in 0.3 M HCl also are soluble in 3 M HNO_3 and aqua regia, but not in water. Those soluble in 3 M HNO_3 also are soluble in HNO_3 but not in water or 0.3 M HCl. Those soluble in bases or in an alkaline sulfide solution generally are not soluble in acids.

8. Describe the acid-base and oxidation-reduction properties of NH_3, N_2H_4, and NH_2OH.

All three of these compounds are weak bases, equations [29] through [32] and are built around an NH_2— core: NH_2—H, NH_2—NH_2, and NH_2—OH.

$$NH_3(aq) + H_2O \rightleftharpoons NH_4^+(aq) + OH^-(aq) \qquad K_b = 1.74 \times 10^{-5} \qquad [29]$$

$$N_2H_4(aq) + H_2O \rightleftharpoons N_2H_5^+(aq) + OH^-(aq) \qquad K_{b1} = 8.5 \times 10^{-5} \qquad [30]$$

$$N_2H_5^+(aq) + H_2O \rightleftharpoons N_2H_6^+(aq) + OH^-(aq) \qquad K_{b2} = 8.9 \times 10^{-16} \qquad [31]$$

$$NH_2OH(aq) + H_2O \rightleftharpoons NH_3OH^+(aq) + OH^-(aq) \qquad K_b = 9.1 \times 10^{-9} \qquad [32]$$

NH_3 always is a reducing agent. NH_2OH acts as either an oxidizing agent ($E° = +1.35$ V for $NH_3OH^+|NH_4^+$ in acidic solution and $E° = +0.42$ V in basic solution for $NH_2OH|NH_3$) or as a reducing agent ($E° = +1.87$ V for $NH_3OH^+|N_2$ in acidic solution and $E° = +3.04$ V for $NH_2OH|N_2$ in basic solution). Likewise, N_2H_4 acts as either an oxidizing agent ($E° = +1.24$ V for $N_2H_5^+|NH_4^+$ in acidic solution and $E° = 0.10$ V for $N_2H_4|NH_3$ in basic solution) or as a reducing agent ($E° = +0.20$ V for $N_2H_5^+|N_2$ in acidic solution and $E° = +1.15$ V for $N_2H_4|N_2$ in basic solution). The ability of N_2H_4, hydrazine, to act as a reducing agent accounts for its use as a rocket fuel in combination with hydrogen peroxide, equation [33].

$$N_2H_4(l) + 2\ H_2O_2(l) \longrightarrow N_2(g) + 4\ H_2O(g) \qquad [33]$$

9. Outline methods of preparing the oxides of nitrogen.

Some methods of preparing the several oxides of nitrogen are summarized in *Table 23-12 in the text*. Notice that two of the nitrogen oxides are produced by thermal decomposition: $N_2O(g)$ from $NH_4NO_3(s)$, and $NO_2(g)$ from the nitrate of an inactive metal: $Pb(NO_3)_2$. Two of the nitrogen oxides are the products of nitric acid reactions: NO(g) from the reaction of moderately concentrated HNO_3 with Cu(s), and $N_2O_5(g)$ as a dehydration product of HNO_3 with $P_4O_{10}(s)$ as a dehydrating agent. Finally, three of the nitrogen oxides are formed by direct combination of two gases: $N_2O_3(g)$ from $NO_2(g)$ and NO(g), $NO_2(g)$ from NO(g) and $O_2(g)$, and $N_2O_4(g)$ from $NO_2(g)$ at low temperature and moderate pressures.

* 10. Describe the simple oxoacids of phosphorus and a number of polyphosphoric acids, metaphosphoric acids, polyphosphates, and metaphosphates by name and formula; and name ortho, meta, and pyro oxoacids according to a scheme based on the loss of H_2O molecules.

The phosphoric acids and their anions are but one example of a series in which the common non-oxygen atom has the same oxidation state in all acids and anions. These compounds can be named by a simple scheme based on H_2O molecule loss summarized in the following rules.

1. Starting with the element E in oxidation state n, form the hydroxy compound $E(OH)_n$.
2. If $E(OH)_n$ (more correctly written H_nEO_n) or its anion exists, it is the ortho acid. Thus H_4SiO_4 is orthosilicic acid. If not, loss of one water molecule yields the ortho acid, with formula $H_{n-2}EO_{n-1}$. Thus, since H_5PO_5 does not exist, H_3PO_4 is orthophosphoric acid.
3. The meta acid is formed by the loss of one water molecule from one molecules of the ortho acid. Thus, H_2SiO_3 is metasilicic acid and HPO_3 is metaphosphoric acid.

TABLE 23-1 Common ortho acids

-ic acids					
Representative elements				*Transition elements*	*-ous acids*
H_3BO_3	H_2CO_3	HNO_3		H_3VO_4	H_2SO_3
H_3AlO_3	H_4SiO_4	H_3PO_4	H_2SO_4	H_3WO_4	H_3PO_3
	H_4GeO_4	H_3AsO_4	H_2SeO_4	H_4ZrO_4	H_3AsO_3
	H_4SnO_4	H_3SbO_4	H_6TeO_6	H_4TiO_4	
		H_3BiO_4		H_3NbO_4	

4. The pyro acid is formed by the loss of one water molecule from *two* molecules of the ortho acid. Thus, $H_4P_2O_7$ is pyrophosphoric acid, and $H_2S_2O_7$ is pyrosulfuric acid.

$$HO{-}SO_2{-}OH + HO{-}SO_2{-}OH \longrightarrow HO{-}SO_2{-}O{-}SO_2{-}OH + H_2O \qquad [34]$$

sulfuric acid pyrosulfuric acid

In order to use these rules, you need to know the ortho acids given in Table 23-1. Very few of these acids actually exist. However, their anions all exist. H_2CO_3, HNO_3, H_2SO_4, and H_2SO_3 rarely are named as ortho acids.

11. Indicate the unique features of chemical bonding in the boron hydrides.

Attempts to draw a Lewis structure of diborane, B_2H_6, fail because there are 12 valence electrons and at least 7 single bonds are needed (2 electrons per bond, totalling 14 electrons). This apparent dilemma has been resolved by recognizing that a three-center two-electron B—H—B bond is stable. Although the valence bond theory (objectives 11-1 through 11-5) explanation of such a bond is somewhat complex, the molecular orbital theory (objectives 10-6 through 10-11) explanation is clearer. When three atoms combine, three molecular orbitals form: a bonding (b) orbital, a nonbonding (n) orbital, and an antibonding (a) orbital. These three orbitals are shown in Figure 23-2. The two electrons of the three atoms fill the bonding orbital and create the bond. A similar explanation applies to the B—B—B bond in the higher boron hydrides, such as B_5H_9, B_9H_{15}, and $B_{10}H_{14}$.

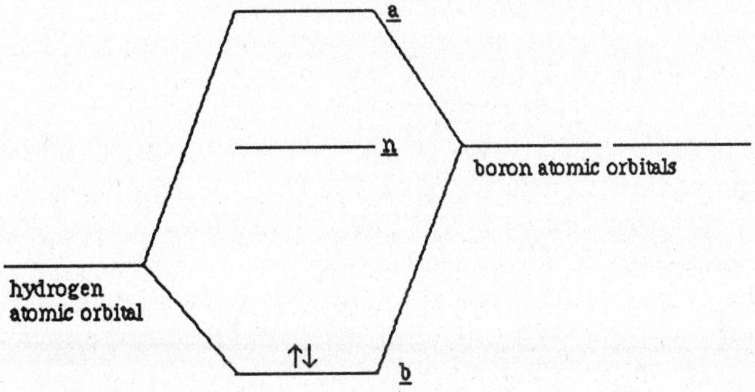

FIGURE 23-2 Molecular Orbital Diagram for the B—H—B Two-center Three-electron Bond
a indicates the antibonding orbital, n the nonbonding orbital, and b the bonding orbital.
The two electrons are indicated as arrows in the molecular orbitals.

* **12. Discuss the oxidizing and/or reducing power of the halogen elements and their oxoacids and oxoanions, oxygen, ozone, hydrogen peroxide, the oxoacids and oxoanions of sulfur and nitrogen, and the noble gas compounds.**

An examination of reduction potential diagrams (*Figures 23-1, 23-6, 23-7, and 23-10 in the text*) reveals that oxoanions are more powerful oxidizing agents in acid than in base. The different half equations reveal why this is so. Consider the reduction of ClO_4^- to ClO_3^- in acidic and basic solutions, equations [35] and [36], respectively.

$$ClO_4^- + 2\,H^+ + 2\,e^- \longrightarrow ClO_3^- + H_2O \qquad [35]$$

$$ClO_4^- + H_2O + 2\,e^- \longrightarrow ClO_3^- + 2\,OH^- \qquad [36]$$

If we consider these two equations in light of LeChâtelier's principle, we notice that in acidic solution, an excess of reactant (H^+) is present, encouraging the forward reaction. In basic solution, on the other hand, an excess of product (OH^-) is present, favoring the reverse reaction and making the forward reaction less likely. A Nernst equation analysis gives a similar result. The Nernst equations for the acidic and basic reactions are given in equations [37] and [38], respectively.

$$E = E^\circ - (0.0592/2)\log([ClO_3^-]/[ClO_4^-][H^+]^2) \qquad [37]$$

$$E = E^\circ - (0.0592/2)\log([ClO_3^-][OH^-]^2/[ClO_4^-]) \qquad [38]$$

Notice that a high $[H^+]$ (or a low $[OH^-]$) increases the value of E, making the oxoanion a better oxidizing agent with a more positive half cell potential.

To compare the oxidizing power of various species, we need to look at their reduction potentials. These reduction potentials must be those of reactions that actually occur (rather than those constructed from half reaction potentials). In addition, it is most helpful if the end product is the same for each series of species that are compared. Such data are presented in Table 23-2. Reduction potentials in acidic solution for the halogens to the halides are presented in the first line, those for the halogen oxoanions to the halogens are in the next three lines, the next line are those of oxygen and hydrogen peroxide, the next two lines are those of the common oxoanions of sulfur and nitrogen, and the last entry is for XeF_2.

TABLE 23-2 Standard Reduction Potentials for Nonmetallic Species and Oxoanions in Acidic Solution.

I_2|I^- +0.535 V	Br_2|Br^- +1.065 V	Cl_2|Cl^- + 1.36 V	F_2|F^- +2.87 V
HClO|Cl_2 +1.62 V	$HClO_2$|Cl_2 +1.63 V	ClO_3^-|Cl_2 +1.47 V	ClO_4^-|Cl_2 +1.38 V
HBrO|Br_2 +1.59 V		BrO_3^-|Br_2 +1.54 V	
HIO|I_2 +1.45 V		IO_3^-|I_2 +1.19 V	
H_2O_2|H_2O + 1.17 V	O_2|H_2O + 1.229 V	O_2|H_2O_2 + 0.682 V	
SO_4^{2-}|SO_2 +0.17 V	SO_2|S +0.45 V		
HNO_2|NO +0.98 V	NO_3^-|HNO_2 +0.94 V	NO_3^-|NO_2 +0.80 V	
XeF_2|Xe,F^- +2.2 V			

A careful examination of the data in this table reveals the following trends.

1. The most electronegative halogens are the strongest oxidizing agents; they have the highest reduction potentials. Thus, oxidizing power decreases in the order $F_2 > Cl_2 > Br_2 > I_2$. XeF_2 is almost as strong an oxidizing agent as F_2.
2. For oxoanions of a given element, the oxoanion with that element in the highest oxidation state is the weakest oxidizing agent; it has the smallest reduction potential. Thus, the oxidizing power of the chlorine oxoanions decrease in the order $HClO \approx HClO_2 > ClO_3^- > ClO_4^-$. In like fashion, O_2 is a better oxidizing agent than is H_2O_2 when both are considered to be reduced to H_2O.
3. For oxoanions of elements in the same period with similar formulas, the oxoanion of the more electronegative nonmetal is the better oxidizing agent. Thus, in terms of oxidizing power, $ClO_4^- > SO_4^{2-}$ and $ClO_3^- > SO_3^{2-}$.

* 13. Cite uses of some of the compounds discussed in this chapter.

Many of the important industrial uses of the representative elements and their compounds are given in Table 14-3. However, uses for some additional compounds have been introduced in this chapter. These compounds, arranged by groups in the periodic table, are given in Table 23-3. That material should be reviewed carefully and committed to memory.

TABLE 23-3 Uses of Some of the Nonmetallic Elements and Some of Their Compounds.

Substance	*Important or Unique Uses*
	Group 7A
ClF_3, BrF_3	fluorination reagents
UF_6	in the separation of uranium isotopes by gaseous diffusion
NaClO	commercial bleach, swimming pool disinfectant
$NaClO_2$	textile bleach
$HClO_4$	metal finishing
$KClO_4$	in detonators and explosives
NH_4ClO_4	in solid propellants
Br_2	to manufacture fire retardants
HBr	used to make inorganic and organic bromides
$C_2H_4Br_2$	gasoline additive, fumigant, pesticide
AgBr	light-sensitive agent in photographic film
I_2	antiseptic; in medications for the thyroid gland; to analyze laboratory reducing agents
ICl	organic iodination reagent
$NaIO_3$	natural source of iodine
AgI	photography
HCN	poison gas; in the preparation of synthetic fibers (Orlon)

	Group 6A
$Na_2S_2O_3$	analytical chemistry reducing agent
$Na_2S_2O_4$	bleach for dyes
$S_2O_8^{2-}$	production of H_2O_2 by hydrolysis
Se	in photoelectric devices and photocopying machines
	Group 5A
N_2H_4	with H_2O_2 as a rocket propellant
$NaNH_2$	dehydrating agent; in synthesis of N-containing organic compounds
NO_2	in the preparation of explosives and organic nitrates
$Pb(N_3)_2$	in detonators
$Na_5P_3O_{10}$	cleaning agent, detergent builder, water softener; in cement manufacturing; in oil well drilling
$(NaPO_3)_n$	water softener

DRILL PROBLEMS

1. Write the balanced half equation and determine the standard reduction potential for each of the following couples. Use the data of *Figures 23-1, 23-6, 23-7, and 23-10 in the text.*
   A. $O_2|H_2O$ (acidic) B. $O_2|OH^-$ (basic) C. $HO_2^-|OH^-$ (basic)
   D. $HClO|Cl^-$ (acidic) E. $ClO_4^-|Cl_2$ (acidic) F. $ClO_4^-|Cl^-$ (acidic)
   G. $HClO_2|Cl^-$ (acidic) H. $ClO_3^-|Cl_2$(basic) I. $ClO_4^-|Cl_2$ (basic)
   J. $ClO_2^-|Cl^-$ (basic) K. $NO_3^-|NO$ (acidic) L. $NO_3^-|NO$ (basic)
   M. $NO_3^-|HNO_2$ (acidic) N. $NO_3^-|NO_2^-$ (basic) O. $NO_3^-|NH_4^+$ (basic)
   P. $HNO_2|N_2$ (acidic) Q. $SO_4^{2-}|S$ (acidic) R. $SO_4^{2-}|S^{2-}$ (basic)
2. Write chemical reactions for the preparation of the following. Be sure to indicate how the final product is separated from the reactions mixture. (For example, if $KClO_3$(s) is requested, and synthesis produces $KClO_3$(aq), indicate that the solution should be gently evaporated to dryness.) Use sodium or potassium halides as the only source of the halogen elements. Use data in the text as necessary.
   A. Cl_2(g) B. Br_2(l) C. I_2(s) D. $KClO_3$(s)
   E. HI(g) F. $HClO_3$(aq) G. $HClO_2$(aq) H. $Ba(ClO_2)_2$(s)
   I. $HClO_4$(l) J. ClF_3(g) K. KI_3(aq) L. F_2(g)
   M. HF(g) N. HCl(g) O. HBr(g)
   P. NaOCl(aq,impure)
3. Give the hybridization on the central atom and the molecular shape of each of the following species.
   A. ICl_5 B. SO_3^{2-} C. IO_6^{5-} D. NO_3^-
   E. BrF_3 F. I_3^- G. ClO_4^- H. BrF_5
   I. IF_7 J. ICl_3 K. NO_2^- L. ClO_2^-
   M. SO_4^{2-} N. BrO_3^- O. ICl_5 P. ClO_3^-
7. Complete and balance each of the following equations. ("No reaction" may be the correct answer.)
   A. CuS + aqua regia $\longrightarrow$ B. SnS + $H_2O \longrightarrow$
   C. $Na_2S + H_2O \longrightarrow$ D. MnS + $H_2O \longrightarrow$
   E. PbS + HCl(aq) $\longrightarrow$ F. CdS + HNO_3(aq) $\longrightarrow$
   G. $As_2S_3 + (NH_4)_2S$(aq) $\longrightarrow$ H. CuS + HCl(aq) $\longrightarrow$
   I. ZnS + aqua regia $\longrightarrow$ J. CaS + HNO_3(aq) $\longrightarrow$
   K. CoS + HNO_3(aq) $\longrightarrow$ L. CuS + KOH(aq) $\longrightarrow$
   M. Na_2S + KOH(aq) $\longrightarrow$
10. Give the correct name or formula of each of the following compounds.
   A. Na_2SiO_3 B. $H_4As_2O_7$ C. $Ca_2B_2O_5$ D. $MgSnO_3$
   E. pyrosilicic acid F. sodium metaaluminate G. calcium pyrophosphate
   H. iron(III) metaborate I. barium orthocarbonate J. magnesium orthobismuthate

12. Use the principles of this objective to rearrange the following species in order of decreasing value (largest to smallest) of reduction potentials.
A. Cl_2, I_2, Br_2, F_2
B. BrO_3^-, BrO_4^-, HBrO
C. ClO_4^-, HPO_4^{2-}, SO_4^{2-}
D. "H_2SO_3", SO_4^{2-}, SO_3^{2-}
E. NO_3^-, HNO_2, $NaNO_2$
F. HPO_4^{2-}, H_3PO_3, $H_2PO_3^-$
G. H_4SiO_4, HSO_4^-, ClO_4^-, HPO_4^{2-}
H. SeO_3^{2-}, H_4GeO_4, BrO_3^-, $H_2AsO_3^-$

13. Give the formulas of at least two substances that are used for each of the following purposes. Use information in Table 23-2 (and Table 14-3 if necessary).
A. photography
B. textile manufacturing
C. analytical reagent
D. detonators
E. explosives
F. water softener
G. medication
H. nitrate synthesis
I. halogenating agents
J. bleach
K. disinfectant
L. poison

QUIZZES (15-20 minutes each) Choose the best answer for each question.

Quiz A

1. Which of the following dissolves in KOH(aq)? (a) HgS; (b) CdS; (c) SnS; (d) CoS; (e) none of these
2. Trigonal pyramid is the molecular shape of which species? (a) ClO_2^-; (b) NO_3^-; (c) IF_3; (d) BrO_3^-; (e) none of these.
3. What is the formula of pyrophosphoric acid? (a) H_3PO_4; (b) HPO_3; (c) H_3PO_3; (d) $H_4P_2O_5$; (e) none of these.
4. In basic solution, $E° = 0.65$ V for $ClO_2^-|OCl^-$ and $E° = 0.40$ V for $OCl^-|Cl_2$. What is the value of $E°$ for $ClO_2^-|Cl_2$? (a) 1.05 V; (b) 0.52 V; (c) 0.48 V; (d) 0.57 V; (e) none of these.
5. Which of the following is the strongest oxidizing agent? (a) Br_2; (b) HNO_2; (c) H_2SO_4; (d) $NaNO_3$; (e) HNO_3.
6. Which compound is used in textile manufacture? (a) HBr; (b) $Na_2S_2O_3$; (c) HCN; (d) N_2H_4; (e) none of these.
7. Which reagent is not used in the preparation of HBr(g)? (a) H_2(g); (b) H_2O; (c) H_2SO_4; (d) H_3PO_4; (e) none of these.
8. The following oxoacids all form pyroacids except (a) H_2SO_4; (b) H_3PO_4; (c) H_4SiO_4; (d) H_3AsO_4; (e) HNO_3.

Quiz B

1. Which of the following is soluble only in aqua regia? (a) HgS; (b) CaS; (c) CdS; (d) As_2S_3; (e) none of these.
2. Which of the following has a linear molecular shape? (a) ClO_2^-; (b) XeF_2; (c) O_3; (d) H_2O; (e) none of these.
3. What is the formula of metasulfurous acid? (a) H_2SO_3; (b) SO_3; (c) $H_2S_2O_5$; (d) $H_2S_2O_7$; (e) none of these.
4. In acidic solution, $E° = 1.229$ V for $O_2|H_2O$ and $E° = 1.77$ V for $H_2O_2|H_2O$. What is the value of $E°$ for $O_2|H_2O_2$? (a) 1.50 V; (b) 0.54 V; (c) 2.31 V; (d) 0.68 V; (e) none of these.
5. Which of the following is the weakest oxidizing agent? (a) $NaClO_3$; (b) $HClO_3$; (c) $HClO_2$; (d) XeF_2; (e) Cl_2.
6. Which of the following is not sensitive to light? (a) AgBr; (b) Se; (c) $H_2 + Cl_2$; (d) NO_2; (e) AgI.
7. Which reagent is not used in a laboratory preparation of O_2(g)? (a) BaO_2; (b) H_2O; (c) $KClO_3$; (d) Ag_2O; (e) $Na_2S_2O_8$.
8. Which central atom produces compounds with the largest number of ligands? (a) F; (b) I; (c) N; (d) Cl; (e) O.

Quiz C

1. Which of the following dissolves in 0.3 M HCl? (a) HgS; (b) SnS; (c) PbS; (d) CdS; (e) none of these.
2. Which of the following has a T-shaped molecular shape? (a) ClO_3^-; (b) ClF_3; (c) NO_3^-; (d) SO_3^{2-}; (e) none of these.
3. What is the formula of pyrophosphorous acid? (a) H_3PO_3; (b) $H_4P_2O_7$; (c) $H_4P_2O_5$; (d) HPO_2; (e) none of these.
4. In acidic solution, $E° = 0.17$ V for $SO_4^{2-}|SO_2$ and $E° = 0.45$ V for $SO_2|S$. What is the value of $E°$ for $SO_4^{2-}|S$? (a) 0.62 V; (b) 0.21 V; (c) 0.36 V; (d) 1.07 V; (e) none of these
5. Which of the following is the strongest oxidizing agent? (a) $HClO_2$; (b) $HClO_3$; (c) $HClO_4$; (d) $KClO_3$; (e) $KClO_2$.
6. Which of the following is *not* used in connection with explosives? (a) HCN; (b) $Pb(N_3)_2$; (c) $KClO_4$; (d) NO_2; (e) none of these.
7. N_2O_3(g) can be prepared from (a) HNO_3; (b) NH_4NO_3; (c) $NO_2 + NO$; (d) $NO + O_2$; (c) NO_2.
8. Which compound forms the weakest hydrogen bonds? (a) HF; (b) H_2O; (c) HCl; (d) NH_3; (e) H_2O_2.

Quiz D

1. Which of the following does *not* dissolve in 3 M HNO_3? (a) PbS; (b) MnS; (c) Sb_2S_3; (d) ZnS; (e) none of these.
2. Which of the following has a square planar molecular shape? (a) ClO_4^-; (b) SO_4^{2-}; (c) ICl_3; (d) ClF_5; (e) none of these.
3. What is the formula of orthoarsenic acid? (a) H_3AsO_4; (b) H_3AsO_3; (c) $HAsO_3$; (d) $H_4As_2O_7$; (e) none of these.
4. In basic solution, $E° = -3.04$ V for $N_2|NH_2OH$ and $E° = 0.42$ V for $NH_2OH|NH_3$. What is the value of $E°$ for $N_2|NH_3$? (a) −2.62 V; (b) −0.87 V; (c) −0.73 V; (d) −1.89 V; (e) none of these.
5. Which of the following is the weakest oxidizing agent? (a) $KClO_4$; (b) $HClO_4$; (c) $KClO_3$; (d) $HClO_3$; (e) $HClO_2$.
6. Which of the following is *not* used for killing (insects, bacteria, and so forth)? (a) Br_2; (b) NaClO; (c) $C_2H_4Br_2$; (d) HCN; (e) I_2.
7. Which reagent is used to prepare H_2O_2? (a) N_2H_4; (b) $Na_2S_2O_8$; (c) $Na_2S_2O_4$; (d) NH_2OH; (e) $H_4P_2O_7$.
8. Which element forms the largest variety of oxoanions? (a) S; (b) Cl; (c) P; (d) Br; (e) I.

SAMPLE TEST (30 minutes)

1. In acidic solution $E° = -1.07$ V for $Br^-|Br_2$, $E° = -1.59$ V for $Br_2|HBrO$, and $E° = -1.49$ V for $HBrO|BrO_3^-$.
   a. Is HBrO spontaneous with respect to disproportionation to Br_2 and BrO_3^-?
   b. What is $E°$ for $Br_2|BrO_3^-$?
2. Write equations for the following processes.
   a. $NaBrO_3$(aq) from Br_2(l), H_2O, and NaOH(aq)
   b. HBr(g) by at least three methods.
   c. O_2(g) by at least three methods.
   d. $NaClO_3$(aq) from NaClO(aq).
   e. H_2S(g) and Cd^{2+}(aq) from CdS.
   f. NO_2(g) by at least two methods.
   g. N_2O_4(g) with HNO_3(aq) as the only source of N.
   h. N_2(g) from a nitrogen-containing compound.
   i. $H_4P_2O_7$ from P_4O_{10} and H_2O
   j. H_2O_2 from a sulfur-containing compound.

24 The Transition Elements

CHAPTER OBJECTIVES

* **1. State ways in which the transition elements differ from the representative elements and know the trends in their properties.**

Transition elements correctly includes those of families 3B through 1B; those elements whose atoms or ions that have partially filled *d* subshells. Often the zinc group (2B) is included in the transition elements, a practice that we will not follow. Sometimes the copper group (1B) and occasionally the scandium group (3B) are excluded from the transition elements, because the chemistry of the elements in these two groups is similar to those of some representative elements. In particular, the chemistry of the elements of group 3B is quite similar to that of the elements of family 3A (Al, Ga, In, Tl). All of the transition elements have high melting and boiling points, and high heats of fusion and vaporization. Most are good conductors of heat and electricity, especially those of group 1B (Cu, Ag, Au). They are denser than the representative elements of the same period. Many transition metal compounds are highly colored and paramagnetic owing to partly filled underlying *d* orbitals.

Except for elements of group 3B (Sc, Y, Ln), the transition elements exhibit more than one oxidation state. Higher oxidation states are common in combination with highly electronegative nonmetals: oxygen, fluorine, and chlorine. The highest oxidation state increases across each period, peaking in family 8B and declining thereafter. Higher oxidation states are more important for heavier members of each family. Thus, Fe(II) and Fe(III) are most important for iron; Os(IV), Os(VI), and Os(VII) for osmium. For representative metals, on the other hand, lower oxidation states are more important for heavier members of a family. Thus, Tl(I) is more common than Tl(III), Pb(II) is more stable than Pb(IV), and Bi(III) is more stable than Bi(V). In all cases, higher oxidation states of a given element form more acidic oxides. Thus CrO is basic, Cr_2O_3 is amphoteric, and CrO_3 is acidic.

One notes the following trends. A general but not constant decrease in atomic radius occurs across a period with the smallest radii in group 8B; group 8B elements are the most dense. The elements are denser but softer toward the bottom of each group. Generally, across each period the electronegativity increases, the reduction potential becomes more positive, and the first ionization potential increases.

2. Describe the lanthanide contraction and explain how it affects the properties of the transition elements.

The lanthanide contraction is the gradual decrease in atomic and ionic radii that occurs from La to Lu (Figure 9-4, objective 9-5.4). Because of the lanthanide contraction, the atoms and ions of each group of the second and third transition series are almost of the same size (Figure 9-3). Because they have the same valence electron configurations, their properties are virtually the same. The similarity is closest at the beginning of the transition series, as shown in Table 24-1.

TABLE 24-1 Periodic Properties of Niobium, Tantalum, Silver, and Gold

	Nb	Ta	Ag	Au
ionic radius, pm	70	73	126	137(+1)
ionization potential, kJ/mol	653	577	732	891
Oxidation states	+5	+5	+1	+3,+1
heat of fusion, kJ/mol	27	28	11.3	12.7
electrical conductance, mho	0.080	0.081	0.616	0.42

* **3. Describe sources and uses of the elements of the first transition series (a 3 × 5 card project).**

Scandium is produced by electrolysis of a molten mixture of $ScCl_3$, LiCl, and KCl. It is too rare (about $150 per gram) for important uses, although its density and strength are similar to those of Al.

Titanium ores are rutile (TiO_2), perovskite ($CaTiO_3$), ilmenite ($FeTiO_3$), and titanite ($TiSiO_4$ and Ca_2SiO_4). Its metallurgy is discussed in objective 24-5. Titanium alloys are strong, lightweight, and heat and corrosion resistant. They are used for aircraft, jet engines, and pipes in chemical plants.

Vanadium is found as V_2O_3, $Pb(VO_4)Cl$, $Pb(VO_3)_2$, and $BiVO_4$. The ores are treated with concentrated HCl, and NH_4Cl is added to produce $(NH_4)_3VO_5$. Roasting $(NH_4)_3VO_5$ yields V_2O_5. Reduction of V_2O_5 with Si, Ca or Mg produces the metal; if this reduction is carried out in the presence of Fe, the resulting mixture is known as ferrovanadium. Vanadium steel is used for automotive springs and axles and high-speed machine tools because of its toughness. Alloyed with Ti, V is both strong and heat resistant and is used in rockets.

Chromium's principal ore is chromite, $Fe(CrO_2)_2$. Reduction with C produces ferrochrome: $Fe(CrO_2)_2 + 4\,C \longrightarrow Fe + 2\,Cr + 4\,CO(g)$. Chromate compounds are produced by roasting with Na_2CO_3 in air: $4\,Fe(CrO_2)_2 + 8\,Na_2CO_3 + 7\,O_2 \longrightarrow 2\,Fe_2O_3 + 8\,Na_2CrO_4 + 8\,CO_2(g)$. Chrome steel is hard and tough. Stainless steel is 14% Cr. Nichrome (80% Ni and 20% Cr) is used for electrical resistance heaters. Chrome plating of steel protects against corrosion.

Manganese's principal ore is pyrolusite, MnO_2. Reduction with Al (or C) produces the metal: $3\,MnO_2 + 4\,Al \longrightarrow 4\,Al_2O_3 + 3\,Mn$. Steel alloy containing 12% Mn is used for armor plate and earth-moving equipment. Manganese also improves the ease of working molten steel.

Iron is found as Fe_2O_3, Fe_3O_4, $FeCO_3$. Its metallurgy is discussed in objective 24-5. Iron and steel are the most important structural metals.

Cobalt's common ores are $CoAs_2$, CoAsS, and Co_3S_4. A series of reactions produces Co_2O_3, which is reduced to the metal by C(s) or $H_2(g)$ at high temperature. Cobalt is alloyed with Fe and other metals to produce heat-resistant metal-cutting tools and surgical instruments. Some alloys are magnetic.

Nickel occurs with Cu and Fe as NiS. To obtain pure nickel, the sulfides are separated by selective flotation. NiS then is roasted to NiO which in turn is reduced with C to Ni: $2\,NiS + 3\,O_2 \longrightarrow 2\,NiO + 2\,SO_2$; $2\,NiO + C \longrightarrow 2\,Ni + CO_2(g)$. The resulting 96% Ni is refined electrolytically. Often the mixed sulfide ore is not separated but roasted to the oxides and reduced with C. The mixture of Ni, Fe, and Cu from the ore is Monel metal, very resistant to chemical attack. Nickel formerly was separated from the mixture with CO(g) at 60°C in the Monel process: $Ni(s) + 4\,CO(g) \rightleftharpoons Ni(CO)_4(g)$. Above 200°C, $Ni(CO)_4(g)$ decomposes, releasing free Ni(s). Nickel plate also makes Fe corrosion resistant.

Copper is found as Cu_2O, $CuFeS_2$, and Cu_3FeS_3, often mixed together. Its metallurgy is discussed in objective 24-5. Electrolytically refined copper is used for electrical wiring. Unrefined copper is used for plumbing and limited structural purposes (rain gutters, roof flashing).

Silver and *gold* often are found free in nature, although rich deposits have been worked. The metallurgies of both metals are described in objective 24-5. Both metals are used in jewelry and electrical circuits.

* **4. Describe uses of some of the important compounds of the transition elements.**

This is a 3 × 5 card project. Use Table 24-2.

TABLE 24-2 Important Compounds of Transition Elements

Sc	No important compounds
TiH_2	removes $O_2(g)$ and $N_2(g)$ from electronic tubes
$TiCl_3$	powerful reducing agent; laundry stain remover
TiO_2	high quality white pigment for paper, plastics, and ceramics
$TiCl_4$	smoke screens as it produces dense clouds of TiO_2 on reaction with H_2O
TiC	edges for cutting tools
$BaTiO_3$	high performance capacitors as a dielectric
VCl_2	mordant in textile dyeing
V_2O_3	industrial catalyst in oxidations
V_2O_5	catalyst for H_2SO_4 production; manufacture of yellow glass; photography; dye mordant
$CrCl_2$	to absorb $O_2(g)$ in gas analysis; organic catalyst
Cr_2O_3	green pigment for paints, glass, and ceramics; abrasive; in semiconductors
Chrome alum	$[KCr(SO_4)_2 \cdot 12\,H_2O]$ dye mordant; waterproofing fabrics

CrO_3	used in chrome plating baths, and in glassware cleaning solutions
$PbCrO_4$, $ZnCrO_4$	yellow paint pigments
Na_2CrO_4	prevents corrosion in boilers and radiators
$FeCrO_4$	lining open-hearth furnaces for steel production
$Na_2Cr_2O_7$	tanning leather; photography; oxidizing agent
$K_2Cr_2O_7$	laboratory oxidizing agent
$MnCl_2$	disinfectant; drying agent in paints
MnO_2	decolorizing glass; drier for paints; oxidant in dry cells; source of Mn compounds
$KMnO_4$	versatile laboratory oxidizing agent
$FeSO_4$	reducing agent; pigment in ink, dye, and paint; weed killer; wood preservative; starting material for Fe compounds
$FeCl_3$	water clarification
$Fe_4[Fe(CN)_6]_3$	ink and paint pigment; blue prints
Fe_2O_3	paint (Venetian red) and cosmetic (rouge) pigment
CoO	blue glass and ceramic pigment
$K_3Co(NO_2)_6$	yellow paint pigment for oil- and water-colors, glass, and porcelain; rubber pigment.
$NiSO_4$	in nickel plating; textile dye mordant
Ni_2O_3	Edison storage battery (nickel-cadmium battery)
Cu_2O	ruby glass pigment; fungicide for seeds
$CuSO_4$	to kill algae in drinking water; calico dyes; starting material from Cu compounds
$Cu_3(AsO_4)_2$	fungicide and insecticide
CuO	insecticide, absorbent for CO
$AgNO_3$	source of Ag compounds
AgX	(silver halides) photography
$HAuCl_4$	("gold chloride") in gold plating; ruby glass manufacture

* **5. Outline the metallurgies of Ti, Fe and steel, Cu, Ag, and Au.**

Titanium is produced mainly from rutile ore, TiO_2. Reaction with HCl or Cl_2 produces the chloride, equation [1]. $TiCl_4(g)$ is reduced with magnesium metal at high temperatures (800 °C) under a He atmosphere, equation [2]

$$TiO_2(s) + 2\,Cl_2(s) + C(s) \longrightarrow TiCl_4(g) + 2\,CO(g) \qquad [1]$$

$$TiCl_4(g) + 2\,Mg(l) \longrightarrow Ti(s) + 2\,MgCl_2(l) \qquad [2]$$

Iron is produced from the oxide ores, Fe_2O_3 and Fe_3O_4. The oxides are reduced by CO(g) and $H_2(g)$, equations [3] and [4]. $H_2(g)$ and CO(g) in turn are produced from coke (produced by heating coal to high temperatures in the absence of air), oxygen, and water, equations [5].

$$Fe_2O_3 + 3\,CO \longrightarrow 2\,Fe + 3\,CO_2 \qquad [3]$$

$$Fe_2O_3 + 3\,H_2 \longrightarrow 2\,Fe + 3\,H_2O \qquad [4]$$

$$2\,C + O_2 \longrightarrow 2\,CO \;\; \textit{and} \;\; C + CO_2 \longrightarrow 2\,CO \;\; \textit{and} \;\; C + H_2O \longrightarrow CO + H_2 \qquad [5]$$

Many impurities are removed in a slag formed with $CaCO_3$, equations [6a] and [6b].

$$CaCO_3 \longrightarrow CaO + CO_2$$

$$CaO + SiO_2 \longrightarrow CaSiO_3 \qquad [6a]$$

$$3\,CaO + P_2O_5 \longrightarrow Ca_3(PO_4)_2 \qquad [6b]$$

The final product, pig iron, contains 4-5% C along with impurities of about 1% each of Si, Mn, and P. It is transformed into steel by melting the mass and injecting $O_2(g)$ in the presence of CaO. The impurities are converted into their oxides which either bubble out of the mixture (CO and CO_2) or combine to form slag [$FeSiO_3$, $MnSiO_3$, $Ca_3(PO_4)_2$]. Alloying metals (Cr, Ni, Mn, V, Mo, and W) are added to the molten iron to produce steel of the desired composition.

Copper ores are concentrated by flotation. Roasting in air at low temperatures produces Cu_2S and converts FeS to FeO, which combines with SiO_2 to form slag, equations [7]. After the slag is separated, air is blown through the molten Cu_2S, yielding free copper, called blister copper because of frozen bubbles of $SO_2(g)$, equation [8]. This impure copper is refined electrolytically.

$$2\ CuFeS_2(s) + 4\ O_2(g) \longrightarrow Cu_2S(l) + 2\ FeO(s) + 3\ SO_2(g)$$

$$FeO(s) + SiO_2(l) \longrightarrow FeSiO_3(l) \qquad [7]$$

$$Cu_2S(l) + O_2(g) \longrightarrow 2\ Cu(l) + SO_2(g) \qquad [8]$$

Both *silver* and *gold* are extracted from their low-grade ores with cyanide-ion containing solutions, equations [9]. The pure metals are recovered from these solutions with the addition of zinc, equations [10]

$$4\ Au(s) + 8\ CN^-(aq) + O_2(g) + 2\ H_2O \longrightarrow 4\ [Au(CN)_2]^-(aq) + 4\ OH^-(aq) \qquad [9a]$$

$$Ag_2S(s) + 4\ CN^-(aq) \longrightarrow 2\ [Ag(CN)_2]^-(aq) + S^{2-}(aq) \qquad [9b]$$

$$2\ [Au(CN)_2]^-(aq) + Zn(s) \longrightarrow 2\ Au(s) + [Zn(CN)_4]^{2-}(aq) \qquad [10a]$$

$$2\ [Ag(CN)_2]^-(aq) + Zn(s) \longrightarrow 2\ Ag(s) + [Zn(CN)_4]^{2-}(aq) \qquad [10b]$$

Ag and Au also are recovered from the anode "muds" produced during the electrolytic refining of Cu and Ni. Ag is recovered when impure Pb is refined.

* **6. State which are the most common oxidation states of the first transition series and group 1B elements, which of these are stable, and which are unstable.**

The common oxidation states of these metals, and an example of a compound of each, are given in Table 24-3. The most stable oxidation state of each element is given in boldface.

TABLE 24-3 Common Oxidation States (and Examples Thereof) of the First Transition Series Elements (**boldface** most stable.)

Scandium			**+3 Sc_2O_3**				
Titanium		+2 TiH_2	+3 $TiCl_3$	**+4 TiO_2**			
Vanadium		+2 VCl_2	+3 V_2O_3	+4 VF_4	**+5 V_2O_5**		
Chromium		+2 $CrCl_2$	+3 Cr_2O_3			**+6 K_2CrO_4**	
Manganese		**+2 $MnCl_2$**	+3 MnF_3	+4 MnO_2		+6 K_2MnO_4	+7 $KMnO_4$
Iron		+2 $FeSO_4$	**+3 $FeCl_3$**				
Cobalt		**+2 CoO**	+3 $K_3[Co(NO_2)_6]$				
Nickel		**+2 $NiSO_4$**	+3 Ni_2O_3				
Copper	+1 Cu_2O	+2 $CuSO_4$					
Silver	**+1 AgX**	+2 Ag_2F					
Gold	+1 AuCl		**+3 $HAuCl_4$**				

* **7. Use electrode potential data to predict the conditions under which certain compounds are likely to disproportionate.**

Electrode potential diagrams for the various oxidation states of vanadium and manganese are given in *Figures 24-3 and 24-8*, respectively, *in the text*. Let us use these data to determine whether $MnO_4^{2-}(aq)$ is stable. We first write the balanced half equations for the oxidation of $MnO_4^{2-}(aq)$ to $MnO_4^-(aq)$ and its reduction to $MnO_2(s)$.

$$2\ e^- + MnO_4^{2-} + 4\ H^+ \longrightarrow MnO_2 + 2\ H_2O \quad E° = 2.26\ V \qquad [11]$$

$$MnO_4^{2-} \longrightarrow MnO_4^- + e^- \quad E° = -0.56\ V \qquad [12]$$

Equation [12] is multiplied by 2 and added to equation [11] to produce the net ionic equation [13]. The positive value of $E°$ indicates that the overall reaction is spontaneous; $MnO_4^{2-}(aq)$ spontaneously disproportionates to $MnO_4^-(aq)$ and MnO_2.

$$3\ MnO_4^{2-} + 4\ H^+ \longrightarrow MnO_2 + 2\ MnO_4^- + 2\ H_2O \quad E° = +1.70\ V \qquad [13]$$

*** 8. Write equations for some important oxidation-reduction reactions, especially those involving permanganate and dichromate ions.**

MnO_4^- is somewhat the stronger oxidizing agent.

$$MnO_4^- + 8\ H^+ + 5\ e^- \longrightarrow Mn^{2+} + 4\ H_2O \quad E° = +1.51\ V \qquad [14]$$

$$Cr_2O_7^{2-} + 14\ H^+ + 6\ e^- \longrightarrow 2\ Cr^{3+} + 7\ H_2O \quad E° = +1.33\ V \qquad [15]$$

Both ions are much better oxidizing agents in acid than in base.

$$MnO_4^- + 2\ H_2O + 3\ e^- \longrightarrow MnO_2 + 4\ OH^- \quad E° = +0.59\ V \qquad [16]$$

$$CrO_4^{2-} + 4\ H_2O + 3\ e^- \longrightarrow Cr(OH)_3 + 5\ OH^- \quad E° = -0.13\ V \qquad [17]$$

9. Discuss the chromate-dichromate equilibrium and the effect of pH on the concentrations of these ions in aqueous solution.

$Cr_2O_7^{2-}$ is a good oxidizing agent (see equation [15]) and a poor precipitating agent (nearly all dichromates are soluble). CrO_4^{2-} is a poor oxidizing agent (see equation [17]) but a good precipitating agent. The two ions are in equilibrium in aqueous solution.

$$2\ CrO_4^{2-} + 2\ H^+ \rightleftharpoons Cr_2O_7^{2-} \quad K_c = 3.2 \times 10^{14} \qquad [19]$$

$$[Cr_2O_7^{2-}]/[CrO_4^{2-}]^2 = 3.2 \times 10^{14}[H^+]^2 \qquad [20]$$

If we know the total chromium(VI) concentration and the pH of the solution, we can determine $[CrO_4^{2-}]$ and $[Cr_2O_7^{2-}]$.

EXAMPLE 24-1 0.200 mol $K_2Cr_2O_7$ is added to 1.00 L of a pH = 4.700 buffer. Determine $[CrO_4^{2-}]$ in the final solution.

$[H^+] = 2.0 \times 10^{-5}$ M. In the final solution, $[Cr_2O_7^{2-}] + 0.5[CrO_4^{2-}] = 0.200$ M.

Rxn:	$2\ CrO_4^{2-}$ +	$2\ H^+$	$\rightleftharpoons$	$Cr_2O_7^{2-}$	$+ H_2O$
Init:		1.00×10^{-5} M		0.200 M	
Change:	$+2x$ M			$-x$ M	
Equil:	$2x$ M	1.00×10^{-5} M		$(0.200 - x)$M	

$[Cr_2O_7^{2-}]/[CrO_4^{2-}] = (0.200 - x)/(2x)^2 = 3.2 \times 10^{14}(2.00 \times 10^{-5}) = 1.3 \times 10^5$

Assuming $x << 0.200$ M, then $0.200/4x^2 \approx 1.3 \times 10^5$ *or* $x = 6.2 \times 10^{-4}$ M.
Thus, $[Cr_2O_7^{2-}] = 1.2 \times 10^{-3}$ M and $[CrO_4^{2-}] = 0.198$ M.

With the same method as Example 24-1, $[Cr_2O_7^{2-}] = 2.0 \times 10^{-3}$ M and $[CrO_4^{2-}] = 0.396$ M in a solution buffered at pH = 9.000

10. Write chemical equations to illustrate the amphoteric nature of certain transition metal oxides and hydroxides, especially those of Cr(III).

An amphoteric substance behaves as a base in the presence of strong acid and as an acid in the presence of strong base. Amphoteric transition metal oxides include: VO_2, V_2O_5, Cr_2O_3, FeO, Fe_2O_3, and CuO. The reactions with acid for the chromium(III) hydroxide and oxide are equations [21]; those with base are equations [22].

$$Cr(OH)_3(s) + 3\ H_3O^+(aq) \rightleftharpoons [Cr(H_2O)_6]^{3+}(aq) \qquad [21a]$$

$$Cr_2O_3(s) + 6\ H_3O^+(aq) + 3\ H_2O \rightleftharpoons 2\ ]Cr(H_2O)_6]^{3}+(aq) \qquad [21b]$$

$$Cr(OH)_3(s) + OH^-(aq) \rightleftharpoons [Cr(OH)_4]^-(aq) \qquad [22a]$$

$$Cr2O_3(s) + 2\ OH^-(aq) + 3\ H_2O \rightleftharpoons 2\ [Cr(OH)_4]^-(aq) \qquad [22b]$$

11. Describe the property of ferromagnetism and the features of atomic structure that lead to it.

Ferromagnetism is the ability of a substance to become strongly and relatively permanently magnetic. For ferromagnetism to be important, the substance must already possess magnetic domains, areas in which the magnetic moments are aligned. If the atoms are small, they will get close enough to pair up (↑↓) and no domains will form.

(You can see this pairing with two small magnets. Place one free on a table and slide the other up to it: side to side and north pole to north pole. When the moving magnet gets close, the free magnet suddenly will swing around.) If the atoms are too large, they will not feel each other's paramagnetism. Under the influence of a strong magnetic field, the domains line up in the same direction. The magnetism of the material is destroyed when this alignment is disrupted by heat or vibration.

12. Describe the formation of metal carbonyls.

The formula of a metal carbonyl can be predicted by beginning with the electron configuration of the element, and adding carbonyl groups with two electrons each until the number of electrons equals that of Kr. If the metal atom has a odd number of electrons, two metal atoms unite with a single bond in the resulting complex. Only $Fe(CO)_5$ and $Ni(CO)_4$ are formed by the direct union of the metal with CO(g). $Fe(CO)_5$ is formed only at high temperature and pressure. Other methods of formation follow.

$$2\ CoCO_3 + 2\ H_2 + 8\ H_2 \longrightarrow Co_2(CO)_8 + 2\ CO_2 + 2\ H_2O$$

$$VCl_3 + CO + Na(excess) + diglyme \xrightarrow[300\ atm]{120°C} [Na(diglyme)_2][V(CO)_6]$$

$$[Na(diglyme)_2][V(CO)_6] \xrightarrow[sublime\ at\ 50°C]{phosphoric\ acid} V(CO)_6 \qquad [24]$$

13. Discuss some of the separations and tests for ions in the ammonium sulfide group of the qualitative analysis scheme.

The tests are outlined in *Figure 24-15 in the text* and described in the accompanying paragraphs. You should be able to reproduce the essential features of *Figure 24-15* from memory. In addition, you should be able to write chemical reactions for each step and explain the observations (formation of a precipitate, development of a color, and so forth). Examples of these reactions are given below. The paragraph numbers are those of the steps in *Figure 24-15*.

1. Treat with NH_3-NH_4Cl buffer and $(NH_4)_2S$.

$$Fe^{3+} + 3\ NH_3 + 3\ H_2O \longrightarrow 3\ Fe(OH)_3(s) + 3\ NH_4+ \qquad [25]$$

$$Fe^{2+} + (NH_4)_2S \longrightarrow FeS(s) + 2\ NH_4^+ \qquad [26]$$

2. Treat precipitate with aqua regia, HCl-HNO_3

$$Fe(OH)_3(s) + 3\ H^+ \longrightarrow Fe^{3+} + 3\ H_2O \qquad [27]$$

$$NiS(s) + 2\ NO_3^- + 8\ H^+ \longrightarrow 3\ Ni^{2+} + 2\ NO(g) + 3\ S(s) + 4\ H_2O \qquad [28]$$

$$MnS(s) + 2\ H^+ \longrightarrow Mn^{2+} + H_2S(g) \qquad [29]$$

3. Treat solution with NaOH(aq)

$$Mn^{2+} + 2\ OH^- \longrightarrow Mn(OH)_2(s) \qquad [30]$$

$$Fe^{3+} + 3\ OH^- \longrightarrow Fe(OH)_3(s) \qquad [31]$$

$$Al^{3+} + 4\ OH^- \longrightarrow [Al(OH)_4]^- \qquad [32]$$

$$Zn^{2+} + 4\ OH^- \longrightarrow [Zn(OH)_4]^{2-} \qquad [33]$$

Then treat with H_2O_2, which forms HO_2^- in alkaline solution.

$$Mn(OH)_2(s) + HO_2^- \longrightarrow MnO_2(s) + OH^- + H_2O \qquad [34]$$

$$2\ Co(OH)_2(s) + HO_2^- + H_2O \longrightarrow 2\ Co(OH)_3(s) + OH^- \qquad [35]$$

$$2\ Cr(OH)_4^{2-} + 3\ HO_2^- \longrightarrow 2\ CrO_4^{2-} + OH^- + 5\ H_2O \qquad [36]$$

4. Treat precipitate from step (3) with HCl(aq), then NH_3(aq).

$$Fe(OH)_3(s) + 3\ H^+ \longrightarrow Fe^{3+} + 3\ H_2O \qquad [37]$$

$$Fe^{3+} + 3\ NH_3 + 3\ H_2O \longrightarrow Fe(OH)_3(s) + 4\ NH_4^+ \qquad [38]$$

$MnO_2(s) + 4\ H^+ + 2\ Cl^- \longrightarrow Mn^{2+} + Cl_2(g) + 2\ H_2O$ [39]

$Ni(OH)_2(s) + 2\ H^+ \longrightarrow Ni^{2+} + 2\ H_2O$ [40]

$Ni^{2+} + 6\ NH_3 \longrightarrow [Ni(NH_3)_6]^{2+}$ [41]

$Co^{3+} + 6\ NH_3 \longrightarrow [Co(NH_3)_6]^{3+}$ [42]

5. Treat the precipitate from step (4) with HCl(aq), and test the solution with KSCN(aq).

$Fe(OH)_3(s) + 3\ H^+ \longrightarrow Fe^{3+} + 3\ H_2O$ [43]

$Fe^{3+} + SCN^- + 5\ H_2O \longrightarrow [Fe(H_2O)_5SCN]^{2+}$(wine red) [44]

6. Treat the solution from step (4) with $(NH_4)_2S$, then treat the solid with HCl(aq) to dissolve only MnS(s), equation [48].

$2\ [Co(NH_3)_6]^{3+} + 3\ (NH_4)_2S \longrightarrow 2\ CoS(s) + S(s) + 6\ NH_4^+ + 12\ NH_3$ [45]

$[Ni(NH_3)_6]^{2+} + (NH_4)_2S \longrightarrow NiS(s) + 2\ NH_4^+ + 6\ NH_3$ [46]

$Mn^{2+} + (NH_4)_2S \longrightarrow MnS(s) + 2\ NH_4^+$ [47]

$MnS(s) + 2\ H^+ \longrightarrow Mn^{2+} + H_2S(g)$ [48]

7. Treat the CoS(s) and NiS(s) with aqua regia (see equation [28]), then NH_3(aq) (equations [41] and [42]).

$CoS(s) + 2\ NO_3^- + 4\ H^+ \longrightarrow Co^{3+} + NO(g) + S(s) + 2\ H_2O$ [49]

Treat half of this solution with dimethylglyoxime (DMG), equation [50], and the other half with, in order: H_2SO_4, NaF, and NH_4SCN.

$[Ni(NH_3)_6]^{2+} + 2\ DMG^- \longrightarrow Ni(DMG)_2$(scarlet,s) $+ 6\ NH_3$ [50]

$[Co(NH_3)_6]^{3+} + 3\ H_2SO_4 \longrightarrow Co^{3+} + 6\ NH_4^+ + 3\ SO_4^{2-}$ [51]

$Co^{3+} + 2\ Cl^- \longrightarrow Co^{2+} + Cl_2(g)$ (Cl^- from aqua regia) [52]

$Co^{2+} + 4\ SCN^- \longrightarrow Co(SCN)_4^{2-}$(blue) [53]

*** 14. Draw conclusions about the presence or absence of ions in a qualitative analysis "unknown," based on the results of laboratory tests.**

The precipitates and solution colors that appear in the course of a qualitative analysis indicate the ions that are present in an unknown. The absence of a precipitate or color indicates the ions that are not present. This uses the data of *Figure 24-15 in the text.*

DRILL PROBLEMS

1. Rewrite each of the following lists in order of increasing value (smallest to largest) of the indicated property.
   - A. atomic radius: Mn, Sc, Fe, Ti
   - B. density: Cr, Sc, Os, Fe
   - C. highest oxidation state: Cr, Sc, V, Mn
   - D. acidity: MnO, Mn_2O_7, MnO_3, MnO_2
   - E. first ionization energy: Zn, Sc, Mn, Fe
   - F. number of unpaired electrons: Fe, V, Sc, Mn
   - G. density: Ti, Pd, Nb, Pt
   - H. acidity: VO, V_2O_3, V_2O_5, VO_2
   - I. highest oxidation state: V, Ti, Mn, Y
   - J. no. of unpaired elns: Cu^{2+}, Fe^{3+}, V^{3+}, Ni^{2+}
   - K. atomic radius: Ti, Cr, Fe, Ca

3. Give the chemical symbol of the transition element or elements with the following properties, sources, or uses.
   - A. Mg is used to produce this metal
   - B. metal produced by electrolysis
   - C. metal refined by electrolysis
   - D. used in electrical wires
   - E. never used in structural alloys
   - F. many ores are sulfides
   - G. metal vaporizes during reduction
   - H. used to coat iron to protect it from corrosion
   - I. present in all steels
   - J. forms heat-resistant alloys
   - K. found in many magnets

4. Give the chemical formula of the transition metal compound or compounds with the following uses.
   A. pigment for paper
   B. fluorescent screens
   C. glass pigment
   D. paint pigment
   E. insecticide
   F. ingredient of burn ointments
   G. used in H_2SO_4 manufacture
   H. oxidizing agent
   I. used in dyeing
   J. corrosion preventative
   K. used in water treatment
   L. used in electrical batteries

5. Write the equation for each of the following processes.
   A. reducing Cu_2S to metallic copper
   B. generating to reducing agent for iron ore
   C. initial treatment of titanium ore.
   D. formation of silicate slag in iron metallurgy
   E. production of titanium metal with C
   F. roasting copper ore
   G. treatment of Ag with $CN^-(aq)$
   H. formation of silicate slag in Cu metallurgy

7. Predict whether or not the following disproportionations will occur. Use the data of *Table 21-1 and Figures 24-3 and 24-8 in the text.* Assume 1.00 M acidic aqueous solution unless otherwise stated.
   A. Cu^+ to Cu^{2+} and Cu°
   B. Sn^{2+} to Sn^{4+} and Sn°
   C. Fe^{2+} to Fe^{3+} and Fe°
   D. H_2O_2 to $O_2(g)$ and H_2O
   E. $MnO_2(s)$ to Mn^{2+} and MnO_4^-
   F. $MnO_2(s)$ to Mn^{2+} & MnO_4^-, 1.0 M OH^-
   G. V^{3+} to V^{2+} and VO^{2+}
   H. MnO_4^{2-} to MnO_4^- & MnO_3^-, 1.0 M OH^-

8. Complete and balance the following equations in aqueous solution.
   A. $K_2Cr_2O_7 + KBr + H_2SO_4 \longrightarrow Br_2(l)$
   B. $KMnO_4 + Na_2C_2O_4 + H_2SO_4 \longrightarrow CO_2(aq)$
   C. $H_2S + KMnO_4 \longrightarrow KOH + SO_2(g)$
   D. $K_2Cr_2O_7 + SnSO_4 + H_2SO_4 \longrightarrow Sn(SO_4)_2$
   E. $KMnO_4 + K_2O_2 \longrightarrow KOH + O_2(g)$
   F. $Cd + KMnO_4 + H_2SO_4 \longrightarrow CdSO_4$
   G. $K_2Cr_2O_7 + HNO_2 \longrightarrow KNO_3$
   H. $NaMnO_4 + NaI \longrightarrow I_2(s)$
   I. $Na_2CrO_4 + SnCl_2 + HCl \longrightarrow SnCl_4$
   J. $HMnO_4 + AsH_3 + H_2SO_4 \longrightarrow H_3AsO_4$
   K. $Cr(OH)_3 + Na_2O_2 \longrightarrow NaOH$
   L. $KMnO_4 + FeSO_4 + H_2SO_4 \longrightarrow Fe_2(SO_4)_3$

12. Each of the following paragraphs gives the results of the qualitative analysis for a different unknown. Use these results to state which ions are definitely present and absent and which are uncertain. All unknowns contain ions from qualitative analysis groups 3 through 5. (See *Figure 5-12 in the text*). (Flame tests are summarized in Table 8-2 of objective 8-3.)
   A. The unknown was treated with NH_4Cl-NH_3 buffer an no precipitated formed. Further treatment with $(NH_4)_2S$ produced a light-colored precipitate that was separated from the solution. When the solution was treated with $(NH_4)_2CO_3$ a white precipitate formed. The remaining solution gave an orange-yellow color to a flame. The white precipitate dissolved in HCl and gave a brick-red color to a flame. The light-colored precipitate dissolved completely in HCl(aq) and did not re-form when NaOH was added.
   B. An unknown was treated with NH_4Cl–NH_3 buffer and a brown-red precipitate formed. Addition of $(NH_4)_2S$ produced no further precipitate. The remaining solution gave no precipitate with $(NH_4)_2CO_3$ but gave a violet color to a flame. The brownish red precipitate dissolved in HCl(aq) but re-formed when the solution was made basic. Treatment of this precipitate with HCl(aq), followed by addition of KSCN(aq) produced a dark red color in the solution.
   C. The unknown gave no precipitate on treatment with NH_4Cl–NH_3 buffer, but a black precipitate when $(NH_4)_2S$ was added. The resulting solution gave a white precipitate on treatment with $(NH_4)_2CO_3$. The solution above the white precipitate gave no positive flame tests. However, when the white precipitate was dissolved in HCl(aq) the solution gave a yellow-green color to a flame. The black precipitate did not dissolve in HCl(aq) but did in aqua regia. Addition of $NH_3(aq)$ produced a faint blue solution, which turned red on the addition of dimethylglyoxime.

QUIZZES (20 minutes each) Choose the best answer for each question.

Quiz A

1. 1. Which element displays ferromagnetism? (a) Co; (b) Cu; (c) Ti; (d) Cr; (e) none of these.
2. Which element has the greatest density? (a) Mn; (b) Ti; (c) Cr; (d) V; (e) Fe.
3. Which is the best oxidizing agent? (a) MnO_4^- in acid; (b) MnO_4^- in base; (c) $Cr_2O_7^{2-}$ in base; (c) $Cr_2O_7^{2-}$ in acid; (e) CrO_4^{2-} in acid.
4. MnO_4^- produces what manganese-containing species when used as an oxidizing agent in base? (a) MnO_4^{2-}; (b) Mn; (c) Mn^{2+}; (d) Mn_2O_3; (e) none of these.
5. Which is the most basic? (a) Cr_2O_3; (b) CrO_3; (c) CrO; (d) CO_2; (e) ClO_2.
6. Which has a tetrahedral shape? (a) $Fe(CO)_5$; (b) $Ni(CO)_4$; (c) $Co_2(CO)_8$; (d) $Mn_2(CO)_{10}$; (e) none of these.
7. The paramagnetism of transition element compounds is due to (a) paired electrons spinning in opposite direction; (b) unpaired electrons in *d* or *f* orbitals; (c) shared valence electrons; (d) unshared valence electron pairs; (e) unpaired electrons in *s* or *p* orbitals.

Quiz B

1. Which metal gives high-temperature strength to steel? (a) Sc; (b) Zn; (c) Cu; (d) Co; (e) none of these.
2. Which element has the lowest first ionization potential? (a) Fe; (b) Ni; (c) Co; (d) Mn; (e) Ti.
3. Which is a green pigment? (a) Cr_2O_3; (b) $PbCrO_4$; (c) ZnO; (d) CoO; (e) none of these.
4. $Cr_2O_7^{2-}$ produces what species when used as an oxidant in base? (a) CrO_3^{2-}; (b) CrO_3; (c) Cr^{3+}; (d) $Cr(OH)_3$; (e) none of these.
5. Which is most acidic? (a) Mn_2O_7; (b) MnO_2; (c) MnO; (d) MgO; (e) Mn_2O_3.
6. Which can be formed by warming the metal in CO(g)? (a) $V(CO)_6$; (b) $Co_2(CO)_8$; (c) $Ni(CO)_4$; (d) $Mn_2(CO)_{10}$; (e) none of these.
7. A property common to all transition metals is that each type of metal (a) is found in many different oxidation states; (b) produces colored compounds; (c) has a high electronegativity; (d) exhibits one oxidation state equal to the group number; (e) reacts with water to produce hydrogen gas.

Quiz C

1. Which metal is very resistant to chemical attack? (a) Fe; (b) Ni; (c) Cu; (d) Zn; (e) none of these.
2. Which has the largest atomic radius? (a) Ti; (b) Fe; (c) Co; (d) Ni; (e) Cr.
3. Which is *not* used in dyeing? (a) $CuSO_4$; (b) ZnO; (c) $FeSO_4$; (d) $KCr(SO_4)_2 \cdot 12H_2O$; (e) none of these.
4. Which will most enhance the oxidizing strength of MnO_4^-? (a) double $[H^+]$; (b) double $[MnO_4^-]$; (c) halve $[H^+]$; (d) halve $[Mn^{2+}]$; (e) halve $[MnO_4^-]$.
5. Which will not dissolve in acidic solution? (a) Al_2O_3; (b) P_2O_5; (c) MgO; (d) Sc_2O_3; (e) all will dissolve.
6. Which is a common test for Fe^{2+}? (a) brownish red oxide; (b) red color with KSCN(aq); (c) white precipitate with NaCl(aq); (d) red color with NH_3(aq); (e) none of these.
7. The last step in the production of copper metal used for electrical conduction is (a) smelting; (b) crushing; (c) roasting; (d) reduction; (e) electrolytic refining.

Quiz D

1. Which metal is produced in the smallest amounts industrially? (a) Fe; (b) Cr; (c) Mn; (d) Cu; (e) Sc.
2. Which has the highest oxidation state in some of its compounds? (a) Mn; (b) Sc; (c) Cu; (d) Zn; (e) V.
3. Which is not used to control insects, fungi, or algae? (a) $CuSO_4$; (b) $ZnCl_2$; (c) CuCl; (d) $Cu_3(AsO_4)_2$; (e) Cu_2O.
4. Under what conditions is Hg^{2+}(aq) mist likely to be formed in a 0.100 M K_2CrO_4, 0.100 M $Hg_2(NO_3)_2$ solution? (a) high temperature; (b) low pH; (c) addition of KOH; (d) none of these will work.
5. Which will dissolve in basic solution? (a) $Ni(OH)_2$; (b) MgO; (c) ZnO; (d) Fe_2O_3; (e) none will dissolve.
6. Which of the following ions precipitates as its hydroxide, rather than its sulfide, when its solution is treated with NH_4Cl-NH_3 and $(NH_4)_2S$ is added? (a) Fe^{2+}; (b) Co^{2+}; (c) Fe^{3+}; (d) Ni^{2+}; (e) Ag^+.
7. The transition elements are *not* characterized by (a) tendency to form complexes; (b) ability to have several different oxidation state; (c) greater reactivity from left to right in a period; (d) tendency to form colored compounds; (e) none of these.

SAMPLE TEST (20 minutes)

1. An alloy is suspected of containing Fe, Cu, and Ni. Describe how you would qualitatively analyze this alloy for these three elements, including how you would dissolve the alloy. Write the chemical equation for each reaction that occurs during the analysis.
2. a. Write equations that demonstrate the production of aluminum chloride and sodium aluminate from aluminum oxide.
   b. What mass of oxide is needed to produce 100.0 g of each of the two products, with the reactions written in the first part of this question?
3. Use the couples $Cr^{2+}|Cr^{3+}$, +1.51 V and $Cr^{3+}|Cr_2O_7^{2-}$, –1.33 V to draw an electrode potential diagram for chromium and determine $E°$ of $Cr_2O_7^{2-}|Cr^{2+}$

25 Complex Ions and Coordination Compounds

CHAPTER OBJECTIVES

* **1. Identify the central ions and the ligands, determine the coordination number and the oxidation state of the central ion, and establish the net charge on a complex ion.**

The formula of a coordination compound has the *complex ion* enclosed in brackets, [ ]. Thus the complex ion of $[Fe(NH_3)_5(NO_3)](NO_3)_2$ is $[Fe(NH_3)_5(NO_3)]^{2+}$, that of $K_2[Zn(CN)_4]$ is $[Zn(CN)_4]^{2-}$, and that of $[Cr(NH_3)_6]Cl_3$ is $[Cr(NH_3)_6]^{3+}$. The *net charge* on the complex ion is that needed to balance the total charge of the simple cations or anions in the compound. The central ion is the metal ion in the complex: Fe^{3+} in $[Fe(NH_3)_5(NO_3)]^{2+}$, Zn^{2+} in $[Zn(CN)_4]^{2-}$, and Cr^{3+} in $[Cr(NH_3)_6]^{3+}$. The *ligands* are the groups attached to the central ion: NH_3 and NO_3^- in $[Fe(NH_3)_5(NO_3)]$, CN^- in $[Zn(CN)_4]^{2-}$, and NH_3 in $[Cr(NH_3)_6]^{3+}$. The ligands are fairly common ions and simple molecules (see Table 25-2). The *oxidation state* of the central ion is determined by the charge on the complex and the total charges on the ligands.

$$\text{complex ion charge} = \text{total charge on ligands} + \text{central ion oxidation state} \quad [1]$$

The *coordination number* is the number of places where ligands are bonded to the central ion. If all the ligands are monodentate—meaning that they each form only one bond to the central ion—the coordination number equals the number of ligands.

* **2. Give the coordination numbers of some common metal ions and write the names and formulas of some common unidentate and multidentate ligands.**

The coordination numbers of common metal ions given in Table 25-1 should be committed to memory. The names, formulas, and charges of the various ligands in Table 25-2 should be memorized also.

TABLE 25-1 Coordination Numbers of Common Metal Ions

Coordination number(s)	*Ion(s)*
2 only	Ag^+
2 or 4	Cu^+, Au^+
4 only	Zn^{2+}, Au^{3+}, Cd^{2+}, Pt^{2+}
4 or 6	Co^{2+}, Ni^{2+}, Cu^{2+}, Al^{3+}
6 only	Ca^{2+}, Fe^{2+}, Sc^{3+}, Cr^{3+}, Fe^{3+}, Co^{3+}, Pt^{4+}

TABLE 25-2 Formulas and Names of Common Ligands

Unidentate ligands—anions					
OH^-	hydroxo	NH_2^-	amido	SO_3^{2-}	sulfito
H^-	hydro	NO_3^-	nitrato	SO_4^{2-}	sulfato
F^-	fluoro	$—NO_2^-$	nitro	$S_2O_3^{2-}$	thiosulfato
Cl^-	chloro	$—ONO^-$	nitrito	O^{2-}	oxo
Br^-	bromo	$—CN^-$	cyano	O_2^{2-}	peroxo
I^-	iodo	$—NC^-$	isocyano	CO_3^{2-}	carbonato
ClO_2^-	chlorito	$—NCS^-$	thiocyanato	Ac^-	acetato ($C_2H_3O_2^-$)
ClO_3^-	chlorato	$—SCN^-$	isothiocyanato		

Unidentate ligands—molecules

NH_3	ammine	H_2O	aqua	CH_3NH_2	methylamine	N_2	nitrogeno
NO	nitrosyl	CO	carbonyl	C_6H_5N	pyridine		

Multidentate ligands[a]

$C_2O_4^{2-}$	oxalato (2) (or ox)	en	ethylenediamine (2)
o-phen	*o*-phenanthroline (2)	dien	diethylenetriamine (3)
EDTA	ethylenediaminetetraacetato (6)	trien	triethylenetetraamine (4)

[a]Numbers in parentheses are the number of monodentate ligands that each multidentate ligand replaces. This is the *denticity* of the multidentate ligand.

* **3. Write distinctive names based on formulas of complexes, and distinctive formulas based on names.**

The following rules are a restatement of those *in the text*.

1. Name the cation first, the anion second.
2. Within the complex ion, name the ligands first (in alphabetical order by ligand name), then the central ion. In writing formulas, this order is reversed except that neutral ligand formulas are written before anion ligand names.
3. Ligand names are those given in Table 25-2.
4. The number of ligands of a given type is indicated by the prefixes: mono- (1, used for emphasis only), di- (2), tri- (3), tetra- (4), penta- (5), and hexa- (6) for unidentate ligands; and bis- (2), tris- (3), and tetrakis- (4) for multidentate ligands. Enclose multidentate ligand names in parentheses.
5. The oxidation state of the central ion is given as a Roman numeral in parentheses, after the name of the complex.
6. Complex *anion* names always end with *-ate*, followed by the appropriate Roman numeral. If the name of the atom ends in *-um*, *-ium*, or *-enum*, the ending is replaced with *-ate* in anion complexes.

Al	aluminate	Cr	chromate	Mo	molybdate

The following metal names are used in anion complexes.

Mn	manganate	Fe	ferrate	Co	cobaltate	Ni	niccolate	Zn	zincate
Cu	cuprate	As	arsenate	Ag	argentate	Pb	plumbate	Au	aurate
Sn	stannate	Sb	antimonate	W	tungstate				

Some examples follow

$[Ag(NH_3)_2]Cl$	diamminesilver(I) chloride
$[Co(NH_3)_3Cl_3]$	triamminetrichlorocobalt(III)
$K_4[Fe(CN)_6]$	potassium hexacyanoferrate(II)
$[Ni(CO)_4]$	tetracarbonylnickel(II)
$[Cu(en)_2]SO_4$	bis(ethylenediamine)copper(II) sulfate
$[Pt(NH_3)_4][PtCl_6]$	tetraammineplatinum(II) hexachloroplatinate(IV)

* **4. Draw plausible structures for complex ions from their names and formulas.**

The coordination number (C.N.) of a metal ion is a guide to the geometry of the ligands around that ion. C.N. = 2 has a linear geometry and that of C.N. = 6 is octahedral. C.N. = 4 can be either square planar, as it is for Ni^{2+}, Cu^{2+}, and Pt^{2+}; or tetrahedral, as it is for Ni^{2+}, Co^{2+}, Zn^{2+}, Cd^{2+}, and Al^{3+}. The square planar geometry is shown by the less active metals that have C.N. = 4, but Ni^{2+} shows both square planar and tetrahedral geometries. The structures of four complexes are drawn in Figure 25-1.

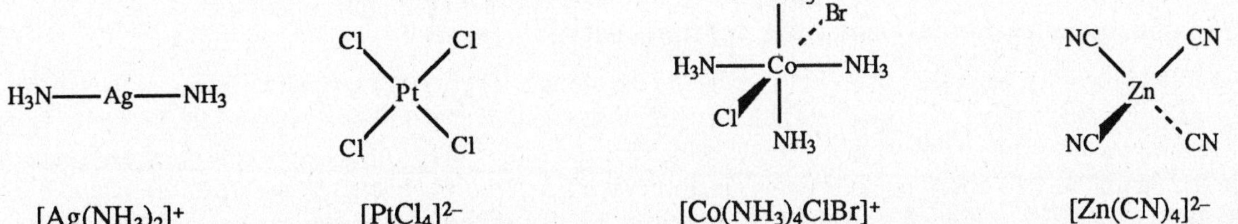

$[Ag(NH_3)_2]^+$ diamminesilver | $[PtCl_4]^{2-}$ tetrachloroplatinate(II) | $[Co(NH_3)_4ClBr]^+$ tetraamminebromochlorocobalt(III) | $[Zn(CN)_4]^{2-}$ tetracyanozincate

FIGURE 25-1 Structures of Four Complex Ions

*** 5. Describe the types of isomerism found among complex ions and identify the possible isomers in specific cases.**

Isomers are compounds with the same formulas but different structures, and thus different chemical or physical properties. Coordination compounds show four types of isomerism (1) Ionization isomers have different ions as ligands. For example $[Co(NH_3)_5SO_4]Br$ is red and gives a positive test for Br^- ion when in solution, but $[Co(NH_3)_5Br]SO_4$ is green and gives a positive sulfate test. (2) Linkage isomerism occurs when the ligand attaches to the central ion in different ways. For example, pentaamminenitrocobalt(III) chloride, $[Co(NH_3)_5NO_2]Cl_2$, is yellow while pentaamminenitritocobalt(III) chloride, $[Co(NH_3)_5ONO]Cl_2$, is red. Other possible linkage isomer ligands are CN^-, SCN^-, and CO. (3) Coordination isomerism occurs when a given ligand is bonded to the cation in one isomer and bonded to the anion in the other. The pair of cobalt(III) complexes $[Co(NH_3)_3(NO_2)_3]$ and $[Co(NH_3)_4(NO_2)_2][Co(NH_3)_2(NO_2)_4]$ are one example. (4) Geometric isomers differ in the arrangement of the ligands around the central ion. This is most easily seen in square planar complexes, such as $[Pt(NH_3)_2Cl_2]$.

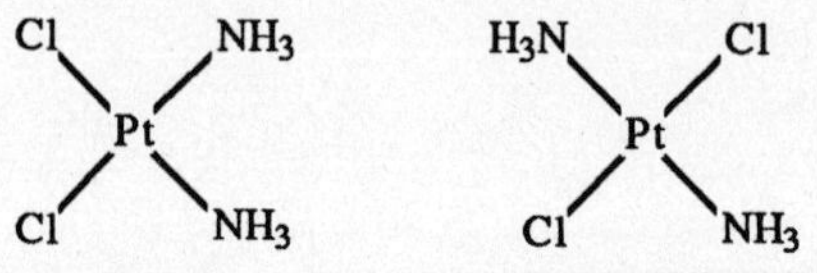

cis isomer *trans* isomer
dichlorodiammineplatinum(II)

In the *trans* isomer the ammines are across the Pt from each other, on opposite sides. In the *cis* isomer they are next to each other. Geometric isomers also occur in octahedral complexes. The *fac* designation arises because the three common ligands are at the corners of a *face* of the octahedron. The *mer* designation signifies that the three common ligands are on the same *mer*idian.

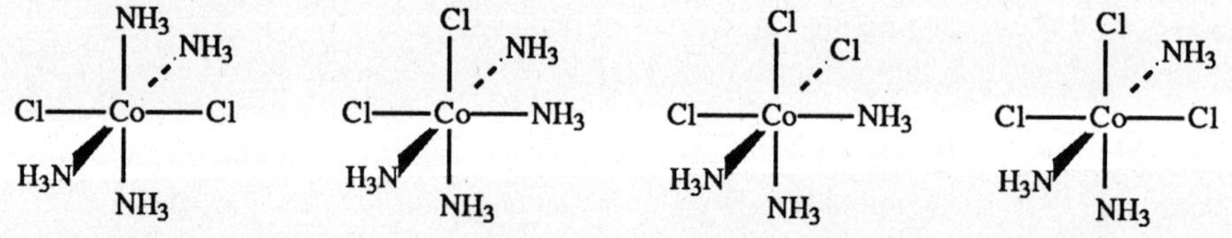

trans isomer *cis* isomer *cis* or *fac* isomer *trans* or *mer* isomer

6. Use valence bond theory to describe the structure and bonding of complex ions.

In valence bond theory a complex is held together by coordinate covalent bonds. The geometry around the central ion corresponds to the hybridization: *sp* for linear, sp^3 for tetrahedral, dsp^2 for square planar, and d^2sp^3 for octahedral. The *s* and *p* orbitals are those of the valence shell of the central atom. Thus, for Ca through Ge, the 4*s* and 4*p* orbitals are used for bonding. Strong bonding ligands use the inner *d* orbitals, 3*d* orbitals for Ca through Ge. These are inner orbital complexes. Weak bonding ligands use the outer *d* orbitals, 4*d* for Ca through Ge. These are called outer orbital complexes. The electron configurations for various complexes, the number of unpaired electrons (encircled) of each, and whether the complex is an inner or outer orbital complex are shown in Figure 24-2. (x's indicate ligand electrons.)

7. Explain the basis of the crystal field theory of bonding in complex ions; and use the spectrochemical series to make predictions about *d*-level splitting and the number of unpaired electrons in complex ions.crystal field splitting energy

In crystal field theory, a complex ion is held together by the attraction between the positive charge of the central ion and the negative electrons of the ligands. When they get close, the ligand electrons repel the valence shell electrons of the central ion. This makes the five *d* orbitals, which are the same energy in the isolated ion, of different energies in the complex. This difference in energies is the *crystal field splitting energy*. It is shown for octahedral and tetrahedral geometries in Figure 25-3. In that figure, the energy of the central ion surrounded by unorganized ligands (the ligands have not yet taken the octahedral or tetrahedral configuration) is called the excited ion's energy. Since the ligands place negatively charged electron pairs in the vicinity of the positively charged central ion, that ion is higher in energy than would be an isolated ion The crystal field splitting energy is symbolized by $\Delta = 10$

Dq. 10 Dq is much smaller for tetrahedral complexes than for octahedral ones. The size of 10 Dq is determined by the ligands. Those that produce a large value of 10 Dq are called *strong field ligands* and are strong Lewis bases. (The complexes that are formed correspond in overall spin to the inner orbital complexes of valance bond theory.)Those that produce a small value of 10 Dq are *weak field ligands* and are weak Lewis bases. The relative strengths of ligands is summarized in the spectrochemical series. (Weak field complexes correspond in overall spin to outer orbital complexes.)

$$CN^- > —NO_2^- > en > py \approx —NH_3 > —NCS^- > H_2O > C_2O_4^{2-} > OH^- > F^- > Cl^- > Br^- > I^- \quad [2]$$

This series should be memorized. Note that nitrogen-bonding ligands are stronger than oxygen-bonding ones. The halides are the weakest ligands.

The strength of the crystal field determines the number of unpaired electrons in d^4, d^5, d^6, and d^7 octahedral complexes and in d^3, d^4, d^5, and d^6 tetrahedral complexes, as shown in Figure 25-4. The dividing line between strong field ligands and weak field ligands is not the same for each cation.

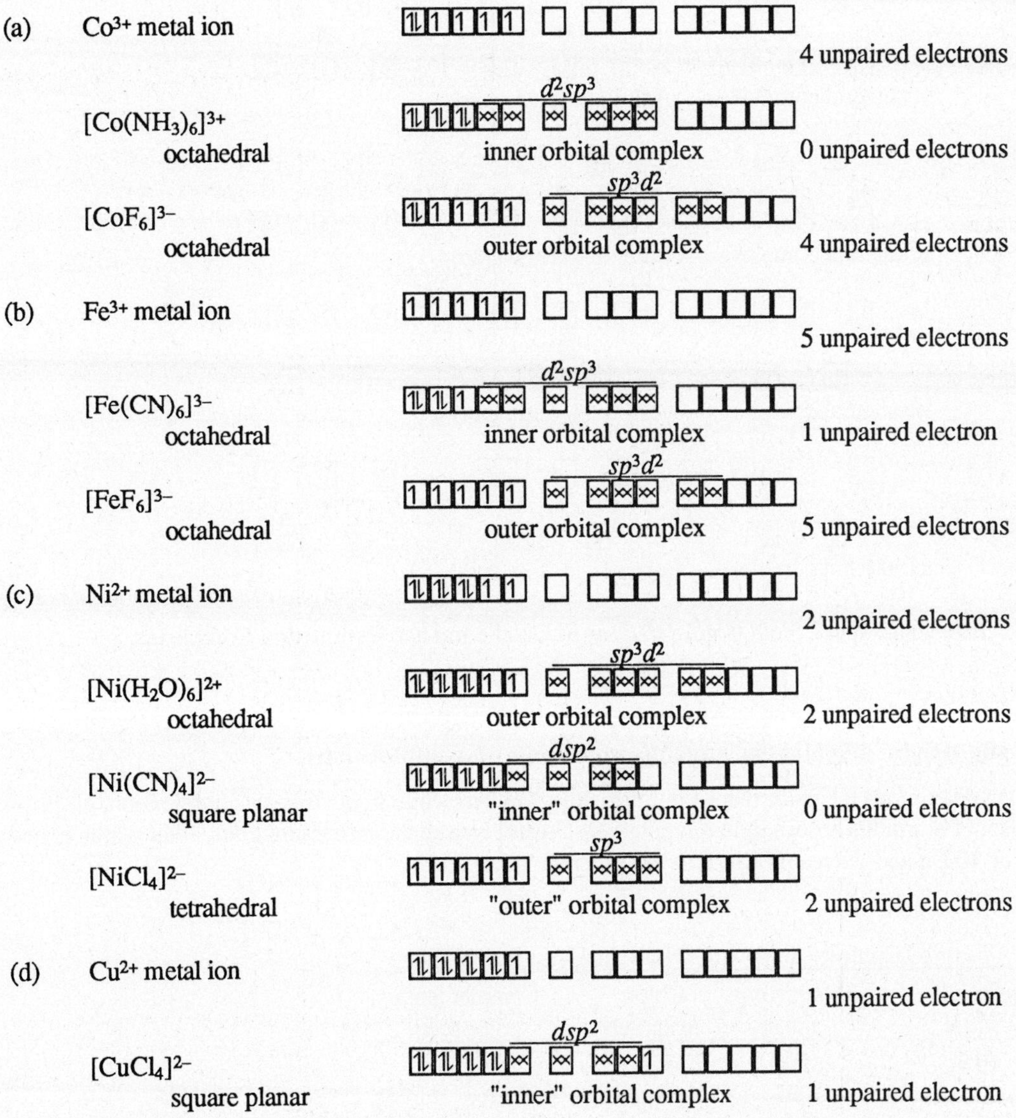

FIGURE 25-2 Orbital Diagrams for Some Inner and Outer Orbital Complexes

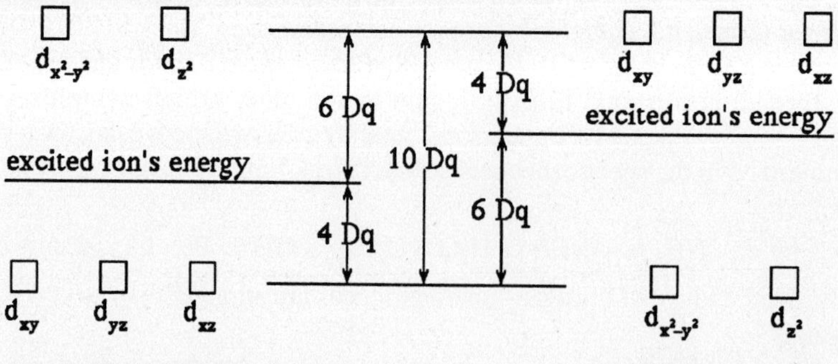

octahedral complex tetrahedral complex

FIGURE 25-3 Ligand Field Splitting in Octahedral and Tetrahedral Complexes

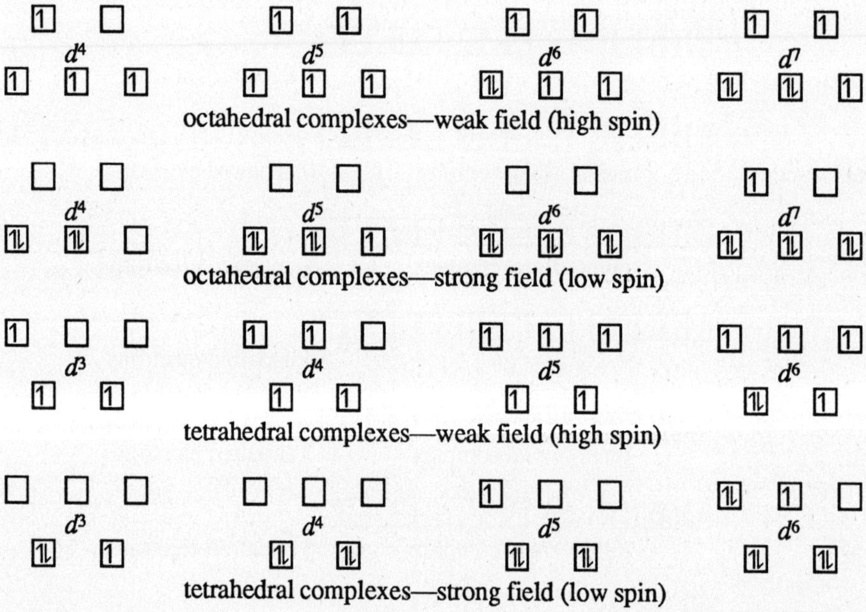

FIGURE 25-4 Strong and Weak Field Configurations in Octahedral and Tetrahedral Complexes.

8. Explain the origin of color in aqueous solutions of complex ions.

The size of 10 Dq determines the color that a complex shows. When white light strikes a complex ion, electrons in the low energy set of *d* orbitals absorb light of energy 10 Dq and we see the remaining light—that which is not absorbed. Values of 10 Dq and the resulting colors are given in Table 25-3.

TABLE 25-3 Colors of Complex Ions with Different Values of 10 Dq

10 Dq = λ (approx) J/molecule	kJ/mol	λ (approx.), nm	Color absorbed	Complementary color seen
3.06×10^{-19}	185	650	red	green
3.31×10^{-19}	200	600	orange	blue
3.43×10^{-19}	206	580	yellow	violet
3.82×10^{-19}	230	520	green	red
4.23×10^{-19}	255	470	blue	orange
4.85×10^{-19}	292	410	violet	yellow

9. Write equations to show the formation of complex ions by the stepwise displacement of H_2O by other ligands, and relate the overall formation constant, K_f, to stepwise formation constants.

In earlier chapters we have written equations for reactions in which complex ions were formed. Many examples are in objective 24-13; for instance

$$Ni^{2+} + 6\ NH_3 \longrightarrow [Ni(NH_3)_6]^{2+} \quad [3]$$

The Ni^{2+} ion in aqueous solution is more properly written $[Ni(H_2O)_6]^{2+}$. A reaction such a [3] is then revealed as one in which one Lewis base, NH_3, is displacing another one, H_2O, as in equation [4]. The complex ion is properly viewed as an adduct between a Lewis acid (the central ion), and Lewis bases (the ligands).

$$[Ni(H_2O)_6]^{2+} + 6\ NH_3 \longrightarrow [Ni(NH_3)_6]^{2+} + 6\ H_2O \quad [4]$$

The exchange of ligands does not take place in one step. In most cases, the equilibrium constant for the replacement of each aqua ligand by an ammine ligand has been measured. For example the reaction of $Ag^+(aq)$ with NH_3 occurs in two steps.

$$[Ag(H_2O)_2]^+ + NH_3 \rightleftharpoons [AgNH_3H_2O]^+ + H_2O \quad K_1 = \frac{[[AgNH_3H_2O]+]}{[[Ag(H_2O)_2]+][NH_3]} = 2.0 \times 10^3 \quad [5]$$

$$[AgNH_3H_2O]^+ + NH_3 \rightleftharpoons [Ag(NH_3)_2]^+ + H_2O \quad K_2 = \frac{[[Ag(NH_3)_2]+]}{[[AgNH_3H_2O]+][NH_3]} = 7.9 \times 10^3 \quad [6]$$

The overall equation is a combination of equations [5] and [6] and its equilibrium constant, K_f, is the product of the equilibrium constants of those two reactions, $K_f = K_1 \times K_2$

$$[Ag(H_2O)_2]^+ + 2\ NH_3 \rightleftharpoons [Ag(NH_3)_2]^+ + 2\ H_2O \quad K_f = \frac{[[Ag(NH_3)_2]^+]}{[[Ag(H_2O)_2]+][NH_3]^2} = 1.6 \times 10^7 \quad [7]$$

Some of the equations from previous chapters appear less artificial when aqua complex ions are written. For example, equation [44] of objective 24-13 becomes

$$[Fe(H_2O)_6]^{3+} + SCN^- \rightleftharpoons [Fe(H_2O)_5SCN]^{2+} + H_2O \quad K_1 = \frac{[[Fe(H_2O)_5SCN]^{2+}]}{[[Fe(H_2O)_6]^{3+}][SCN^-]} = 890 \quad [8]$$

10. Describe how an aqua complex ion ionizes as an acid, and explain amphoterism from this standpoint.

In addition to $Fe^{3+}(aq)$ mentioned *in the text,* $Co^{3+}(aq)$ also ionizes as an acid. In fact, the first ionization constant for $Co^{3+}(aq)$ is stronger than that for any of the weak acids listed in *Table 17-2 in the text.*

$$[Co(H_2O)_6]^{3+} + H_2O \rightleftharpoons [Co(H_2O)_5OH]^{2+} + H_3O^+ \quad K = 0.018 \quad [9]$$

Amphoterism also can be understood from the standpoint of the addition and removal of protons of aqua ligands. Consider the reactions from $Cr(OH)_3$ with strong acid and with strong base.

$$[Cr(OH)_3(H_2O)_3](s) + 3\ H_3O^+ \rightleftharpoons [Cr(H_2O)_6]^{3+} + 3\ H_2O \quad [10]$$

$$[Cr(OH)_3(H_2O)_3](s) + 3\ OH^- \rightleftharpoons [Cr(OH)_6]^{3-} + 3\ H_2O \quad [11]$$

11. Explain how complex ion formation can be used to stabilize oxidation states.

We normally think of copper(I) as unstable in aqueous solution. In fact, copper(I) spontaneously disproportionates to copper(II) and elemental copper.

$$2\ Cu^+ \rightleftharpoons Cu^{2+} + Cu \quad E° = +0.37\ V \quad [12]$$

From this value of $E°$ we can compute a value of $\Delta G°$ ($= -n\mathcal{F}E°$) and from that a value of K_{eq} ($\Delta G° = -RT \ln K_{eq}$).

$$K_{eq} = [Cu^{2+}]/[Cu^+]^2 = 2 \times 10^6 \quad [13]$$

This ratio indicates that the value of $[Cu^+]$ must be quite small. For instance, if $[Cu^{2+}] = 1.0$ M, then $[Cu^+] = 7 \times 10^{-4}$ M. We can achieve this small $[Cu^+]$ if we add cyanide ion to the solution and form $[Cu(CN)_4]^{3-}$.

$$Cu^+ + 4\ CN^- \rightleftharpoons [Cu(CN)_4]^{3-} \quad K_f = 2.0 \times 10^{30} \quad [14]$$

12. Cite ways in which complex ion equilibria are used in the qualitative analysis scheme.

Complex ion equilibria are often used to separate one ion or a group of ions from a larger group of ions or to confirm the presence of an ion by displaying a distinctive color. Some separations include: (1) Often Ag^+ is separated from the mixed precipitate of the chloride qualitative analysis group by adding $NH_3(aq)$. Neither Hg_2^{2+} nor Pb^{2+} forms a very stable ammine complex, but Ag^+ does, equation [7]. (2) In the ammonium sulfide cation group, Al^{3+}, Cr^{3+}, and Zn^{2+} are kept in solution by adding OH^-; at the same time $Mn(OH)_2$, $Co(OH)_2$, $Ni(OH)_2$, and $Fe(OH)_3$ precipitate. (Examples are equations [30] through [33] of objective 24-13.) (3) $NH_3(aq)$ keeps $[Ni(NH_3)_6]^{2+}$ and $[Co(NH_3)_6]^{3+}$ in solution, while $Fe(OH)_3(s)$ precipitates (equations [38], [41], and [42] of objective 24-13).

Some qualitative analysis confirmations are: (1) The formation of $[Fe(H_2O)_5SCN]^{2+}$ confirms Fe^{3+}, equation [8]. (2) The formation of $Ni(DMG)_2$ confirms Ni^{2+}.

$$Ni[(NH_3)_6]^{2+} + 2\ DMG^- \longrightarrow [Ni(DMG)_2] + 6\ NH_3 \qquad [15]$$

(3) $[Pb(C_2H_3O_2)_2]$ formation begins the confirmation of Pb.

$$PbSO_4(s) + 2\ C_2H_3O_2^-(aq) \longrightarrow [Pb(C_2H_3O_2)_2] + SO_4^{2-}(aq) \qquad [16]$$

13. Describe applications of complex ion formation in the photographic process, in electroplating, and in water treatment.

In developing *photographic* film, the unexposed AgBr is removed by forming a thiosulfate complex ion.

$$AgBr(s) + 2\ S_2O_3^{2-} \longrightarrow [Ag(S_2O_3)_2]^{3-} + Br^- \qquad [17]$$

In *electroplating*, the cation of the element to be plated often is in solution as a complex ion. This maintains a low concentration of free metal ion in solution, which produces a smoother, more adherent plating. Frequently even low concentrations of some metal ions in *water* have undesired effects on chemical reactions and manufacturing processes. Often these metal ions can be incorporated by chelation with a multidentate ligand into a stable complex.

DRILL PROBLEMS

1. For each of the following complexes, give the central atom and its oxidation state, the coordination number, and the formulas of the ligands.

A. $[Ag(S_2O_3)_2]^{3-}$ B. $[AlH_4]^-$ C. $[Co(H_2O)_2(CN)_4]^-$
D. $[Pt(NH_3)_2Cl_4]$ E. $[Ni(NH_3)_4]^{2+}$ F. $[Cr(NH_3)_2(SCN)_4]^-$
G. $[Fe(en)_3]^{3+}$ H. $[Pt(NO_2)_4]^{2-}$ I. $[Co(NH_3)_5Cl]^{2+}$
J. $[Al(H_2O)_5OH]^{2+}$ K. $[Pt(NH_3)_3Cl_3]^+$ L. $[Fe(CO)_4]$
M. $[Mn(NO)(CN)_5]^{3-}$ N. $[AuF_4]$ O. $[Co(NH_3)_4(NH_2)_2]^+$
P. $[SbCl_5]^{2-}$ Q. $[NiCl_2(en)_2]$ R. $[Fe(CN)_6]^{4-}$
S. $[Ni(NH_3)_2Br_2]$ T. $[Fe(H_2O)_4(OH)_2]^+$ U. $[Ni(H_2O)_2(NH_3)_4]^{2+}$

2. (1) List all the ions for each coordination number

A. two (2) B. six (6) C. four (4) or six (6)

(2) Give the coordination number(s) of the following ions.

D. Cu^{2+} E. Fe^{2+} F. Fe^{3+} G. Ca^{2+} H. Co^{2+}
I. Au^+ J. Al^{3+} K. Co^{3+} L. Ag^+

(3) Give the name or formula, as appropriate, and the denticity (1 for unidentate, 2 for bidentate, and so forth) of the following ligands. For example: Cl^-, chloro (1).

M. $—NO_3^-$ N. EDTA O. triethylenetetraamine
P. ammine Q. hydroxo R. carbonyl
S. $—NO_2^-$ T. thiocyanato U. NO
V. N_2 W. fluoro X. sulfato
Y. cyano Z. $—ONO^-$

3. Name all of the ions listed in the drill problems of objective 25-1.

V. $Na_3[Au(CN)_4]$ W. sodium dithiosulfatoargentate
X. $K_3[Co(NO_2)_6]$ Y. diamminedibromoplatinum(II)
Z. $[Co(NH_3)_4Cl_2]Br$ Γ. tetraaquadichlorochromium(III) bromide
Δ. diaquatetrachloroplatinum(IV) Θ. $K_2[PtCl_6]$
Λ. diamminesilver chloride Ξ. hexaamminenickel(II) sulfate

Π. $Na[Cu(en)_2(SO_4)_2]$ | Σ. $[Al(H_2O)_6][Co(CN)_6]$
ϒ. $Ca[Cr(NH_3)_2(SCN)_4]_2$ | Φ. $[Co(NH_3)_6]ClSO_4$
Ψ. sodium trisoxalatochromate(III) | Ω. bisethylenediaminecopper(II) chloride

4. Draw a reasonable structure for each of the complexes given in the drill problems of objective 25-1.

5. Use these ten coordination compounds in answering the questions below.

1. $Li[Al(CN)_4]$	2. $[Co(NH_3)_6][ClSO_4]$	3. $[Co(H_2O)_4(CN)_2]Cl$
4. $[Co(en)_2(NH_3)_2]SO_4$	5. $[Al(H_2O)_6][Co(CN)_6]$	6. $[Pt(H_2O)_2(NO_2)_2]$
7. $[Cr(H_2O)_4Cl_2]Br$	8. $[Pt(H_2O)_4][PtCl_6]$	9. $[Co(NH_3)_5Cl]SO_4$
10. $[Mn(CO)_3(CN)_3]$		

A. Identify all of the compounds that have ionization isomers and give the names and formulas of two ionization isomers of each one of them.
B. Identify all of the compounds that have linkage isomers and give the names of two linkage isomers of each compound.
C. Identify all of the compounds that have coordination isomers and give the formulas of two coordination isomers of each compound.
D. Identify all of the compounds that can display geometric isomerism. Draw the structures and designate each geometric isomer as *cis* or *trans*.

7. Predict whether each of the following complex ions will have a strong field or a weak field and predict the number of unpaired electrons in each complex.

A. $[Co(NH_3)_6]^{3+}$	B. $[Cr(en)_3]^{3+}$	C. $[PtCl_6]^{2-}$	D. $[Co(ox)_3]^{4-}$
E. $[Fe(CN)_6]^{4-}$	F. $[CoCl_4]^{2-}$	G. $[CoI_6]^{3-}$	H. $[Ni(C_2O_4)_3]^{4-}$
I. $[Sc(H_2O)_3Cl_3]$	J. $[Co(NH_3)_4]^{2+}$		

QUIZZES (20 minutes each) Choose the best answer or fill in the blank.

Quiz A

1. $[Fe(C_2O_4)_3]^{2-}$ has a coordination number of (a) 2; (b) 3; (c) 4; (d) 6; (e) none of these.
2. Au^{3+} shows a coordination number of (a) 2 only; (b) 2 or 4; (c) 4 only; (d) 4 or 6; (e) 6 only.
3. The nitrito ligand has the formula (a) $—NO_2^-$; (b) $—NO_3^-$; (c) $—ONO^-$; (d) NO; (e) none of these.
4. The name of $[Au(NH_3)_2Cl_2]Br$ is ______________________________.
5. The formula of tri(ethylenediamine)cobalt(III) acetate is ______________________________.
6. $[Fe(CN)_6]^{4-}$ has how many unpaired electrons? (a) 2; (b) 0; (c) 4; (d) 3; (e) none of these.
7. Coordination isomerism could be shown by (a) $Li[AlH_4]$; (b) $[Ag(NH_3)_2][CuCl_2]$; (c) $[Co(NH_3)_4Cl_2]Br$; (d) $[Pt(H_2O)_4Cl_2]$; (e) none of these.
8. The formation constant for the complex $[Zn(NH_3)_4]^{2+}$ is the equilibrium constant of the reaction represented by (a) $Zn^{2+}(aq) + 4\ NH_3(aq) \rightleftharpoons [Zn(NH_3)_4]^{2+}(aq)$; (b) $[Zn(NH_3)_3(H_2O)]^{2+}(aq) + NH_3(aq) \rightleftharpoons [Zn(NH_3)_4]^{2+}(aq) + H_2O$; (c) $Zn(s) + 4\ NH_3(aq) \rightleftharpoons [Zn(NH_3)_4]^{2+}(aq) + 2\ e^-$; (d) $[Zn(H_2O)_4]^{2+}(aq) + 4\ NH_3(aq) \rightleftharpoons [Zn(NH_3)_4]^{2+}(aq) + 4\ H_2O$

Quiz B

1. $[Co(en)_2Cl_2]^+$ has a coordination number of (a) 2; (b) 4; (c) 6; (d) 3; (e) none of these.
2. Cu^{2+} shows a coordination number of (a) 2 only; (b) 2 or 4; (c) 4 only; (d) 4 or 6; (e) 6 only.
3. The carbonyl ligand has the formula (a) CO; (b) CO_3^{2-}; (c) SCN^-; (d) CN^-; (e) none of these.
4. The name of $Ca[Cu(CN)_4]$ is ______________________________.
5. The formula of potassium pentachloroantimonate(III) is ______________________________.
6. $[CoCl_6]^{4-}$ has how many unpaired electrons? (a) 6; (b) 3; (c) 1; (d) 0; (e) none of these.
7. Ionization isomerism could be shown by (a) $[Co(en)_2Cl_2]$; (b) $Na[Ag(CN)_2]$; (c) $[Cr(NH_3)_5Cl]I_2$; (d) $[PtCl_4][Pt(NH_3)_4Cl_2]$; (e) none of these.
8. The $[Fe(CN)_6]^{3-}$ complex ion (a) exhibits square planar geometry; (b) is diamagnetic; (c) involves d^2sp^3 hybridization of Fe(III); (d) has two unpaired electrons; (e) none of these.

Quiz C

1. $[Zn(en)Cl_2]$ has a coordination number of (a) 2; (b) 3; (c) 4; (d) 6; (e) none of these.
2. Au^+ shows a coordination number of (a) 2 only; (b) 2 or 4; (c) 4 only; (d) 4 or 6; (e) 6 only.

3. The oxalato ligand has the formula (a) $C_2O_4^{2-}$; (b) O^{2-}; (c) O_2^-; (d) CO_3^{2-}; (e) none of these.
4. The name of $K_2[Pt(NH_3)_2Cl_2]$ is ______________________.
5. The formula of calcium diamminetetrachlorochromate(III) is ______________________.
6. $[Pt(NH_3)_6]^{4+}$ has how many unpaired electrons (a) 0; (b) 3; (c) 2; (d) 1; (e) none of these.
7. Linkage isomerism can be shown by (a) $[Pt(H_2O)_4Cl_2]$; (b) $[Co(NH_3)_4Cl_2]$; (c) $[Zn(H_2O)_4][CdCl_4]$; (d) $K[Ag(CN)_2]$; (e) none of these.
8. Which of the following is a chelating agent? (a) H_2O; (b) $H_2NCH_2CH_2NH_2$; (c) HCl; (d) NH_3; (e) $S_2O_3^{2-}$.

Quiz D

1. $[Cr(en)(NH_3)(H_2O)_2]^{3+}$ has a coordination number of (a) 3; (b) 4; (c) 5; (d) 6; (e) none of these.
2. Al^{3+} shows a coordination number of (a) 2 only; (b) 2 or 4; (c) 4 only; (d) 4 or 6; (e) 6 only.
3. The hydroxo ligand has the formula (a) OH^-; (b) H_2O; (c) O^{2-}; (d) $C_2O_4^{2-}$; (e) none of these.
4. The name of $Na_2[Co(NO)_2Cl_4]$ is ______________________.
5. The formula of potassium tetracyanodimethylamineferrate(II) is ______________________.
6. A $[Co(H_2O)_4(OH)_2]^+$ ion has how many unpaired electrons? (a) 0; (b) 4; (c) 1; (d) 3; (e) none of these.
7. Geometric isomerism can be shown by (a) $[Ag(NH_3)(CN)]$; (b) $Na_2[Cd(NO_2)_4]$; (c) $[PtCl_4I_2]$; (d) $[Au(CN)_4][Pt(NH_3)_3Cl]$; (e) none of these.
8. One mole of a compound with empirical formula $CoCl_3 \cdot 4NH_3$ yields one mole of AgCl on treatment with excess $AgNO_3(aq)$. Ammonia is not removed by treatment with concentrated H_2SO_4. The formula of this compound is (a) $Co(NH_3)_4Cl_3$; (b) $[Co(NH_3)_4]Cl_3$; (c) $[Co(NH_3)_3Cl_3]NH_3$; (d) $[Co(NH_3)_4Cl_2]Cl$; (e) none of these.

SAMPLE TEST

1. For each of the following complexes, draw the *d* orbital splitting diagram and predict how many unpaired electrons are in the complex.
   a. $[Co(CN)_4]^{2-}$ b. $W(CO)_6$ c. $[MnI_4]^{2-}$ d. $[Fe(CN)_6]^{3-}$
2. A compound contains 24.5% Na, 34.2% Cu, and 41.0% F. A 0.010 M solution of this compound has an osmotic pressure at 298 K of 559 mmHg. What is the formula and the name of this compound?
3. 350.0 mL of 0.200 M $AgNO_3(aq)$ is added to 250.0 mL of 0.240 M $Na_2SO_4(aq)$ and Ag_2SO_4 ($K_{sp} = 1.4 \times 10^{-5}$) precipitates. 400.0 mL of 0.500 M $Na_2S_2O_3(aq)$ is added to this mixture. What mass of $Ag_2SO_4(s)$ remains? $K_f = 1.7 \times 10^{13}$ for $[Ag(S_2O_3)_2]^{3-}$.

26 Nuclear Chemistry

CHAPTER OBJECTIVES

1. Name the different types of radioactive decay processes and describe the characteristics of their radiation.

Alpha (α) *decay* involves nuclides with an atomic number larger than 83 ($Z > 83$) and a mass number larger than 200 ($A > 200$). It often leaves the nucleus in an excited state. The alpha particle is a ^4He nucleus, $^4He^{2+}$. Thus the atomic number decreases by two and the mass number by four units. The alpha particle has great ionizing power but poor penetrating power; it can be stopped by a sheet of paper.

Beta (β) *decay*occurs when electrons are emitted from the nucleus. The nuclide's mass number is unchanged, but its atomic number increases by one. b particles are less ionizing but more penetrating than a particles. They are stopped by about 0.5 mm of Al, whereas an a particle is stopped by 0.015 mm of Al.

Gamma (γ) *decay*is the emission of high-energy photons. This is how the nucleus gets rid of excess energy. It occurs within one nanosecond of some other decay process. If the high-energy (excited) nucleus survives for more than a nanosecond, it is called an "isomer" and its gamma decay is an *isomeric transition* (IT). Gamma rays have relatively little ionizing power but great penetrating power. They are stopped by 5 to 11 mm of Al.

Positron (β^+) *emission* occurs only in artificial nuclides. A positron is a positively-charged electron. Positron emission leaves the mass number unchanged and decreases the atomic number by one. A positron has almost no penetrating power; it is destroyed as soon as it encounters an electron.

Electron capture occurs when a nucleus captures or absorbs one of its own electrons, usually one in the *K* or *L* shell. An "isomer' is formed that emits x rays or g rays to get rid of its excess energy.

* **2. Complete nuclear equations for radioactive decay processes.**

Make sure that the sum of the mass numbers and the sum of the atomic numbers are the same on each side of the equation. Recall (objective 2-8) that the mass number is the preceding superscript of the atomic symbol and the atomic number is the preceding subscript. A decay process has one nuclide on the left side (the "parent" nucleus) and another on the right side (the "daughter" nucleus) along with the emitted particle: alpha (^4_2He), beta ($^{\ 0}_{-1}e$), or positron (^0_1e).

parent nucleus $\longrightarrow$ daughter nucleus + emitted particle [1]

EXAMPLE 26-1 Alpha decay of ^{210}Po yields what nuclide?

The incomplete decay equation is $^{210}_{84}Po \longrightarrow ? + ^4_2He$. The daughter has an atomic number of 82 ($84 = 82 + 2$) and a mass number of 206 ($210 = 206 + 4$). An atomic number of 82 corresponds to Pb. Thus, the product is $^{206}_{82}Pb$.

3. Describe the three natural radioactive decay series,using the uranium series as an example.

$^{238}_{92}U$, $^{232}_{90}Th$, and $^{235}_{92}U$ are the naturally occurring parents of three decay series: the uranium series, the thorium series, and the actinium series, respectively. These three series are shown in Figure 26-1. Note that the atomic numbers (on the horizontal axis) are offset so that the series are not plotted on top of each other. The thorium series is the $4n$ series (the mass numbers of its nuclides are multiples of 4). The uranium series is the $4n + 2$ series, and the actinium series is the $4n + 3$ series. Half-lives and isotope masses of the nuclides in these three series are given in Table 26-1.

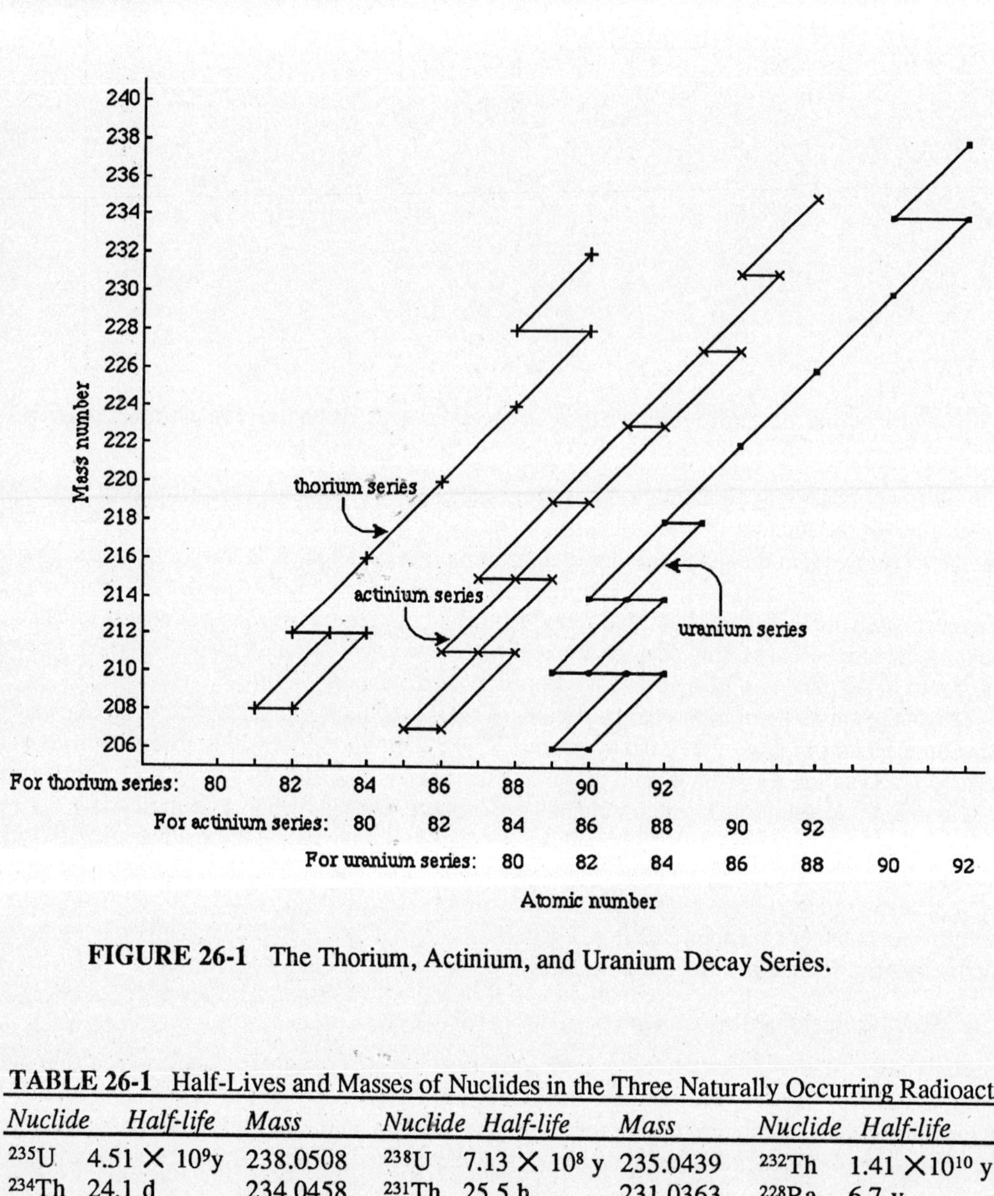

FIGURE 26-1 The Thorium, Actinium, and Uranium Decay Series.

TABLE 26-1 Half-Lives and Masses of Nuclides in the Three Naturally Occurring Radioactive Series

Nuclide	*Half-life*	*Mass*	*Nuclide*	*Half-life*	*Mass*	*Nuclide*	*Half-life*	*Mass*
^{235}U	4.51×10^9y	238.0508	^{238}U	7.13×10^8 y	235.0439	^{232}Th	1.41×10^{10} y	232.0381
^{234}Th	24.1 d	234.0458	^{231}Th	25.5 h	231.0363	^{228}Ra	6.7 y	228.0311
^{234}Pa	6.75 h	234.0433	^{231}Pa	3.25×10^4 y	231.0359	^{228}Ac	6.13 h	228.0310
^{234}U	2.47×10^5 y	234.0409	^{227}Ac	21.6 y	227.0278	^{228}Th	1.91 y	228.0287
^{230}Th	8.0×10^4 y	230.0331	^{227}Th	18.17 d	227.0278	^{224}Ra	3.64 d	224.0202
^{226}Ra	1602 y	226.0254	^{223}Fr	22 min	233.0198	^{220}Rn	53.3 s	220.0114
^{222}Rn	3.82 d	222.0175	^{223}Ra	11.4 d	223.1086	^{216}Po	145 ms	216.0019
^{218}Po	3.05 min	218.0089	^{219}At	54 s	219.0113	^{212}Pb	10.64 h	211.9919
^{218}At	2.0 s	218.0086	^{219}Rn	4.0 s	219.0095	^{212}Bi	60.6 min	211.9913
^{214}Pb	26.8 min	213.9998	^{215}Bi	8 min	215.0018	^{212}Po	0.30 μs	211.9889
^{214}Bi	19.7 min	213.9987	^{215}Po	1.8 ms	214.9995	^{208}Tl	3.10 min	207.9820
^{214}Po	150 μs	213.9952	^{215}Pb	100 μs	214.9987	^{208}Pb	stable	207.9766
^{210}Tl	1.32 min	209.9901	^{211}Pb	36.1 min	210.9887			
^{210}Pb	20.4 y	209.9842	^{211}Bi	2.15 min	210.9873			
^{210}Po	138 d	209.9829	^{207}Tl	4.78	206.9775			
^{206}Tl	4.19 min	205.9761	^{207}Pb	stable	206.9757			
^{206}Pb	stable	205.9745						

* **4. Write equations for the artificial production of nuclides.**

Many of these equations look like equation [2].

target nucleus + bombarding particle ⟶ emitted particle + product nucleus [2]

Occasionally more than one particle and/or product results. A shorthand notation for equation [2] is equation [3].

target nucleus (bombarding particle, emitted particle) product nucleus [3]

Thus, the bombardment of ^{242}Cm by alpha particles is equation [4] or equation [5].

$$^{242}_{96}Cm + ^{4}_{2}He \longrightarrow ^{245}_{98}Cf + ^{1}_{0}n \quad [4]$$

$$^{242}_{96}Cm(\alpha,n)^{245}_{98}Cf \quad [5]$$

In the shorthand notation, we use the abbreviated symbols of the particles given in Table 26-2. Predicting products is not expected of general chemistry students. But, by balancing mass numbers and atomic numbers, we can predict one of the four species in equation [2], given the other three.

TABLE 26-2 Particles Involved in Nuclear Reactions

Name	*Complete symbol*	*Abbreviated symbol*	*Mass, amu*
alpha particle	$^{4}_{2}He^{2+}$	α	4.001507
beta particle *or* electron	$^{0}_{-1}e$	β *or* β^-	0.000548
neutron	$^{1}_{0}n$	n	1.008665
proton	$^{1}_{1}H^+$	p	1.007277
deuteron	$^{2}_{1}H^+$	d	2.013554
helium atom	$^{4}_{2}He^0$		4.002603
positron	$^{0}_{1}e$	β^+	
hydrogen atom	$^{1}_{1}H^0$		1.007825
deuterium atom	$^{2}_{1}H^0$		2.014102

5. Name some of the transuranium elements and describe how they are made.

The transuranium elements are those with atomic number greater than 92 and are made by bombarding lighter nuclei with other particles. The equations for their production are given in Table 26-3. The names and discoverers of elements with atomic numbers higher than 103 have yet to be officially decided.In the interim, their symbols and names are based on their atomic numbers, by stringing together the syllables: nil (0), uni (1), bi (2), tri (3), quad (4), pent (5), hex (6), sept (7), oct (8). and enn (9), and ending with -ium. Thus, element 108, which has been reported, is Uniniloctium

Neptunium, plutonium, americium, curium, berkelium, and californium are available commercially in kilogram quantities. Where two reactions are given in Table 26-3 with only one set of discoverers, the first is the reaction of discovery. The second reaction is that of commercial preparation. You should know the commercial preparations for neptunium through californium.

6. Describe how a charged-particle accelerator operates.

A charged-particle accelerator produces beams of charged particles having very high energies per particle. Often these high-energy particles are atomic nuclei and thus are positively charged. Charged particles are accelerated by an electrical potential that exerts a force on them. In a circular accelerator, the polarity of the electric potential (whether it is positive or negative) in each half of the accelerator alternates as the particles move in a nearly circular path. The field alternates so that it always accelerates the particles. Such circular accelerators also are called cyclotrons.

In a synchrotron, the beam of particles travels through a donut-shaped tube. Magnetic fields are used to keep the beam in the center of the tube (focusing magnets) and to bend the beam into a circular path (deflection magnets). The particles are accelerated by many electrical fields spaced around the ring (or donut). The voltage of each field varies so that the particles are accelerated as they pass. As the particles move around the ring more rapidly, the field must vary more rapidly. (Think of a steel ball traveling around the outer track in a roulette wheel.

If you want to speed up the ball by hitting it each time it passes, you have to hit it more often as it goes faster.) The strength of the deflection magnets also must increase to keep the particles in their circular path.

Linear accelerators also use electrical fields of increasing strength to accelerate particles. However, particles go around a synchrotron many times, being accelerated constantly. Thus a synchrotron can accelerate a particle to a much higher speed than can a linear accelerator.

TABLE 26-3 Production of Transuranium Elements

Element	*Equation*	*Discoverers and date*
neptunium	$^{238}U(n,\beta)^{239}Np$	McMillan and Abelson
	$^{235}U(n,\gamma)^{236}U(n,\gamma)^{237}U(\ ,\beta)^{237}Np$	Berkeley, CA (1940)
plutonium	$^{238}U(d,2n)^{238}Np(\ ,\beta)^{238}Pu$	Seaborg, McMillan, Kennedy, Wahl
	$^{238}U(n,\gamma)^{239}U(\ ,\beta)^{239}Np(\ ,\beta)^{239}Pu$	Berkeley, CA (1940)
americium	$^{239}Pu(n,\gamma)^{240}Pu(n,\gamma)^{241}Pu(\ ,\beta)^{241}Am$	Seaborg, James, Morgan, Ghiorso
		Chicago, Il (1944)
curium	$^{239}Pu(\alpha,n)^{242}Cm$	Seaborg, James, Ghiorso
	$^{241}Am(n,\gamma)^{242m}Am(\ ,\beta)^{242}Cm$	Berkeley, CA (1944)
berkelium	$^{241}Am(\alpha,2n)^{243}Bk$	Thomson, Ghiorso, Seaborg
		Berkeley, CA (Dec 1949)
californium	$^{242}Cm(\alpha,n)^{245}Cf$	Thomson, Street, Ghiorso, Seaborg
	$^{249}Bk(n,\gamma)^{250}Bk(\ ,\beta)^{250}Cf$	Berkeley, CA (1950)
		In nuclear reactors
einsteinium	$^{238}U(15\ n,7\ \beta)^{253}Es$	Ghiorso and colleagues
	products of first hydrogen bomb	Berkeley, CA (Dec 1952)
fermium	$^{238}U(^{16}O,4\ n)^{250}Fm$	Ghiorso and colleagues
	products of first hydrogen bomb	Berkeley, CA (Dec 1952)
mendelevium	$^{253}Es(\alpha,n)^{256}Md$	Ghiorso, Harvey, Choppin, Thompson,
		Seaborg Berkeley, CA (1955)
nobelium	$^{246}Cm(^{12}C,4\ n)^{254}No$	Ghiorso, Sikkeland, Walton, Seaborg
		Berkeley, CA (Apr 1958)
lawrencium	$^{252}Cf(^{11}B,5\ n)^{258}Lw$	Ghiorso, Sikkeland, Larsh, Latimer
		Berkeley, CA (Mar 1961)
element 104	$^{242}Pu(^{22}Ne,4\ n)^{260}Ku$	Joint Nuclear Research Institute
(kurchatovium)		Dubna, USSR (1964)
(rutherfordium)	$^{249}Cf(^{12}C,4\ n)^{257}Ru$	Ghiorso and colleagues
		Berkeley, CA (1969)
element 105	$^{243}Am(^{22}Ne,\ 4n)^{261}Unp$	Joint Nuclear Research Institute
	(no name proposed)	Dubna, USSR (early 1970)
(hanium)	$^{249}Cf(^{15}N,4\ n)^{250}Ha$	Ghiorso and colleagues
		Berkeley, CA (Mar 1970)
element 106	$^{249}Cf(^{18}O,4\ n)^{263}Unh$	Lawrence Berkeley Laboratories
	(no name proposed)	Berkeley, CA (Sept 1974)
	$^{206\text{-}208}Pb(^{54}Cr,?)^{?}Unh$	Joint Nuclear Research Institute
	(no name proposed)	Dubna, USSR (June, 1974)
element 107	$^{204}Bi(^{54}Cr,?)^{?}Uns$	Joint Nuclear Research Institute
	(no name proposed)	Dubna, USSR (1976)
element 109	$^{209}Bi(^{58}Fe,n)^{266}Une$	Heavy Ion Research Laboratory
	(no name proposed)	Darmstadt, West Germany

* **7. Given two of the quantities: rate of radioactive decay, half-life, and number of atoms in a sample of a radioactive nuclide, calculate the third.**

It will be very helpful for you to realize that the kinetics of radioactive decay are first-order kinetics. Problems involving radioactive decay therefore can be solved with the techniques of Chapter 15. The three quantities mentioned are related as follows.

$$t_{1/2} = 0.693/\lambda \qquad [6]$$

decay rate = λN [7]

In these equations, $t_{1/2}$ is the half-life of the nuclide, λ is the decay constant of the nuclide (a first-order rate constant), and N is the number of atoms in the sample. Decay rate also is known as activity.

EXAMPLE 26-2 A 1.00×10^{-6} g sample of a nuclide with an atomic weight of 231 g/mol decays at a rate of 1.00×10^5 atoms/min. What are the half-life and the identity of the nuclide (Table 26-1)?

$N = 1.00 \times 10^{-6}$ g $\times$ mol/231 g $\times$ 6.022×10^{23} atoms/mol

$= 2.61 \times 10^{15}$ atoms

Now we determine the decay constant and the half-life.

1.00×10^5 atoms/min $= \lambda\,(2.61 \times 10^{15}$ atoms$)$ *or* $\lambda = 3.83 \times 10^{-11}$/min

$t_{1/2} = 0.693/3.83 \times 10^{-11}$/min $= 1.81 \times 10^{10}$ min $\times$ h/60 min $\times$ d/24 h $\times$ y/365 d

$= 3.44 \times 10^4$ y

This is quite close to the half-life of ^{231}Pa

* **8. Determine the ages of rocks from the measured ratio of a stable nuclide to a radioactive one and the ages of carbon-containing materials from the decay rate of ^{14}C.**

The age of a rock can be found from the relative amounts of the parent and the final stable product of one of the three decay series.

$\ln (N_t/N_0) = \ln N_t - \ln N_0 = -\lambda t$ [8]

N_t is the number of atoms of parent nuclide present now and N_0 is the number of atoms of parent present when the rock was formed.

EXAMPLE 26-3 A rock has 18.05 g of ^{235}U for every 1.00 g of ^{207}Pb. Assume that all the ^{207}Pb was produced by the decay of ^{235}U and estimate the age of the rock.

The half-life of ^{235}U is 7.13×10^8 y, and thus

$\lambda = 0.693/t_{1/2} = 0.693 / 7.13 \times 10^8$ y $= 9.72 \times 10^{-10}$/y

N_t is found from the mass of ^{235}U.

18.05 g ^{235}U $\times$ mol ^{235}U/235 g $\times$ 6.022×10^{23} atoms/mol $= 4.63 \times 10^{22}$ atoms ^{235}U

We also determine the number of ^{207}Pb atoms.

1.00 g ^{207}Pb $\times$ mol ^{207}Pb/207 g $\times$ 6.022×10^{23} atoms/mol $= 0.291 \times 10^{22}$ atoms ^{207}Pb

Each atom of ^{235}U produces one atom of ^{207}Pb (Figure 26-1). The overall equation is

$^{235}U \longrightarrow {}^{207}Pb + 7\,{}^{4}He + 4\,\beta^-$

Thus, $N_0 = N_t$ + number of ^{207}Pb atoms present now

$= 4.63 \times 10^{22} + 0.291 \times 10^{22}$ atoms

$= 4.92 \times 10^{22}$ atoms of ^{235}U

$\ln (N_t/N_0) = \ln (4.63 \times 10^{22}/4.92 \times 10^{22}) = -0.0608$

$= -\lambda t = -(9.72 \times 10^{-10}/\text{y})t$ *or* $t = 6.25 \times 10^7$ y

Equation [8] also can be used when only the decay rate (λN) is known at two different times.

EXAMPLE 26-4 A sample of radioactive nuclide has a decay rate of 1.000×10^6 β emissions/min. 100 min later, the decay rate is 14.7% of the initial rate. What is the half-life of the nuclide? Use the data of Table 26-1 and Figure 26-1 to identify the nuclide.

rate at 100 min $= (1.000 \times 10^6/\text{min})(0.147) = 0.147 \times 10^6$/min

$\ln(N_t/N_0) = \ln (0.147 \times 10^6/\text{min} \div 1.000 \times 10^6/\text{min}) = -1.92 = -\lambda(100 \text{ min})$

or $\lambda = 0.0192$/min

$t_{1/2} = 0.693/\lambda = 0.693 \div 0.0192/\text{min} = 36.1$ min

^{211}Pb is the only nuclide in Table 26-1 with a 36.1-min half-life, and ^{211}Pb is a beta emitter (Figure 26-1).

* 9. Calculate the energies associated with nuclear reactions.

In spontaneous nuclear reactions, the products have less mass than the reactants. The mass difference appears as energy and is determined by equation [9]. Energies are often expressed in millions of electron volts (MeV = mega electron volts) where 1.000 MeV = 1.602×10^{-13} J.

$E = mc^2$ ($c = 3.00 \times 10^8$ m/s, the speed of light) [9]

EXAMPLE 26-5 How much energy in MeV is produced by the alpha (4.0026 amu) decay of each ^{235}U atom (235.0439 amu)? The daughter is ^{231}Th (231.0363 amu).

Mass difference = 235.0429 – (231.0363 + 4.0026) = 0.0050 amu

$E = mc^2 = 0.0050 \text{ amu} \times 1.660 \times 10^{-27} \text{ kg/amu} \times (3.00 \times 10^8 \text{ m/s})^2$

$= 7.47 \times 10^{-13} \text{ J} \times 1 \text{ MeV}/1.602 \times 10^{-13} \text{ J} = 4.7 \text{ MeV}$

In calculations of this sort, we must make sure that the masses of the electrons are taken into account. Although ^{235}U emits an alpha particle ($^4He^{2+}$), the calculation is performed as if it emits a helium-4 atom ($^4He^0$).

* 10. Calculate the average nuclear binding energy per nucleon for a nuclide.

A nuclide always weighs less than the sum of its particle masses. Consider a ^{12}C atom with a mass of 12.000 amu.

total particle mass = (6 × electron mass) + (6 × neutron mass) + (6 × proton mass)

= (6 × 0.00548 amu) + (6 × 1.008665 amu) + (6 × 1.007277 amu)

= 12.09894 amu

The mass defect is the difference between the total particle mass and the mass of the nuclide, and the binding energy is the mass defect expressed in units of energy, J or MeV.

mass defect = 12.09894 amu – 12.00000 amu = 0.09894 amu

$$\text{binding energy} = 0.09894 \text{ amu} \times \frac{1.000 \times 10^{-3} \text{ kg}}{6.022 \times 10^{23} \text{ amu}} \times (3.00 \times 10^8 \text{ m/s})^2$$

$$= 1.478 \times 10^{-11} \text{ J} \times 1 \text{ MeV} / 1.602 \times 10^{-13} \text{ J} = 92.26 \text{ MeV}$$

Often binding energy is expressed as energy per nucleon.

binding energy/nucleon = 92.29 MeV/12 nucleons = 7.690 MeV/nucleon

The higher the binding energy per nucleon, the more stable the nucleus is. A heavy nucleus of low stability can break apart or fission to produce more stable light nuclei. On the other hand, two very light nuclei can join together or fuse to produce a more stable heavy nucleus.

* 11. Describe the factors that determine nuclear stability, establish whether a nuclide is likely to be stable or radioactive, and predict the type of decay process expected for a radioactive nuclide.

Nuclei tend to be stable if they have a "magic number" of protons or neutrons (Table 26-4). Nuclei with an even number of neutrons and/or an even number of neutrons also tend to be stable. All nuclei with $Z > 83$ are unstable.

TABLE 26-4 Magic Numbers of Protons (Z) and Neutrons (A)

Z	2	8	20	28	50	82	114		
A	2	8	20	28	50	82	126	184	196

The unstable nuclei decay by α, β, or β^+ emission, or electron capture. We can predict the type of decay: α decay if in the belt of stability with $Z > 83$, β decay if above and to the left of the belt of stability, and β^+ decay or electron capture if below and to the right of the belt of stability.

12. Describe the processes of nuclear fission and nuclear fusion, including the problems with using them as energy sources.

Nuclear fission is the splitting apart of large nuclei into more stable ones, often initiated by neutrons. The fission reaction of ^{235}U is not a simple process. Two possibilities are given in equations [10] and [11].

$$^{235}U + {}^{1}n \longrightarrow {}^{144}Xe + {}^{90}Sr + 2\ {}^{1}n \qquad [10]$$

$$^{235}U + {}^{1}n \longrightarrow {}^{140}Ba + {}^{93}Kr + 3\ {}^{1}n \qquad [11]$$

These reactions also occur in the atomic bomb. About 3.20×10^{-11} J or 200 MeV is produced by each ^{235}U fission. Once started by a stray neutron, the reaction continues because more neutrons are produced than are consumed. But the neutrons given off are moving too fast to efficiently split ^{235}U nuclei. They are slowed down by the graphite moderator in the reactor. The ^{1}H nuclei of the reactor's cooling water also serves as moderators. Boron control rods absorb neutrons to stop the reaction.

$$^{10}B + {}^{1}n \longrightarrow {}^{7}Li + {}^{4}He \qquad [12]$$

The ^{1}H nuclei of the cooling water also absorb neutrons. Because of this, ^{235}U fuel is enriched to 1-4% from a natural 0.7%. If heavy water is used as cooling water, however, the fuel need not be enriched. ^{1}H nuclei absorb nuclei 600 times better than do the ^{2}H nuclei of heavy water. The CANDU reactor of Canada uses heavy water as a coolant. The savings are considerable; enriched fuel is about five times as expensive to produce as the fuel used by the CANDU reactor.

In breeder reactors, ^{239}Pu is produced from ^{238}U (natural abundance = 99.3%). See Table 26-3 for the reactions. Plutonium has the benefits of an abundant raw material (^{238}U) and no need for moderation as fast neutrons are more efficient than slow ones. Two disadvantages are plutonium's toxicity and its use in nuclear weapons.

In a fusion reaction, light nuclei bind together to form heavier, more stable ones. The overall reaction in the sun is

$$4\ {}^{1}H \longrightarrow {}^{4}He + 2\ \beta^{-} + 26.7\ \text{MeV} \qquad [13]$$

Two fusion reactions are being considered for commercial use.

$$^{6}Li + {}^{2}H \longrightarrow 2\ {}^{4}He + 22.4\ \text{MeV} \qquad [14]$$

$$5\ {}^{2}H \longrightarrow {}^{4}He + {}^{3}He + {}^{1}H + 2\ {}^{1}n + 24.9\ \text{MeV} \qquad [15]$$

Although the products of reactions [14] and [15] are not radioactive, high gamma and neutron radiation cause the coolant and the reactor vessel to become radioactive.

All nuclear reactors have the advantage of producing large quantities of energy from small quantities of fuel. In addition, they do not contribute to pollution or to the CO_2 burden of the atmosphere, as do fossil fuel power plants. On the other hand, the reactors produce nuclear radiation while they are operating and leave the reactor and the spent fuel radioactive. The disposal of the spent fuel is a problem for which no final solution has been adopted.

13. Explain the effects of ionizing radiation on matter and describe several radiation-detection devices based on these effects.

When ionizing radiation passes through matter, it leaves a trail of ions in its path. It does this by knocking electrons off the atoms and molecules that it encounters. Of course, the bonds in molecules are severely disrupted by these ionizations. In the cloud chamber and the bubble chamber, the ions promote the formation of droplets of liquid and bubbles of vapor, respectively. In a Geiger-Muller counter, ions create a pathway for an electric spark. Because of their lack of charge, neutrons penetrate matter as far as gamma rays, but they produce few ions. They damage matter by combining with stable nuclei to produce new ones that often are radioactive, a process known as transmutation.

14. Discuss methods of expressing radiation dosages, some biological hazards of ionizing radiation, and sources of radiation to which the general population is exposed.

The units of radiation dosage are in *Table 26-6 in the text*. Rem abbreviates "radiation equivalent—man" and depends on the amount of biological damage. Radiation can kill directly and quickly by ionizing molecules to stop an organism from functioning. Radiation can disrupt individual cells, causing some of them to reproduce without limit. Cancer is one result of such uncontrolled growth. Radiation can change genetic material causing mutations in offspring. Radiation also can destroy only part of an organism as a serious burn from fire or chemicals will destroy tissue.

The isotopes of the naturally occurring decay series are spread throughout the environment. Starting with the rock (such as granite) in which uranium and thorium commonly are found, the parents and daughters are spread by weathering processes. Certain plants concentrate these elements and thus are a source of these isotopes. Mining activities of all kinds speed up the dispersion of these elements by bringing rocks to the surface where they are exposed to the weather and erode faster. In addition a considerable fraction of the total radiation enters from outer space. All of these sources produce an annual background of 130 mrem per person. X rays for dentistry and medicine also are a source of radiation, as are nuclear power plants.

15. Discuss some practical, beneficial uses of radioisotopes.

Radioisotopes are widely used in medicine to deliver radiation to the cancer cells to be destroyed without irradiating many normal cells. This is done by injecting the radioisotope or by using natural body processes. (For example, iodine will concentrate in the thyroid gland.) Radioisotopes are used as tracers, to detect the fate of certain substances during plant and animal growth, and in industrial processes and chemical reactions. Certain radioisotopes concentrate in specific organs of the body. The organ then can be readily seen by a radiation-detecting camera, allowing us to see abnormalities. Radioisotopes also are used industrially in quality control, as in the production of aluminum foil or plastic film. The isotope is placed below the moving sheet and a detector above. The radiation reaching the detector depends on the thickness of the aluminum or plastic.

TABLE 26-5 Decay Modes, Half-lives, and Masses of Some Nuclides

			Nuclide mass, amu	
Nuclide	Decay mode	Half-life	Parent	Daughter
^{3}H	β^-	12.26 y	3.016050	3.016030
^{6}He	β^-	0.81 s	6.018893	6.015125
^{7}Be	EC	53.37 d	7.016929	7.016004
^{8}Be	α	2×10^{-16} s	8.0053	4.002603
^{12}B	β^-	0.02 s	12.0143	8.022487
^{13}N	β^+	9.96 m	13.005738	13.003354
^{37}Ar	EC	35 d	36.966772	36.965898
^{61}Co	β^-	1.65 h	60.932440	60.931056
^{90}Sr	β^-	28 y	89.907747	89.907163
^{98}Tc	β^-	1.5×10^{6} y	97.907110	97.905289
^{109}In	β^+	4.3 h	108.907096	108.904928
^{161}Tm	EC	30 m	160.933730	160.929950
^{194}Au	β^+	39.5 h	193.965418	193.962725

DRILL PROBLEMS

2. (1) Complete each of the following nuclear equations.

A. $^{210}Pb \longrightarrow$ ___ $+ ^{210}Bi$ B. $^{8}Be \longrightarrow ^{4}He +$ ___ C. $^{205}Pb +$ ___ $\longrightarrow ^{205}Tl$
D. $^{227}Pa \longrightarrow$ ___ $+ ^{223}Ac$ E. ___ $\longrightarrow ^{4}He + ^{228}Th$ F. $^{194}Hg +$ ___ $\longrightarrow ^{194}Au$
G. ___ $\longrightarrow \beta^- + ^{200}Hg$ H. $^{6}He \longrightarrow \beta^- +$ ___ I. $^{191}Os \longrightarrow$ ___ $+ ^{191}Ir$
J. ___ $\longrightarrow \beta^- + ^{17}O$ K. $^{199}Pb \longrightarrow$ ___ $+ ^{199}Tl$ L. ___ $+ \beta^- \longrightarrow ^{37}Cl$
M. $^{210}Pb \longrightarrow ^{4}He +$ ___ N. $^{191}Hg \longrightarrow \beta^- +$ ___ O. ___ $\longrightarrow ^{4}He + ^{218}Po$

(2) Write a nuclear equation for the reaction that occurs when each nuclide below decays in the manner indicated.

P. ^{7}Be (EC) Q. ^{191}Hg (β^+) R. ^{177}Pt (α) S. ^{3}H (β^-) T. ^{177}W (EC)
U. ^{109}In (β^+) V. ^{186}Re (β^-) W. ^{235}U (α) X. ^{12}B (α) Y. ^{161}Tm (EC)
Z. ^{98}Tc (β^-)

4. (1) Write full and abbreviated equations in each of the following cases.

A. ^{59}Co (p, n) ___ B. ^{9}Be (^{6}Li, ___) ^{14}N C. ^{14}N (n, ___) ^{9}Be
D. ___ (p, α) ^{12}C E. ___ (p, d) ^{4}He F. ___ (d, α) ^{21}Ne
G. ^{7}Li (α, ___) ^{10}B H. ^{9}Be (α, n) ___ I. ^{35}Cl (n, p) ___
J. ___ (p, n) ^{44}Sc K. ^{27}Al (d, α) ___ L. ___ (α, p) ^{28}Al

(2) Expand each of the abbreviated equations given in equation Table 26-3.

M. neptunium discovery N. neptunium in reactors O. plutonium discovery
P. americium Q. curium discovery R. curium in reactors

S. berkelium T. californium U. mendelevium
V. nobelium W. lawrencium

7. (1) Find $t_{1/2}$ and λ of each nuclide and the activity of the given mass. Use the data of Tables 26-1 and 26-5 as needed.
A. ^{223}Fr, 1.00 μg B. ^{61}Co, 4.00 pg C. ^{98}Tc, 7.00 g
D. ^{3}H, 1.31 mg E. ^{194}Au, 1.00 g F. ^{228}Ra, 1.32 g
(2) Find $t_{1/2}$ and λ of each nuclide and the mass needed to produce the given activity.
G. ^{6}He, 3.3 × 10^{15}/s H. ^{7}Be, 5.1 × 10^{15}/d I. ^{194}Au, 19.4 × 10^{16}/s
J. ^{3}H, 187 × 10^{20}/m K. ^{90}Sr, 8.46 × 10^{13}/m L. ^{212}Bi, 34.5 × 10^{15}/s
(3) Find $t_{1/2}$ and λ of each nuclide based on mass number, mass, and activity.
M. $A = 214$, 45.3 μg, 3.3 × 10^{15}/s N. $A = 109$, 7.56 pg, 1.87 × 10^{9}/s
O. $A = 212$, 0.165 g, 8.46 × 10^{15}/s P. $A = 37$, 4.20 μg, 9.41 × 10^{11}/m
Q. $A = 215$, 25.2 g, 1.02 × 10^{20}/s R. $A = 210$, 37.2 μg, 145 × 10^{10}/m

8. (1) Given the initial decay rate, the elapsed time, and the final decay rate (in that order), determine the half-life of each nuclide.
A. 1.43 × 10^{8}/s, 20.4 d, 6.73 × 10^{7}/s B. 1.94 × 10^{7}/m, 12.1 h, 8.43 × 10^{5}/m
C. 7.93 × 10^{12}/s, 1.41 M, 5.73 × 10^{11}/s D. 3.14 × 10^{5}/s, 8.75 m, 1.72 × 10^{5}/s
E. 1.77 ×10^{12}/m, 7.63 d, 1.82 × 10^{11}/m F. 9.32 × 10^{6}/s, 1.32 d, 4.14 × 10^{5}/s
G. 4.17 ×10^{10}/m, 4.00 d, 1.00 × 10^{10}/m H. 8.00 × 10^{12}/s, 10.0 d, 9.00 × 10^{10}/s
(2) In each part below, the mass number of the radioactive isotope is followed by the mass number of the stable isotope. Then, the weight ratio of the two isotopes (radioactive/stable) in a sample is followed by the half-life of the first isotope. Assume that none of the stable isotope was present when the sample was formed, and determine the age of the sample.
I. 231, 227, 8.00/0.625, 3.4 × 10^{4} y J. 209, 205, 6.50/1.25, 103 y
K. 233, 209, 0.265/1.00,1.6 × 10^{5} L. 210, 210, 1.13, 0.132, 19.4 y
M. 41, 41, 7.30/1.00, 8.0 × 10^{4} y N. 14, 14, 1.26/5.00, 5700 y
(3) In each part below, the initial rate of decay of a sample is followed by the decay rate at some later time, and the half-life (all assumed to be three significant figures) of the nuclide that is decaying. Determine the age of each sample.
O. 3.12 × 10^{4}/s, 164/s, 7.20 y P. 6.64 × 10^{9}/m, 7.30 × 10^{5}/m, 5700 y
Q. 1.72 × 10^{5}/s, 4.44 × 10^{4}/d, 80000 y R. 8.52 × 10^{6}/h, 1.66 × 10^{4}/h, 5.27 y
S. 1.96 × 10^{4}/h, 127/d, 16.0 y T. 8.95 × 10^{6}/h, 47.5/s, 2.60 y

9. Use the data of Tables 26-2 and 26-5 to determine the energy associated with the decay of each of the following nuclides. Determine each energy both in MeV/atom and in kJ/mol
A. ^{3}H B. ^{6}He C. ^{7}Be D. ^{8}Be E. ^{12}B
F. ^{13}N G. ^{37}Ar H. ^{61}Co I. ^{90}Sr J. ^{98}Tc
K. ^{109}In L. ^{161}Tm M. ^{194}Au

10. The nuclide mass in amu of the most abundant stable isotope of an element is given in each part below. Compute the average binding energy per nucleon for each nuclide in MeV. Then plot the energies you have computed against atomic number.
A. ^{1}H, 1.007825 B. ^{56}Fe, 55.934936 C. ^{4}He, 4.002603
D. ^{58}Ni, 57.935342 E. ^{9}Be, 9.012186 F. ^{64}Zn, 63.929146
G. ^{12}C, 12.000000 H. ^{80}Se, 79.916527 I. ^{16}O, 15.994915
J. ^{84}Kr, 83.911503 K. ^{20}Ne, 19.992440 L. ^{90}Zr, 89.904700
M. ^{24}Mg, 23.985042 N. ^{102}Ru, 101.904348 O. ^{28}Si, 27.976928
P. ^{114}Cd, 113.903360 Q. ^{32}S, 31.972074 R. ^{130}Te, 129.906238
S. ^{40}Ar, 39.962384 T. ^{138}Ba, 137.905000 U. ^{40}Ca, 39.962589
V. ^{142}Nd, 141.907663 W. ^{48}Ti, 47.947960 X. ^{158}Gd, 157.924178
Y. ^{52}Cr, 51.940513 Z. ^{166}Er, 165.932060

11. Predict whether each of the following nuclides is likely to be stable or radioactive. If radioactive, predict the type of decay process that will occur.

A. ^{8}B	B. ^{234}U	C. ^{19}F	D. ^{119}Cd	E. ^{50}Sc
F. ^{212}Po	G. ^{88}Sr	H. ^{17}F	I. ^{226}Ra	J. ^{50}Mn
K. ^{106}In	L. ^{197}Au	M. ^{63}Co		

QUIZZES (20 minutes each) Choose the best answer for each question.

Quiz A

1. The alpha decay of ^{226}Ra produces (a) ^{3}He; (b) ^{4}Li; (c) ^{224}Rn; (d) ^{224}Po; (e) none of these.
2. Which of the following could be a member of the thorium series, which begins with ^{232}Th? (a) ^{223}Fr; (b) ^{216}Po; (c) ^{221}Rn; (d) ^{214}Pb; (e) none of these.
3. Neutron bombardment of ^{23}Na produces an isotope that is a beta emitter. After beta emission, the final product is (a) ^{24}Na; (b) ^{24}Mg; (c) ^{23}Ar; (d) ^{24}Ar; (e) none of these.
4. A nuclide has a decay rate of 2.00×10^{10}/s. 25.0 days layer, its decay rate is 6.25×10^{8}/s. What is the nuclide's half-life? (a) 25.0 d; (b) 12.5 d; (c) 50.0 d; (d) 5.00 d; (e) none of these.
5. A nuclide has a half-life of 35.0 h. What is the value of its decay constant? (a) 0.0285; (b) 0.0198; (c) 24.3; (d) 0.0412; (e) none of these.
6. The binding energy per nucleon is largest for (a) ^{3}He; (b) ^{59}Co; (c) ^{235}U; (d) ^{98}Tc; (e) ^{31}P.
7. The most massive particle is (a) alpha; (b) beta; (c) gamma; (d) positron; (e) neutron.
8. Based on magic numbers, which nuclide is the most stable? (a) ^{3}He; (b) ^{16}O; (c) ^{15}N; (d) ^{119}Sn; (e) ^{206}Pb.

Quiz B

1. Electron capture by ^{41}Ca produces (a) ^{41}K; (b) ^{40}Ca; (c) ^{42}Ca; (d) ^{41}Sc; (e) none of these.
2. Which could be a member of the actinium series, which begins with ^{235}U? (a) ^{223}Ra; (b) ^{221}Ra; (c) ^{214}Ba; (d) ^{216}Po; (e) none of these.
3. Proton bombardment of ^{230}Th followed by emission of two alpha particles produces (a) ^{222}Rn; (b) ^{223}Fr; (c) ^{223}Ra; (d) ^{222}Fr; (e) none of these.
4. A nuclide has a half-life of 1.91 y. Its decay constant has a numerical value of (a) 1.32; (b) 2.76; (c) 0.363; (d) 0.524; (e) none of these.
5. The activity of a radioactive sample declines to 1.00% of its original value in 300. days. What is the half-life of the nuclide in this sample? (a) 3.35×10^{-5} d; (b) 0.0145 d; (c) 0.00667 d; (d) 3.00 d; (e) none of these.
6. The binding energy per nucleon is smallest for (a) ^{13}C; (b) ^{3}He; (c) ^{52}Cr; (d) ^{56}Fe; (e) none of these.
7. The most highly charged particle of the following is (a) alpha; (b) beta; (c) gamma; (d) positron; (e) neutron.
8. Based on magic numbers, which nuclide is the least stable? (a) ^{96}Nb; (b) ^{119}Sn; (c) ^{40}K; (d) ^{15}O; (e) ^{40}Ca.

Quiz C

1. The beta decay of ^{90}Sr produces (a) ^{90}Sr; (b) ^{91}Sr; (c) ^{89}Sr; (d) ^{89}Rb; (e) none of these.
2. Which could be a member of the uranium series, which begins with ^{238}U? (a) ^{215}Po; (b) ^{213}Po; (c) ^{210}Po; (d) ^{212}Po; (e) none of these.
3. Alpha particle bombardment of ^{27}Al followed by neutron emission produces (a) ^{30}P; (b) ^{30}Si; (c) ^{31}Si; (d) ^{29}P; (e) none of these.
4. A nuclide's activity decreases 75% in 4.00 days. Its half-life is (a) 2.00 d; (b) 1.00 d; (c) 4.00 d; (d) 0.347 d; (e) none of these.
5. A nuclide is a beta emitter with a 28-y half-life. The ratio of its mass to the mass of its stable product in a sample is 1.00/8.25. If none of the stable product was present initially, what is the age of the sample?
(a) 89.9 y; (b) 85.3 y; (c) 231 y; (d) 259 y; (e) none of these.
6. The binding energy per nucleon is largest for isotopes of (a) U; (b) Co; (c) Cs; (d) He; (e) Hg.
7. The most ionizing radiation is of what type? (a) alpha; (b) beta; (c) gamma; (d) positron; (e) neutron.
8. Based on magic numbers, which nuclide is the most stable? (a) ^{91}Nb; (b) ^{91}Zr; (c) ^{58}Co; (d) ^{13}C; (e) ^{20}Ne.

Quiz D

1. The beta decay of ^{45}Ca produces (a) ^{45}K; (b) ^{45}Sc; (c) ^{44}K; (d) ^{44}Ca; (e) none of these.
2. Which could be a member of the uranium series, which begins with ^{238}U? (a) ^{236}U; (b) ^{217}At; (c) ^{216}Po; (d) ^{219}At; (e) none of these.
3. Deuteron (^2H) bombardment of ^{96}Mo followed by neutron emission produces (a) ^{98}Tc; (b) ^{97}Tc; (c) ^{97}Nb; (d) ^{98}Mo; (e) none of these.
4. A nuclide has a decay constant of 4.28×10^{-4}/h. If the activity of a sample is 3.14×10^5/s, how many atoms of the nuclide are present in the sample? (a) 2.64×10^{12}; (b) 7.34×10^8; (c) 2.04×10^6; (d) 4.40×10^{10}; (e) none of these.
5. The activity of a nuclide declines to 10.0 % of its original value in 145 d. What is the decay constant of this nuclide? (a) 0.0159/d; (b) 63.0 d; (c) 0.00690/d; (d) 0.00478/d; (e) none of these.
6. The binding energy per nucleon is smallest for isotopes of (a) Li; (b) Co; (c) Ge; (d) Sc; (e) S.
7. The least penetrating radiation is of what type? (a) alpha; (b) beta; (c) gamma; (d) positron; (e) neutron.
8. Based on magic numbers, which nuclide is the least stable? (a) ^{59}Ni; (b) ^{51}V; (c) ^{122}Sb; (d) ^{16}O; (e) ^{12}C.

SAMPLE TEST (30 minutes)

1. If there were no nuclear explosions, would the ratio of ^{12}C to ^{14}C in living material increase, decrease, or remain the same? Briefly explain why.
2. An atom of ^{253}Es decays to one of ^{237}Np. What kinds of particles are given off in this process, and how many of each are given off?
   - a. alpha emission by ^{243}Cm
   - b. positron emission by ^{18}F
   - c. electron capture by ^{88}Zr
   - d. ^{10}B (α, n) ____
   - e. ^{45}Sc (α, p) ____
   - f. ____ + $^2H \longrightarrow 2\ ^1n + ^{51}Cr$
3. Write or complete, as appropriate, and balance each of the following nuclear equations.
4. ^{11}Be decays via the reaction $^{11}Be \longrightarrow \beta^- + ^{11}B + g$. What is the maximum energy of the gamma ray produced if atomic masses are ^{11}B = 11.00931 amu, ^{11}Be = 11.0216 amu, β^-= 0.00055 amu (1.000 amu = 932.8 MeV)?
5. How many ^{238}U atoms would decay in a mole of ^{238}U during 995 million years? The half-life of ^{238}U is 4.51×10^9 y.

27 Organic Chemistry

CHAPTER OBJECTIVES

1. Give examples of alkane, alkene, alkyne, and aromatic hydrocarbons.

Hydrocarbons are compounds that contain only carbon and hydrogen. Alkanes have only single bonds between the carbons; alkenes have at least one carbon-carbon double bond (C=C); and alkynes have at least one carbon-carbon triple bond (C≡C). Aromatic hydrocarbons have a benzene ring.

* **2. Draw structural and condensed formulas for hydrocarbons, given systematic (IUPAC) names; and name hydrocarbon molecules, given structural or condensed formulas.**

Condensed formulas are simply a way of writing structural formulas all on one line. We have been writing condensed formula for some time. For example, CH_4 is the condensed formul for methane and its Lewis structure is its structural formula. The condensed formula of a straight chain hydrocarbon is fairly easy to write. Simply write down the carbons and, after each carbon, write the formula of the hydrogen atoms tattached to it. Double and triple bonds are written between a carbon and the preceding carbon or hydrogen. Thus $CH_3CH_2CH_3$ is propane, while $CH_2{=}CHCH_3$ is the structural formula of propene. If a multi atom group replaces a hydrogen atom, tits formula is enclosed in parentheses. Thus, if a methyl group $—CH_3$ replaces the hydrogen on the central carbon of propene, the structural formula of the resulting compound is $CH_2{=}C(CH_3)CH_3$

TABLE 27-1 The First Ten Alkanes

Chain length	*Formula*	*Name*
1	CH_4	methane
2	CH_3CH_3	ethane
3	$CH_3CH_2CH_3$	propane
4	$CH_3(CH_2)_2CH_3$	butane
5	$CH_3(CH_2)_3CH_3$	pentane
6	$CH_3(CH_2)_4CH_3$	hexane
7	$CH_3(CH_2)_5CH_3$	heptane
8	$CH_3(CH_2)_6CH_3$	octane
9	$CH_3(CH_2)_7CH_3$	nonane
10	$CH_3(CH_2)_8CH_3$	decane

For *noncyclic hydrocarbons* (those with no a ring of carbon atoms), the IUPAC rules for naming follow.

1. The longest hydrocarbon chain determines the base name (see Table 27-1).
2. The longest chain must contain any double or triple bonds. After that requirement is satisfied, and there is more than one possible longest chain, choose the one that has the greatest number of side chains. (This ensures that the side chains—the substituents—are as simple as possible,)
3. Add-*ane* to the base name if no multiple bonds are present, *-ene* for a C=C bond, and *-yne* for a C≡C bond.
4. The position of the multiple bond is shown by a number that precedes the base name and is separated from it by a hyphen. The number must be as small as possible. There are always two ways to number the longest carbon chain, depending on the end you start at.
5. *Alkyl groups* (side chains or branches) are named by changing their alkane names from *-ane* to *-yl* (see Figure 27-1). Their positions on the chain are indicated by the number (2, 3, 4, and so on) of the carbon to

which they are attached. This number is separated from the alkyl group name by a hyphen. The alkyl group numbers must be as small as possible, after rule (4) is satisfied. There must be a number prefix for *each* side chain even if the same number is repeated.

6. The number of alkyl groups of one type is given as a prefix (di-, tri-, tetra-, penta-, hexa-, hepta-, and so on) to the alkyl group name. (This prefix follows the numbers that indicate positions on the main chain. Thus, one would say 2,2,3,4-tetramethylpentane)

7. The side chains are named in alphabetical order.

8. Numbers are separated from each other by commas and from letters by hyphens, with no spaces in a name.

$$\begin{array}{cccc}
\overset{\displaystyle CH_3}{\underset{\displaystyle |}{}} & \overset{\displaystyle CH_3}{\underset{\displaystyle |}{}} & \overset{\displaystyle CH_3}{\underset{\displaystyle |}{}} & \overset{\displaystyle CH_3}{\underset{\displaystyle |}{}} \\
CH_3-CH- & CH_3-CH_2-CH- & CH_3-CH-CH_2- & CH_3-\overset{|}{\underset{|}{C}}-CH_3 \\
\text{isopropyl} & sec\text{-butyl} & \text{isobutyl} & tert\text{-butyl}
\end{array}$$

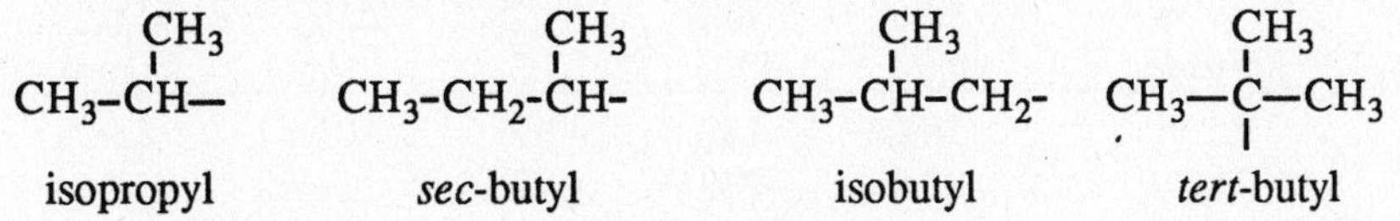

FIGURE 27-1   Names and Structures of Some Alkyl Groups

For *cyclic hydrocarbons* the IUPAC rules for naming are

9. The number of carbons in the ring determines the base name, which always begins with *cyclo-* for nonaromatic hydrocarbons.

10. A double or triple bond in the ring is assumed to be between carbons 1 and 2. Numbering proceeds around the ring so that the side chains are attached at the lowest possible numbers.

11. Cyclic hydrocarbons names then follow rules 5, 6, 7, and 8 above.

Aromatic hydrocarbons are named as derivatives of benzene.

12. Side-chain positions are designated by the smallest possible numbers, the largest side chain is on carbon 1.

13. For two groups on the ring, the position of the second group can be named. Positions 2 and 6 are ortho (*o-*), 3 and 5 are meta (*m-*), and 4 is para (*p-*).

14. If aromatic hydrocarbons are named as derivatives of toluene, the methyl group is on carbon 1.

15. Aromatic hydrocarbon names then follow rules 5, 6, 7, and 8 above.

* **3.   Determine all the possible skeletal isomers of simple hydrocarbons of given formulas.**

An abbreviated way of drawing hydrocarbon molecules is to draw only the bonds. All bonds are shown except those to hydrogen atoms. Abbreviated skeletons are drawn in Figure 27-2.

ethane   propane   propene   propyne   butane   1-butene   2-butene   1-butyne

2-methylpropane   2-methylbutane   2,3-dimethylbutane   cyclopentane   methylcyclopropane

1,1-dimethylcyclopropane   1,2-dimethylcyclopropane   1-pentene   2-pentene

2-methyl-1-butene   3-methyl-1-butene   2-methyl-2-butene

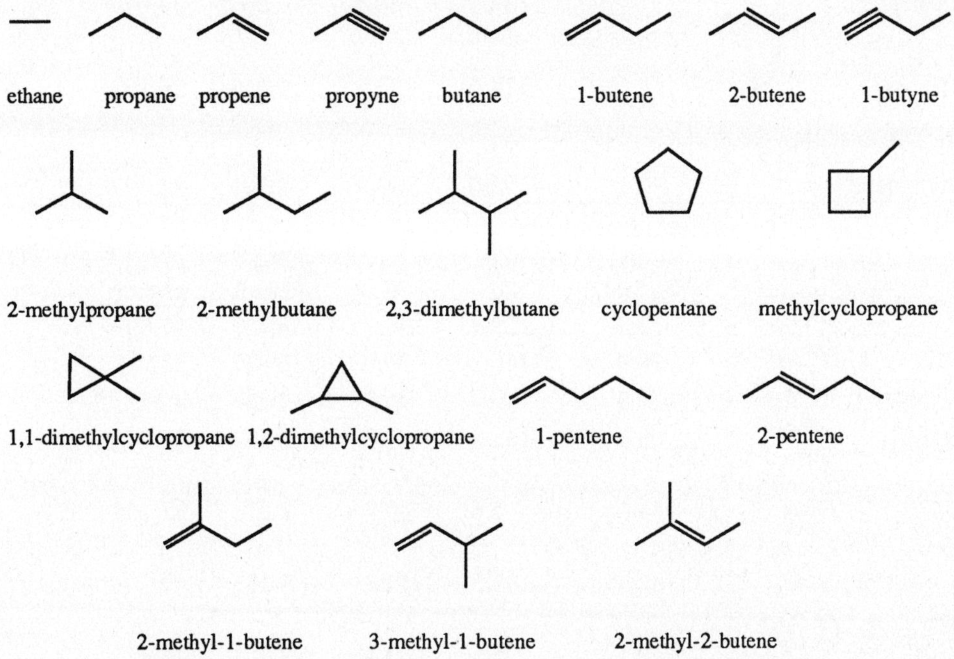

FIGURE 27-2   Abbreviated Skeletons of Some Hydrocarbons

EXAMPLE 27-1 Use the abbreviated skeletal notation to draw and name all the isomers of formula C_5H_{10}. These isomers are the last nine in Figure 27-2.

TABLE 27-2 Physical Properties of Hydrocarbons

		Effect of molecular weight—straight-chain alkanes				
Name	*Formula*	$\mathcal{M}$, g/mol	m.p., °C	b.p., °C	ΔH_{fus} kJ/mol	ΔH_{vap} kJ/mol
methane	CH_4	16.04	−182.5	−164	0.937	8.908
ethane	C_2H_6	30.1	−183.3	− 88.6	2.862	15.648
propane	C_3H_8	44.1	−181.7	− 42.1	3.527	20.133
butane	C_4H_{10}	58.1	−138.3	− 0.5	4.661	24.271
pentane	C_5H_{12}	72.2	−129.7	36.1	8.427	27.593
hexane	C_6H_{14}	86.2	− 95.3	69.0	13.079	31.911
heptane	C_7H_{16}	100.2	− 90.6	98.4	14.163	35.187
octane	C_8H_{18}	114.2	− 56.8	125.7	20.652	38.581
		Effect of branching—heptanes				
Name	*Structure*		m.p., °C	b.p., °C	ΔH_{fus} kJ/mol	ΔH_{vap} kJ/mol
n-heptane			− 90.6	98.4	14.163	35.187
2-methylhexane	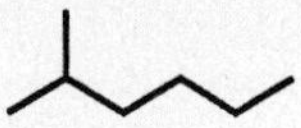		−118.2	90.0	8.870	35.727
2,2-dimethylpentane	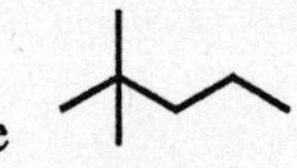		−123.8	79.2	5.862	33.920
2,4-dimethylpentane	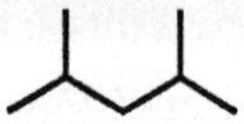		−119.9	89.8	6.686	34.171
3,3-dimethylpentane	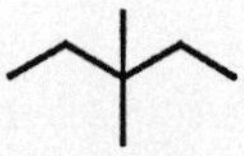		−134.9	80.5	7.067	34.079
3-ethylpentane	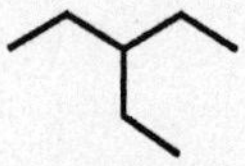		−118.6	93.5	9.552	36.162
2,2,3-trimethylbutane	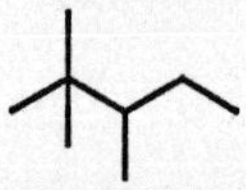		− 25.0	80.9	2.201	32.497
		Effect of π bonding				
Name	*Structure*	$\mathcal{M}$, g/mol	m.p., °C	b.p., °C	ΔH_{fus} kJ/mol	ΔH_{vap} kJ/mol
n-pentane		72.2	−129.7	36.1	8.427	27.593
1-pentene		70.4	−166.2	30.0	5.837	28.999
2-pentene		70.4	−140.2	36.4	8.389	n.a.[a]
1,3-pentadiene		68.1	−148.8	42.0	6.142	30.548
cyclohexane	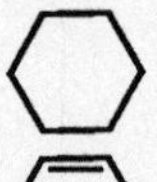	84.16	6.6	80.7	2.632	32.765
cyclohexene		82.15	−103.5	82.9	3.293	n.a[a]
benzene		78.12	5.53	80.1	9.954	42.903

[a]n.a. indicates data are not available.

* 4.   Discuss the physical properties of aliphatic and aromatic hydrocarbons in relation to their bonding, structures, and molecular weights.

Within a series of hydrocarbons, such as the alkanes, boiling points increase with molecular weight. London forces between larger molecules are stronger, and it is more difficult to disrupt these stronger intermolecular attractions. The melting points and the molar heats of fusion and vaporization increase with mole weight for the same reason. A branched molecule boils at a lower temperature than does a straight-chain one with the same number of carbons. The branched molecule has less surface area and hence less area over which intermolecular forces can act. The melting points and the molar heats of fusion and vaporization also decrease as branching increases. Finally, we notice that the number of π electrons in a molecule increases the intermolecular forces. For molecules with the same number of carbons, those with more π electrons have larger molar heats of fusion and vaporization, although there is no clear trend in melting and boiling points. Data illustrating these predictions are given in Table 27-2. Predictions based on molecular weight are the best ones. Boiling points and molar heats of vaporization are predicted better than are melting points and molar heats of fusion.

* 5.   Write equations for several reactions used in the preparation of alkanes, alkenes, and alkynes.

These equations are summarized below and should be memorized. R represents any alkyl group.

(a)   Hydrogenation of multiple bonds.

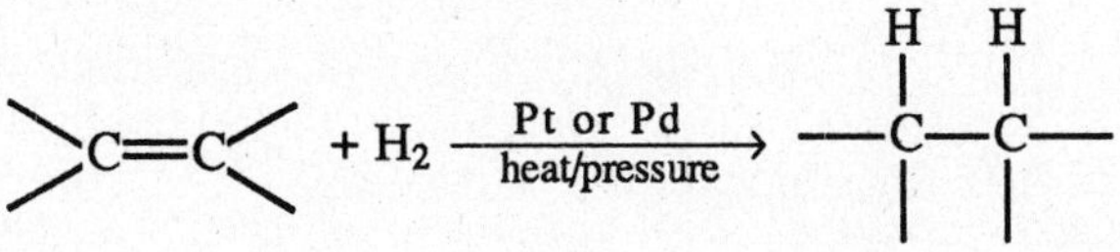

$$\begin{array}{c} \diagdown \\ \diagup \end{array} C{=}C \begin{array}{c} \diagup \\ \diagdown \end{array} + H_2 \xrightarrow[\text{heat/pressure}]{\text{Pt or Pd}} \quad \overset{\displaystyle H}{\underset{\displaystyle |}{-}}\overset{\displaystyle |}{\underset{\displaystyle |}{C}}\overset{\displaystyle H}{\underset{\displaystyle |}{-}}\overset{}{\underset{}{C}}\overset{}{\underset{}{-}} \qquad\qquad [1]$$

$$-C{\equiv}C- + H_2 \xrightarrow[\text{heat/pressure}]{\text{Pt or Pd}} \quad -\overset{H}{\underset{H}{C}}-\overset{H}{\underset{H}{C}}- \qquad\qquad [2]$$

(b)   Wurtz reaction: chain length is doubled.

$$R{-}Br + 2\,Na \longrightarrow NaBr + R{-}R \qquad\qquad [3]$$

(c)   Alkali carbonates (such as Na_2CO_3, K_2CO_3) fused with alkali hydroxides (such as NaOH, KOH).

$$R{-}\overset{\displaystyle O}{\overset{\displaystyle \|}{C}}{-}ONa + NaOH \xrightarrow{\text{heat}} Na_2CO_3 + R{-}H \qquad\qquad [4]$$

(d)   Dehydration of (removing the elements of water from) an alcohol.

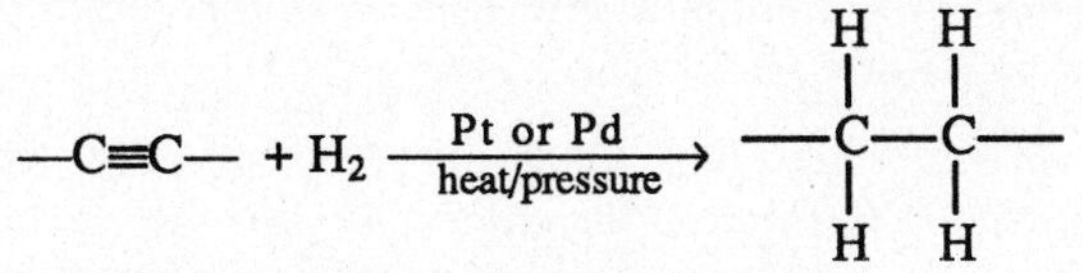

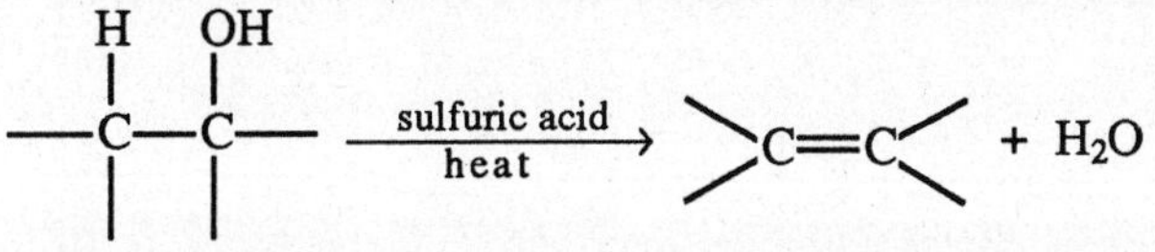

$$-\overset{H}{\underset{|}{C}}-\overset{OH}{\underset{|}{C}}- \xrightarrow[\text{heat}]{\text{sulfuric acid}} \begin{array}{c}\diagdown \\ \diagup\end{array}C{=}C\begin{array}{c}\diagup \\ \diagdown\end{array} + H_2O \qquad\qquad [5]$$

(e)   Dehydrohalogenation of (removing the elements of HX from) an alkyl halide.

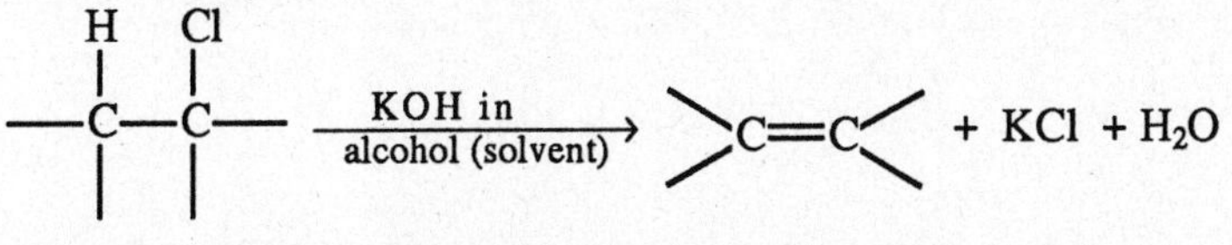

$$-\overset{H}{\underset{|}{C}}-\overset{Cl}{\underset{|}{C}}- \xrightarrow[\text{alcohol (solvent)}]{\text{KOH in}} \begin{array}{c}\diagdown \\ \diagup\end{array}C{=}C\begin{array}{c}\diagup \\ \diagdown\end{array} + KCl + H_2O \qquad\qquad [6]$$

(f)   Extension of alkyne chain length with sodium amide.

$$H{-}C{\equiv}C{-}H + NaNH_2 \longrightarrow NH_3 + H{-}C{\equiv}C^- \, Na^+$$

$$H{-}C{\equiv}C^- \, Na^+ + R{-}CH_2Br \longrightarrow H{-}C{\equiv}C{-}CH_2R + NaBr \qquad\qquad [7]$$

(g)   double dehydrohalogenation of an alkyl halide.

$$-\overset{Cl}{\underset{H}{C}}-\overset{Cl}{\underset{H}{C}}- \xrightarrow[\text{alcohol (solvent)}]{\text{KOH in}} -\overset{H}{\underset{|}{C}}{=}\overset{Cl}{\underset{|}{C}}- \xrightarrow[\text{alcohol (solvent)}]{\text{KOH in}} -C{\equiv}C- \qquad\qquad [8]$$

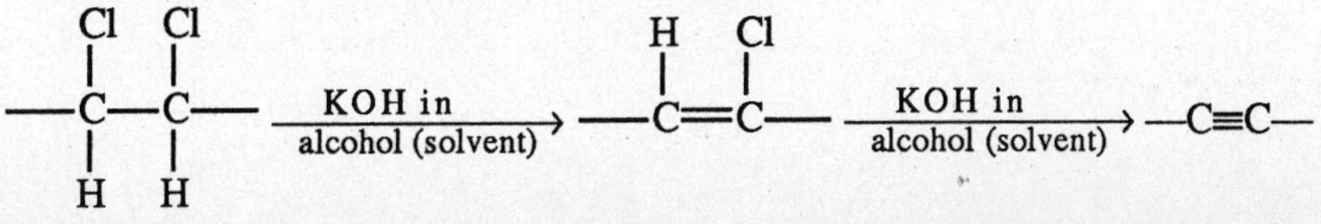

6.   Explain why alkanes and aromatic hydrocarbons react by substitution, and alkenes and alkynes react by addition.

For alkanes, substitution is the only possible type of reaction, as carbon is capable of forming only four bonds. For alkenes and alkynes, addition reactions produce more energy than do substitution reactions. Consider the hydrogenation of a double bond.

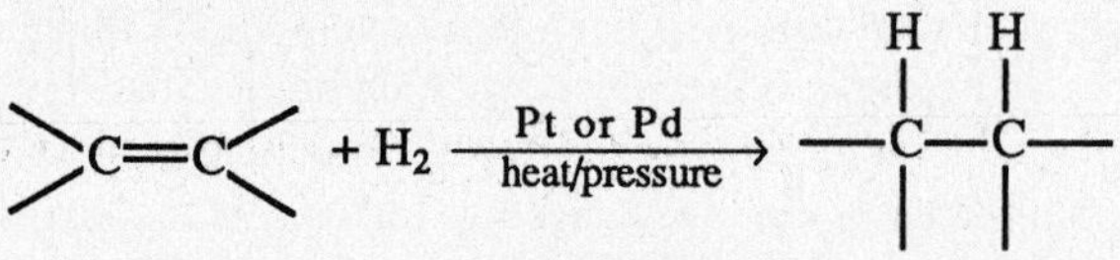

$$>C=C< \;+\; H_2 \;\xrightarrow[\text{heat/pressure}]{\text{Pt or Pd}}\; \overset{\displaystyle H \;\;\; H}{\underset{\displaystyle | \;\;\; |}{-C-C-}} \tag{1}$$

The net result is that a C=C bond (611 kJ/mol) and an H—H bond (435 kJ/mol) are broken, and a C—C bond (347 kJ/mol) and two C—H bonds (414 kJ/mol) are formed. Thus the enthalpy change for hydrogenation of a double bond is –127 kJ/mol [= 611 + 435 – 347 – 2(414)].

Aromatic hydrocarbons are considerably more stable than we might expect based on the number of single and double bonds present (see objectives 11-10 and 11-11). This extra stability is the resonance stabilization energy. If an aromatic hydrocarbon participates in an addition reaction, the resonance stabilization energy is lost. For example, when benzene is hydrogenated to cyclohexane, the enthalpy change is –206.1 kJ/mol.

$$C_6H_6(g) + 3\,H_2(g) \xrightarrow[\text{heat/pressure}]{\text{Pt or Pd}} C_6H_{12}(g) \tag{9}$$

This –206.1 kJ/mol is far short of the –381 kJ/mol (3 × –127 kJ/mol) that would result from hydrogenating three moles of C=C bonds. The difference, 175 kJ/mol, is the resonance stabilization energy, the energy by which benzene is more stable than a non-resonance structure of C_6H_6 that contains three double bonds.

7.   Write equations for the reactions of alkanes with halogens and with oxygen, emphasizing the radical chain nature of the alkane-halogen reaction.

When an alkane reacts with a halogen (F_2, Cl_2, Br_2, or I_2, symbolized as X_2), halogen atoms substitute for some of the hydrogens, producing the halogenated hydrocarbon and HX. The reaction proceeds by by a free-radical chain mechanism consisting of initiation, propagation, and termination steps. A free radical is one of the species produced when a bond breaks and the two electrons in the bond are split between the two fragments, one electron going with each. This is homolytic cleavage of the bond. (Heterolytic cleavage produces a cation and an anion.) Free-radical halogenation of an alkane follows the following steps.

Initiation: $X—X \xrightarrow{\text{heat or light}} 2\,X\cdot$ $\tag{10}$

Propagation: $R—H + X\cdot \longrightarrow R\cdot + H—X$

$R\cdot + X—X \longrightarrow R—X + X\cdot$ $\tag{11}$

Termination: $X\cdot + \cdot X \longrightarrow X—X$ *and* $R\cdot + \cdot X \longrightarrow R—X$ *and* $R\cdot + \cdot R \longrightarrow R—R$ $\tag{12}$

The relative strengths of the R—X and H—X bonds compared to R—H and X—X determine the energy and speed of the reaction. In general, reactions of F_2 are faster and more energetic than those of Cl_2, Br_2, or I_2. Any number of hydrogens may be substituted, producing a mixture of products.

Of course, all hydrocarbons react with oxygen to produce carbon dioxide and water.

* 8.   Predict the products of an addition reaction at a multiple bond.

The more positive fragment (usually a hydrogen atom) of an unsymmetrical addition reagent attaches to the carbon atom of a multiple bond that has the largest number of hydrogen atoms, For alkynes with the triple bond at the end of the molecule, the last carbon has the most hydrogen atoms (one). For alkenes, there are three possibilities. They are shown in Figure 27-3. The carbon to which the positive fragment would add is shown by an arrow. R is some group other than hydrogen.

$$R—C{\equiv}C—H \qquad \overset{\displaystyle H \;\;\; H}{\underset{}{R—C{=}C—H}} \qquad \overset{\displaystyle R \;\;\; H}{\underset{}{R—C{=}C—H}} \qquad \overset{\displaystyle R \;\;\; R}{\underset{}{R—C{=}C—H}}$$

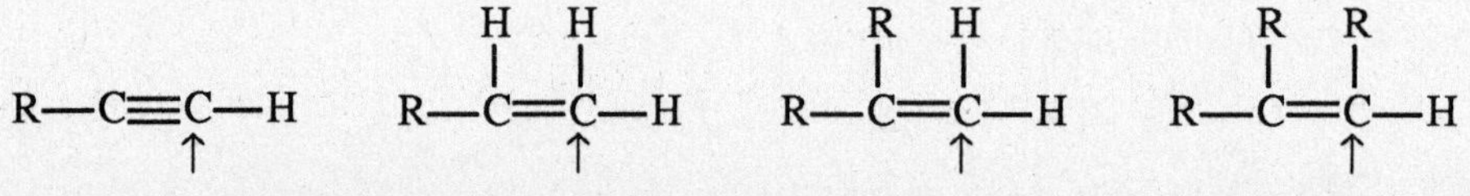

FIGURE 27-3 Addition Sites for Positive Fragments

9. Name the common functional groups and give examples of compounds containing them.

While the descriptive chemistry of inorganic compounds is based on the periodic table, organic chemistry is based on functional groups. Organic compounds with the same functional group react in similar ways. Some of the common classes of organic compounds are given in Table 27-3. R is any alkyl group.

TABLE 27-3 Some Common Classes of Organic Compounds

Class of compound	*General formula*	*Examples and names*		
alkane	R—H	CH_3CH_3 ethane	$CH_3\overset{CH_3}{\overset{\vert}{C}}CH_3$ 2-methylpropane	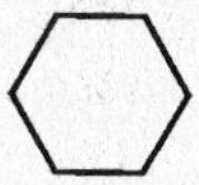 cyclohexane
alkene	$R_2C{=}CR_2$	$H_2C{=}CHCH_3$ 1-propene	$(CH_3)_2C{=}C(CH_3)_2$ 2,3-dimethyl-2-butene	 cyclopentene
alkyne	R—C≡C—R	$CH_3C{\equiv}CCH_3$ 2-butyne	$CH_3\text{-}\overset{CH_3}{\overset{\vert}{CH}}\text{-}C{\equiv}CCH_2CH_3$ 2-methyl-3-hexyne	
arene (aromatic hydrocarbon)	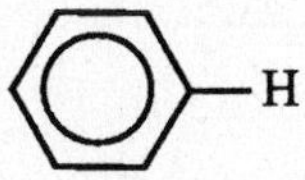 or Ar—H	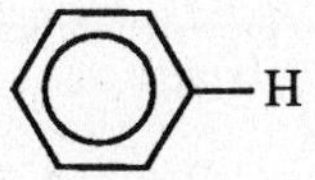 benzene	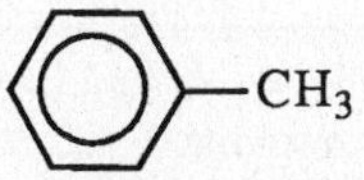 methylbenzene *or* toluene	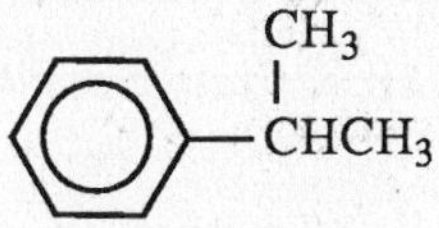 isopropylbenzene
alkyl halide	R—X	CH_3CH_2Br bromoethane	$CH_3CF_2CH_2CH_3$ 2,2-difluorobutane	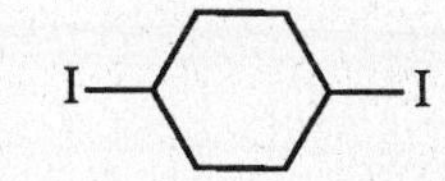 1,4-diiodocyclohexane
alcohol	R—OH	$CH_3CH_2\overset{OH}{\overset{\vert}{C}}HCH_2CH_3$ 3-pentanol	$CH_3-\overset{CH_3}{\underset{OH}{C}}-CH_2CH_3$ 2-methyl-2-butanol	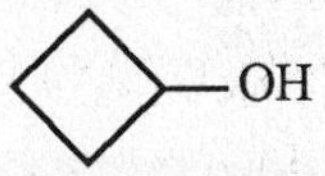 cyclobutanol
ether	R—O—R'	$CH_3—O—CH_3$ dimethyl ether	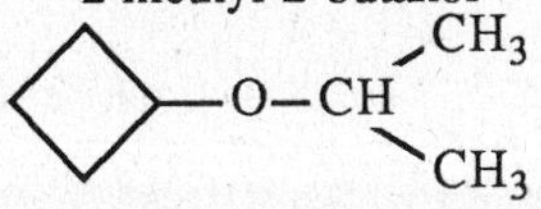 cyclobutyl isopropyl ether	
aldehyde	$R—\overset{O}{\overset{\Vert}{C}}—H$	$CH_3\overset{O}{\overset{\Vert}{C}}-H$ ethanal *or* acetaldehyde	$CH_3CH_2CH_2\overset{O}{\overset{\Vert}{C}}-H$ butanal *or* butryraldehyde	
ketone	$R—\overset{O}{\overset{\Vert}{C}}—R'$	$CH_3\overset{O}{\overset{\Vert}{C}}CH_3$ propanone	$CH_3\text{-}\overset{CH_3}{\overset{\vert}{CH}}\text{-}CH_2\overset{O}{\overset{\Vert}{C}}CH_3$ 4-methyl-2-pentanone	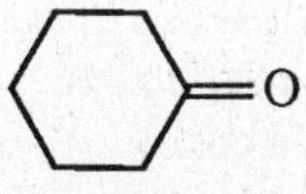 cyclohexanone
acid	$R—\overset{O}{\overset{\Vert}{C}}—OH$	$CH_3\overset{O}{\overset{\Vert}{C}}-OH$ ethanoic acid *or* acetic acid	$CH_3\text{-}\overset{CH_3}{\overset{\vert}{CH}}\text{-}CH_2\overset{O}{\overset{\Vert}{C}}-OH$ 3-methylbutanoic acid	

ester	$R-\overset{O}{\overset{\|\|}{C}}-OR'$	$CH_3\overset{O}{\overset{\|\|}{C}}-OCH_3$ methyl ethanoate *or* methyl acetate	$CH_3 \cdot \overset{CH_3}{\overset{\|}{CH}}-CH_2\overset{O}{\overset{\|\|}{C}}-OCH_2CH_3$ ethyl 3-methylbutanoate	
amine	$R-NH_2$	$CH_3CH_2NH_2$ ethylamine	$CH_3CH_2\overset{CH_3}{\overset{\|}{C}}-NH_2$ 2-aminobutane	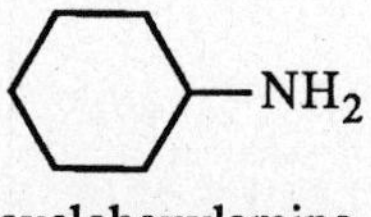 cyclohexylamine

* **10. Write structural formulas, identify possible isomers, and name organic compounds containing functional groups.**

First one needs to memorize the structural formulas in Table 27-3. The summary in Table 27-4 may help in writing structural formulas. Different structural isomers are found by drawing all the possible carbon skeletons and then placing the functional group in all possible positions. Nomenclature depends on the class of the compound. Alcohols, alkyl halides, ketones, and amines have names obtained from the corresponding alkane, with the position of the group on the main chain indicated with the smallest possible number. This number precedes the base name and is separated from it by a dash. The *-ane* ending of the base name is replaced by the ending given in Table 27-5. For both aldehydes and carboxylic acids, the functional group is at one end of the longest chain and the carbon of the functional group is numbered 1.

TABLE 27-4 Characteristics of Common Classes of Organic Compounds

Class of compounds	*Structural group present*
alkane	only C—H and C—C single bonds
alkene	at least one C=C bond
alkyne	at least one C≡C bond
arene	an aromatic ring, 
alkyl halide	an alkyl group bonded to a F, Cl, Br, or I atom
alcohol	an —OH group bonded to a C atom
ether	two alkyl group bonded to an O atom
aldehyde	an alkyl group bonded to a $-\overset{O}{\overset{\|\|}{C}}-H$ group
ketone	two alkyl groups bonded to a $-\overset{O}{\overset{\|\|}{C}}-$ group
acid	an alkyl group bonded to a $-\overset{O}{\overset{\|\|}{C}}-OH$ group
ester	two alkyl groups bonded to each end of a $-\overset{O}{\overset{\|\|}{C}}-O-$ group
amine	an alkyl group bonded to an $-NH_2$ group

TABLE 27-5 Name Endings for Classes of Organic Compounds

Class of compounds	*Ending*
alkyl halide	-yl halide (as -yl iodide)
alcohol	-ol
ether	-yl ether
aldehyde	-anal
ketone	-one
acid	-anoic acid
ester	-anoate
amine	-yl amine

Esters are derived from carboxylic acids by replacing the H atom in the —COOH group with an alkyl group. The alkyl group name precedes that of the acid and the *-oic acid* ending is replaced by *-oate*. In ethers the two alkyl group names are followed by *ether*. In order to determine the numbering of substituents on alkyl groups in ethers, the oxygen-bonded carbon atoms are numbered 1.

11.   Describe methods of preparing alcohols, ethers, aldehydes, ketones, acids, and amines; and write equations for some typical reactions of these classes of organic compounds

Many methods of preparation are summarized below. Following them are some of the reactions that compounds in each class of compounds undergoes.

a.   *Alcohols*
Hydration (adding water to) of alkenes:

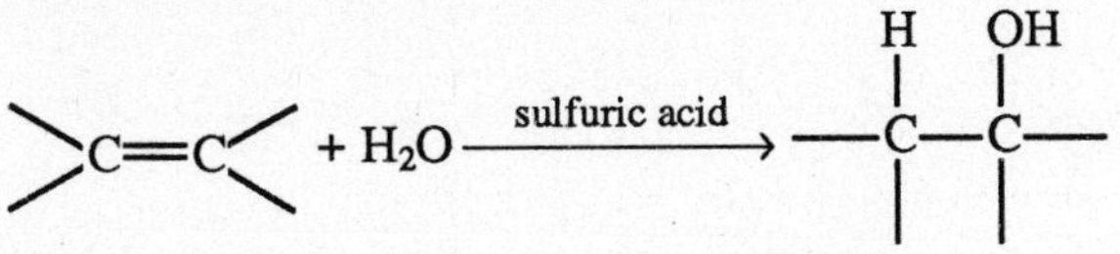

$$>C=C< \quad + H_2O \xrightarrow{\text{sulfuric acid}} \quad -\overset{H}{\underset{|}{C}}-\overset{OH}{\underset{|}{C}}- \tag{13}$$

Hydrolysis (attack by water) of alkyl halides:

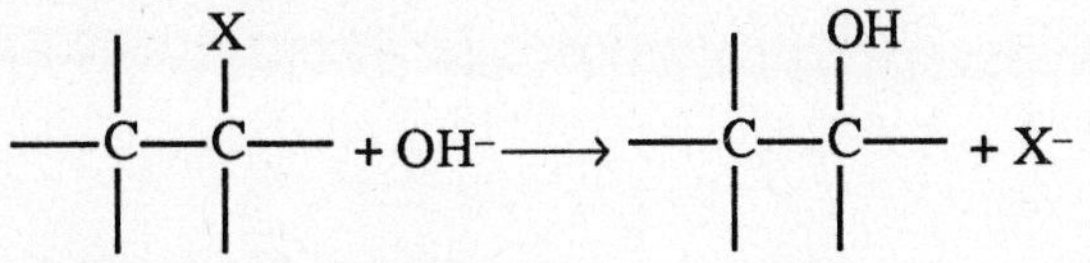

$$-\overset{|}{\underset{|}{C}}-\overset{X}{\underset{|}{C}}- \quad + OH^- \longrightarrow \quad -\overset{|}{\underset{|}{C}}-\overset{OH}{\underset{|}{C}}- \quad + X^- \tag{14}$$

Production of methanol:

$$CO(g) + 2 H_2(g) \xrightarrow[\text{ZnO, Cr2O3}]{\text{350°C, 200 atm}} CH_3OH(g) \tag{15}$$

b.   *Ethers*
Elimination of water between two alcohol molecules, a condensation reaction:

$$R—OH + HO—R' \xrightarrow{\text{conc.sulfuric acid}} R—O—R' + H_2O \tag{16}$$

c.   *Aldehydes and ketones.* Produced by oxidation of alcohols.
Primary alcohols form aldehydes:

$$RCH_2OH \xrightarrow{Cr_2O_7^{2-},H^+} R—\overset{O}{\overset{||}{C}}—H \xrightarrow{Cr_2O_7^{2-},H^+} R—\overset{O}{\overset{||}{C}}—OH \tag{17}$$

$$RCH_2OH \xrightarrow{Cu, 200\text{-}300°C} R—\overset{O}{\overset{||}{C}}—H \tag{18}$$

Secondary alcohols form ketones:

$$R—\overset{OH}{\underset{|}{CH}}—R' \xrightarrow[\text{or Cu, 200-300°C}]{KMnO_4 \text{ or } K_2Cr_2O_7} R—\overset{O}{\overset{||}{C}}—R' \tag{19}$$

d.   *Carboxylic acids*
Oxidation of a primary alcohol:

$$RCH_2OH \xrightarrow{KMnO_4, OH^- \text{ or } Cr_2O_7^{2-}, H^+} R—\overset{O}{\overset{||}{C}}—OH \tag{20}$$

Oxidation of an aldehyde:

$$R—\overset{O}{\overset{||}{C}}—H \xrightarrow{KMnO_4, OH^- \text{ or } Cr_2O_7^{2-}, H^+} R—\overset{O}{\overset{||}{C}}—OH \tag{21}$$

e.   *Amines*
Reduction of nitro compounds:

$$R—NO_2 \xrightarrow{Fe, HCl} R—NH_3^+Cl^- \xrightarrow{NaOH} R—NH_2 \tag{22}$$

Reactions of these classes of compounds are summarized in the following general equations.

f. *Alcohols*

1. Dehydration (see equation [5]).
2. Reactions with acid halides.

$$R'{-}OH + R{-}\overset{\overset{O}{\|}}{C}{-}X \longrightarrow R{-}\overset{\overset{O}{\|}}{C}{-}OR' + HX \qquad [23]$$

3. Reactions with an carboxylic acids.

$$R'{-}OH + R{-}\overset{\overset{O}{\|}}{C}{-}OH \overset{H^+}{\rightleftharpoons} R{-}\overset{\overset{O}{\|}}{C}{-}OR' + H_2O \qquad [24]$$

4. Reactions with nitric acid.

$$R{-}OH + HONO_2 \longrightarrow R{-}ONO_2 + H_2O \qquad [25]$$

5. Reactions with alkali hydroxides.

$$R{-}OH + NaOH \longrightarrow R{-}C^- + Na^+ + H_2O \qquad [26]$$

6. Reaction with sodium.

$$R{-}OH + Na^0 \longrightarrow R{-}O^- Na^+ + \tfrac{1}{2}H_2(g) \qquad [27]$$

g. *Ethers*

1. Stable when attacked by most oxidizing and reducing agents and dilute acids and alkalis.
2. Cleavage by strong acids.

$$R{-}O{-}R' + 2\ HCl \longrightarrow R{-}Cl + R'{-}Cl + H_2O \qquad [28]$$

h. *Aldehydes*

1. Oxidation to acids (see equation [17]).
2. Reduction with hydrogen.

$$R{-}\overset{\overset{O}{\|}}{C}{-}H \xrightarrow{H_2,\ Ni} R{-}CH_2{-}OH \qquad [29]$$

i. *Ketones*

Reduction with hydrogen.

$$R{-}\overset{\overset{O}{\|}}{C}{-}R' \xrightarrow{H_2,\ Ni} R{-}\overset{\overset{OH}{|}}{CH}{-}R' \qquad [30]$$

j. *Acids*

1. Reaction with ammonia, followed by heating.

$$R{-}\overset{\overset{O}{\|}}{C}{-}OH + NH_3 \longrightarrow R{-}\overset{\overset{O}{\|}}{C}{-}O^-NH_4^+ \xrightarrow{\Delta,\ P_2O_5} R{-}C{\equiv}N + H_2O \qquad [31]$$

2. Reaction with an alcohol to give an ester (see equation [24]).
3. Elimination of water between two acid molecules.

$$2\ R{-}\overset{\overset{O}{\|}}{C}{-}OH \xrightarrow{\Delta,\ P_2O_5} R{-}\overset{\overset{O}{\|}}{C}{-}O{-}\overset{\overset{O}{\|}}{C}{-}R \qquad [32]$$

k. *Amines*

Reaction with an alkyl halide.

$$RNH_2 + R'X \longrightarrow R{-}NH{-}R' + HX \qquad [33]$$

* **12. Propose schemes for synthesizing some simple organic compounds.**

You should be able to combine the reactions given in equations [1] through [8], and [13] through [33] to synthesize simple organic compounds. Often it is easier to work backward: determine what reactants are needed to produce the desired compound. Then figure out how to make those reactants, always keeping in mind what the starting materials are.

EXAMPLE 27-2 Produce propyl butanoate from simple alcohols.

$$CH_3CH_2CH_2CH_2OH \xrightarrow{KMnO_4} CH_3CH_2CH_2COOH$$

$$CH_3CH_2CH_2COOH + HOCH_2CH_2CH_3 \xrightarrow{H^+} CH_3CH_2CH_2COOCH_2CH_2CH_3 + H_2O$$

13. Describe the processes that are used to increase the yield of gasoline from petroleum and to improve the performance of gasoline in internal combustion engines.

Gasoline consists of hydrocarbons with six to nine carbon atoms per molecule. Most of the hydrocarbons in petroleum are quite a bit larger. The molecules of different chain lengths, and thus different boiling points, are separated by fractional distillation (see objective 13-10). When heated with a catalyst, the long-chain hydrocarbons break down—a process known as cracking. Other catalysts form double bonds and rearrange straight chains into branched ones—reforming. Very short two- to four-carbon chains combine together to form branched hydrocarbon chains—alkylation. Finally, toluene and other aromatic hydrocarbons are added.

14. Illustrate the principal methods of polymer formation—free-radical addition and condensation—through structural formulas of monomers and polymers.

Free-radical polymerization consists of initiation, propagation, and termination steps, as does free-radical halogenation (objective 27-7). There are, however, two major differences. First, there is very little initiator present. In halogenation, there are about equal amounts of halogen and alkane, while here the mixture is principally alkene. Second, propagation consists of one reaction rather than two. The propagating free radical is an ever-lengthening molecule. Organic peroxides easily form free radicals and often are used as chain initiators, equation [34]. Propagation depends on the alkene monomer; three examples are given in equations [35] through [37]. Termination occurs when any two free radicals combine.

Initiation: $RO{-}OR \longrightarrow 2\ RO\cdot$ [34]

Polyethylene propagation (ethylene, $CH_2{=}CH_2$, monomer).

$RO\cdot + CH_2{=}CH_2 \longrightarrow ROCH_2CH_2\cdot$

$ROCH_2CH_2 + CH_2{=}CH_2 \longrightarrow RO(CH_2)_3CH_2\cdot$

$RO(CH_2)_3CH_2\cdot + CH_2{=}CH_2 \longrightarrow RO(CH_2)_5CH_2\cdot$ [35]

Polyvinyl chloride propagation (vinyl chloride, $CH_2{=}CHCl$, monomer).

$RO\cdot + CH_2{=}CHCl \longrightarrow ROCH_2CHCl\cdot$

$ROCH_2CHCl\cdot + CH_2{=}CHCl \longrightarrow RO(CH_2CHCl)_2\cdot$

$RO(CH_2CHCl)_2\cdot + CH_2{=}CHCl \longrightarrow RO(CH_2CHCl)_3\cdot$ [36]

Polystyrene propagation (styrene, $CH_2{=}CH\phi$, monomer; $\phi = C_6H_5$, a phenyl group).

$RO\cdot + CH_2{=}CH\phi \longrightarrow ROCH_2CH\phi\cdot$

$ROCH_2CH\phi\cdot + CH_2{=}CH\phi \longrightarrow RO(CH_2CH\phi)_2\cdot$

$RO(CH_2CH\phi)_2\cdot + CH_2{=}CH\phi \longrightarrow RO(CH_2CH\phi)3\cdot$ [37]

Condensation polymerization occurs when a small molecule such as water is eliminated between the two growing molecules. This reaction is promoted by the presence of a substance that consumes the small molecule as it is formed. For the removal of water, often an acid anhydride is used to increase the yield, for example, acetic anhydride [$(CH_3CO)_2O + H_2O \longrightarrow 2\ CH_3COOH$]. This also speeds up the reaction, since many of these polymerization reactions are acid-catalyzed. The reaction can be represented by the generalized equation [38]. Note that there must be two functional groups per molecule of monomer, one on each end.

$HO{-}R{-}OH + HO{-}R{-}OH \longrightarrow HO{-}R{-}O{-}R{-}OH + H_2O$ [38]

15. Distinguish between natural and synthetic polymers and among elastomers, fibers, and plastics.

Many natural polymers are discussed in Chapter 28. These include polysaccharides (objectives 28-5 and 28-7), which have monosaccharides as monomers; proteins (objective 28-9), which have amino acids as monomers; and nucleic acids (objective 28-14), which have nucleotides as monomers. In addition, rubber also is a natural polymer. Rubber is composed of isoprene monomers, as shown in Figure 27-4. Synthetic polymers include polyethylene, polystyrene, polyvinyl chloride, dacron, nylon, neoprene, teflon, and lucite. These are classified and summarized in *Table 10-4 in the text.* Elastomers are polymers that can deform or stretch under stress. Fibers are long single- or double-stranded chains. Plastics are crosslinked chains that form a somewhat rigid two- or three-dimensional structure. Thus they can be formed into films such as mylar (backing for magnetic tapes for audio casettes and VCR's) or solids such as polyvinyl chloride (plumbing pipe and phonograph records).

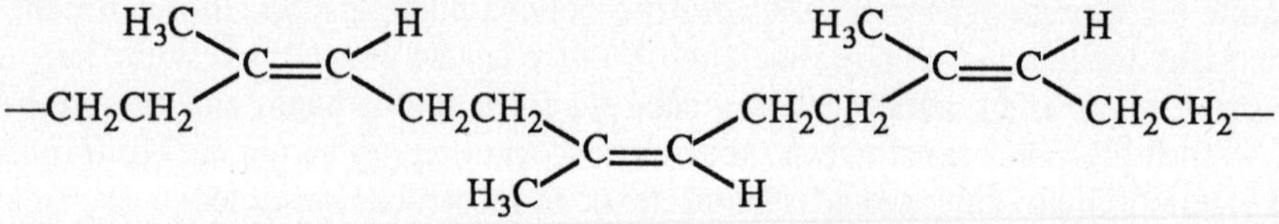

FIGURE 27-4 Rubber—Polyisoprene

DRILL PROBLEMS

2. Draw structural formulas for each of the following compounds. (You may draw either the *cis* or the *trans* isomer for those compounds that exhibit geometric isomerism.)

A. 3-methyl-3-hexene
B. 2-pentene
C. 2-methylbutane
D. 2,2,4-trimethylpentane
E. 2-butyne
F. 2-methylethylbenzene
G. toluene
H. *m*-xylene
I. 3-ethylcyclopentene
J. 2,3,4-trimethylpentane
K. 2,3-dimethylbutane
L. 2,4-dimethyl-4-ethylheptane
M. 2,3-dimethyl-2-butene
N. 3-methyl-4-ethyl-3-hexene
O. 2-pentyne
P. 1,3-dimethylcyclohexane
Q. 3,5-dimethylcyclopentene
R. 3-methylcyclopentene
S. cyclohexylcyclohexane
T. 1,1-dimethyl-4-ethylcyclopentane
U. 1-methylcyclopentene
V. 1,2,3,4-tetramethylcyclobutene
W. 1,3,5-trimethylbenzene

(2) Give the systematic (IUPAC) name of each of the following compounds.

A. 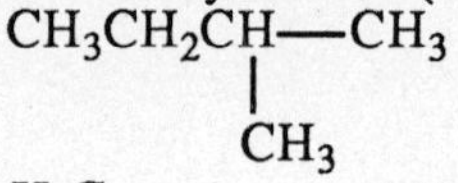

B. 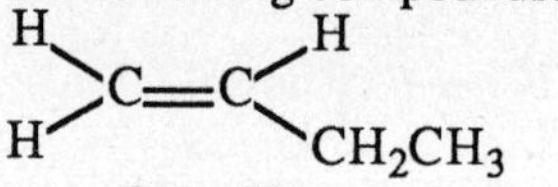

C. 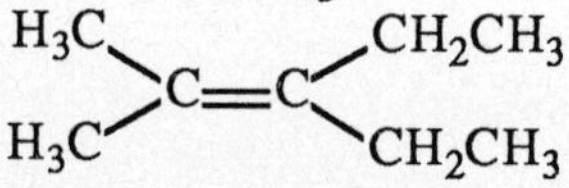

D. 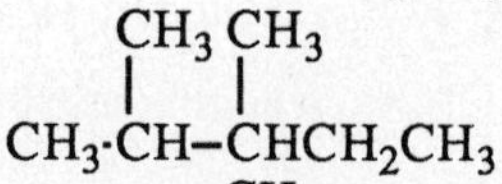

E. 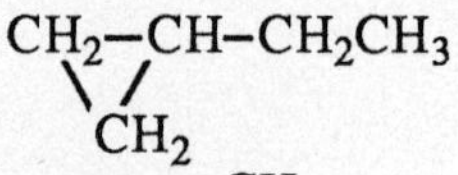

F. 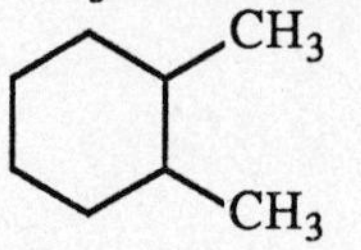

G. 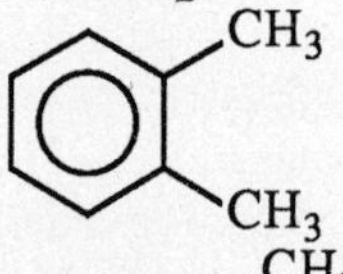

H. 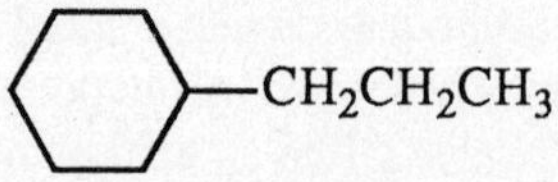

I. 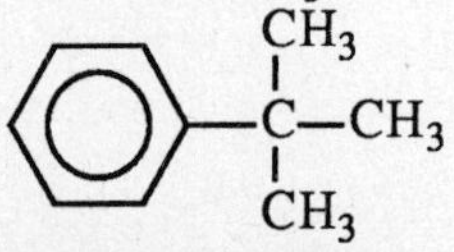

J. 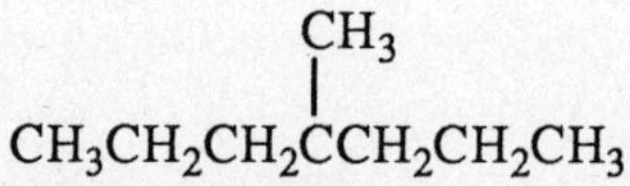

K. $CH_3CH_2C{\equiv}CH$

L. $CH_3CH{=}CHCH_3$

M. $CH_3CH_2CH_2CH=CH_2$

N. $CH\equiv CCH_3$

O. C_3H_8

P.
$$CH_3$$
$$|$$
$$CH_3CHCH_3$$

Q.
$$CH_3\ CH_3$$
$$|\ \ \ \ |$$
$$CH_3-C-CHCH_3$$
$$|$$
$$CH_3$$

R.
$$CH_3\ CH_2CH_3$$
$$|\ \ \ \ \ |$$
$$CH_3CH-CCH_2CH_2CH_3$$
$$|$$
$$CH_2CH_3$$

S.
$$CH_3\ CH_2CH_3$$
$$|\ \ \ \ \ |$$
$$CH_3CH-CHCH_2CH_3$$

T.
$$CH_3$$
$$|$$
$$CH_3CHCH_2CH_2CH_3$$

U.
$$CH_3\ CH_3$$
$$|\ \ \ \ |$$
$$CH_3CH_2CH-CCH_3$$
$$|$$
$$CH_3$$

V.
$$CH_3$$
$$|$$
$$CH_3CH_2CH_2CH_2CCH_2CH_2CH_3$$
$$|$$
$$CH_3$$

W.
—CH_3

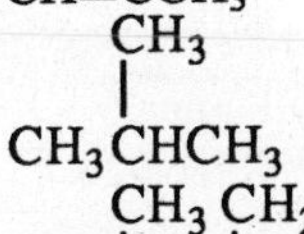

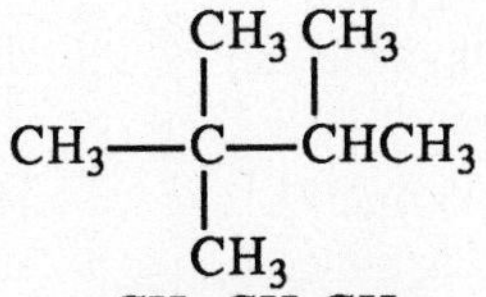

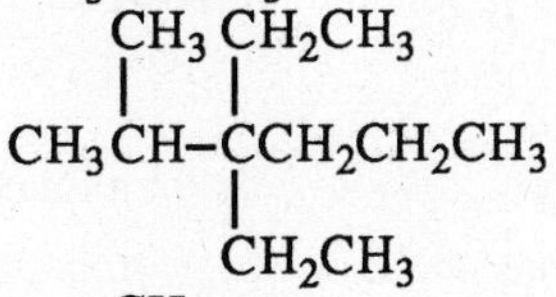

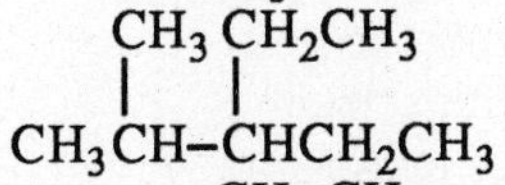

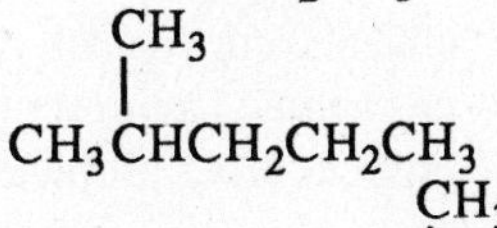

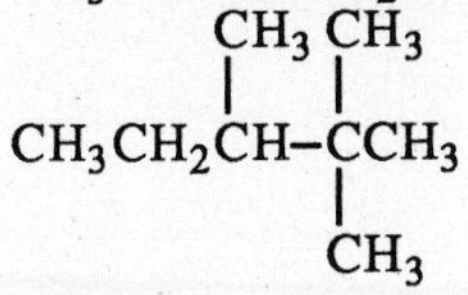

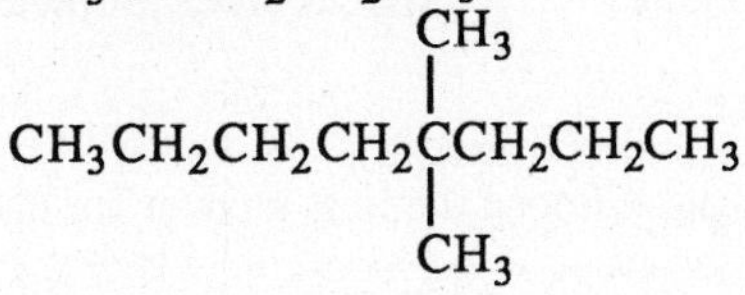

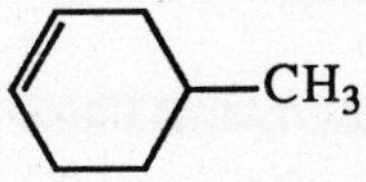

3. Draw all the possible skeletal isomers of each of the following hydrocarbons. Then name each isomer.
  A. aromatic compounds of formula C_9H_{12} B. C_4H_8
  C. cyclic compounds of formula C_6H_{10} D. C_6H_{14}
  E. C_4H_{10} F. C_5H_{10} G. hydrocarbons of molecular weight 86

4. Predict which compound in each group below has the highest value of the listed property and which one has the lowest value.
  Melting point
  A. cyclohexane; hexane; benzene   B. pentane; 2,2-dimethylpropane; 2-methylbutane
  C. ethyne; ethane; ethene   D. benzene; 1,4-dimethylbenzene; toluene
  Boiling point
  E. butane; propane; hexane   F. 3,3-dimethylpentane; 2-methylhexane; n-heptane
  G. cyclohexane; benzene; cyclohexene   H. butane; cyclobutane; methylcyclopropane
  Molar heat of fusion
  I. methane; ethane; ethene   J. ethyne; propyne; butyne
  K. cyclohexane; hexane; benzene   L. n-hexane; 2,2-dimethylbutane; 3-methylpentane
  Molar heat of vaporization
  M. xylene; benzene; toluene   N. 3,3-dimethylpentane; 2-methylhexane; n-heptane
  O. cyclohexane; hexane; benzene   P. butane; pentane; propane

5. Write the complete equation for each of the following reactions.
  A. 1-bromohexane with KOH in alcohol   B. 1-propanol $\longrightarrow$ propylene
  C. ethanol with H_2SO_4 and heat   D. chloroethane $\longrightarrow$ butane
  E. 1,1-dichloropropane $\longrightarrow$ propyne F. 2-chloropropane $\longrightarrow$ propylene
  G. 2-chlorobutane $\longrightarrow$ NaCl H. 1-butene $\longrightarrow$ butane
  I. 2-bromo-2,3-dimethylpentene with KOH in alcohol

8. Predict the product of each of the following reactions.
  A. 1-butene + H_2O $\xrightarrow{H_2SO_4,\ HgSO_4}$ B. 3-methyl-2-pentene + H_2O $\xrightarrow{H_2SO_4}$
  C. 1-butene + HI $\longrightarrow$ D. 2-methyl-2-butene + HBr $\longrightarrow$
  E. 2-methyl-2-pentene + HCl $\longrightarrow$ F. 2-methylcyclohexene + HCl $\longrightarrow$

10. (1) Give the name of each compound
  A. $CH_3CH_2CHBrCH_2$ B. $(CH_3)_2CH_2CHOH(CH_3)_3$
  C. $CH_3CH_2CH_2COCH_2CH_3$ D. $CH_3CH_2CH_2CH_2COOH$
  E. $(CH_3)_3CCHO$ F. $CH_3OCH_2CH_2CH_3$
  G. $CH_3CH_2CH_2CH_2CH_2NH_2$ H. HCOOH
  I. $CH_3CH_2COOCH_2CH_3$ J. $(CH_3)_3COH$
  K. $(CH_3)_3COCH_2CH_3$ L. $CH_3CCl_2CHClCH_2CH_3$
  M. $CH_3CH_2COCH_2CH_3$ N. $CH_3CH_2COOCH_2CH_2CH_3$

O. $CH_3CH_2CH_2CH_2CHO$
(2) Give the condensed structural formula of each compound.
P. butanone Q. 2-methylbutanal R. 2-methylpentanoic acid
S. 3-bromopentane T. methyl butanoate U. propyl isopropyl ether
V. 3-methylpentanal W. 2,2-dichloropropane X. 2-amino-3-methylpentane
Y. 3-hexanol Z. 2-propanol Γ. 3-aminopentane
Δ. 2-methylpropanoic acid Θ. 1-iodo-2-methylbutane Λ. 2-pentanone

12. Detail all the steps in synthesizing the following compounds from the given starting material.
A. propyne from ethyne B. ethene from ethanal C. cyclohexane from hexane
D. 2-butene from ethane E. propanone from propane F. diethyl ether from ethane
G. propanal from propane H. propene from propane I. ethyl butanoate from ethane

QUIZZES (20 minutes each) Choose the best answer for each question.

Quiz A

1. The IUPAC name of $(CH_3)_2CHCH_2CH_3$ is (a) isopentane; (b) *sec*-pentane; (c) 3-methylbutane; (d) 2-methylbutane; (e) none of these.
2. The formula of 2-methyl-1-butanol is (a) $CH_3CH_2COH(CH_3)_2$; (b) $CH_3CH_2CH(CH_3)CH_2OH$; (c) $CH_3CH(CH_3)CH_2OH$; (d) $CH_3COH(CH_3)_2$; (e) none of these.
3. C_4H_8 has how many skeletal isomers, including geometric isomers? (a) 2; (b) 3; (c) 4; (d) 5; (e) none of these.
4. Which compound is the lowest boiling? (a) methane; (b) ethene; (c) ethane; (d) cyclopropane; (e) ethyne.
5. Which monomer would be suitable for condensation polymerization? (a) CH_3CH_2OH; (b) $CH_3CH{=}CH_2$; (c) $CH_2{=}CHCl$; (d) $HOCH_2CH_2OH$; (e) none of these.
6. What reactant could *not* be used to produce an alkene in fewer than three steps? (a) bromoethane; (b) bromobutane; (c) methane; (d) propyne; (e) none of these.
7. Which class of compounds is the least reactive? (a) ethers; (b) alcohols; (c) ketones; (d) alkyl halides; (e) aldehydes.
8. To produce an organic ester, one uses a carboxylic acid and an (a) alkyl halide; (b) alkane; (c) alcohol; (d) aldehyde; (e) none of these.

Quiz B

1. The IUPAC name of $H_2C{=}C(CH_3)CH(Cl)CH_3$ is (a) 4-chloro-2-methylbutene; (b) 2-chloro-2-methyl-4-butene; (c) 4-chloro-2-methylpentene; (d) 2-chloro-4-methyl-4-butene; (e) none of these.
2. The formula of 2-methylbutanal is (a) $CH_3CH_2CH(CH_3)CHOH$; (b) $CH_3CH(CH_3)CHOH$; (c) $CH_3CH_2CH(CH_3)CHO$; (d) $CH_3CH(CH_3)CHO$; (e) none of these.
3. C_5H_{12} has how many skeletal isomers? (a) 2; (b) 3; (c) 4; (d) 5; (e) none of these.
4. Which compound has the lowest molar heat of vaporization? (a) hexane; (b) cyclohexane; (c) benzene; (d) heptane; (d) 2-heptene.
5. Which process is not used in refining gasoline from petroleum? (a) distillation; (b) cracking; (c) reforming; (d) polymerization; (e) alkylation.
6. What could be used to produce an alkane in one step? (a) an alkene; (b) an alcohol; (c) an acid; (d) an alkali metal carbonate; (e) none of these.
7. A free-radical chain reaction is used to prepare (a) methanol; (b) methanal; (c) methyl chloride; (d) dimethyl ether; (e) none of these.
8. To produce an alcohol from a ketone, one used (a) Cu, heat; (b) HCl; (c) H_2, Ni; (d) NH_3; (e) none of these.

Quiz C

1. The formula of 2-chloro-1-butene is (a) $CH_2{=}CClCH_2CH_3$; (b) $CH_3CCl{=}CHCH_3$; (c) $CH_2{=}CClCH_3$; (d) $CH_3CCl{=}CH_2$; (e) none of these.
2. The name of $CH_3CH_2COCH_2CH_3$ is (a) diethyl ether; (b) ethyl propyl ether; (c) butanone; (d) 3-pentanone; (e) none of these.
3. C_4H_9Cl has how many skeletal isomers (a) 1; (b) 2; (c) 3; (d) 4; (e) none of these.

4. Which compound is the highest melting? (a) propanone; (b) propene; (c) cyclopropane; (d) propyne; (e) methylpropane.
5. Which of the following is the formula of a straight chain alkene? (a) C_2H_6; (b) C_2H_2; (c) C_6H_{10}; (d) C_6H_{14}; (e) none of these.
6. To add to a double bond, which group of reactants is *not* used? (a) HCl; (b) H_2, Pt; (c) Na; (d) H_2O, H_2SO_4; (e) none of these.
7. A free-radical chain reaction is *not* used to produce (a) bromoethane; (b) polystyrene; (c) polyethylene; (d) ethanol; (e) none of these.
8. To produce an ether from methanol, one uses (a) heat; (b) concentrated H_2SO_4; (c) KOH in alcohol; (d) Cu, heat; (e) none of these.

Quiz D

1. The name of $CH_3CCl{=}CBr_2$ is (a) 1,1-dibromo-2-chloropropane; (b) 3,3-dibromo-2-chlorethane; (c) 1,1-dibromo-2-chloropropene; (d) 3,3-dibromo-2-chloropropene; (e) none of these.
2. The formula of 2-ethylbutanal is (a) $CH_3CH_2CH(CH_3)CHO$; (b) $(CH_3)_2CHCH_2CHO$; (c) $(CH_3)_2CHCHO$; (d) $CH_3CH(CH_3)CHO$; (e) none of these.
3. C_3H_8O has how many skeletal isomers? (a) 1; (b) 2; (c) 3; (d) 4; (e) none of these.
4. Which compound has the highest molar enthalpy of fusion? (a) butane; (b) 2-butane; (c) cyclobutane; (d) methylcyclopropane; (e) methylcyclobutane.
5. Which of the following could be used as a reactant with HBr to produce 1-bromobutane? (a) $CH_2{=}CHCH_2CH_3$; (b) $CH{\equiv}CCH_2CH_3$; (c) $CH_3CH_2CH_2CH_3$; (d) $CH_3CH{=}CHCH_3$; (e) none of these.
6. Which could produce an alkene in one step? (a) an amine; (b) an alkyl halide; (c) an alkyl dihalide; (d) an alkane; (e) all of these could be used.
7. Which class of compounds is the most oxidized? (a) alkanes; (b) alcohols; (c) acids; (d) aldehydes; (e) ketones.
8. To produce an aldehyde from 2-butanol, what reactant is used? (a) MnO_4^-; (b) Cu, heat; (c) $Cr_2O_7^{2-}$; (d) H_2SO_4; (e) impossible with any of these reagents.

SAMPLE TEST (20 minutes)

1. Draw all of the ethers of formula $C_4H_{10}O$. Name each compound.
2. Explain why propanol is soluble in water but hexanol is not.
3. Consider the dehydrohalogenation of 2-bromobutane in the presence of KOH. Draw the structures of all the products and name them.
4. Give the names and structures of compounds A, B, C, and D formed in the following sequence of reactions.

$$\text{2-methylpropanal} \xrightarrow{H_2,Pt} A \xrightarrow{H_2SO_4, 200°C} B \xrightarrow{HBr} C \xrightarrow{KOH, alcohol} D$$

28 Chemistry of the Living State

CHAPTER OBJECTIVES

1. List the four principal types of substances found in cells—lipids, polysaccharides, proteins, and nucleic acids—and describe the chemical composition of each.

Lipids readily dissolve in nonpolar solvents. They are nonpolar or of very low polarity. Lipids are subdivided into triglycerides (fats and oils), phosphatides or phospholipids, and waxes. See objectives 28-2, 28-3, and 28-4.

Polysaccharides are carbohydrate polymers composed of more than ten monosaccharides units. A monosaccharide is an aldehyde or a ketone with hydroxyl groups (—OH) on all other carbons. Aldehyde monosaccharides are aldoses; ketone ones are ketoses. See objectives 28-5 and 28-7.

Proteins are polymers of amino acids. An α-amino acid is a carboxylic acid with an amino group (—NH_2) attached to the α-carbon, the carbon bonded to the carboxylic group (—COOH). These acids have the general formula shown in Figure 28-1. R groups of common amino acids are given in *Table 28-3 in the text*. See objectives 28-9 and 28-10.

Nucleic acids are composed of phosphoric acid, pentose sugars (ribose and 2-deoxyribose), and the purine and pyrimidine bases adenine, guanine, thymine, cytosine, and uracil that are shown in *Figure 28-20 in the text*. See objective 27-14.

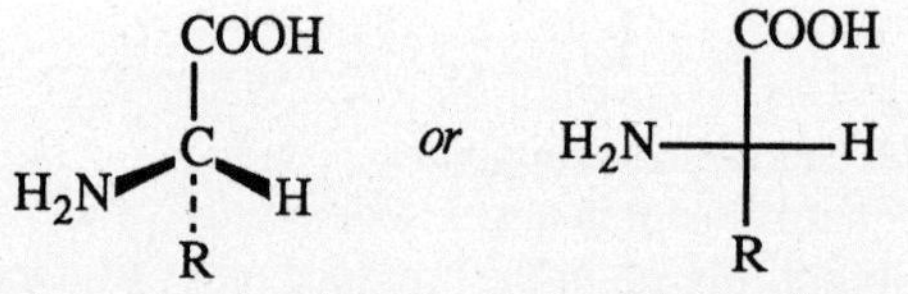

FIGURE 28-1 General Formula for Amino Acids

2. Write structural formulas and names of triglycerides and indicate whether their constituent fatty acids are saturated or unsaturated.

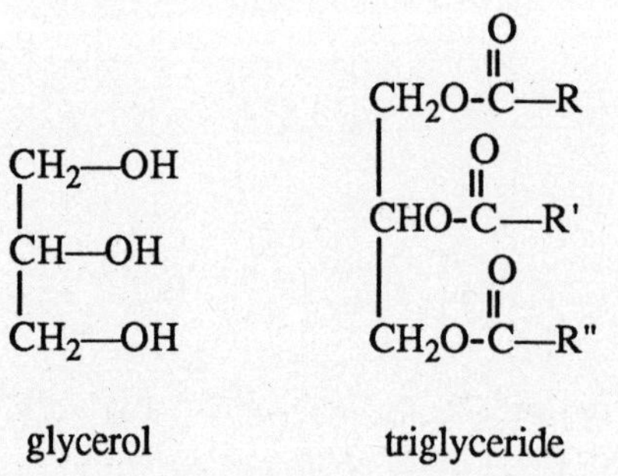

FIGURE 28-2 General Formula of Triglycerides

Triglycerides are esters of glycerol and carboxylic acids with long hydrocarbon chains (Figure 28-2). When the carboxylic acids are saturated, meaning they contain no double (C=C) bonds, the triglyceride is a saturated fat. Saturated fats are waxy solids, such as lard. When the R groups are unsaturated, containing one or more double bonds, the triglyceride is an unsaturated fat. Unsaturated fats are liquids, commonly called oils. Because of the reactivity of the double bonds, oils react with oxygen in the air to become rancid.

Fats are names beginning with "glyceryl," followed by the names of the fatty acid anions. (The anion of stearic acid, $C_{17}H_{35}COOH$, is the stearate anion, $C_{17}H_{35}COO^-$.) A prefix indicates when there is more than one anion of a given type, such as "distearate." In a mixed triglyceride—where the three fatty acids are not the same—the anions are named in alphabetical order and the *-ate* endings of the first two are changed to *-o*. All three names are written together with no spaces. The names of some common fatty acids are given in *Table 28-1 in the text.*

3. Relate the structures, saponification values, and iodine numbers of triglycerides.

The saponification value is the mass in milligrams of KOH that reacts with one gram of triglyceride. We know that 3 moles of KOH reacts with 1 mole of triglyceride, as shown in equation [1]. Thus, we can use the saponification value to determine the mole weight of a triglyceride, as in Example 28-1.

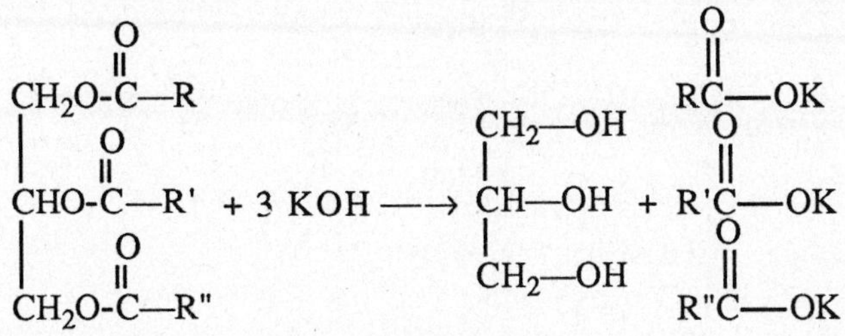

$$
\begin{array}{c}
\text{CH}_2\text{O-C}\overset{\overset{\displaystyle O}{\|}}{}\text{—R} \\
\text{CHO-C}\overset{\overset{\displaystyle O}{\|}}{}\text{—R'} + 3\ \text{KOH} \longrightarrow \text{CH—OH} + \\
\text{CH}_2\text{O-C}\overset{\overset{\displaystyle O}{\|}}{}\text{—R''}
\end{array}
\quad
\begin{array}{c}
\text{CH}_2\text{—OH} \\
\text{CH—OH} \\
\text{CH}_2\text{—OH}
\end{array}
+
\begin{array}{c}
\text{RC}\overset{\overset{\displaystyle O}{\|}}{}\text{—OK} \\
\text{R'C}\overset{\overset{\displaystyle O}{\|}}{}\text{—OK} \\
\text{R''C}\overset{\overset{\displaystyle O}{\|}}{}\text{—OK}
\end{array}
\qquad [1]
$$

EXAMPLE 28-1 What is the mole weight of triglyceride with a saponification value of 185.7?

$$185.7\ \text{mg KOH} \times \frac{1.000\ \text{g}}{1000\ \text{mg}} \times \frac{\text{mol KOH}}{56.11\ \text{g}} \times \frac{\text{mol triglyceride}}{3\ \text{mol KOH}} = 1.103 \times 10^{-3}\ \text{mol}$$

MW = 1.00 g/ 1.103 $\times$ 10^{-3} mol = 906.3 g/mol

The iodine number is the mass in grams of I_2 that reacts with 100 g of triglyceride by addition of I_2 across the double bonds. Thus, the iodine number can be used to establish the number of double bonds in a triglyceride molecule.

EXAMPLE 28-2 How many double bonds are there in the triglyceride of Example 28-1 it its iodine number is 168 g of I_2

$$\frac{168\ \text{g}\ I_2}{100\ \text{g triglyceride}} \times \frac{906.3\ \text{g triglyceride}}{\text{mol triglyceride}} \times \frac{\text{mol}\ I_2}{253.8\ \text{g}\ I_2} = 6.00\ \frac{\text{mol}\ I_2}{\text{mol triglyceride}}$$

There are six double bonds in this triglyceride molecule.

$$
\begin{array}{l}
\text{CH}_2\text{O-C}\overset{\overset{\displaystyle O}{\|}}{}\text{—R} \\
\text{CHO-C}\overset{\overset{\displaystyle O}{\|}}{}\text{—R'} \\
\text{CH}_2\text{O-PO—A} \\
\qquad\ \ \text{O}^- \\
\text{phospholipid}
\end{array}
\qquad
\begin{array}{c}
\text{RC}\overset{\overset{\displaystyle O}{\|}}{}\text{—OR'} \\
\text{wax} \\[1em]
\text{CH}_3(\text{CH}_2)_{24}\text{-C}\overset{\overset{\displaystyle O}{\|}}{}\text{O(CH}_2)_{30}\text{CH}_3 \\
\text{carnauba wax}
\end{array}
\qquad
\begin{array}{c}
\text{CH}_3(\text{CH}_2)_{12}\text{-C}\overset{\overset{\displaystyle O}{\|}}{}\text{O(CH}_2)_{25}\text{CH}_3 \\
\text{beeswax} \\[1em]
\text{CH}_3(\text{CH}_2)_{14}\text{-C}\overset{\overset{\displaystyle O}{\|}}{}\text{O(CH}_2)_{15}\text{CH}_3 \\
\text{spermaceti wax}
\end{array}
$$

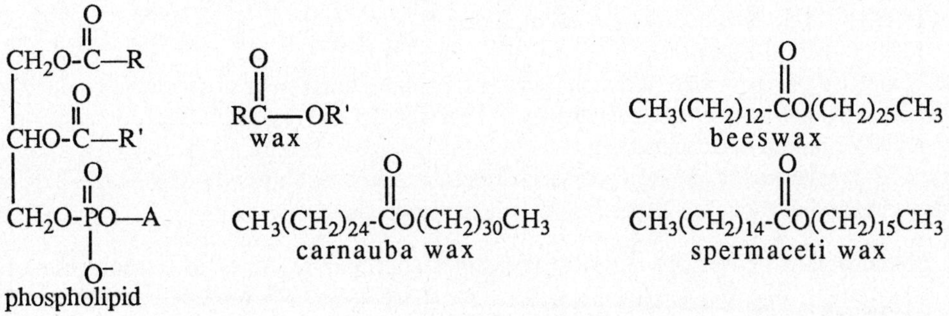

FIGURE 28-3 Structures of Phospholipids and Waxes

4.   Explain how phosphatides and waxes, which also are lipids, differ from triglycerides.

Phosphatides (or phospholipids) are esters of glycerol with two fatty acids and a derivative of phosphoric acids (Figure 28-3). There are two major classes of phosphatides: lecithins, in which —A is —$CH_2CH_2N(CH_3)_3^+$; and cephalins, in which —A is —$CH_2CH_2NH_3^+$. A wax is an ester of a long-chain *mono*hydric alcohol (an alcohol with *one* OH group) with a fatty acid.

5.   Classify carbohydrates as monosaccharides, oligosaccharides, and polysaccharides, and give examples of each.

The general formula of a monosaccharide is $(CH_2O)_n$ where $n = 3, 4, 5, 6$, etc. The total number of carbon atoms in the molecule determines its name: triose ($n = 3$), tetrose ($n = 4$), pentose ($n = 5$), hexose ($n = 6$), and so on. Whether the monosaccharide is an aldose or a ketose is indicated by a prefix: aldo- or keto-. In addition, the position of the C=O in a ketose is indicated by a preceding number (as in Figure 28-4). With the exception of ketotriose, there are several optical isomers of each monosaccharide (see objective 28-6); the number of optical isomers is given in Figure 28-4.

CHO | CHOH | CH₂OH	CH₂OH | C=O | CH₂OH	CHO | CHOH | CHOH | CH₂OH	CH₂OH | C=O | CHOH | CH₂OH	CHO | CHOH | CHOH | CHOH | CH₂OH
aldotriose (1)	ketotriose (1)	aldotetrose (4)	ketotetrose (2)	aldopentose (8)

CH₂OH | C=O | CHOH | CHOH | CH₂OH	CH₂OH | CHOH | C=O | CHOH | CH₂OH	CHO | CHOH | CHOH | CHOH | CHOH | CH₂OH	CH₂OH | C=O | CHOH | CHOH | CH₂OH	CH₂OH | CHOH | C=O | CHOH | CHOH | CH₂OH
2-ketopentose (4)	3-ketopentose (3)	aldohexose (16)	2-ketohexose (8)	3-ketohexose (8)

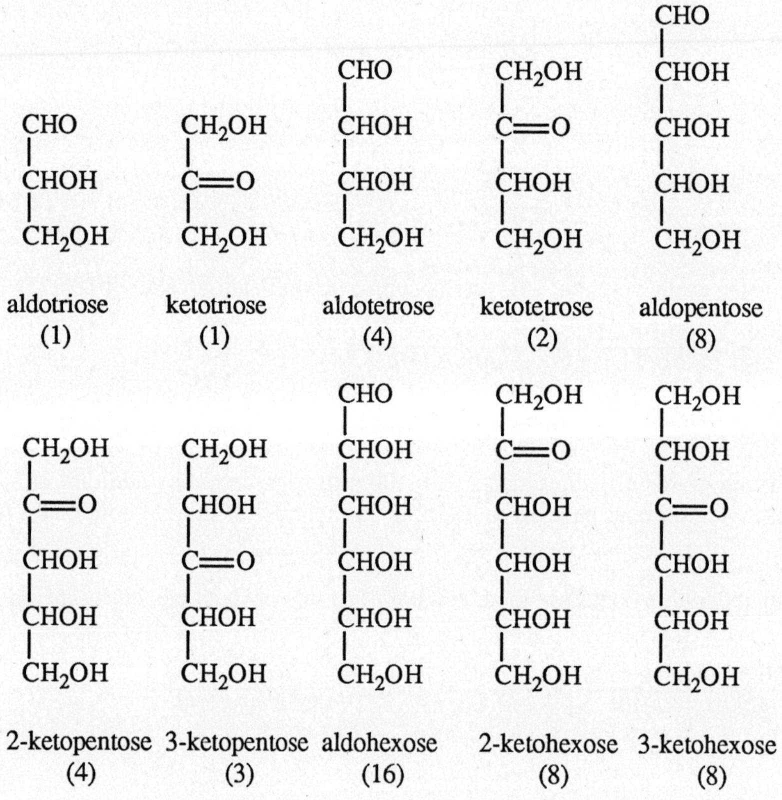

FIGURE 28-4   Names of Monosaccharides.

Oligosaccharides are composed of from two to ten monosaccharide units. The number of saccharide units is given by a numerical prefix: disaccharide, trisaccharide, tetrasaccharide, and so forth. Polysaccharides contain more than ten monosaccharide units. Common disaccharides are sucrose (cane sugar), lactose (milk sugar), maltose, and cellobiose. (See *Figure 28-8 in the text.*) Common polysaccharides are starch, cellulose (both shown *in Figure 28-9 in the text*), glycogen and amylopectin. All of these are polymers of D-glucose.

* **6.   Describe the phenomenon of optical activity and the structural features of a molecule that produces it.**

An optically active molecule is one that rotates the plane of polarized light passed through it. Light is accompanied by both an electric field and a magnetic field, which vibrate at right angles to the direction of travel. If we could see this vibration, light traveling at us would look like Figure 28-5a. When light is plane polarized, the electric or magnetic field vibrates in one plane (Figure 28-5b). Light can be polarized in two ways: by passing it through polarizing material or by reflecting it off a flat surface. You probably have seen light polarized by a flat surface. It looks like shimmering water (a mirage) at a distance and often is seen on the highway on bright days.

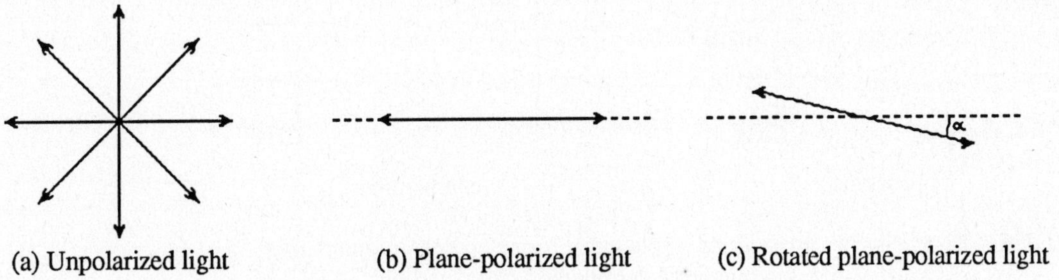

FIGURE 28-5 The Polarization of Light

Optically active molecules rotate the plane of polarized light, as shown in Figure 28-5c, where the plane is rotated by an angle α. A molecule is optically active if it cannot be superimposed on its mirror image. If there are four different groups bonded to a carbon atom in a molecule, that carbon atom is termed a chiral carbon or a chiral center. (A chiral carbon formerly was referred to as an asymmetric carbon.) A chiral carbon often is present in an optically active molecule. A molecule with only one chiral carbon must be optically active. The easiest way to see optical activity is to build a model of the molecule and of its mirror image and see if they can be superimposed. After some practice with models, you can used sketches. There are two types of sketches that you can draw: wedge and dash sketches, and Fischer projections. Both types are shown in Figure 28-6 for CHFClBr. In Fischer projections, the groups above and below the chiral carbon are assumed to point away from you and the groups on either side point toward you. Two compounds that differ only in their optical activity are called enantiomers.

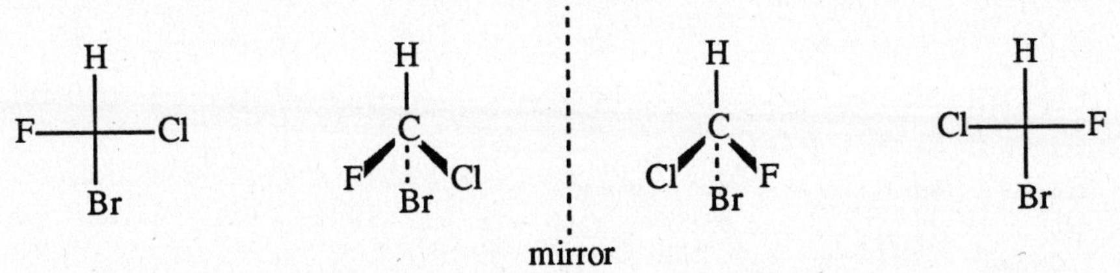

FIGURE 28-6 Wedge and dash sketches (in center) and Fischer projections (on outside) of the two enantiomers of CHFClBr.

The presence of a chiral carbon does not go hand-in-hand with optical activity. A molecule may *lack* a chiral carbon and yet be optically active, that is, possess a nonsuperimposable mirror image. An example is 1-bromo-1-chloro-2-fluoro-1,2-propadiene, shown in Figure 28-7.

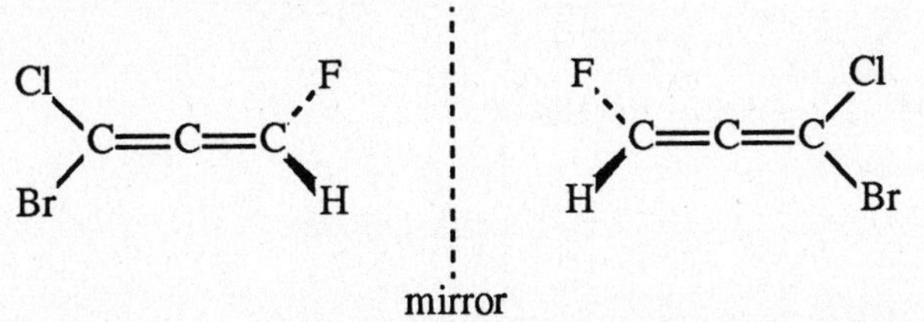

FIGURE 28-7 Wedge and dash drawings of the two enantiomers of 1-bromo-1-chloro-2-fluoro-1,2-propadiene.

On the other hand, a molecule may posses two or more chiral carbons and yet its mirror image *is* superimposable, as exemplified by 2,3-dichlorobutane in Figure 28-8.

mirror

CH₃ | H—*—Cl | H—*—Cl | CH₃   CH₃ | Cl—*—H | Cl—*—H | CH₃

meso forms

mirror

CH₃ | Cl—*—H | H—*—Cl | CH₃   CH₃ | H—*—Cl | Cl—*—H | CH₃

enantiomers

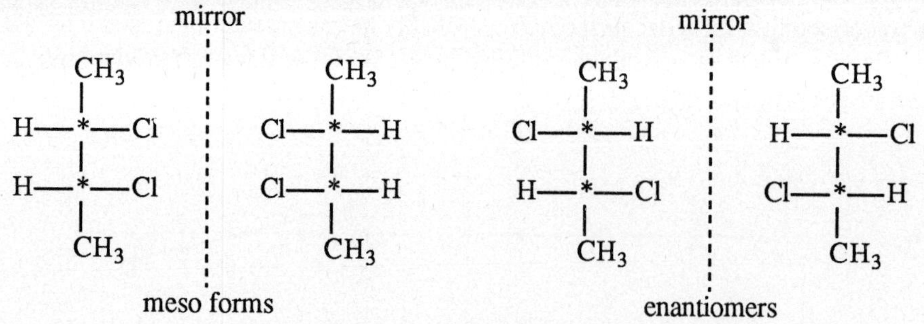

FIGURE 28-8 Fischer projections of the meso forms and enantiomers of
2,3-dichlorobutane (* designates a chiral carbon).

7. Describe the conversion of a straight-chain monosaccharide into its cyclic form, and the joining of monosaccharide units into polysaccharides.

The aldohexoses do not exist in the straight-chain form shown in Figure 28-4 but rather as six-membered rings. The principal evidence for the ring is that two forms of D-glucose exist: α-D-glucose and β-D-glucose. Whereas the straight-chain form of glucose has four chiral carbons (numbered 2-5 in Figure 28-9), the ring form has five (1-5 in Figure 28-9). α-D-glucose and β-D-glucose also are known as pyranoses, because of their similarity to the pyran ring structure. The glucose ring opens and closes readily in aqueous solution and thus the α and β forms of D-glucose interconvert. At 25°C, the equilibrium mixture contains about two molecules of β-D-glucose for every α-D-glucose molecule. The chair conformations of α-D-glucose and β-D-glucose in Figure 28-9 show that the —OH group on carbon 1 is axial in α-D-glucose and equatorial in β-D-glucose. The equatorial position is favored for a large group (since there is more room available there than at the axial position), and thus β-D-glucose is more stable.

α-D-glucose   pyran   β-D-glucose

α-D-glucose   β-D-glucose

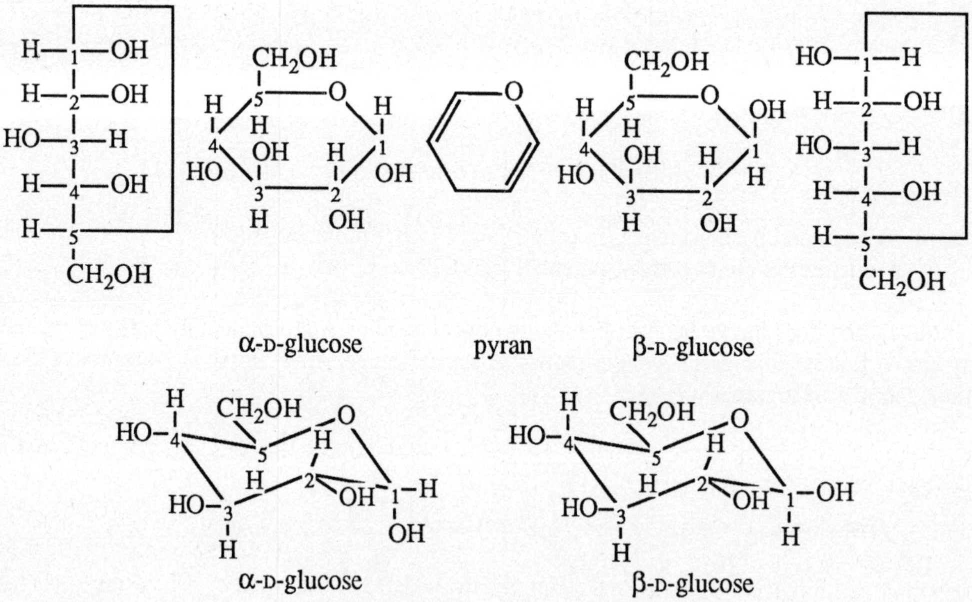

FIGURE 28-9 α- and β-D-glucose

Monosaccharides join together to form disaccharides, other oligosaccharides, and polysaccharides, with the loss of one water molecule for each bond that is formed. The linkage reaction is given in equation [2]. Note that it is the —OH groups on the two molecules that react in forming the bond.

$$>C\diagdown_{H}^{OH} \; + \; \diagup_{C}^{HO}< \; \rightleftharpoons \; >C\diagdown_{H} \diagup^{O} C< \; + \; H_2O \qquad [2]$$

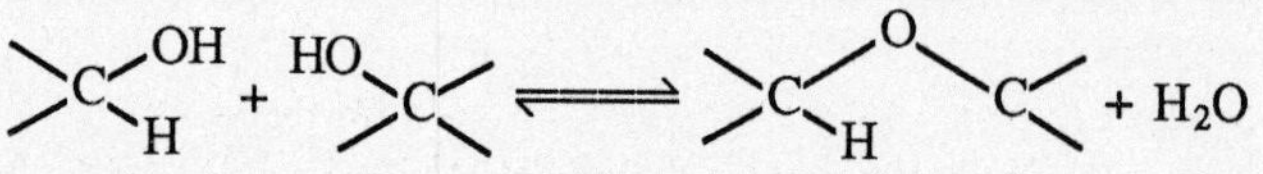

Hexoses can join between carbon 1 on one molecule, a (1⟶4) linkage, or they can join by a (1⟶6) linkage. These linkages may be either α or β (axial or equatorial from carbon 1. The monomers (or subunits) and the linkages for common disaccharides are given in Table 28-1.

TABLE 28-1 Joining of Monosaccharides

Carbohydrate name	*Name of subunit(s)*	*Linkage(s)*
maltose	glucose	(1⟶4)
lactose	galactose, glucose	(1⟶4)
cellobiose	glucose	(1⟶4)
α-amylose (starch)	glucose	(1⟶4)
amylopectin (a component of starch)	glucose	(1⟶4) backbone α(1⟶6) side chain branches
glycogen (in liver and muscle)	glucose	same as amylopectin, but more highly branched
cellulose	glucose	β(1⟶4)

8. Show how the form adopted by an amino acid in solution varies with the pH.

Two of the groups on an amino acid change form depending on acidity: the amino group (—NH_2) and the carboxylic acid group (—COOH). We have shown both of these groups as uncharged (Figure 28-1). Since the forces between uncharged molecules are relatively low, we would expect solid amino acids to be low melting. Yet most crystalline amino acids melt above 200°C, a temperature more characteristic of ionic compounds. ($LiNO_3$ melts at 264°C.) This is one of the many data that indicate that solid amino acids exist as dipolar ions called zwitterions, which are depicted in Figure 28-10. (The name comes from the German word for two, *zwei*.)The zwitterion also exists in neutral solution or, more correctly, at the isoelectric point, a pH intermediate between acidic and basic. The isoelectric point depends on the acidic character of the specific amino acid. In basic solution, a proton is donated by (and thus removed from) the nitrogen of the ammonium group (—NH_3^+), yielding an anion. In acidic solution, a proton is accepted by (and thus added to) an oxygen of the carboxylate group (—COO^-), yielding a cation.

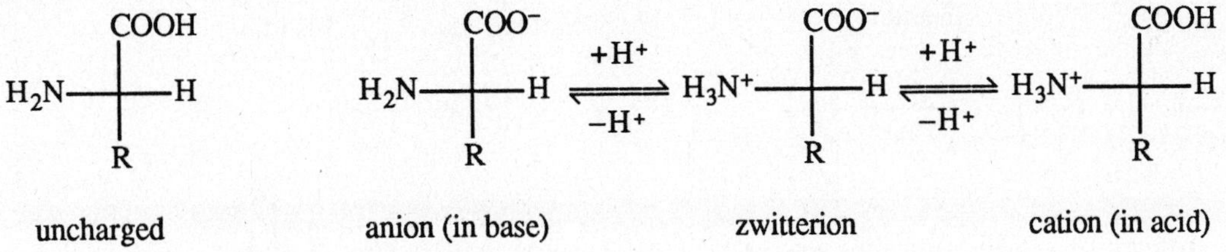

FIGURE 28-10 Different Forms of Amino Acids

9. Describe the formation of a peptide bond between two amino acids, and assign structures and systematic names to polypeptides.

We have already seen several cases of a bond being formed between two molecules by the elimination of water between them. Among these are the formation of pyrophosphates and pyrosulfates in objective 23-10, and the formation of ethers and esters in equations [16] and [23] of objective 27-11. An analogous reaction occurs between two amino acids.

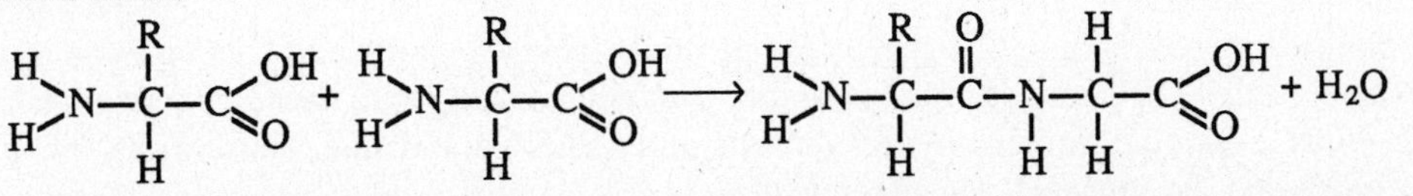

A polypeptide is named by naming the amino acids in order starting at the N-terminal end. Except for the last amino acid (the one at the C-terminal end) the *-an* or *-ine* ending of each amino acid is changed to *-yl*. For large polypeptides, more than ten amino acids, only the peptide sequence is given. This also starts the the N-terminal end, and uses the three-letter symbols given in *Table 28-3 in the text*. The symbols, names, and structures in that table should be memorized.

*** 10. Determine the sequence of amino acids in a polypeptide chain from information acquired in the hydrolysis of the polypeptide.**

With regard to polypeptide analysis, degradation is the process of breaking the molecule apart, identifying the fragments, and then deducing the structure. In determining the structure—the sequence of amino acids—a large "marker" group is attached to the N-terminal end (the end with the amino group, —NH_2). The polypeptide is gently hydrolyzed by heating gently in a mildly acidic or alkaline solution. (A more precise method of hydrolysis is treatment with selective enzymes. Some of these enzymes are so specific that they only hydrolyze the peptide joining two distinct amino acids as, for example, an alanine-glycine bond.) Note that hydrolysis is the reverse of peptide bond formation. After the initial hydrolysis, the N-terminal marker is placed on each fragment. This process—of gentle hydrolysis followed by marking the N-terminal group—is repeated until the chain has been degraded to di-, tri-, and tetrapeptides. The identities of these small molecules are determined by conventional means: melting point, solubility, spectral analysis, and so forth. Putting together the sequence of the polypeptide is a matter of piecing together these fragments.

*** 11. Explain the meaning of primary, secondary, tertiary, and quaternary structure of a protein.**

The *primary* structure of a protein is the sequence of amino acids in the molecule, starting at the N-terminal end.

The *secondary* structure refers to how the backbone of the polypeptide chain is arranged in space. Most often the polypeptide chain has the form of an α-helix. Hydrogen bonds are formed between the carboxyl oxygen atoms of one amino acid and the amide nitrogen of the amino acid four positions further along the chain. These hydrogen bonds are shown as dashed lines in *Figure 28-12 in the text.*

The *tertiary* structure describes how different parts of the α-helix are held together. The R groups of the amino acids point outward from the helix and therefore can attract each other. Thus the helix will bend into a three-dimensional coil, rather like a twisted worm. The forces that hold the tertiary structure together are those between the atoms of the R groups. They are of four types, and are illustrated in Figure 28-11.

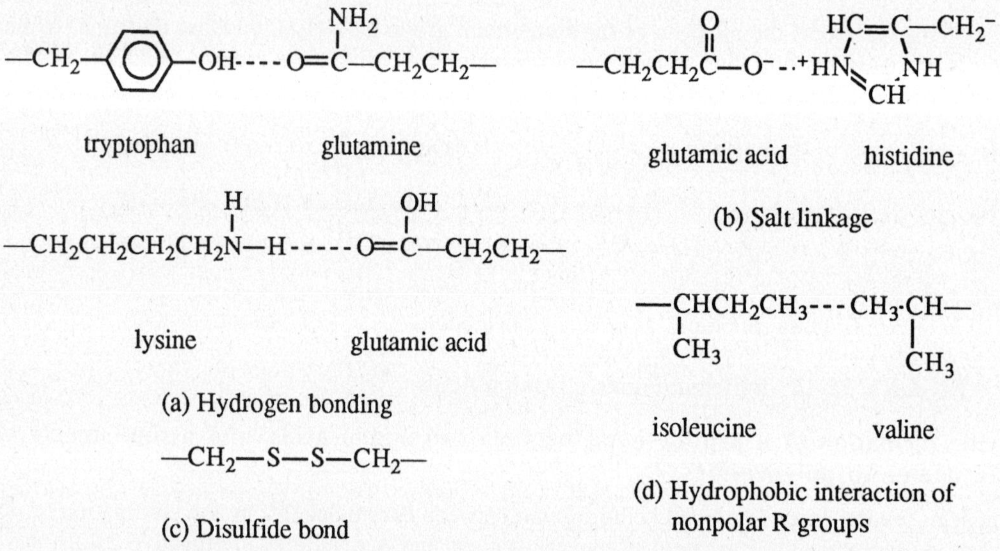

FIGURE 28-11 Interactions that Influence Tertiary Structure

(a) *Hydrogen bonds* link —NH_2 or —OH groups of one amino acid's R group and C=O groups of another's. The amino acids with —NH_2 groups are asparagine, glutamine, tryptophan, lysine, arginine, and histidine. Those with —OH groups are serine, 4-hydroxyproline, threonine, and tyrosine. Those with C=O groups are asparagine, glutamine, aspartic acid, and glutamic acid.

(b) *Salt linkages* occur between acidic and basic R groups. Acidic amino acids are aspartic acid and glutamic acid. Basic ones are lysine, arginine, and histidine.

(c) *Disulfide bonds* join the two "halves" of cystine.

(d) *Hydrophobic interactions* occur because nonpolar R groups, such as those in glycine, alanine, valine, leucine, phenylalanine, and proline, tend to point inward in the molecule, away from the polar solvent.

In addition to these attractions between R groups, the protein may be wrapped around and bound to another type of molecule. Thus in hemoglobin the proteins (called globins) are bound to a heme molecule.

The quaternary structure of a protein is how two or more individual polypeptide chains are packed together to form the final protein. Many proteins possess only one polypeptide chain and thus have no quaternary structure. But in hemoglobin, for example, four globin chains and four heme molecules combine to produce the final protein.

12. In general terms, explain how lipids, polysaccharides, and proteins are metabolized in the human body and how the energy released by metabolism is stored.

The body breaks down lipids, polysaccharides, and proteins by hydrolysis of the bond between the parts. Thus, lipids break down into glycerol and fatty acids, polysaccharides into monosaccharides, and proteins into amino acids.

Lipids, polysaccharides, and proteins are not rapidly hydrolyzed. The human body uses protein catalysts, called enzymes, to speed up these reactions. Both fatty acids and the monosaccharides are further broken down within the body, resulting ultimately in carbon dioxide and water. Some are stored as fat deposits in the body, and some as carbohydrate deposits, principally glycogen in liver and muscle tissue. Amino acids can be further metabolized to provide energy, but they also are used as raw material for protein synthesis.

13. Explain enzyme action in terms of the "lock-and-key" model.

Enzymes are protein catalysts. Each enzyme speeds up a certain type of chemical reaction. For example, hydrolases speed up hydrolysis reactions. Other enzymes control the rate of peptide bond formation, disulfide bond formation or breakage, and so forth. Enzymes can be quite specific. There are enzymes for triglyceride hydrolysis that will not affect the rate of polysaccharide hydrolysis. Enzymes also are stereospecific. An enzyme may catalyze a reaction involving α-D-glucose but not affect a reaction involving α-L-glucose.

The molecule on which the enzyme acts is known as the substrate. The involved structure of an enzyme and its high specificity are strong evidence that the substrate fits almost perfectly into a pocket in the enzyme, called the active site. The active site often is only a very small portion of the large enzyme molecule. The substrate fits into the active site much as does a key into a lock. While the substrate is held in place, the enzyme causes the desired reaction.

14. Name the principal constituents of nucleic acids and indicate how these constituents are linked together into chains and, in the case of DNA, a double helix.

Nucleic acids consist of phosphate groups, the purine and pyrimidine bases, and the pentose sugars (shown in *Figure 28-21 in the text*). The purine and pyrimidine bases bond to the 1' carbon of the pentose sugar, thus forming *nucleosides*.. Guanine bonds to the pentose through the same nitrogen as does adenine. Cytosine and uracil bond through the same nitrogen as does thymine. A *nucleotide* consists of a phosphate group bound to a nucleoside. The phosphate group can bond to the 5' or the 3' carbon of either pentose, and also to the 2' carbon of ribose. However, all naturally occurring nucleotides have the phosphate group bound to the 5' carbon.

Nucleotides for chains by a bond between the phosphate group on one nucleotide and the 3' carbon on the next nucleotide. This is known as a 3'-5' linkage. A portion of a nucleic acid chain is show in *Figure 28-21 in the text*.

The single-stranded DNA forms a double-stranded helix by formation of hydrogen bonds. A purine (the larger molecules: adenine and guanine) always is hydrogen-bonded to a pyrimidine (the smaller molecules: thymine and cytosine). This base pairing holds the two strands of the chain together. The base pairing and a model of the double helix are shown in *Figure 28-22 in the text.*

15. Explain how DNA replicates itself during cell division; and outline the process of protein synthesis, indicating the roles of DNA, mRNA, and tRNA.

In the replication of DNA, the molecule unwinds gradually. As it does so, nucleotides in the cell attach to the exposed base pairs of the unwinding chains. The phosphate-pentose bonds are formed through the action of enzymes. The exact details are still being worked out.

In one-celled organisms, protein synthesis begins with transcription, the production of an RNA copy of part of the DNA strand. This close copy is messenger RNA or mRNA, and is similar to a copy of a master blueprint. The mRNA molecule, which is single stranded, leaves the nucleus for the cytoplasm of the cell, where it attaches to one or more ribosomes. This grouping of an mRNA molecule and several ribosomes is a polysome. Ribosomes are made of protein and ribosomal RNA or rRNA. Molecules of transfer RNA (or tRNA) are attracted to the polysomes. There a portion of the tRNA molecule, called an anticodon, matches up with a codon on mRNA. Each codon specifies a particular amino acid. An amino acid is bound to the end of the tRNA molecule away from the anticodon and has been carried along to the polysome with the tRNA. An enzyme in the cell then forms the peptide bond between adjacent amino acids.

DRILL PROBLEMS

2. (1) Name the following triglycerides.

A.
$$\begin{array}{l} CH_2OOC(CH_2)_7CH{=}CH(CH_2)_7CH_3 \\ | \\ CHOOC(CH_2)_7CH{=}CH(CH_2)_7CH_3 \\ | \\ CH_2OOC(CH_2)_7CH{=}CH(CH_2)_7CH_3 \end{array}$$

B.
$$\begin{array}{l} CH_2OOC(CH_2)_8CH_3 \\ | \\ CHOOC(CH_2)_8CH_3 \\ | \\ CH_2OOC(CH_2)_8CH_3 \end{array}$$

C.
$$\begin{array}{l} CH_2OOC(CH_2)_{10}CH_3 \\ | \\ CHOOC(CH_2)_7CH{=}CHCH_2CH_3 \\ | \\ CH_2OOC(CH_2)_8CH_3 \end{array}$$

D.
$$\begin{array}{l} CH_2OOC(CH_2)_7(CH{=}CH)_3(CH_2)_3CH_3 \\ | \\ CHOOC(CH_2)_7(CH{=}CH)_3(CH_2)_3CH_3 \\ | \\ CH_2OOC(CH_2)_7(CH{=}CH)_3(CH_2)_3CH_3 \end{array}$$

(2) Draw the condensed formula of the following triglycerides.

E. glyceryl dipalmitoleate
F. glyceryl lauromyristopalmitate
G. glyceryl dilaurolinoleate
H. glyceryl trilinolenate
I. glyceryl triricinoleate
J. glyceryl dioleolaurate

3. (1) Determine the average mole weights and the average number of double bonds per molecule for each of the following oil by using the given average value of the saponification value and the iodine number (in that order in parentheses).

A. butter (220, 27)
B. coconut oil (256, 8.8)
C. soybean oil (192, 132)
D. safflower oil (191, 148)

(2) Determine the saponification value and the iodine number of each of the following triglycerides.

E. glyceryl trioleate
F. glyceryl eleosterate
G. glyceryl tristearate
H. glyceryl dilaurolinoleate

6. Draw the enantiomers in Fischer projection for each of the following molecules which is optically active.

A. 2-chloropentane
B. 3-chloro-1-pentene
C. $CH_3CH(NH_2)COOH$
D. $CH_3CH_2CHOHCH_3$
E. 2-chlorobutane
F. 1-chloro-3-methylpentane
G. 1,2-dichlorobutane
H. 1,4-dichlorobutane
I. 2,2-dichloropentane
J. 3-chloropentanol
K. 1,3-dibromopentane
L. 1-bromo-3-chloro-1-butene
M. 3-chloro-4-methyl-1-pentane

10. (1) Based on the hydrolysis fragments that follow, determine the sequence of each polypeptide, starting with the N-terminal end. Each fragment starts with the N-terminal end.

A. Lys-Asp-Gly, Glu-Ser-Gly, Ala-Ala-Glu, Asp-Gly-Ala, Ala-Glu-Ser
B. Lys-Phe-Ile, Arg-Glu-Lys, Ala-Ala-His, Glu-Lys-Phe, Ala-His-Arg-Glu
C. Cys-Lys-Ala, Tyr-Cys-Lys, Arg-Arg-Gly, Ala-Arg-Arg
D. Phe-Ala, Ser-Ala-Gly, Glu-Ser-Ala, Ala-Glu
E. Val-Ala-Lys-Glu, Met-Gly-Gly-Phe, Val-Met-Tyr-Cys, Glu-Glu-Phe-Val, Tyr-Cys-Glu, Glu-Phe-Val, Glu-Trp-Met, Gly-Gly-Phe

(2) Name the results of A through D above.

QUIZZES (15-20 minutes each) Choose the best answer for each question.

Quiz A

1. Which function is not regulated by a hormone? (a) pregnancy; (b) vision; (c) metabolism; (d) growth; (e) reproduction.
2. The mole weight of a triglyceride can be determined from its (a) iodine number; (b) oxygen content; (c) unsaturation; (d) saponification value; (e) none of these.
3. All lipids contain (a) nitrogen atoms; (b) long-chain alcohols; (c) glycerol; (d) a phosphate group; (e) none of these.

4. Which of the following is not a monosaccharide? (a) glucose; (b) ribose; (c) glyceraldehyde; (d) sucrose; (e) fructose.
5. Which molecule is not optically active? (a) 1,2-dichlorobutane; (b) 1,4-dichlorobutane; (c) 1,3-dichlorobutane; (d) 1,2-dichloropropane; (e) none of these.
6. Which symbol does not specify how a compound rotates polarized light? (a) d; (b) +; (c) *dl*; (d) D; (e) none of these.
7. Amino acids are joined into proteins by (a) a 3'-5' linkage; (b) a β(1⟶4) linkage; (c) DNA; (d) an ester; (e) none of these.
8. The R group $NH_2CH_2CH_2$— is that of what amino acid? (a) alanine; (b) serine; (c) threonine; (d) arginine; (e) none of these.
9. Glycine is (a) a sugar; (b) an amino acid; (c) a nucleoside; (d) a lipid; (e) none of these.

Quiz B

1. Which of the following is not present in DNA? (a) purine; (b) pentose; (c) phosphate; (d) heme; (e) pyrimidine.
2. The number of double bonds in a triglyceride molecule can be determined from its (a) mole weight; (b) saponification value; (c) iodine number; (d) oxygen content; (e) none of these.
3. Lipids always contain (a) glycerol; (b) phosphate groups; (c) double bonds; (d) fatty acids; (e) none of these.
4. Which of the following is a monosaccharide? (a) sucrose; (b) glycogen; (c) starch; (d) cellulose; (e) none of these.
5. Which molecule is optically active? (a) 1,2-dichlorobutane; (b) 1,4-dichlorobutane; (c) *cis*-1,2-dichlorocyclobutane; (d) *cis*-2-butene; (e) none of these.
6. D refers to (a) rotation of plane polarized light; (b) a meso form; (c) configuration of sugars; (d) racemic mixtures; (e) none of these.
7. What is present in all amino acid molecules? (a) a chiral carbon; (b) two nitrogen atoms; (c) a peptide bond; (d) a —COOH or $—COO^-$ group; (e) none of these.
8. Glycine has what R group? (a) $—CH_3$; (b) —H; (c) $—CH_2CH_2COOH$; (d) $—CH_2SH$; (e) none of these.
9. 2'-Deoxyribose is a (a) nucleoside; (b) sugar; (c) wax; (d) nucleotide; (e) purine.

Quiz C

1. All of the following are major component of every cell except (a) triglycerides; (b) carbohydrates; (c) proteins; (d) nucleic acids; (e) hormones.
2. Which of the following fatty acids is unsaturated? (a) palmitic acid; (b) oleic acid; (c) lauric acid; (d) stearic acid; (e) none of these.
3. Biological waxes are (a) the same as triglycerides; (b) the same as lecithins; (c) soluble in water; (d) esters of long-chain alcohols; (e) none of these.
4. Which of the following does *not* contain six carbon atoms? (a) deoxyribose; (b) glucose; (c) aldohexose; (d) methylcyclopentane; (e) benzene.
5. Which molecule is optically active? (a) 2-chlorobutane; (b) 1-chlorobutane; (c) *trans*-propene; (d) *cis*-propene; (e) none of these.
6. Which of the following is optically active? (a) racemate; (b) *dl*; (c) an enantiomer; (c) DL; (e) none of these.
7. In alkaline solution, an amino acid has what charges? (a) + and –; (b) – only; (c) + only; (d) no charges; (e) none of these.
8. The —R group of serine is (a) $—CH_2OH$; (b) $—CH_2COOH$; (c) $—CH_2CH_2CH_3$; (d) $—CH_2CH_2CH_2CH_2NH_2$; (e) none of these.
9. Cytosine is (a) an amino acid; (b) a nucleoside; (c) a purine; (d) a sugar; (e) none of these.

Quiz D

1. All of the following contain both carbon and nitrogen atoms except (a) nucleosides; (b) nucleotides; (c) pyrimidines; (d) purines; (e) none of these.
2. A fatty acid usually has (a) many branched chains; (b) many double bonds; (c) no double bonds; (d) an odd number of carbon atoms; (e) none of these.
3. Which of the following is *false* about lipids? (a) some are triglycerides; (b) they are soluble in water; (c) they are soluble in hexane; (d) some are lecithins; (e) all are true.
4. Which of the following is *not* a monosaccharide? (a) ribose; (b) fructose; (c) glucose; (d) glycogen;

(e) pentose.

5. Enantiomers *always* (a) have an asymmetric carbon; (b) have different physical properties; (c) change the color of light; (d) rotate polarized light; (e) none of these.
6. Which of the following is *not* broken by hydrolysis with dilute acid? (a) DNA; (b) peptide bonds; (c) polysaccharides; (d) ring forms of hexoses; (e) triglycerides.
7. At the isoelectric point, an amino acid has what charges? (a) + and –; (b) none; (c) + only; (d) – only; (e) none of these.
8. 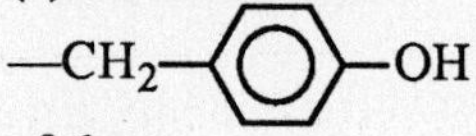 is the R group of (a) tryptophan; (b) tyrosine; (c) serine; (d) phenylalanine; (e) none of these.
9. Which of the following is a pyrimidine? (a) adenine; (b) glycine; (c) phenylalanine; (d) thymine; (e) none of these.

SAMPLE TEST (20 minutes)

1. Draw the cyclic forms of threose, $OHC(CHOH)_2CH_2OH$, remembering that five- or six-membered rings are more stable than larger or smaller ones.
2. Describe three structural features of proteins which influence their tertiary structures. Give an example of each feature.
3. On gentle hydrolysis, a protein gives the following polypeptides, with the N-terminal end written first for each fragment. Give the amino acid sequence of this protein.
   Gly-His-Leu, Phe-Val-Asp, Cys-Gly-Ser-His, His-Leu-Cys-Gly, Asp-Gly-His, Gly-Ser-His, Phe-Val

Appendix I

Answers and Solutions

CHAPTER 1

Drill Problems

8. (1) A. 4 B. 6 C 4 D. 4 E. 2 F. 4-6 G. 3 H. 4 I. 8 J. 5 K. 6 L. 7 M. 6 N. 4 O. 5 P. 5 Q. 2-4 R. 1-5 S. 4 T. 6 U. 6 V. 3 W. 1-4 X. 3
(2) Y. 6.168 Z. 213.2 Γ. 1.200×10^3 Δ. 3136 Θ. 6.196 Λ. 14.16 Ξ. 0.002246 Π. 152.0 Σ. 0.001987 Υ. 3.024×10^5 Φ. 14.16 Ψ. 3.141×10^4
9. A. 0.592 B. 1.701×10^{-23} C. 56 D. 2×10^4 or 1.7×10^4 or 1.75×10^4 E. 8.9×10^2 F. 3.0×10^1 G. 13.6 H. 0.21 or 0.208 or 0.2078 I. 15.3 J. 3.60×10^7 K. 122 L. 87.6 M. −91 N. 108.8 O. 4.362 P. 696 Q. 1033.7 R. 75.9 S. 1.4×10^3 T. 412 U. −1.33 V. 3.8 W. 1.98
10. A. 3.74×10^{-3} B. 1.2×10^3 C. 4.06389×10^3 D. 1.751×10^5 E. 6.4604×10^{10} F. 6.627×10^{-27} G. 9.475×10^{-3} H. 3.74×10^4 I. 1.42×10^{-4} J. 1.7645×10^4 K 2.12×10^8 L. 2.66×10^{-3} M. 8.43×10^7 N. 9.400×10^7 O. 4.963×10^{-8} P. 8.43214×10^{-1} Q. 2.12×10^8 R. 8.39×10^{-1} S. 8.9413×10^2 T. 8.314×10^{-7} U. 4.90006×10^4 V. 9.204×10^8 W. 8.7012×10^{27} X. 1.413×10^{-7} Y. 1.7645×10^{-11} Z. 2.66 Γ. 8.134×10^{-5}
11. (1) A. 88 in. B. 205 fl oz C. 7.244 mi D. 2.346 lb E. 165.4 mm F. 3.74×10^4 μm G. 0.146 L H. 0.03754 m^3 I. 2.72×10^7 g J. 0.454 lb K. 8.46 fl oz L. 2.23 m M 0.355 L N. 85.0 kg O. 22 cm P. 3.78 L Q. 236 cm^3 R. 14.2 g S. 57.7 kg T. 5.92 gal U. 0.998 oz V. 3.38 fl oz W. 16.4 yd X. 157 cm
(2) Y. 76.5 m^3 Z. 1.31 yd^3 Γ. 1728 $in.^3$ Δ. 6.45 cm^3 Θ. 929 cm^3 Λ. 202 gal Ξ. 7.35 ft^3 Π. 4.10 L Σ. 1.804 $in.^3$
12. A. 5.00 g, 0.176 oz B. 96.5 g, 3.40 oz C. 56.7 g, 2.00 oz D. 197 g E. 250 g F. 4.41 cm^3 G. 5.60 cm^3 H. 57.7 cm^3 I. 23.5 cm^3 J. K. 208 lb L. 746 g M. 17.6 $in.^3$ N. 7.17 $in.^3$
13. A. 2.86 g B. 0.13 ton Al, 87 L Fe C. 3.7×10^5 L N_2, 3.2×103 L air D. 71.4 lb solder E. 20 lb bleach, 2.3 gal bleach F. 1.22 qt whiskey G. 1.99 L vinegar H. 2.27×10^3 g sulfuric acid I. \$0.50, 2.0×10^3 pennies J. \$33.60, 3.72 g silver nitrate
14. A. 0°C B. 37.8°C C. 209°F D. −3.9°C E. 212°F F. 39°F G. −26°F H. 260°C I. −76°C J. −459.67°F
15. A. $x = 0.602$ B. $r = (3V/4\pi)^{1/3}$ C. $x = 14/15$ D. $h = 1.15H/13.6$ E. $T = pvt/PV$ F. $R = PV/nT$ G. $n/V = P/RT$ H. $\mathcal{M} = mRT/PV$ I. $m/V = P\,\mathcal{M}/RT$ J. $u = (3RT/\mathcal{M})^{1/2}$ K. $n = \{1/[(1/4) - (v/Rc)]\}^{1/2}$ L. $v = h/m\lambda$ M. $x = 4/5$

Quizzes

A: 1. (d) 2. (a) 3. (e) 4. (e) 5. (d) 6. (d) 7. (b) 8. (b) 9. (a) 10. (a) 90°F 11. (c) 12. (d)
B: 1. (a) 2. (a) 3. (a) 4. (e) 5. (a) 6. (d) 7. (c) 8. (c) 9. (e) 10. (a) 11. (a) 12. (d)
C: 1. (c) 2. (b) 3. (c) 4. (d) 5. (a) 6. (c) 7. (e) 8. (c) 9. (e) 10. (e) 11. (e) 12. (b)
D: 1. (b) 2. (d) 3. (b) 4. (b) 5. (d) 6. (a) 7. (c) 8. (b) 9. (a) 10. (c) 11. (c) 12. (b)

Sample Test

1. $10.72\ \text{g} \times \frac{0.417\ \text{g Au}}{\text{g ring}} \times \frac{\text{lb}}{454\ \text{g}} \times \frac{16\ \text{oz}}{\text{lb}} \times \frac{\$652.50}{\text{oz}}$
$= \$103$

2. a. physical—no change of identity, only of physical state b. physical—separation of a mixture c. chemical—two new substances are produced d. chemical—a compound is produced from elements e. physical—separation of a mixture Air is a mixture that later is separated physically; pressure and temperature are not forms of matter; natural gas is possibly a mixture, more likely a compound, since it is later used in a reaction; steam and carbon dioxide are compounds; hydrogen is one of the elements; and ammonia gas is a compound.

3. a. $12.5 \frac{\text{lb}}{\text{in.}^2} \times \frac{16\ \text{oz}}{\text{lb}} \times \left(\frac{\text{in.}}{2.54\ \text{cm}}\right)^2 = 31.0\ \text{oz/cm}^2$

$$31.0 \frac{\text{oz}}{\text{cm}^2} \times \frac{\text{lb}}{16\ \text{oz}} \times \frac{\text{kg}}{2.20\ \text{lb}} \times \left(\frac{100\ \text{cm}}{\text{m}}\right)^2$$
$$= 8.81 \times 10^3\ \text{kg/m}^2$$

b. $(77.4°\text{F} - 32.0)\frac{5}{9} = 25.2°\text{C}$ c. $5.2\ \text{m}^3 \times \frac{1000\ \text{L}}{\text{m}^3}$

$$= 5.2 \times 10^3\ \text{L} \times \frac{\text{qt}}{0.946\ \text{L}} = 5.5 \times 10^3\ \text{qt}$$

CHAPTER 2

Drill Problems

1. (1) A. 8.21 g products, 1.80 g water B. 8.15 g products, 6.35 g Cu, 79.8% Cu C. 10.8 g water, 60.0% water D. 2.23 g Fe, 69.9% Fe, 0.592 g carbon dioxide, 0.592 ton carbon dioxide E. 14.36 g products, 7.96 g copper(II) oxide F. 10.8 g water, 19.2 g oxygen, 4.27 lb oxygen
(2) G. 28.6 g S, 28.5 g oxygen H. 27.3 g C, 72.7 g oxygen I. 8.32 g hydrogen, 66.7 g oxygen J. 21.1 g oxygen, 53.0 g Ca K. 23.1 g table salt, 9.1 g Na L. 182 g ammonia, 150 g nitrogen
(3) Mass ratios are given in the same order as the compounds are listed, followed by the whole number ratios.
A. g K/g O = 4.88:2.45:1.63 or 1.99:1.50:1.00 or about 6:3:2
B. g C/g H = 12.0:3.00:4.00:5.99 or about 12:3:4:6
C. g Cl/g O = 4.43:2.22:0.634 or 6.99:3.50:1.00 or about 14:7:2
D. g Cl/g S= 1.11:2.22:4.43 or 1.00:2.00:3.99 or about 1:2:4
E. g N/g H = 4.68:7.00:42.5:13.9 or 1.00:1.50:9.08:2.97 or about 2:3:18:6 F. g P/g H = 10.2:10.2:10.4 or about 1:1:1; g O/g P = 2.07:1.54:1.03 or 2.01:1.50:1.00 or about 4:3: G. g S/g H = 16:16:16 or 1:1:1; g O/g S = 1.50:1.99:2.49 or about 3:4:5
8. A. $^{81}Br^-$, −1, 46 n B. $^{59}Co^{3+}$, Z = 27, 24 e^- C. +2, A = 43, Z = 20, 18 e^-, 23 n D. $^{15}N^{3-}$, A = 15, 10 e^-

E. ^{20}Ne, 10 e^-, 10 n F. $^{127}I^-$, −1, 74 n G. $^{23}Na^+$, $Z = 11$, 10 e^-
H. 4+, $A = 192$, $Z = 76$, 72 e^-, 116 n
I. impossible, since $A < A - Z$ J. $^{52}Cr^{2+}$, 22 e^-, 28 n
K. $^{60}Co^{2+}$, +2, 33 n L. $^{17}F^-$, $Z = 9$, 10 e^-
M. −2, $A = 80$, $Z = 34$, 36 e^-, 46 n N. $^{14}C^{4-}$, $Z = 6$, 10 e^-
O. $^{118}Sn^{4+}$, 46 e^-, 68 n

9. A. 246.5314 B. 1.11435 C. 5.06364 D. 3.62797 E. 162.219 F.254.293 G. 249.140 H. 159.868 I. 10.6212 J. 41.6073 K. 54.2572 L. 0.383873

10. A. Li, 6.941, 92.58% B. 10.81, 20.0%, 80.0% C.13.0, 1.11% D. Ne, 20.18, 9.6% E. Cu, 69.5%, 30.5% F. 35.453, 36.947, 24.47% G. K, 41, 6.9% H. Ga, 69.7, 39.6% I. 79.904, 50.7%, 49.3% J. 85.47, 86.91, 27.85%

11. A. 1.01, 0.416 g, 2.48×10^{23} atoms B. S, 2.72 mol, 1.64×10^{24} atoms C. 16.0, 25.4 g, 9.58×10^{23} atoms D. 12.0, 6.02 g, 0.501 mol E. Cl, 1.52 g, 2.57×10^{22} atoms F. N, 0.153 mol, 9.21×10^{22} atoms G. 24.3, 40.8 mol, 2.46×10^{24} atoms H. P, 155 g, 5.00 mol I. 79.9, 4.89×10^{-3} g, 3.69×10^{19} atoms J. 39.1, 5.64×10^{-3}, 1.44×10^{-4} mol K. 9.01, 4.52×10^{11} g, 3.02×10^{34} atoms L. F, 1.59×10^{-5} mol, 9.57×10^{18} atoms

Quizzes

A: 1. (a) 2. (c) 3. (a) 4. (b) 5. (c) 6. (c) 7. (c) 8. (c)
B: 1. (c) 2. (b) 3. (c) 4. (a) 5. (e) 6. (b) 7. (a) 8. (a)
C: 1. (a) 2. (e) 3. (b) 4. (d) 5. (d) 6. (d) 7. (b) 8. (d)
D. 1. (e) 2. (a) 3. (e) 4. (b) 5. (b) 6. (d) 7. (e) 8. (e)

Sample Test

1. $\frac{16.00 \text{ g O}}{\text{mol O}} \times \frac{\text{mol O}}{\text{mol X}} \times \frac{46.7 \text{ g X}}{53.5 \text{ g O}} = 14.0 \frac{\text{g X}}{\text{mol}}$

2. a. isotope b. radioactivity c. atomic number d. Dalton e. cathode tube emission f. Millikan g. Rutherford h. fundamental particles i. Thomson

3. $[10.013 + (11.009 \times 4)]/5 = 10.810$ amu (boron)

CHAPTER 3

Drill Problems

2. (1) A. 472.2 g/mol B. 58.5 g/mol C. 260.7 g/mol D. 106.3 g/mol E. 84.0 g/mol F. 213.0 g/mol G. 208.3 g/mol H. 34.0 g/mol I. 27.0 g/mol J. 301.2 g/mol K. 152.0 g/mol L. 27.6 g/mol
(2) M. 152.0 g/mol, 1.14 mol N. 93.9 g/mol, 232 g O. 58.5 g/mol, 3.98 mol P. 159.4 g/mol, 11.1 g Q. 162.3 g/mol, 0.346 mol R. 34.0 g/mol, 66.7 mol S. 40.0 g/mol, 58.8 g T. 76.3 g/mol, 0.638 g U. 324.6 g/mol, 0.0127 mol V. 99.0 g/mol, 789 g W. 85.0 g/mol, 7.64 mol X. 144.9 g/mol,316 g

3. & 4. A. 1.05×10^{23} C atoms B. 4.01×10^{-3} g NH_3 C. 19.8 kg $CsIO_3$ D. 1.57×10^{24} O atoms E. 2.70×10^{24} HCl molecules F. 5.83×10^{24} atoms G. 1.21 Mg H_2SO_4 H. 6.88 kg $(NH_4)_2Cr_2O_7$ I. 2.82×10^{24} atoms J. 0.0266 g $BaCl_2$ K. 2.06×10^{25} O atoms L. 7.36 mol C M. 0.500 mol H N. 111 mol C O. 2.70 mol Al

5. & 6. A. H_2SO_4, 82.06 g/mol, 58,54% O B. 40.0 g/mol, 57.5% Na, 40.0% O, 2.5% H C. $MgSO_3$, 84.3 g/mol, 28.9% Mg D. 213.0 g/mol, 12.7% Al, 19.7% N, 67.6% O E. $LiClO_4$, 106.4 g/mol, 33.3% Cl F. 153.4 g/mol, 42.6% Zn, 15.6% C, 41.7% O G. K_2MnO_4, 197.1 g/mol, 32.5% O H. 76.3 g/mol, 31.8% Mg, 31.4% C, 36.7% N I. $CuCO_3$, 123.5 g/mol, 9.7% C J. 98.0 g/mol, 3.1% H, 31.6% P, 65.3% O K. $Hg(BrO_3)_2$, 456.4 g/mol, 44.0% Hg

7. A. 0.406 g C, 0.102 g H, 0.812 g O; $C_2H_6O_3$ B. 61.72 g C, 12.87 g H, 0 g O; C_2H_5 C. 3.05 g C, 1.03 g H, 4.06 g O; CH_4O D. 0.140 g C, 0.00118 g H, 0 g O, CH E. 0.658 g C, 0.110 g H, 0.175 g O; $C_5H_{10}O$ F. 6.11 g C, 1.53 g H, 4.06 g O; C_2H_6O G. 7.77 g C, 1.30 g H, 10.33 g O; CH_2O H. 3.23 g C, 0.6501 g H, 3.44 g O; $C_5H_{12}O_4$ I. 0.00665 g C, 0.00112 g H, 0.0177 g O; CH_2O_2 J. 0.106 g C, 0.0178 g H, 0.432 g O; CH_2O_3 K. 60.6 g C, 5.04 g H, 80.4 g O; CHO L. 86.62 g C, 5.82 g H, 0 g O; C_5H_4

8. A. 67.49% Ba B. 10.92% Mg C. 73.20% Pb D. 0.668% Ca E. 29.42% Al F. 24.7% Cr G. 4.27% Li H. 52.20% F I. 15.44% S J. 95.45% C

9. A. 58.71 g/mol B. 55.87 g/mol C. 15.6 g/mol D. 23.00 g/mol E. 140.1 g/mol F. 26.97 g/mol G. 39.11 g/mol H. 144.2 g/mol

10. (1) A. Hg = +1, Cl = −1 B. Na = +1, Cl = −1 C. Sn = +4, Cl = −1 D. Ca = +2, H = +1, S = −2 E. Na = +1, H = +1, O = −2, C = +4 F. Al = +3, O = −2, N = + 5 G. Ba = +2, Cl = −1 H. H = +1, O = −1 I. H = +1, N = −3, C = +2 J. Cl = −1, S = −2, C = +5 K. H = +1, N = −3, O = −2, Cr = +6 L. H = +1, B = −3
(2) M. H = +1, N = −3, O = −2, Cr = +6 N. Li = +1, O = −2, S = + 4 O. Na = +1, Cl = −1, P. Ag = +1, O = −2, Cl = +1 Q. Cl = −1, Fe = +3 R. H = +1, O = −1 S. Na = +1, H = +1, O = −2 T. Mg = +2, N = −3, C = +2 U. Hg = +2, O = −2, N = +5 V. Cl = −1, Cu = +1 W. Na = +1, O = −2 , N = +5 X. H = +1, O = −2, Br = + 7

11. A. sodium chloride B. aluminum sulfate C. strontium chloride D. chromium(II) hydrogen phosphate E. sodium thiosulfate F. copper(I) chloride G. iron(III) phosphate H. lithium sulfate I. ammonium chromate J. potassium hydrogen sulfate K. calcium acetate L. copper(II) hydrogen sulfide M. magnesium carbonate N. chromium(III) cyanide O. barium sulfate P. $KMnO_4$ Q. $NaC_2H_3O_2$ R. $FeCl_3$ S. $Sr(NO_2)_2$ T. $AgClO_3$ U. Cu_2O V. $Hg_2(NO_3)_2$ W. $Fe(NO_2)_3$ X. $Sr(HCO_3)_2$ Y. $Zn_3(PO_4)_2$ Z. K_2CrO_4 Γ. $Mg(CN)_2$ Δ. Al_2O_3 Θ. NaOH Λ. $Ca(OCl)_2$

12. A. water B. hydrogen sulfide C. carbon dioxide D. dinitrogen trioxide E. diphosphorus pentoxide F. iodine pentachloride G. nitrogen trichloride H. sulfur tetrachloride I. chlorine dioxide J. sulfur trioxide K. NO L. P_2O_5 M. NH_3 N. HF O. SiF_4 P. XeF_6 Q. PBr_3 R. SiS_2 S. CH_4 T. BCl_3 U. N_2O_3 V. Sb_2S_5

13. A. sodium perchlorate: Na +1, Cl +7, O −2 B. H_2SeO_4: H +1, Se +6, O −1 C. ammonium sulfide: H +1, N −3, S −2 D. copper(I) iodate: Cu +1, I +5, O −2 E. carbonic acid: H +1, C +4, O −2 F. H_3PO_4: H +1, P +5, O −2 G. Na_3BO_3: Na +1, B +3, O −2 H. sodium nitrate: Na +1, N +5, O −2 I. lithium hypochlorite: Li +1, Cl +1, O −2 J. H_2SO_2: H +1, S +2, O −2 K. aluminum phosphate: Al +3, P +5, O −2 L. H_3BO_3: H +1, B +3, O −2 M. perchloric acid: H +1, Cl +7, O −2 N. HNO_2: H +1, N +3, O −2 O. $Mg(ClO_3)_2$: Mg +2, Cl +5, O −2 P. scandium aluminate: Sc +3, Al +3, O −2 Q. magnesium germinate: Mg +2, Ge +4, O −2 R. ammonium nitrate: N −3, H +1, N +5, O −2 S. hypofluorous acid: H +1, F +1, O −2 T. HNO: H +1, N +1, O −2

14. A. 9.38 B. 0.113 C. 0.116 D. 0.122 E. 0.832 F. 360. G. 0.884 H. 100. I. 1.58 J. 331 K. 0.498 L. 0.173

Quizzes

A: 1. (d) 2. (d) 3. (b) 4. (c) 5. (e) 6. (b) 7. (c) 8. (d) 9. (d) 10. (b) 11. (c)
B: 1. (a) 2. (a) 3. (c) 4. (e) 5. (d) 6. (b) 7. (b) 8. (b) 9. (b) 10. N_2O_3 11. (b)
C: 1. (d) 2. (d) 3. (d) 4. (b) 5. (c) 6. (c) 7. (d) 8. (c) 9. (a) 10. dichlorine heptoxide 11. (d)
D: 1. (a) 2. (a) 3. (b) 4. (e) 5. (d) 6. (e) 7. (a) 8. (a) 9. (d) 10. P_2O_5 11. (b)

Sample Test

1. $16.88\text{ g }CO_2 \times \frac{12.01\text{ g C}}{44.01\text{ g }CO_2} = 4.606\text{ g C} \times \frac{\text{mol C}}{12.01\text{ g}}$
$= 0.3835\text{ mol C}$
$9.21\text{ g }H_2O \times \frac{2.016\text{ g H}}{18.01\text{ g }H_2O} = 1.03\text{ g H} \times \frac{\text{mol H}}{1.008\text{ g H}}$
$= 1.023\text{ mol H}$
mass O = 9.72 g − 4.606 g C − 1.03 g H
$= 4.08\text{ g O} \times \frac{\text{mol O}}{16.0\text{ g O}} = 0.255\text{ mol O}$
0.384 mol C/0.255 = 1.51 mol C
1.023 mol H/0.0255 = 4.01 mol H
0.255 mol O/0.255 = 1.00 mol O $C_3H_8O_2$

2. $$\frac{2.9\text{ g Fe} \times \frac{\text{mol Fe}}{55.8\text{ g}} \times \frac{6.02 \times 10^{23}\text{ atoms}}{\text{mole Fe}}}{2.6 \times 10^{13}\text{ red blood cells}}$$
$= 1.2 \times 10^{9}\ \frac{\text{Fe atoms}}{\text{red blood cell}}$

3. a. $CuSO_4$: Cu +2, S +6, O −2 b. sodium thiosulfate: Na +1, S +2, O −2 c. $Mg(ClO_2)_2$: Mg +2, Cl +3, O −2 d. sodium perborate: Na +1, B +5, O −2 e. $(NH_4)_2O$: N −3, H +1, O −2 f. nitrogen triiodide: N +3, I −1 g. calcium xenate: Ca +2, Xe +6, O −2 h. Ba +2, O −1 i. Sb_2S_3 Sb +3, S −2

CHAPTER 4

Drill Problems

1. A. $2\ Al + 3\ MgO \longrightarrow 3\ Mg + Al_2O_3$
B. $AlCl_3 + 3\ NaOH \longrightarrow Al(OH)_3 + 3\ NaCl$
C. $3\ AgNO_3 + Na_3PO_4 \longrightarrow Ag_3PO_4 + 3\ NaNO_3$
D. $Cl_2 + 2\ KI \longrightarrow I_2 + 2\ KCl$
E. $Fe_2(SO_4)_3 + 3\ Ca(OH)_2 \longrightarrow 2\ Fe(OH)_3 + 3\ CaSO_4$
F. $Ba(NO_3)_2 + (NH_4)_2CO_3 \longrightarrow BaCO_3 + 2\ NH_4NO_3$
G. $2\ KOH + H_2SO_4 \longrightarrow K_2SO_4 + 2\ H_2O$
H. $Fe(OH)_2 + 2\ HCl \longrightarrow FeCl_2 + 2\ H_2O$
I. $CuSO_4 + H_2S \longrightarrow CuS + H_2SO_4$
J. $2\ AgNO_3 + K_2CrO_4 \longrightarrow Ag_2CrO_4 + 2\ KNO_3$

1a. (1) A. potassium dichromate B. silver bromate C. magnesium bisulfite D. potassium hypoiodite E. sodium perchlorate F. potassium thiocyanate G. lead(IV) carbonate H. copper(II) oxalate I. potassium oxide J. thallium(I) sulfide K. tin(IV) manganate *or* tin(II) permanganate L. manganese(II) iodate M. nickel(II) cyanide N. manganese(II) cyanate O. cadmium sulfate P. manganese(II) sulfite Q. ammonium dichromate R. cobalt(II) manganate S. silver thiocyanate T. lithium peroxide U. mercury(I) acetate V. thallium(III) sulfide W. silver dichromate X. gold(III) sulfate Y. magnesium iodate Z. cadmium bisulfite

(2) A. SrO_2 B. $MnSO_3$ C. PbC_2O_4 D. $RbBrO_3$ E. $Mg(CNO)_2$ F. NaHS G. $Ca(IO)_2$ H. FeO_2 I. Ag_2S J. $Ba(IO_3)_2$ K. $CdSO_3$ L. K_2MnO_4 M. $AsCl_3$ N. $Cr(BrO_2)_2$ O. $Sn(CO_3)_2$ P. As_2S_3 Q. $Mn(HCO_3)_2$ R. $Pb(CO_3)_2$ S. $Au_2Cr_2O_7$ T. $K_2C_2O_4$ U. MnS_2O_3 V. H_2O_2 W. $Au(IO_3)_3$ X. NaCN Y. $Cu(CNO)_2$ Z. SnF_2

2. A. $P_4 + 5\ O_2 \longrightarrow 2\ P_2O_5$ B. $2\ Na + O_2 \longrightarrow Na_2O_2$
C. $2\ Al + 6\ HCl \longrightarrow 2\ AlCl_3 + 3\ H_2$
D. $Ca + 2\ H_2O \longrightarrow Ca(OH)_2 + H_2$
E. $2\ FeCl_3 + 3\ Ca(OH)_2 \longrightarrow 2\ Fe(OH)_3 + 3\ CaCl_2$
F. $2\ Al + N_2 \longrightarrow 2\ AlN$
G. $6\ HCl + Fe_2O_3 \longrightarrow 2\ FeCl_3 + 3\ H_2O$
H. $3\ Cl_2 + 3\ H_2O \longrightarrow 5\ HCl + HClO_3$
I. $Al(OH)_3 + 3\ HCl \longrightarrow AlCl_3 + 3\ H_2O$
J. $CaSO_3 + H_2SO_4 \longrightarrow CaSO_4 + H_2O + SO_2$
K. $2\ NaCl + H_2SO_4 \longrightarrow Na_2SO_4 + 2\ HCl$
L. $2\ Pb(NO_3)_2 \longrightarrow 2\ PbO + 4\ NO + 3\ O_2$
M. $3\ HNO_2 \longrightarrow HNO_3 + 2\ NO + H_2O$
N. $3\ Ca(OH)_2 + 2\ H_3PO_4 \longrightarrow Ca_3(PO_4)_2 + 6\ H_2O$
O. $SiF_4 + 2\ H_2O \longrightarrow 4\ HF + SiO_2$

2a. A. $Cu + 2\ Fe^{3+} \longrightarrow Cu^{2+} + 2\ Fe^{2+}$
B. $Zn + 2\ H^{+} \longrightarrow Zn^{2+} + H_2$ C. $2\ H^{+} + S^{2-} \longrightarrow H_2S(g)$
D. $Pb^{2+} + 2\ Hg^{2+} \longrightarrow Pb^{4+} + Hg_2^{2+}$
E. $2\ Fe^{3+} + 3\ Sn^{2+} \longrightarrow 2\ Fe + 3\ Sn^{4+}$
F. $3\ H^{+} + N^{3-} \longrightarrow NH_3(g)$ G. $6\ CN^{-} + Fe^{2+} \longrightarrow Fe(CN)_6^{4-}$
H. $2\ Bi^{3+} + 3\ S^{2-} \longrightarrow Bi_2S_3(s)$

3. A. $2\ CH_3OH + 3\ O_2 \longrightarrow 2\ CO_2 + 4\ H_2O$
B. $2\ C_6H_6 + 15\ O_2 \longrightarrow 12\ CO_2 + 6\ H_2O$
C. $C_5H_{12} + 8\ O_2 \longrightarrow 5\ CO_2 + 6\ H_2O$
D. $C_{12}H_{22}O_{11} + 12\ O_2 \longrightarrow 12\ CO_2 + 11\ H_2O$
E. $2\ CO + O_2 \longrightarrow 2\ CO_2$
F. $HC_2H_3O_2 + 2\ O_2 \longrightarrow 2\ CO_2 + 2\ H_2O$
G. $2\ C_7H_6O_2 + 15\ O_2 \longrightarrow 14\ CO_2 + 6\ H_2O$
H. $2\ C_9H_6O_4 + 17\ O_2 \longrightarrow 18\ CO_2 + 6\ H_2O$
I. $C_{10}H_8 + 12\ O_2 \longrightarrow 10\ CO_2 + 4\ H_2O$
J. $C_{19}H_{28}O_2 + 25\ O_2 \longrightarrow 19\ CO_2 + 14\ H_2O$

4. C = combination, De = decomposition, Di = displacement, M = metathesis
A. C: $2\ Al + 3\ Br_2 \longrightarrow 2\ AlBr_3$
B. Di: $2\ Al + Fe_2O_3 \longrightarrow 2\ Fe + Al_2O_3$
C. M: $AlCl_3 + 3\ KOH \longrightarrow Al(OH)_3 + 3\ KCl$
D. M: $BaCl_2 + Na_2CO_3 \longrightarrow BaCO_3 + 2\ NaCl$
E. M: $BaCl_2 + 2\ AgNO_3 \longrightarrow 2\ AgCl + Ba(NO_3)_2$
F. C: $BaO + SO_3 \longrightarrow BaSO_4$
G. M: $Ba(OH)_2 + H_2CO_3 \longrightarrow BaCO_3 + 2\ H_2O$
H. C: $Br_2 + H_2 \longrightarrow 2\ HBr$
I. M: $Cd(NO_3)_2 + (NH_4)_2S \longrightarrow CdS + 2\ NH_4NO_3$
J. Di: $Ca + 2\ H_2O \longrightarrow Ca(OH)_2 + H_2$
K. Di: $6\ HCl + 2\ Al \longrightarrow 2\ AlCl_3 + 3\ H_2$
L. C: $H_2O + Li_2O \longrightarrow 2\ LiOH$
M. M: $3\ Ba(OH)_2 + 2\ H_3PO_4 \longrightarrow Ba_3(PO_4)_2 + 6\ H_2O$
N. C: $BaO + H_2O \longrightarrow Ba(OH)_2$
O. M: $CaO + 2\ HCl \longrightarrow CaCl_2 + H_2O$
P. C: $CaO + CO_2 \longrightarrow CaCO_3$
Q. M: $Ca(OH)_2 + 2\ HNO_3 \longrightarrow Ca(NO_3)_2 + 2\ H_2O$
R. M: $CaO + 2\ HNO_3 \longrightarrow Ca(NO_3)_2 + H_2O$
S. C: $NH_3(g) + HCl(g) \longrightarrow NH_4Cl(s)$
T. C: $2\ SO_2 + O_2 \longrightarrow 2\ SO_3$
U. Di: $2\ AgNO_3 + Cu \longrightarrow 2\ Ag + Cu(NO_3)_2$
V. C: $Cl_2 + 2\ FeCl_2 \longrightarrow 2\ FeCl_3$
W. Di: $Cl_2 + 2\ KI \longrightarrow 2\ KCl + I_2$
X. C: $N_2 + 3\ H_2 \longrightarrow 2\ NH_3$
Y. M: $2\ Al(OH)_3 + 3\ H_2SO_4 \longrightarrow Al_2(SO_4)_3 + 6\ H_2O$

5. & 5a. A. 35.4 g HCl, 52.4 g $FeCl_3$, 17.4 cm^3 H_2O B. 40.5 g $Fe(OH)_3$, 20.5 g H_2O C. 17.8 g $Fe(OH)_3$, 27.0 g $FeCl_3$ D. 50.8 g O_2, 43.7 g CO_2, 21.5 cm^3 H_2O E. 10.4 g C_5H_{12}, 15.6 g H_2O F. 0.410 g C_5H_{12}, 0.640 L CO_2, 0.616 cm^3 H_2O G. 0.314 g H_2, 1.77 g NH_3 H. 11.0 g N_2, 2.37 g H_2 I. 90.1 g N_2, 175 L NH_3 J. 10.6 g HNO_3, 6.05 g H_2O, 3.76 L NO K. 13.3 g HNO_3, 6.34 g NO, 22.5 g HCl L. 12.2 L HCl, 6.56 g H_2O, 4.08 L NO M. 8.30 g HNO_3, 0.988 g NO N. 9.20 g Cu,27.1 g $Cu(NO_3)_2$ O. 460 g HNO_3, 513 g $Cu(NO_3)_2$, 40.8 L NO P. 0.802 cm^3 Cu, 1.69 L NO, 2.71 g H_2O

6. A. 51.26 g. 0.8535 mol B. 46.30 g, 3.175 M C. 73.2 g, 200 mL (0.200 L) D. 50.9 g, 0.244 mol E. 0.244 mol, 167 mL F. 0.492 mol, 2.83 M G. 21.30 g, 0.2128 mol H. 0.1155 mol, 0.3301 M I. 781 g, 2.06 L (2.06 × 10^3 mL) J. 33.7 g, 166 mL K. 511.0 g, 5.210 mol L. 0.223 mol, 0.745 M

7. A. $[KNO_3]$ = 0.0526 M, 570 mL B. 240 mL HCl soln, 360 mL H_2O C. 64.8 mL water, 152 mL total D. [NaCl] = 0.271 M, 620 mL total E. 31.3 mL $MgSO_4$ soln, 269 mL water F. 840 mL total, 700 mL water G. $[K_2SO_4]$ = 0.117 M, 720 mL total H. 44.4 mL AlI_3 soln, 356 mL water I. [NaCl] = 0.228 M, 300 mL total J. 135 mL total, 85.0 mL KCl soln K. $[NaNO_3]$ = 0.456 M, 300 mL total L. 305 mL soln A, 195 mL soln B M. 300 mL HCl soln, [HCl] = 1.90 M N. $[NaNO_3]$ = 3.00 M, 500 mL total

8. A. 4.19 g Ag_3PO_4, $[NaNO_3]$ = 0.300 M B. 31.5 mL Na_3PO_4 soln, $[NaNO_3]$ = 0.473 M C. 131 mL $AgNO_3$ soln, 36.6 g Ag_3PO_4 D. 0.178 M Na_3PO_4 soln, 32.0 mL $NaNO_3$ soln, E. 9.48 g HCN, 176 mL $MgSO_4$ soln F. 13.4 g $Mg(CN)_2$, 29.3 mL H_2SO_4 soln G. 45.9 g $Mg(CN)_2$, 29.2 mL H_2SO_4 soln H. 26.7 g $Mg(CN)_2$, 32.5 g HCN I. 11.3 mL H_2SO_4 soln, 1.22 g H_2O J. 55.4 g Al_2O_3, 29.3 g H_2O K. $[Al_2(SO_4)_2]$ = 0.333 M, 3.53 g H_2O L. 14.0 g Al_2O_3, 412 mL H_2SO_4 soln M. [KOH] = 7.06 M, 42.5 g CO_2, $[K_2CO_3]$ = 3.53 M N. 4.27 g CO_2, 194 mL K_2CO_3 soln, 1.75 g H_2O O. [KOH] = 0.429 M, 0.548 g H_2O P. 3.81 g CO_2, 1.56 g H_2O Q. $[H_2SO_4]$ = 1.82 M, 121 mL K_2SO_4 soln, 3.28 g H_2O R. 217 mL KOH soln, 5.15 g H_2O S. 88.3 mL H_2SO_3 soln, 124 mL K_2SO_3 soln

9. A. 64.8 g Ag_3PO_4 produced, 23.1 g $AgNO_3$ left B. 22.1 g H_2O produced, 23.0 g H_2SO_4 left C. 6.79 g CO produced, 9.7 g C left D. 0.220 g H_2 produced, 6.2 g HCl left E. 34.1 g MgS produced, 6.6 g Mg left F. 68.8 g CuCl produced, 6.7 g HCl left G. 32.8 g H_2O produced, 29.5 g SiO_2 left H. 46.6 g H_2O produced, 175 g $Al(OH)_3$ left I. 57.8 g H_2SO_4 produced, 121 g HNO_3 left J. 212 g $FeCl_3$ produced, 100 g Fe_2O_3 left K. 70.3 g H_2O produced, 64.1 g H_2S left L. 92.0 g NaCl produced, 5.8 g HCl left M. 219 g $ZnCl_2$ produced, 137 g H_3AsO_4 left, 11 g HCl left N. 42.0 g K_2SO_4 produced, 7 g H_2SO_4 left, 48 g Hg left O. 19.9 g Cl_2 produced, 48.2 g C_2H_3OCl left, 45 g H_2SO_4 left

10. A. 91.6% B. 63.6 g PCl_3 C. 29.0 g P D. 79.7% E. 59.4 g CaO F. 98.1 g $CaCO_3$ G. 94.4% H. 249 g O_2 I. 20.8 g H_2O J. 64.0% K. 330 g Fe L. 2.01 g H_2 M. 89.8% N. 5.93 g H_2O O. 3.29 g H_2S

11. A. 5.03 g NaCl, 172 mL NaCl soln B. 8.45 g CO_2, 2.70 × 10^3 mL HCl soln C. 61.6 g CO_2, 110.6 g solid D. 1.23 L KOH soln, 13.8 g H_2O E. 0.563 g H_2, 2.79 L HCl soln F. 33.4 g H_2O, 535 mL H_2SO_4 soln G. 20.4 g $NaNO_3$, 41.9 g solid

12. A. 19.9 g H_2SO_4 B. 17.9 g O_2 C. 153 g NaCl D. 152 mL H_2SO_4 mL H_2SO_4 soln E. 15.3 g Cu F. 255 g HNO_3 G. 1.03 g H_2 H. 21.6 g $CaCl_2$ I. 25.1 g NH_3 J. 120 g $CaCO_3$ K. 301 g AgCl L. 283 mL HNO_3 soln M. 1.97 g P, 6.78 g Cl_2

13. $CuS + 2\,O_2 + H_2O \longrightarrow CuO + H_2SO_4$

$6\,NaCl + 3\,H_2SO_4 + 2\,Fe + 3\,CuO \longrightarrow 3\,Na_2SO_4 + 2\,FeCl_3 + 3\,Cu + 3\,H_2O$

$6\,N_2 + 18\,H_2 + 21\,O_2 \longrightarrow 14\,H_2O + 8\,HNO_3 + 4\,NO$

$CaCO_3 + 2\,NH_4Cl \longrightarrow CO_2 + CaCl_2 + 2\,NH_3 + H_2O$

$2\,P + 3\,Cl_2 + 6\,H_2O + 6\,AgNO_3 \longrightarrow 2\,H_3PO_3 + 6\,AgCl + 6\,HNO_3$

Quizzes

A: 1. sodium nitrite 2. H_2SO_4 3. (d) 4. (e) (sum = 9) 5. (a) 6. (b) 7. $Na_2S + 2\,HCl \longrightarrow 2\,NaCl + H_2S(g)$ 8. (b) 9. (d) 10. (a)

B: 1. magnesium chlorate 2. NaFO 3. (b) 4. (b) 5. (b) 6. (d) 7. $Al_2O_3 + 3\,H_2SO_4 \longrightarrow Al_2(SO_4)_3 + 3\,H_2O$ 8. (c) 9. (e) 10. (c)

C: 1. rubidium perbromate 2. HNO_3 3. (c) 4. (c) 5. (b) 6. (b) 7. $2\,NaOH + H_2SO_4 \longrightarrow Na_2SO_4 + 2\,H_2O$ 8. (d) 9. (a) 10. (b)

D: 1. sulfuric acid 2. $NaHCO_3$ 3. (c) 4. (e) (sum = 9) 5. (c) 6. (b) 7. $CaCO_3 + 2\,HCl \longrightarrow CaCl_2 + CO_2 + H_2O$ 8. (d) 9. (b) 10. (a)

Sample Test

1. a. lead(II) thiocyanate b. $Pb(S_2O_3)_2$ c. As_2S_3 d. manganese(II) hypobromite e. nitrous acid f. HIO_4 g. SnS h. NH_4HCO_3

2. 200.0 mL × 0.240 M × 2 mmol K^+/mmol K_2SO_4 = 96.0 mmol K^+

Let V be the volume of KNO_3 solution added, in mL.

$$0.400\ M = \frac{96.0\ mmol + (01.60\ M \times V)}{200.0 + V}$$

or V = 66.7 mL

3. a. $BaCl_2 + 2\,AgNO_3 \longrightarrow Ba(NO_3)_2 + 2\,AgCl$
b. $Ca(OH)_2 \xrightarrow{heat} CaO + H_2O$ c. $NH_3 + HCl \longrightarrow NH_4Cl$
d. $Zn + CuSO_4 \longrightarrow ZnSO_4 + Cu$
e. $2\,C_2H_6O_2 + 5\,O_2 \longrightarrow 4\,CO_2 + 6\,H_2O$

4. $$55.2\ g\ CuO \times \frac{mol\ CuO}{79.5\ g\ CuO} \times \frac{1 mol\ H_2O}{2\ mol\ HCl} = 0.213\ mol\ H_2O$$

0.213 mol H_2O × 18.0 g H_2O/mol = 3.84 g H_2O

5. $$26.4\ g \times \frac{0.400\ g\ CaO}{g\ mixture} \times \frac{mol\ CaO}{56.1\ g\ CaO} \times \frac{mol\ CaCl_2}{mol\ CaO} \times \frac{111.1\ g\ CaCl_2}{mol\ CaCl_2} = 20.9\ g\ CaCl_2$$

$$26.4\ g \times \frac{0.600\ g\ NaOH}{g\ mixture} \times \frac{mol\ NaOH}{40.0\ g\ NaOH} \times \frac{mol\ NaCl}{mol\ NaOH} \times \frac{58.5\ g\ NaCl}{mol\ NaCl} = 23.2\ g\ NaCl$$

Total mass of solid = 20.9 g $CaCl_2$ + 23.2 g NaCl = 44.1 g

CHAPTER 5

Drill Problems

1. Electrolytes are designated as follows: N = nonelectrolyte, W = weak electrolyte, and S = strong electrolyte. Also a = acid, b = base, and s = salt. Of course, for instance, a weak electrolyte that is a base is also a weak base.
A. Sa B. N C. Wb D. Ss E. Sa F. Wa G. Ss H. N I. Ss J. Ss K. Ss L. Sb M. Ss N. Wa O. N P. Ss Q. Ss R. Ss

S. Ss T. Wb U. N V. Ss W. Ss X. Wa Y. Ss Z. Ss Γ. N Δ. Ss Θ. Ss Λ. Wa Ξ. N Π. N Σ. Ss Υ. Ss Φ. Ss Ψ. N

1a. A. sulfuric acid B. HNO_3 C. H_2SO_4 D. acetic acid E. $HC_2H_3O_2$ F hydrochloric acid G. phosphoric acid H. perchloric acid I. iodic acid J. HF K. HNO_2 L. hypofluorous acid M. oxalic acid N. HIO_4 O. manganic acid P. HBrO Q. HCN R. hypoiodous acid S. H_2CO_3 T. $HMnO_4$ U. $HClO_4$ V. chlorous acid

2. (1) A. 6.48 M B. 1.22 M C. 0.466 M D. 0.630 M E. 4.80 M F. 2.23 M G. 0.269 M H. 0.214 M I. 0.169 M J. 0.203 M K. 0.324 M L. 0.665 M
(2) M. 31.3 mL $MgSO_4$ soln, 269 mL water N. 840 mL total, 700 mL water O. $[K^+] = 0.117$ M, 720 mL total P. 44.7 mL $Al(NO_3)_3$ soln, 356 mL water Q. $[Na^+] = 0.228$ M, 300 mL total R. $[Na^+] = 0.456$ M, 300 mL total S. 305 mL $MgSO_4$ soln, 195 mL Na_2SO_4 soln T. 300 mL KCl soln, $[KCl] = 1.90$ M U. $[Cl^-] = 3.00$ M, 500 mL total

3. i = insoluble, s = soluble, m = moderately soluble, p = sparingly soluble A. i B. i C. s D. i E. s F. s G. s H. m I. s J. s K. i L. s M. i N. i O. s P. s Q. s R. s S. s T. i U. s V. s W. s X. i Y. s Z. m Γ i Δ. s Θ. i Λ. s Ξ. s Π. i Σ. m Υ. i Φ. i Ψ. s

3. & 4. C. E. L. and P are the same: $OH^- + H^+ \longrightarrow H_2O$
A. $2\,Cl^- + Pb^{2+} \longrightarrow PbCl_2(s)$ B. $3\,Ca^{2+} + 2\,PO_4^{3-} \longrightarrow Ca_3(PO_4)_3(s)$ D. $Cl^- + H^+ \longrightarrow HCl(g)$ F. $S^{2-} + 2\,H^+ \longrightarrow H_2S(g)$ G. $Ba^{2+} + SO_4^{2-}$ H. $2\,Ag^+ + SO_4^{2-} \longrightarrow Ag_2SO_4(s)$ I. $CO_3^{2-} + 2\,H^+ \longrightarrow H_2O + CO_2(g)$ J. $NH_4^+ + OH^- \longrightarrow H_2O + NH_3(g)$ K. $Cd^{2+} + S^{2-} \longrightarrow CdS(s)$ M. $Fe^{3+} + 3\,OH^- \longrightarrow Fe(OH)_3(s)$ N. $3\,Ag^+ + PO_4^{3-} \longrightarrow Ag_3PO_4(s)$ O. $3\,Sr^{2+} + 2\,PO_4^{3-} \longrightarrow Sr_3(PO_4)_2(s)$ and $H^+ + OH^- \longrightarrow H_2O$

5. If the reaction goes to completion, the reason will be included in brackets: [p] = precipitate forms, [n] = nonelectrolyte forms, [g] = gas escapes from solution, [eq] reaction does not go to completion. A. [p] $3\,Hg_2(NO_3)_2(aq) + 2\,H_3PO_4(aq) \longrightarrow Hg_6(PO_4)_2(s) + 6\,HNO_3(aq)$
B. [p] $Pb(NO_3)_2(aq) + 2\,HCl(aq) \longrightarrow PbCl_2(s) + 2\,HNO_3(aq)$
C. [p] $Sr(NO_3)_2(aq) + H_2SO_4(aq) \longrightarrow SrSO_4(s) + 2\,HNO_3(aq)$
D. [n] $2\,RbOH(aq) + H_2SO_3(aq) \longrightarrow Rb_2SO_3(aq) + 2\,H_2O(l)$
E. [eq] $2\,AsCl_3(aq) + 3\,H_2S(aq) \longrightarrow As_2S_3(s) + 6\,HCl(aq)$
F. [eq] $MnCl_2(aq) + H_2S(aq) \longrightarrow MnS(s) + 2\,HCl(aq)$
G. [n,p] $Ba(OH)_2(aq) + H_2SO_4(aq) \longrightarrow BaSO_4(s) + 2\,H_2O(l)$
H. [n] $Zn(OH)_2(aq) + 2\,HCl(aq) \longrightarrow ZnCl_2(aq) + 2\,H_2O(l)$
I. [n] $Al_2O_3(aq) + 6\,HNO_3(aq) \longrightarrow 2\,Al(NO_3)_3(aq) + 3\,H_2O(l)$
J. [n] $Cu_2O(s) + 2\,HCl(aq) \longrightarrow 2\,CuCl(s) + H_2O(l)$
K. [n] $Cr_2O_3(s) + 3\,H_2SO_4(aq) \longrightarrow Cr_2(SO_4)_3(aq) + 3\,H_2O(l)$
L. [n] $2\,KOH(aq) + CO_2(g) \longrightarrow K_2CO_3(aq) + H_2O(l)$
M. [eq] $CuSO_4(aq) + H_2S(aq) \longrightarrow CuS(s) + H_2SO_4(aq)$
N. [eq] $ZnCl_2(aq) + H_2S(aq) \longrightarrow ZnS(s) + 2\,HCl(aq)$
O. [n] $3\,ZnO(s) + 2\,H_3PO_4(aq) \longrightarrow Zn_3(PO_4)_2(aq) + 3\,H_2O(l)$
P. [n] $Al_2O_3(s) + 3\,H_2SO_4(aq) \longrightarrow Al_2(SO_4)_3(aq) + 3\,H_2O(l)$
Q. [g] $2\,HCl(aq) + Na_2S(s) \longrightarrow 2\,NaCl(aq) + H_2S(g)$
R. [g] $K_2CO_3(aq) + H_2SO_4(aq) \rightarrow K_2SO_4(aq) + H_2O(l) + CO_2(g)$
S. [g] $NH_4Cl(aq) + NaOH(aq) \rightarrow NaCl(aq) + H_2O(l) + NH_3(g)$
T. [g] $CuSO_3(s) + 2\,HBr(aq) \longrightarrow CuBr_2(aq) + H_2O(l) + SO_2(g)$

6. Oxidizing agent given first, followed by reducing agent. A. HNO_3 or NO_3^-, Ag B. PbO_2, HCl or Cl^- C. HBr or H^+, Al D. $BaSO_4$ or SO_4^{2-}, C E. Br_2, SO_2 F. Br_2, Br_2 G. HNO_3 or NO_3^-, Cu H. $Ca_3(PO_3)_2$ or PO_3^{3-}, C I. H_2O_2, ClO_2 J. HNO_3 or NO_3^-, Cu K. $HClO_3$, $HClO_3$ L. I_2, H_3AsO_3 M. Na_2O_2 or O_2^{2-}, $Cr(OH)_3$ N. H_2SO_4 or SO_4^{2-}, FeI_2 O. HNO_3 or NO_3^-, FeS

7. A. $3\,Ag + 4\,HNO_3 \longrightarrow 3\,AgNO_3 + 2\,H_2O + NO$
B. $4\,HCl + PbO_2 \longrightarrow 2\,H_2O + Cl_2 + PbCl_2$
C. $2\,Al + 6\,HBr \longrightarrow 2\,AlBr_3 + 3\,H_2$
D. $BaSO_4 + 4\,C \longrightarrow BaS + 4\,CO$
E. $Br_2 + 2\,H_2O + SO_2 \longrightarrow 2\,HBr + H_2SO_4$
F. $3\,Br_2 + 6\,KOH \longrightarrow 5\,KBr + KBrO_3 + 3\,H_2O$
G. $C + 4\,HNO_3 \longrightarrow CO_2 + 2\,H_2O + 4\,NO_2$
H. $3\,Ca(PO_3)_2 + 10\,C \longrightarrow Ca_3(PO_4)_2 + 10\,Co + P_4$
I. $2\,ClO_2 + H_2O_2 + 2\,KOH \longrightarrow 2\,H_2O + 2\,KClO_2 + O_2$
J. $Cu + 4\,HNO_3 \longrightarrow Cu(NO_3)_2 + 2\,NO_2 + 2\,H_2O$
K. $3\,HClO_3 \longrightarrow HClO_4 + 2\,ClO_2 + H_2O$
L. $H_3AsO_3 + H_2O + I_2 \longrightarrow 2\,HI + H_3AsO_4$
M. $2\,Cr(OH)_3 + 3\,Na_2O_2 \longrightarrow 2\,Na_2CrO_4 + 2\,NaOH + 2\,H_2O$
N. $2\,FeI_2 + 6\,H_2SO_4 \longrightarrow Fe_2(SO_4)_3 + 2\,I_2 + 3\,SO_2\ 6\,H_2O$
O. $FeS + 6\,HNO_3 \longrightarrow Fe(NO_3)_3 + S + 3\,NO_2 + 3\,H_2O$

8. A. $Ag \rightarrow Ag^+ + e^-$; B. $NO_3^- + 4\,H^+ + 3\,e^- \rightarrow NO + 2\,H_2O$; C. $2\,Cl^- \rightarrow Cl_2\ 2\,e^-$; D. $PbO_2 + 4\,H^+ + 2\,e^- \rightarrow Pb^{2+} + 2\,H_2O$; E. $Br_2 + 6\,H_2O \rightarrow 2\,BrO_3^- + 12\,H^+ + 10\,e^-$; F. $Br_2 + 12\,OH^- \rightarrow 2\,BrO_3^- + 6\,H_2O + 10\,e^-$; G. $NO_3^- + 2\,H^+ + e^- \rightarrow NO_2 + H_2O$; H. $H_2O_2 + 2\,OH^- \rightarrow O_2 + 2\,H_2O + 2\,e^-$; I. $ClO_3^- + 2\,H^+ + e^- \rightarrow ClO_2 + H_2O$; J. $ClO_3^- + H_2O \rightarrow ClO_4^- + 2\,H^+ + 2\,e^-$; K. $Cr(OH)_3 + 5\,OH^- \rightarrow CrO_4^{2-} + 4\,H_2O + 3\,e^-$; L. $O_2^{2-} + 2\,H_2O + 2\,e^- \rightarrow 4\,OH^-$; M. $SO_4^{2-} + 4\,H^+ + 2\,e^- \rightarrow SO_2 + 2\,H_2$ N. $NO_2^- + 2\,OH^- \rightarrow NO_3^- + H_2O + 2\,e^-$; O. $MnO_4^{2-} + 2\,H_2O + 2\,e^- \rightarrow MnO_2 + 4\,OH^-$
(2) A. $H_2O_2 \rightarrow 2\,H^+ + O_2 + 2\,e^-$; $Cu^{2+} + 2\,e^- \rightarrow Cu$; $CuSO_4 + H_2O_2 \longrightarrow Cu + O_2 + H_2SO_4$ B. $Mg \rightarrow Mg^{2+} + 2\,e^-$; $Ag^+ + e^- \rightarrow Ag$; $Mg + 2\,AgNO_3 \longrightarrow 2\,Ag + Mg(NO_3)_2$ C. $Al \rightarrow Al^{3+} + 3\,e^-$; $2\,H^+ + 2\,e^- \rightarrow H_2$; $2\,Al + 3\,H_2SO_4(aq) \longrightarrow Al_2(SO_4)_3(aq) + 3\,H_2(g)$ D. $Pb \rightarrow Pb^{2+} + 2\,e^-$; $Fe^{3+} + 3\,e^- \rightarrow Fe$; $Fe_2(SO_4)_3 + 3\,Pb \longrightarrow 3\,PbSO_4 + 2\,Fe$ E. $2\,I^- \rightarrow I_2 + 2\,e^-$; $F_2 + 2\,e^- \rightarrow 2\,F^-$; $2\,NaI + F_2 \longrightarrow I_2 + 2\,NaF$ F. $H_2 + 2\,OH^- \rightarrow 2\,H_2O + 2\,e^-$; $ClO^- + H_2O + 2\,e^- \rightarrow Cl^- + 2\,OH^-$; $H_2 + KClO \longrightarrow KCl + H_2O$ G. $2\,Cl^- \rightarrow Cl_2 + 2\,e^-$; $MnO_2 + 4\,H^+ + 2\,e^- \rightarrow Mn^{2+} + 2\,H_2O$; $MnO_2 + 4\,HCl \longrightarrow MnCl_2 + 2\,H_2O + Cl_2$ H. $I_2 + 6\,H_2O \rightarrow 2\,IO_3^- + 12\,H^+ + 10\,e^-$; $Br_2 + 2\,e^- \rightarrow 2\,Br^-$; $I_2 + 6\,H_2O + 5\,Br_2 \longrightarrow 10\,HBr + 2\,HIO_3$ I. $Ag \rightarrow Ag^+ + e^-$; $NO_3^- + 4\,H^+ + 3\,e^- \rightarrow NO + 2\,H_2O$; $3\,Ag + 4\,HNO_3 \longrightarrow 3\,AgNO_3 + NO + 2\,H_2O$ J. $S + 3\,H_2O \rightarrow SO_3^{2-} + 6\,H^+ + 4\,e^-$; $Pb^{2+} + 2\,e^- \rightarrow Pb$; $S + 3\,H_2O + 2\,Pb(NO_3)_2 \longrightarrow 2\,Pb + H_2SO_3 + 4\,HNO_3$; K. $O_2 + H_2O \rightarrow O_3 + 2\,H^+ + 2\,e^-$; $O_2 + 2\,H^+ + 2\,e^- \rightarrow H_2O_2$; $2\,O_2 + H_2O \longrightarrow O_3 + H_2O_2$; L. $H_2S \rightarrow S + 2\,H^+ + 2\,e^-$; $Br_2 + 2\,e^- \rightarrow 2\,Br^-$; $H_2S + Br_2 \longrightarrow S + 2\,HBr$; M. $Fe^{2+} \rightarrow Fe^{3+} + e^-$; $Sn^{2+} + 2\,e^- \rightarrow Sn$; $SnSO_4 + 2\,FeSO_4 \longrightarrow Sn + Fe_2(SO_4)_3$; N. $S + 3\,H_2O \rightarrow H_2SO_3 + 4\,H^+ + 4\,e^-$; $SO_4^{2-} + 4\,H^+ + 2\,e^- \rightarrow H_2SO_3 + H_2O$; $2\,H_2SO_4 + S + H_2O \longrightarrow 3\,H_2SO_3$; O. $I_2 + 6\,H_2O \rightarrow 2\,IO_3^- + 12\,H^+ + 10\,e^-$; $I_2 + 2\,e^- \rightarrow 2\,I^-$; $3\,I_2 + 3\,H_2O \longrightarrow 5\,HI + HIO_3$

9. (4-2) A. c, o B. c, o C. di, o D. di, o E. m, p F. c, o G. m, a H. ?, a I. m, a J. m, p K. m, g L. de, o M. de, o N. m, (p, a) O. m, (p [$SiO_2(s)$], g)
(4-4) A. c, o B. di, o C. m, p D. m, p E. m, p F. c, a G. m, (a,p) H. c, o I. m, (p,g) J. di, o K. di, o L. c, ? M. m, (p, a) N. c, a O. m, a P. c, a Q. m, a R. m, a S. c, a T. c, o U. di, o V. c, o W. di, o X. c, o Y. m, a Z. c, o

10. A. 0.0750 M B. 0.165 M C. 1.25 M D. 1.18 M E. 0.0510 M F. 3.10 M G. 0.0388 M H. 19.4% I. 0.0697 M J. 0.384 M K. 0.0919 M L. 41.3 % M. 4.44% N. 62.5% O. 0.142 M

12. A. NH_4NO_3 B. $FeSO_4$ C. KCl D. $AgClO_3$ E. $MgSO_4$ F. FeS G. $FeSO_4$ H. $CoCl_2$ I. MgS J. $NaC_2H_3O_2$ K. Na_2CO_3 L. $MgCO_3$

Quizzes

A: 1. (d) 2. (e) 3. (a) 4. (e) 5. (b) 6. (a) 7. (e) (5) 8. (c)
9. $2\ AgClO_3(aq) + H_2SO_4(aq) \longrightarrow Ag_2SO_4(s) + 2\ HClO_3(aq)$
10. $2\ LiOH(aq) + H_2C_2O_4(aq) \longrightarrow Li_2C_2O_4(aq) + 2\ H_2O(l)$
B: 1. (c) 2. (a) 3. (d) 4. (b) 5. (b) 6. (b) 7. (c) 8. (c)
9. $MnSO_3(aq) + 2\ HCl(aq) \longrightarrow MnCl_2(aq) + H_2O + SO_2(g)$
10. $ZnSO_4(aq) + BaS(aq) \longrightarrow ZnS(s) + BaSO_4(s)$
C: 1. (c) 2. (b) 3. (e) 4. (b) 5. (d) 6. (a) 7. (d) 8. (a)
9. $FeS(s) + H_2SO_4(aq) \longrightarrow FeSO_4(aq) + H_2S(g)$
10. $2\ Al(OH)_3(s) + 3\ H_2SO_4(aq) \longrightarrow Al_2(SO_4)_3(aq) + 6\ H_2O(l)$
D: 1. (b) 2. (d) 3. (b) 4. (c) 5. (d) 6. (d) 7. (d) 8. (c)
9. $BaCO_3(s) + 2\ HNO_3(aq) \rightarrow Ba(NO_3)_2(aq) + H_2O(l) + CO_2(g)$
10. $Ba(OH)_2(aq) + H_2SO_4(aq) \longrightarrow BaSO_4(s) + 2\ H_2O(l)$

Sample Test

1. The abbreviations are the same as those for objective 5-9.
a. $Sr(NO_3)_2(aq) + H_2SO_4(aq) \rightarrow SrSO_4(s) + 2\ HNO_3(aq)$ [m, p]
b. $Na_2CO_3(s) + 2\ HCl(aq) \longrightarrow 2\ NaCl(aq) + H_2O(l) + CO_2(g)$ [m, g] c. $Zn(s) + CuSO_4(aq) \longrightarrow ZnSO_4(aq) + Cu(s)$ [di, o]
d. $CaO(s) + 2\ HNO_3(aq) \longrightarrow Ca(NO_3)_2(aq) + H_2O$ [m, a]
e. $2\ NaI(aq) + Cl_2(aq) \longrightarrow 2\ NaCl(aq) + I_2(s)$ [di, o]
f. $BaBr_2(aq) + 2\ AgNO_3(aq) \longrightarrow Ba(NO_3)_2(aq) + 2\ AgBr(s)$ [m, p] g. $CaO(s) + CO_2(g) \longrightarrow CaCO_3(s)$ [c, ?]
h. $2\ Al(s) + 3\ Cl_2(g) \longrightarrow 2\ AlCl_3(s)$ [c, o]
2. a. $3\ Ag(s) + 4\ HNO_3(aq) \rightarrow 3\ AgNO_3(aq) + NO(g) + 2\ H_2O$
b. $Cl_2(aq) + 2\ KOH(aq) \longrightarrow KCl(aq) + KClO(aq) + H_2O(l)$
c. $Fe_2(SO_4)_3(aq) + SnSO_4(aq) \longrightarrow Sn(SO_4)_2(aq) + 2\ FeSO_4(aq)$
d. $5\ K_2C_2O_4(aq) + 2\ KMnO_4(aq) + 8\ H_2SO_4(aq) \longrightarrow$
$2\ MnSO_4(aq) + 6\ K_2SO_4(aq) + 10\ CO_2(g) + 8\ H_2O(l)$
3. The pungent odor with NaOH(aq) indicates the presence of NH_4^+ ion. The compound thus is expected to be soluble. The absence of a precipitate with $Cu^{2+}(aq)$ eliminates the generally insoluble ions: carbonate, chromate, phosphate, oxalate, sulfite, hydroxide, oxide, and sulfide anions. The formation of a precipitate with $Ba^{2+}(aq)$ eliminates chloride, bromide, and iodide anions. This leaves sulfate and thiosulfate anions, of which the latter forms $SO_2(g)$ on the addition of acid. The unknown is ammonium sulfate, $(NH_4)_2SO_4$.

CHAPTER 6

Drill Problems

1. A. 441 lb/in.^{2} = 3.04 × 10^3 kPa B. 2.75 lb/in.^{2} = 1.89 × 10^4 Pa C. 16.5 atm = 1.67 × 10^3 kPa D. 3.80 × 10^5 torr = 5.07 × 10^7 Pa E. 0.658 atm = 6.67 × 10^4 Pa F. 4.93 × 10^{-3} atm = 3.75 torr G. 34.0 atm = 3.45 × 10^6 Pa H. 0.511 lb/in.^{2} = 2.64 mmHg
3. A. 6.13 L B. 39.5 atm C 1.16 × 10^{-4} L D.23.0 atm E. 12.6 L F. 0.334 atm G. 1.51 × 10^3 ft^3 H. 257 kPa I. 187 L J. 756 mmHg
4. A. 8.73 L B. 1.59 × 10^3 K C. 170 L D. –24.5°C E. 24.5 L F. 1.22 × 10^3 K G. 13.6 L H. –214°C
8. A. 49.7 L B. 187 K C. 0.277 mol D. 22.4 atm E. 31.9 qt F. –72.0°C G. 19.9 mol H. 6.57 L I. 122 torr J. 8.50 L K. 2.43 g H_2O L. 0.337 lb/in.^{2} M. 284°F N. 0.418 L O. 7.22 g CH_4 P. 44.5 torr
9. (1) A. 215 K B. 21.9 L C. 2.65 atm D. 1.30 × 10^3 K E. 1.32 L F. 0.342 atm G. 559 K H. 143 L I. 0.688 atm J. 715 K K. 7.27 L L. 12.1 atm
(2) M. 10.7 mol N. 25.1 atm O. 0.634 mol P. 685°C Q. 2.90 mol R. 67.8 L
10. A. 2.03 mol, 15.4 g/mol, 31.3 g B. 3.02 L, 32.3 g/mol, 24.8 g/L C. 19.4 g/mol,307 K, 0.474 g/L D. 0.511 atm, 14.5 g, 0.418 g/L E. 1.11 × 10^3 K, 42.0 K, 42.0 g, 4.04 g/L F. 422 mol, 33.4 kg, 249 g/L G. 13.2 L, 181 g, 13.7 g/L H. 15.5 L, 40.0 g/mol, 4.92 g/L I. 24.2 L, 5.15 mol, 17.2 g/mol J. 95.9 L, 1.08 atm, 205 g K. 333 L, 0.0853 atm, 82.5 g/mol L. 95.2 L, 2.72 atm, 9.00 mol M. 0.0881 L, 0.0116 mol, 409 g/mol N. 0.546 mol, 24.5 g/mol, 13.4 g O. 99.7 g/mol, 203 K, 542 g P. 0.770 mol, 204 K, 649 g
11. 0.459 mol, 42.1 g/mol, C_3H_6 B. 0.788 mol, 128 g/mol, $C_4H_4F_4$ C. 0.394 mol, 186 g/mol, C_6F_6 D. 0.0548 mol, 32.0 g/mol, N_2H_4 E. 0.109 mol, 84.4 g/mol, C_6H_{12} F. 3.50 mol, 48.0 g/mol, O_3 G. 0.0244 mol 52.1 g/mol, C_4H_4 H. 21.5 mol, 46.0 g/mol, NO_2 I. 0.237 mol, 40.0 g/mol, C_2H_2N J. 2.72 mol, 58.0 g/mol, $C_2H_2O_2$ K. 0.723 mol, 60.0 g/mol, $C_2H_4O_2$ L. 0.570 mol, 30.0 g/mol, CH_2O_2
12. (1) A. 59.3 L O_2, 37.1 L CO_2 B. 141 L O_2, 105 L H_2O C. C_5H_{12} limiting: 22.8 L CO_2 , 40.5 L H_2O D. 81.6 L H_2, 54.4 L NH_3 E. H_2 limiting: 16.8 L NH_3 F. 10.7 L N_2, 32.1 L H_2 G. 6.09 L H_2O, 3.05 L NO H. 3.26 L HCl, 1.09 L NO, 1.63 L Cl_2 I. 0.973 L H_2O, 0.487 L NO J. 4.61 L NO, 6.92 L H_2O K. O_2 limiting: 33.7 L NO, 50.5 L H_2O L. 3.43 L O_2, 4.11 L H_2O M. 38.4 L NH_3, 28.8 L O_2 N. O_2 limiting: 12.4 L N_2, 37.2 L H_2O O. 9.73 L NH_3, 7.30 L O_2, 4.87 L N_2
(2) P. 12.9 L CO_2 Q. O_2 limiting: 22.6 L H_2O R. 101 L O_2, 67.1 L CO_2 S. 7.15 L O_2, 4.29 L CO_2 T. 2.82 g O_2, 1.59 g H_2O, 8.16 L CO_2 U. C_6H_6 limiting: 3.07 L H_2O V. CH_4 limiting: 12.3 L CO_2 W. $C_2H_6O_2$ limiting: 2.07 L CO_2 X. 8.96 L CO_2 Y. 14.8 L CO_2, 12.2 g H_2O
13. A. 48.2 L B. 2.53 atm C. 308 K D. 584 atm E. 139 K F. 7.41 atm G. 15.5 L H. 6.13 atm I. 160 K
17. A. He 1356 m/s, Ne 608.8 m/s, H_2 1927 m/s, N_2 516.8 m/s, O_2 483.6 m/s, CO_2 412.3 m/s, Ar 432.8 m/s, H_2O 644.5 m/s, NH_3 662.8 m/s, CH_4 682.9 m/s B. He 14.4 K, Ne 72.8 K, H_2 7.3 K, N_2 101.1 K, O_2 115.5 K, CO_2 158.8 K, Ar 144.1 K, CH_4 57.9 K, NH_3 61.5 K, H_2O 65.0 K C. 2.65 times faster D. 1.069 times faster E. 25.3 hr F. 536 hr

Quizzes

A. 1. (b) 2. (a) 3. (d) 4. (b) 5. (a) 6. (c) 7. (c) 8. (e) 9. (c) 10. (c)
B. 1. (c) 2. (d) 3. (b) 4. (d) 5. (a) 6. (a) 7. (c) 8. (a) 9. (b) 10. (c)
C. 1. (d) 2. (a) 3. (e) 4. (a) 5. (b) 6. (b) 7. (e) 8. (c) 9. (b) 10. (d)
D. 1. (b) 2. (a) 3. (d) 4. (b) 5. (d) 6. (b) 7. (a) 8. (a) 9. (e) 10. (e)

Sample Test

1. a. 1.80 atm = 182 kPa b. 0.0566 m^3 = 56.6 L c. 33.3°C = 306.5 K

2. $n = PV/RT = \dfrac{(1.80\ \text{atm})(56.6\ \text{L})}{[(0.08206\ \text{L atm mol}^{-1}\ \text{K}^{-1})(306.5\ \text{K})]} = 4.05\ \text{mol} \quad \mathcal{M} = \dfrac{(1.305\ \text{lb} \times 453.5\ \text{g/lb})}{4.05\ \text{mol}} = 146\ \text{g/mol}$

3. $\dfrac{0.08206\ \text{L atm}}{\text{mol K}} \times \dfrac{1000\ \text{mL}}{\text{L}} \times \dfrac{760\ \text{mmHg}}{\text{atm}} \times \dfrac{\text{mol}}{1000\ \text{mmol}} = 62.37\ \text{mL mmHg mmol}^{-1}\ \text{K}^{-1}$

4. $n_A = PV/RT = \dfrac{(1.067\ \text{atm})(14.20\ \text{L})}{[(0.08206\ \text{L atm mol}^{-1}\ \text{K}^{-1})(303.1\ \text{K})]} = 0.6092\ \text{mol}$

$n_B = PV/RT = \dfrac{(26.42 \text{ atm})(1.251 \text{ L})}{[(0.08206 \text{ L atm mol}^{-1} \text{ K}^{-1})(327.5 \text{ K})]}$

$= 1.230$ mol $n_{total} = 1.839$ mol

a. $P = \dfrac{(1.839 \text{ mol})(0.08206 \text{ L atm mol}^{-1} \text{ K}^{-1})(291.0 \text{ K})}{(8.78 \text{ atm})}$

$= 2.91$ atm

b. $P_A = (n_A/n_{total})P = (0.6092 \text{ mol}/1.839 \text{ mol})(8.78 \text{ atm})$ $= 2.91$ atm $V_B = (n_B/n_{total})V = (1.230 \text{ mol}/1.839 \text{ mol})(5.00 \text{ L})$

5. a. Ne b. He c. He d. equal pressures

CHAPTER 7

Drill Problems

2. (1) A. 30.5°C B. 441 J C 0.62 J g^{-1} °C^{-1} D. 117 g E. 29.9°C F. 1.18 kJ G. 0.0862 J g^{-1} °C^{-1} H.157 g I. 21.2°C J. 1.5 mol K. 24.2 J mol^{-1} deg^{-1} L. 0.27 kJ M. 20.0°C (2) N. 1.8 J mol^{-1} deg^{-1} O. 20.4°C P. 153.5°C Q. 522 g R. 6.53 g S. 0.208 J g^{-1} °C^{-1} T. 73.7°C U. 46.0°C V. 22.3°C

3. (1) Values in calories: A. 4.216 C. 586.8 E. 74.66 G. 150.0 I. 1619 K. 956.5 Values in joules: B. 2394 D. 2268 F. 306.3 H. 5.288 J. 10.88 L. 144.4 (2) M. –68 J N. –1207 J O. + 55 J P. +1663 J Q. +1.96 kJ R. –3.88 kJ S. –0.68 kJ T. +5.28 kJ U. –0.314 kJ V. 989 J W. –1382 J X. 272 J

7. A. 0.47 kJ/°C B. 40.7 kJ/g C. 21.3°C D. 0.79 g E. 0.716 kJ/°C F. 48 kJ/g G. 1.09 g H. 63.2°F

9. A. + 39.7 kJ/mol B. 21.8°C C. 244 g H_2O D. +24.6 kJ/mol E. 40.3°C F. 18.4°C G. + 25.1 kJ/mol

10. All values are in kJ/mol A. +107.6 B. –406.8 C. –112.3 D. –411.0 E. –167.4 F. –217.5 G. +406 H. –529 I. –25.5 J. +2126.8 K. –155.2

10a. A. 2 C(graphite) + 3 H_2(g) + $\frac{1}{2}O_2$(g) ⟶ C_2H_5OH(l)

B. N_2(g) + 4 H_2(g) + Sb(c,III) + $\frac{3}{2}Cl_2$(g) ⟶ $(NH_4)_2SbCl_5$(s)

C. 2 Hg(l) + Cl_2(g) ⟶ Hg_2Cl_2(s)

D. $\frac{3}{2}H_2$(g) + P(s,white) + 2 O_2(g) ⟶ H_3PO_4(s)

E. Na(s) + $\frac{1}{2}I_2$(g) + $\frac{3}{2}O_2$(g) ⟶ $NaIO_3$(s)

F. Sn(s,white) + F_2(g) ⟶ SnF_2(s)

G. Xe(g) + $\frac{3}{2}O_2$(g) ⟶ XeO_3(g)

H. $\frac{1}{2}H_2$(g) + $\frac{1}{2}N_2$(g) + $\frac{3}{2}O_2$(g) ⟶ HNO_3(l)

I. K(s) + $\frac{1}{2}Br_2$(l) + $\frac{3}{2}O_2$(g) J. $\frac{1}{2}I_2$(s) + $\frac{3}{2}Cl_2$(g) ⟶ ICl_3(g)

12. All values are in kJ/mol A. +177.9 B. –176.9 C. –126.7 D. –260.8 E. +226.3 F. –92.8 G. –1530.6 H. –1169.2 I. –113.04 J. –71.7 K. –308.4 L. –323.2 M. –115.2 N. +873.1 O. –65.2

Quizzes

A. 1. (a) 2. (c) 3. (d) 4. (d) 5. (e) 6. (b) 7. (e)(–103 kJ/mol) 8. (b)

B. 1. (c) 2. (b) 3. (c) 4. (a) 5. (c) 6. (b) 7. (e)(108.8 kJ/mol) 8. (d)

C. 1. (b) 2. (b) 3. (a) 4. (c) 5. (b) 6. (c) 7. (a) 8. (e)(–1125.6 kJ/mol)

D. 1. (d) 2. (d) 3. (e)(+24.8 J) 4. (b) 5. (e) 6. (d) 7. (d) 8. (b)

Sample Test

1. a. Sn(s,white) + 2 Cl_2(g) → $SnCl_4$(s)
b. 7 C(graphite) + 3 H_2(g) + O_2(g) ⟶ C_6H_5COOH(s)
c. C(graphite) + $\frac{1}{2}O_2$(g) + Cl_2(g) ⟶ $COCl_2$(g)
2. a. –185 kJ/mol b. –807 kJ/mol
3. $5.00 \text{ g} \times \dfrac{\text{mol LiCl}}{42.4 \text{ g}} \times \dfrac{37.2 \text{ kJ}}{\text{mol}} = 4.39 \text{ kJ} \times \dfrac{1000 \text{ J}}{\text{kJ}}$ $\times \dfrac{\text{g °C}}{4.00 \text{ J}} = 1097$ g °C

$\Delta t = 1097$ g °C/115.0 g = 9.54°C. Thus $t_f = 29.5$°C.

CHAPTER 8

Drill Problems

1. & 2. A. 1.1 × 10^3 nm, ir B. 1.9 × 10^{18} Hz, γ ray C. 12 m, TV D. 200 kHz, radio E. 1.71 × 10^{-5} nm, γ ray F. 4.13 × 10^{10} Hz, microwave G. 5.83 km, radio H. 242 MHz, radar I. 4.04 × 10^{-8} cm, x ray J. 5.95 × 10^{11} kHz, visible K. 95.2 nm, UV L. 2.410 × 10^7 Hz, TV M. 3.55 × 10^{-15} m x ray N. 2.21 × 10^5 MHz, microwave O. 5.98 × 10^3 Å, visible P. 3.64 × 10^9 MHz, UV Q. 2.05 × 10^{-5} km, microwave R. 2.27 × 10^{21} Hz, x ray S. 0.355 cm, microwave T. 7.43 × 10^{19} MHz, x ray U. 1.58 × 10^6 m, radio

4. A. 2, Balmer, visible B. 3, Lyman, UV C. 379.9 nm, Balmer, visible D. 1875 nm, Paschen, ir E. 2, Balmer, visible F. 5, Balmer, visible G. 1945 nm, Brackett, ir H. 12372 nm, Humphreys, ir I. 5, Pfund, ir J. 2, Lyman, UV K. 389.0 nm, Balmer, visible L. 92.3 nm, Lyman, UV

5. A. 1.95 × 10^{18} Hz, 1.29 × 10^{-8} erg, 7.78 × 10^5 kJ/mol B. 6.94 × 10^{-12} erg, 1.05 × 10^{15} Hz, 286 nm C. 2.86 m, 6.96 × 10^{-19} erg, 4.19 × 10^{-5} kJ/mol D. 0.0644 kJ/mol, 1.61 × 10^{11} Hz, 1.85 × 10^6 nm E. 5.69 × 10^{14} Hz, 3.77 × 10^{-12} erg, 227 kJ/mol F. 4.73 × 10^{-13} erg, 7.14 × 10^{13} Hz, 4.20 × 10^{-4} cm G. 4.90 × 10^4 nm, 4.06 × 10^{-14} erg, 2.44 kJ/mol H. 4.90 × 10^{-6} kJ/mol, 1.23 × 10^7 Hz, 24.5 m I. 2.24 × 10^5 Hz, 1.48 × 10^{-21} erg, 8.94 × 10^{-8} kJ/mol J. 8.73 × 10^{-15} erg, 1.32 × 10^{12} Hz, 0.0228 cm K. 0.586 km, 3.39 × 10^{-21} erg, 2.04 × 10^{-7} kJ/mol L. 1.89 × 10^4 kJ/mol, 4.74 × 10^{16} Hz, 6.33 × 10^{-3} pm M. 4.88 × 10^9 MHz, 3.23 × 10^{-17} erg, 1.95 × 10^{-3} kJ/mol N. 1.20 × 10^{-14} erg, 1.82 × 10^{12} Hz, 0.165 mm O. 8.77 nm, 2.27 × 10^{-10} erg, 1.37 × 10^4 kJ/mol P. 7.47 × 10^7 kJ/mol, 1.87 × 10^{20} Hz, 1.60 pm

9. (1) A. $m_l = 0$ B. $3p_x$, $3p_y$, $3p_z$; $m_l = -1, 0, +1$

C. $4d_{xy}$, $4d_{yx}$, $4d_{xz}$, $4d_{x^2-y^2}$, $4d_{z^2}$; $m_l = -2, -1, 0, +1, +2$

D. $m_l = 0$ E. $m_l = -3, -2, -1, 0, +1, +2, +3$

F. $4p_x$, $4p_y$, $4p_z$; $m_l = -1, 0, +1$ G. $m_l = 0$

H. $2p_x$, $2p_y$, $2p_z$; $m_l = -1, 0, +1$

I. $5d_{xy}$, $5d_{yx}$, $5d_{xz}$, $5d_{x^2-y^2}$, $5d_{z^2}$; $m_l = -2, -1, 0, +1, +2$

J. $3d_{xy}$, $3d_{yx}$, $3d_{xz}$, $3d_{x^2-y^2}$, $3d_{z^2}$; $m_l = -2, -1, 0, +1, +2$

(2) K. 1*s*; 2 electrons L. 3*s*, 3*p*, 3*d*; 18 electrons M. 2*s*, 2*p*; 8 electrons N. 4*s*, 4*p*, 4*d*, 4*f*; 32 electrons O. 5*s*, 5*p*, 5*d*, 5*f*, 5*g*; 50 electrons

10. See Figure A-1.

12. A. (3); $1s^22s^22p_x^12p_y^12p_z^1$ B. (5); $1s^22s^22p^63s^23p^1$

C. (5); $1s^22s^22p^1$ D. (5); $1s^22s^22p^63s^23p^3$ E. (4); [Ar]$4s^13d^{10}$

F. (5); $1s^22s^2$ G. (2); [Ne] [↑↓] H. (1); $1s^22s^22p_x^12p_y^1$

I. (1); $1s^22s^22p^63s^23p^64s^23d^3$ J (3); [Ne]$3s^23p_x^23p_y^13p_z^1$

K. (4); [Ar]$4s^23d^5$ L. (5); $1s^22s^22p^3$ M. (4); [Kr]$5s^14d^{10}$

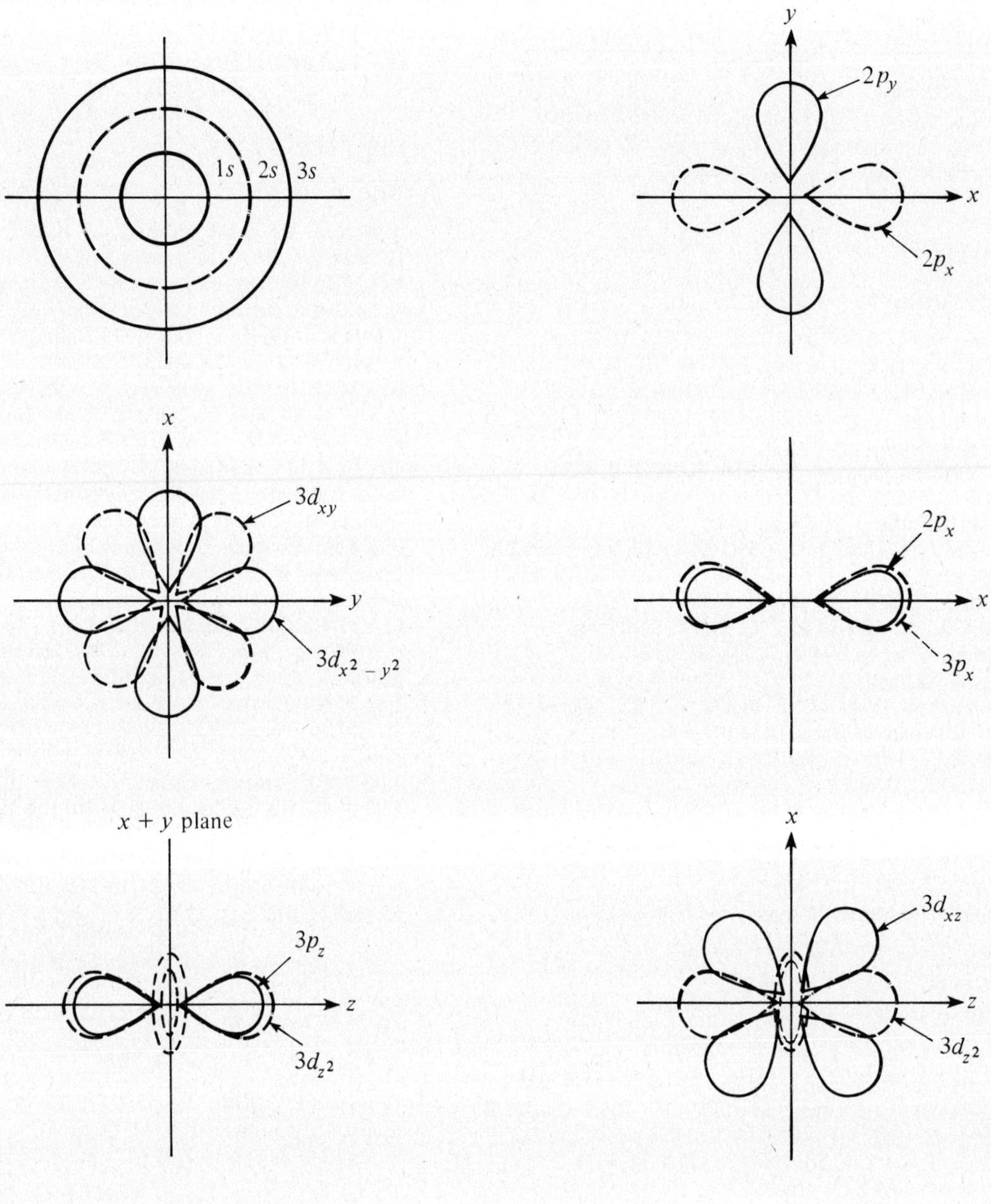

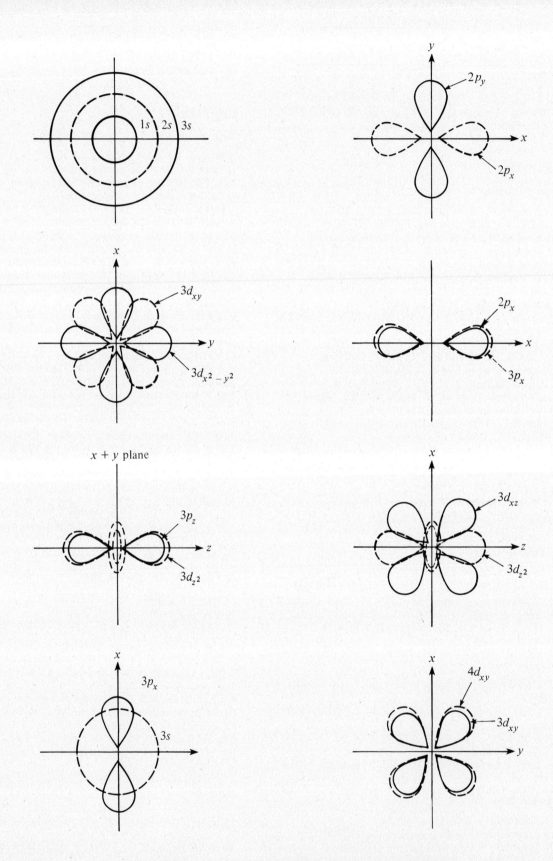

FIGURE A-1 Answers to Objective 8-10 Drill Problems

N. (3); $[Kr]5s^24d_{xy}^2 4d_{xz}^2 4d_{yz}^1 4d_{z^2}^1 4d_{x^2-y^2}^1$ O. (5); $1s^22s^22p^63s^1$

P. (1); $[Ne]3s^23p^64s^23d^1$ Q. (5); $1s^22s^22p^63s^23p^5$

R. (3); $1s^22s^22p_x^12p_y^1$ S. (5); [Ne] [↑↓] [↑↓] [↑↓] [↑↓] T. (1); $1s^22s^22p_x^1$

U. When two subshells are very close in energy, an electron can move from one to the other if that moves results in the increase in the number of empty, half-filled, or full electron shells. Cr is $[Ar]4s^13d^5$; Ag is $[Kr]5s^14d^{10}$. V. Each orbital of a subshell must contain one electron before electrons are paired in the orbitals of that subshell. N is $1s^22s^22p_x^12p_y^12p_z^1$; Si is $[Ne]3s^23p_x^13p_y^1$. Electrons fill orbitals in the order given in expression [15], Figure 8-4, and Figure 8-5. K is $1s^22s^22p^63s^23p^64s^1$; Fr is $1s^22s^22p^63s^23p^64s^23d^{10}4p^65s^24d^{10}5p^66s^24f^{14}5d^{10}6p^67s^1$. X. No two electrons may have the same set of four quantum numbers. C is $1s^22s^2$ [↑] [↑] [ ], not $1s^22s^2$ [↑↓] [ ] [ ] and not $1s^32s^3$.

13. (1) A. Sc $[Ar]4s^23d^1$ B. C $1s^22s^22p_x^12p_y^1$

C. F $1s^22s^22p_x^22p_y^22p(1,z)$ D. Ti [Ar}$4s^23d^2$ E. Al $[Ne]3s^23p^1$

F. Cl $1s^22s^22p^63s^23p^5$ G. Cr $[Ar]4s^13d^5$
H. Y $[Ar]4s^23d^{10}4p^65d^24d^1$ I. Sc $1s^22s^22p^63s^23p^64s^23d^1$
J. B $1s^22s^22p^1$ K. N $1s^22s^22p_x^12p_y^12p(1,z)$

(2) L. S M. Sc N. Zr O. Tl P. Ce Q. Mo, Tc R. V S. Se T. Sr

(3) U. $[He]2s^22p_x^22p_y^22p(1,z)$; [Ne] $_{3s}$[↑↓] $_{3p}$[↑] [↑] [↑];

V. $[Ne]3s^23p_x^23p_y^13p(1,z)$; [Ne] $_{3s}$[↑↓] $_{3p}$[↑↓] [↑] [↑];

W. $[Ne]3s^23p_x^23p_y^23p(1,z)$; [Ne] $_{3s}$[↑↓] $_{3p}$[↑↓] [↑↓] [↑];

X. $[Ne]3s^23p_x^13p_y^1$; [Ne] $_{3s}$[↑↓] $_{3p}$[↑] [↑]; Y. $[Ar]4s^13d^{10}$

[Ar] $_{4s}$[↑] $_{3d}$[↑↓] [↑↓] [↑↓] [↑↓] [↑↓] Z. $[Ne]3s^1$; [Ne] $_{3s}$[↑] Γ. $[Xe]6s^1$;
[Xe] $_{6s}$[↑] Δ. $[Xe]6s^25d^14f^1$;
[Xe] $_{6s}$[↑↓] $_{5d}$[↑] [ ] [ ] [ ] [ ] $_{4f}$[↑] [ ] [ ] [ ] [ ] [ ] [ ] Θ. $[Ar]4s^23d^1$;
[Ar] $_{4s}$[↑↓] $_{3d}$[↑] [ ] [ ] [ ] [ ] Λ. $[He]2s^22p_x^22p_y^12p(1,z)$;

[He] $_{2s}$[↑↓] $_{2p}$[↑↓] [↑] [↑]; Π. $[Ar]4s^23d^6$ [Ar] $_{4s}$[↑↓] $_{3d}$[↑↓] [↑] [↑] [↑] [↑] [↑]
S. $[Kr]6s^2$; [Kr] $_{6s}$[↑↑]

Further Questions A. $[Ar]4s^1$ B. $[He]2s^22p^2$ C. $[Ar]4s^23d^5$ D. $[Ar]4s^23d^1$ E. $[Ar]4s^13d^5$ F. $[Xe]6s^14f^{14}4d^{10}$ G. $[Ne]3s^23p^2$ H. $[Ar]4s^23d^{10}$ I. $[Ne]3s^2$ J. $[He]2s^22p^5$ K. $[Ne]3s^23p^3$ L. $[Ar]4s^23d^6$ M. $[Rn]7s^26d^15f^3$ N. $[He]2s^22p^6$ O. $[Xe]6s^25d^14f^2$, but actually $[Xe]6s^25d^04f^3$ P. $[Kr]5s^14d^{10}$ Q. $[Ne]3s^23p^1$ R. $[Kr]5s^24d^{10}5p^2$ S. $[Ar]4s^23d^{10}4p^2$ T. $[Kr]5s^14d^5$ U. $[Ar]4s^23d^{10}4p^1$ V. $[Ar]4s^2$ W. $[Kr]5s^14d^{10}5p^5$ X. $[Xe]6s^14f^{14}5d^5$, but actually $[Xe]6s^24f^{14}5d^4$ Y. $[He]2s^22p^1$ Z. $[Ar]4s^13d^{10}$

Quizzes

A: 1. (a) 2. (c) 3. (b) 4. (a) 5. (c) 6. (d) 7. (c) 8. (a) 9. (a) 10. (b) 11. (c) 12. (e)(Sc)
B: 1. (c) 2. (d) 3. (e) 4. (d) 5. (c) 6. (d) 7. (b) 8. (c) 9. (c) 10. (d) 11. (b) 12. (d)
C: 1. (d) 2. (c) 3. (d) 4. (c) 5. (b) 6. (e) 7. (c) 8. (d) 9. (b) 10. (c) 11. (b) 12. (a)
D: 1. (b) 2. (b) 3. (a) 4. (b) 5. (d) 6. (b) 7. (c) 8. (a) 9. (d) 10. (d) 11. (c) 12. (b)

Sample Test

1. $$\varepsilon = \frac{hc}{\lambda} = \frac{(6.63 \times 10^{-34}\ \text{J·s})(3.00 \times 10^8\ \text{m/s})}{3.28\ \text{m}}$$
$$= 6.06 \times 10^{-26}\ \text{J/photon}$$
$$E = \frac{6.06 \times 10^{-26}\ \text{J}}{\text{photon}} \times \frac{6.022 \times 10^{23}\ \text{photon}}{\text{mole}} \times \frac{\text{kJ}}{1000\ \text{J}}$$
$$= 3.65 \times 10^{-5}\ \text{kJ/mol}$$
2. See Figures 8-2 and 8-3.
3. a. $1s^22s^22p^63p_x^13p_y^13p_z^1$ b. $1s^22s^22p^63s^23p^64s^23d^5$
c. $1s^22s^22p^63s^23p_x^13p_y^1$
4. a. [Ar] $_{4s}$[↑↓] $_{3d}$[↑↓] [↑] [↑] [↑] [↑] b. [Ne] $_{3s}$[↑↓] $_{3p}$[↑] [ ] [ ]
c. [He] $_{2s}$[↑↓] $_{2p}$[↑↓] [↑] [↑]

CHAPTER 9

Drill Problems

3. (1) A. $ns^2(n-1)d^x$ B. ns^2 C. ns^2np^5 D. ns^2np^2 E. ns^2np^1 F. ns^2np^4 G. ns^1 H. ns^2np^6 I. ns^2np^4 J. $ns^2(n-1)d^1$ K. $ns^2(n-1)d^2$ L. ns^2np^3
(2) M. s^2p^3 N. s^2p^2 O. s^2p^6 P. s^2p^3 Q. s^1 R. s^2p^4 S. s^2p^1 T. s^2p^5 U. s^1 V. s^2p^6 W. s^2 X. s^2p^1 Y. s^2p^4 Z. s^2p^2 Γ. s^2p^5 Δ. s^2
(3) Θ. $[Ar]4s^13d^{10}$ Λ. $[Kr]5s^14d^5$ Ξ. $[Xe]6s^14f^{14}5d^{10}$ Π. $[Xe]6s^25d^04f^7$ Σ. $[Xe]6s^25d^1$ Υ. $[Xe]6s^24f^{14}5d^6$ Φ. $[Ar]4s^23d^6$ Ψ. $[Ar]4s^23d^3$ Ω. $[Kr]5s^24d^3$ Ø. $[Kr]5s^24d^1$

5. A. Be<Mg<Ca<Sr B. O<S<Te<Po C. He<Ar<Kr<Rn D. Si<Ge<Sn,Pb E. B<Al<Ga<In F. O<C<B<Li G. Bi<Tl<Ba<Cs H. Te<Sn<In<Rb I. Br<As<Ge<K J. Cl<P<Al,Mg K. Fe<V<Ti<K L. Os<Hf<La<Ba M. $W^{3+}<Tm^{3+}<Eu^{3+}<Ce^{3+}$ N. Os<W<U<Ac O. $Ca^{2+}<Ar<P^{3-}<Si^{4-}$ P. $Na^+<Ne<O^{2-}<C^{4-}$ Q. $Sc^{3+}<K^+<Cl^-<S^{2-}$ R. $Y^{3+}<Sr^{2+}<Br^-<Ga^{5-}$ S. Ne<O<Si<Ge T. $Be^{2+}<Mg^{2+}<F^-<C^{4-}$ U. N<Al<Na<K V. $Na^+<Ne<N<Li$ W. Ge<Ga<Tl<Ba

6. A. Cs<Rb<K<Na B. Sb<As<P<N C. Tl<In<Al<B D. Sr<Ca<Mg<Be E. Na<Al<Cl<Ar F. Ca<Ge<As<Kr G. Rb<In<Te<I H. Li<B<N<Ne I. Ba<Ga<Se<F J. Sr<Sb<P<Ne K. Cs<Sn<As<Cl L. Rb<In<Al<F

7. A. F<Cl<Br<I B. N<P<Sb<Bi C. Li<Na<K<Rb D. Br<Se<Ge<K E. I<Sb<In<Rb F. F<P<In<Cs G. Cl<Se<Sb<Tl H. B < Al < In < Tl I. Cl < As < In < Sr

8. (1) A. C<N<O<F B. K<Ca<As<Br C. I<Br<Cl<F D. Cs<Rb<K<Li E. Rb<Ru<In<Se F. Na<Li<B<F G. Cs<As<Cl<F H. Ge < Si < N < O I. Cs < Ba < Tl < Bi
(2) J. Na < K < Rb < Cs K. F < P < In < Cs L. Br < As < Ca < K M. F < N < O < C N. Cl < Al < Mg < Na O. Se < Ge < Fe < Ce P. Be < Mg < Ca < Sr Q. C < B < Be < Li R. Al < Ga < Sr < Rb

9. A. $K^+<Ce^{3+}<C<P$ B. $(He,Cs^+)<Sc<S$ C. B<(Si,S)<P D. Ca<(K,Sc)<Ti E. Cu<V<Mn<Cr F. (Zn,Ca)<Ti<As G. $Na^+<(Na,S^-)<S$ H. Be<(Li,B)<C I. $Cu^+ < Cu^{2+} < Fe^{2+} < Fe^{3+}$

Quizzes

A: 1. (c) 2. (d) 3. (d) 4. (c) 5. (b) 6. (b) 7. (e) 8. (c)
B: 1. (c) 2. (c) 3. (e) 4. (a) 5. (b) 6. (c) 7. (a) 8. (a)
C: 1. (e) 2. (e) 3. (c) 4. (b) 5. (c) 6. (b) 7. (a) 8. (a)
D: 1. (a) 2. (b) 3. (d) 4. (b) 5. (b) 6. (a) 7. (c) 8. (a)

Sample Test

1. Se_8 or Se(metal), both allotropes exist. 2. 34 3. $4s^24p^4$ 4. 78.96 (vertical trends predict 78.4) 5. 2 6. 116 pm (horizontal trends predict 116 pm or 117 pm) 7. 16.5 mL/mol (period 5 trend predicts 16.1 mL/mol, halogen group trend predicts 18.9 mL/mol) 8. 2.4, same as predicted by period trend 9. 9.8 eV/atom, same as predicted by group trend 10. 0.35 J g^{-1} °C^{-1} (group trends predict 0.30 J g^{-1} °C^{-1})

CHAPTER 10

Drill Problems

2. A. He: B. Li· C. :Ne: D. ·In· E. :I: F. :Sb· G. ·Sn· H. Sr: I. ·S· J. :F: K. :C· L. Ca: M. K· N. :Al· O. ·P· P. :Se·

3. A. Li^+ [H:]$^-$ B. Na^+ [|S|]$^{2-}$ C. Ca^{2+} [|Br|]$^-$ D. Sr^{2+} [|O|]$^{2-}$ E. Mg^{2+} [|N|]$^{3-}$ F. Sc^{3+} [|S|]$^{2-}$ G. Na^+ [|F|]$^-$ H. Ti^{4+} [|O|]$^{2-}$ I. Mg^{2+} [|S|]$^{2-}$ J. La^{3+} [|Cl|]$^-$ K. K^+ [|Se|]$^{2-}$ L. Mg^{2+} [|C|]$^{4-}$ M. Ce^{4+} [|Cl|]$^-$ N. Al^{3+} [|N|]$^{3-}$ O. Cs^+ [|Te|]$^{2-}$ P. Rb^+ [|P|]$^{3-}$

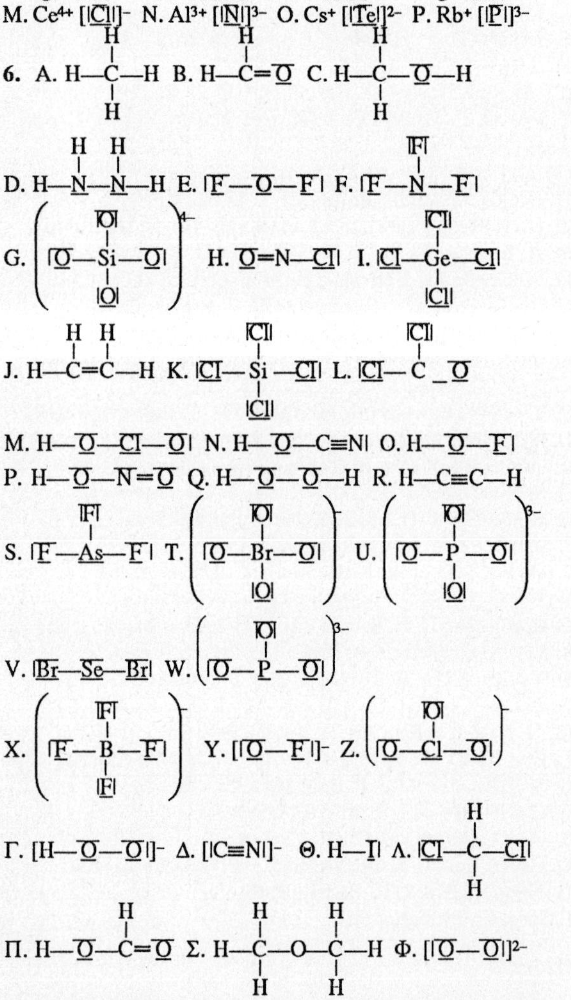

7. Formal charges are given in parentheses after each structure, when they are different from zero. The "best" structure is drawn first in each case.
A. H—O—C≡N|; H—O—N≡C| (N, +1; C, −1)
B. H—C≡N|; H—N≡C| (N, +1; C, −1)
C. 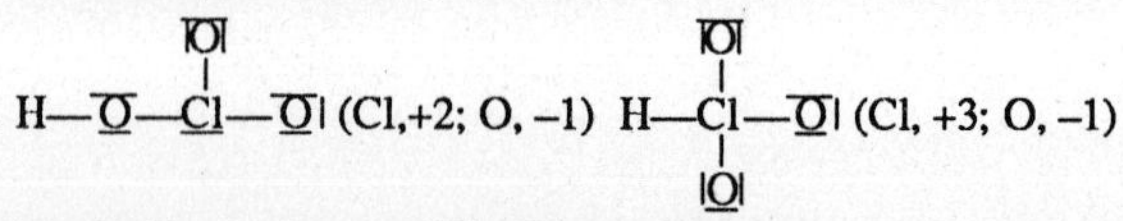
D. |C≡O| (C, −1; O, +1); |C=O (C, −2; O, +2)
E. O=N—Cl|; N=O—Cl| (N, −1; O, +1)
F. [|N=C=S|]$^-$ (N, −1); [|C=N=S|]$^-$ (C, −2; N, +1)
G. H—N=C=S|; H—C=N=S| (C, −1; N, +1)
H. 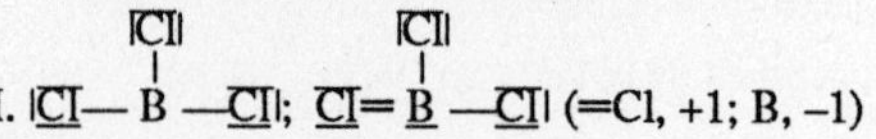
I. H—N=O|; H—O=N| (O, +1; N, −1)
J. |Cl—Be—Cl|; Cl=Be—Cl| (=Cl, +1; Be, −1)
K. 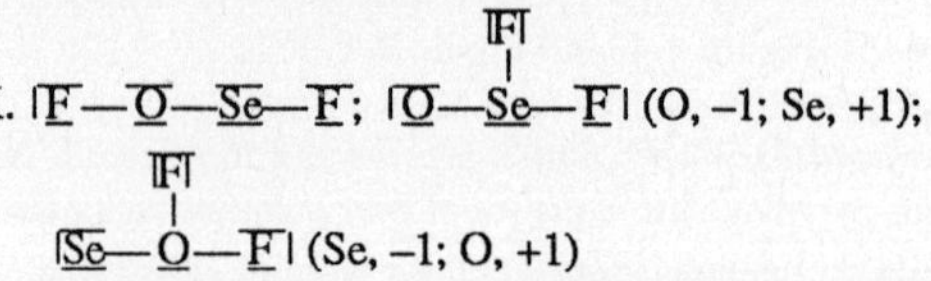
L. [|N=C=N|]$^{2-}$ (both N,−1); [|C=N=N|]$^{2-}$ (C,−2; N,+1; N, −1)
M. 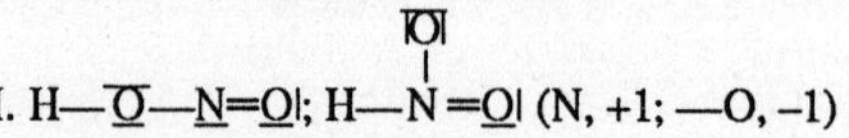
N. 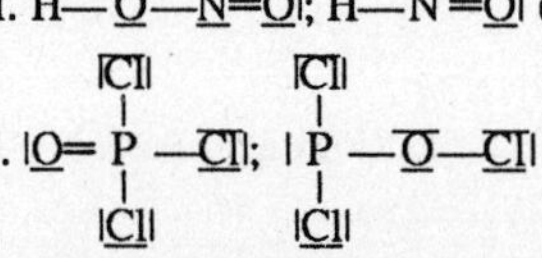
O. |Cl—C≡N|; |Cl—N≡C| (N, +1; C, −1)
P. H—N=N=N (=N=, +1; =N, −1); H—N≡N—N| (both —N≡, +1; —N|, −2)
Q. 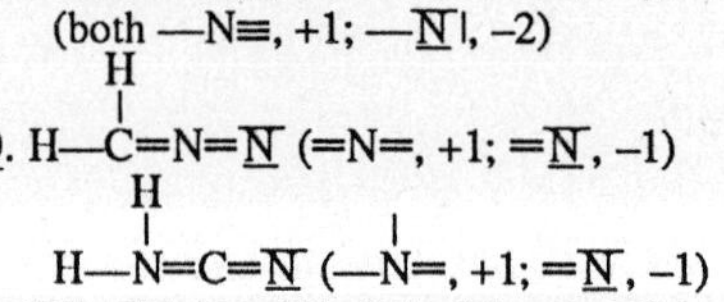
R. |Cl—O—N=O; |Cl—O—O=N (—O=, +1; N, −1)

8. A. 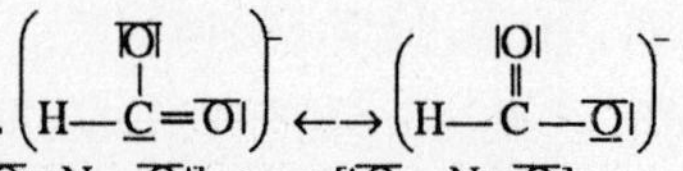
B. [O=N—O|]$^-$ ⟷ [|O—N=O]$^-$
C. [N=N=N]$^-$ ⟷ [|N≡N—N|]$^-$ ⟷ [|N—N≡N|]$^-$
D. O=S—O| ⟷ |O—S=O
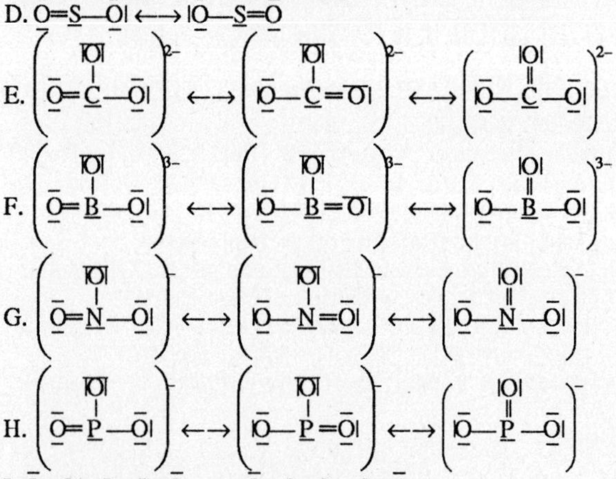
I. O=C—O=C=O ⟷ O=C=O—C=O
J. 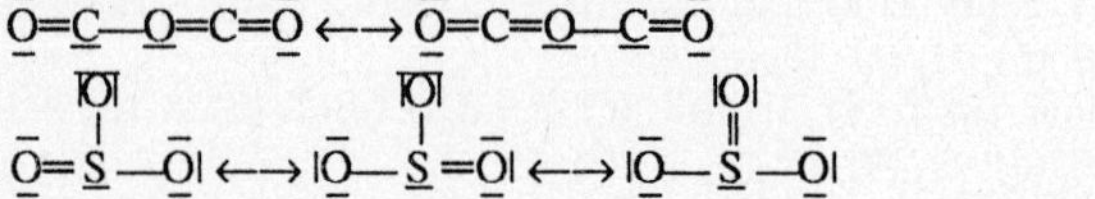

K. 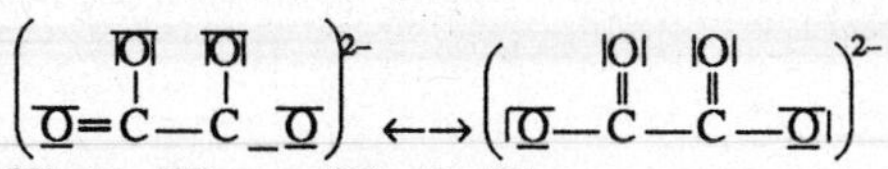

L. O=O—O ⟷ O—O=O

9. All formal charges equal zero unless otherwise indicated.

A. N̈_O B. O—Ṅ=O ⟷ O=Ṅ—O (—O, −1; N, +1)

C. Br—Sn—Br

D. O—Ċl=O ⟷ O=Ċl—O (—O, −1; Cl, +1)

E. I—Pb—I F. Cl—Be—Cl

G. 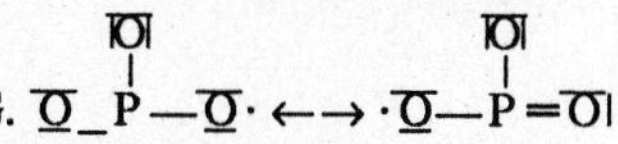 (O—, −1; P, +1) and four other resonance forms.

H. H—Be—H I. [N≡C—N·]⁻ ⟷ [·N—C≡N]⁻ [N, −1]

J. B≡N K. 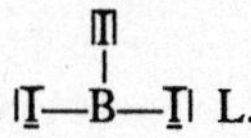 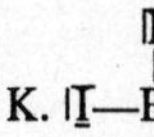 L. 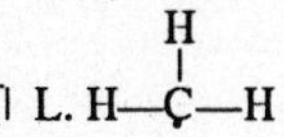

M. [·O—S=O]⁺ ⟷ [O=S—O·]⁺ (S, +1)

N. 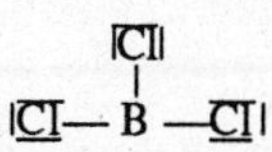 O. [·O—O]⁻ ⟷ [O—O·]⁻ (O—, −1)

10. All formal charges equal zero unless otherwise indicated.

A. 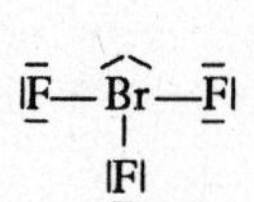 B. 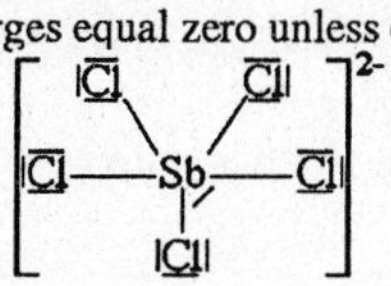 (Sb, −2)

C. 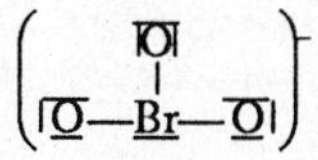 (all O, −1; Br, +2)

D. [I—I—I]⁻ (central I, −1) E. 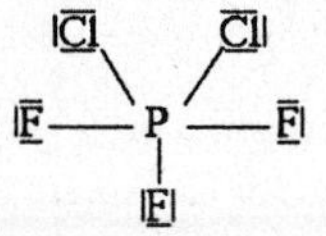

F. Cl—Pb—Cl G. 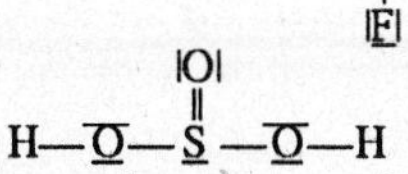

H. 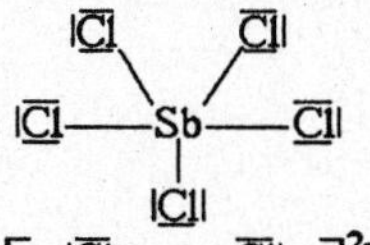 I. 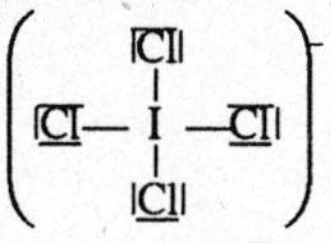 (I, −1)

J. 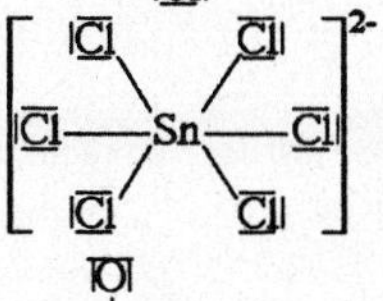 (Sn, −2) K. 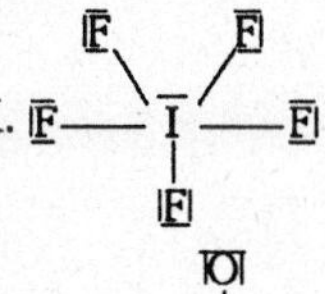

L.  (—O, −1)

M. 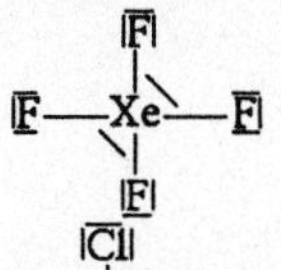 N. 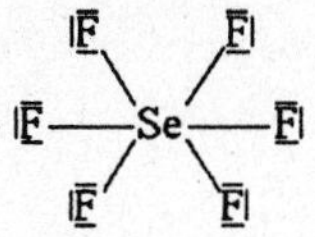 O. F—Ar—F

P. 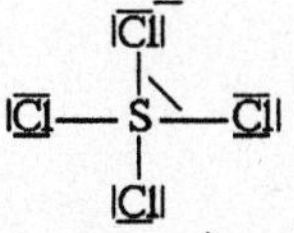 Q. 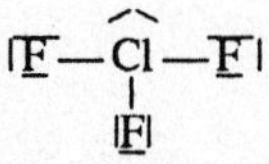 R. 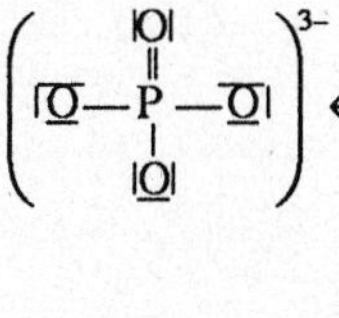 ⟷

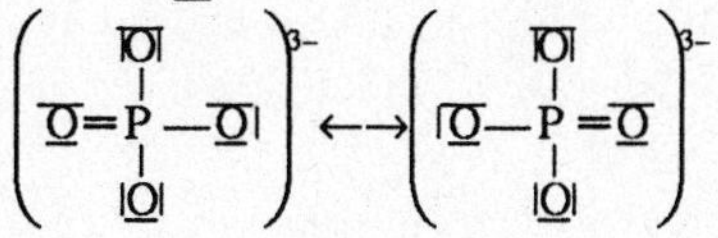

⟷ 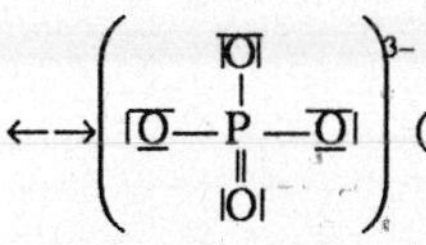 (—O, −1)

S. [S—I—S]⁻ (S, −1; I, +1) T. F—Xe—F

U. 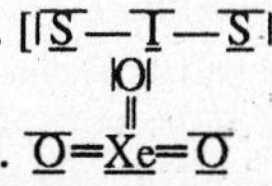

11. All values are in kJ/mol

A. −56 B. −618 C. −309 D. −821 E. −374 F. −83 G. −858 H. −1694 I. +121 J. −2628 K. −158 L. −814 M. −35 N. −8 O. +24 P. −1458

12. Electron pair geometry is given first, and molecular shape is given second, abbreviated as follows: lin, linear; tripl, trigonal planar; tet, tetrahedron; tripyr, trigonal pyramid; tbp, trigonal bipyramid; irtet, irregular tetrahedron; T, T-shaped; oct, octahedron; sqpyr, square pyramid; sqpl, square planar.

A. tripl, tripl B. tet, tet C. tet, bent D. tbp, T

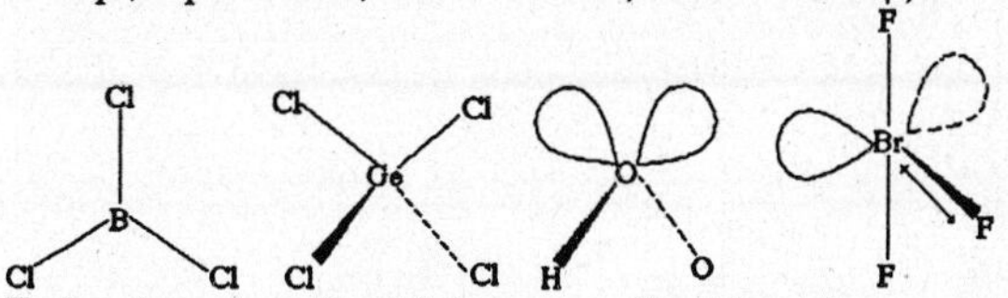

E. tbp, tbp F. tripl, bent G. tripl, tripl

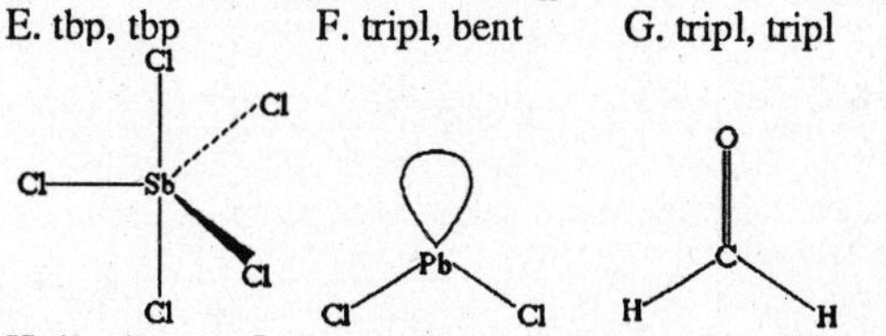

H. lin, lin I. oct, sqpl J. tbp, lin K. tbp, irtet

Cl—Be—Cl 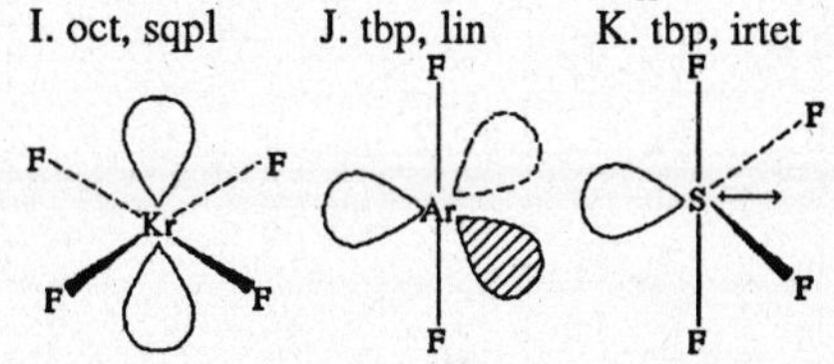

L. tripl, bent M. tet, tripyr N. tbp, T O. oct, sqpyr

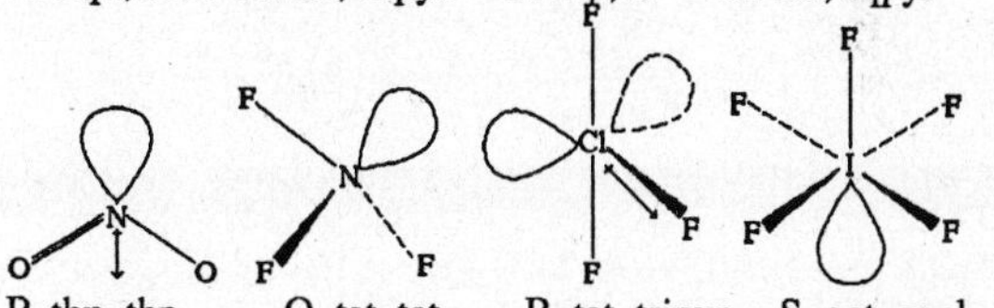

P. tbp, tbp Q. tet, tet R. tet, tripyr S. oct, sqpl

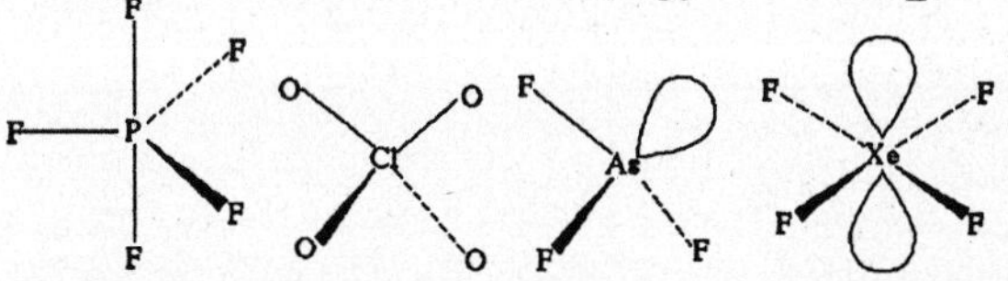

T. tripl, tripl U. tet, bent V. tet, tet W. tbp, irtet

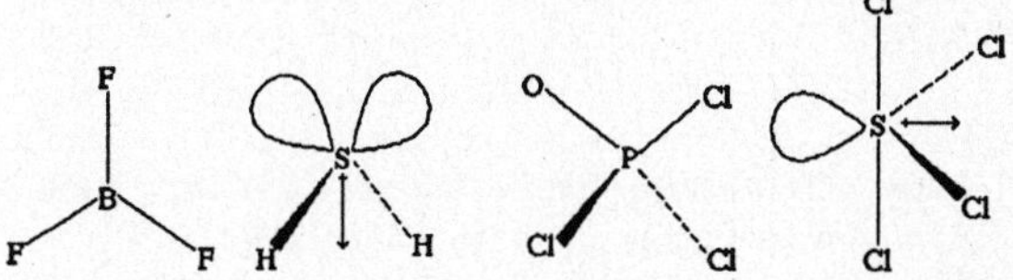

X. tripl, tripl Y. tripl, tripl Z. lin, lin Γ. tet, tet

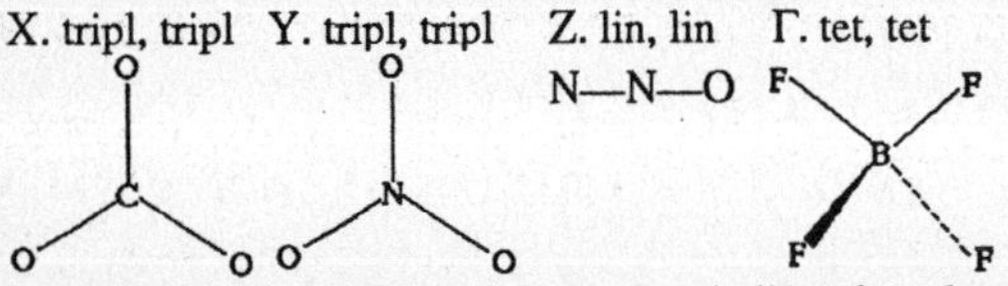

13. Most of the molecular polarities are indicated on the sketches that are part of the answers to the drill problems of objective 10-12. The following molecular polarities are described because they do not appear clearly on a sketch. M. from N to the center of the 3-F triangle R. from As to the center of the 3-F triangle V. along the P—O bond Z. from N to O.

Quizzes

A: 1. (d) 2. (a) 3. (d) 4. (e) 5. (c) 6. (b) 7. (c) 8. (d) 9. (c) 10. (a)
B: 1. (b) 2. (d) 3. (c) 4. (e) 5. (d) 6. (c) 7. (a) 8. (a) 9. (a) 10. (c)
C: 1. (b) 2. (b) 3. (b) 4. (d) 5. (a) 6. (d) 7. (e) (linear) 8. (b) 9. (b) 10. (e) (–420)
D: 1. (c) 2. (c) 3. (d) 4. (b) 5. (d) 6. (d) 7. (b) 8. (a) 9. (d) 10. (d)

Sample Test

1. Formal charge is given only when not zero.

a. (O=P—O with O above)⁻ ⟷ (O—P=O with O above)⁻ ⟷ (O—P—O with O double-bonded above)⁻

b. [O=N=O]⁺ (N, +1) c. H—C(H)(H)—Cl d. F—I(F)—F

2. Electron pair geometry is given first, followed by molecular shape, abbreviated as in the answers for objective 10-10.

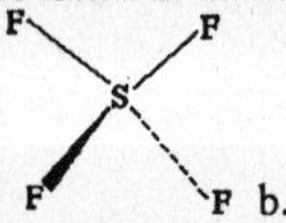

a. tet, tet b. tet, tripyr, M c. tbp, irtet, W d. oct, sqpyr, O (For questions b, c, and d, the sketches are given in the answers to objectives 10-12 and 10-13, to which the capital letters [M, W, O] refer.)

3. $CH_4(g) + 2\ O_2(g) \longrightarrow CO_2(g) + 2\ H_2O(g)$ (C O.S. = –4 in CH_4) Bonds broken: 4 C—H + 2 O=O = 4(414) + 2(498) = 2642 kJ Bonds formed: 2 C=O + 4 H—O = 2(707) + 4 (464) = 3270 kJ. Energy/g = (2642 – 3270)kJ/16 g = –38.6 kJ/g. $2\ CH_3OH(g) + 3\ O_2(g) \longrightarrow 2\ CO_2(g) + 4\ H_2O(g)$ (C O.S. = –2 in CH_3OH) Bonds broken: 6 C—H + 2 C—O + 2 O—H + 3 O=O = 6(414) + 2(360) + 2(464) + 3(498) = 5626 kJ/mol. Bonds formed: 4 C=O + 8 H—O = 4(707) + 8(464) = 6540 kJ/mol. Energy/g = (5626 – 6540)kJ/64.0 g = –14.3 kJ/g. $H_2CO(g) + O_2(g) \longrightarrow CO_2(g) + H_2O(g)$ (C O.S. = 0 in H_2CO) Bonds broken: 2 C—H + 1 C=O + 1 O=O = 2(414) + 1(707) + 1(498) = 2033 kJ. Bonds formed: : 2 C_O + 2 H—O = 2 (707) + 2 (464) = 2342 kJ. Energy/g = (2033 – 2342)kJ/30.0 g = –10.3 kJ/g. $2\ HCOOH(g) + O_2(g) \longrightarrow 2\ CO_2(g) + 2\ H_2O(g)$ (C O.S. = +2 in HCOOH) Bonds broken: 2 H—O + 2 H—C + 2 C=O + 2 C—O + O=O = 2(464) + 2(414) + 2(707) + 2(360) + 1(498) + 4388 kJ Bonds formed: 4 C=O + 4 H—O = 4(707) + 4(464) = 4684 kJ. Energy/g = (4388 – 4684)kJ/92.0 g = –3.22 kJ/g. The more negative the oxidation state of carbon, the more energy given off per gram of combusted material.

CHAPTER 11

Drill Problems

Abbreviations are the same as used in answers for objective 10-10.

2. (1) A. *sp*, lin, lin B. sp^3, tet, tet C. sp^2, tripl, bent D. sp^3, tet, tripyr E. sp^3, tet, tripyr F. sp^3, tet, bent G. sp^3, tet, bent H. sp^3, tet, bent I. sp^2, tripl, tripl J. sp^2, tripl, bent K. sp^3, tet, tet L. sp^3, tet, tet M. sp^3, tet, tripyr N. sp^3, tet, tet O. sp^2, tripl, tripl P. sp^3, tet, bent
(2) Q. sp^3d, tbp, irtet R. sp^3d^2, oct, oct S. sp^3d, tbp, T T. sp^3d, tbp, tbp U. sp^3d, tbp, T V. sp^3d, tbp, lin W. sp^3d^2, oct, sqpyr X. sp^3d, tbp, lin Y. sp^3d, tbp, tbp Z. sp^3d, tbp, lin Γ. sp^3d, tbp, tbp Δ. sp^3d^2, oct, sqpyr Θ. sp^3d^2, oct, sqpl Λ. sp^3d, tbp, irtet Ξ. sp^3d^2, oct, sqpyr Π. sp^3d, tbp, irtet Σ. sp^3d^2, oct, sqpl Υ. sp^3d, tbp, T Φ. sp^3d^2, oct, oct Ψ. sp^3d^2, oct, sqpl Ω. sp^3d^2, oct, oct

3. All bonds are σ bonds unless specified as π bonds. The number of bonds of a given type is given before the abbreviated description of the bond. "l.p." means lone pair; "l.e." means lone or unpaired electron. Only the lone pairs and lone electrons of the central atom are indicated. In the cases where there is some doubt, "hyb" precedes the description that assumes the central atom to be hybridized, and "unhyb" precedes the description in which the central atom is not assumed to be hybridized.
A. 2 σ Be(*sp*)–H(1*s*) B. 3 σ C(sp^3)–H(1*s*), l.e. in C(sp^3) C. 4 σ Kr(d^2sp^3)–F(2*p*), 1 l.p. in Kr(d^2sp^3) D. 4 σ B(sp^3)–F(2*p*) E. hyb: 2 σ Sn(sp^2)–Cl(3*p*), 1 l.p. in Sn(sp^2); unhyb: 2 σ Sn(5*p*)–Cl(3*p*), 1 l.p. in Sn(5*s*) F. 4 σ N(sp^3)–H(1*s*) G. 4 σ I(sp^3d^2)–Cl(3*p*), 2 l.p. in I(sp^3d^2) H. 3 σ C(sp^2)–H(1*s*) I. hyb: 2 σ S(sp^3)–I(5*p*), 2 l.p. in S(sp^3); unhyb: 2 σ S(3*p*)–I(5*p*), l.p. in S(3*p*), l.p. in S(3*s*) J. 4 σ Se(sp^3d)–F(2*p*), l.p. in Se(sp^3d) K. 2 σ Xe(sp^3d)–F(2*p*), 3 l.p. in Xe(sp^3d) L. hyb: 2 σ C(sp^2)–Cl(3*p*), l.p. in C(sp^2); unhyb: 2 σ C(2*p*)–Cl(3*p*), l.p. in C(2*s*) M. 2 σ N(sp^3)–H(1*s*), 2 l.p. in N(sp^3) N. 3 σ Br(sp^3d)–F(2p), l.p. in Br(sp^3d) O. 5 σ I(sp^3d^2)–F(2*p*), 1 l.p. in I(sp^3d^2) P. 2 σ Cl(sp^3)–O(2p), 2 l.p. in Cl(sp^3) Q. 3 σ B(sp^2)–Cl(3*p*) R. Sb(sp^3d)–Cl(3*p*) S. 6 σ Se(sp^3d^2)–F(2*p*) T. 3 σ O(sp^3)–H(1s), 1 l.p. in O(sp^3)

4. Refer to the paragraph at the beginning of the answers for objective 11-3. These bonding descriptions are based on the Lewis structure that is drawn first in each part.
A. $[N^a{=}C{=}N^b]^{2-}$ 2 σ: C(*sp*)–N(2*p*), π C($2p_y$)–N^a($2p_y$), π C($2p_z$)–Nb($2p_z$)
B. H^b—C(H^a)=S 3 σ: C(sp^2)–H^a(1*s*), C(sp^2)–H^b(1*s*), C(sp^2)–S($3p_y$); π C($2p_z$)–S($3p_z$)
C. Cl^b—C^a(Cl^a)=C(H^a)—H^b 5 σ: C^a(sp^2)–Cl^a(3*p*), C^a(sp^2)–Cl^b(3*p*), C^b(sp^2)–H^a(1*s*), C^b(sp^2)–H^b(1*s*), C^a(sp^2)–C^b(sp^2); π C^a($2p_z$)–C^b($2p_z$)
D. Cl^b—C(Cl^a)_S 3 σ: Cl^a(3*p*)–C(sp^2), Cl^b(3*p*)–C(sp^2), C(sp^2)–S(3*p*), π C($2p_z$)–S($3p_z$)
E. $[S{=}C{=}N]^-$ σ C(*sp*)–S($3p_y$), σ C(*sp*)–N($2p_z$), π C($2p_y$)–N($2p_y$), π C($2p_z$)–S($3p_z$)
F. N^a≡N^b—O σ N^a($2p_x$)–N^b(*sp*), σ N^b(*sp*)–O($2p_z$), π N^a($2p_y$)–N^b($2p_y$), π N^a($2p_z$)–N^b($2p_z$)

G. $\overline{O}^a{=}C{=}\overline{O}^b$ σ $O^a(2p_y)$–$C(sp)$, σ $O^b(2p_z)$–$C(sp)$, π $O^a(2p_z)$–$C(2p_z)$, π $O^b(2p_y)$–$C(2p_y)$
H. H—$\overline{O}$—C≡N| 3 σ: $H(1s)$–$O(sp^3)$, $O(sp^3)$–$C(sp)$, $C(sp)$–$N(2p_x)$; 2 l.p. in $O(sp^3)$; π $C(2p_y)$–$N(2p_y)$, π $C(2p_z)$–$N(2p_z)$
I. H—$\underline{N}{=}\overline{O}$ σ $H(1s)$–$N(sp^2)$, σ $N(sp^2)$–$O(2p_y)$, l.p. in $N(sp^2)$, π $N(2p_z)$–$O(2p_z)$
J. H—C≡N| σ $H(1s)$–$C(sp)$, σ $C(sp)$–$N(2p_x)$, π $C(2p_y)$–$N(2p_y)$, π $C(2p_z)$–$N(2p_z)$
K. H^a—C^a≡C^b—H^b 3 σ: $H^a(1s)$–$C^a(sp)$, $C^a(sp)$–$C^b(sp)$, $C^b(sp)$–$H^b(1s)$; π $C^a(2p_y)$–$C^b(2p_y)$, $C^a(2p_z)$–$C^b(2p_z)$
L. $[\overline{O}^a{=}N{=}\overline{O}^b]^+$ σ $O^a(2p_y)$–$N(sp)$, σ $N(sp)$–$O^b(2p_z)$, π $O^a(2p_z)$–$N(2p_z)$, π $N(2p_y)$–$O^b(2p_y)$
M. $\overline{O}{=}C{=}\overline{S}$ σ $O(2p_y)$–$C(sp)$, σ $C(sp)$–$S(2p_z)$, π $O(2p_z)$–$C(2p_z)$, π $C(2p_y)$–$S(2p_y)$
N. $[\overline{F}$—$\underline{P}{=}\overline{O}$ σ $F(2p_z)$–$P(sp^2)$, σ $P(sp^2)$–$O(2p_y)$, l.p. in $P(sp^2)$, π $P(3p_z)$–$O(2p_z)$

O. 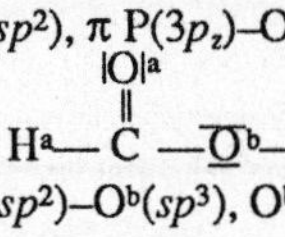 4 σ: $H^a(1s)$–$C(sp^2)$, $C(sp^2)$–$O^a(2p_y)$, $C(sp^2)$–$O^b(sp^3)$, $O^b(sp^3)$–$H^b(1s)$; 2 l.p. in $O^b(sp^3)$, π $C(2p_z)$–$O^a(2p_z)$

8. Simplified molecular orbital diagram is given first (in the order σ^b_{2s}, σ^*_{2s}, π^b_{2p}, σ^b_{2p}, π^*_{2p}, σ^*_{2p}), followed by answers (1)–(6).

		σ^b_{2s}	σ^*_{2s}	π^b_{2p}	σ^b_{2p}	π^*_{2p}	σ^*_{2p}	
A.	KK	[⇅]	[⇅]	[⇅][⇅]	[⇅]	[ ][ ]	[ ]	8, 2, 3, 6, 4, 0
B.	KK	[⇅]	[⇅]	[⇅][⇅]	[⇅]	[⇅][⇅]	[ ]	8, 6, 1, 6, 8, 0
C.	KK	[⇅]	[⇅]	[⇅][⇅]	[⇅]	[↑][⇅]	[ ]	8, 3, 5/2, 6, 5, 1
D.	KK	[⇅]	[⇅]	[⇅][⇅]	[ ]	[ ][ ]	[ ]	6, 2, 2, 4, 4, 0
E.	KK	[⇅]	[⇅]	[⇅][⇅]	[ ]	[ ][ ]	[ ]	6, 2, 2, 4, 4, 0
F.	KK	[⇅]	[⇅]	[⇅][⇅]	[ ]	[ ][ ]	[ ]	6, 2, 2, 4, 4, 0
G.	KK	[⇅]	[↑]	[ ][ ]	[ ]	[ ][ ]	[ ]	2, 1, 1/2, 3, 0, 1
H.	KK	[⇅]	[⇅]	[⇅][⇅]	[↑]	[ ][ ]	[ ]	7, 2, 5/2, 5, 4, 1
I.	KK	[⇅]	[⇅]	[⇅][⇅]	[⇅]	[⇅][⇅]	[↑]	8, 7, 1/2, 7, 8, 1
J.	KK	[⇅]	[⇅]	[⇅][⇅]	[⇅]	[⇅][⇅]	[↑]	8, 7, 1/2, 7, 8, 1
K.	KK	[⇅]	[⇅]	[⇅][⇅]	[⇅]	[⇅][↑]	[ ]	8, 5, 3/2, 6, 7, 1
L.	KK	[⇅]	[⇅]	[⇅][⇅]	[⇅]	[↑][↑]	[ ]	8, 4, 2, 6, 6, 2
M.	KK	[⇅]	[⇅]	[ ][ ]	[ ]	[ ][ ]	[ ]	2, 2, 0, 4, 0, 0

10. Refer to the paragraph at the beginning of objective 11-3. These descriptions are based on the resonance structure drawn first in each part.
A. $\overline{O}^a{=}\underline{N}$—$\overline{O}\cdot^b$ ←→ ·$\overline{O}$—$\underline{N}{=}\overline{O}$ σ $O^a(2p_y)$–$N(sp^2)$, σ $N(sp^2)$–$O^b(2p_y)$, l.p. in $N(sp^2)$, π $O^a(2p_z)$–$N(2p_z)$–$O^b(2p_z)$ (3–eln, 3–center π bond)

B. 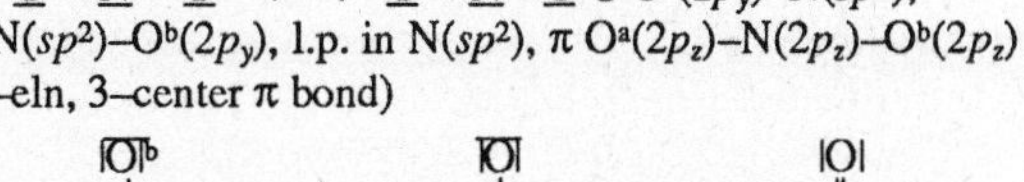
3 σ: $S(sp^2)$–$O^a(2p_y)$, $S(sp^2)$–$O^b(2p_y)$, $S(sp^2)$–$O^c(2p_y)$, π $S(3p_z)$–$O^a(2p_z)$–$O^b(2p_z)$–$O^c(2p_z)$ (6–electron, 4–center π bond)

C. 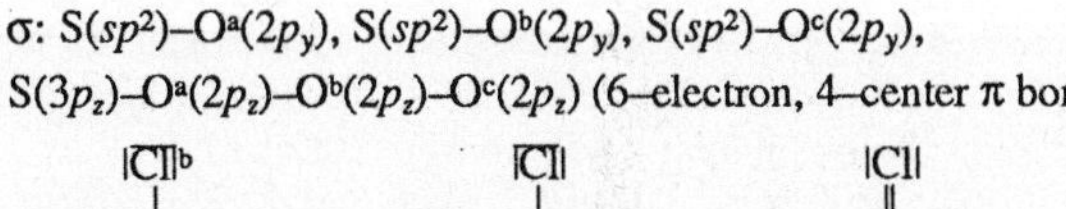
3 σ: $B(sp^2)$–$Cl^a(3p_y)$, $B(sp^2)$–$Cl^b(3p_y)$, $B(sp^2)$–$Cl^c(3p_y)$, π $B(2p_z)$–$Cl^a(3p_z)$–$Cl^b(3p_z)$–$Cl^c(3p_z)$ (6–electron, 4–center π bond)

D. $\overline{O}^a{=}O^b$—$\overline{O}|^c$ ←→ $|\overline{O}$—$O{=}\overline{O}$ σ $O^a(2p_y)$–$O^b(sp^2)$, σ $O^b(sp^2)$–$O^c(2p_y)$, l.p. in $O^b(sp^2)$, π $O^a(2p_z)$–$O^b(2p_z)$–$O^c(2p_z)$ (4–center, 3– electron π bond)

E. 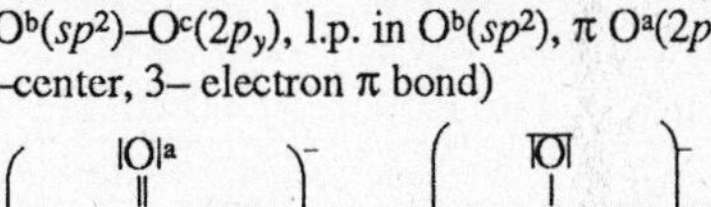 3 σ: $H(1s)$–$C(sp^2)$, $C(sp^2)$–$O^a(2p_y)$, $C(sp^2)$–$O^b(2p_y)$; π $C(2p_z)$–$O^a(2p_z)$–$O^b(2p_z)$ (4–center, 3–electron π bond)

F. $\overline{O}{=}\underline{C}$—$\overline{F}|^c$ ←→ $|\overline{O}$—$\underline{C}{=}\overline{F}$ σ $O(2p_y)$–$C(sp^2)$, σ $C(sp^2)$–$F(2p_y)$, l.p. in $C(sp^2)$, π $O(2p_z)$–$C(2p_z)$–$F(2p_z)$ (4–center, 3–electron π bond)

G. 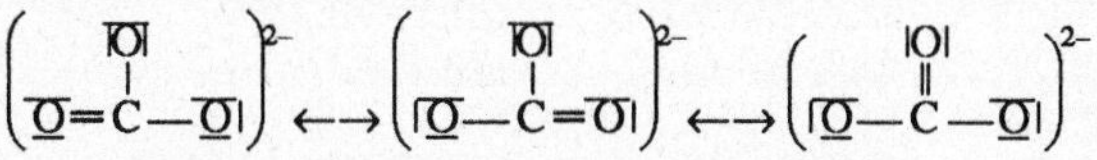
σ $O^a(2p_y)$–$C(sp^2)$, σ $O^b(2p_y)$–$C(sp^2)$, σ $O^c(2p_y)$–$C(sp^2)$
H. Same as BCl_3 except $2p$ orbitals used on ligands.

Quizzes

A: 1. (c) 2. (c) 3. (b) 4. (a) 5. (b) 6. (a) 7. (c) 8. (e)
B: 1. (a) 2. (b) 3. (b) 4. (b) 5. (b) 6. (b) 7. (a) 8. (c)
C: 1. (a) 2. (b) 3. (d) 4. (b) 5. (b) 6. (b) 7. (d) 8. (a)
D: 1. (a) 2. (a) 3. (c) 4. (d) 5. (c) 6. (c) 7. (d) 8. (c)

Sample Test

1. a. 2, 3, 2 b. 0, 0, 2 c. linear, triangular planar, tetrahedral d. linear, triangular planar, bent e. none, none sp^3 f. sp, sp^2, sp^3 g. $N(2p_x)$–$C^a(sp)$, $C^a(sp)$–$C^b(sp^2)$, $C^b(sp^2)$–$S(3p_x)$, $C^b(sp^2)$–$O(sp^3)$, $O(sp^3)$–$H(1s)$ h. $N(2p_y)$–$C^a(2p_y)$, $N(2p_z)$–$C^a(2p_z)$, $C^b(2p_z)$–$S(3p_y)$ i. 180° j. 120° k. 120° l. 109.5°
2. a. 5/2 b. NO^+ as it has a bond order of 3 compared to 5/2 for O_2^+ c. 1, 2, 0

CHAPTER 12

Drill Problems

2. A. 200. g × mol/32.0 g × 0.0816 kJ mol^{-1} K^{-1} × (64.96°C – 50.0°C) = 7.63 kJ; (184.7 kJ – 7.63 kJ) × mol/39.2 kJ × 32.0 g/mol = 145 g vaporizes; m_l = 55 g; t_f = 64.96°C B. q = 204 g × mol/46.0 g × [0.113 kJ mol^{-1} K^{-1} (78.5°C – 20.2°C) + 40.5 kJ/mol + 0.0657 kJ mol^{-1} K^{-1} (100.0°C – 78.5°C)] = 215 kJ. C. 186 g × mol/64.1 g × 0.0852 kJ mol^{-1} K^{-1} (–10.0°C + 51.0°C) + (186 g – 46 g) mol/64.1 g × 26.8 kJ/mol = 68.7 kJ; D. 20.2 kJ = m × mol/27.0 g [0.0707 kJ mol^{-1} K^{-1} (26.0°C + 14.0°C) + 30.7 kJ/mol + 0.0360 kJ mol^{-1} K^{-1} (36.5°C – 26.0°C)]; m = 16.1 g E. same method as A; t_f = 118.5°C, m_l = 177 g F. same method as B; q = 126 g G. same method as A; t_f = 34.6°C,.m_l =54.5 g H. q = 14.0 g × mol/154.0 g × 0.133 kJ mol^{-1} K^{-1} (76.8°C – 25.0°C) = 0.626 kJ; (16.4 – 0.626 kJ) × mol/{[0.133 kJ(76.8°C – 25.0°C)] + 34.4 kJ} × 154.0 g/mol = 58.9 g; total mass = 58.9 g + 72.9 g I. 314 g × mol/119.5 g × 0.115 kJ mol^{-1} K^{-1}(t_f – 0.0°C) = 14.3 kJ; t_f = 47.3°C J. Same method as B; 92.1 kJ K. Same method as C; 230 kJ; L. t_f= 58.78°C: q = 124 g × mol/159.8 g × 0.0757 kJ mol^{-1} K^{-1}(58.78°C – 20.40°C) + 100. g × mol/159.8 g × 33.9 kJ/mol = 23.5 kJ M. Same method as I; t_f = –34.35°C

4. (1) A. 8.43 mmHg B. 35.0 mmHg C. 94.6 mmHg D. 45.0 mmHg E. 101 mmHg F. 8.01 mmHg G. 1.10 mmHg (2) H. 450 mmHg, 6.8 g liquid I. 190 mmHg, 2.3 g liquid J. 370 mmHg, no liquid K. 110 mmHg, 2.4 g liquid L. 570 mmHg, no liquid M. 400 mmHg, 4.6 g liquid N. 650 mmHg, no liquid

5. (1) A. 31.4 kJ/mol, 41.7°C B. 27.4 kJ/mol, 5.0°C C. 50.2 kJ/mol, 149°C D. 46.0 kJ/mol, 139.7°C E. 32.1 kJ/mol, 70.3°C F. 19.6 kJ/mol, –98.9°C G. 78.5 kJ/mol, 678°C
(2) H. 113 mmHg, –9.3°C I. 1386 mmHg, 31°C J. 143 mmHg. –15°C K. 315 mmHg, 36°C L. 6.24 mmHg, 47°C M. 25.9 mmHg, 37°C N. 237 mmHg, –54°C O. 1087 mmHg, 17°C

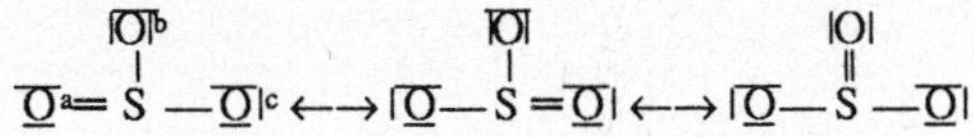

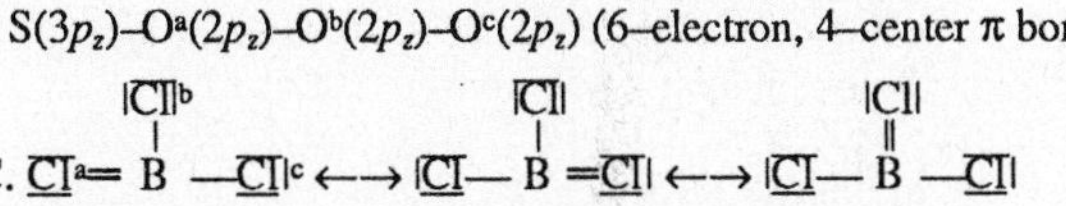

7. The enthalpies (in kJ/mol) are, in order ΔH_{fus}, ΔH_{vap}, ΔH_{sub}, ΔH_{sol}, ΔH_{cond}, ΔH_{dep}.
A. 6.01, 44.0, 50.0, –6.01, –44.0, –50.0
B. 10.7, 232.8, 243.5, –10.7, –232.8, –243.5
C. 27.2, 176.9, 204.1, –27.2, –176.9, –204.1
D. 47.2, 181.8, 229.0, –47.2, –181.8, –229.0
E. 201., 158.6, 178.7, –20.1, –158.6, –178.7
F. 19.0, 41.4, 60.4, –19.0, –41.4, –60.4
G. 56.6, 668.5, 725.1, –56.6, –668.5, –725.1

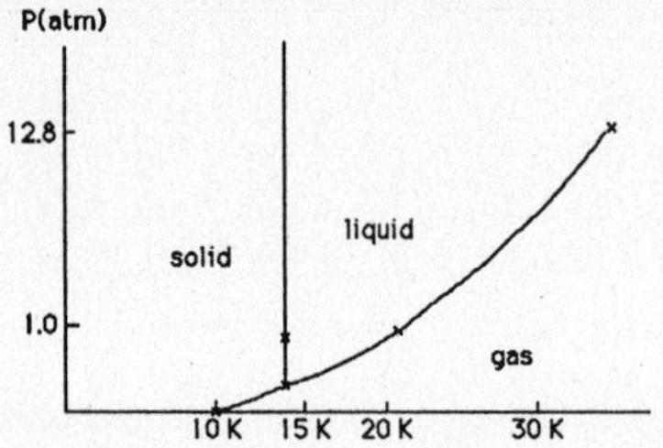

8. A.
B. Yes, at about 14.00 K C. No; the critical pressure is 12.8 atm. D. No; the solid-liquid line slopes upward to the right.

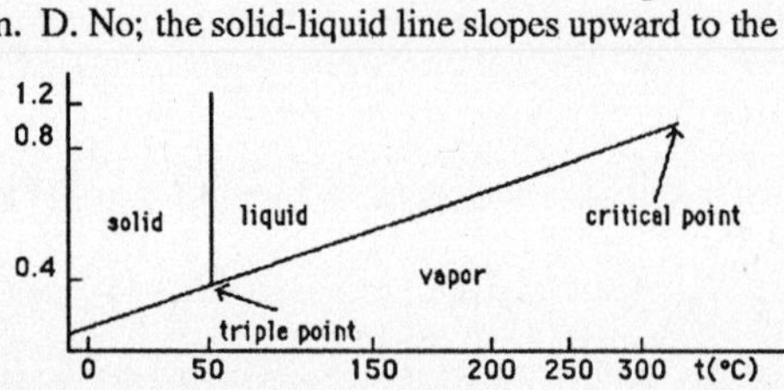

E.
F. Yes, about 49.6°C G. Yes, about 218°C H. Yes; the solid-liquid line slopes upward to the left. I. Begins as a vapor; turns solid at about 5 mmHg, turns liquid between 300 and 400 mmHg and remains liquid. J. Begins as a solid; turns liquid at about 273.15 K; vaporizes at 131°C (from the Clausius-Clapeyron equation). K. Begins as a solid; sublimes at about –70°C. L. Begins as a solid; melts at about –56.7°C; vaporizes at perhaps 0°C. M. Begins as a solid; melts at about –55°C; never vaporizes as 80.0 atm is above the critical pressure. N. Begins as a solid; melts at 14.01 K; boils at 20.38 K. O. Starts as a vapor; becomes solid at 0.07 atm; turns liquid at about 0.8 atm and remains so. P. Begins as a vapor; becomes a solid at 0.40 atm; turns liquid at about 1.40 atm and remains so. Q. Begins as a solid; melts at about 49.6°C; vaporizes at about 225°C.
9. A. HI, Xe B. PF_3, BCl_3 C. HBr, HCl D. CH_4, GeH_4 E. PF_3, AlF_3 F. CH_3OH, C_3H_8OH G. H_2O, HF H. I_2, Cl_2 I. CH_4, CCl_4 J. CCl_3F, Br_2
12. A. SrS, MgS B. NaH, NaBr C. SrI_2, $CaCl_2$ D. MgS, Na_2S E. $CaCl_2$, CaO F. $Al_2(CO_3)_3$, Na_2CO_3 G. LiF, LiI H. MgS, K_2S I. $Pb(NO_3)_2$, $Mg(NO_3)_2$ J. RbCl, $InCl_3$

Quizzes

A: 1. (e) 2. (b) 3. (e) 4. (b) 5. (b) 6. (d) 7. (a) 8. (e)
B: 1. (b) 2. (e) 3. (d) 4. (d) 5. (a) 6. (e) 7. (c) 8. (b)
C: 1. (c) 2. (b) 3. (a) 4. (d) 5. (e) 6. (a) 7. (a) 8. (b)
D: 1. (c) 2. (b) 3. (b) 4. (e) 5. (c) 6. (d) 7. (e) 8. (e)

Sample Test

1. $-8.314 \frac{J}{mol \cdot K} \ln \frac{50.0}{760.0} = 32.0 \times 10^3 \left(\frac{1}{T} - \frac{1}{329.4}\right)$;
$T = -6.0°C$
2. See Figure 12-1.
3. Review objective 12-9.
4. MgO. The lattice energy depends on $Z_A Z_C/(r_A + r_C)$. Large lattice energy means highly charged ions with the smallest radii.

CHAPTER 13

Drill Problems

2. A. 30.00 cm³, 8.91 cm³, 21.80 g, 31.16%, 29.70% B. 50.00 g, 0.9912 g/cm³, 5.00%, 4.96%, C. 4.29 cm³, 40.13 g, 9.11%, 9.13% D. 91.22 cm³, 48.00 g, 120.00 g, 60.00%, 67.81% E. 6.995 cm³, 0.870 cm³, 6.18 g, 9.81%, 12.43% F. 60.00 g, 1.0094 g/cm³, 9.00%, 9.08%, 8.19% G. 4.35 cm³, 120.46 g, 4.00%, 4.02% H. 39.34 cm³, 102.00 g, 150.00 g, 32.00%, 28.51% I. 36.00 cm³, 7.50 cm³, 28.96 g, 25.41%, 20.83% J. 100.00 g, 1.044 g/cm³, 19.00%, 19.84%, 15.73% K. 31.99 cm³, 84.07 g, 48.00%, 53.80% L. 74.30 cm³, 11.20 g, 70.00 g, 84.00%, 88.80% M. 71.40 cm³, 53.75 cm³, 20.02 g, 59.58%, 75.29%
3. A. 0.8535 mol, 4.163 *m*, 19.99% B. 53.5 g/mol, 0.667 mol, 1.406 *m* C. 0.224 mol, 1.156 g/mL, 0.9391 *m* D. 3.175 M, 3.383 *m*, 10.98% E. 42.40 g/mol, 500.0 mL, 2.620 *m* F. 200. mL, 1.17 *m*, 28.0% G. 42.5 g/mol, 0.963 M, 0.981 *m* H. 0.2128 mol, 1.065 g/mL, 1.110 *m* I. 0.9768 mol, 1.302 M, 1.368 *m* J. 74.54 g/mol, 0.3084 mol, 0.9326 *m* K. 0.2007 mol, 0.7434 M, 0.7598 *m* L. 165.7 mL, 1.29 *m*, 20.0% M. 3.782 mol, 1.261 g/mL, 6.060 *m* N. 5.210 mol, 32.31 *m*, 76.06% O. 420. mL, 14.3 M, 16.7 *m* P. 58.44 g/mol, 6.012 *m*, 1.598 mol Q. 400.0 mL, 6.803 *m*, 40.00% R. 0.205 mol, 0.8548 M, 0.8873 *m*
4. A. 0.06977, 93.023% B. 0.0247, 97.53% C. 0.0166, 98.34% D. 0.05744, 94.256% E. 0.04507, 95.493% F. 0.0208, 97.92% G. 0.0172, 98.28% H. 0.01959, 98.041% I. 0.02404, 97.596% J. 0.0165, 98.35% K. 0.0135, 98.65% L. 0.255, 77.5% M. 0.09846, 90.154% N. 0.3677, 63.23% O. 0.231, 76.9% P. 0.09769, 90.231 Q. 0.1092, 89.08% R. 0.01573, 98.427%
6.. A. 0.0355 *m* B. 7.67 atm C. 0.000234 *m* D. 0.493 atm E. 0.0144 *m* F. 0.457 atm G. 0.00278 *m* H. 4.58 atm I. 0.0854 *m* J. 1.17 atm
8. A. $\chi_B = 0.25$, $P = 87.5$ B. $\chi_A = 0.542$, $\chi_B = 0.458$ C. $P_A = 340$, $\chi_B = 0.75$ D. $P_B = 254$, $\chi_B = 0.70$ E. $P_A = 125$, $\chi_A = 0.86$ F. $\chi_A = 0.65$, $P = 218$ G. $\chi_A = 0.125$, $\chi_B = 0.875$ H. $\chi_A = 0.200$, $\chi_B = 0.800$
11. A. 106 g/mol, 15.83°C B. 4.88°C/*m*, 0.217 *m*, 80.64°C C. 7.00°C/*m*, 23.1 g, 212.09°C D. 3.56°C/*m*, 83.8 g/mol, 40.91°C E. 0.325 *m*, 100.17°C F. 14.2 g, 0.369 *m*, –0.69°C
12. A. 0.0398 M, 0.749 g B. 0.00740 M, 512 g/mol C. 488 mmHg, 0.0269 M D. 0.0808 M, 7.75 g E. 0.0258 M, 1214 g/mol F. 579 mmHg, 0.0322 M G. 0.0968 M, 8.02 g H. 0.0275 M, 99.8 g/mol I. 455 mmHg, 0.0241 M J. 0.0256 M, 1.73 g K. 0.0414 M, 164 g/mol L. 351 mmHg, 0.0187 M M. 0.0524 M, 1.04 g N. 0.0350 M, 154 g/mol O. 47.2 mmHg, 0.00250 M
13. A. $[Li^+] = 0.0709$ M, $[Ca^{2+}] = 0.436$ M, $[NO_3^-] = 0.943$ M
B. $[Cl^-] = 0.0788$ M, $[SO_4^{2-}] = 0.0966$ M, $[H^+] = 0.2720$ M
C. $[Na^+] = 0.124$ M, $[Al^{3+}] = 0.305$ M, $[Br^-] = 1.040$ M
D. $[Cl^-] = 0.0611$ M, $[PO_4^{3-}] = 0.0916$ M, $[Na^+] = 0.3360$ M
E. $[H^+] = 0.829$ M, $[Na^+] = 0.457$ M, $[NO_3^-] = 1.286$ M
F. $[Ca^{2+}] = 0.0894$ M, $[Mg^{2+}] = 0.120$ M, $[Cl^-] = 0.420$ M
G. $[Na^+] = 0.112$ M, $[Al^{3+}] = 0.144$ M, $[SO_4^{2-}] = 0.272$ M
H. $[SO_4^{2-}] = 0.0946$ M, $[PO_4^{3-}] = 0.184$ M, $[Na^+] = 0.741$ M
I. $[Br^-] = 0.308$ M, $[Cl^-] = 0.336$ M, $[Na^+] = 0.644$ M
J. $[Cl^-] = 0.536$ M, $[SO_4^{2-}] = 0.0848$ M, $[Li^+] = 0.706$ M

Quizzes

A: 1. (e) 2. (e) (concentrated) 3. (a) 4. (a) 5. (d) 6. (d) 7. (e) 8. (a)
B: 1. (b) 2. (a) 3. (c) 4. (e) 5. (e) (8/100) 6. (c) 7. (e) 8. (b)
C: 1. (e) 2. (d) 3. (d) 4. (c) 5. (e) 6. (e) 7. (a) 8. (c)
D: 1. (b) 2. (c) 3. (d) 4. (e) 5. (c) 6. (d) 7. (d) 8. (a)

Sample Test

1. a. (750 g + 85 g)/810 mL = 1.03 g/mL
b. [86 g/(750 g + 85 g)] × 100% = 10.2%
c. $\dfrac{85.0\text{ g} \times \text{mol}/180.0\text{ g}}{985.0\text{ g} \times \text{mol}/180.0\text{ g}} + (750\text{ g} \times \text{mol}/46.0\text{ g})) = 0.0281$
$= \chi_{sucrose}$
d. $\dfrac{875.0\text{ g} \times \text{mol}/180.0\text{g}}{750\text{ g} \times \text{kg}/1000\text{ g}} = 0.630\ m$
e. $\dfrac{85.0\text{ g} \times \text{mol}/180.0\text{ g}}{810\text{ mL} \times \text{L}/1000\text{ mL}} = 0.583\text{ M}$
2. 914.0°F – 32)(5/9) = –10.°C; $\Delta t_f = 10.0°C = K_F m$;
$m = 10.0°C/(1.86\ °C/m) = 5.38\ m$
mols of solute = 20.0 L H_2O × (1.00 kg/L) (5.38 m) = 108 mol
$108\text{ mol} \times \dfrac{62.0\text{ g}}{\text{mol}} \times \dfrac{\text{cm}^3}{1.12\text{ g}} \times \dfrac{\text{L}}{1000\text{ cm}^3}$
= 5.95 L ethylene glycol

CHAPTER 14

Drill Problems

1. A. Ba<Ca<Mg<C B. Br<Se<Ca<C C. $Be^{2+}<Li^{+}<K^{+}<Cs^{+}$ D. $Cs^{+}<Ca^{2+}<Mg^{2+}<Al^{3+}$ E. Cs<Sr<Al<F F. Na<Mg<S<I G. Cs<Rb<Ca<Al H. Cs<Ca<Mg<Al predicted, Cs<Mg<Al<Ca actual I. Cs<Rb<Na<Mg J. $Si^{4+}<Al^{3+}<Mg^{2+}<Li^{+}$ K. Br<I<Sr<Cs L. $Al^{3+}<Mg^{2+}<Na^{+}<K^{+}$ M. Cs<Rb<Na<Li predicted, Cs<Rb<Li<Na actual N. Cs<Sr<Ca<Mg predicted, Cs<Mg<Sr<Ca actual O. C<P<S<I predicted, P<S<C<I actual (Notice how often predictions are incorrect because a second period element is "out of place.")
2. The predicted value is given first, followed by the actual value in parentheses. (1) A. 74°C (–129°C) B. yellow-green (yellow-green) C. 330°C (302°C) D. 411°C (337°C) E. 1000°C (unknown) F. black (unknown) G. –170°C (–185°C) H. 15.6 kJ/mol (15.1 kJ/mol) I. –64°C (–52°C) J. –70°C (100°C) K. –97°C (0°C) L. –150.5°C (–157.3°C) M. –146.5°C (–152°C)
(2) N. 14.7°C (28.7°C) O. (602°C (690°C) P. 2.1 g/cm³ (1.90 g/cm³) Q. 4.7 g/cm³ (4.25 g/cm³)
6. A. Mg, B, Al, Be B. $Na_6P_6O_{18}$, Na_2CO_3, Na_4SiO_4, borax, silicates C. Na, Cs, Be, Al, Si, Li D. NaOH, Na_2SO_4, KOH E. Li_2CO_3, $NaHCO_3$, MgO, H_3BO_3, I_2, $MgSO_4$, N_2O, H_2O_2 F. Na_4SiO_4, $MgSO_4$, $Al_2(SO_4)_3$, alum G. KNO_3, HNO_3, NH_4NO_3, S, P H. NaCl, SO_2, NH_3 I. KCl, $NaNO_3$, $Ca(NO_3)_2{\cdot}4H_2O$, $Ca(H_2PO_4)_2$, NH_3, NH_4NO_3, HNO_3 J. MgO, Al_2O_3, CaO K. KNO_3, Sr compounds, S, $KClO_3$ L. $CaCO_3$, $BaSO_4$, Al M. $LiAlH_4$, Na, Al N. alum, O_3, NaF, Cl_2 O. KNO_3, $KMnO_4$, $KClO_3$, bromates, O_3, HNO_3 P. NaCN, H_2O_2, NaF, Cl_2 Q. Na_2SO_4, alum, HNO_3, H_3PO_4, KNO_3 R. H_3BO_3, Br_2, alum S. NaOH, Na_2SO_4, Na_2O_2, sulfates, $Na_2S_2O_3$, Cl_2, $CaCO_3$ T. CaO, $CaSO_4{\cdot}2H_2O$, SiO_2 U. Na_2CO_3, Na_2SO_4, K_2CO_3, B_2O_3, SiO_2, $CaCO_3$ V. B_4C_3, Al_2O_3, SiO_2< SiC W. Sizing is a material used to fill in the pores of surfaces of paper, textiles, or leather.
6. & 7. A. $K(s) + O_2(g) \longrightarrow KO_2(g)$ B. $2\,Al(s) + 3\,F_2(g) \longrightarrow 2\,AlF_3(s)$ C. $Ca(s) + 2\,H_2O(l) \longrightarrow Ca(OH)_2(aq) + H_2(g)$ D. $Mg(s) + I_2(s) \longrightarrow MgI_2(s)$ E. $CaO(s) + H_2O(l) \longrightarrow Ca(OH)_2$(aq *or* s) F. $2\,Na(s) + O_2(g) \longrightarrow Na_2O_2(s)$ G. $2\,Na(s) + 2\,H_2O(l) \longrightarrow 2\,NaOH(aq) + H_2(g)$ H. $Mg(s) + 2\,HCl(aq) \longrightarrow MgCl_2(aq) + H_2(g)$ I. $Mg(s) + H_2O(l, hot) \longrightarrow MgO(s) + H_2(g)$ J. $Na(s) + N_2(g) \longrightarrow$ no reaction K. $2\,Ca(s) + O_2(g) \longrightarrow 2\,CaO(s)$ L. $NaF(aq) + Cl_2(g) \longrightarrow$ no reaction M. $H_2(g) + F_2(g) \longrightarrow 2\,HF(g)$ N. $Cl_2(g) + H_2O(l) \longrightarrow HCl(aq) + HClO(aq)$ O. $Mg(s) + H_2(g) \longrightarrow$ no reaction P. $6\,Li(s) + N_2(g) \longrightarrow 2\,Li_3N(s)$ Q. $4\,Cl_2(g) + S_8(s) \longrightarrow 4\,S_2Cl_2(g)$ R. $4\,Li(s) + O_2(g) \longrightarrow 2\,Li_2O(s)$ S. $Be(s) + H_2(g) \longrightarrow$ no reaction T. $8\,Cl_2(g) + 8\,H_2S(g) \longrightarrow 16\,HCl(g) + S_8(s)$ U. $2\,F_2(g) + 2\,H_2O(l) \longrightarrow 4\,HF(aq) + O_2(g)$ V. $2\,Na(s) + Cl_2(g) \longrightarrow 2\,NaCl(s)$ W. $Li(s) + H_2(g) \longrightarrow$ no reaction.

Quizzes

A: 1. (c) 2. (d) 3. (a) 4. (a) 5. (e) 6. (c) 7. (c) 8. (b)
B: 1. (c) 2. (a) 3. (a) 4. (d) 5. (c) 6. (a) 7. (a) 8. (a)
C: 1. (b) 2. (a) 3. (a) 4. (c) 5. (c) 6. (a) 7. (c) 8. (e)
D: 1. (c) 2. (c) 3. (e) 4. (e) 5. (e) 6. (c) 7. (d) 8. (d)

Syntheses

Only one possible sequence of reactions is given, except for the second question in which several are provided to give some indication of the possible diversity of answers. ESTD means "evaporate solution to dryness.
1. $2\,Na(s) + O_2(g) \longrightarrow Na_2O_2(s)$ **2.** $2\,Na(s) + 2\,H_2O(l) \longrightarrow 2\,NaOH(aq) + H_2(g)$; $NaOH(aq) + HCl(aq) \longrightarrow NaCl(aq) + H_2O(l)$ ESTD. *or* $2\,Na(s) + 2\,HCl(aq) \longrightarrow 2\,NaCl(aq) + H_2(g)$, but this would be a very violent reaction. *or* $2\,HCl(aq) \xrightarrow{\text{electricity}} H_2(g) + Cl_2(g)$; gases separate naturally. Cl_2 at anode, H_2 at cathode; then $2\,Na(s) + Cl_2(g) \longrightarrow 2\,NaCl(s)$ (pretty violent) **3.** $2\,Li(s) + 2\,H_2O(l) \longrightarrow 2\,LiOH(s) + H_2(g)$ ESTD carefully; $2\,LiOH(s) \longrightarrow Li_2O(s) + H_2O(g)$; $Li_2O(s) + CO_2(g) \longrightarrow Li_2CO_3(s)$ **4.** $Ca(s) + 2\,H_2O(l) \longrightarrow Ca(OH)_2(aq) + H_2(g)$; $Ca(OH)_2(aq) + H_2SO_4(aq) \longrightarrow CaSO_4(s) + 2\,H_2O(l)$ filter off solid **5.** $2\,Al(s) + 3\,H_2SO_4(aq) \longrightarrow Al_2(SO_4)_3(aq) + 3\,H_2(g)$ ESTD **6.** $Mg(s) + H_2O(l) \xrightarrow{\Delta} MgO(s) + H_2(g)$ **7.** $Li_2CO_3(s) \xrightarrow{\Delta} Li_2O(s) + CO_2(g)$; $Na_2O(s) + CO_2(g) \longrightarrow Na_2CO_3(s)$ **8.** $CaCO_3(s) \xrightarrow{\Delta} CaO(s) + CO_2(g)$; $CaO(s) + H_2O(l) \longrightarrow Ca(OH)_2(aq)$ ESTD carefully **9.** $2\,Ca(s) + O_2(g) \longrightarrow 2\,CaO(s)$ at low temperatures; $C(s) + O_2(g) \longrightarrow CO_2(g)$; $CaO(s) + CO_2(g) \longrightarrow CaCO_3(s)$
10. $2\,Ca(s) + O_2(g) \longrightarrow 2\,CaO(s)$ at low temperatures; $S(s) + O_2(g) \longrightarrow SO_2(g)$; $CaO(s) + SO_2(g) \longrightarrow CaSO_3(s)$
11. $2\,Na(s) + Cl_2(g) \longrightarrow 2\,NaCl(s)$; $3\,NaCl(s) + H_3PO_4(aq) \longrightarrow Na_3PO_4(aq) + 3\,HCl(g)$. $H_2SO_4(aq)$ also could be used as the strong acid. **12.** $2\,H_2(g) + O_2(g) \longrightarrow 2\,H_2O(l)$; $P_4(s) + 5\,O_2(g) \longrightarrow P_4O_{10}(s)$; $P_4O_{10}(s) + 6\,H_2O(l) \longrightarrow 4\,H_3PO_4(aq)$ **13.** $6\,Li(s) + N_2(g) \longrightarrow 2\,Li_3N(s)$; $Li_3N(s) + 3\,H_2O(l) \longrightarrow 3\,LiOH(aq) + NH_3(g)$
14. $Ca(s) + 2\,H_2O(l) \longrightarrow Ca(OH)_2(aq) + H_2(g)$
15. $Ca(s) + 2\,H_2O(l) \longrightarrow Ca(OH)_2(aq) + H_2(g)$ ESTD; $Ca(OH)_2(aq) \xrightarrow{\Delta} CaO(s) + H_2O(g)$; $C(s) + O_2(g) \longrightarrow CO_2(g)$; $CaO(s) + CO_2(g) \longrightarrow CaCO_3(s)$ **16.** $Ba(s) + O_2(g) \xrightarrow{600°C} BaO_2(s)$; $BaO_2(s) + H_2SO_4(aq) \longrightarrow BaSO_4(s) + H_2O_2(aq)$
17. $2\,H_2(g) + O_2(g) \longrightarrow 2\,H_2O(l)$; $2\,Li(s) + 2\,H_2O(l) \longrightarrow 2\,LiOH(aq) + H_2(g)$; $LiOH(aq) + HNO_3(aq) \longrightarrow$

$LiNO_3(aq) + H_2O(l)$ ESTD **18.** $Ca(s) + S(s) \longrightarrow CaS(s)$; $CaS(s) + 2\ HCl(aq) \longrightarrow CaCl_2(aq) + H_2S(g)$
19. $Mg(s) + Cl_2(g) \longrightarrow MgCl_2(s)$; $3\ MgCl_2(s) + 2\ H_3PO_4(aq) \longrightarrow Mg_3(PO_4)_2(aq) + 6\ HCl(g)$ **20.** $P_4(s) + 6\ Br_2(l) \longrightarrow 4\ PBr_3(l)$; $2\ H_2(g) + O_2(g) \longrightarrow 2\ H_2O(l)$; $PBr_3(l) + 3\ H_2O(l) \longrightarrow H_3PO_3(aq) + 3\ HBr(g)$. Use water sparingly.

CHAPTER 15

Drill Problems

1. A. rate = $\Delta[O_2]/\Delta t = \Delta[NO_2]/4\Delta t = -\Delta[N_2O_5]/2\Delta t$ B. rate = $\Delta[CH_4]/\Delta t = \Delta[CO]/\Delta t = -\Delta[CH_3CHO]/\Delta t$ C. rate = $\Delta[CO_2]/\Delta t = \Delta[NO]/\Delta t = -\Delta[CO]/\Delta t = -\Delta[NO_2]/\Delta t$ D. rate = $\Delta[N_2]/\Delta t = \Delta[H_2O]/2\Delta t = -\Delta[NO]/2\Delta t = -\Delta[H_2]/2\Delta t$ E. rate = $\Delta[Cl_2]/\Delta t = \Delta[NO]/2\Delta t = -\Delta[NOCl]/\Delta t$ F. rate = $\Delta[FClO_2]/\Delta t = -\Delta[F_2]/\Delta t = -\Delta[ClO_2]/2\Delta t$ G. rate = $\Delta[Cl^-]/\Delta t = \Delta[CH_3F]/\Delta t = -\Delta[F^-]/\Delta t = -\Delta[CH_3Cl]/\Delta t$ H. rate = $\Delta[Br_2]/3\Delta t = \Delta[H_2O]/3\Delta t = -\Delta[BrO_3^-]/\Delta t = -\Delta[Br^-]/5\Delta t = -\Delta[H^+]/5\Delta t$ I. rate = $\Delta[N_2]/\Delta t = \Delta[H_2O]/2\Delta t = -\Delta[NH_4^+]/\Delta t = -\Delta[NO_2^-]/\Delta t$ J. rate = $\Delta[HPO_3^-]/\Delta t = \Delta[H_2]/\Delta t = -\Delta[OH^-]/\Delta t$ K. rate = $\Delta[N_2O]/\Delta t = \Delta[H_2O]/\Delta t = -\Delta[NO]/2\Delta t = -\Delta[H_2]/\Delta t$
3. (1) A. 2.9×10^{-3} mol L^{-1} min^{-1} B. 9.3×10^{-4} mol L^{-1} min^{-1} C. 3.4×10^{-4} mol L^{-1} min^{-1} D. 2.1×10^{-3} mol L^{-1} min^{-1} E. 7.0×10^{-4} mol L^{-1} min^{-1} F 1.4×10^{-3} mol L^{-1} min^{-1} G. 2.2×10^{-4} mol L^{-1} min^{-1} H. 0.14 mol L^{-1} hr^{-1} I. 0.23 mol L^{-1} hr^{-1} J. 0.051 mol L^{-1} hr^{-1} K. 0.010 mol L^{-1} hr^{-1} L. 0.031 mol L^{-1} hr^{-1} M. 0.018 mol L^{-1} hr^{-1} N. 0.028 mol L^{-1} hr^{-1}
(2) O. 44 min, 2.3×10^{-3} mol L^{-1} min^{-1} P. 320 min, 1.1×10^{-3} mol L^{-1} min^{-1} Q. 484 min, 7.6×10^{-4} mol L^{-1} min^{-1} R. 688 min, 5.1×10^{-4} mol L^{-1} min^{-1} S. 1.6 hr, 0.28 mol L^{-1} hr^{-1} T. 21.2 hr, 0.044 mol L^{-1} hr^{-1} U. 60.6 hr, 0.010 mol L^{-1} hr^{-1} V. 40.7 hr, 0.018 mol L^{-1} hr^{-1}
5. A. $k[N_2O_5]$, first B. $k[CH_3CHO]^2$, second C. $k[CO][NO_2]$, second D. $k[NO]^2[H_2]$, third E. $k[NOCl]^2$, second F. $k[F_2][ClO_2]$, second G. $k[CH_3Cl][F^-]$, second H. $k[BrO_3^-][Br^-][H^+]^2$, fourth I. $k[NH_4^+][NO_2^-]$, second J. $k[H_2PO_2^-][OH^-]^2$, third K. $k[NO]^2[H_2]$, third
6. A. 0.294/hr, 0.0588 mol L^{-1} hr^{-1}, 0.294 mol L^{-1} hr^{-1} B. 9.00×10^{-5} L mol^{-1} hr^{-1}, 2.03×10^{-6} mol L^{-1} s^{-1}, 1.02×10^{-5} mol L^{-1} s^{-1} C 1.2 L mol^{-1} s^{-1}, 0.30 mol L^{-1} s^{-1}, 0.86 mol L^{-1} s^{-1} D. 1.35 L^2 mol^{-2} s^{-1}, 7.15×10^{-3} mol L^{-1} s^{-1}, 0.0335 mol L^{-1} s^{-1} E. 4.0×10^{-8} L mol^{-1} s^{-1}, 1.0×10^{-8} mol L^{-1} s^{-1}, 2.1×10^{-9} mol L^{-1} s^{-1} F. 2.4 L^2 mol^{-2} s^{-1}, 0.084 mol L^{-1} s^{-1}, 0.077 mol L^{-1} s^{-1} G. 4.8×10^{11} L mol^{-1} s^{-1}, 0.39 mol L^{-1} s^{-1}, 0.21 mol L^{-1} s^{-1} H. 2.11 L^3 mol^{-3} s^{-1}, 0.00601 mol L^{-1} s^{-1}, 2.32×10^{-4} mol L^{-1} s^{-1}, I. 3.9×10^{-4} L mol^{-1} s^{-1}, 7.7×10^{-9} mol L^{-1} s^{-1} J. 5.3×10^{-5} L^2 mol^{-2} s^{-1}, 4.6×10^{-6} mol L^{-1} s^{-1}, 6.2×10^{-6} mol L^{-1} s^{-1} K. 2.2 atm^{-2} min^{-1}, 4.3×10^{-3} atm/min, 0.260 atm/min
7. A. second 2.0×10^{-3} mol L^{-1} s^{-1} B. first, 0.0207/min C. first, 0.0902/hr D. second, 0.0581 L mol^{-1} min^{-1} E. second, 3.33×10^{-3} L mol^{-1} min^{-1} F. first, 5.86×10^{-3}/min G. first, 0.0523/min H. first, 2.01×10^{-3}/min I. first, 0.0103/min J. 1.974×10^{-2}/min
8. A. 2.9×10^2 s B. 33.5 min C. 7.68 hr D. 43.0 min E. 2.63×10^3 min F. 118 hr G. 13.3 min H. 345 min I. 67.3 min J. 35.11 min
12. A. 24.91 kJ/mol, 0.479/s B. 134.01 kJ/mol, 6.0×10^{-14}/s C. 97.10 kJ/mol, 4.9×10^{-39}/s D. 269.36 kJ/mol, 8.09/s E. 80.96 kJ/mol, 0.797 L mol^{-1} s^{-1}
15. (1) A. rate = $k[CO][NO_2]$ B. rate = $k[NO_2]^2[CO]^2$ C. rate = $k[NO]^2$ D. rate = $k[NO_2]^2$ Since C involves an unlikely fast termolecular step, D is more plausible. They both give the same total reaction.
(2) E. rate = $k[Cl_2]^{3/2}[CO]$ F. rate = $k[Cl_2]^{3/2}[CO]$ G. rate = $k[Cl_2][CO]$ H. rate = $k[Cl_2]^{3/2}[CO]$ E, F, and H all sum to the total reaction and give the correct rate law, but H does so in an unlikely way, and E contains Cl_3, a very unlikely intermediate. Thus F seems the most plausible.
(3) I. rate = $k[H_2][Br_2]$ J. rate = $k[Br_2][H_2]^{1/2}$ K. rate = $k[H_2][Br_2]^{1/2}$ if the first step in J and K is disregarded as occurring to only a very small extent, all three sum to the overall reaction. Only K gives the correct rate law, and it is the most plausible.
(4) L. rate = $k[NO]^2[O_2]$ M. rate = $k[NO]^2[O_2]$ N. rate = $k[NO]^2[O_2]$ All three sum to the overall reaction and give the correct rate law. However, both L and N have unlikely termolecular steps. Thus M is the most plausible.
(5) O. rate = $k[C_4H_9Br][OH^-]/[H_2O]$ P. rate = $k[C_4H_9Br]$ Q. rate = $k[C_4H_9Br][OH^-]$ All sum to the overall reaction, but only Q gives the correct rate law.
(6) R sums to the overall reaction if the first step is disregarded as proceeding only to a minor extent. No termolecular collisions. rate = $k[CH_3Cl][Cl]^{1/2}$
(7) S sums to the overall reaction and contains no termolecular collisions. rate = $k[O_3][NO]$

Quizzes

A: 1. (d) 2. (d) 3. (b) 4. (d) 5. (e) (all change the *rate*) 6. (e) (zero order) 7. (b) 8. (d)
B: 1. (d) 2. (e) 3. (e) (0.012) 4. (b) 5. (a) 6. (e) (L mol^{-1} s^{-1}) 7. (c) 8. (e)
C: 1. (b) 2., (b) 3. (a) 4. (d) 5. (b) 6. (d) 7. (c) 8. (d)
D: 1. (c) 2. (c) 3. (e) 4. (c) 5. (d) 6. (a) 7. (d) 8. (e) (changes the *rate*)

Sample Test

1. -8.314 J mol^{-1} K^{-1} $\ln[(7.00 \times 10^{-9})/(1.00 \times 10^{-4})] = E_a(1/400 - 1/500)$; $E_a = 159.1$ kJ/mol
2. Inspection of the data reveals a nearly constant half life of 280 seconds. This suggests a first -order reaction. Check this by calculating several values of k. $kt = \ln([B]_0/[B]_t)$ $k = \ln(3.116/2.291) \div 120$ s $= 2.56 \times 10^{-3}$ s^{-1} and $k = \ln(2.542/1.223) \div 320 = 2.29 \times 10^{-3}$ s^{-1}. We use the latter value to determine the rate. Rate = $k[B] = 2.29 \times 10^{-3}$ s^{-1} $\times$ 0.00504 M = 1.15×10^{-5} M s^{-1}.

CHAPTER 16

Drill Problems

2. A. $[CH_4] = 0.0300$ M, $[CCl_4] = 0.138$ M, $[CH_2Cl_2] = 0.0199$ M B. $[CH_4] = [CCl_4] = 0.0390$ M, $[CH_2Cl_2] = 0.0121$ M C. $[CH_4] = 0.1604$ M, $[CCl_4] = 0.1004$ M, $[CH_2Cl_2] = 0.0392$ M D. $[CH_4] = [CCl_4] = 0.5154$ M, $[CH_2Cl_2] = 0.01874$ M E. $[PCl_5] = 0.0038$ M, $[PCl_3] = [Cl_2] = 0.0462$ M F. $[PCl_5] = 0.0229$ M, $[PCl_3] = 0.1381$ M, $[Cl_2] = 0.1211$ M G. $[PCl_5] = 0.00100$ M, $[PCl_3] = 0.0340$ M, $[Cl_2] = 0.0170$ M H. $[HCHO] = 0.600$ M, $[H_2] = [CO] = 0.400$ M I. $[HCHO] = 0.750$ M, $[H_2] = 1.300$ M, $[CO] = 0.250$ M J. $[HCHO] = 0.167$ M, $[H_2] = 0.133$ M, $[CO] = 0.333$ M K. $[NO] = 1.00$ M, $[Br_2] = 2.50$ M, $[NOBr] = 4.00$ M L. $[NO] = 0.188$ M, $[Br_2] = 2.50$ M, $[NOBr] = 0.750$ M M. $[NO] = 0.0781$ M, $[Br_2] =$

1.41 M, [NOBr] = 0.234 M N. $[H_2]$ = 2.152 M, $[N_2]$ = 0.0185 M, $[NH_3]$ = 0.698 M O. $[H_2]$ = 0.0697 M, $[N_2]$ = 0.0640 M, $[NH_3]$ = 1.742 M

3. (1) A. $[NO]^2[Br_2]/[NOBr]^2$ B. $[H_2O]^2[Cl_2]^2/[HCl]^4[O_2]$ C. $[NO]^2/[N_2][O_2]$ D. $[CH_2Cl_2]^2/[CH_4][CCl_4]$ E. $[PCl_3][Cl_2]/[PCl_5]$ F. $[H_2][CO]/[HCHO]$ G. $[NH_3]^2/[N_2][H_2]^3$ H. $[H_2O]^2[SO_2]^2/[H_2S]^2[O_2]^3$ I. $[NF_3]^2[HF]^6/[NH_3]^2[F_2]^6$
(2) J. 0.0263 M K. 1.42 M L. 0.348 M M. 6.59 M N. 0.769 M O. 0.157 M P. 0.00155 M Q. 0.0386 M R. 0.0560 M S. 0.199 M T. 0.775 U. 0.0396 M

4. A. $[HCHO]/[H_2][CO]$ =3.75 B. $[PCl_5]/[PCl_3][Cl_2]$ = 1.784 C. $[N_2][H_2]^3/[NH_3]^2 = 7.14 \times 10^{-6}$ D. $[Br_2][Cl_2]/[BrCl]^2$ = 0.159 E. $[CH_4][CCl_4]/[CH_2Cl_2]^2$ = 10.48 F. $\sqrt{[CH_4][CCl_4]}/[CH_2Cl_2]$ G. $[NOBr]/[NO]\sqrt{[Br_2]}$ = 2.529 H. $[NH_3]/[N_2]^{1/2}[H_2]^{3/2}$ = 374 I. $[BrCl]/\sqrt{([Br_2][Cl_2])}$ = 2.51 J. $\sqrt{[N_2][O_2]}/[NO] = 3.96 \times 10^7$ K. $([HCl]^2/[H_2O])\sqrt{[O_2]}/[Cl_2]$ = 0.0259 L. $[H_2O][SO_2]/[H_2S][O_2]^{3/2} = 4.17 \times 10^{22}$

6. A. $P(PCl_5)/P(PCl_3)P(Cl_2)$, $K_p = 0.0362$
B. $P(BrCl)^2/P(Br_2)P(Cl_2)$, $K_c = 6.30$
C. $P(H_2O)^2P(SO_2)^2/P(H_2S)^2P(O_2)^3$, $K_p = 2.12 \times 10^{43}$
D. $P(NH_3)^2/P(N_2)P(H_2)^3$, $K_p = 92.9$ E. $P(NO_2)^2/P(N_2O_4)$, $K_c = 4.62 \times 10^{-3}$ F. $P(NO)^2P(Br_2)/P(NOBr)^2$, $K_p = 4.489$
G. $P(CH_3OH)/P(H_2)^2P(CO)$, $K_c = 1.23 \times 10^7$
H. $P(CH_2Cl_2)^2/P(CH_4)P(CCl_4)$, $K_p = 0.0956$
I. $P(NO)^2P(Cl_2)/P(NOCl)^2$, $K_c = 2.69 \times 10^{-3}$
J. $P(H_2O)^2P(Cl_2)^2/P(HCl)^4P(Cl_2)$, $K_p = 0.0956$
K. $P(H_2)^2P(S_2)/P(H_2S)^2 = 4.64 \times 10^{-5}$ L. $P(NO)^2/P(N_2)P(O_2) = 6.37 \times 10^{-16}$

7. A. $1/[CO_2][NH_3]^2$ B. $[SO_2]/[CO]^2$ C. $[NH_3][HCl]$ D. $[H_2]/[NH_3]$ E. $[SO_2][Cl_2]$ F. $1/[H_2]^2[O_2]$ G. $[N_2O][H_2O]^2$ H. $[NaHS]/[H_2S][NaOH]$ I. $[H_3PO_3][HCl]^3/[PCl_3]$ J. $[AgNO_3]^3[NO]/[HNO_3]^4$ K. $[H_2SO_4][NO_2]^6/[HNO_3]^6$ L. $[CO_2]$ M. $[NH_3]^2$

8. (1) A. 0.0954 B. 0.562 C. 0.267 D. 6.4 E. 1.40×10^5 (2) F. 1.16×10^{-3} G. 2.05×10^4 H. 1.95×10^{-5} I. 1.21×10^{-4}

9. A. $Q_c = 1.49 \times 10^3 = K_c$, equilibrium B. $Q_c = 162 < K_c$, right C. $Q_c = 0.423 > K_c$, left D. $Q_c = 1.189 > K_c$, left E. $Q_c = 0.0956 = K_c$, equilibrium F. $Q_c = 0.0624 < K_c$, right G. $Q_c = 0.0844 < K_c$, right H. $Q_c = 0.346 > K_c$, left I. $Q_c = 7.87 \times 10^4 > K_c$, left J. $Q_c = 7.87 < K_c$, right K. $Q_c = 17.3 > K_c$, left L. $Q_c = 1.67 < K_c$, right M. $Q_p = 0.0889 < K_p$, right N. $Q_p = 0.398 > K_p$, left O. $Q_p = 1.98 < K_p$, right P. $Q_p = 8.01 > K_p$, left Q. $Q_p = 0.168 < K_p$, right R. $Q_p = 0.957 > K_p$, left S. $Q_p = 0.306 > K_p$, left T. $3.18 \times 10^{-3} < K_p$, right U. $Q_p = 0.744 < K_p$, right V. $Q_p = 2.89 > K_p$, left

10. A. $K_p = 0.112$ B. $K_p = 0.334$ C. $K_p = 91.0$ D. $K_p = 0.556$ E. $K_p = 2.69$ F. $K_p = 8.4 \times 10^{-3}$ G. $K_p = 0.675$

11. A. R B. L C. R D. L E. U F. U G. L H. R I. R J. L K. R L. L M. R N. R O. U P. L

12. C. $[CH_3ONO]$ = 0.1203 M, [HCl] = 0.2623 M, $[CH_3OH]$ = 0.0987 M, [NOCl] = 0.0778 M D. $[CH_3ONO]$ = 0.418 M, [HCl] = 0.174 M, $[CH_3OH]$ = 0.140 M, [NOCl] = 0.126 M E. is at equilibrium F. $[CH_4]$ = 0.2057 M, $[CCl_4]$ = 0.0143 M, $[CH_2Cl_2]$ = 0.0168 M K $[PCl_3] = [Cl_2]$ = 0.201 M, $[PCl_5]$ = 0.0702 M L. $[PCl_5]$ = 0.1079 M, $[PCl_3]$ = 0.2001 M, $[Cl_2]$ = 0.3021 M M. $P(N_2O_4)$ = 0.245 atm, $P(NO_2)$ = 0.167 atm N. $P(N_2O_4)$ = 0.185 atm, $P(NO_2)$ = 0.144 atm O. $P(Br_2) = P(Cl_2)$ = 0.163 atm, $P(BrCl)$ = 0.410 atm P. $P(Br_2)$ = 0.157 atm, $P(Cl_2)$ = 0.171 atm, $P(BrCl)$ = 0.410 atm Q. $P(PCl_5)$ = 0.089 atm, $P(PCl_3)$ = 0.319 atm, $P(Cl_2)$ = 0.190 atm R. $P(PCl_5)$ = 0.058 atm, $P(PCl_3)$ = 0.290 atm, $P(Cl_2)$ = 0.135 atm S. $P(H_2) = P(CO_2)$ = 0.293 atm, $P(H_2O) = P(CO)$ = 0.027 atm T. $P(H_2)$ = 0.795 atm, $P(CO_2)$ = 0.495 atm, $P(H_2O)$ = 0.081 atm, $P(CO)$ = 0.040 atm U. $P(SO_2)$ = 2.44 atm, $P(Cl_2)$ = 1.10 atm V. $P(SO_2) = P(Cl_2)$ = 1.64 atm W. $P(HCHO)$ = 0.121 atm, $P(H_2) = P(CO)$ = 0.179 atm X. $P(HCHO)$ = 0.180 atm, $P(H_2) = P(CO)$ = 0.220 atm

13. Since the stoichiometric coefficients are the same on the left- and right-hand side of each equation, there is no change in the position of equilibrium by changing the volume in parts J, K, L, M, R, or S. T. shift left, $[Cl_2]$ = 0.220 M U. shift right, $[PCl_3]$ = 0.505 M.

Quizzes

A: 1. $[NaClO_3]^2[SO_2]/[Na_2SO_3][HClO_3]^2$ 2. (e) (1.40) 3. (b) 4. (e) 5. (a) 6. (b) 7. (d)
B: 1. $P(SO_2)P(Cl_2)^2/P(O_2)$ 2. (c) 3. (c) 4. (b) 5. (a) 6. (e) 7. (a)
C: 1. $[Cu(NO_3)_2]^3[NO]^2/[HNO_3]^8$ 2. (d) 3. (b) 4. (c) 5. (e) 6. (e) 7. (d)
D: 1. $P(CO_2)/P(SO_2)$ 2. (c) 3. (b) 4. (a) 5. (d) 6. (c) 7. (e)

Sample Test

1. $PCl_5(g) \rightleftharpoons PCl_3(g) + Cl_2(g)$

	2.00 atm	1.00 atm
x	$-x$	$-x$
x	$2.00 - x$	$1.00 - x$

$1.19 = (2.00 - x)(1.00 - x)/x$
$x = (4.19 \pm \sqrt{17.56 - 8.00})/2 = 0.549 \text{ atm} = P(PCl_5)$

2. Reaction: $HCHO(g) \rightleftharpoons H_2(g) + CO(g)$

Reaction:	$HCHO(g)$	$H_2(g)$	$CO(g)$
Initial:	0.900	0.000	0.600
Changes:	−0.400	+0.400	+0.400
1st Eq:	0.500	0.400	1.000
V ÷ 3	1.50	1.20	3.00
Changes	+ x	− x	− x
2nd Eq:	$1.50 + x$	$1.20 - x$	$3.00 - x$

The information in the first three lines are used to establish the value of the equilibrium constant:

$$K_c = \frac{(0.400 \text{ mol}/3.00 \text{ L})(1.000 \text{ mol}/3.00 \text{ L})}{1.500 \text{ mol}/3.00 \text{ L}} = 0.0889$$

Then, the set-up in the last three lines is used to determine $[H_2]$ once equilibrium is re-established:

$$K_c = \frac{(1.20 - x)(3.00 - x)}{1.50 + x} = 0.0889$$

$3.60 - 4.20\,x + x^2 = 0.0889\,x + 0.133$
or $x^2 - 4.29 + 3.47 = 0$

$$x = \frac{4.29 \pm \sqrt{18.4 - 13.9}}{2} = 1.08$$

Thus, $[H_2]$ = 1.20 − 1.08 = 0.12 M

3..a. left b. no effect c. right d. right e. right f. left g. right h. right i. no effect j. no effect

CHAPTER 17

Drill Problems

2. (1) In these equations, the species appear in the order: acid + base $\rightleftharpoons$ acid + base.
(2) The products are followed by one of the conjugate pairs in parentheses, conjugate base written first.
K. $Cl^- + Na^+$ (NH_3, NH_4^+) L. $H_2O + H_3PO_4$ (H_2O, H_3O^+) M. $C_2H_3O_2^- + H_2S$ (HS^-, H_2S) N. $N_3^- + CH_3NH_3^+$ (N_3^-, HN_3) O. $CH_3NH_2 + H_2O$ (OH^-, H_2O) P. $NO_2^- + HN_3$ (NO_2^-, HNO_2)

3. The base is drawn first in each case.

A. $H-\overset{..}{\underset{..}{O}}-H$, $\overset{..}{O}=\overset{..}{\underset{}{S}}-\overset{..}{\underset{..}{O}}| \leftrightarrow |\overset{..}{\underset{..}{O}}-\overset{..}{\underset{}{S}}=\overset{..}{O}$ B. $[|\overset{..}{\underset{..}{F}}|]^-$, $|\overset{..}{\underset{..}{F}}-Be-\overset{..}{\underset{..}{F}}|$

C. $H-\overset{\overset{\displaystyle H}{|}}{\underset{..}{N}}-H$, H^+ from $H-\overset{..}{\underset{..}{C}l}|$ D. $[H-\overset{..}{\underset{..}{O}}|]^-$, $\overset{..}{O}=C=\overset{..}{O}$

E. $H-\overset{\overset{\displaystyle H}{|}}{\underset{..}{N}}-H$, Ag^+ F. $[|\overset{..}{\underset{..}{O}}|]^{2-}$, H^+ from $H-\overset{..}{\underset{..}{O}}-H$

G. $[|\overset{..}{\underset{..}{F}}|]^-$, $|\overset{..}{\underset{..}{F}}-\overset{\overset{\displaystyle |\overset{..}{\underset{..}{F}}|}{|}}{B}-\overset{..}{\underset{..}{F}}|$ H. $H-\overset{..}{\underset{..}{O}}-H$, H^+ from $H-\overset{..}{\underset{..}{C}l}|$

I. $H-\overset{..}{\underset{..}{O}}-H$, $|\overset{..}{\underset{..}{F}}-\overset{\overset{\displaystyle |\overset{..}{\underset{..}{F}}|}{|}}{B}-\overset{..}{\underset{..}{F}}|$ J. $[|\overset{..}{\underset{..}{S}}|]^{2-}$,

$\overset{..}{O}=\overset{\overset{\displaystyle |\overset{..}{O}|}{|}}{\underset{..}{S}}-\overset{..}{\underset{..}{O}}| \longleftrightarrow |\overset{..}{\underset{..}{O}}-\overset{\overset{\displaystyle |\overset{..}{O}|}{|}}{\underset{..}{S}}=\overset{..}{O}| \longleftrightarrow |\overset{..}{\underset{..}{O}}-\overset{\overset{\displaystyle |\overset{..}{O}||}{|}}{\underset{..}{S}}-\overset{..}{\underset{..}{O}}|$

K. $\left(\overset{..}{O}=\overset{\overset{\displaystyle |\overset{..}{O}|}{|}}{C}-\overset{..}{\underset{..}{O}}|\right)^{2-} \longleftrightarrow \left(|\overset{..}{\underset{..}{O}}-\overset{\overset{\displaystyle |\overset{..}{O}|}{|}}{C}=\overset{..}{O}|\right)^{2-} \longleftrightarrow \left(|\overset{..}{\underset{..}{O}}-\overset{\overset{\displaystyle |\overset{..}{O}||}{|}}{C}-\overset{..}{\underset{..}{O}}|\right)^{2-}$

H^+ L. $H-\overset{..}{\underset{..}{O}}-H$, $\overset{..}{O}=C=\overset{..}{O}$ M. $H-\overset{\overset{\displaystyle H}{|}}{\underset{\underset{\displaystyle H}{|}}{O}}-\overset{}{C}-H$, $|\overset{..}{\underset{..}{F}}-\overset{\overset{\displaystyle |\overset{..}{\underset{..}{F}}|}{|}}{B}-\overset{..}{\underset{..}{F}}|$

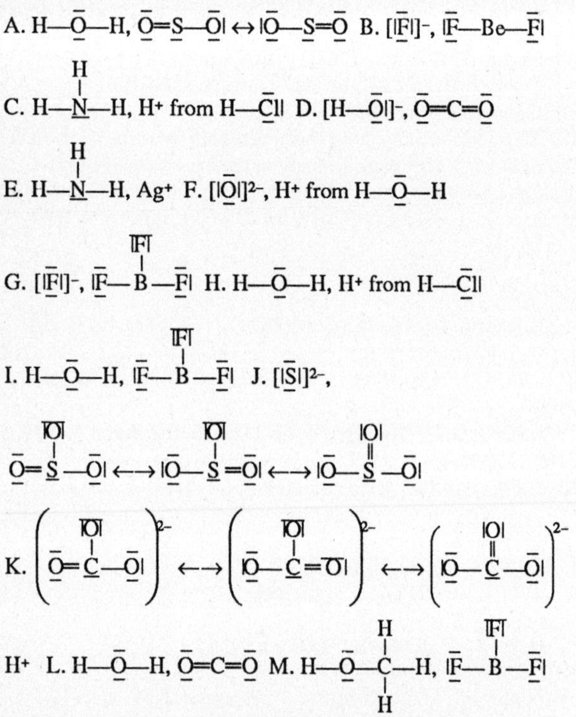

5. & 6. A. 1.6 g/L, 2.5×10^{-13} M, 0.040 M, 12.60, 1.40
B. 6.25×10^{-4} M, 0.0613 g/L, 8.00×10^{-12} M, 2.903, 11.097
C. 1.67×10^{-2} M, 1.23 g/L, 3.00×10^{-13} M, 12.522, 1.478
D. 0.0375 M, 3.84 g/L, 2.67×10^{-13} M, 12.574, 1.426
E. 0.0382 M, 0.0382 M, 2.62×10^{-13} M, 1.418, 12.582
F. 2.995×10^{-3} M, 1.67×10^{-12} M, 5.99×10^{-3} M, 11.777, 2.223 G. 0.788 g/L, 0.0125 M, 8.00×10^{-13} M, 1.903, 12.097 H. 0.045 M, 2.5 g/L, 2.2×10^{-13} M, 12.65, 1.35
I. 6.7×10^{-6} M, 8.2×10^{-4} g/L, 7.4×10^{-10} M,
1.3×10^{-5} M, 4.87 J. 2.2×10^{-5} M, 3.4×10^{-3} g/L,
4.5×10^{-10} M, 2.2×10^{-5} M, 9.35 K. 7.1×10^{-4} M,
0.057 g/L, 7.1×10^{-4} M, 1.4×10^{-11} M, 10.85 L. 0.056 M,
7.2 g/L, 0.056 M, 1.8×10^{-13} M, 1.25
7. A. $HOBr + H_2O \rightleftharpoons H_3O^+ + OBr^-$;
$K_a = [H_3O^+][OBr^-]/[HOBr]$ B. $(CH_3)_2NH + H_2O \rightleftharpoons$
$(CH_3)_2NH_2^+ + OH^-$; $K_b = [(CH_3)_2NH_2^+][OH^-]/[(CH_3)_2NH]$
C. $HC_6H_7O_2 + H_2O \rightleftharpoons C_6H_7O_2^- + H_3O^+$;
$K_a = [C_6H_7O_2^-][H_3O^+]/[HC_6H_7O_2]$ D. $N_2H_4 + H_2O \rightleftharpoons$
$N_2H_5^+ + OH^-$; $K_b = [N_2H_5^+][OH^-]/[N_2H_5^+]$ E. $HC_3H_5O_3 + H_2O$
$\rightleftharpoons H_3O^+ + C_3H_5O_3^-$; $K_a = [H_3O^+][C_3H_5O_3^-]/[HC_3H_5O_3]$
F. $(CH_3)_3N + H_2O \rightleftharpoons (CH_3)_3NH^+ + OH^-$; $K_b =$
$[(CH_3)_3NH^+][OH^-]/[(CH_3)_3N]$ G. $HC_3H_5O_2 + H_2O \rightleftharpoons$
$H_3O^+ + C_3H_5O_2^-$; $K_a = [H_3O^+][C_3H_5O_2^-]/[HC_3H_5O_2]$
H. $HC_2HCl_2O_2 + H_2O \rightleftharpoons H_3O^+ + C_2HCl_2O_2^-$; $K_a =$
$[H_3O^+][C_2HCl_2O_2^-]/[HC_2HCl_2O_2]$ I. $H_3BO_3 + H_2O \rightleftharpoons$
$H_3O^+ + H_2BO_3^-$; $K_a = [H_3O^+][H_2BO_3^-]/[H_3BO_3]$
J. $HC_2Cl_3O_2 + H_2O \rightleftharpoons H_3O^+ + C_2Cl_3O_2^-$; $K_a =$
$[H_3O^+][C_2Cl_3O_2^-]/[HC_2Cl_3O_2]$
8. (1) A. 2.10×10^{-9} B. 7.10×10^{-4} C. 8.02×10^{-5}
D. 9.92×10^{-7} E. 8.38×10^{-4} F. 7.41×10^{-5}
G. 1.40×10^{-5} H. 3.32×10^{-2} I. 5.81×10^{-10} J. 0.210
(2) K. 2.68 L. 2.06 M. 1.92 N. 1.84 O. 1.92 P. 0.98
Q. 11.66 R. 12.56 S. 12.67 T. 2.03 U. 9.83 V. 8.90
9. A. pH = 4.36, $[H_2CO_3] = 4.6 \times 10^{-3}$ M, $[HCO_3^-] =$

4.4×10^{-5} M, $[CO_3^{2-}] = 5.6 \times 10^{-11}$ M B. pH = 4.48, $[H_2S] =$
1.00×10^{-2} M, $[HS^-] = 3.3 \times 10^{-5}$ M, $[S^{2-}] = 1.0 \times 10^{-14}$
C. pH = 0.64, $[H_2C_2O_4] = 0.97$ M, $[HC_2O_4^-] = 0.23$ M,
$[C_2O_4^{2-}] = 5.4 \times 10^{-5}$ M D. pH = 1.20 , $[H_3PO_4] = 0.686$ M,
$[H_2PO_4^-] = 0.064$ M, $[HPO_4^{2-}] = 6.2 \times 10^{-8}$, $[PO_4^{3-}] =$
4.7×10^{-19} M E. pH = 0.72, $[H_3PO_3] = 0.711$ M; $[H_2PO_3^-] =$
0.189 M; $[HPO_3^{2-}] = 2.5 \times 10^{-7}$ M F. pH = 0.97, $[H_2SO_3] =$
0.89 M; $[HSO_3^-] = 0.11$ M; $[SO_3^{2-}] = 6.3 \times 10^{-8}$ M
10. Only the products of the dissolving reactions are written. All
ions are (aq). A. Na^+ + Cl^-; neutral; no hydrolysis
B. NH_4^+ + Cl^-; acidic; $NH_4^+ + H_2O \rightleftharpoons NH_3(aq) + H_3O^+$
C. $N_2H_5^+$ + Br^-; acidic; $N_2H_5^+ + H_2O \rightleftharpoons N_2H_4(aq) + H_3O^+$
D. K^+ + NO_3^-; neutral; no hydrolysis E. $C_2H_5NH_3^+ + H_2O$;
acidic; $C_2H_5NH_3^+ + H_2O \rightleftharpoons C_2H_5NH_2(aq) + H_3O^+$
F. 2 Na^+ + CO_3^{2-}; alkaline; $CO_3^{2-} + H_2O \rightleftharpoons HCO_3^- + OH^-$
 G. $C_6H_5NH_3^+$ + I^-; acidic; $C_6H_5NH_3^+ + H_2O \rightleftharpoons$
$C_6H_5NH_2(aq) + OH^-$ H. Ca^{2+} + 2 ClO^-; alkaline; $ClO^- + H_2O$
$\rightleftharpoons HClO(aq) + OH^-$ I. Rb^+ + ClO_2^-; alkaline; $ClO_2^- + H_2O$
$\rightleftharpoons HClO_2(aq) + OH^-$ J. 2 $HONH_3^+$ + SO_4^{2-}; acidic;
$HONH_3^+ + H_2O \rightleftharpoons HONH_2(aq) + H_3O^+$ K. Sr^{2+} + 2 F^-;
alkaline; $F^- + H_2O \rightleftharpoons HF(aq) + OH^-$ L. Cs^+ + $C_2H_3O_2^-$;
alkaline; $C_2H_3O_2^- + H_2O \rightleftharpoons HC_2H_3O_2(aq) + OH^-$
M. NH_4^+ + CN^-; basic; $NH_4^+ + H_2O \rightleftharpoons NH_3(aq) + H_3O^+$;
$CN^- + H_2O \rightleftharpoons HCN(aq) + OH^-$ N. $HONH_3^+$ + F^-; acidic;
$HONH_3^+ + H_2O \rightleftharpoons HONH_2(aq) + H_3O^+$; $F^- + H_2O \rightleftharpoons$
$HF(aq) + OH^-$ O. $CH_3NH_3^+$ + ClO_2^-; acidic; $CH_3NH_3^+ + H_2O$
$\rightleftharpoons CH_3NH_2(aq) + H_3O^+$; $ClO_2^- + H_2O \rightleftharpoons HClO_2(aq) + OH^-$
11. (1) A. 7.00 B. 4.62 C. 4.00 D. 7.00 E. 9.38 F. 10.19
G. 2.32 H. 10.92 I. 7.96 J. 2.83 K. 8.74 L. 5.32
(2) M. 9.03 N. 8.75 O. 8.87 P. 8.45 Q. 8.06 R. 8.00
S. 4.76 T. 5.48 U. 5.50 V. 4.00 W. 3.348 X. 5.051
12. Acids: HCl > H_3O^+ > HSO_4^- > H_3PO_4 > HNO_2 > HN_3 >
$HC_2H_3O_2$ > H_2S > NH_4^+ > HCN > HCO_3^- > $CH_3NH_3^+$ > $H_2O \cong$
HS^-
Bases: S^{2-} > OH^- > CH_3NH_2 > CO_3^{2-} > CN^- > NH_3 > HS^- >
$C_2H_3O_2^-$ > N_3^- > NO_2^- > $H_2PO_4^-$ > SO_4^{2-} > H_2O > Cl^-
The products of each reaction follow, acids first.
A. HN_3 + OH^-, left B. H_2O + NO_2^-, right C. $HCN + H_2O$,
right D. HCN + NH_2^-, left E. H_3PO_4 + CN^-, left
F. NH_3 + $H_2PO_4^-$, right G. H_2O + $H_2PO_4^-$, right
H. NH_3 + H_2O, right I. $HC_2H_3O_2$ + CO_3^{2-}, left
J. H_2S + SO_4^{2-}, right *or* H_2SO_4 + S^{2-}, left K. H_2O + HS^-, right
L. $HC_2H_3O_2$ + OH^-, left M. HSO_4^- + CO_3^{2-}, left
N. H_2O + SO_4^{2-}, right
13. (1) A. basic B. acidic C. amphoteric D. acidic E. acidic
F. acidic G. amphoteric H. acidic I. amphoteric J. basic
K. amphoteric L. amphoteric M. basic N. acidic O. basic
(2) P. $CaO + SiO_2 \longrightarrow CaSiO_3$ Q. $Al_2O_3 + 3 SO_2 \longrightarrow$
$Al_2(SO_3)_3$ R. no reaction S. no reaction T. $BaO + SO_2 \longrightarrow$
$BaSO_3$ U. 3 $PbO + P_2O_5 \longrightarrow Pb_3(PO_4)_2$ V. 3 $Na_2O + Al_2O_3$
$\longrightarrow 2 Na_3AlO_3$
14. A. H_2CO_3 more electronegative central atom B. H_2SO_4
higher oxidation state of sulfur C. H_3PO_4 more electronegative
central atom D. $HClO_4$ higher oxidation state of chlorine
E. HClO more electronegative central atom F. HNO_3 higher
oxidation state of nitrogen G. H_2SO_4 more electronegative
central atom H. CF_3OH greater inductive effect by F than by H
I. FC_6H_4OH greater inductive effect by F than by Cl J. $HBrO_3$
more double-bonded oxygens on central atom K. $HClO_4$ more
double-bonded oxygens on central atom L. H_2CO_3 more
electronegative central atom

Quizzes

A: 1. (c) 2. (d) 3. (b) 4. (e) 5. (d) 6. (c) 7. (d) 8. (c) 9. (a)
B: 1. (b) 2. (a) 3. (a) 4. (c) 5. (d) 6. (d) 7. (c) 8. (b) 9. (b)
C: 1. (d) 2. (e) 3. (e) 4. (a) 5. (c) 6. (a) 7. (c) 8. (e) 9. (e)
D: 1. (b) 2. (a) 3. (c) 4. (b) 5. (a) 6. (d) 7. (d) 8. (d) 9. (e)

Sample Test

1. $[HA] = \frac{3.050\ g}{0.250\ L} \times \frac{mol}{122.0\ g} = 0.100\ M$
or $[H_3O^+] = 2.5 \times 10^{-3}$ M and pH = 2.60
2. a. Brønsted-Lowry; acids: HF, $N_2H_5^+$; bases: N_2H_4, F^-
b. Brønsted-Lowry; acids: H_2O, OH^-; bases O^{2-}, OH^- c. Lewis; acid: SOI_2; base: $BaSO_3$ d. Lewis; acid: $HgCl_3^-$; base: Cl^-
3. a. $H_2SO_3 > HF > N_2H_5^+ > CH_3NH_3^+ > H_2O$ b. $OH^- > CH_3NH_2 > N_2H_4 > F^- > HSO_3^-$ (c) (i) to the right; (ii) to the left
4. $[ClO_2^-] = 2 \times 3.00$ mol/2.50 L = 2.40 M;
$K_b = 1.00 \times 10^{-14}/1.2 \times 10^{-2} = 8.3 \times 10^{-13}$;
$[OH^-] = 1.4 \times 10^{-6}$ M *or* pH = 8.15

CHAPTER 18

Drill Problems

1. A. 2.68, 4.44 B. 2.68, 0.85 C. 2.06, 3.54 D. 1.92, 0.62 E. 1.84, 2.67 F. 1.98, 3.48 G. 0.98, 1.20 H. 11.66, 13.51 I. 12.06, 11.00 J. 11.89, 13.33 K. 11.77, 9.62 L. 11.47, 8.76 M. 11.84, 12.31 N. 12.22, 10.16
5. A. pH = 4.03 B. $[HC_7H_5O_2] = 0.660$ M $[C_7H_5O_2^-] = 0.640$ M; pH = 4.19 C. $[HCHO_2] = 0.105$ M, $[CHO_2^-] = 0.710$ M, pH = 4.57 D. pH = 9.72 E. $[C_2H_5NH_2] = 1.29$ M, $[C_2H_5NH_3^+] = 1.02$, pH = 10.74 F. $[CH_3NH_2] = 0.111$ M, $[CH_3NH_3^+] = 0.441$ M, pH = 10.10 G. pH = 3.08 H. $[HC_2H_2ClO_2] = 0.79$ M, $[C_2H_2ClO_2^-] = 2.14$ M, pH = 3.30 I. $[HNO_2] = 1.302$ M, $[NO_2^-] = 0.550$ M, pH = 2.916 J. pH = 11.05 K. $[(CH_3)_3N] = 0.922$ M, $[(CH_3)_3NH^+] = 0.314$ M, pH = 10.34 L. $[NH_3] = 0.0731$ M, $[NH_4^+] = 0.1549$ M, pH = 8.915 M. $[CHO_2^-] = 0.208$ M, pH = 2.77 N. $[HC_2H_3O_2] = 0.0337$ M, $[C_2H_3O_2^-] = 0.0224$ M, pH = 4.58 O. $[HC_7H_5O_2] = 0.512$ M, $[C_7H_5O_2^-] = 0.231$ M, pH = 3.86 P. = $[(CH_3)_3N] = 0.218$ M, $[(CH_3)_3NH^+] = 0.613$ M, pH = 9.42 Q. pH = 10.84 R. $[C_2H_5NH_2] = 0.148$ M, $[C_2H_5NH_3^+] = 1.001$ M, pH = 9.80
6. The pH after acid addition is given first, followed by the pH after addition of base. A. 3.88, 4.15 B. 4.17, 4.21 C. 4.54, 4.61 D. 9.52, 9.96 E. 10.53, 10.95 F. 9.99, 10.20 G. 2.91, 3.23 H. 3.22, 3.39 I. 2.783, 3.034 J. 11.00, 11.10, K. 10.06, 10.,74 L. 8.781, 9.035 M. 2.67, 2.86 N. 4.52, 4.64 O. 3.60, 4.01 P. 9.30, 9.563 Q. 10.23, 11.66 R. 9.27, 10.07
7. Answers are given in the order: buffer range, acid capacity (mol H^+), base capacity (mol OH^-). A. 3.76-5.76, 0.0852, 0.970 B. 3.20-5.20, 0.417, 0.433 C. 2.75-4.75, 0.636, 0.0309 D. 8.24, 10.24, 0.0798, 0.0193 E. 9.63-11.63, 2.20, 1.65 F. 9.70-11.70, 0.376, 2.31 G. 2.17-4.17, 0.999, 1.30 H. 1.87-3.87, 21.4, 5.97 I. 2.29-4.29, 3.32, 9.87 J. 9.70-11.70, 0.519, 0.189 K. 8.87-10.87, 0.721, 0.179 L. 8.24-10.24, 3.98, 106 M. 2.75-4.75, 0.00683, 0.982 N. 3.76-5.76, 1.78, 2.95 O. 3.20-5.20. 0.248, 0.676 P. 8.87-10.87, 0.534, 2.02 Q. 9.70-11.70, 10.7, 7.29 R. 9.63-11.63, 0.0435, 0.897

10. & 11. Given in the order: pH at 0%, 10%, 50%, 90%, 100%, and 110% titration.
A. 0.61, 0.71, 1.12, 1.93, 7.00, 12.02
B. 13.40, 13.26, 12.74, 11.89, 7.00, 2.18
C. 14.40, 14.26, 13.74, 12.89, 7.00, 1.18
D. 0.92, 1.02, 1.43, 2.24, 7.00, 11.71
E. 0.60, 0.74, 1.25, 2.11, 7.00, 11.82
F. 13.80, 13.63, 13.08, 12.21, 7.00, 1.85
G. 12.30, 12.24, 11.85, 11.07, 7.00, 2.97
H. 1.95, 2.34, 3.29, 4.24, 8.17, 12.02
I. 12.05, 11.65, 10.70, 9.75, 5.92, 2.18
J. 12.52, 11.58, 10.63, 9.68, 5.39, 1.18
K. 2.06, 2.22, 3.17, 4.12, 7.96, 11.71
L. 2.17, 2.80, 3.75, 4.70, 8.30, 11.82
M. 12.25, 11.65, 10.70, 9.75, 5.75, 1.85
N. 8.80, 6.25, 5.30, 4.35, 3.63, 2.97
12. Suitable indicators: A. though G. phenol red for all seven titrations. H. phenophthalien I. methyl red J. methyl red K. phenophthalein L. phenophthalein M. methyl red N. indistinct equivalence point; no indicator possible.
11. A. 8.32 B. 11.48 C. 4.72 D. 2.77 E. 4.60 F. 2.76 G. 9.24 H. 4.18 I. 9.2 J. 10.4 K. 10.7 L. 4.52
15. A. HCl: 36.5 g/eq, 0.104 N; $Ca(OH)_2$: 37.1 g/eq, 0.104 N B. H_2SO_4: 98.1 g/eq, 0.704 N; KOH: 56.1 g/eq, 0.197 N C. HNO_3: 63.0 g/eq, 0.146 N; LiOH: 23.9 g/eq, 0.412 N D. HBr: 80.9 g/eq, 0.904 N; $Al(OH)_3$: 26.0 g/eq, 0.666 N E. H_2SO_4: 49.1 g/eq, 1.64 N; $Ca(OH)_2$: 37.1 g/eq, 1.05 N F. H_3PO_4: 32.7 g/eq, 0.600 N; $Al(OH)_3$: 26.0 g/eq, 0.996 N G. H_3PO_4: 49.0 g/eq, 0.330 N; $Sr(OH)_2$: 60.8 g/eq, 0.454 N H. $H_2C_2O_4$: 90.0 g/eq, 0.774 N; NaOH: 40.0 g/eq, 0.146 N
16. A. 11.30 mL B. 0.07784 N C. 35.03 mL D. 0.3480 N E. 20.47 mL F. 0.09358 N G. 118.3 mL H. 0.2462 N

Quizzes

A: 1. (b) 2. (e) (2.0) 3. (d) 4. (c) 5. (b) 6. (e) (pH = pK_a) 7. (d)
B: 1. (a) 2. (c) 3. (e) 4. (a) 5. (b) 6. (b) 7. (e) (no OH^- present)
C: 1. (c) 2. (c) 3. (d) 4. (a) 5. (e) [(c) = endpoint] 6. (b) 7. (e) (8.79)
D: 1. (b) 2. (b) 3. (e) 4. (d) 5. (d) 6. (a) 7. (e) (pOH = pK_b)

Sample Test

1. $pK_a = 4.602$ is $K_a = 2.50 \times 10^{-5}$; pH = 2.716 is $[H^+] = 1.92 \times 10^{-3}$ M. $(1.92 \times 10^{-3})^2/[HA] = 2.50 \times 10^{-5}$ *or* [HA] = 0.148 M. Thus $[HA]_i = [HA] + [H^+] = 0.150$ M.
0.500 L × 0.150 M = 0.0750 mol HA = 7.500 g *and* $\mathcal{M}$ = 7.500 g/0.0750 mol = 100. g/mol
2. a. weak base
b. 50 mL HCl × L/1000 mL × 0.100 mol/L = 5.0×10^{-3} mol HCl $\mathcal{M}$ = 0.400 g/5.0×10^{-3} mol = 80. g/mol
c. pH = 8.0 at midpoint *or* pOH = 6.0 = pK_b. Thus $K_b = 1 \times 10^{-6}$
d. pH = 5 to pH = 3, methyl orange

CHAPTER 19

Drill Problems

1. A. $AgCN(s) \rightleftharpoons Ag^+(aq) + CN^-(aq)$; $K_{sp} = [Ag^+][CN^-]$
B. $SrF_2(s) \rightleftharpoons Sr^{2+}(aq) + 2\,F^-(aq)$; $K_{sp} = [Sr^{2+}][F^-]^2$
C. $TlBr(s) \rightleftharpoons Tl^+(aq) + Br^-(aq)$; $K_{sp} = [Tl^+][Br^-]$
D. $Zn_3(AsO_4)_2(s) \rightleftharpoons 3\,Zn^{2+}(aq) + 2\,AsO_4^{3-}(aq)$;

$K_{sp} = [Zn^{2+}]^3[AsO_4^{3-}]^2$
E. $Hg_2Br_2(s) \rightleftharpoons Hg_2^{2+}(aq) + 2\ Br^-(aq)$; $K_{sp} = [Hg_2^{2+}][Br^-]^2$
F. $Ag_3AsO_4(s) \rightleftharpoons 3\ Ag^+(aq) + AsO_4^{3-}(aq)$;
$K_{sp} = [Ag^+]^3[AsO_4^{3-}]$
G. $Li_2CO_3(s) \rightleftharpoons 2\ Li^+(aq) + CO_3^{2-}(aq)$; $K_{sp} = [Li^+]^2[CO_3^{2-}]$
H. $PbC_2O_4(s) \rightleftharpoons Pb^{2+}(aq) + C_2O_4^{2-}(aq)$; $K_{sp} = [Pb^{2+}][C_2O_4^{2-}]$
I. $La(IO_3)_3(s) \rightleftharpoons La^{3+}(aq) + 3\ IO_3^-(aq)$; $K_{sp} = [La^{3+}][IO_3^-]^3$
J. $Ce_2(C_2O_4)_3(s) \rightleftharpoons 2\ Ce^{3+}(aq) + 3\ C_2O_4^{2-}(aq)$;
$K_{sp} = [Ce^{3+}]^2[C_2O_4^{2-}]^3$
K. $CdCO_3(s) \rightleftharpoons Cd^{2+}(aq) + CO_3^{2-}(aq)$; $K_{sp} = [Cd^{2+}][CO_3^{2-}]$
L. $AlPO_4(s) \rightleftharpoons Al^{3+}(aq) + PO_4^{3-}(aq)$; $K_{sp} = [Al^{3+}][PO_4^{3-}]$
M. $Al_2S_3(s) \rightleftharpoons 2\ Al^{3+}(aq) + 3\ S^{2-}(aq)$; $K_{sp} = [Al^{3+}]^2[S^{2-}]^3$
N. $Ca(IO_3)_2(s) \rightleftharpoons Ca^{2+}(aq) + 2\ IO_3^-(aq)$; $K_{sp} = [Ca^{2+}][IO_3^-]^2$
O. $CoS(s) \rightleftharpoons Co^{2+}(aq) + S^{2-}(aq)$; $K_{sp} = [Co^{2+}][S^{2-}]$
P. $FeAsO_4(s) \rightleftharpoons Fe^{3+}(aq) + AsO_4^{3-}(aq)$; $K_{sp} = [Fe^{3+}][AsO_4^{3-}]$

2. A. 1.16×10^{-12}, 1.08×10^{-6} M, 1.79×10^{-5} g
B. 2.2×10^{-5}, 4.7×10^{-3} M, 0.095 g
C. 0.031, 0.18 M, 0.18 M
D. 4.5×10^{-5} M, 4.5×10^{-5} M, 2.3×10^{-3} g
E. 6.3×10^{-19}, 7.9×10^{-10} M, 9.7×10^{-7} g
F. 3.2×10^{-6}, 9.3×10^{-3} M, 0.37 g
G. 1.6×10^{-16}, 3.4×10^{-6} M, 6.8×10^{-6} M
H. 5.6×10^{-3} M, 1.1×10^{-2} M, 0.22 g
I. 2.5×10^{-48}, 8.5×10^{-17} M, 1.4×10^{-15} g
J. 1.0×10^{-12}, 1.3×10^{-4} M, 3.3×10^{-3} g
K. 3.4×10^{-8}, 4.1×10^{-3} M, 2.0×10^{-3} M
L. 1.3×10^{-5} M, 3.9×10^{-5} M, 7.8×10^{-4} g
M. 1.3×10^{-16}, 4.8×10^{-5} M, 2.0×10^{-3} g
N. 6.5×10^{-12}, 6.9×10^{-4} M, 0.046 g
O. 1.9×10^{-19}, 2.1×10^{-4} M, 1.4×10^{-6} M
P. 0.036 M, 0.045 M, 0.27 g

3. A. 2.6×10^{-6} M B. 1.2×10^{-11} M C. 3.2×10^{-24} M D. 2.1×10^{-4} M E. 1.4×10^{-6} M F. 6.2×10^{-10} M G. 4.5×10^{-13} M H. 6.1×10^{-10} M I. 7.4×10^{-4} M J. 1.7×10^{-9} M

5. (1) A. 4.3×10^{-23} M B. 4.2×10^{-7} M C. 7.9×10^{-8} M D. 6.8×10^{-4} M E. 1.8×10^{-6} M F. 0.033 M G. 0.56 M H. 1.1×10^{-3} M I. 4.2×10^{-7} M J. 1.3×10^{-3} M
(2) Values of Q are given, followed by *yes* if precipitation should occur. K. 5.12×10^{-17}, yes L. 3.3×10^{-8}, yes M. 1.10×10^{-12}, yes N. 1.1×10^{-6}, yes O. 2.5×10^{-11}, no P. 1.5×10^{-8}, yes Q. 1.3×10^{-8}, yes R. 2.9×10^{-10}, no S. 1.7×10^{-2}, yes T. 1.7×10^{-25}, no

6. The concentrations of the two ions are followed by the formula(s) of the ion(s) for which precipitation is complete (or by neither). (1) K. $[Co^{2+}] = 3.2 \times 10^{-5}$ M, $[S^{2-}] = 1.3 \times 10^{-16}$ M, S^{2-} L. $[Pb^{2+}] = 4.6 \times 10^{-3}$ M, $[SO_4^{2-}] = 3.5 \times 10^{-6}$ M, neither M. $[Ag^+] = 1.2 \times 10^{-5}$ M, $[I^-] = 7.1 \times 10^{-12}$ M, I^- N. $[Ba^{2+}] = 6.4 \times 10^{-2}$ M, $[SO_4^{2-}] = 1.7 \times 10^{-9}$ M, SO_4^{2-} O. no ppt. P. $[Pb^{2+}] = 1.7 \times 10^{-10}$ M, $[CrO_4^{2-}] = 1.7 \times 10^{-3}$ M, Pb^{2+} Q. $[Ag^+] = 0.017$ M, $[CrO_4^{2-}] = 8.4 \times 10^{-9}$ M, CrO_4^{2-} R. no ppt. S. $[Li^+] = 1{,}6$ M, $[PO_4^{3-}] = 3.2 \times 10^{-9}$ M, PO_4^{3-} T. no ppt.
(2) A. $[Zn^{2+}] = 1.1 \times 10^{-5}$ M, $[S^{2-}] = 9.1 \times 10^{-17}$ M, S^{2-} B. $[Hg_2^{2+}] = 3.4 \times 10^{-4}$ M, $[Cl^-] = 6.2 \times 10^{-8}$ M, neither C. $[Al^{3+}] = 6.6 \times 10^{-23}$ M, $[OH^-] = 2.7 \times 10^{-4}$ M, Al^{3+} D. $[Pb^{2+}] = 1.0$ M, $[Cl^-] = 4.0 \times 10^{-3}$ M, neither E. $[Ag^+] = 1.5 \times 10^{-10}$ M, $[Br^-] = 3.3 \times 10^{-3}$ M, Ag^+ F. $[Bi^{3+}] = 2.7 \times 10^{-6}$ M, $[S^{2-}] = 2.4 \times 10^{-29}$ M, S^{2-}

7. The formula of the precipitating ion is given first, followed by its concentrations needed: to just cause precipitation, and for complete precipitation of each ion; "yes" means that the two ions can be separated by fractional precipitation.
A. Ag^+: Cl^- 1.3×10^{-4} M, 0.13 M; Br^- 5.6×10^{-10} M, 5.6×10^{-7} M; yes B. SO_4^{2-}; Ba^{2+} 6.9×10^{-6}M, 6.9×10^{-3} M; Pb^{2+} 1.2×10^{-3} M, 1.2 M; no (barely) C. S^{2-}: Pb^{2+} 2.1×10^{-24} M, 2.1×10^{-21} M; Hg_2^{2+} 4.1×10^{-49} M, 4.1×10^{-46} M; yes (Cl^- also can be used as a precipitating ion.) D. SO_4^{2-}: Sr^{2+} 20 M, 2.0×10^4 M; Pb^{2+} 4.7×10^{-4} M, 0.47 M; yes E. Mg^{2+}: OH^- 2.3×10^{-5} M, 23 M; F^- 2.2×10^{-6} M, 2.2 M; no F. CrO_4^{2-}: Ag^+ 1.1×10^{-7} M, 0.11 M; Pb^{2+} 1.2×10^{-11} M, 1.2×10^{-8} M; yes G. S^{2-}: Zn^{2+} 1.3×10^{-8} M, 1.3×10^{-5} M; Sn^{2+} 3.4×10^{-14} M, 3.4×10^{-11} M; yes H. PO_4^{3-}; Li 1.6×10^{-8} M, 11 M; Mg^{2+} 1.6×10^{-8} M, 5.0×10^{-4} M; no

8. Concentrations are given at pH = 2, pH = 7, and pH = 10, in that order. A. 2.6×10^{-9} M, 1.3×10^{-12} M, 1.3×10^{-21} M
B. 1.2×10^{-3} M, 1.2×10^{-3} M, 1.2×10^{-3} M
C. 1.7×10^{-4} M, 1.7×10^{-4} M, 1.4×10^{-4} M
D. 3.4×10^{-8} M, 1.1×10^{-8} M, 1.6×10^{-14} M
E. 0.108 M, 0.108 M, 0.108 M
F. 2×10^{-10} M, 4×10^{-17} M, 4×10^{-26} M
G. 0.011 M, 0.011 M, 0.011 M
H. 1.4×10^{-4} M, 1.4×10^{-4} M, 1.0×10^{-4} M

9. A. $Ag^+ + 2\ S_2O_3^{2-} \rightleftharpoons [Ag(S_2O_3)_2]^{3-}$ $S_2O_3^{2-} + H_2O \rightleftharpoons HS_2O_3^- + OH^-$; $Ag^+ + Cl^- \rightleftharpoons AgCl(s)$ decreases precipitate B. $Ag^+ + I^- \rightleftharpoons AgI(s)$; $Hg^{2+} + 4\ I^- \rightleftharpoons [HgI_4]^{2-}$; $AgI(s) + I^- \rightleftharpoons [AgI_2]^-$ decreases precipitate C. $Zn^{2+} + S^{2-} \rightleftharpoons ZnS(s)$; $Zn^{2+} + 4\ CN^- \rightleftharpoons [Zn(CN)_4]^{2-}$; $CN^- + H_2O \rightleftharpoons HCN(aq) + OH^-$ decreases precipitate D. $Fe^{2+} + 2\ OH^- \rightleftharpoons Fe(OH)_2(s)$; $Fe^{2+} + 6\ CN^- \rightleftharpoons [Fe(CN)_6]^{4-}$; $CN^- + H_2O \rightleftharpoons HCN(aq) + OH^-$ decreases precipitate E. $Al^{3+} + 3\ F^- \rightleftharpoons AlF_3(s)$; $Ca^{2+} + 2\ F^- \rightleftharpoons CaF_2(s)$; $AlF_3(s) + 3\ F^- \rightleftharpoons [AlF_6]^{3-}$ $F^- + H_2O \rightleftharpoons HF(aq) + OH^-$ decreases precipitate F. $Cu^+ + Cl^- \rightleftharpoons CuCl(s)$; $Ag^+ + Cl^- \rightleftharpoons AgCl(s)$ $Cu^+ + 2\ CN^- \rightleftharpoons [Cu(CN)_2]^-$ $Ag^+ + 2\ CN^- \rightleftharpoons [Ag(CN)_2]^-$; $CN^- + H_2O \rightleftharpoons HCN(aq) + OH^-$ decreases precipitate G. $Al^{3+} + 3\ OH^- \rightleftharpoons Al(OH)_3(s)$; $Al^{3+} + 6\ F^- \rightleftharpoons [AlF_6]^{3-}$; $F^- + H_2O \rightleftharpoons HF(aq) + OH^-$ decreases precipitate H. $2\ Co^{3+} + 3\ S^{2-} \rightleftharpoons Co_2S_3(s)$; $Co^{3+} + 3\ OH^- \rightleftharpoons Co(OH)_3(s)$; $Co^{3+} + 6\ NH_3 \rightleftharpoons [Co(NH_3)_6]^{3+}$; $S^{2-} + H_2O \rightleftharpoons HS^- + OH^-$; $NH_3 + H_2O \rightleftharpoons NH_4^+ + OH^-$ decreases precipitate I. $Cu^{2+} + 2\ OH^- \rightleftharpoons Cu(OH)_2(s)$; $Cu^{2+} + 4\ NH_3 \rightleftharpoons [Cu(NH_3)_4]^{2+}$; $NH_3 + H_2O \rightleftharpoons NH_4^+ + OH^-$ decreases precipitate

10. Free ion concentration given first, followed by that of complex ion. A. 6.4×10^{-17} M, 0.002 M B. 3.3×10^{-6} M, 7.9×10^{-5} M C. 4.5×10^{-15} M, 0.0067 M D. 9.7×10^{-22} M, 5.0×10^{-4} M E. 4.6×10^{-31} M, 5.0×10^{-4} M F. 1.4×10^{-40}, 5.0×10^{-4} M G. 1.4×10^{-45}, 5.0×10^{-4} M H. 2.5×10^{-40}, 1.4×10^{-5} M I. 2.0×10^{-10} M, 1.3×10^{-4} M J. 1.3×10^{-36} M, 2.4×10^{-6} M K. 2.0×10^{-21} M, 2.4×10^{-6} M

11. A. $[Ag^+] = 2.5 \times 10^{-11}$ M, $Q = 2.5 \times 10^{-13}$, precipitate forms B. $[Ag^+] = 2.5 \times 10^{-11}$ M, $Q = 2.5 \times 10^{-13}$, no precipitate C. $[Zn^{2+}] = 2.6 \times 10^{-18}$ M, $Q = 2.6 \times 10^{-20}$, precipitate forms D. $[Ag^+] = 2.7 \times 10^{-14}$ M, $Q = 2.7 \times 10^{-16}$, no precipitate E. $[Hg^{2+}] = 3.1 \times 10^{-31}$ M, $Q = 6.3 \times 10^{-33}$, no precipitate F. $[Cu^{2+}] = 1.7 \times 10^{-17}$ M, $Q = 3.3 \times 10^{-19}$, precipitate forms G. $[Ag^+] = 1.9 \times 10^{-21}$ M, $Q = 9.3 \times 10^{-24}$, precipitate forms

14. A. 0.71 B. 2.81 C. 2.07 D. 1.32 E. 3.14 F. –9.92 G. –9.31 (The last two sulfides cannot be dissolved in the solution described by adding acid.

Quizzes

A: 1. (a) 2. (d) 3. (e) 4. (d) 5. (d) 6. (c) 7. (e) 8. (a)
B: 1. (a) 2. (e) 3. (a) 4. (d) 5. (d) 6. (b) 7. (c) 8. (c)
C: 1. (d) 2. (b) 3. (b) 4. (d) 5. (a) 6. (c) 7. (c) 8. (b)
D: 1. (c) 2. (c) 3. (a) 4. (e) 5. (c) 6. (c) 7. (b) 8. (e)

Sample Test

1. The concentration of PO_4^{3-} removed is $\Delta[PO_4^{3-}] = 8.06 \times 10^{-4}$ M $- 1.00 \times 10^{-12}$ M $= 8.06 \times 10^{-4}$ M The volume of the pond is 100. m × 200. m × 8.00 m × 1000 L/m³ = 1.60×10^8 L The $[Ca^{2+}]$ that must be present in solution when precipitation is complete is given by $[Ca^{2+}]^3[PO_4^{3-}]^2 = [Ca^{2+}]^3(1.00 \times 10^{-12})^2 = 1.30 \times 10^{-32}$ *or* $[Ca^{2+}] = 2.35 \times 10^{-3}$ M The $Ca(NO_3)_2$ is needed both (1) to form precipitate and (2) to maintain the needed $[Ca^{2+}]$.
(1) 1.60×10^8 L × (3 mol Ca^{2+}/2 mol PO_4^{3-}) × 8.06×10^{-4} M $= 1.93 \times 10^5$ mol $Ca(NO_3)_2$
(2) 1.60×10^8 L × 2.35×10^{-3} M $= 3.76 \times 10^5$ mol $Ca(NO_3)_2$
total $= 5.69 \times 10^5$ mol $Ca(NO_3)_2$ × 164.1 g/mol × lb/453.5 g × ton/2000 lbs = 103 tons
2. $[Ag^+]^2(2.0 \times 10^{-3}\text{ M}) = 2.4 \times 10^{-12}$
or $[Ag^+] = 3.5 \times 10^{-5}$ M for CrO_4^{2-} pptn
$[Ag^+](1.0 \times 10^{-5}\text{ M}) = 1.6 \times 10^{-10}$
or $[Ag^+] = 1.6 \times 10^{-5}$ M for Cl^- pptn
AgCl precipitates first. When Ag_2CrO_4 begins to precipitate, $(3.5 \times 10^{-5}\text{ M})[Cl^-] = 1.6 \times 10^{-10}$ *or* $[Cl^- = 4.6 \times 10^{-6}$ M, 46% of its initial value.
3. Initially: $[Ag^+] = 0.0700$ M, $[SO_4^{2-}] = 0.0600$ M, $[S_2O_3^{2-}] = 0.200$ M

$$\frac{[[Ag(S_2O_3)_2]^{3-}]}{[Ag^+][S_2O_3^{2-}]^2} = \frac{0.0700 - x}{x(0.0600 + 2x)^2} = 1.7 \times 10^{13}$$

or $x = 1.1 \times 10^{-12}$ M $= [Ag^+]$
$Q = [Ag^+]^2[SO_4^{2-}] = 7.8 \times 10^{-26} < 1.4 \times 10^{-5} = K_{sp}$
It all dissolves

CHAPTER 20

Drill Problems

2. In order, the values are q, w, and ΔE. A. –196 J, +123 J, –319 J B. –189 J, –247 J, +58 J C. +647 J, +92.4 J, + 555 J D. +1362 J, +321 J, +1041 J E. –142 J, –254 J, +112 J F. +802 J, –897 J, +1699 J G. –1432 J, –937 J, –495 J H. –972 J, +432 J, –1404 J
3. ΔE is given first, followed by ΔH (all in kJ/mol). A. –2667, –2672 B. – 3061, –3063 C. –96.91, –91.83 D. –547.0, –563.5 E. –72.47, –72.47 F. –8981.2, –9015.9 G. –333.2, –338.1 H. +177.20, +184.73
6. + for $\Delta S > 0$, – for $\Delta S < 0$, u for unclear. A. – B. + C. + D. + E. + F. u G. – H. – I. u J. – K. – L. – M. u N. + O. u P. – Q. – R. + S. – T. + U. – V. u W. u X. +
7a. & 7b. In order, the numbers are $\Delta G°$ (in kJ/mol), $\Delta H°$ (in kJ/mol), and $\Delta S°$ (in J mol⁻¹ K⁻¹) A. –91.98, –176.89, –284.6, S, (2) B. +177.5, +226.3, +161.1, N (3) C. +11.9, +126.7, +335.2, N (2) D. +5.40, +58.03, +176.7. N (3) E. +255.8, +177.8, +160.7, N (3) F. –215.5, –224.1, –28.8, S (2) G. –38.3, –92.5, –182, S (2) H. –140.0, –196.6, –189.6, S (2) I. –4.90, –1.3, +11.5, S (1) J. –1009.2, –1125.3, –389.4, S (2) K. –33.28, –92.38, –198.3, S (2) L. –24.6, –90.7, –221, S (2) M. –2.4, +17.5, –27.2, S (4) N. +75.56, +75.36, +116, N (3) O. +173.38, +180.74, +24.7, N (3) P. –6.7, –133.2, –424.0, S (2) Q. –25.8, –75.9, –167.8, S (2) R. +14, +92, +265, B (3) S. + 8.2, –19.4, –92.3, N (2) T. –326.8, –284, +139, S (1) U. –76.1, –114.4, –128.4, S (2) V. +41.2, +28.5, +42.4, S (3) W. +16.8, +10.5, –21.4, N (4) X. –27.3, +5.4, +109.8, S (3)
8. Value for $\Delta G°$ (in kJ/mol) given first, followed by K_{eq}.200 K values given first, followed by 400 K values, separated by a semicolon. A. –119.97, 2.2×10^{31}; –63.05, 1.7×10^8
B. +34.98, 7.3×10^{-10}; +65.76, 2.6×10^{-9}
C. –58.85, 2.3×10^{15}; –52.23, 6.6×10^6
D. +22.69, 1.2×10^{-6}; –12.65, 44.9
E. +145.8, 8.5×10^{-39}; +113.62, 1.5×10^{-15}
F. –218.3, 1.1×10^{57}; –212.5, 5.8×10^{27}
G. –56.1, 4.5×10^{14}; –19.7, 374
H. –158.7, 2.8×10^{41}; –120.8, 5.9×10^{15}
I. –3.6, 8.71; –5.9, 5.9
J. –1047.42, 10^{274}; –969.54, 3.9×10^{126}
K. –52.72, 5.9×10^{13}; –13.06, 50.8
L. –46.5, 1.4×10^{12}; –2.3, 2.00
M. + 22.9, 1.0×10^{-6}; +28.4, 1.97×10^{-4}
N. +52.16, 2.4×10^{-14}; +29.86, 1.7×10^{-4}
O. +175.8, 1.2×10^{-46}; +170.86, 4.9×10^{-23}
P. –48.4, 4.4×10^{12}; +36.4, 1.8×10^{-5}
Q. –42.3, 1.1×10^{11}; –8.78, 14
R. + 39, 6.5×10^{-11}; –14, 67
S. –0.94, 1.76; +17.5, 5.15×10^{-3}
T. –312, 2.7×10^{81}; –340, 2.2×10^{44}
U. –88.7, 1.5×10^{23}; –63.0, 1.7×10^8
V. + 20.0, 5.9×10^{-6}; 11.5, 0.031
W. +14.8, 1.4×10^{-4}; +19.1, 3.2×10^{-3}
X. –16.6, 2.1×10^4; –38.5, 1.1×10^5
9. Transitions are all from the first to the second substance. Values given are ΔH°_{tr} (in kJ/mol), ΔS°_{tr} (in J mol⁻¹ K⁻¹), and T_{tr} in order unless otherwise indicated. A. –67.2, –152, 442 K B. –22, –96, +231 J mol⁻¹ K⁻¹ = $S°$ of $AsF_3(l)$ C. + 660.3, +183.3, 3602 K D. +23.4, +80.9, 289 K E. –214.8, –93.3, 24.8 kJ/mol = ΔH°_f of $Br_2(g)$ F. –12, –5, 2400 K G. –62.25, –143.9, 432.6 K H. 1.55, 0.5, 3100 K I. 0.50, 1.25, 71.95 J mol⁻¹ K⁻¹ = $S°$ for HgO (yellow).
11. (1) Expression for K_{eq} is followed by value of K_{eq} at 298 K. A. $1/P(NH_3)P(HCl) = 1.3 \times 10^{16}$
B. $P(CO_2) = 8 \times 10^{-32}$ C. $P(NH_3)P(CO_2) = 8.2 \times 10^{-3}$
D. $P(NO_2)^2/P(N_2O_4) = 0.113$ E. $P(CO_2) = 1.4 \times 10^{-45}$
F. $P(CO_2)/P(SO_3) = 6.0 \times 10^{37}$ G. $P(PCl_5)/P(PCl_3)(Cl_2) = 5.2 \times 10^6$ H. $P(SO_2)^2/P(SO_2)^2P(O_2) = 3.5 \times 10^{24}$
I. $P(BrCl)^2/P(Br_2)P(Cl_2) = 7.23$ J. $P(SO_2)^2/P(H_2S)^2P(O_2)^3 = 7 \times 10^{176}$ K. $P(NH_3)^2/P(N_2)P(H_2)^3 = 6.8 \times 10^5$
L. $P(CH_3OH)/P(CO)P(H_2)^2 = 2.1 \times 10^4$
M. $P(CH_2Cl_2)^2/P(CH_4)P(CCl_4) = 2.63$
N. $P(NO)^2P(Cl_2)/P(NOCl)^2 = 5.7 \times 10^{-14}$
O. $P(NO)^2/P(N_2)P(O_2) = 4.1 \times 10^{-31}$ P. $1/P(CO_2)P(NH_3)^2 = 15$ Q. $P(SO_2)/P(CO)^2 = 3.3 \times 10^4$ R. $P(SO_2)P(Cl_2) = 3.5 \times 10^{-3}$ S. $[CO_2]/P(CO_2) = 0.037$ T. $P(O_2)^3/P(O_3)^2 = 1.9 \times 10^{57}$ U. $P(H_2O)^2P(Cl_2)^2/P(HCl)^4P(O_2) = 2.2 \times 10^{13}$
V. $P(H_2O)P(CO)/P(H_2)P(CO_2) = 6.0 \times 10^{-8}$
W. $P(H_2)P(I_2)/P(HI)^2 = 1.1 \times 10^{-3}$
X. $P(H_2)P(CO)/P(HCHO) = 6.1 \times 10^4$
(2) Values of $\Delta G°$ (in kJ/mol): Y. 0.00 Z. 0.00 Γ. 17.1 Δ. 11.5 Θ. 28.7 Λ. –11.5 Ξ. –17.1 Π. – 28.7
13. A. 216.9 kJ/mol B. 0.147, 355 K C. 3.57×10^4, 385 K D. –17.32 kJ/mol E. 92.2 kJ/mol F. 22.2, 297 K G. 7.81×10^{-72}, 509 K

Quizzes

A: 1. (a) 2. (b) 3. (e) (–4.0) 4. (d) 5. (a) 6. (b) 7. (d) 8. (a)
B: 1. (e) (–80.66) 2. (c) 3. (c) 4. (e) 5. (d) 6. (c) 7. (a) 8. (c)
C: 1. (d) 2. (b) 3. (e) (–1796.9) 4. (a) 5. (a) 6. (a) 7. (b) 8. (b)
D: 1. (c) 2. (a) 3. (d) 4. (a) 5. (d) 6. (c) 7. (b) 8. (a)

Sample Test

1. a. $P(NH_3)^2/[Mg(OH)_2]^3$ b. $P(SO_2Cl_2)/P(SO_2)P(Cl_2)$
c. $1/P(CO)P(H_2)^2$ d. $1/[H_2SO_4]$
2. a. $\Delta G° = -RT \ln K_{eq}$
$= -(8.314 \text{ J mol}^{-1} \text{ K}^{-1})(298 \text{ K})\ln(1.87 \times 10^{-7})$
$= 38.38 \times 10^3$ J/mol. The reaction is non-spontaneous since $\Delta G° > 0$ and $K_{eq} < 1$.
b. $\Delta G° = \Delta H° - T\Delta S° = 38.38$ kJ/mol
$= \Delta H° - 298 \text{ K}(0.18192 \text{ kJ mol}^{-1} \text{ K}^{-1})$ or $\Delta H° = 92.60$ kJ/mol
c. As temperature decreases, the $-T\Delta S$ factor becomes less negative. Thus $\Delta G°$ becomes more positive, making K_{eq} smaller. Let's check.
$\Delta G° = 92.60 \text{ kJ/mol} - 200 \text{ K } (0.18192 \text{ kJ mol}^{-1} \text{ K}^{-1})$
$= 56.211 \text{ kJ/mol} = -RT \ln K_{eq}$ or $K_{eq} = 2.08 \times 10^{-15}$
This is in accord with LeChâtelier's principle. Lowering the temperature of an endothermic reaction shifts the equilibrium left.
d. $\Delta G° = \Delta H° - T\Delta S° = 92.60$ kJ/mol
$= 320 \text{ K}(0.18192 \text{ kJ mol}^{-1} \text{ K}^{-1}) = 34.39$ kJ/mol. We assumed that $\Delta H°$ and $\Delta S°$ remain constant with temperature.
3. CaO(s) is a solid and thus more ordered than any gas. The remaining gases have more and more atoms. There are increasingly many mays in which their molecules can bend and stretch, and thus their molecules are more disordered.
4. $\ln(1.456/14.2) = -(\Delta H°/8.314 \text{ J mol}^{-1} \text{ K}^{-1})(1/273 - 1/298)$;
$\Delta H° = +61.6$ kJ/mol

CHAPTER 21

Drill Problems

1. Sketches of the cells are in Figure A-2. The anode is the left half cell of each sketch. Cell diagrams follow.
A. $Pt(s)|H_2O_2(aq)|H(aq)||Cu^{2+}(aq)|Cu(s)$
B. $Mg(s)|Mg^{2+}(aq)||Ag^+(aq)|Ag(s)$
C. $Al(s)|Al^{3+}(aq)||H^+(aq)|H_2(g)|Pt(s)$
D. $Pb(s)|Pb^{2+}(aq)||Fe^{3+}(aq)|Fe(s)$
E. $Pt(s)|I_2(s)|I^-(aq)||F^-(aq)|F_2(g)|Pt(s)$
F. $Pt(s)|H_2(g)|OH^-(aq)||Cl^-(aq)|ClO^-(aq)|Pt(s)$
G. $Pt(s)|Cl_2(g)|Cl+\!-(aq)||Mn^{2+}(aq)|MnO_2(s)|Pt$
H. $Pt(s)|I_2(s)|IO_3^-(aq)||Br^-(aq)|Br_2(l)|Pt(s)$
I. $Ag(s)|Ag^+(aq)||NO_3^-(aq)|NO(g)|Pt(s)$
J. $Pt(s)|S(s)|SO_3^{2-}(aq)||Pb^{2+}(aq)|Pb(s)$
3. A. –0.345 V B. +3.17 V C. +1.66 V D. +0.090 V E. +2.33 V F. +1.718 V G. –0.13 V H. –0.130 V I. +0.16 V J. –0.58 V
4. A. –0.307 V B. +3.11 V C. +1.71 V D. +0.235 V E. +2.28 V F. +1.737 V G. +0.02 V H. –0.059 V I. +0.19 V J. –0.61 V
5. In order $\Delta G°$, K_{eq}, ΔG for each part. Free energies are in kJ/mol. A. +66.6, 2.2×10^{-12}, + 59.2
B. –612, 1×10^{107}, –598 C. –961, 2×10^{168}, –990.
D. –52, 1.3×10^9, –136 E. –450, 6×10^{78}, –440
F. –331.6, 1.3×10^{58}, – 335.2 G. 13, 6×10^{-3}, –2
H. 125, 1.1×10^{-22}, 57 I. –46, 1.3×10^8, –55
J. –224, 1.8×10^{39}, –235
9. A. Cl_2, H_2, – 1.746 V B. Cl_2, Cu, +1.064 V
C. Cl_2, H_2, –0.972 V D. O_2, H_2, –1.229 V
E. Cl_2, Cu, –1.002 V F. I_2, Fe, –1.04 V
10. A. 0.560 $\mathfrak{F}$ B. 226 g Sn C. 0.3174 g Cu D. 49.3 g Cu E. 1.28×10^3 g Hg F. 1.01×10^5 s = 28.0 hr G. 701 g Au H. 28.0 L O_2 I. 82.1 g Al J. 5.85 g Cr K. 61.1 g Au L. 40.2 g Ag
11. A. $CuSO_4$: 79.80 g/eq, H_2O_2: 17.01 g/eq, 0.320 N $CuSO_4(aq)$ B. $AgNO_3$: 169.87 g/eq, Mg: 12.15 g/eq, 0.152 N $AgNO_3(aq)$ C. H_2SO_4: 49.04 g/eq, Al: 8.99 g/eq, 0.0410 N $H_2SO_4(aq)$ D. $Fe_2(SO_4)_3$: 66.64 g/eq, Pb: 103.6 g/eq, 0.00473 N $Fe_2(SO_4)_3(aq)$ E. NaI: 149.89 g/eq, F_2: 19.00 g/eq, 0.0378 N NaI(aq) F. KClO: 45.28 g/eq, H_2: 1.01 g/eq, 0.0628 N KClO(aq) G. MnO_2: 43.47 g/eq, HCl, 36.46 g/eq, 0.0521 N HCl(aq) H. Br_2: 79.90 g/eq, I_2: 25.38 g/eq
I. HNO_3: 21.00 g/eq, Ag: 108.87 g/eq, 0.0817 N $HNO_3(aq)$
J. $Pb(NO_3)_2$: 165.5 g/eq, S: 8.02 g/eq K. O_2: 16.00 g/eq, H_2O: 9.01 g/eq L. Br_2:79.90 g/eq, H_2S: 17.04 g/eq
M. $SnSO_4$: 107.37 g/eq, $FeSO_4$: 151.90 g/eq, 0.177 N $SnSO_4(aq)$ N. H_2SO_4: 49.04 g/eq, S: 8.02 g/eq
O. I_2: 126.90 g/eq, I_2: 25.38 g/eq

Quizzes

A: 1. (e) (5) 2. (c) 3. (d) 4. (d) 5. (b) 6. (a) 7. (b) 8. (c) 9. (d) 10. (a)
B: 1. (c) 2. (c) 3. (e) (9) 4. (b) 5. (a) 6. (d) 7. (a) 8. (e) (Nernst) 9. (c) 10. (c)
C: 1. (d) 2. (a) 3. (b) 4. (d) 5. (e) 6., (b) 7. (d) 8. (a) 9. (e) (1.04 M) 10. (d)
D: 1. (d) 2. (c) 3. (a) 4. (a) 5. (c) 6. (e) 7. (b) 8. (e) (11.7) 9. (b) 10. (a)

Sample Test

1. a. $Al(s)|Al^{3+}(aq)||Pb^{2+}(aq)|Pb(s)$
b. $3\,(Pb^{2+} + 2\,e^- \longrightarrow Pb°)$ $E° = -0.126$ V
$\underline{2(Al \longrightarrow Al^{3+} + 3\,e^-)}$ $\underline{E° = +1.66 \text{ V}}$
$3\,Pb^{2+} + 2\,Al \longrightarrow 2\,Al^{3+} + 3\,Pb$ $E° = +1.53$ V
c. $E_{cell} = E°_{cell} - \dfrac{0.0592}{n} \log Q = 1.53 - \dfrac{0.0592}{6} \log \dfrac{(0.0200)^2}{(4.00)^3}$
$= 1.53 + 0.05 = 1.58$ V
2. $6.00 \times 10^{-4}\ \mathfrak{F}$ used
a. $Cu^{2+} + 2\,e^- \longrightarrow Cu$ $6.00 \times 10^{-4}\ \mathfrak{F} \times (\text{mol Cu}/2\ \mathfrak{F}) \times$ (1000 mmol Cu/mol Cu) = 0.300 mmol Cu
50.0 mL $\times$ 0.100 M = 5.00 mmol Cu^{2+}
(5.00 – 0.30) mmol = 4.70 mmol Cu^{2+} left in solution
$[Cu^{2+}]$ = 4.70 mmol/50.0 mL = 0.094 M
b. $2\,H_2O \longrightarrow O_2 + 4\,H^+ + 4\,e^-$
$6.00 \times 10^{-4}\ \mathfrak{F} \times (\text{mol } H^+/\mathfrak{F}) \times (1000 \text{ mmol } H^+/\text{mol } H^+)$
= 0.600 mmol H^+
$[H^+]$ = 0.600 mmol/50.0 mL = 0.0120 M or pH = 1.921
c. $2\,KMnO_4 + 3\,H_2S \longrightarrow 2\,MnO_2 + 3\,S + 2\,KOH + 2\,H_2O$
$3\,NiS + KClO_3 + 6\,HCl \longrightarrow 3\,NiCl_2 + KCl + 3\,S + 3\,H_2O$

CHAPTER 22

Drill Problems

2. A. $MgCl_2(l) \xrightarrow{\text{electrolysis}} Mg(l) + Cl_2(g)$
B. $2\,NaCl(l) \xrightarrow{\text{electrolysis}} 2\,Na(l) + Cl_2(g)$
C. $2\,Al_2O_3(l) + 3\,C(s) \xrightarrow{\text{electrolysis}} 4\,Al(l) + 3\,CO_2(g)$
D. $Al_2O_3(s) + 2\,OH^-(aq) + 3\,H_2O \longrightarrow 2\,Al(OH)_4^-(aq)$;
$Al(OH)_4^-(aq) + H_3O^+ \longrightarrow Al(OH)_3(s) + 2\,H_2O$

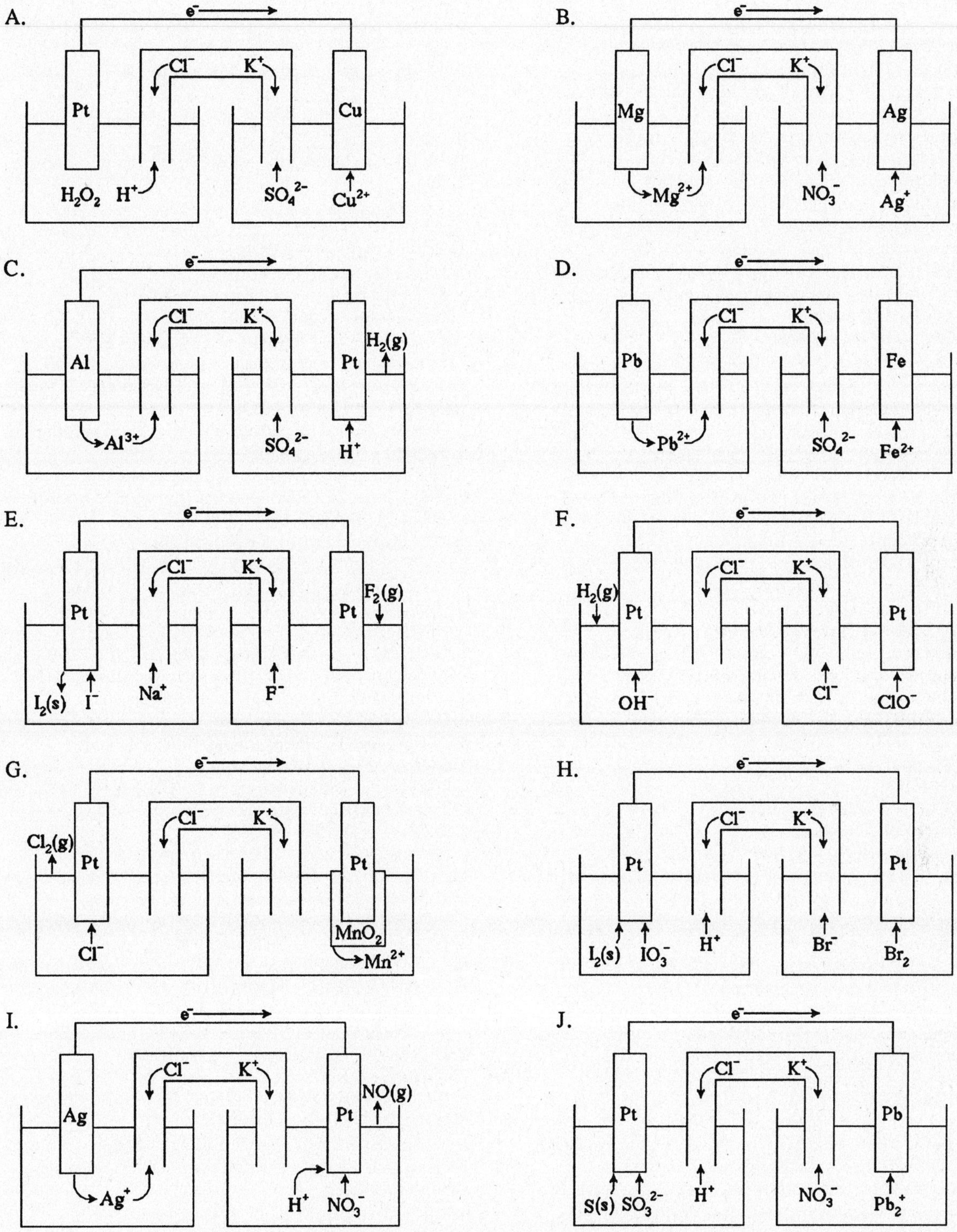

FIGURE A-2 Answers to Objective 21-1 Drill Problems

E. $Mg^{2+}(aq) + 2\ OH^-(aq) \longrightarrow Mg(OH)_2(s)$
F. $CaCO_3(s) \xrightarrow{\Delta} CaO(s) + CO_2(g)$
G. from question F plus $CaO(s) + H_2O \longrightarrow Ca(OH)_2(aq)$
5. $NaAl(SO_4)_2 \cdot 12\ H_2O$, sodium alum; $KAl(SO_4)_2 \cdot 12\ H_2O$, alum; $NH_4Al(SO_4)_2 \cdot 12\ H_2O$, ammonium alum; $TlAl(SO_4)_2 \cdot 12\ H_2O$, thallous alum; $NaV(SO_4)_2 \cdot 12\ H_2O$, sodium vanadium(III) alum; $KV(SO_4)_2 \cdot 12\ H_2O$, vanadium(III) alum; $NH_4V(SO_4)_2 \cdot 12\ H_2O$, ammonium vanadium(III) alum; $TlV(SO_4)_2 \cdot 12\ H_2O$, thallous vanadium(III) alum; $NaCr(SO_4)_2 \cdot 12\ H_2O$, sodium chromium(III) alum; $KCr(SO_4)_2 \cdot 12\ H_2O$, chromium(III) alum; $NH_4Cr(SO_4)_2 \cdot 12\ H_2O$, ammonium chromium(III) alum; $TlCr(SO_4)_2 \cdot 12\ H_2O$, thallous chromium(III) alum; $NaFe(SO_4)_2 \cdot 12\ H_2O$, sodium iron(III) alum; $KFe(SO_4)_2 \cdot 12\ H_2O$, iron(III) alum; $NH_4Fe(SO_4)_2 \cdot 12\ H_2O$, ammonium iron(III) alum; $TlFe(SO_4)_2 \cdot 12\ H_2O$, thallous iron(III) alum
6. A. $ZnCO_3(s) \xrightarrow{\Delta} ZnO(s) + CO_2(g)$
B. $2\ ZnS(s) + 3\ O_2(g) \xrightarrow{\Delta} 2\ ZnO(s) + 2\ SO_2(g)$
C. $NiO(s) + H_2(g) \xrightarrow{\Delta} Ni(s) + H_2O(g)$ D. $HgS(s) + O_2(g) \xrightarrow{\Delta} Hg(l) + SO_2(g)$ E. $Fe_2O_3(s) + 3\ H_2(g) \xrightarrow{\Delta} 2\ Fe(s) + 3\ H_2O(g)$ F. $SO_2(g) + 2\ C(s) \xrightarrow{\Delta} S(s) + 2\ CO(g)$
G. $SnO_2(s) + 2\ C(s) \xrightarrow{\Delta} Sn(l) + 2\ CO(g)$
H. $PbO(s) + CO(g) \xrightarrow{\Delta} Pb(l) + CO_2(g)$ I. $PbS(s) + 3\ O_2(g) \xrightarrow{\Delta} 2\ PbO(s) + 2\ SO_2(g)$ J. $CuO(s) + H_2(g) \xrightarrow{\Delta} Cu(s) + H_2O(g)$ K. $2\ CdS(s) + 3\ O_2(g) \xrightarrow{\Delta} 2\ CdO(s) + SO_2(g)$
7. A. no reaction B. $2\ K(s) + 2\ HCl(aq) \longrightarrow 2\ KCl(aq) + H_2(g)$ C. $Zn(s) + 2\ HCl(aq) \longrightarrow ZnCl_2(aq) + H_2(g)$ D. no reaction
E. $2\ Al(s) + 2\ NaOH(aq) + 6\ H_2O \longrightarrow 2\ NaAl(OH)_4(aq) + 3\ H_2(g)$ F. $Sn(s) + 2\ H^+(aq) + 4\ Cl^-(aq) \longrightarrow H_2(g) + SnCl_4^{2-}(aq)$ G. $Pb(s) + 2\ H_2SO_4(conc.) \longrightarrow PbSO_4(s) + SO_2(g) + 2\ H_2O$ H. $2\ Al(s) + 6\ HCl(aq) \longrightarrow 2\ AlCl_3(aq) + 3\ H_2(g)$ I. $Cd(s) + H_2SO_4(aq) \longrightarrow CdSO_4(aq) + H_2(g)$ J. $6\ Hg(l) + 8\ HNO_3(dil.) \longrightarrow 3\ Hg_2(NO_3)_2(aq) + 2\ NO(g) + 4\ H_2O$
K. $Pb(s) + 4\ HNO_3(conc.) \longrightarrow Pb(NO_3)_2(aq) + 2\ NO_2(aq) + 2\ H_2O$
L. $5\ Be(s) + 12\ HNO_3(conc.) \longrightarrow 5\ Be(NO_3)_2(aq) + N_2(g) + 6\ H_2O$
M. $Be(s) + 2\ NaOH(aq) + 2\ H_2O \longrightarrow Na_2Be(OH)_4(aq) + H_2(g)$
N. no reaction
8. A. $ZnO(s) + 2\ HCl(aq) \longrightarrow ZnCl_2(aq) + H_2O$
B. $SnO_2(s) + 2\ H_2SO_4(aq) \longrightarrow Sn(SO_4)_2(aq) + 2\ H_2O$
C. $ZnO(s) + H_2O + 2\ NaOH \longrightarrow Na_2Zn(OH)_4(aq)$
D. $SnO(s) + 2\ HCl \longrightarrow SnCl_2(aq) + H_2O$
E. $Zn(OH)_2(s) + H_2SO_4(aq) \longrightarrow ZnSO_4(aq) + 2\ H_2O$
F. $Sn(OH)_2(s) + 2\ H_2O \longrightarrow Na_2Sn(OH)_4$
G. $SnO_2(s) + 2\ NaOH(aq) + 2\ H_2O \longrightarrow Na_2Sn(OH)_6$
H. $Sn(OH)_4(s) + 2\ H_2SO_4 \longrightarrow Sn(SO_4)_2(aq) + 4\ H_2O$
I. $Sn(OH)_4(s) + CaO(s) + H_2O \longrightarrow CaSn(OH)_6$
J. $Al_2O_3(s) + CaO(s) + 4\ H_2O \longrightarrow Ca[Al(OH)_4]_2$
10. A. K_2CO_3, CdS, PbO, $CaCO_3$ B. Cd, In, Pb C. KOH, K_2CO_3, CdS D. KNO_3, $Ca(NO_3)_2 \cdot 2\ H_2O$, $Ca_3(PO_4)_2$
E. $KClO_3$, $SrCl_2$, $Sr(NO_3)_2$, $Ba(NO_3)_2$, $Ba(ClO_3)_2$ F. Rb, Ba
G. Cd, Zn, $ZnCrO_4$, Sn, Pb_3O_4 H. Na_3AlF_6, $CaCO_3$, Zn
I. Rb, Cs, ZnO, CdS J. PbO_2, $Hg(CNO)_2$, $KClO_3$
K. $BiONO_3$, $(BiO)_2CO_3$, SnF_2, Hg_2Cl_2, ZnO L. Bi_2O_3, $ZnCl_2$
M. $HgCl_2$, $ZnSO_4$, $ZnCl_2$ N. ZnS, CdS O. $Al_2(SO_4)_3$, $CaCO_3$, $BaSO_4$ P. $SnCl_2$, Ca Q. KNO_3, $KClO_3$, $Pb(C_2H_3O_2)_4$ R. CdO, Ca, Cd, $CdSO_4$, HgO, Pb S. CdO, Cs_2SO_4 T. K_2CO_3, $ZnSO_4$, CdS U. ZnO, BiOCl

Quizzes

A: 1. (e) 2. (e) 3. (c) 4. (d) 5. (c) 6. (b) 7. (c) 8. (c)
B: 1. (d) 2. (e) 3. (b) 4. (c) 5. (d) 6. (b) 7. (e) 8. (b)
C: 1. (d) 2. (e) 3. (d) 4. (d) 5. (c) 6. (d) 7. (b) 8. (c)
D: 1. (a) 2. (b) 3. (d) 4. (a) 5. (b) 6. (e) 7. (e) 8. (e)

Sample Test

1. $2\ C(s) + O_2(g) \longrightarrow 2\ CO(g)$
$2\ FeO(s) \longrightarrow 2\ Fe(s) + O_2(g)$
$2\ C(s) + 2\ FeO(s) \longrightarrow 2\ Fe(s) + 2\ CO(g)$
$\Delta H°_{rxn} = -221 - (-544) = +323$ kJ/mol
$\Delta S°_{rxn} = +179.5 - (-138.1) = +317.6$ J mol^{-1} K^{-1}
The reaction becomes spontaneous where $\Delta G° = 0 = \Delta H° - T\Delta S°$. We solve for $T = \Delta H°/\Delta S°$ = (323 kJ/mol)/(0.3176 kJ mol^{-1} K^{-1}) = 1017 K = 744°C.
2. Only one method (of perhaps several) is given in each part.
a. $3\ PbS(s) + 8\ HNO_3(aq,dil) \longrightarrow 3\ Pb(NO_3)_2(aq) + 2\ NO(g) + 3\ S(s)$ [filter solution]; $Pb(NO_3)_2(aq) + 2\ HCl(aq) \longrightarrow PbCl_2(s) + 2\ HNO_3(aq)$ [filter solution] b. $HgS(s) + O_2(g) \xrightarrow{\Delta} Hg(l) + SO_2(g)$; $6\ Hg(l) + 8\ HNO_3(aq,dil) \longrightarrow 3\ Hg_2(NO_3)_2(aq) + 2\ NO(g) + 4\ H_2O$; $2\ NaOH(aq) + H_2SO_4(aq) \longrightarrow Na_2SO_4(aq) + 2\ H_2O$; $Hg_2(NO_3)_2(aq) + Na_2SO_4(aq) \longrightarrow Hg_2SO_4(s) + 2\ NaNO_3(aq)$ c. $Al_2O_3(s) + 3\ H_2O \longrightarrow 2\ Al(OH)_3(aq)$; $CaCO_3(s) \xrightarrow{\Delta} CaO(s) + CO_2(g)$; $CaO(s) + H_2O \longrightarrow Ca(OH)_2(aq)$; $2\ Al(OH)_3(s) + 3\ Ca(OH)_2(aq) \longrightarrow Ca_3(AlO_3)_2(s) + 6\ H_2O$ [filter solutions; all solutions are saturated solutions of sparingly soluble compounds];
d. $CdS(s) + 8\ HNO_3(aq,dil) \longrightarrow 3\ Cd(NO_3)_2(aq) + 2\ NO(g) + 3\ S(s)$ [filter solution] $CaCO_3(s) \xrightarrow{\Delta} CaO(s) + CO_2(g)$; $Cd(NO_3)_2(aq) + 2\ NaOH(aq) \longrightarrow Cd(OH)_2(s) + 2\ NaNO_3(aq)$ [filter solution] $Cd(OH)_2$(s, slurry in water) + $CO_2(g) \longrightarrow CdCO_3(s) + H_2O$ [filter solution]
e. Mg^{2+}(aq, seawater) + 2 OH^- (aq, from NaOH) $\longrightarrow Mg(OH)_2(s)$ [filter off solid]; $Mg(OH)_2(s) + H_2SO_4(aq) \longrightarrow MgSO_4(aq) + 2\ H_2O$ [evaporate solution to dryness]
f. $SnO_2(s) + 2\ C(s) \xrightarrow{\Delta} Sn(l) + 2\ CO(g)$ [solidify metal]; $Sn(s) + 2\ HCl(aq) \longrightarrow SnCl_2(aq) + H_2(g)$; $SnCl_2(aq) + 2\ NaOH(aq) \longrightarrow 2\ NaCl(aq) + Sn(OH)_2(s)$ [filter off solid]; $Sn(OH)_2(s) + NaOH(aq) \longrightarrow NaSn(OH)_3(aq)$ [evaporate solution to dryness]; $NaSn(OH)_3(s) + NaOH(s) \xrightarrow{\Delta} Na_2SnO_2(s) + 2\ H_2O(g)$ g. $ZnS(s) + 2\ HCl(aq) \longrightarrow ZnCl_2(aq) + H_2S(g)$; $ZnCl_2(aq) + 2\ NaOH(aq) \longrightarrow Zn(OH)_2(s) + 2\ NaCl(aq)$; $Zn(OH)_2(s) \xrightarrow{\Delta} ZnO(s) + H_2O(g)$
h. $BaCO_3(s) + H_2SO_4(aq) \longrightarrow BaSO_4(s) + H_2O + CO_2(g)$ [filter off solid] i. $2\ KCl(l) \xrightarrow{electrolysis} 2\ K(l) + Cl_2(g)$ [solidify metal]; $2\ K(s) + 2\ H_2O \longrightarrow 2\ KOH(aq) + H_2(g)$; $KOH(aq) + HNO_3(aq) \longrightarrow KNO_3(aq) + H_2O$ [evaporate solution to dryness]

CHAPTER 23

Drill Problems

1. A. $O_2 + 4\ H^+ + 4\ e^- \longrightarrow 2\ H_2O$, + 1.229 V
B. $O_2 + 2\ H_2O + 4\ e^- \longrightarrow 4\ OH^-$, +0.401 V
C. $HO_2^- + H_2O + 2\ e^- \longrightarrow 3\ OH^-$, + 0.878 V

D. $HClO + H^+ + 2\ e^- \longrightarrow Cl^- + H_2O$, +1.49 V
E. $2\ ClO_4^- + 16\ H^+ + 14\ e^- \longrightarrow Cl_2 + 8\ H_2O$, +1.39 V
F. $ClO_4^- + 8\ H^+ + 8\ e^- \longrightarrow Cl^- + 4\ H_2O$, +1.39 V
G. $HClO_2 + 3\ H^+ + 4\ e^- \longrightarrow Cl^- + 2\ H_2O$, +1.56 V
H. $2\ ClO_3^- + 6\ H_2O + 10\ e^- \longrightarrow Cl_2 + 12\ OH^-$, +0.28 V
I. $2\ ClO_4^- + 8\ H_2O + 14\ e^- \longrightarrow Cl_2 + 16\ OH^-$, +0.45 V
J. $ClO_2^- + 2\ H_2O + 4\ e^- \longrightarrow Cl^- + 4\ OH^-$
K. $NO_3^- + 4\ H^+ + 3\ e^- \longrightarrow NO + 2\ H_2O$, +0.96 V
L. $NO_3^- + 2\ H_2O + 3\ e^- \longrightarrow NO + 4\ OH^-$, –0.14 V
M. $NO_3^- + 3\ H^+ + 2\ e^- \longrightarrow HNO_2 + H_2O$
N. $NO_3^- + H_2O + 2\ e^- \longrightarrow NO_2^- + 2\ OH^-$, +0.02 V
O. $NO_3^- + 10\ H^+ + 8\ e^- \longrightarrow NH_4^+ + 3\ H_2O$, +0.88 V
P. 2. $HNO_2 + 6\ H^+ + 6\ e^- \longrightarrow N_2 + 4\ H_2O$, +1.45 V
Q. $SO_4^{2-} + 8\ H^+ + 6\ e^- \longrightarrow S + 4\ H_2O$, +0.36 V
R. $SO_4^{2-} + 4\ H_2O + 8\ e^- \longrightarrow S^{2-} + 8\ OH^-$, –0.68 V

2. A. $2\ NaCl(aq) + 2\ H_2O \xrightarrow{\text{electrolysis}}$ $2\ NaOH(aq) + H_2(g) + Cl_2(g)$ [in a Downs cell] B. $2\ KBr(aq) + Cl_2(g$, from A) $\longrightarrow 2\ KCl(aq) + Br_2(l)$ [separate denser $Br_2(l)$] C. $2\ KI(s) + Cl_2(g$, from A) $\xrightarrow{\Delta}$ $2\ KCl(s) + I_2(g)$ [deposit $I_2(s)$ on a cool surface] D. $6\ Cl_2(g$, from A) $+ 6\ Ba(OH)_2(aq) \xrightarrow{\Delta}$ $5\ BaCl_2(aq) + Ba(ClO_3)_2(aq) + 6\ H_2O$ [separate salts by fractional crystallization]; $Ba(ClO_3)_2(aq) + H_2SO_4(aq) \longrightarrow$ $BaSO_4(s) + 2\ HClO_3(aq)$ [filter off solid]; $HClO_3(aq) + KOH(aq) \longrightarrow KClO_3(aq) + H_2O$ [carefully evaporate solution to dryness] E. $NaI(s) + H_3PO_4(aq) \longrightarrow NaH_2PO_4(aq) + HI(g)$ F. first two reactions of question D G. $2\ HClO_3(aq$, from F) $+ H_2C_2O_4(s) \longrightarrow 2\ ClO_2(g) + 2\ CO_2(g) + 2\ H_2O$; $4\ ClO_2(g) + 2\ Ba(OH)_2(aq) \longrightarrow Ba(ClO_2)_2(s) + Ba(ClO_3)_2(aq) + 2\ H_2O$ [filter off solid]; $Ba(ClO_2)_2(s) + H_2SO_4(aq) \longrightarrow BaSO_4(s) + 2\ HClO_2(aq)$ [filter off solid] H. First two equations of question G I. $ClO_3^-(aq$, from F) $+ H_2O \xrightarrow{\text{electrolysis}} ClO_4^-(aq) + H_2(g)$ [in acidic solution] J. $2\ NaF(l) \xrightarrow{\text{electrolysis}} 2\ Na(l) + F_2(g)$ [in a Downs-type cell]; $Cl_2(g$, from A] $+ 3\ F_2(g) \longrightarrow$ $2\ ClF_3(g)$ [controlled ratio of reactants]; $KI(aq) + I_2(s$, from C) $\longrightarrow KI_3(aq)$ L. First reaction of question J M. $NaF(s) + H_2SO_4(aq) \longrightarrow NaHSO_4(aq) + HF(g)$ N. $NaCl(s) + H_2SO_4(aq) \longrightarrow NaHSO_4(aq) + HCl(aq)$ O. $KBr(s) + H_3PO_4(aq) \longrightarrow NaH_2PO_4(aq) + HBr(g)$ P. $Cl_2(g) + 2\ NaOH(aq) \longrightarrow NaCl(aq) + NaOCl(aq) + H_2O$

3. Shape abbreviations are as in the answers for objective 10-12. A. d^2sp^3, sqpyr B. sp^3, tet C. d^2sp^3, oct D. sp^2, tripl E. dsp^3, T F. dsp^3, lin G. sp^3, tet H. d^2sp^3, sqpyr I. d^3sp^3, pentagonal bipyramid J. dsp^3, T K. sp^2, bent L. sp^3, bent M. sp^3, tet N. sp^3, tripyr O. d^2sp^3, sqpyr P. sp^3, tripyr

7. A. $CuS + 8\ HNO_3 \longrightarrow 3\ Cu(NO_3)_2 + 3\ S + 2\ NO + 4\ H_2O$ B. no reaction C. $Na_2S + H_2O \longrightarrow$ $2\ Na^+(aq) + HS^-(aq) + OH^-(aq)$ D. no reaction E. no reaction F. $3\ CdS + 8\ HNO_3 \longrightarrow 3\ Cu(NO_3)_2 + 3\ S + 2\ NO + 4\ H_2O$ G. $As_2S_3 + (NH_4)_2S \longrightarrow 2\ AsS_2^- + 2\ NH_4^+$ H. no reaction I. $ZnS + H^+(aq) \longrightarrow Zn^{2+}(aq) + HS^-(aq)$ J. $CaS + H_2O \longrightarrow$ $Ca^{2+}(aq) + HS^-(aq) + OH^-(aq)$ K. $CoS + H^+(aq) \longrightarrow$ $Co^{2+}(aq) + HS^-(aq)$ L. no reaction M. no reaction

10. A. sodium metasilicate B. pyroarsenic acid C. calcium pyroborate D. magnesium metastannate E. $H_6Si_2O_7$ F. $NaAlO_2$ G. $Ca_2P_2O_7$ H. $Fe(BO_2)_2$ I. $BaCO_3$ J. $Mg_3(BiO_4)_2$

12. A. F2 > $Cl_2 > Br_2 > I_2$ B. $HBrO > BrO_3^- > BrO_4^-$ C. $ClO_4^- > SO_4^{2-} > HPO_4^{2-}$ D. "H_2SO_3" > $Na_2SO_3 > SO_4^{2-}$ E. $HNO_2 > NaNO_2 > NO_3^-$ F. $H_3PO_3 > H_2PO_3^- > HPO_4^{2-}$ G. $ClO_4^- > HSO_4^- > HPO_4^{2-} > H_4SiO_4$ H. $BrO_3^- > SeO_3^{2-} > H_2AsO_3^- > H_4GeO_4$

13. A. AgBr, AgI B. $NaClO_2$, HCN C. I2, $Na_2S_2O_3$ D. $KClO_4$, $Pb(N_3)_2$ E. $KClO_4$, NO_2 F. $(NaPO_3)_n$, Na_2CO_3 G. I_2, Li_2CO_3 H. NO_2, HNO_3 I. ClF_3, BrF_3, HBr, ICl J. NaClO, $NaClO_2$, $Na_2S_2O_4$ K. NaClO, I2 L. HCN, NaF

Quizzes

A: 1. (c) 2. (d) 3. (e) 4. (d) 5. (e) 6. (c) 7. (c) 8. (e)
B: 1. (a) 2. (b) 3. (e) 4. (d) 5. (a) 6. (d) 7. (e) 8. (b)
C: 1. (e) 2. (d) 3. (c) 4. (c) 5. (a) 6. (a) 7. (c) 8. (c)
D: 1. (c) 2. (e) 3. (c) 4. (c) 5. (a) 6. (a) 7. (b) 8. (a)

Sample Test

1. a. $2\ HBrO + 2\ H^+ + 2\ e^- \longrightarrow Br_2 + 2\ H_2O$ $E° = +1.59$ V
$HBrO + 2\ H_2O \longrightarrow BrO_3^- + 5\ H^+ + 4\ e^-$ $E° = +1.49$ V
$5\ HBrO \longrightarrow 2\ Br_2 + BrO_3^- + 2\ H_2O + 2\ H^+$ $E° = +3.08$ V
Yes, it is spontaneous
b. $Br_2 + 2\ H_2O \longrightarrow 2\ HBrO + 2\ H^+ + 2\ e^-$ $E° = -1.59$ V
$2\ HBrO + 4\ H_2O \longrightarrow 2\ BrO_3^- + 10\ H^+ + 8\ e^-$ $E° = +1.49$ V
$Br_2 + 6\ H_2O \longrightarrow 2\ BrO_3^- + 12\ H^+ + 10\ e^-$
$E° = [2(-1.59) + 8(1.49)]/10 = +0.874$ V

2. a. $Br_2(l) + 6\ NaOH(aq) \xrightarrow{\Delta}$ $2\ NaBrO_3(aq) + 5\ NaBr(aq) + 3\ H_2O$ b. $H_2(g) + Br_2(g) \longrightarrow$ $2\ HBr(g)$, $PBr_3(g) + 3\ H_2O \longrightarrow H_3PO_3(aq) + 3\ HBr(g)$, $NaBr(s) + H_3PO_4(aq) \longrightarrow NaH_2PO_4(aq) + HBr(g)$
c. $2\ H_2O \xrightarrow{\text{electrolysis}} 2\ H_2(g) + O_2(g)$, $2\ H_2O_2$(l *or* aq) $\longrightarrow$ $2\ H_2O + O_2(g)$, $2\ HgO(s) \xrightarrow{\Delta} 2\ Hg(l) + O_2(g)$
d. $3\ NaClO(s) \xrightarrow{\Delta} 2\ NaCl(s) + NaClO_3(s)$
e. $CdS(s) + 2\ HCl(aq) \longrightarrow CdCl_2(aq) + H_2S(g)$
f. $2\ NO(g) + O_2(g) \longrightarrow 2\ NO_2(g)$, $N_2O_4(g) \rightleftharpoons 2\ NO_2(g)$, $2\ Pb(NO_3)_2(s) \xrightarrow{\Delta} 2\ PbO(s) + 4\ NO_2(g) + O_2(g)$
g. $2\ Cu(s) + 8\ HNO_3(aq) + 2\ NO(g) + 4\ H_2O$, $2\ NO(g) + O_2(g) \longrightarrow 2\ NO_2(g) \rightleftharpoons N_2O_4(g)$ [favored by high P and low T]
h. $N_2H_4(l) + 2\ H_2O_2(l) \longrightarrow N_2(g) + 4\ H_2O(g)$
i. $P_4O_{10}(s) + 6\ H_2O \longrightarrow 4\ H_3PO_4(l) \xrightarrow{\Delta}$ $2\ H_4P_2O_7(l) + 2\ H_2O(g)$ j. $Na_2S_2O_8(s) + 2\ H_2O \longrightarrow$ $2\ NaHSO_4(aq) + H_2O_2(aq)$

CHAPTER 24

Drill Problems

1. A. Fe < Mn < Ti < Sc B. Sc < Cr < Fe < Os C. Sc < Ti < Cr < Fe D. $MnO < MnO_2 < MnO_3 < Mn_2O_7$ E. Sc < Mn < Fe < Zn F. Sc < V < Fe < Mn G. Ti < Nb < Pb < Pt H. $VO < V_2O_3 < VO_2 < V_2O_5$ I. Y < Ti < V < Mn J. $Cu^+ < V^{3+} < Ni^{2+} < Fe^{3+}$ K. Fe < Cr < Ti < Ca

3. A. Ti, V B. Sc C. Cu D. Cu E. Sc F. Cu, Co, Ni G. Zn H. Zn, Ni, Cr I. Fe J. Ti, V K. Fe, Co, Ni

4. A. TiO_2 B. ZnS C. CoO, Cu_2O, CuO, Cr_2O_3 D. Cr_2O_3, $PbCrO_4$, $FeSO_4$, $Fe_4[Fe(CN)_6]_3$, Fe_2O_3, $ZnCrO_4$ E. CuCl, $Cu_3(AsO_4)_2$ F. ZnO G. V_2O_5 H. $KMnO_4$, $Na_2Cr_2O_7$ I. $KCr(SO_4)_2 \cdot 12H_2O$, $FeSO_4$, $CuSO_4$ J. Na_2CrO_4, $ZnCrO_4$ K. $FeCl_3$, $CuSO_4$ L. MnO_2, Ni_2O_3

5. A. $Cu_2S(l) + O_2(g) \longrightarrow 2\ Cu(l) + SO_2(g)$ B. $2\ C(s) + O_2(g) \longrightarrow 2\ CO(g)$ C. $TiO_2(s) + 2\ Cl_2(g) + C(s) \longrightarrow TiCl_4(g) + 2\ CO(g)$ D. $CaO(s) + SiO_2(l) \longrightarrow CaSiO_3(l)$ E. $TiO_2(s) + C(s) \longrightarrow$ $Ti(l) + CO_2(g)$ F. $2\ CuFeS_2(s) + 4\ O_2(g) \longrightarrow$ $Cu_2S(l) + 2\ FeO(s) + 3\ SO_2(g)$ G. $4\ Ag(s) + 8\ CN^-(aq) + O_2(g) + 2\ H_2O \longrightarrow$ $4\ [Ag(CN)_2^-](aq) + 4\ OH^-(aq)$ H. $FeO(s) + SiO_2(l) \longrightarrow$ $FeSiO_3(l)$

7. Standard cell potentials follow; spontaneous reactions have positive $E°$ values. A. +0.37 V B. +0.01 V C. –1.211 V

D. +1.09 V E. −0.47 V F. −0.63 V G. −0.616 V H. −0.3 V

8. A. $K_2Cr_2O_7 + 6\ KBr + 7\ H_2SO_4 \longrightarrow$
$3\ Br_2 + 4\ K_2SO_4 + Cr_2(SO_4)_3 + 7\ H_2O$
B. $2\ KMnO_4 + 5\ Na_2C_2O_4 + 8\ H_2SO_4 \longrightarrow$
$2\ MnSO_4 + 10\ CO_2 + 5\ Na_2SO_4 + K_2SO_4 + 8\ H_2O$
C. $H_2S + 2\ KMnO_4 \longrightarrow 2\ KOH + SO_2 + 2\ MnO_2$
D. $K_2Cr_2O_7 + 3\ SnSO_4 + 7\ H_2SO_4 \longrightarrow$
$3\ Sn(SO_4)_2 + K_2SO_4 + Cr_2(SO_4)_3 + 7\ H_2O$
E. $2\ KMnO_4 + 3\ H_2O_2 \longrightarrow 3\ O_2 + 2\ MnO_2 + 2\ H_2O + 2\ KOH$
F. $5\ Cd + 2\ KMnO_4 + 8H_2SO_4 \longrightarrow$
$5\ CdSO_4 + 2\ MnSO_4 + K_2SO_4 + 8\ H_2O$
G. $K_2Cr_2O_7 + 3\ HNO_2 + 5\ HCl \longrightarrow$
$2\ KNO_3 + 4\ H_2O + CrCl_3 + CrCl_2NO_3$
H. $2\ NaMnO_4 + 6\ NaI + 4\ H_2O \longrightarrow$
$3\ I_2(s) + 2\ MnO_2 + 8\ NaOH$
I. $2\ Na_2CrO_4 + 3\ SnCl_2 + 16\ HCl \longrightarrow$
$3\ SnCl_4 + 2\ CrCl_3 + 4\ NaCl + 8\ H_2O$
J. $8\ HMnO_4 + 5\ AsH_3 + 8\ H_2SO_4 \longrightarrow$
$5\ H_3AsO_4 + 8\ MnSO_4 + 12\ H_2O$ K. $2\ Cr(OH)_3 + 3\ Na_2O_2 \longrightarrow$
$2\ Na_2CrO_4 + 2\ NaOH + 2\ H_2O$
L. $2\ KMnO_4 + 10\ FeSO_4 + 8\ H_2SO_4 \longrightarrow$
$5\ Fe_2(SO_4)_3 + 2\ MnSO_4 + K_2SO_4 + 8\ H_2O$

13. Ions that are present: A. Na^+, Sr^{2+}, Zn^{2+} B. K^+, Fe^{3+} C. Ba^{2+}, Ni^{2+}

Quizzes

A: 1. (a) 2. (e) 3. (a) 4. (e) (MnO_2) 5. (c) 6. (b) 7. (b)
B: 1. (d) 2. (e) 3. (a) 4. (d) 5. (a) 6. (c) 7. (d)
C: 1. (b) 2. (a) 3. (b) 4. (a) 5. (b) 6. (e) 7. (e)
D: 1. (e) 2. (a) 3. (b) 4. (b) 5. (c) 6. (c) 7. (c)

Sample Test

1. a. Dissolve in nitric acid. $Fe(s) + 2\ H^+ \longrightarrow Fe^{2+} + H_2(g)$ $Ni(s) + 2\ H^+ \longrightarrow Ni^{2+} + H_2(g)$ $3\ Cu(s) + 2\ NO_3^- + 8\ H^+ \longrightarrow 3\ Cu^{2+} + 2\ NO(g) + 4\ H_2O$
b. Treat with NH_3-NH_4Cl, $(NH_4)_2S$. $Cu^{2+} + 4\ NH_3(aq) \longrightarrow [Cu(NH_3)_4]^{2+}$; $Fe^{2+} + (NH_4)_2S(aq) \longrightarrow FeS(s) + 2\ NH_4^+$; $Ni^{2+} + (NH_4)_2S(aq) \longrightarrow NiS(s) + 2\ NH_4^+$
c. Treat remaining solution with Zn(s). $Zn(s) + [Cu(NH_3)_4]^{2+} \longrightarrow Cu(s) + [Zn(NH_3)_4]^{2+}$
d. Treat precipitate from step b with aqua regia. $FeS(s) + NO_3^- + 4\ H^+ \longrightarrow Fe^{3+} + NO(g) + S(s) + 2\ H_2O$; $3\ NiS(s) + 2\ NO_3^- + 8\ H^+ \longrightarrow 3\ Ni^{2+} + 2\ NO(g) + 3\ S(s) + 4\ H_2O$
e. Treat solution with $NH_3(aq)$ $Fe^{3+} + 3\ NH_3(aq) + 3\ H_2O \longrightarrow Fe(OH)_3(s) + 3\ NH_4^+$; $Ni^{2+}(aq) + 6\ NH_3(aq) \longrightarrow [Ni(NH_3)_6]^{2+}$
f. Test solution from e with dimethylglyoxime. $[Ni(NH_3)_6]^{2+} + 2\ DMG^- \longrightarrow Ni(DMG)_2$(scarlet, s) $+ 6\ NH_3(aq)$
g. Dissolve solid from e in HCl(aq), then test with KSCN(aq). $Fe(OH)_3(s) + 3\ H^+ \longrightarrow Fe^{3+} + 3\ H_2O$ $Fe^{3+} + SCN^- + 5\ H_2O \longrightarrow [Fe(H_2O)_5SCN]^{2+}$(wine red)

2. a. $Al_2O_3 + 6\ HCl(aq) \longrightarrow 3\ H_2O + 2\ AlCl_3(aq)$; $Al_2O_3(s) + 6\ NaOH(aq) \longrightarrow 2\ Na_3AlO_3(aq) + 3\ H_2O$
b. $$100.0\ g\ AlCl_3 \times \frac{mol\ AlCl_3}{133.5\ g} \times \frac{mol\ Al_2O_3}{2\ mol\ AlCl_3} \times \frac{102.0\ g}{mol\ Al_2O_3} = 38.20\ g\ Al_2O_3$$
$$100.0\ g\ Na_3AlO_3 \times \frac{mol\ Na_3AlO_3}{144.0\ g} \times \frac{mol\ Al_2O_3}{2\ mol\ Na_3AlO_3} \times \frac{102.0\ g}{mol\ Na_3AlO_3} = 35.42\ g\ Al_2O_3$$

3. $Cr_2O_7^{2-} \xrightarrow{+1.33\ V} Cr^{3+} \xrightarrow{-1.51\ V} Cr^{2+}$ (overall: +0.62 V)
$E° = [3(1.33) - 1.51] \div 4 = +0.62\ V$

CHAPTER 25

Drill Problems

1. A. Ag, +1, 2, $S_2O_3^{2-}$ B. Al, +3, 4, H^- C. Co, +3, 6, H_2O & CN^- D. Pt, +4, 6, NH_3 & Cl^- E. Ni, +2, 4, NH_3 F. Cr, +3, 6, NH_3 & SCN^- G. Fe, +3, 6, en H. Pt, +2, 4, NO_2^- I. Co, +3, 6, NH_3 Cl^- J. Al, +3, 6, H_2O & OH^- K. Pt, +4, 6, NH_3 & Cl^- L. Fe, 0, 4, CO M. Mn, +2, 6, NO & CN^- N. Au, +3, 4, F^- O. Co, +3, 6, NH_3 & NH_2^- P. Sb, +3, 5, Cl^- Q. Ni, +2, 6, Cl^- & en R. Fe, +2, 6, CN^- S. Ni, +2, 4, NH_3 & Br^- T. Fe, +3, 6, H_2O & OH^- U. Ni, +4, 4, H_2O & NH_3.

2. (1) A. Ag^+ B. Ca^{2+}, Fe^{2+}, Sc^{3+}, Cr^{3+}, Fe^{3+}, Co^{3+}, Pt^{4+} C. Co^{2+}, Ni^{2+}, Cu^{2+}, Al^{3+}
(2) D. 4,6 E. 6 F. 6 G. 6 H. 4,6 I. 2,4 J. 4,6 K. 6 L. 2
(3) M. nitrato (1) N. ethylenediamminetetraacetato (6) O. trien (4) P. NH_3 (1) Q. OH^- (1) R. CO (1) S. nitro (1) T. —NCS^- (1) U. nitrosyl (1) V. nitrogeno (1) W. F^- (1) X . SO_4^{2-} (aq) Y. CN^- (1) Z. nitrito (1)

3. (1) A. dithiosulfatoargentate(I) ion B. tetrahydroaluminate(III) ion C. diaquatetracyanocobaltate(III) ion D. diamminetetrachloroplatinum(IV) E. tetraamminenickel(II) ion F. diamminetetrathiocyanatochromate(III) ion G. tris(ethylenediamine)iron(III) ion H. tetranitroplatinate(II) ion I. pentaamminechlorocobalt(III) ion J. pentaaquahydroxoaluminum(III) ion K. triamminetrichloroplatinum(IV) ion L. tetracarbonyliron(0) M. pentacyanonitrosylmanganate(II) ion N. tetrafluoroaurate(III) ion O. diamidotetraamminecobalt(III) ion P. pentachloroantimonate(III) ion Q. dichlorobis(ethylenediamine)nickel(II) R . hexacyanoferrate(II) ion S. diamminedibromonickel(II) T. tetraaquadihydroxoiron(III) ion U. tetraamminediaquanickel(II) ion
(2) V. sodium tetracyano aurate(I) W. $Na_3[Ag(S_2O_3)]$ X. potassium hexanitrocobaltate(III) ion Y. $[Pt(NH_3)_2Br_2]$ Z. tetraamminedichlorocobalt(III) bromide Γ. $[Cr(H_2O)_4Cl_2]Br$ Δ. $[Pt(H_2O)_2]Cl_4$ Θ. potassium hexachloroplatinate(IV) Λ. $[Ag(NH_3)_2]Cl$ Ξ. $[Ni(NH_3)_6SO_4$ Π. sodium bis(ethylenediamine)disulfatocuprate(II) Σ. hexaaquaaluminum hexacyanocobaltate(III) Υ. calcium diamminetetraisothiocyanatochromate(III) Φ. hexaamminecobalt(III) chloride sulfate Ψ. $Na_3[Cr(C_2O_4)_3]$ W. $[Cu(en)_2]Cl_2$

4. A. $[S_2O_3$—Ag—$S_2O_3]^{3-}$
B.
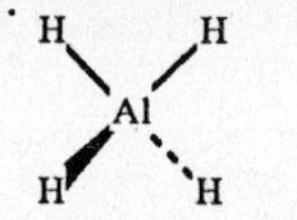

C.
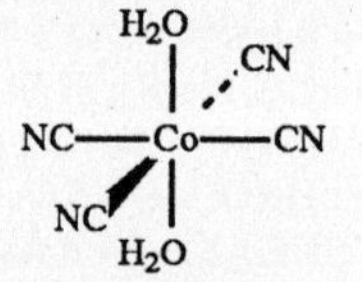

D.
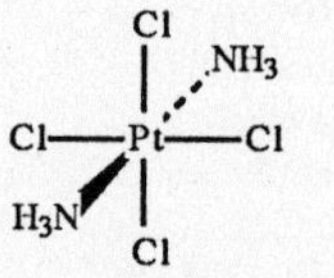

E.
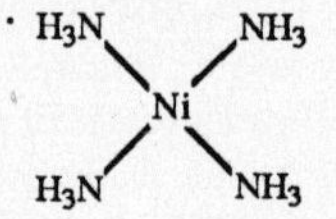

F.
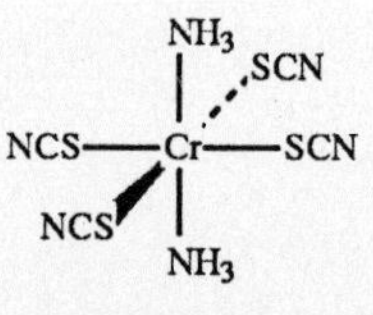

G.
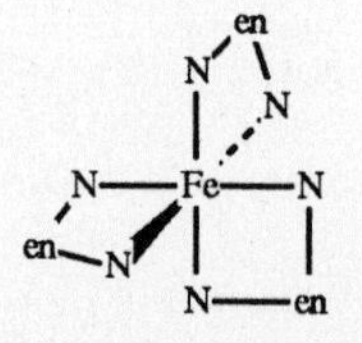

H. 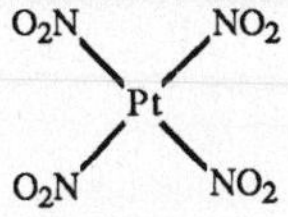

I. 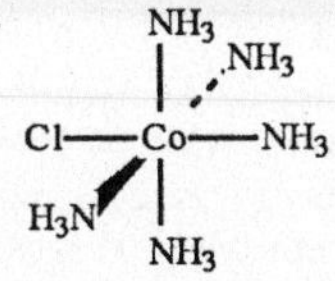

J. 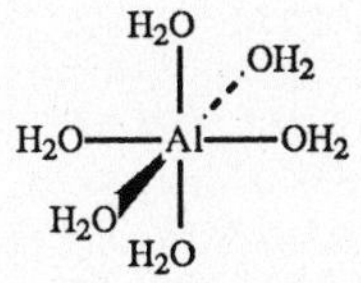

K. 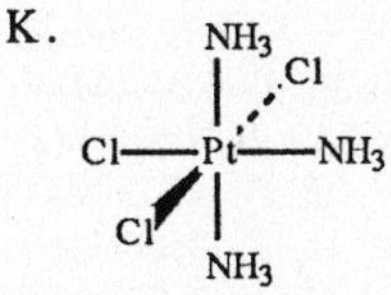

L. 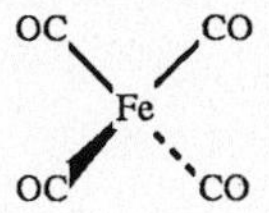

M. 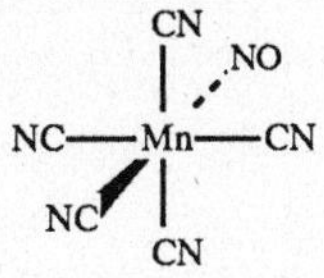

N. 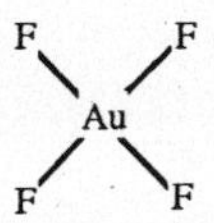

O. 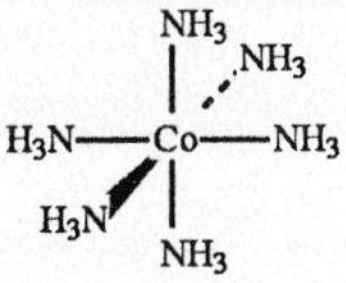

P. 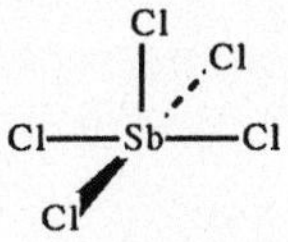

Q. 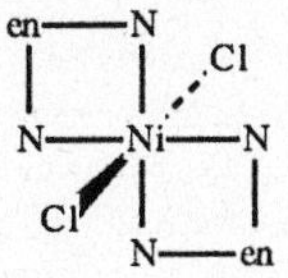

R. 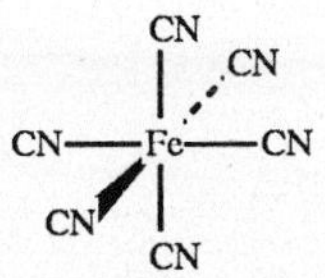

S. 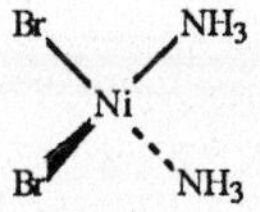

T. 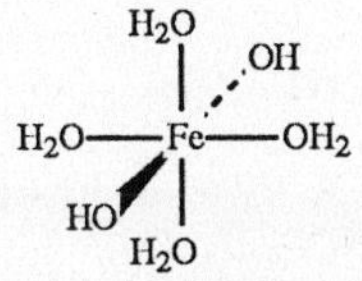

U. 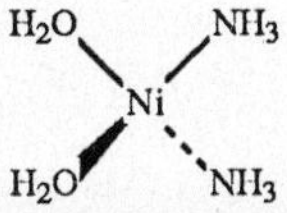

5. A. 3: $[Co(H_2O)_4(CN)_2]Cl$, tetraaquadicyanocobalt(III) chloride; $[Co(H_2O)_4(CN)Cl]CN$, tetraaquachlorocyanocobalt(III) cyanide 7: $[Cr(H_2O)_4Cl_2]Br$, tetraaquadichlorochromium(III) bromide; $[Cr(H_2O)_4ClBr]Cl$, tetraaquabromochlorochromium(III) chloride 9: $[Co(NH_3)_5Cl]SO_4$, pentaamminechlorocobalt(III) sulfate; $[Co(NH_3)_5SO_4]Cl$, pentaamminesulfatocobalt(III) chloride B. 1: lithium tetracyanoaluminate, lithium tetraisocyanoaluminate 3: tetraaquadicyanocobalt(III) chloride, tetraaquadiisocyanocobalt(III) chloride 5: hexaaquaaluminum hexacyanocobaltate(III), hexaaquaaluminum hexaisocyanocobaltate(III) 6: diaquadinitroplatinum(II), diaquadinitritoplatinum(II) 10: tricyanotrinitrosylmanganese(III), triisocyanotrinitrosylmanganese(III)
C. 5: $[Al(H_2O)_6][Co(CN)_6]$, $[Co(H_2O)_6][Al(CN)_6]$
8: $[Pt(H_2O)_4Cl_2][PtCl_4]$

D. 3: *trans* 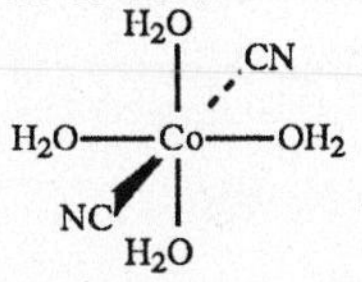 *cis* 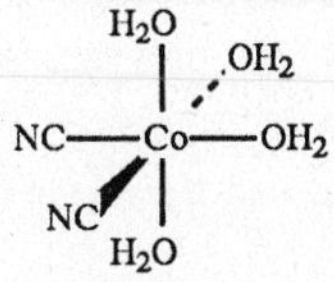

4: *trans* 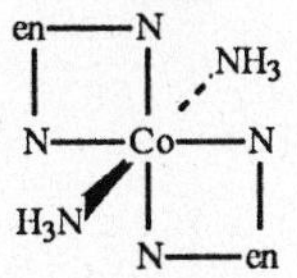 *cis* 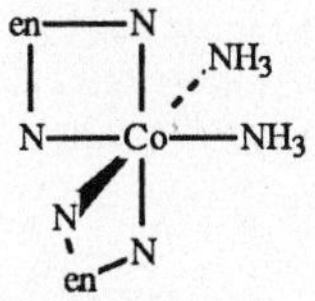

6: *trans* 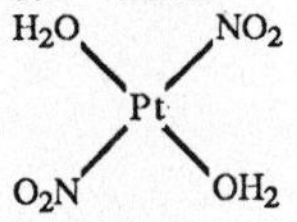 *cis* 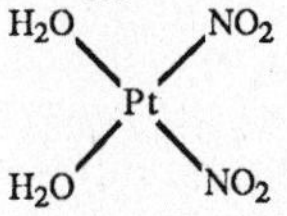

7: *trans* 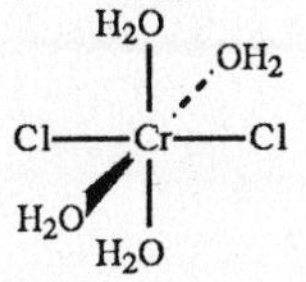 *cis* 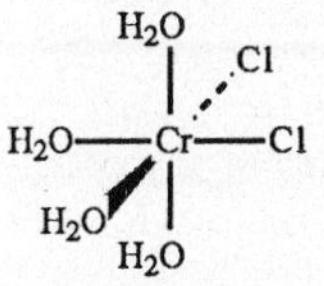

10: *cis* 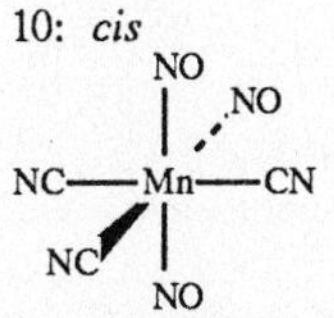 *trans* 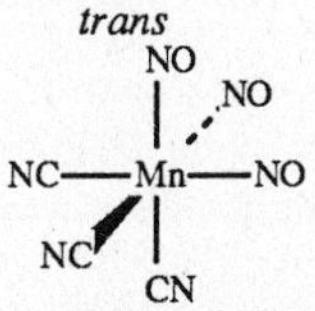

7. A. strong, 0 B. strong, 3 C. weak, 4 D. weak, 1 E. strong, 0 F. weak, 3 G. weak 4 H. weak, 2 I. weak, 0 J. strong, 3

Quizzes

A: 1. (d) 2. (c) 3. (c) 4. diamminedichlorogold(III) bromide 5. $[Co(en)_3](C_2H_3O_2)_3$ 6. (b) 7. (b) 8. (a)
B: 1. (c) 2. (d) 3. (a) 4. calcium tetrachlorocuprate(II) 5. $K_2[SbCl_5]$ 6. (b) 7. (c) 8. (c)
C: 1. (c) 2. (b) 3. (a) 4. potassium diamminedichloroplatinate(II) 5. $Ca[Cr(NH_3)_2Cl_4]_2$ 6. (a) 7. (d) 8. (b)
D: 1. (c) 2. (d) 3. (a) 4. sodium tetrachlorodinitrosylcobaltate(II) 5. $K_2[Fe(CH_3NH_2)_2(CN)_4]$ 6. (b) 7. (c) 8. (d)

Sample Test

1. Number of unpaired electrons is given in parentheses.

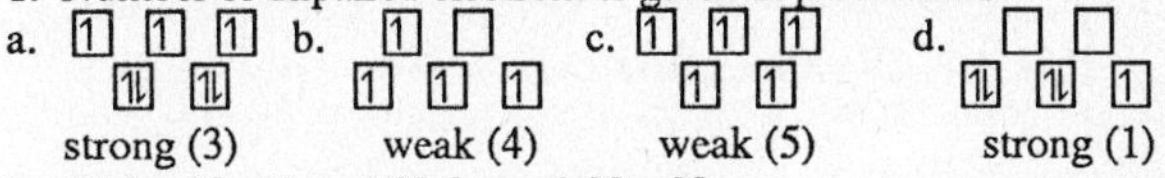

a. b. c. d.

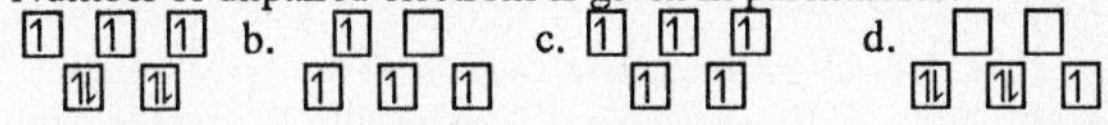

strong (3) weak (4) weak (5) strong (1)

2. 24.8 g Na × mol/23.0 g = 1.08 g Na
34.2 g Cu × mol/63.5 g = 0.539 mol Cu
41.0 g F × mol/19.0 g = 2.16 mol F
$i(0.010\text{ M})(0.0821\text{ L atm mol}^{-1}\text{ K}^{-1})(298\text{ K})$
$= 559\text{ mmHg} \times \text{atm}/760\text{ mmHg}$
$i = 3.00$. Thus, the compound is $Na_2[CuF_4]$, sodium tetrafluorocuprate(II)

CHAPTER 26

Drill Problems

2. (1) A. β^- B. ^4He C. β^- D. ^4He E. ^{232}U F. β^- G. ^{200}Au H. ^6Li I. β^- J. ^{17}F K. β^+ L. ^{37}Ar M. ^{206}Hg N. ^{191}Au O. ^{222}Rn
(2) P. $^7Be + \beta^- \longrightarrow {}^7Li$ Q. $^{191}Hg \longrightarrow \beta^+ + {}^{191}Au$
R. $^{177}Pt \longrightarrow {}^4He + {}^{173}Os$ S. $^3H \longrightarrow \beta^- + {}^3He$
T. $^{177}W + \beta^- \longrightarrow {}^{177}Ta$ U. $^{109}In \longrightarrow \beta^+ + {}^{109}Cd$
V. $^{186}Re \longrightarrow \beta^- + {}^{186}Os$ W. $^{235}U \longrightarrow {}^4He + {}^{231}Th$
X. $^{12}B \longrightarrow {}^4He + {}^8Li$ Y. $^{161}Tm + \beta^- \longrightarrow {}^{161}Er$
Z. $^{98}Tc \longrightarrow \beta^- + {}^{98}Ru$
4. (1) A. $^{59}Co + {}^1H \longrightarrow {}^1n + {}^{59}Ni$; $^{59}Co(p,n)^{59}Ni$
B. $^9Be + {}^6Li \longrightarrow {}^1n + {}^{14}N$; $^9Be(^6Li,n)^{14}N$
C. $^{14}N + {}^1n \longrightarrow {}^9Be + {}^6Li$; $^{14}N(n,^6Li)^9Be$
D. $^{15}N + {}^1H \longrightarrow {}^4He + {}^{12}C$; $^{15}N(p,\alpha)^{12}C$
E. $^5He + {}^1H \longrightarrow {}^2H + {}^4He$; $^5He(p,d)^4He$
F. $^{23}Na + {}^2H \longrightarrow {}^4He + {}^{21}Ne$; $^{23}Na(d,\alpha)^{21}Ne$
G. $^7Li + {}^4He \longrightarrow {}^1n\ {}^{10}B$; $^7Li(\alpha,n)^{10}B$
H. $^9Be + {}^4He \longrightarrow {}^1n + {}^{12}C$; $^9Be(\alpha,n)^{12}C$
I. $^{35}Cl + {}^1n \longrightarrow {}^1H + {}^{35}S$; $^{35}Cl(n,p)^{35}S$
J. $^{44}Ca + {}^1H \longrightarrow {}^1n + {}^{44}Sc$; $^{44}Ca(p,n)^{44}Sc$
K. $^{27}Al + {}^2H \longrightarrow {}^4He + {}^{25}Mg$; $^{27}Al(p,\alpha)^{25}Mg$
L. $^{25}Mg + {}^4He \longrightarrow {}^1H + {}^{28}Al$; $^{25}Mg(\alpha,p)^{28}Al$
(2) M. $^{238}U + {}^1n \longrightarrow \beta^- + {}^{239}Np$ N. $^{235}U + {}^1n \longrightarrow {}^{236}U + \gamma$; $^{236}U + {}^1n \longrightarrow {}^{237}U + \gamma$; $^{237}U \longrightarrow {}^{237}Np + \beta^-$
O. $^{238}U + {}^2H \longrightarrow {}^{238}Np + 2\ {}^1n$; $^{238}Np \longrightarrow {}^{238}Pu + \beta^-$
P. $^{239}Pu + {}^1n \longrightarrow {}^{240}Pu + \gamma$; $^{240}Pu + {}^1n \longrightarrow {}^{241}Pu + \gamma$; $^{241}Pu \longrightarrow {}^{241}Am + \beta^-$ Q. $^{239}Pu + {}^4He \longrightarrow {}^{242}Cm + {}^1n$
R. $^{241}Am + {}^1n \longrightarrow {}^{242}Am + \gamma$; $^{242}Am \longrightarrow {}^{242}Cm + \beta^-$
S. $^{241}Am + {}^4He \longrightarrow {}^{243}Bk + 2\ {}^1n$
T. $^{242}Cm + {}^4He \longrightarrow {}^{245}Cf + {}^1n$
U. $^{253}Es + {}^4He \longrightarrow {}^{256}Md + {}^1n$
V. $^{246}Cm + {}^{12}C \longrightarrow {}^{254}No + 4\ {}^1n$
W. $^{252}Cf + {}^{11}B \longrightarrow {}^{258}Lw + 5\ {}^1n$
7. (1) Given in the order: $t_{1/2}$, λ, activity. A. 22 min, 0.0315/min, 8.5×10^{13}/min B. 1.65 h, 0.42/h, 1.7×10^{10}/h C. 1.5×10^6 y, 4.6×10^{-7}/y, 2.0×10^{16}/y D. 12.26 y, 0.05653/y, 1.49×10^{19}/y E. 39.5 h, 0.0175/h, 5.45×10^{19}/h F. 6.7 y, 0.10/y, 3.6×10^{20}/y
(2) Given in the order: $t_{1/2}$, λ, mass. G. 0.81 s, 0.86/s, 38 ng H. 53.37 d, 0.01298/d, 4.6 μg I. 39.5 h, 0.0175/h, 12.9 g J. 12.26 y, 0.05653/y, 866 kg K. 28 y, 0.025/y, 268 mg L. 60.6 min, 0.0114/min, 63.9 mg
(3) Given in the order: $t_{1/2}$, λ. M. 27 s, 0.026/s N. 15.5 s, 0.45/s O. 10.7 h, 1.81×10^{-5}/s P. 14.0 h, 1.38×10^{-6}/s Q. 8.00 min, 1.44×10^{-3}/s R. 14.2 h, 1.36×10^{-5}/s
8. (1) A. 18.8 d B. 2.67 h C. 0.372 min D. 10.1 min E. 2.32 d F. 0.293 d G. 1.94 d H. 1.54 d
(2) I. 3.8×10^3 y J. 26.6 y K. 3.81×10^5 y L. 3.09 y M. 1.5×10^4 y N. 1.32×10^4 y
(3) O. 54.5 y P. 7.50×10^4 y Q. 1.56×10^5 y R. 12.4 y S. 19.0 y T. 14.9 y
9. A. 2.0×10^{-5} amu, 1.8×10^9 kJ/mol, 0.019 MeV
B. 0.003768 amu, 3.162×10^{11} kJ/mol, 3.514 MeV
C. 0.000377 amu, 3.16×10^{10} kJ/mol, 0.352 MeV
D. 0.000094 amu, 7.9×10^9 kJ/mol, 0.088 MeV
E. 0.0143 amu, 1.20×10^{12} kJ/mol, 13.3 MeV
F. 0.001288 amu, 1.081×10^{11} kJ/mol, 1.201 MeV
G. 0.000326 amu, 2.74×10^{10} kJ/mol, 0.304 MeV
H. 0.001384 amu, 1.161×10^{11} kJ/mol, 1.291 MeV
I. 0.000548 amu, 4.90×10^{10} kJ/mol, 0.545 MeV
J. 0.001821 amu, 1.528×10^{11} kJ/mol, 1.698 MeV
K. 0.001072 amu, 8.996×10^{10} kJ/mol, 0.9997 MeV
L. 0.003232 amu, 2.712×10^{11} kJ/mol, 3.014 MeV
M. 0.001597 amu, 1.340×10^{11} kJ/mol, 1.489 MeV
10. Binding energy per nucleon is given first, then atomic number. A. 0.000 MeV, 1 B. 8.801 MeV, 26 C. 7.082 MeV, 2 D. 8.742 MeV, 28 E. 6.470 MeV, 4 F. 8.746 MeV, 30 G. 7.689 MeV, 6 H. 8.721 MeV, 34 I. 7.985 MeV, 8 J. 8.728 MeV, 36 K. 8.042 MeV, 10 L. 8.720 MeV, 40 M. 8.270 MeV, 12 N. 8.617 MeV, 44 O. 8.458 MeV, 14 P. 8.542 MeV, 48 Q. 8.503 MeV, 16 R. 8.440 MeV, 56 S. 8.605 MeV, 18 T. 8.405 MeV, 56 U. 8.561 MeV, 20 V. 8.356 MeV, 60 W. 8.733 MeV, 22 X. 8.211 MeV, 64 Y. 8.786 MeV, 24 Z. 8.142 MeV, 68 Results (except for parts A and E) are plotted in Figure A-3.

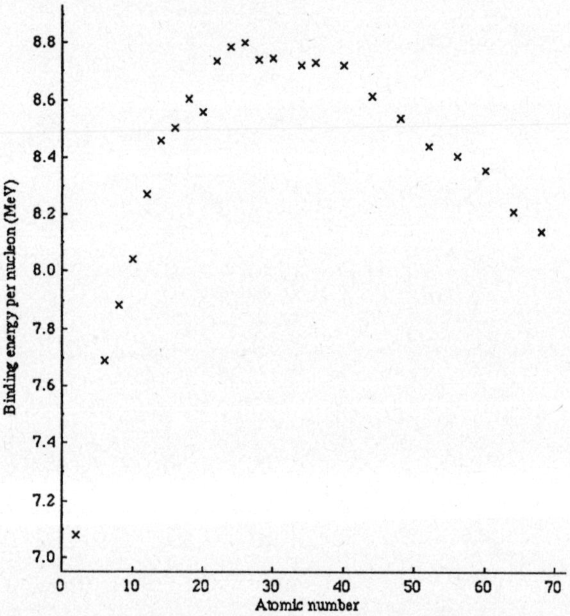

11. A. $^8B \longrightarrow {}^8Be + \beta^+$ B. $^{238}U \longrightarrow {}^{234}Th + {}^4He$
C. ^{19}F stable D. $^{119}Cd \longrightarrow {}^{119}In + \beta^-$ E. $^{50}Sc \longrightarrow {}^{50}Ti + \beta^-$
F. $^{212}Po \longrightarrow {}^{208}Pb + {}^4He$ G. ^{88}Sr stable H. $^{17}F \longrightarrow {}^{17}O + \beta^+$
I. $^{226}Ra \longrightarrow {}^{222}Rn + {}^4He$ J. $^{50}Mn \longrightarrow {}^{50}Cr + \beta^+$
K. $^{106}In \longrightarrow {}^{106}Cd + \beta^+$ L. ^{197}Au stable
M. $^{63}Co \longrightarrow {}^{63}Ni + \beta^-$

Quizzes

A: 1. (e) (^{222}Rn) 2. (b) 3. (b) 4. (d) 5. (b) 6. (b) 7. (a) 8. (b)
B: 1. (a) 2. (a) 3. (b) 4. (c) 5. (e) (45 d) 6. (b) 7. (a) 8. (a)
C: 1. (e) (^{90}Y) 2. (c) 3. (a) 4. (a) 5. (a) 5. (a) 6. (b) 7. (a) 8. (e)
D: 1. (b) 2. (e) 3. (b) 4. (a) 5. (a) 6. (a) 7. (d) 8. (b)

Sample Test

1. Remain the same. ^{14}C is not produced by nuclear explosions but by cosmic rays in the upper atmosphere:
$$^{14}N + {}^1n \longrightarrow {}^{14}C + {}^1H$$
2. Only α (^4He) and β (e^-) particle are given off; $^{253}Es \longrightarrow {}^{237}Np + 4\ {}^4He + 2\ e^-$. Thus four α particles and two β particles are given off.
3. a. $^{243}Cm \longrightarrow {}^{239}Pu + {}^4He$ b. $^{18}F \longrightarrow b^+ + {}^{18}O$
c. $^{88}Zr + b^- \longrightarrow {}^{88}Y$ d. $^{10}B + {}^4He \longrightarrow {}^1n + {}^{13}N$
e. $^{45}Sc + {}^4He \longrightarrow {}^1H + {}^{48}Ti$ f. $^{51}V + {}^2H \longrightarrow 2\ {}^1n + {}^{51}Cr$

4. $(11.0216 - 11.00931)$ amu $\times$ 932.8 MeV/amu = 11.46 MeV

5. $\ln(N_t/N_0) =$

$-0.693t/t_{1/2} -0.693\,(995 \times 10^6\,y) \div (4.51 \times 10^9\,y) = -0.1529$

$N_t = N_0 e^{-0.1529} = 0.858 \times 6.022 \times 10^{23}$ atoms number decayed

$= N_0 - N_t = (6.022 - 5.17) \times 10^{23} = 8.5 \times 10^{22}$ atoms

CHAPTER 27

Drill Problems

2. (1) A.

$$CH_3CH_2\overset{\overset{\displaystyle CH_3}{|}}{C}=CHCH_2CH_3$$

B. $CH_3CH=CHCH_2CH_3$ C.

$$CH_3\overset{\overset{\displaystyle CH_3}{|}}{C}HCH_2CH_3$$

D.

$$CH_3\text{-}\overset{\overset{\displaystyle CH_3}{|}}{\underset{\underset{\displaystyle CH_3}{|}}{C}}CH_2\text{-}\overset{\overset{\displaystyle CH_3}{|}}{C}HCH_3$$

E. $CH_3C{\equiv}CCH_3$

F. (benzene with CH₂CH₃ and CH₃)

G. (benzene with CH₃)

H. H_3C (benzene) CH_3

I. (cyclopentene with CH₂CH₃)

J.

$$CH_3\text{-}\overset{\overset{\displaystyle CH_3}{|}}{C}H\text{-}\overset{\overset{\displaystyle CH_3}{|}}{C}H\text{-}\overset{\overset{\displaystyle CH_3}{|}}{C}H\text{-}CH_3$$

K.

$$CH_3\overset{\overset{\displaystyle CH_3}{|}}{C}H\text{—}\overset{\overset{\displaystyle CH_3}{|}}{C}HCH_3$$

L.

$$CH_3\text{-}\overset{\overset{\displaystyle CH_3}{|}}{C}HCH_2\text{-}\overset{\overset{\displaystyle CH_3}{|}}{\underset{\underset{\displaystyle CH_2CH_3}{|}}{C}}CH_2CH_2CH_3$$

M.

$$CH_3\overset{\overset{\displaystyle CH_3}{|}}{C}=\overset{\overset{\displaystyle CH_3}{|}}{C}CH_3$$

N.

$$CH_3CH_2\overset{\overset{\displaystyle CH_3}{|}}{C}=\overset{\overset{\displaystyle CH_2CH_3}{|}}{C}CH_2CH_3$$

O. $CH_3C{\equiv}CCH_2CH_3$

P. H_3C (cyclohexane) CH_3

Q. H_3C (cyclopentene) CH_3

R. H_3C (cyclopentene)

S. (two cyclohexane rings)

T. H_3C / H_3C (cycloheptane) CH_2CH_3

U. (cyclopentane) CH_3

V. H_3C / CH_3 (cyclobutane) H_3C / CH_3

W. (benzene with CH₃ and two CH₃) H_3C CH_3

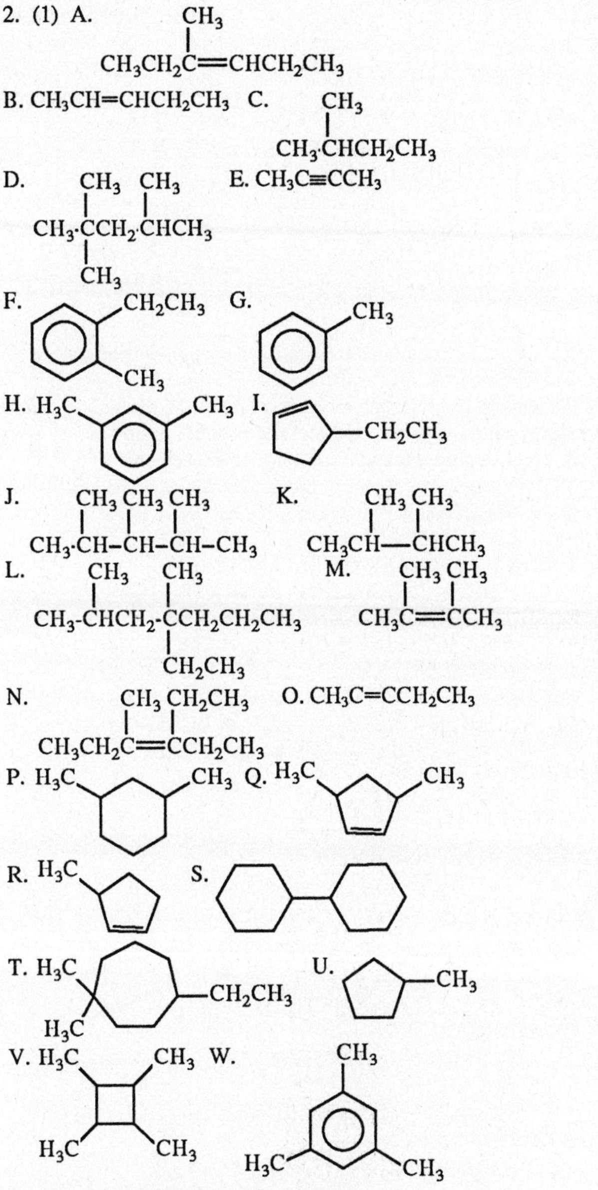

2. (2) A. 2-methylbutane B. 1-butene
C. 3-ethyl-2-methyl-2-pentene D. 2,3-dimethylpentane
E. ethylcyclopropane F. 1,2-dimethylcyclohexane G. 1-pentene
H. 4-n-propylcyclohexane I. t-butylbenzene J. 4-methylheptane
K. 1-butyne L. 2-butene M. 1-pentane N. 1-propyne *or*
propyne O. propane P. 2-methylpropane
Q. 2,2,3-trimethylbutane R. 3,3-diethyl-2-methylhexane
S. 3-ethyl-2-methylpentane T. 2-methylpentane
U. 2,2,3-trimethylpentane V. 4,4-dimethyloctane
W. 4-methylcyclohexane

3. A.

n-propylbenzene

isopropylbenzene

o-ethyltoluene

m-ethyltoluene

p-ethyltoluene

1,2,3-trimethylbenzene

1,2,4-trimethylbenzene

1,3,5-trimethylbenzene

B. $H_2C=CHCH_2CH_3$ $CH_3CH=CHCH_3$
1-butene   2-butene

cyclobutane   methylcyclopropane   2-methyl-1-propene

C. cyclohexene   1-methylcyclopentene

3-methylcyclopentene   4-methylcyclopentene

2-ethylcyclobutene   3-ethylcyclobutene

1,2-dimethylcyclobutene   2,3-dimethylcyclobutene

1,3-dimethylcyclobutene
and other unusual ones such as: cyclobutylethene

methylenecyclopentane

3,4-dimethylcyclobutene   1-cyclopropylpropene

3-cyclopropylpropene   1-propylcyclopropene

3-propylcyclopropene   3-ethyl-1-methylcyclopropene

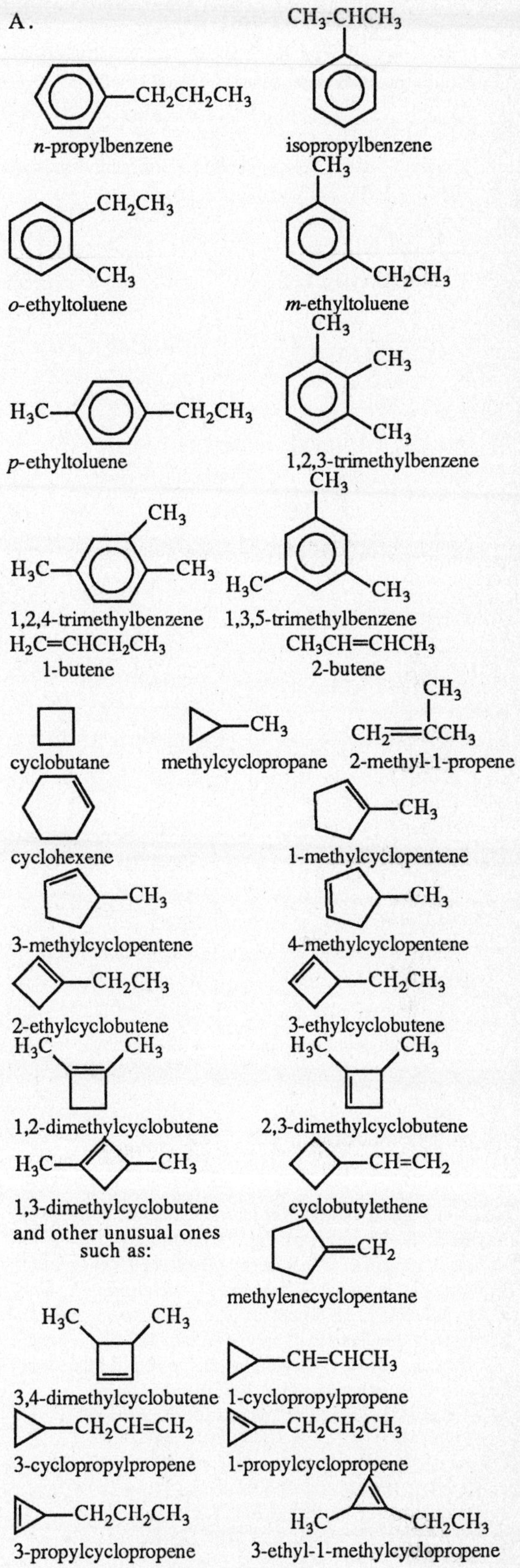

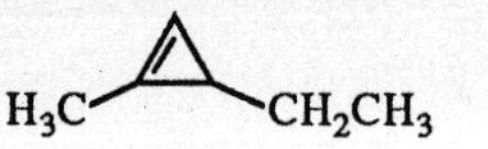
1-ethyl-3-methylcyclopropene
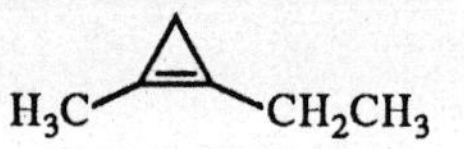
1-ethyl-2-methylcyclopropene
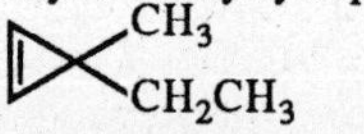
3-methyl-3-ethylcyclopropene
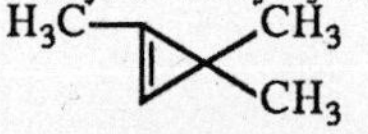
1,3,3-trimethylcyclopropene
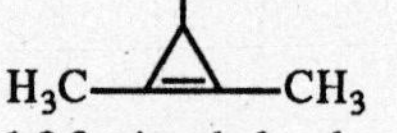
1,2,3-trimethylcyclopropene
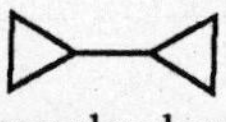
cyclopropylcyclopropane

D.

$CH_3CH_2CH_2CH_2CH_2CH_3$
hexane

CH_3
|
$CH_3{\cdot}CHCH_2CH_2CH_3$
2-methylpentane

CH_3
|
$CH_3CH_2CHCH_2CH_3$
3-methylpentane

CH_3 CH_3
| |
$CH_3{\cdot}CH{-}CHCH_3$
2,3-dimethylbutane

CH_3
|
$CH_3{\cdot}CCH_2CH_3$
|
CH_3
3,3-dimethylbutane

E.

$CH_3CH_2CH_2CH_3$
butane

CH_3
|
$CH_3{\text -}CHCH_3$
2-methylpropane

F. $CH_2{=}CHCH_2CH_2CH_3$
1-pentene

$CH_3CH{=}CHCH_2CH_3$
2-pentene

CH_3
|
$CH_2{=}CH{-}CHCH_3$
3-methyl-1-butene

CH_3
|
$CH_2{=}CCH_2CH_3$
2-methyl-1-butene

CH_3
|
$CH_3C{=}CHCH_3$
2-methyl-2-butene

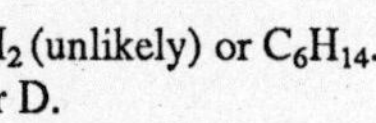
ethylcyclopropane

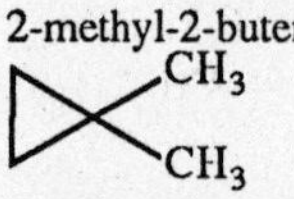
1,1-dimethylcyclopropane

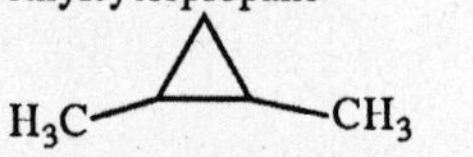
1,2-dimethylcyclopropane

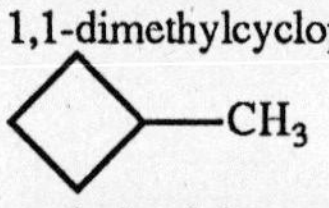
methylcyclobutane

G. Such a hydrocarbon could be C_7H_2 (unlikely) or C_6H_{14}. Thus, the answers are the same as for D.

4. H for highest value; L for lowest value. A. benzene H, hexane L B. ethyne H, ethane L C. pentane H, 2,2-dimethylpropane L D. 1,4-dimethylbenzene H, benzene L E. hexane H, butane L F. benzene H, cyclohexane L G. butane H, cyclobutane L H. *n*-heptane H, 3,3-dimethylpentane L I. ethene H, methane L J. butyne H, ethyne L K. *n*-hexane H, 2,2-dimethylbutane L L. benzene H, hexane L M. xylene H, benzene L N. *n*-heptane H, 3,3-dimethylpentane L O. hexane H, benzene L P. pentane H, propane L

5. A. $BrCH_2CH_2CH_2CH_2CH_2CH_3 \xrightarrow{\text{KOH, alcohol}}$
$CH_2{=}CHCH_2CH_2CH_2CH_3 + KBr + H_2O$
B. $HOCH_2CH_2CH_3 \xrightarrow{\text{sulfuric acid, }\Delta} CH_2{=}CHCH_3 + H_2O$
C. $HOCH_2CH_3 \xrightarrow{\text{sulfuric acid, }\Delta} H_2C{=}CH_2 + H_2O$
D. $2\ ClCH_2CH_3 + 2\ Na \longrightarrow CH_3CH_2CH_2CH_3 + 2\ NaCl$
E. $Cl_2CHCH_2CH_3 \xrightarrow{\text{KOH, alcohol}} HC{\equiv}CCH_3 + 2\ KCl + 2\ H_2O$
F. $CH_3CHClCH_3 \xrightarrow{\text{KOH, alcohol}} CH_2{=}CHCH_3 + KCl + H_2O$

G.
CH_3 CH_3
| |
$CH_3CH{-}CClCH_2CH_3 \xrightarrow{\text{KOH, alcohol}}$

CH_3 CH_3
| |
$CH_3{\cdot}C{=}CCH_2CH_3 + 2\ KCl + 2\ H_2O$ and some

CH_3 CH_3
| |
$CH_3{\cdot}CH{-}C{=}CHCH_3$

H. $CH_2{=}CHCH_2CH_3 \xrightarrow{H_2,\ Pt} CH_3CH_2CH_2CH_3$

I.
$CH_3{\cdot}CHCH_2CH_3$
|
$2\ CH_3CHClCH_2CH_3 + 2\ Na \longrightarrow CH_3{\cdot}CHCH_2CH_3$
(3,4-dimethylhexane)

8. A.
CH_3
|
$CH_3{\cdot}CClCH_2CH_2CH_3$

B.
CH_3
|
$CH_3CH_2{-}CClCH_2CH_3$

C.
I
|
$CH_3{\cdot}CHCH_2CH_3$

D.
CH_3
|
$CH_3{\cdot}CBrCH_2CH_3$

E.
OH
|
$CH_3{\cdot}CHCH_2CH_3$

F.
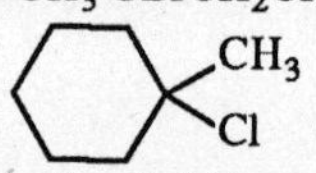

10. (1) A. 2-bromobutane B. 2,2,4-trimethyl-3-pentanol C. 3-hexanone D. pentanoic acid E. 2,2-dimethylpropanol F. methyl *n*-propyl ether G. *n*-pentylamine H. methanoic acid *or* formic acid I. ethyl propanoate J. 2-methyl-2-propanol K. ethyl isopropyl ether. L. 2,2,3-trichloropentane M. 3-pentanone N. *n*-propyl propanoate O. pentanal
(2) P. $CH_3COCH_2CH_3$ Q. $CH_3CH_2CH(CH_3)CHO$
R. $CH_3CH_2CH_2CH(CH_3)COOH$ S. $CH_3CH_2CHBrCH_2CH_3$
T. $CH_3CH_2CH_2COOH$ U. $CH_3CH_2OCH(CH_3)_2$
V. $CH_3CH_2CH(CH_3)CH_2CHO$ W. $CH_3CCl_2CH_3$
X. $CH_3CH(NH_2)CH(CH_3)CH_2CH_3$
Y. $CH_3CH_2CHOHCH_2CH_2CH_3$ Z. $CH_3CHOHCH_3$
Γ. $CH_3CH_2CH(NH_2)CH_2CH_3$ Δ. $CH_3CH(CH_3)COOH$
Θ. $CH_2ICH(CH_3)CH_2CH_3$ Λ. $CH_3COCH_2CH_2CH_3$

12. A. $CH{\equiv}CH + NaNH_2 \longrightarrow NH_3 + HC{\equiv}C^-\ Na^+$;
$HC{\equiv}C^-\ Na^+ + CH_3Br \longrightarrow HC{\equiv}CCH_3 + NaBr$
B.
O
‖
$CH_3C{-}H \xrightarrow{H_2/Ni,\ H_2SO_4} CH_3CH_2OH$;
$CH_3CH_2OH \xrightarrow{H_2SO_4,\ \text{heat}} CH_2{=}CH_2$
C. $CH_3CH_2CH_3 \xrightarrow{Cl_2} CH_3CH_2CH_2Cl + CH_3CHClCH_3$;
either chloropropane $\xrightarrow{\text{KOH, alcohol}} CH_3CH{=}CH_2 + KCl + H_2O$
D. $CH_3CH_3 + Cl_2 \longrightarrow CH_3CH_2Cl + HCl$;
$2\ CH_3CH_2Cl + 2\ Na \longrightarrow CH_3CH_2CH_2CH_3 + 2\ NaCl$;
$CH_3CH_2CH_2CH_3 + Cl_2 \longrightarrow$
$CH_3CH_2CH_2CH_2Cl + CH_3CH_2CHClCH_3 + HCl$;
$CH_3CH_2CHClCH_3 \xrightarrow{\text{KOH, alcohol}}$
$CH_3CH_2CH{=}CH_2 + CH_3CH{=}CHCH_3 + KCl + H_2O$
E. $CH_3CH_2CH_3 + Cl_2 \longrightarrow$
$CH_3CHClCH_3 + CH_3CH_2CH_2Cl + HCl$;

OH
|
$CH_3CHClCH_3 + OH^- \longrightarrow CH_3{\text -}CHCH_3 + Cl^-$;

OH O
| ‖
$CH_3{\text -}CHCH_3 \xrightarrow{K_2Cr_2O_7} CH_3CCH_3$

F. $CH_3CH_3 + Cl_2 \longrightarrow CH_3CH_2Cl + HCl$;
$CH_3CH_2Cl + OH^- \longrightarrow CH_3CH_2OH + Cl^-$;
$CH_3CH_2OH \xrightarrow{H_2SO_4,\ \text{conc}} CH_3CH_2OCH_2CH_3$

G. $CH_3CH_2CH_3 + Cl_2 \longrightarrow CH_3CHClCH_3 + CH_3CH_2CH_2Cl$

$CH_3CH_2CH_2Cl + OH^- \longrightarrow CH_3CH_2CH_2OH + Cl^-$

$CH_3CH_2CH_2OH \xrightarrow{\text{Cu, 200-300°C}} CH_3CH_2CHO + H_2$

H. $CH_3CH_3 + Cl_2 \longrightarrow CH_3CH_2Cl$;

$CH_3CH_2Cl + OH^- \longrightarrow CH_3CH_2OH + Cl^-$;

$CH_3CH_2Cl + Na \longrightarrow CH_3CH_2CH_2CH_3 + NaCl$;

$CH_3CH_2CH_2CH_3 + Cl_2 \longrightarrow$
$CH_3CH_2CHClCH_3 + CH_3CH_2CH_2CH_2Cl + HCl$;

$CH_3CH_2CH_2CH_2Cl + OH^- \longrightarrow CH_3CH_2CH_2CH_2OH + Cl$;

$CH_3CH_2CH_2CH_2OH \xrightarrow{KMnO_4} CH_3CH_2CH_2COOH$;

$CH_3CH_2CH_2COOH + CH_3CH_2OH \longrightarrow$

$CH_3CH_2CH_2C(=O)—OCH_2CH_3$

I. $CH_3CH_2CH_2CH_2CH_2CH_3 + Cl_2 \longrightarrow$ many isomers including
$ClCH_2CH_2CH_2CH_2CH_2CH_2Cl$;

$ClCH_2CH_2CH_2CH_2CH_2CH_2Cl + Na \longrightarrow$ + NaCl

Quizzes

A: 1. (d) 2. (b) 3. (c) 4. (a) 5. (d) 6. (c) 7. (a) 8. (c)
B: 1. (e) 2. (a) 3. (b) 4. (a) 5. (d) 6. (a) 7. (c) 8. (c)
C: 1. (a) 2. (d) 3. (d) 4. (e) 5. (e) 6. (c) 7. (d) 8. (b)
D: 1. (c) 2. (a) 3. (c) 4. (e) 5. (a) 6. (b) 7. (c) 8. (b)

Sample Test

1. $CH_3CH_2OCH_2CH_3$ diethyl ether; $CH_3OCH_2CH_2CH_3$ methyl propyl ether; $CH_3OCH(CH_3)CH_3$ methyl isopropyl ether

2. Propanol contains an —OH group in a three-carbon molecule. This —OH group will hydrogen bond to water molecules, and the remaining $CH_3CH_2CH_2$— propyl group is small enough that this hydrogen bonding is not significantly disrupted. Hence, a solution forms. In the case of hexanol, however, the relatively long six-carbon hexyl chain will disrupt the hydrogen bonds significantly and no solution forms.

3. $CH_2{=}CHCH_2CH_3$ 1-butene

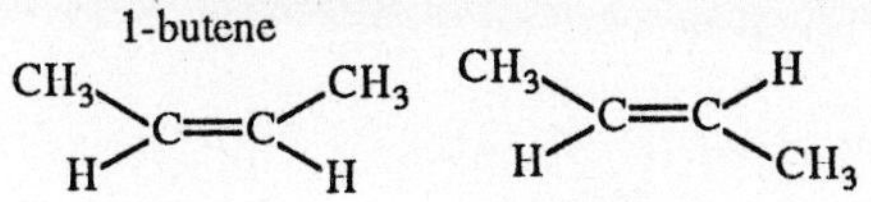

cis-2-butene *trans*-2-butene

4. $CH_3CH(CH_3)C(H){=}O \xrightarrow{H_2,\ Pt} CH_3CH(CH_3)CH_2CH_3$ 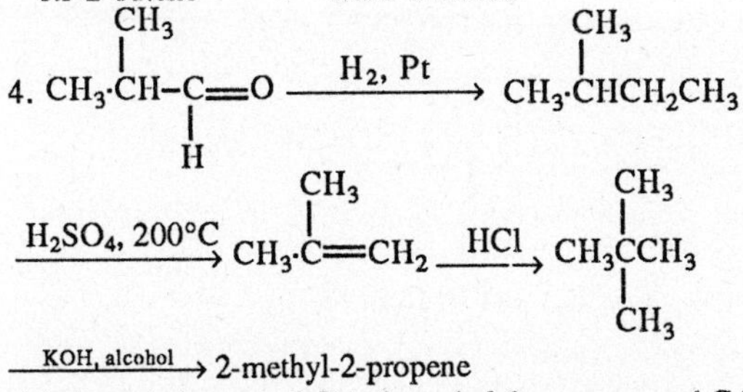

$\xrightarrow{H_2SO_4,\ 200°C} CH_3C(CH_3){=}CH_2 \xrightarrow{HCl} CH_3C(CH_3)_2CH_3$

$\xrightarrow{\text{KOH, alcohol}}$ 2-methyl-2-propene

A is 2-methylpropanol, B is 2-methyl-2-propene, and C is 2-bromo-2-methylpropane.

CHAPTER 28

Drill Problems

2. (1) A. glyceryl trioleate B. glyceryl tricaprate C. glyceryl caprolaurolinolenate D. glyceryl trieleostearate

(2) E. $CH_2—OOC(CH_2)_{14}CH_3$
$CH—OOC(CH_2)_7CH{=}CH(CH_2)_7CH_3$
$CH_2—OOC(CH_2)_{14}CH_3$

F. $CH_2—OOC(CH_2)_{10}CH_3$
$CH—OOC(CH_2)_{12}CH_3$
$CH_2—OOC(CH_2)_{14}CH_3$

G. $CH_2—OOC(CH_2)_{10}CH_3$
$CH—OOC(CH_2)_{10}CH_3$
$CH_2—OOC(CH_2)_7(CH{=}CHCH_2)_2(CH_2)_3CH_3$

H. $CH_2—OOC(CH_2)_7(CH{=}CHCH_2)_3CH_3$
$CH—OOC(CH_2)_7(CH{=}CHCH_2)_3CH_3$
$CH_2—OOC(CH_2)_7(CH{=}CHCH_2)_3CH_3$

I. $CH_2OOC(CH_2)_7CH{=}CHCH_2CHOH(CH_2)_5CH_3$
$CHOOC(CH_2)_7CH{=}CHCH_2CHOH(CH_2)_5CH_3$
$CH_2OOC(CH_2)_7CH{=}CHCH_2CHOH(CH_2)_5CH_3$

J. $CH_2OOC(CH_2)_7CH{=}CH(CH_2)_7CH_3$
$CHOOC(CH_2)_7CH{=}CH(CH_2)_7CH_3$
$CH_2OOC(CH_2)_{10}CH_3$

3. (1) A. 765.0 g/mol; 0.81 double bond/triglyceride B. 657.4 g/mol; 0.23 double bond/triglyceride C. 876.6 g/mol; 4.56 double bonds/triglyceride D. 881.2 g/mol; 5.14 double bonds/triglyceride

(2) E. 885.40 g/mol; SV =190.08; 3 C=C/molecule; IN = 86.0
F. 873.31 g/mol; SV = 192.72; 9 C=C/molecule; IN = 261.56
G. 891.46 g/mol; SV = 188.79; no C=C bonds; IN = 0.00
H. 719.11 g/mol; SV =234.04; 2C=C/molecule; IN = 70.59

6.

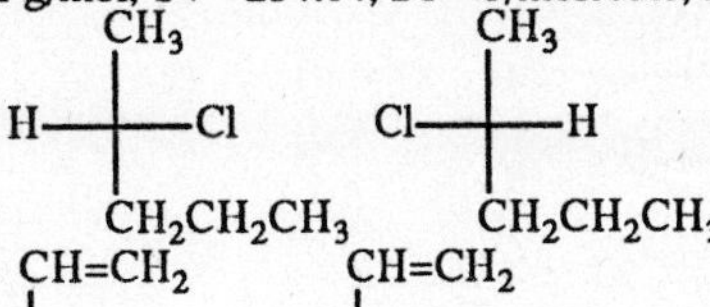

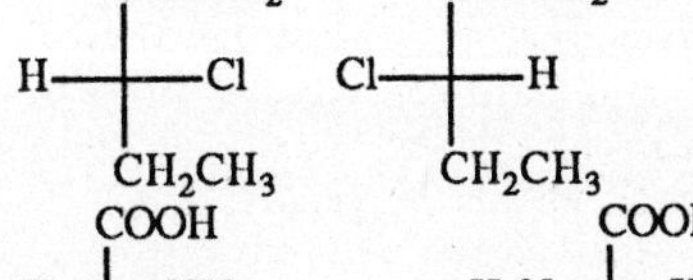

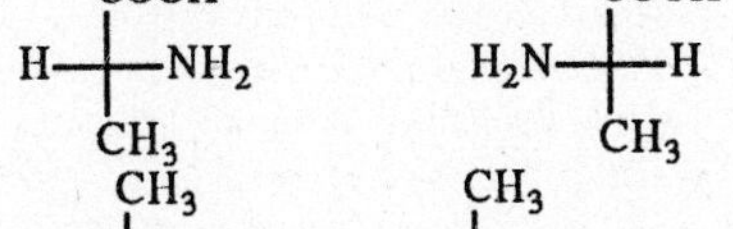

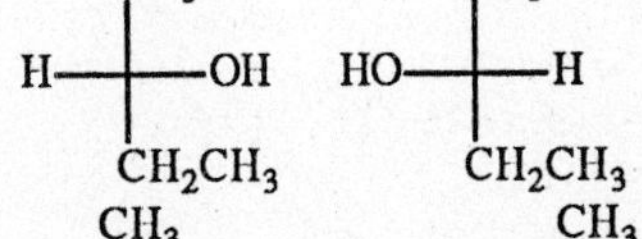

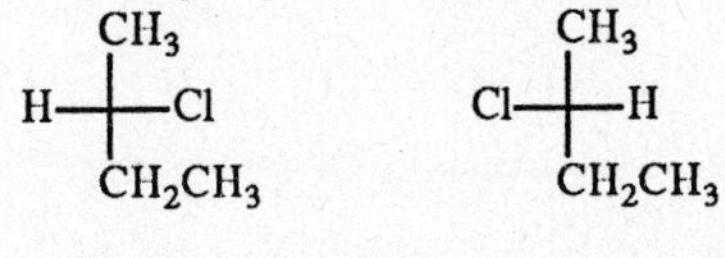

F.
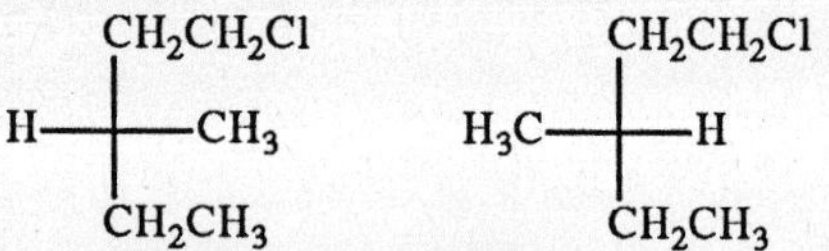

G.
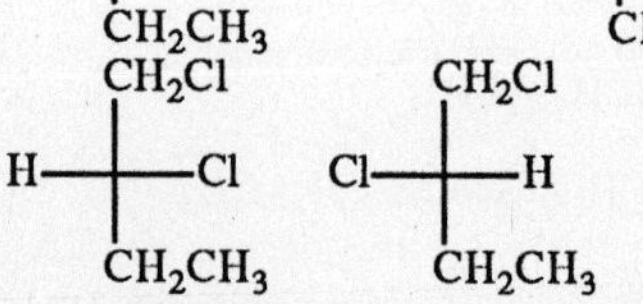

H., I., J. not optically active

K.
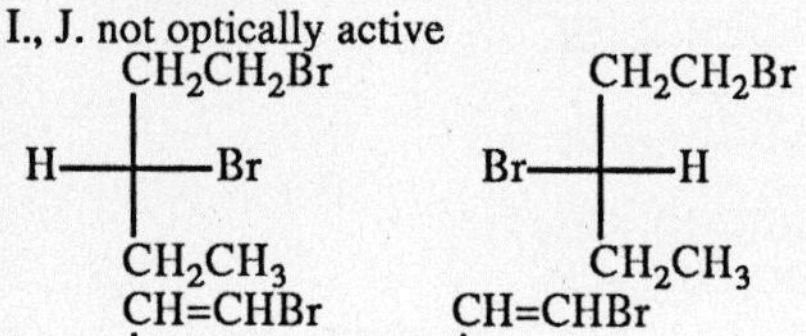

L.
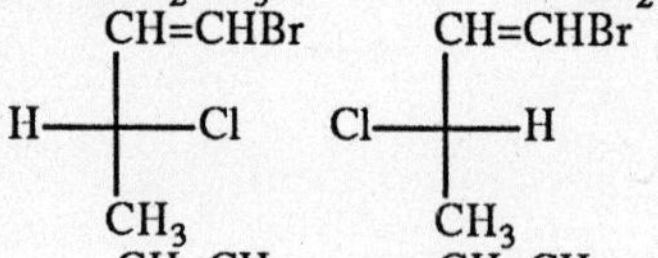

M.
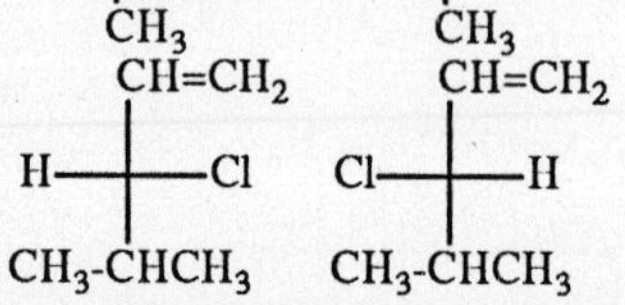

10. (1) A. Lys-Asp-Gly-Ala-Ala-Glu-Ser-Gly B. Ala-Ala-His-Arg-Glu-Lys-Phe-Ile C. Tyr-Cys-Lys-Ala-Arg-Arg-Gly D. Phe-Ala-Glu-Ser-Ala-Gly E. Val-Ala-Lys-Glu-Glu-Phe-Val-Met-Tyr-Cys-Glu-Trp-Met-Gly-Phe

(2) A. Lysylaspartylglycylalanylalanylglutylserylglycine
B. Alanylalanylhistidylarginylglutyllysylphenylalanylisoleucine
C. Tyrosylcysteyllysylalanylarginylarginylglycine
D. Phenylalanylglutylserylalanylglycine

Quizzes

A: 1. (b) 2. (d) 3. (c) 4. (d) 5. (b) 6. (d) 7. (e) 8. (e) 9. (b)
B: 1. (d) 2. (c) 3. (a) 4. (e) 5. (a) 6. (c) 7. (d) 8. (b) 9. (b)
C: 1. (e) 2. (b) 3. (d) 4. (a) 5. (a) 6. (c) 7. (b) 8. (a) 9. (e) (a pyrimidine)
D: 1. (e) 2. (e) 3. (b) 4. (d) 5. (d) 6. (d) 7. (a) 8. (b) 9. (d)

Sample Test

1.
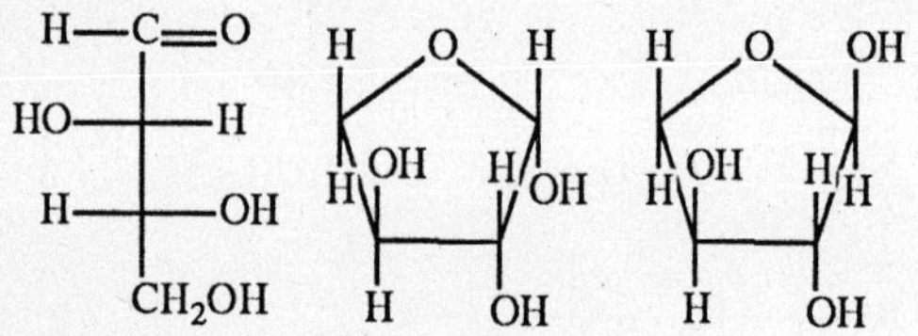

2. Refer to objective 28-11 and Figure 28-11
3. Phe-Val-Asp-Gly-His-Leu-Cys-Gly-Ser-His

Appendix II

Selected Tables and Figures from Petrucci's
General Chemistry

TABLE 10-4
Some synthetic carbon-chain polymers

Name of polymer	Repeating unit	Some uses
Elastomers neoprene [polychloroprene]	$\left(\begin{array}{c} H \\ | \\ -C- \\ | \\ H \end{array}\begin{array}{c} H \\ \\ -C= \\ \\ \end{array}\begin{array}{c} Cl \\ \\ C- \\ \\ \end{array}\begin{array}{c} H \\ | \\ C- \\ | \\ H \end{array}\right)_n$	wire and cable insulators, industrial hoses and belts, shoe soles and heels, gloves
silicone rubber	$\left(\begin{array}{c} CH_3 \\ | \\ -Si-O- \\ | \\ CH_3 \end{array}\right)_n$	gaskets, electrical insulation, surgical membranes, medical devices for use in the body
Fibers nylon 66	$\left(\begin{array}{c} O \\ \parallel \\ -C-(CH_2)_4-C-N-(CH_2)_6-N- \end{array}\right)_n$ with O, H, H substituents	hosiery, rope, tire cord, fish line, parachutes, artificial blood vessels
Acrilan, Orlon	$\left(\begin{array}{c} H \\ | \\ -C- \\ | \\ H \end{array}\begin{array}{c} C\equiv N \\ | \\ C- \\ | \\ H \end{array}\right)_n$	fabrics, carpets, drapes, upholstery, electrical insulation
Plastics polyethylene	$\left(\begin{array}{c} H \\ | \\ -C- \\ | \\ H \end{array}\begin{array}{c} H \\ | \\ C- \\ | \\ H \end{array}\right)_n$	bags, bottles, tubing, packaging film, paper coating
polypropylene	$\left(\begin{array}{c} H \\ | \\ -C- \\ | \\ H \end{array}\begin{array}{c} CH_3 \\ | \\ C- \\ | \\ H \end{array}\right)_n$	laboratory and household ware, storage battery cases, artificial turf, surgical casts, toys
PVC, "vinyl" [poly(vinyl chloride)]	$\left(\begin{array}{c} H \\ | \\ -C- \\ | \\ H \end{array}\begin{array}{c} Cl \\ | \\ C- \\ | \\ H \end{array}\right)_n$	bottles, records, floor tile, food wrap, piping, hoses, linings for ponds and reservoirs
Teflon [poly(tetrafluoroethylene)]	$\left(\begin{array}{c} F \\ | \\ -C- \\ | \\ F \end{array}\begin{array}{c} F \\ | \\ C- \\ | \\ F \end{array}\right)_n$	bearings, insulation, gaskets, nonstick surfaces (ovenware, frying pans), heat-resistant industrial plastics

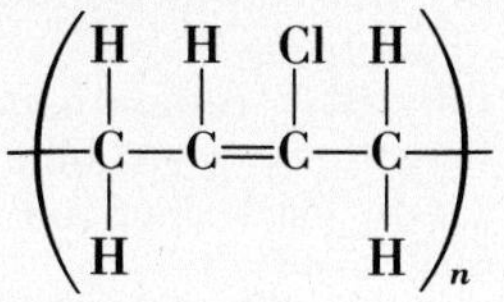

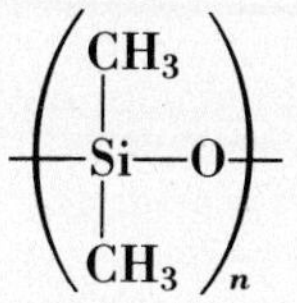

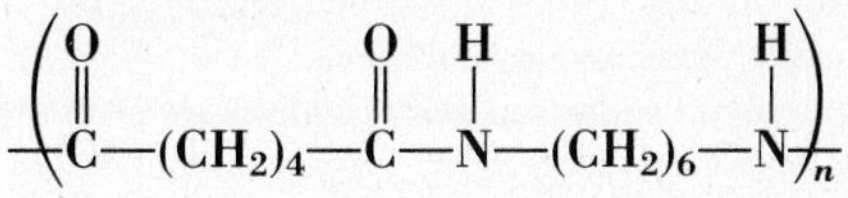

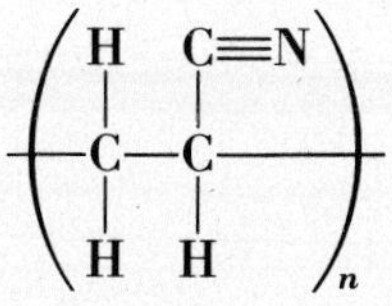

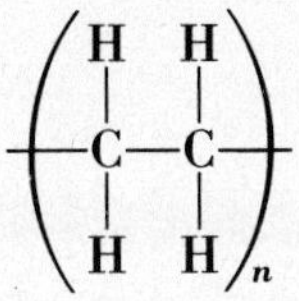

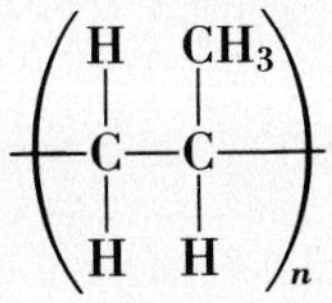

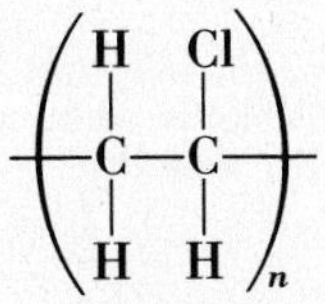

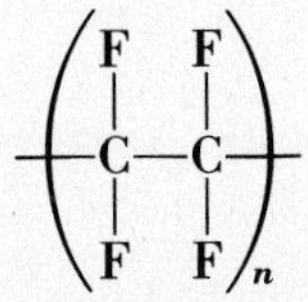

TABLE 13-3
Some common types of colloids

Dispersed phase	Dispersion medium	Type	Examples
solid	liquid	sol	clay sols, colloidal gold
liquid	liquid	emulsion	oil in water, milk, mayonnaise
gas	liquid	foam	soap and detergent suds, whipped cream, meringues
solid	gas	aerosol	smoke, dust-laden air
liquid	gas	aerosol	fog, mist (as in aerosol products)
solid	solid	solid sol	ruby glass, certain natural and synthetic gems, blue rock salt, black diamond
liquid	solid	solid emulsion	opal, pearl
gas	solid	solid foam	pumice, lava, volcanic ash

[a] In water purification it is sometimes necessary to precipitate clay particles or other suspended colloidal materials. This is often done by treating the water with an appropriate electrolyte. Clay sols are also suspected of adsorbing organic substances, such as pesticides, and distributing them in the environment.

[b] Smogs are complex materials that are at least partly colloidal. The suspended particles are both solid (smoke) and liquid (fog): smoke + fog = smog. Other constituents of smog are molecular, such as sulfur dioxide, carbon monoxide, nitric oxide, and ozone.

[c] The bluish haze of tobacco smoke and the brilliant sunsets in desert regions are both attributable to the scattering of light by colloidal particles suspended in air.

TABLE 9-8
Selected reactions of fluorine and chlorine (X = F or Cl)

Reaction with	Reaction equation
alkali metals	$2\ M + X_2 \longrightarrow 2\ MX$
alkaline earth metals	$M + X_2 \longrightarrow MX_2$
other metals (e.g., Fe)	$2\ Fe + 3\ X_2 \longrightarrow 2\ FeX_3$
hydrogen	$H_2 + X_2 \longrightarrow 2\ HX$
sulfur	$S_8 + 4\ Cl_2 \longrightarrow 4\ S_2Cl_2$
	$S_8 + 24\ F_2 \longrightarrow 8\ SF_6$
phosphorus[a]	$P_4 + 6\ X_2 \longrightarrow 4\ PX_3$
other halogens[b]	$Cl_2 + F_2 \longrightarrow 2\ ClF$
	$Cl_2 + Br_2 \longrightarrow 2\ BrCl$
water	$2\ X_2 + 2\ H_2O \longrightarrow 4\ HX + O_2$
	also: $Cl_2 + H_2O \longrightarrow HCl + HOCl$

[a] PF_5 and PCl_5 are also formed.

[b] Other diatomic interhalogen compounds that are known are BrF, IF, ICl, and IBr. Depending on reaction conditions, more complex interhalogen compounds can be obtained; most of these contain fluorine, e.g., ClF_3, BrF_3, IF_3, ICl_3, ClF_5, BrF_5, IF_5, and IF_7.

TABLE 17-2
Ionization constants for some weak acids and weak bases in water at 25°C

	Ionization equilibrium	Ionization constant, K	pK
Acid		$K_a =$	p$K_a =$
acetic	$HC_2H_3O_2 + H_2O \rightleftharpoons H_3O^+ + C_2H_3O_2^-$	1.74×10^{-5}	4.76
benzoic	$HC_7H_5O_2 + H_2O \rightleftharpoons H_3O^+ + C_7H_5O_2^-$	6.3×10^{-5}	4.20
chlorous	$HClO_2 + H_2O \rightleftharpoons H_3O^+ + ClO_2^-$	1.2×10^{-2}	1.92
formic	$HCHO_2 + H_2O \rightleftharpoons H_3O^+ + CHO_2^-$	1.8×10^{-4}	3.74
hydrocyanic	$HCN + H_2O \rightleftharpoons H_3O^+ + CN^-$	4.0×10^{-10}	9.40
hydrofluoric	$HF + H_2O \rightleftharpoons H_3O^+ + F^-$	6.7×10^{-4}	3.17
hypochlorous	$HOCl + H_2O \rightleftharpoons H_3O^+ + OCl^-$	2.95×10^{-8}	7.53
monochloroacetic	$HC_2H_2ClO_2 + H_2O \rightleftharpoons H_3O^+ + C_2H_2ClO_2^-$	1.35×10^{-3}	2.87
nitrous	$HNO_2 + H_2O \rightleftharpoons H_3O^+ + NO_2^-$	5.13×10^{-4}	3.29
phenol	$HOC_6H_5 + H_2O \rightleftharpoons H_3O^+ + C_6H_5O^-$	1.6×10^{-10}	9.80
Base		$K_b =$	p$K_b =$
ammonia	$NH_3 + H_2O \rightleftharpoons NH_4^+ + OH^-$	1.74×10^{-5}	4.76
aniline	$C_6H_5NH_2 + H_2O \rightleftharpoons C_6H_5NH_3^+ + OH^-$	4.30×10^{-10}	9.37
ethylamine	$C_2H_5NH_2 + H_2O \rightleftharpoons C_2H_5NH_3^+ + OH^-$	4.4×10^{-4}	3.36
hydroxylamine	$HONH_2 + H_2O \rightleftharpoons HONH_3^+ + OH^-$	9.1×10^{-9}	8.04
methylamine	$CH_3NH_2 + H_2O \rightleftharpoons CH_3NH_3^+ + OH^-$	4.2×10^{-4}	3.38
pyridine	$C_5H_5N + H_2O \rightleftharpoons C_5H_5NH^+ + OH^-$	2.0×10^{-9}	8.70

[a] Although some of these pK values could be expressed with an additional significant figure, the circumstances of a calculation often do not warrant this.

TABLE 17-3
Ionization constants of some common polyprotic acids

Acid	Ionization equilibria	Ionization constants, K	pK
carbonic[a]	$H_2CO_3 + H_2O \rightleftharpoons H_3O^+ + HCO_3^-$	$K_{a_1} = 4.2 \times 10^{-7}$	p$K_{a_1} = 6.38$
	$HCO_3^- + H_2O \rightleftharpoons H_3O^+ + CO_3^{2-}$	$K_{a_2} = 5.6 \times 10^{-11}$	p$K_{a_2} = 10.25$
hydrosulfuric[b]	$H_2S + H_2O \rightleftharpoons H_3O^+ + HS^-$	$K_{a_1} = 1.1 \times 10^{-7}$	p$K_{a_1} = 6.96$
	$HS^- + H_2O \rightleftharpoons H_3O^+ + S^{2-}$	$K_{a_2} = 1.0 \times 10^{-14}$	p$K_{a_2} = 14.00$
phosphoric	$H_3PO_4 + H_2O \rightleftharpoons H_3O^+ + H_2PO_4^-$	$K_{a_1} = 7.11 \times 10^{-3}$	p$K_{a_1} = 2.15$
	$H_2PO_4^- + H_2O \rightleftharpoons H_3O^+ + HPO_4^{2-}$	$K_{a_2} = 6.34 \times 10^{-8}$	p$K_{a_2} = 7.20$
	$HPO_4^{2-} + H_2O \rightleftharpoons H_3O^+ + PO_4^{3-}$	$K_{a_3} = 4.22 \times 10^{-13}$	p$K_{a_3} = 12.38$
phosphorous	$H_3PO_3 + H_2O \rightleftharpoons H_3O^+ + H_2PO_3^-$	$K_{a_1} = 5.0 \times 10^{-2}$	p$K_{a_1} = 1.30$
	$H_2PO_3^- + H_2O \rightleftharpoons H_3O^+ + HPO_3^{2-}$	$K_{a_2} = 2.5 \times 10^{-7}$	p$K_{a_2} = 6.60$
sulfurous[c]	$H_2SO_3 + H_2O \rightleftharpoons H_3O^+ + HSO_3^-$	$K_{a_1} = 1.3 \times 10^{-2}$	p$K_{a_1} = 1.89$
	$HSO_3^- + H_2O \rightleftharpoons H_3O^+ + SO_3^{2-}$	$K_{a_2} = 6.3 \times 10^{-8}$	p$K_{a_2} = 7.20$
sulfuric[d]	$H_2SO_4 + H_2O \longrightarrow H_3O^+ + HSO_4^-$	K_{a_1} = very large	p$K_{a_1} < 0$
	$HSO_4^- + H_2O \rightleftharpoons H_3O^+ + SO_4^{2-}$	$K_{a_2} = 1.29 \times 10^{-2}$	p$K_{a_2} = 1.89$

[a] H_2CO_3 cannot be isolated. It is in equilibrium with H_2O and dissolved CO_2. The value for K_{a_1} is actually for the reaction: $CO_2(aq) + H_2O \rightleftharpoons H_3O^+ + HCO_3^-$

[b] The value of K_{a_2} for H_2S has always been subject to doubt. Latest evidence seems to suggest that the best value may be as low as 10^{-19}. [See, Myers, R. J., "The New Low Value for the Second Ionization Constant for H_2S," *J. Chem. Educ.*, **63**, 687 (1986).]

[c] H_2SO_3 is a hypothetical, nonisolable species produced in the reaction $SO_2(aq) + H_2O \rightleftharpoons H_2SO_3(aq)$

[d] H_2SO_4 is completely ionized in the first step.

TABLE 19-1
Solubility product constants at 25°C

Solute	Solubility equilibrium	K_{sp}
aluminum hydroxide	$Al(OH)_3(s) \rightleftharpoons Al^{3+}(aq) + 3\,OH^-(aq)$	1.3×10^{-33}
barium carbonate	$BaCO_3(s) \rightleftharpoons Ba^{2+}(aq) + CO_3{}^{2-}(aq)$	5.1×10^{-9}
barium hydroxide	$Ba(OH)_2(s) \rightleftharpoons Ba^{2+}(aq) + 2\,OH^-(aq)$	5×10^{-3}
barium sulfate	$BaSO_4(s) \rightleftharpoons Ba^{2+}(aq) + SO_4{}^{2-}(aq)$	1.1×10^{-10}
bismuth(III) sulfide	$Bi_2S_3(s) \rightleftharpoons 2\,Bi^{3+}(aq) + 3\,S^{2-}(aq)$	1×10^{-97}
cadmium sulfide	$CdS(s) \rightleftharpoons Cd^{2+}(aq) + S^{2-}(aq)$	8×10^{-27}
calcium carbonate	$CaCO_3(s) \rightleftharpoons Ca^{2+}(aq) + CO_3{}^{2-}(aq)$	2.8×10^{-9}
calcium fluoride	$CaF_2(s) \rightleftharpoons Ca^{2+}(aq) + 2\,F^-(aq)$	5.3×10^{-9}
calcium hydroxide	$Ca(OH)_2(s) \rightleftharpoons Ca^{2+}(aq) + 2\,OH^-(aq)$	5.5×10^{-6}
calcium sulfate	$CaSO_4(s) \rightleftharpoons Ca^{2+}(aq) + SO_4{}^{2-}(aq)$	9.1×10^{-6}
chromium(III) hydroxide	$Cr(OH)_3(s) \rightleftharpoons Cr^{3+}(aq) + 3\,OH^-(aq)$	6.3×10^{-31}
cobalt(II) sulfide	$CoS(s) \rightleftharpoons Co^{2+}(aq) + S^{2-}(aq)$	4×10^{-21}
copper(II) sulfide	$CuS(s) \rightleftharpoons Cu^{2+}(aq) + S^{2-}(aq)$	6×10^{-36}
iron(II) sulfide	$FeS(s) \rightleftharpoons Fe^{2+}(aq) + S^{2-}(aq)$	6×10^{-18}
iron(III) hydroxide	$Fe(OH)_3(s) \rightleftharpoons Fe^{3+}(aq) + 3\,OH^-(aq)$	4×10^{-38}
lead(II) chloride	$PbCl_2(s) \rightleftharpoons Pb^{2+}(aq) + 2\,Cl^-(aq)$	1.6×10^{-5}
lead(II) chromate	$PbCrO_4(s) \rightleftharpoons Pb^{2+}(aq) + CrO_4{}^{2-}(aq)$	2.8×10^{-13}
lead(II) iodide	$PbI_2(s) \rightleftharpoons Pb^{2+}(aq) + 2\,I^-(aq)$	7.1×10^{-9}
lead(II) sulfate	$PbSO_4(s) \rightleftharpoons Pb^{2+}(aq) + SO_4{}^{2-}(aq)$	1.6×10^{-8}
lead(II) sulfide	$PbS(s) \rightleftharpoons Pb^{2+}(aq) + S^{2-}(aq)$	8×10^{-28}
lithium phosphate	$Li_3PO_4(s) \rightleftharpoons 3\,Li^+(aq) + PO_4{}^{3-}(aq)$	3.2×10^{-9}
magnesium carbonate	$MgCO_3(s) \rightleftharpoons Mg^{2+}(aq) + CO_3{}^{2-}(aq)$	3.5×10^{-8}
magnesium fluoride	$MgF_2(s) \rightleftharpoons Mg^{2+}(aq) + 2\,F^-(aq)$	3.7×10^{-8}
magnesium hydroxide	$Mg(OH)_2(s) \rightleftharpoons Mg^{2+}(aq) + 2\,OH^-(aq)$	1.8×10^{-11}
magnesium phosphate	$Mg_3(PO_4)_2(s) \rightleftharpoons 3\,Mg^{2+}(aq) + 2\,PO_4{}^{3-}(aq)$	1×10^{-25}
manganese(II) sulfide	$MnS(s) \rightleftharpoons Mn^{2+}(aq) + S^{2-}(aq)$	2×10^{-13}
mercury(I) chloride	$Hg_2Cl_2(s) \rightleftharpoons Hg_2{}^{2+}(aq) + 2\,Cl^-(aq)$	1.3×10^{-18}
mercury(II) sulfide	$HgS(s) \rightleftharpoons Hg^{2+}(aq) + S^{2-}(aq)$	2×10^{-52}
nickel(II) sulfide	$NiS(s) \rightleftharpoons Ni^{2+}(aq) + S^{2-}(aq)$	3×10^{-19}
silver bromide	$AgBr(s) \rightleftharpoons Ag^+(aq) + Br^-(aq)$	5.0×10^{-13}
silver carbonate	$Ag_2CO_3(s) \rightleftharpoons 2\,Ag^+(aq) + CO_3{}^{2-}(aq)$	8.1×10^{-12}
silver chloride	$AgCl(s) \rightleftharpoons Ag^+(aq) + Cl^-(aq)$	1.8×10^{-10}
silver chromate	$Ag_2CrO_4(s) \rightleftharpoons 2\,Ag^+(aq) + CrO_4{}^{2-}(aq)$	2.4×10^{-12}
silver iodide	$AgI(s) \rightleftharpoons Ag^+(aq) + I^-(aq)$	8.5×10^{-17}
silver sulfate	$Ag_2SO_4(s) \rightleftharpoons 2\,Ag^+(aq) + SO_4{}^{2-}(aq)$	1.4×10^{-5}
silver sulfide	$Ag_2S(s) \rightleftharpoons 2\,Ag^+(aq) + S^{2-}(aq)$	6×10^{-50}
strontium carbonate	$SrCO_3(s) \rightleftharpoons Sr^{2+}(aq) + CO_3{}^{2-}(aq)$	1.1×10^{-10}
strontium sulfate	$SrSO_4(s) \rightleftharpoons Sr^{2+}(aq) + SO_4{}^{2-}(aq)$	3.2×10^{-7}
tin(II) sulfide	$SnS(s) \rightleftharpoons Sn^{2+}(aq) + S^{2-}(aq)$	1×10^{-25}
zinc sulfide	$ZnS(s) \rightleftharpoons Zn^{2+}(aq) + S^{2-}(aq)$	1×10^{-21}

TABLE 19-2
Formation constants for some complex ions

Complex ion	Equilibrium reaction	K_f
$[AlF_6]^{3-}$	$Al^{3+} + 6\,F^- \rightleftharpoons [AlF_6]^{3-}$	6.7×10^{19}
$[Cd(CN)_4]^{2-}$	$Cd^{2+} + 4\,CN^- \rightleftharpoons [Cd(CN)_4]^{2-}$	7.1×10^{18}
$[Co(NH_3)_6]^{3+}$	$Co^{3+} + 6\,NH_3 \rightleftharpoons [Co(NH_3)_6]^{3+}$	4.5×10^{33}
$[Cu(CN)_3]^{2-}$	$Cu^+ + 3\,CN^- \rightleftharpoons [Cu(CN)_3]^{2-}$	2×10^{27}
$[Cu(NH_3)_4]^{2+}$	$Cu^{2+} + 4\,NH_3 \rightleftharpoons [Cu(NH_3)_4]^{2+}$	1.1×10^{13}
$[Fe(CN)_6]^{4-}$	$Fe^{2+} + 6\,CN^- \rightleftharpoons [Fe(CN)_6]^{4-}$	1×10^{37}
$[Fe(CN)_6]^{3-}$	$Fe^{3+} + 6\,CN^- \rightleftharpoons [Fe(CN)_6]^{3-}$	1×10^{42}
$[PbCl_3]^-$	$Pb^{2+} + 3\,Cl^- \rightleftharpoons [PbCl_3]^-$	2.4×10^{1}
$[HgCl_4]^{2-}$	$Hg^{2+} + 4\,Cl^- \rightleftharpoons [HgCl_4]^{2-}$	1.2×10^{15}
$[HgI_4]^{2-}$	$Hg^{2+} + 4\,I^- \rightleftharpoons [HgI_4]^{2-}$	1.9×10^{30}
$[Ni(CN)_4]^{2-}$	$Ni^{2+} + 4\,CN^- \rightleftharpoons [Ni(CN)_4]^{2-}$	1×10^{22}
$[Ag(NH_3)_2]^+$	$Ag^+ + 2\,NH_3 \rightleftharpoons [Ag(NH_3)_2]^+$	1.6×10^{7}
$[Ag(CN)_2]^-$	$Ag^+ + 2\,CN^- \rightleftharpoons [Ag(CN)_2]^-$	5.6×10^{18}
$[Ag(S_2O_3)_2]^{3-}$	$Ag^+ + 2\,S_2O_3^{2-} \rightleftharpoons [Ag(S_2O_3)_2]^{3-}$	1.7×10^{13}
$[Zn(NH_3)_4]^{2+}$	$Zn^{2+} + 4\,NH_3 \rightleftharpoons [Zn(NH_3)_4]^{2+}$	4.1×10^{8}
$[Zn(CN)_4]^{2-}$	$Zn^{2+} + 4\,CN^- \rightleftharpoons [Zn(CN)_4]^{2-}$	1×10^{18}
$[Zn(OH)_4]^{2-}$	$Zn^{2+} + 4\,OH^- \rightleftharpoons [Zn(OH)_4]^{2-}$	4.6×10^{17}

TABLE 21-1
Some selected standard electrode potentials

Reduction half-reaction	$E°$, V
Acidic solution	
$F_2(g) + 2\,e^- \longrightarrow 2\,F^-(aq)$	+2.87
$O_3(g) + 2\,H^+(aq) + 2\,e^- \longrightarrow O_2(g) + H_2O$	+2.07
$S_2O_8^{2-}(aq) + 2\,e^- \longrightarrow 2\,SO_4^{2-}(aq)$	+2.01
$H_2O_2(aq) + 2\,H^+(aq) + 2\,e^- \longrightarrow 2\,H_2O$	+1.77
$MnO_4^-(aq) + 8\,H^+(aq) + 5\,e^- \longrightarrow Mn^{2+}(aq) + 4\,H_2O$	+1.51
$PbO_2(s) + 4\,H^+(aq) + 2\,e^- \longrightarrow Pb^{2+}(aq) + 2\,H_2O$	+1.455
$Cl_2(g) + 2\,e^- \longrightarrow 2\,Cl^-(aq)$	+1.360
$Cr_2O_7^{2-}(aq) + 14\,H^+(aq) + 6\,e^- \longrightarrow 2\,Cr^{3+}(aq) + 7\,H_2O$	+1.33
$MnO_2(s) + 4\,H^+(aq) + 2\,e^- \longrightarrow Mn^{2+}(aq) + 2\,H_2O$	+1.23
$O_2(g) + 4\,H^+(aq) + 4\,e^- \longrightarrow 2\,H_2O$	+1.229
$2\,IO_3^-(aq) + 12\,H^+(aq) + 10\,e^- \longrightarrow I_2(s) + 6\,H_2O$	+1.195
$Br_2(l) + 2\,e^- \longrightarrow 2\,Br^-(aq)$	+1.065
$NO_3^-(aq) + 4\,H^+(aq) + 3\,e^- \longrightarrow NO(g) + 2\,H_2O$	+0.96
$Ag^+(aq) + e^- \longrightarrow Ag(s)$	+0.800
$Fe^{3+}(aq) + e^- \longrightarrow Fe^{2+}(aq)$	+0.771
$O_2(g) + 2\,H^+(aq) + 2\,e^- \longrightarrow H_2O_2(aq)$	+0.682
$I_2(s) + 2\,e^- \longrightarrow 2\,I^-(aq)$	+0.535
$Cu^+(aq) + e^- \longrightarrow Cu(s)$	+0.52
$H_2SO_3(aq) + 4\,H^+(aq) + 4\,e^- \longrightarrow S(s) + 3\,H_2O$	+0.45
$Cu^{2+}(aq) + 2\,e^- \longrightarrow Cu(s)$	+0.337
$SO_4^{2-}(aq) + 4\,H^+(aq) + 2\,e^- \longrightarrow 2\,H_2O + SO_2(g)$	+0.17
$Sn^{4+}(aq) + 2\,e^- \longrightarrow Sn^{2+}(aq)$	+0.154
$S(s) + 2\,H^+(aq) + 2\,e^- \longrightarrow H_2S(g)$	+0.141
$2\,H^+(aq) + 2\,e^- \longrightarrow H_2(g)$	0.0000
$Pb^{2+}(aq) + 2\,e^- \longrightarrow Pb(s)$	−0.126
$Sn^{2+}(aq) + 2\,e^- \longrightarrow Sn(s)$	−0.136
$Cr^{3+}(aq) + e^- \longrightarrow Cr^{2+}(aq)$	−0.41
$Fe^{2+}(aq) + 2\,e^- \longrightarrow Fe(s)$	−0.440
$Zn^{2+}(aq) + 2\,e^- \longrightarrow Zn(s)$	−0.763
$Al^{3+}(aq) + 3\,e^- \longrightarrow Al(s)$	−1.66
$Mg^{2+}(aq) + 2\,e^- \longrightarrow Mg(s)$	−2.375
$Na^+(aq) + e^- \longrightarrow Na(s)$	−2.714
$Ca^{2+}(aq) + 2\,e^- \longrightarrow Ca(s)$	−2.76
$K^+(aq) + e^- \longrightarrow K(s)$	−2.925
$Li^+(aq) + e^- \longrightarrow Li(s)$	−3.045

continues

TABLE 21-1 (continued)

Reduction half-reaction	$E°$, V
Basic solution	
$O_3(g) + H_2O + 2\,e^- \longrightarrow O_2(g) + 2\,OH^-$	+1.24
$OCl^-(aq) + H_2O + 2\,e^- \longrightarrow Cl^- + 2\,OH^-$	+0.89
$O_2(g) + 2\,H_2O + 4\,e^- \longrightarrow 4\,OH^-(aq)$	+0.401
$CrO_4^{2-}(aq) + 4\,H_2O + 3\,e^- \longrightarrow Cr(OH)_3(s) + 5\,OH^-$	−0.13
$S(s) + 2\,e^- \longrightarrow S^{2-}(aq)$	−0.48
$2\,H_2O + 2\,e^- \longrightarrow H_2(g) + 2\,OH^-(aq)$	−0.828
$SO_4^{2-}(aq) + H_2O + 2\,e^- \longrightarrow SO_3^{2-}(aq) + 2\,OH^-(aq)$	−0.93

TABLE 23-7
Solubilities of some metal sulfides

	Soluble in			
H_2O	0.3 M HCl (K_{sp})	3 M HNO_3 (K_{sp})	Aqua regia (K_{sp})	KOH(aq) or
K_2S	MnS (2.5×10^{-13})	CdS (8.0×10^{-27})	HgS (1.6×10^{-52})	SnS
Na_2S	FeS (6.3×10^{-18})	PbS (8×10^{-28})		As_2S_3
CaS	CoS (4.0×10^{-21})	CuS (6.3×10^{-36})		Sb_2S_3
	ZnS (1.0×10^{-21})			

TABLE 23-12
Preparation of oxides of nitrogen

Oxide	A method of preparation
N_2O	$NH_4NO_3(s) \xrightarrow{\Delta} N_2O(g) + 2\,H_2O(l)$
NO	$3\,Cu(s) + 8\,H^+ + 2\,NO_3^- \longrightarrow 3\,Cu^{2+} + 2\,NO(g) + 4\,H_2O$
N_2O_3	$NO(g) + NO_2(g) \rightleftharpoons N_2O_3(g)$ at 298 K: $K_p = 0.48$
NO_2	$2\,Pb(NO_3)_2(s) \xrightarrow{\Delta} 2\,PbO(s) + 4\,NO_2(g) + O_2(g)$
	$2\,NO(g) + O_2(g) \rightleftharpoons 2\,NO_2(g)$ at 298 K: $\Delta\overline{H}° = -113$ kJ/mol; $K_p = 1.6 \times 10^{12}$
N_2O_4	$2\,NO_2(g) \rightleftharpoons N_2O_4(g)$ at 298 K: $\Delta\overline{H}° = -58$ kJ/mol; $K_p = 8.84$
N_2O_5	$4\,HNO_3(l) + P_4O_{10}(s) \longrightarrow 4\,HPO_3 + 2\,N_2O_5(s)$

TABLE 26-6
Units of radiation dosage[a]

Unit	Definition
Curie	An amount of radioactive material decaying at the same rate as 1 g of radium (3.7×10^{10} dis/s).
Rad	A dosage of radiation able to deposit 1×10^{-2} J of energy per kilogram of matter.
Rem	A unit related to the rad, but taking into account the varying effects of different types of radiation of the same energy on biological matter. This relationship is through a "quality factor," which may be taken as equal to one for x rays, γ rays, and β particles. For protons and slow neutrons the factor has a value of 5; and for α particles, 10. Thus, an exposure to 1 rad of x rays is about equal to 1 rem, but 1 rad of α particles is about equal to 10 rem.

[a] Sources of α radiation are relatively harmless when external to the body and extremely hazardous when taken internally, as in the lungs or stomach. Other forms of radiation (x rays, γ rays), because they are highly penetrating, are hazardous even when external to the body.

TABLE 28-1
Some common fatty acids

Common name	IUPAC name	Formula
	Saturated acids	
lauric acid	dodecanoic acid	$C_{11}H_{23}CO_2H$
myristic acid	tetradecanoic acid	$C_{13}H_{27}CO_2H$
palmitic acid	hexadecanoic acid	$C_{15}H_{31}CO_2H$
stearic acid	octadecanoic acid	$C_{17}H_{35}CO_2H$
	Unsaturated acids	
oleic acid	9-octadecenoic acid	$C_{17}H_{33}CO_2H$
linoleic acid	9,12-octadecadienoic acid	$C_{17}H_{31}CO_2H$
linolenic acid	9,12,15-octadecatrienoic acid	$C_{17}H_{29}CO_2H$
eleostearic acid	9,11,13-octadecatrienoic acid	$C_{17}H_{29}CO_2H$

TABLE 28-3
Some common amino acids

Name	Symbol	Formula	p*I*
		Neutral amino acids	
glycine	Gly	$HCH(NH_2)CO_2H$	5.97
alanine	Ala	$CH_3CH(NH_2)CO_2H$	6.00
valine	Val	$(CH_3)_2CHCH(NH_2)CO_2H$	5.96
leucine	Leu	$(CH_3)_2CHCH_2CH(NH_2)CO_2H$	6.02
isoleucine	Ileu or Ile	$CH_3CH_2CH(CH_3)CH(NH_2)CO_2H$	5.98
serine	Ser	$HOCH_2CH(NH_2)CO_2H$	5.68
threonine	Thr	$CH_3CHOHCH(NH_2)CO_2H$	5.6
phenylalanine	Phe	$C_6H_5CH_2CH(NH_2)CO_2H$	5.48
methionine	Met	$CH_3SCH_2CH_2CH(NH_2)CO_2H$	5.74
cysteine	Cys	$HSCH_2CH(NH_2)CO_2H$	5.05
cystine	$(Cys)_2$	$[SCH_2CH(NH_2)CO_2H]_2$	4.8
tyrosine	Tyr	4-$HOC_6H_4CH_2CH(NH_2)CO_2H$	5.66
tryptophan	Trp	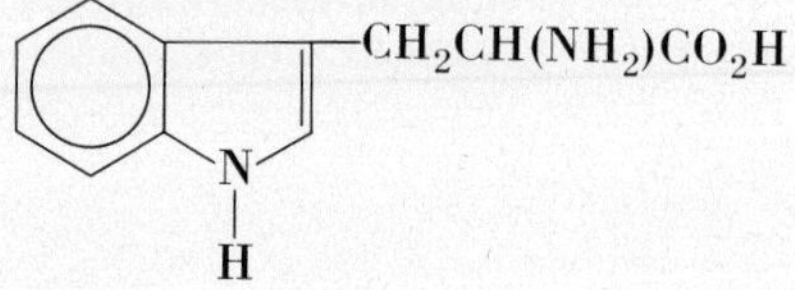	5.89
proline	Pro	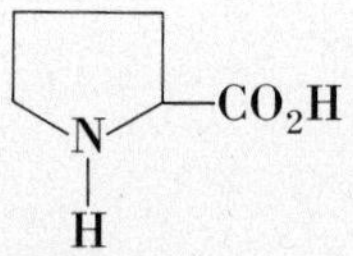	6.30
hydroxyproline	Hyp	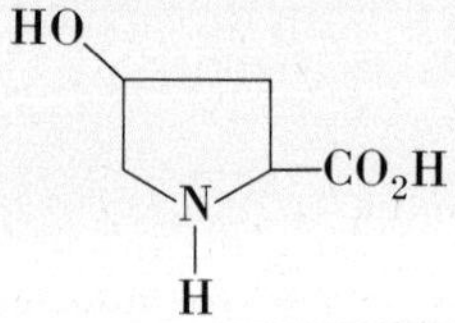	
		Acidic amino acids	
aspartic acid	Asp	$HO_2CCH_2CH(NH_2)CO_2H$	2.77
glutamic acid	Glu	$HO_2CCH_2CH_2CH(NH_2)CO_2H$	3.22
		Basic amino acids	
lysine[a]	Lys	$H_2N(CH_2)_4CH(NH_2)CO_2H$	9.74
arginine	Arg	$H_2NCNHNH(CH_2)_3CH(NH_2)CO_2H$	10.76
histidine	His	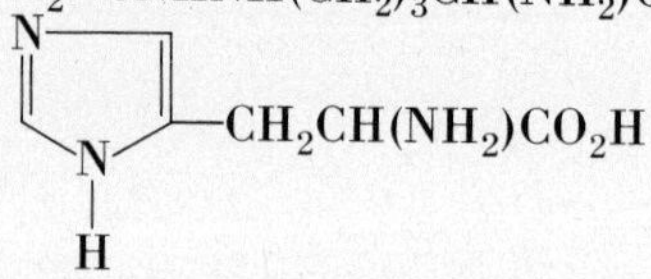	

FIGURE 2-12
A mass spectometer

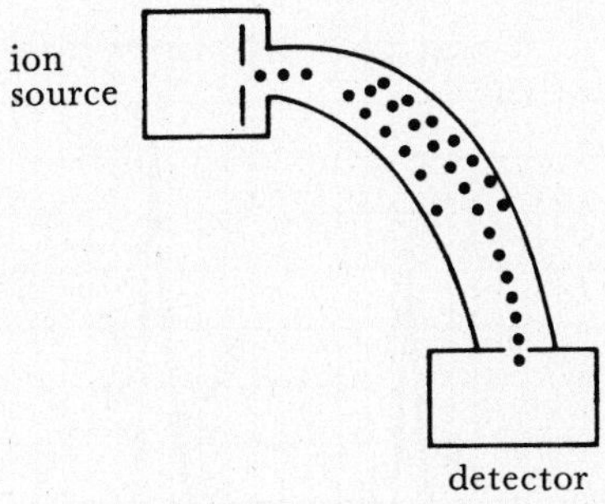

FIGURE 8-5
The electromagnetic spectrum.

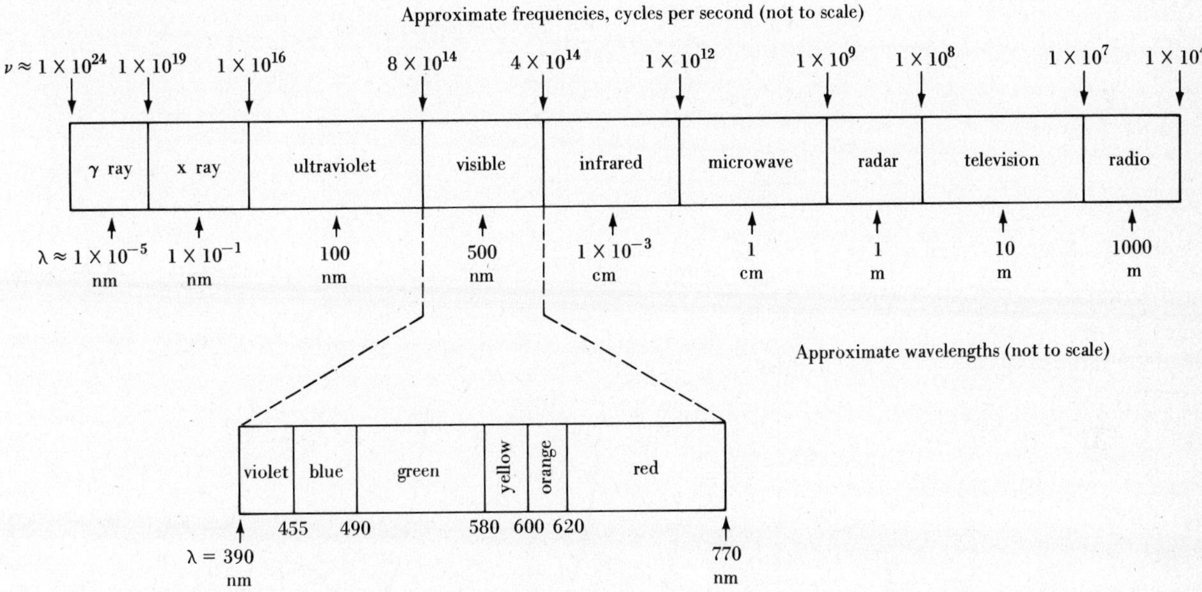

FIGURE 8-17
Energy-level diagram for the hydrogen atom.

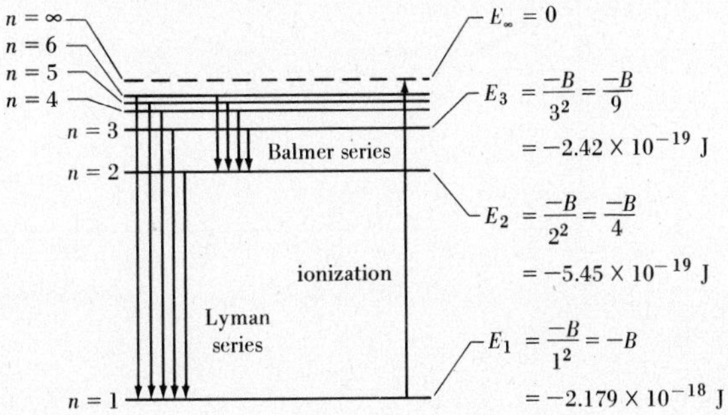

FIGURE 9-1
An illustration of the periodic law—atomic volume as a function of atomic number.

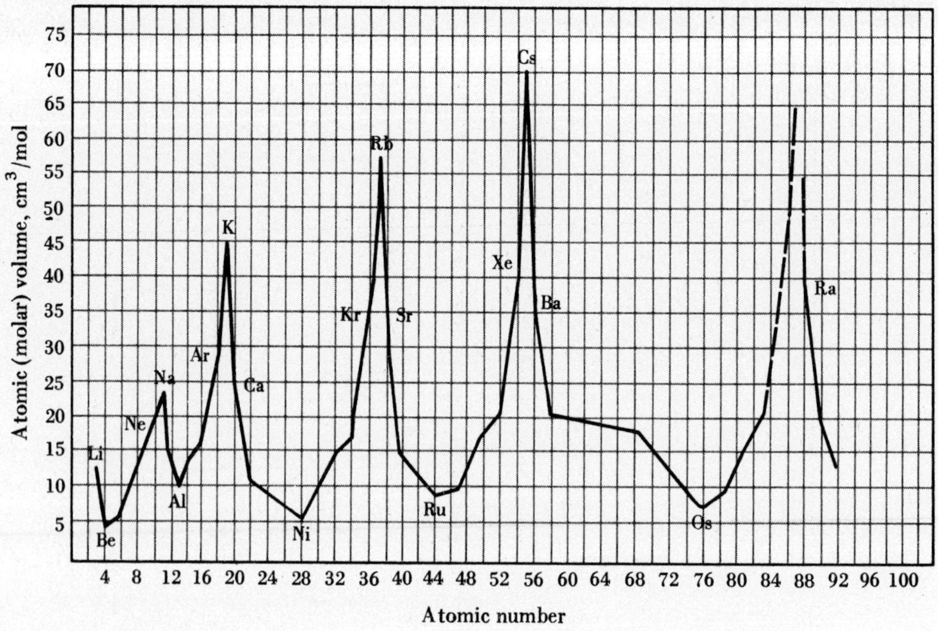

FIGURE 9-8
Covalent radii of atoms.

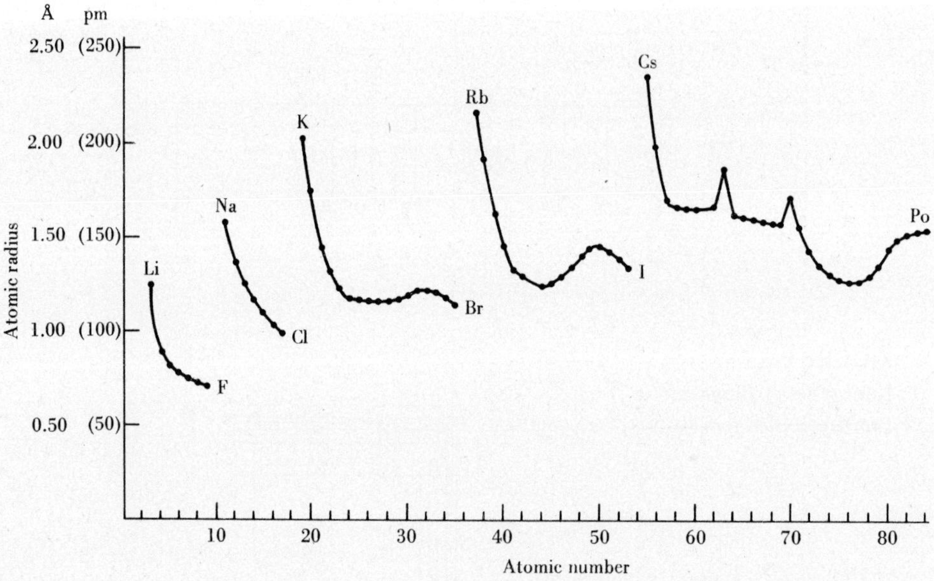

FIGURE 9-13
First ionization energies as a function of atomic number.

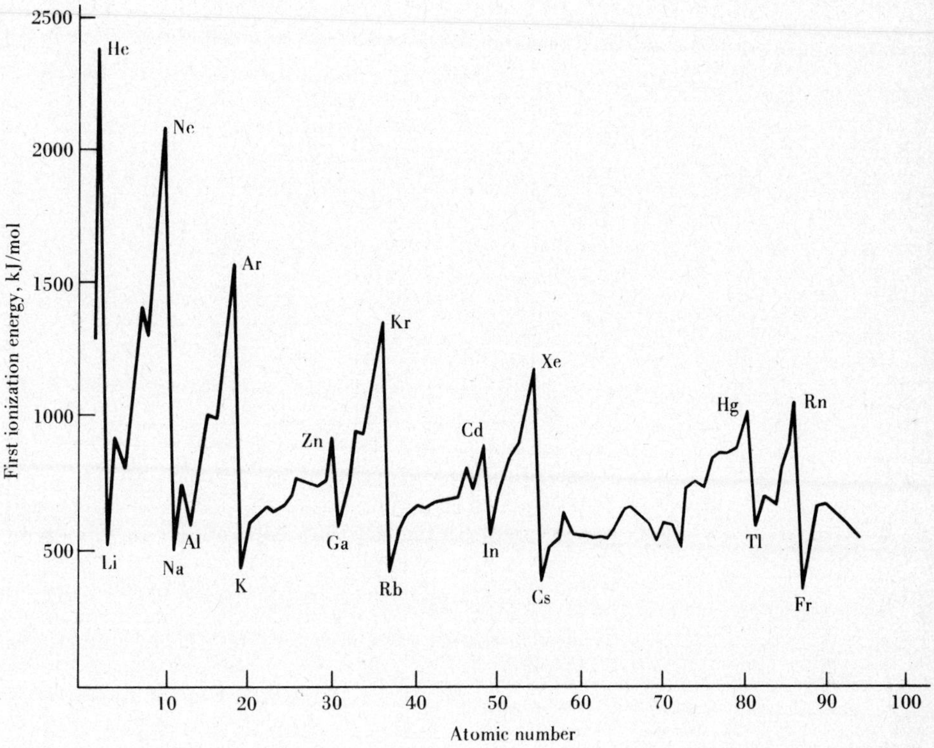

FIGURE 12-11
Vapor pressure curves of several liquids.

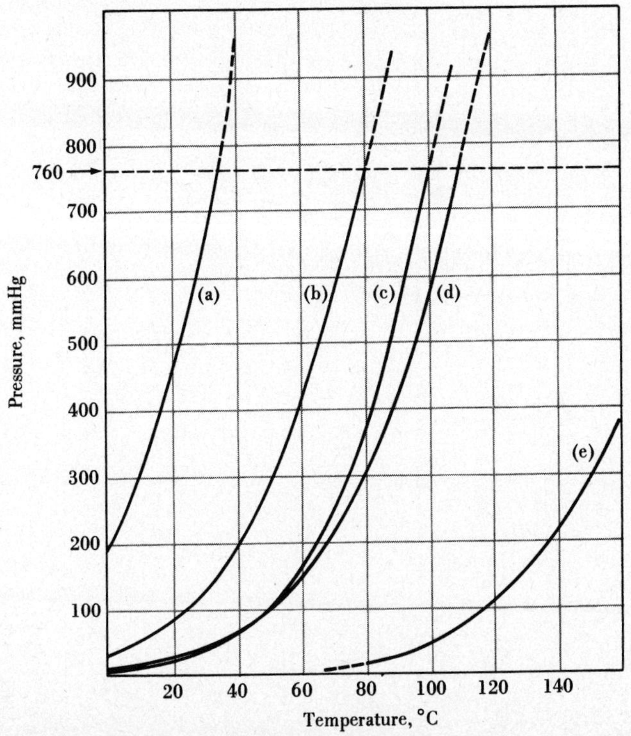

(a) Diethyl ether, $C_4H_{10}O$; (b) benzene, C_6H_6;
(c) water, H_2O; toluene, C_7H_8; (e) aniline, C_6H_7N.

FIGURE 12-20
Phase diagram for carbon dioxide.

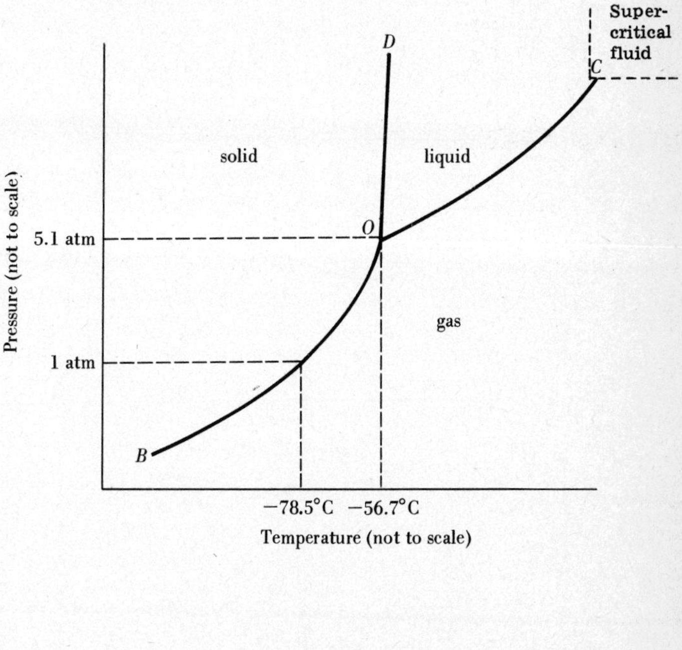

FIGURE 12-21
Phase diagram for water.

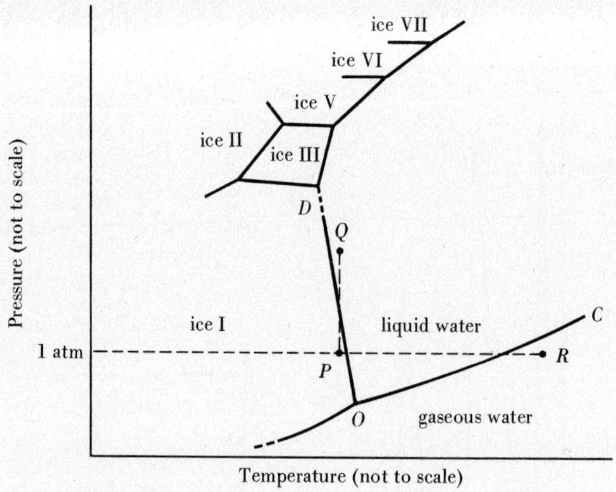

Point O, the triple point, is at $+0.0098°C$ and 4.58 mmHg. The critical point, C, is at 374.1°C and 218.2 atm. The negative slope of the fusion curve OD is exaggerated in this diagram. An increase in pressure of about 125 atm is required to produce a decrease of 1°C in the melting point of ice I. All of the high pressure forms of ice are more dense than liquid water. Ice I, ice III, and liquid water are at equilibrium (point D) at $-22.0°C$ and 2045 atm. Ice VII can be maintained at temperatures approximating the normal boiling point of water (100°C), but only at pressures in excess of 25,000 atm.

FIGURE 13-15
Vapor pressure lowering by a nonvolatile solute.

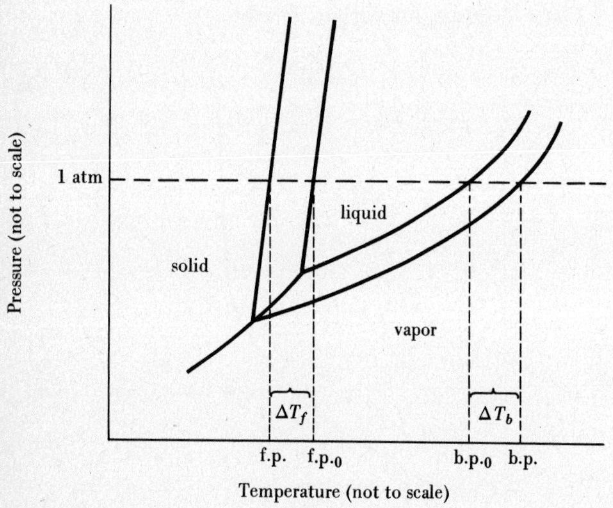

FIGURE 14-23
The Frasch process for mining sulfur.

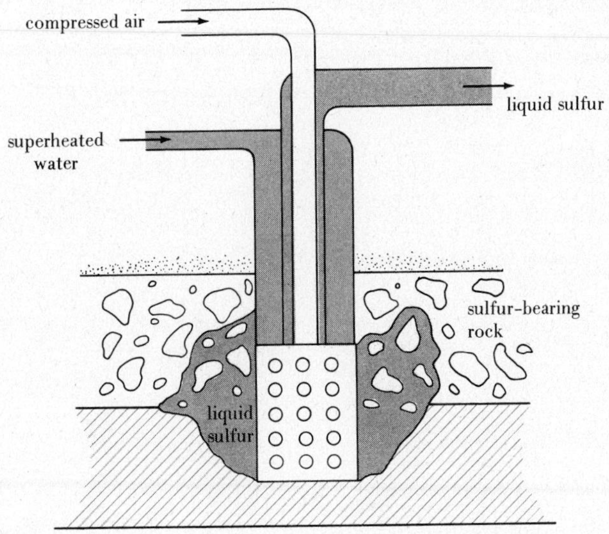

FIGURE 5-12
A qualitative analysis scheme for cations.

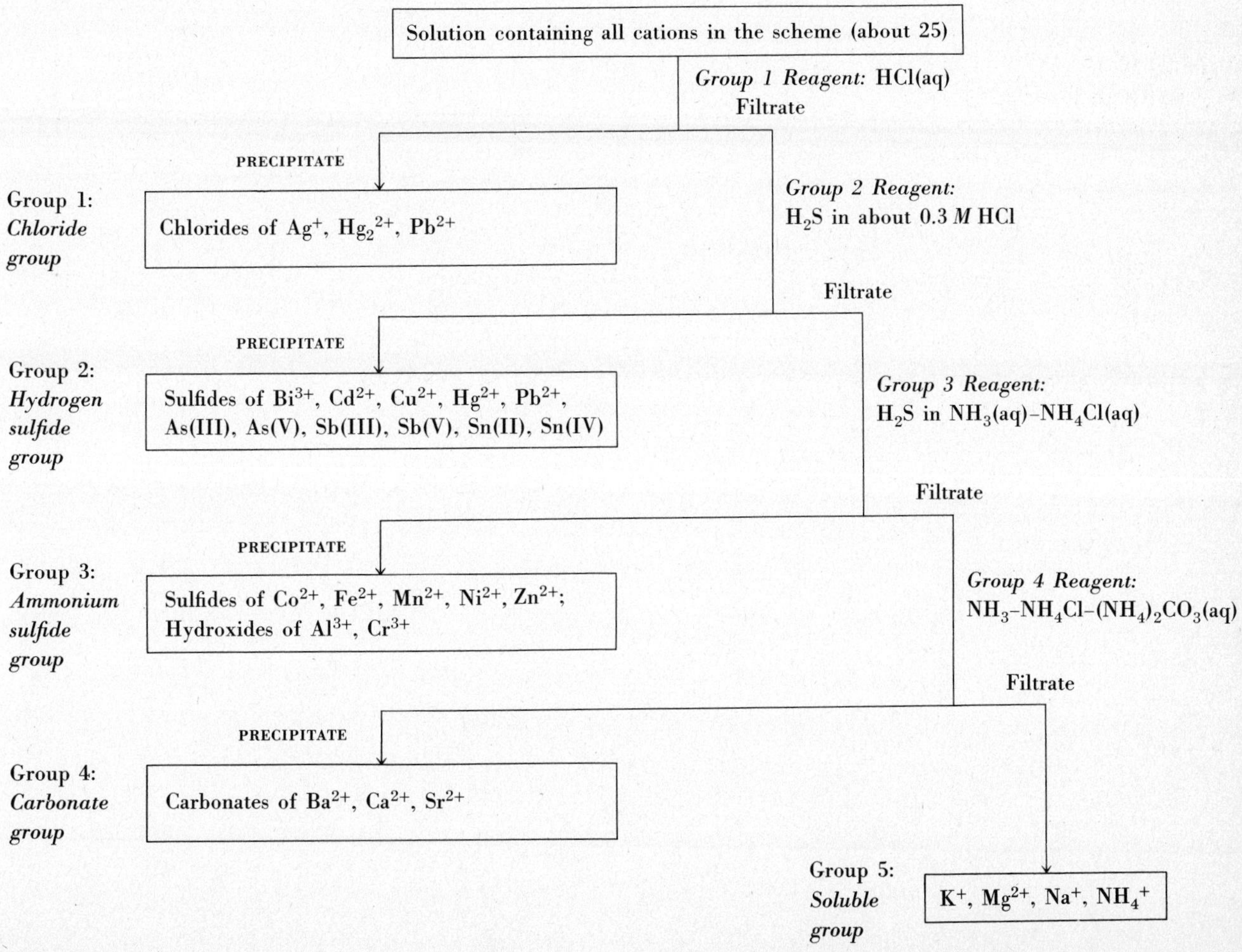

FIGURE 23-1
Standard electrode potential diagram for chlorine.

Acidic solution ($[H^+] = 1\ M$):

+7 | +5 | +3 | +1 | 0 | −1

ClO_4^- —1.19 V— ClO_3^- —1.21 V— $HClO_2$ —1.63 V— $HOCl$ —1.62 V— Cl_2 —1.36 V— Cl^-

ClO_3^- to Cl_2: 1.47 V

Basic solution ($[OH^-] = 1\ M$):

+7 | +5 | +3 | +1 | 0 | −1

ClO_4^- —0.36 V— ClO_3^- —0.35 V— ClO_2^- —0.65 V— OCl^- —0.40 V— Cl_2 —1.36 V— Cl^-

OCl^- to Cl^-: 0.88 V

FIGURE 23-6
Electrode potential diagram for oxygen.

Acidic solution ($[H^+] = 1\ M$):

O_3 —2.07 V— O_2 —0.682 V— H_2O_2 —1.77 V— H_2O

O_2 to H_2O: 1.229 V

Basic solution ($[OH^-] = 1\ M$):

O_3 —1.24 V— O_2 —0.076 V— HO_2^- —0.878 V— OH^-

O_2 to OH^-: 0.401 V

FIGURE 23-7
Electrode potential diagram for sulfur.

Acidic solution ($[H^+] = 1\ M$):

+6 | +5 | +4 | +2.5 | +2 | 0 | −2

SO_4^{2-} —−0.22 V— $S_2O_6^{2-}$ —0.57 V— SO_2 —0.51 V— $S_4O_6^{2-}$ —0.08 V— $S_2O_3^{2-}$ —0.50 V— S —0.14 V— H_2S

SO_4^{2-} to SO_2: 0.17 V; SO_2 to S: 0.45 V

Basic solution ($[OH^-] = 1\ M$):

+6 | +4 | +2.5 | +2 | 0 | −2

SO_4^{2-} —−0.93 V— SO_3^{2-} —−0.79 V— $S_4O_6^{2-}$ —0.08 V— $S_2O_3^{2-}$ —−0.74 V— S —−0.48 V— S^{2-}

FIGURE 23-10
Electrode potential diagram for nitrogen.

Acidic solution ($[H^+] = 1\ M$):

NO_3^- —+0.81 V— NO_2 —+1.07 V— HNO_2 —+0.99 V— NO —+1.59 V— N_2O —+1.77 V— N_2 —−1.87 V— NH_3OH^+ —+1.46 V— $N_2H_5^+$ —+1.24 V— NH_4^+

Basic solution ($[OH^-] = 1\ M$):

NO_3^- —−0.85 V— NO_2 —+0.88 V— NO_2^- —−0.46 V— NO —+0.76 V— N_2O —+0.94 V— N_2 —−3.04 V— NH_2OH —+0.74 V— N_2H_4 —+0.10 V— NH_3

FIGURE 24-3
Electrode potential diagram for vanadium.

$VO_2^+(aq)$ (yellow) —+1.00 V— $VO^{2+}(aq)$ (blue) —+0.361 V— $V^{3+}(aq)$ (green) —−0.255 V— $V^{2+}(aq)$ (violet) —−1.18 V— $V(s)$

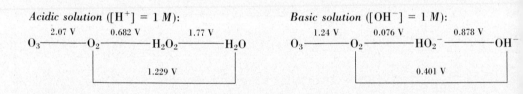

FIGURE 24-8
Electrode potential diagram for manganese.

Acidic solution ([H^+] = 1 *M*):

$$MnO_4^- \xrightarrow{0.56\ V} MnO_4^{2-} \xrightarrow{2.26\ V} MnO_2 \xrightarrow{0.95\ V} Mn^{3+} \xrightarrow{1.49\ V} Mn^{2+} \xrightarrow{-1.18\ V} Mn$$

(purple)   (green)   (black)   (red)   (pale pink)

1.70 V   1.23 V

Basic solution ([OH^-] = 1 *M*):

$$MnO_4^- \xrightarrow{0.56\ V} MnO_4^{2-} \xrightarrow{0.3\ V} MnO_3^- \xrightarrow{0.8\ V} MnO_2 \xrightarrow{-0.2\ V} Mn(OH)_3 \xrightarrow{0.1\ V} Mn(OH)_2 \xrightarrow{-1.55\ V} Mn$$

(purple) (green) (blue) (black) (brown) (pink)

0.60 V   −0.04 V

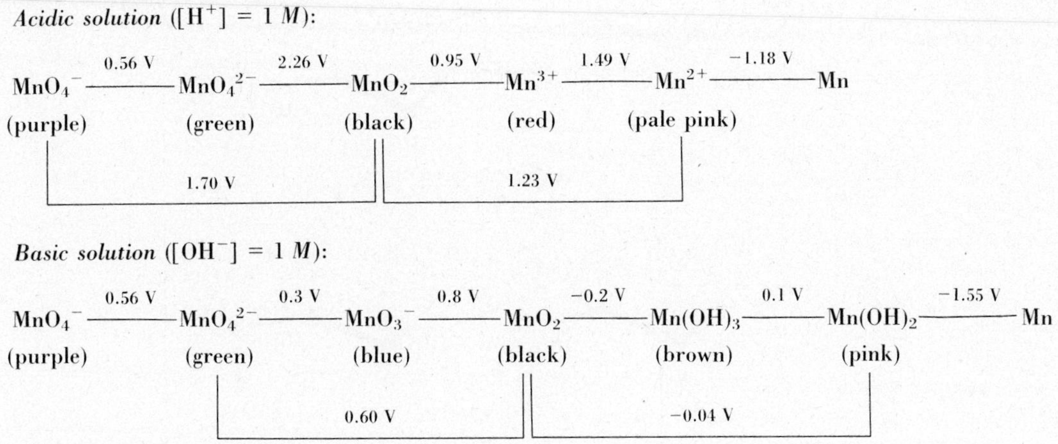

FIGURE 24-15
Qualitative analysis of the cation group 3.

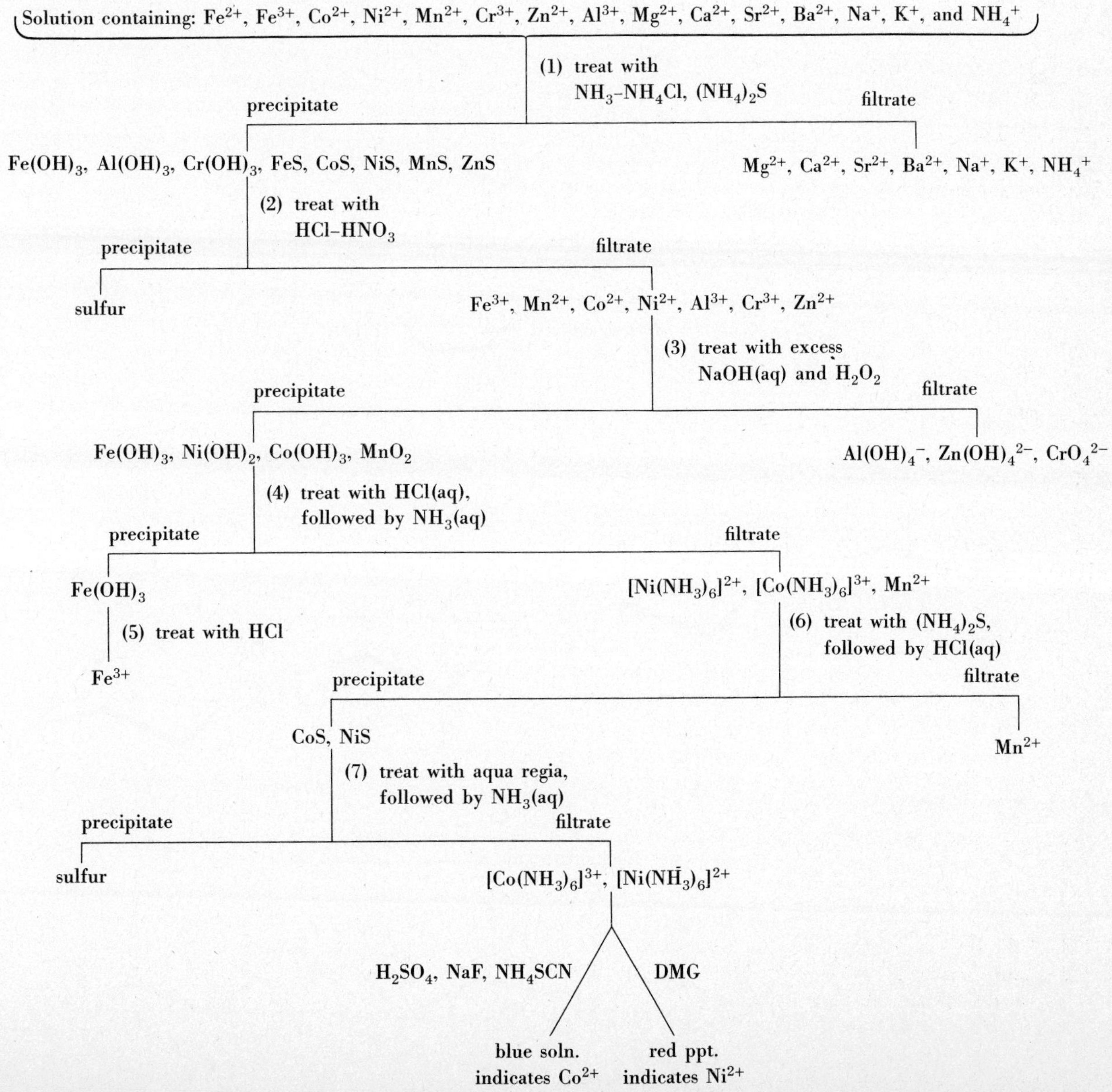

FIGURE 28-8
Some common disaccharides.

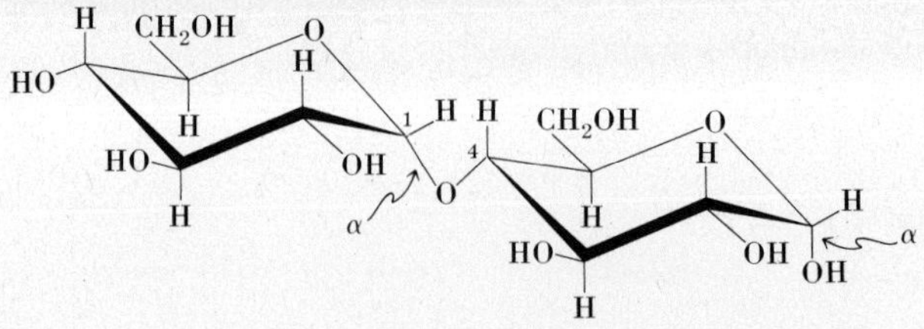

maltose (α-form)

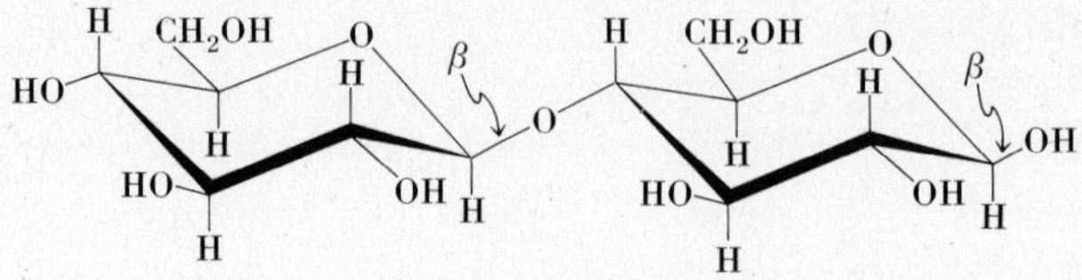

cellobiose

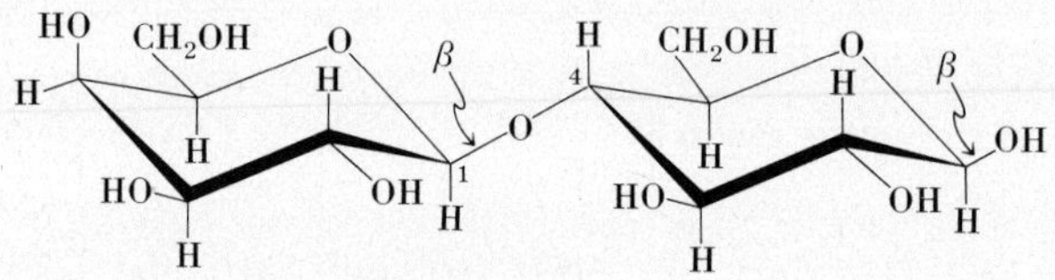

lactose (β-form)

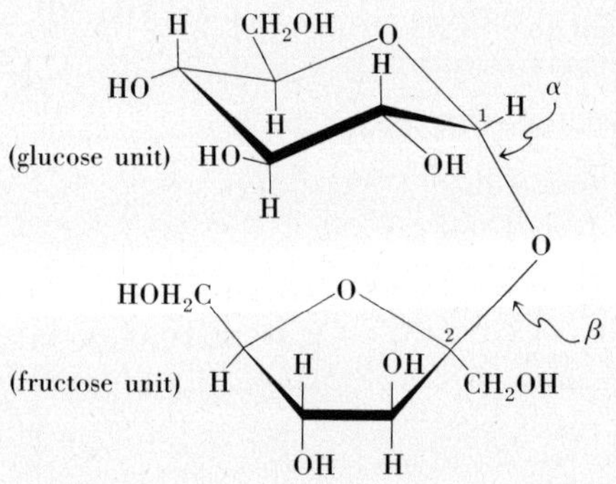

sucrose

FIGURE 28-9
Two common polysaccharides.

starch

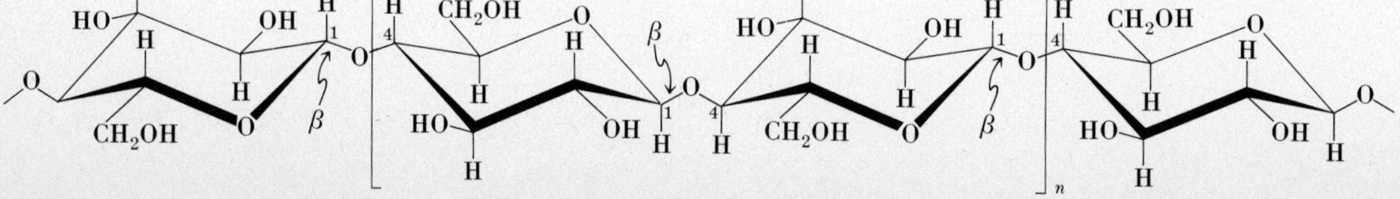

cellulose

FIGURE 28-12
An alpha helix—secondary structure of a protein.

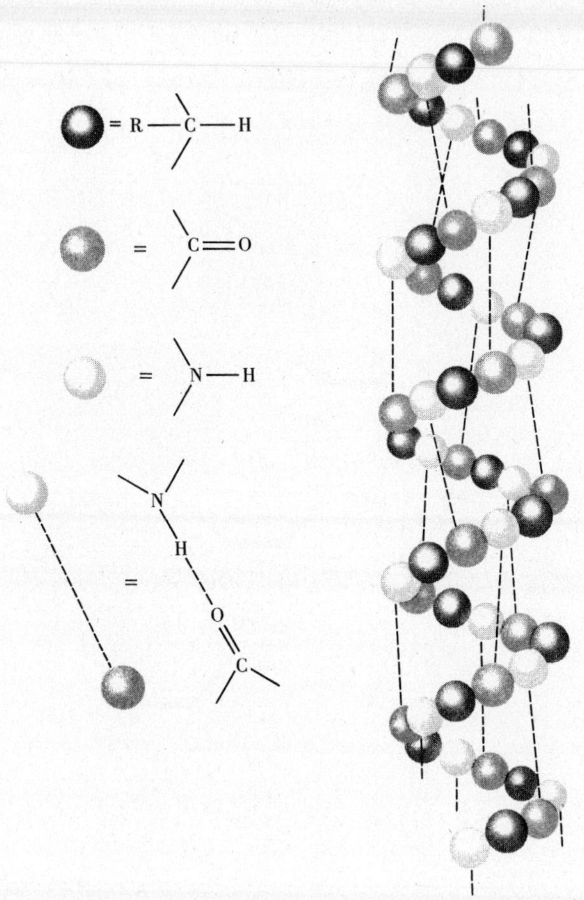

FIGURE 28-20
Hydrolysis products of nucleic acids.

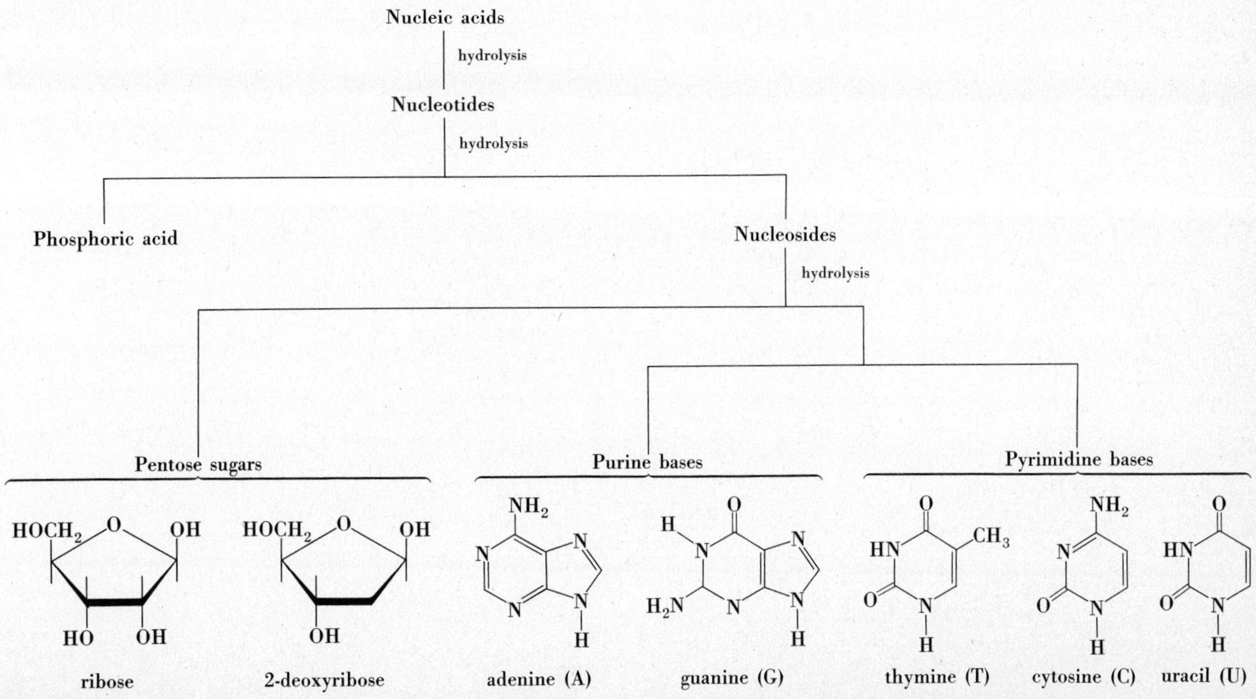

FIGURE 28-21
A portion of a nucleic acid chain.

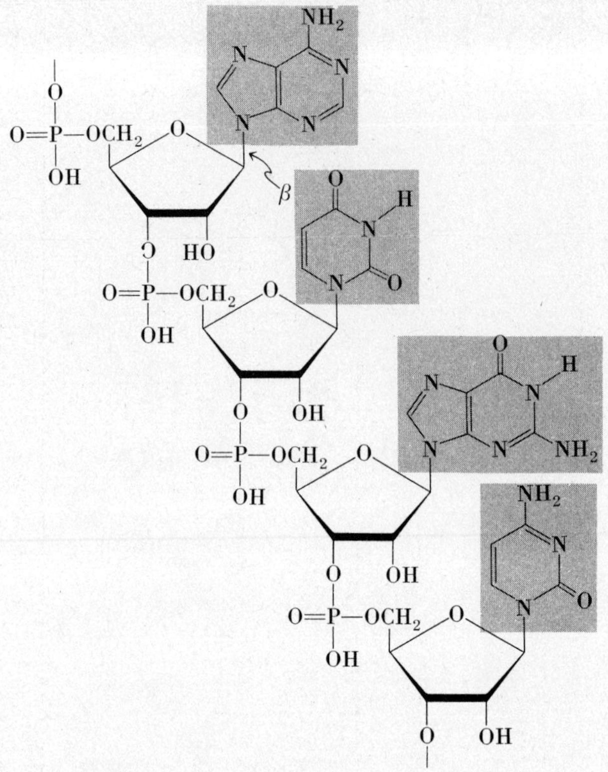

FIGURE 28-22
DNA model.

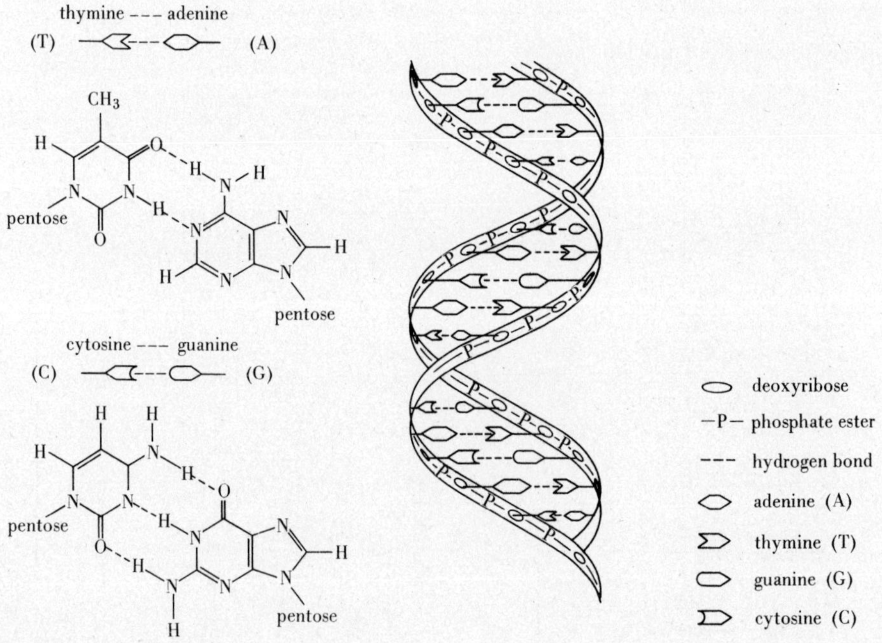

Index